10th Anniversary of Biomedicines—Translational Laboratory and Experimental Medicine for the Sake of Neurological Diseases and Mental Illnesses

10th Anniversary of Biomedicines—Translational Laboratory and Experimental Medicine for the Sake of Neurological Diseases and Mental Illnesses

Editor

Masaru Tanaka

MDPI

Basel • Beijing • Wuhan • Barcelona • Belgrade • Novi Sad • Cluj • Manchester

Editor
Masaru Tanaka
HUN-REN-SZTE
Neuroscience Research
Group, Hungarian Research
Network, University of
Szeged (HUN-REN-SZTE)
Szeged
Hungary

Editorial Office
MDPI
St. Alban-Anlage 66
4052 Basel, Switzerland

This is a reprint of articles from the Special Issue published online in the open access journal *Biomedicines* (ISSN 2227-9059) (available at: https://www.mdpi.com/journal/biomedicines/special_issues/10th_anniversary_neuro).

For citation purposes, cite each article independently as indicated on the article page online and as indicated below:

Lastname, A.A.; Lastname, B.B. Article Title. *Journal Name* **Year**, *Volume Number*, Page Range.

ISBN 978-3-7258-1267-7 (Hbk)
ISBN 978-3-7258-1268-4 (PDF)
doi.org/10.3390/books978-3-7258-1268-4

Contents

About the Editor

Masaru Tanaka

Masaru Tanaka, M.D., Ph.D. is a Senior Research Fellow in the Danube Neuroscience Research Laboratory, HUN-REN-SZTE Neuroscience Research Group, Hungarian Research Network, University of Szeged (HUN-REN-SZTE). He obtained his PhD in Medicine and MD in general Medicine from the University of Szeged, and his bachelors' degree in Biophysics from the University of Illinois, Urbana-Champaign. His scientific interests include depression, anxiety, dementia, pain, their comorbid nature, and translational research on neurological diseases and psychiatric disorders. He is the author of over 80 papers in peer-reviewed journals and was placed in D1 in the global publication rank. His articles have been ranked in the top 99.9 percentile for citations in various fields. He is an Editorial Board Member of *Biomedicines, Biomolecules & Biomedicine, and Frontiers in Neuroscience*, among others.

biomedicines

MDPI

Editorial

A Decade of Dedication: Pioneering Perspectives on Neurological Diseases and Mental Illnesses

Masaru Tanaka [1,*,†] and László Vécsei [1,2,†]

[1] HUN-REN-SZTE Neuroscience Research Group, Hungarian Research Network, University of Szeged, Danube Neuroscience Research Laboratory, Tisza Lajos krt. 113, H-6725 Szeged, Hungary; vecsei.laszlo@med.u-szeged.hu
[2] Department of Neurology, Albert Szent-Györgyi Medical School, University of Szeged, Semmelweis u. 6, H-6725 Szeged, Hungary
* Correspondence: tanaka.masaru.1@med.u-szeged.hu; Tel.: +36-62-342-847
† These authors contributed equally to this work.

1. Introduction

Welcome to *Biomedicines'* 10th Anniversary Special Issue, a journey through the human mind's labyrinth and complex neurological pathways. This edition, focused on "Translational Laboratory and Experimental Medicine for Neurological Diseases and Mental Illnesses", presents 21 pioneering papers that explore the enigmas of the brain and its remarkable ability to heal and adjust. We investigate the various impacts of time on neural circuits and cognitive responses. Our research spans studying how the brain can adjust and recuperate following a stroke, a process known as neuroplasticity, to exploring the intricate link between age and behavior [1–6].

We are particularly interested in the neural mechanisms that underpin these mechanisms, such as the role of neural circuits and their plasticity in cognitive responses [7–11]. By investigating neural activity and connectivity, we hope to gain insight into brain adaptation [12–14]. This entails investigating how these changes affect cognitive functions such as memory and decision-making, as well as their implications for cognitive development and disorders [15–21]. Hence, we may be able to uncover the complex mechanisms that underpin neurodegenerative disorders and investigate potential therapeutic strategies that hold promise for novel treatments [22–28].

Preclinical research plays a crucial role in understanding neuropsychiatric conditions [29–32]. By conducting studies in vitro and in vivo, researchers gather valuable data that would be impractical to obtain directly from humans [33–38]. These preclinical findings, combined with ongoing clinical studies, help us better understand the behavioral aspects of neuropsychiatric disorders [39–41]. Computational and inferential methods also contribute to new approaches to treating neurological and psychiatric disorders by helping to unravel the underlying pathology [42–48]. Integrating interdisciplinary methods further optimizes drug development research, leading to the evaluation of potential lead compounds [27,49–52]. Promising interventions, such as brain stimulation, have the potential to transform treatment and pave the way for new and more effective drugs for neurological and psychiatric conditions [53–57].

In our quest to break barriers and unveil unknowns, we also delve into the realm of mental health, exploring the biochemical basis of suicidal thoughts and the relationship between mental illness and pain. Whether you are a clinician, researcher, or simply curious about the complexities of the human mind, this collection of articles promises to challenge conventional wisdom and expand your horizons [58–60]. Join us in commemorating ten years of groundbreaking exploration and advancement in the realm of biomedicine.

Citation: Tanaka, M.; Vécsei, L. A Decade of Dedication: Pioneering Perspectives on Neurological Diseases and Mental Illnesses. *Biomedicines* **2024**, *12*, 1083. https://doi.org/10.3390/biomedicines12051083

Received: 2 May 2024
Accepted: 11 May 2024
Published: 13 May 2024

2. Special Issue Articles

2.1. Stroke and Neuroplasticity: Unraveling the Brain's Resilience

The complex interaction between stroke and neuroplasticity is central to the process of post-stroke recovery [61–63]. When a stroke happens, it disrupts the complex neural pathways, leading to a sequence of impairments [64–66]. Nonetheless, the brain, as the master of adaptation, makes use of its hidden asset: neuroplasticity. This remarkable phenomenon facilitates neuroplasticity, which involves the brain's ability to reorganize itself by forming new synaptic connections and redirecting functions to unaffected areas [67–69]. The present section explores five fascinating studies that shed light on the interaction between stroke consequences and the brain's extraordinary capacity for recovery. The authors of the referenced articles investigate brain legion prediction via dizziness, cognitive symptoms caused by subcortical damage, the gut microbiota in stroke patients, the effect of alcohol on neurogenesis, and the potential of virtual reality in cognitive rehabilitation. These articles provide a promising and resilient perspective on overcoming neurological challenges (Table 1).

Table 1. Major subjects covered in the Special Issue "10th Anniversary of *Biomedicines*—Translational Laboratory and Experimental Medicine for the Sake of Neurological Diseases and Mental Illnesses".

	Subjects	Ref.
1. Stroke and neuroplasticity		
	Vertigo and stroke	[70]
	Pontine stroke effects	[71]
	Gut dysbiosis and stroke	[72]
	Alcohol and neurogenesis	[73]
	VR cognitive training	[74]
2. Age and behavioral studies		
	Ketamine vulnerability	[75]
	Melatonin and anxiety	[76]
	Platelet mitochondrial changes	[77]
	Guanfacine and behavior	[78]
3. Neuropsychiatric disorders and treatments		
	Enteric nervous system and PD	[79]
	UBL3 and alpha-synuclein	[80]
	NLRP3 inflammasome in brain diseases	[81]
	Metabolism and MS	[82]
	Stem cells and febrile seizures	[83]
	Stem cells and ALS	[84]
	Rehabilitation and spinal cord injury	[85]
4. Mental health and disorders		
	Intentional forgetting	[86]
	Gaze perception	[87]
	BDNF and major depression	[88]
	Autism and suicidal thoughts	[89]
	AI and mental illness	[90]

Abbreviations: AI: artificial intelligence, ALS: amyotrophic lateral sclerosis, BDNF: brain-derived neurotrophic factor, MS: multiple sclerosis, NLRP3: NOD-, LRR-, and pyrin domain-containing protein 3, PD: Parkinson's disease, UBL3: ubiquitin-like 3, VR: virtual reality.

Vertigo is a rare symptom in people who have recently suffered from a stroke, and there is currently a limited understanding of its significance [91]. d'Annunzio et al. make a notable contribution by showing that vertigo in patients with acute stroke can be used as an indicator of the stroke's location, specifically in the cerebellum and/or brainstem [70]. However, it does not have an impact on early outcomes or increase the risk of mortality during hospitalization. Cortical damage is commonly associated with cognitive dysfunction, while subcortical damage is an aspect of cognitive dysfunction that is frequently overlooked in research [92]. Shimmyo and Obayashi improve our understanding of cognitive deterioration following pontine stroke, a frequently overlooked condition due to the incorrect belief that subcortical damage is less likely to induce cognitive dysfunction [71]. The study employs two neuroimaging techniques to better understand the neurophysiology that underpins cognitive decline. The study results suggest that the degree of cognitive decline may be related to the responses observed in the supplementary motor area. This phenomenon may be attributed to the breakdown of hierarchical cognitive processing in the fronto–ponto–cerebellar–thalamic loop.

A growing body of evidence suggests that disorders of the central nervous system (CNS) can be linked to peripheral body regions [93,94]. Park et al. investigated the gut microbiota in individuals who have suffered from strokes, uncovering significant imbalances in both the taxonomic composition and functional characteristics of the microbiota when compared to a group of healthy individuals. Patients who have experienced a stroke exhibit changes in their gut microbiota, which may be a sign of malnutrition. Adjusting their diet could help restore a healthy balance of gut bacteria, leading to better outcomes and a decrease in disability and death rates in stroke patients.

Alcohol consumption is well known to affect the risk and prognosis of ischemic stroke [95,96]. Li et al. examined the effects of light alcohol consumption (LAC) on the growth of new neurons in the brain in the context of ischemic stroke [73]. The findings of their study indicate that LAC can considerably enhance neurogenesis in both normal conditions and after an ischemic stroke. This process has the potential to minimize brain damage and enhance locomotor activity, suggesting that LAC may have a protective effect against ischemic stroke. There is an increasing need for more objective outcome measures in cognitive rehabilitation (CR) for stroke patients [97–99]. Gangemi et al. contribute to the field of CR by showing that a virtual reality-based approach has the potential to effectively promote neuroplastic changes in patients with chronic ischemic stroke [74]. This is supported by significant improvements in electroencephalogram (EEG)-related neural activity and variations in power spectral density in the alpha and beta band powers (Table 1).

2.2. *Age and Behavioral Studies: Unraveling the Complexities of Lifespan Influence*

We gain more insight into the complex relationship between age and behavior as we investigate the various ways that aging affects brain networks and cognitive processes [100–102]. This section launches a journey through five illuminating studies, each shedding light on the dynamic relationship between age and behavioral outcomes. The section showcases various aspects of scientific research, including the vulnerability of mice to ketamine, the complex relationship between the dosage and effects of ketamine, the connection between melatonin and anxiety in C57/B6J mice, the role of platelet mitochondria, and the involvement of noradrenaline in regulating learned and innate behaviors in rats lacking the dopamine transporter. These articles invite us to reflect on the complex interplay of age, behavior, and the constantly changing brain.

Ketamine is frequently abused as a psychedelic substance [103–105]. Chen et al. examined the impact of ketamine on glutamatergic neurotransmission, which plays a vital role in memory retention, addiction, and psychosis [75]. The authors of the study investigate the varying sensitivity to ketamine in mice of different ages and strains. The results indicate that an individual's response to ketamine, as observed through their locomotor behavior, is determined by biological factors and can differ depending on dosage and age.

The production of melatonin decreases as one ages, and its effectiveness may vary depending on age [106–108]. The study conducted by Nasini et al. examines the impact of melatonin on anxiety-related behavior and the circuit connecting the medial prefrontal cortex and dorsal hippocampus in both adolescent and adult mice [76]. The results emphasize the variations in the effects of melatonin based on age, indicating that age can have a substantial influence on outcomes.

Mitochondrial dysfunction, characterized by a decline in mitochondrial respiratory function and an increase in reactive oxygen species production, is a key cellular hallmark of aging and neurodegenerative diseases [52,109,110]. Fišar et al. utilized platelets as a model to assess age-related mitochondrial parameters and the influence of cognitive impairment on these parameters [77]. The study shows age-dependent changes in mitochondrial function in platelets but no significant difference between individuals with and without cognitive impairment. Platelet mitochondrial respiration may serve as a promising biomarker for aging and a target for interventions aimed at combating aging and neurodegenerative processes.

Developing focused treatment approaches for attention-deficit hyperactivity disorder requires investigation of the underlying mechanisms involving dopamine dysregulation [111–113]. Volnova et al. examined the impacts of guanfacine, an α2A-adrenoceptor agonist, on the behavior and brain activity of dopamine transporter knockout rats [78]. Guanfacine has been shown to improve spatial working memory and pre-pulse inhibition in dopamine transporter knockout rats. This supports the role of noradrenergic modulation in attention regulation and suggests potential combined treatments to maintain dopamine–norepinephrine balance (Table 1).

2.3. *Neuropsychiatric Disorders and Treatments: Unraveling Pathways and Novel Approaches*

Neurodegenerative disorders pose a substantial and increasing public health issue, impacting a considerable population worldwide [114–116]. These conditions, including Alzheimer's disease, Parkinson's disease (PD), multiple sclerosis (MS), and amyotrophic lateral sclerosis (ALS), are defined by the gradual deterioration and depletion of nerve cells in the brain and spinal cord [117–120]. This section is focused on recent advances in understanding these diseases, their causes, and potential treatments including neuroprotection, highlighting innovative research that offers hope for new therapies [121]. The six articles we feature cover a broad spectrum of neurodegeneration, each focusing on a different aspect of these complex conditions.

The enteric nervous system (ENS) is intricately linked to the CNS and plays an important role in the pathophysiology of PD [122–124]. Montanari et al. discuss the early involvement of the ENS in PD pathogenesis, with α-synuclein (α-syn) aggregation occurring before CNS symptoms [79]. By proposing the ENS as a target for potential new PD therapies, this could provide insights into brain health and advance the development of novel therapeutic options.

Little is known about the interactions between ubiquitin-like 3 (UBL3) and alpha-synuclein (α-syn), and their modulation by drugs, which are relevant for understanding and treating α-synucleinopathies [125–127]. Chen et al. examined the interaction between UBL3 and α-syn in order to comprehend its function in α-synucleinopathies [80]. UBL3 interacts with α-syn, and this interaction is modulated by osimertinib, an inhibitor of the epidermal growth factor receptor pathway. This study advances the field by identifying the UBL3 pathway as a potential new therapeutic target for α-synucleinopathies.

Neuroinflammation is increasingly recognized as a significant factor in a variety of brain diseases, with microglia and monocytes playing an important role in the robust activation of the NOD-, LRR- and pyrin domain-containing protein 3 (NLRP3) inflammasome [128–130]. Chiarini et al. discuss the regulation of NLRP3 and its involvement in diverse neurological disorders [81]. The authors acknowledge the absence of proof regarding the impact of NLRP3 inhibition on human diseases and emphasize the possibility of other inflammasomes stepping in to fill the gap. They advocate for the use of human neural cell-based models to gain a deeper understanding of these diseases and develop more effective treatment strategies.

Anomalies in the tryptophan (Trp)-kynurenine (KYN) metabolic system have been detected in individuals with MS [131–133]. However, the specific profile of KYN metabolites in progressive MS is still uncertain. Polyák et al. examined KYN metabolite levels in a cuprizone-induced mouse model of demyelination [82]. The authors show significant reductions in specific KYN metabolites, suggesting that these metabolites are potential biomarkers for personalized MS treatment.

Sodium voltage-gated channel alpha subunit 1 (SCN1A) gene mutations cause cellular immaturity in neurons, resulting in delayed maturation and reduced excitability, both of which contribute to the development of febrile seizures [134–136]. Scalise et al. investigated the effect of SCN1A gene mutations from a well-characterized Italian family on neurons derived from induced pluripotent stem cell-derived neurons [83]. The mutations cause reduced excitability in neurons as well as intrinsic cellular immaturity. The authors provide strong evidence that SCN1A gene mutations play a role in the development of febrile seizures, highlighting the potential of diseased neurons for personalized therapy and ex vivo drug screening for human epileptic disorders.

The secretome of dental pulp stem cells (DPSCs) on motoneurons in ALS demonstrated neuroprotective effects; however, the mechanism of action remains unknown [137–139]. Younes et al. investigated the effects of the DPSC secretome on the survival, axonal length, and electrical activity of cultured wild-type and superoxide dismutase 1 (SOD1) G93A motoneurons, as well as the roles of two DPSC-secreted factors, growth/differentiation factor 15 (GDF15) and heparin-binding epidermal growth factor-like growth factor (HB-EGF) [84]. The secretome of DPSCs has neuroprotective effects on motoneurons and could be a therapeutic candidate for ALS, highlighting the roles of GDF15 and HB-EGF, two DPSC-secreted factors that protect motor neurons from nitric oxide-induced death.

Individuals with spinal cord injury (SCI) experience rapid and debilitating muscle and bone loss, necessitating the development of effective bone mass preservation and maintenance strategies to reduce the risk of fragility and fracture in these vulnerable populations [85,140–142]. Leone et al. investigated the pathophysiology and risk factors of muscle and bone loss after SCI, the mechanisms that contribute to this loss, and current and future pharmacological and non-pharmacological therapies for reducing or eliminating neurogenic bone loss after SCI [85]. Pharmacological and non-pharmacological treatments can lessen or completely prevent neurogenic bone loss following SCI. Additionally, people with SCI have more rapid and severe bone and muscle loss because of a number of different factors (Table 1).

2.4. Mental Health and Disorders: Breaking Barriers and Unveiling Secrets

Mental health is a fundamental component of our general state of being, impacting our cognitive processes, emotions, and social interactions [143–145]. This section explores a wide range of research articles that provide insights into different aspects of mental health and disorders. These studies provide valuable insights into the relationship between mental illness and pain, as well as the biochemical basis of suicidal thoughts. Let us delve into the complex neural pathways, biological indicators, and psychological phenomena that influence our comprehension of mental health [146].

Despite the importance of intentional forgetting (IF) in daily performance, psychological well-being, and memory functioning, the neuropsychological mechanisms underlying successful IF are unknown [147–149]. Gamboa et al. investigated the neural correlates of IF using two meta-analytic algorithms, activation likelihood estimation, and latent Dirichlet allocation, and evaluated the proposed neurobiological models' compatibility with existing brain imaging data [86]. IF involves the interaction of two networks: a primarily right-lateralized frontal–parietal circuit and a less constrained supportive network that includes frontal–hippocampal interactions. In support of the inhibitory or thought suppression hypothesis, the study also discovered a neural signature of IF that is consistent across various experimental paradigms and may open new avenues for developing effective clinical interventions.

Gaze cueing plays an important role in the reflexive orientation of attention and its susceptibility to context [149–151]. However, the distinct functional roles of the amygdala and the superior temporal lobe, particularly the superior temporal sulcus (STS), in gaze processing remain unknown, as does the interaction of contextual factors with gaze-cueing [152–154]. Battaglia et al. investigated the neural bases of gaze cueing and gaze direction perception, how contextual factors interact with the gaze shift of attention, and the distinct functional roles of the amygdala and STS in gaze perception [87]. The amygdala and the STS are important components in gaze perception, and gaze-cueing is influenced by a variety of context-specific factors. The idea of invariant representation is a useful framework for further research, highlighting the disparities in attempts to characterize the distinct functional roles of these regions in the processing of gaze. Th authors emphasize the role of the amygdala and the STS in gaze perception and introduce the concept of invariant representation as a valuable conceptual framework for future research on the perceptual processing of gaze within the STS.

The differences in serum brain-derived neurotrophic factor (BDNF) levels during pharmacotherapy in major depressive disorder (MDD) patients, particularly between treatment-response and treatment-nonresponse groups, remain unclear [155–157]. Yoshimura et al. studied changes in serum BDNF concentrations in first-episode, drug-naive MDD patients during antidepressant treatment and compared them to treatment-response and treatment-nonresponsive groups [88]. In first-episode, drug-naive MDD patients, serum BDNF levels did not differ significantly between treatment-response and treatment-nonresponse groups. However, the responder group showed statistically significant changes in serum BDNF, implying that the changes in serum BDNF may differ between the two groups and that measuring serum BDNF has the potential to be a useful predictor of pharmacotherapy in these patients. The authors demonstrate that serum BDNF measurement has the potential to be a useful predictor of pharmacotherapy in first-episode, drug-naive MDD patients.

Despite the emphasis on neurobiological underpinnings and the poor predictive accuracy of many sociodemographic risk factors and prognostic markers, understanding and predicting suicide remain significant challenges [158–160]. Cremone et al. examined the relationship between blood levels of serotonin, BDNF, Trp and its metabolites, interleukin-6 (IL-6), and homocysteine levels and suicidality in adults with autism and explored how these biochemical parameters may be linked to an elevated risk of suicide [89]. There is a link between suicidality and autism, and suicidality is associated with elevated homocysteine and IL-6 levels, as well as decreased Trp and KYNA levels. The authors show a possible transnosographic link between these biochemical parameters and increased suicide risk, which potentially improves our understanding and prediction of suicide.

Despite the known association between psychological events and pain intensity, there is no comprehensive mathematical model that accurately captures the multidimensional nature of pain, particularly low back pain, and its relationship with psychological factors [161–163]. Parolini et al. investigated the development of a mathematical representation of the International Association for the Study of Pain (IASP) pain model, using an artificial neural network to identify patterns in the relationship between various variables related to low back pain, as well as how these patterns differ between groups with altered patterns in the context of low back pain [90]. The authors show a direct correlation between psychological and pain events in the context of low back pain, suggesting that mental illness can exacerbate pain episodes and impact functionality. They also found that the developed artificial neural network model was able to identify patterns and relationships between variables and differentiate groups with altered patterns (Table 1).

3. Conclusions

This 10th Anniversary Special Issue of *Biomedicines* has thoroughly examined the field of translational laboratory and experimental medicine in relation to neurological diseases and mental illnesses. Within this compilation, scholars have extensively examined the intricate mechanisms that form the basis of these conditions, offering novel perspectives on possible therapeutic strategies and interventions. In addition to improving our understanding of the human mind, this Special Issue has facilitated groundbreaking advances in the diagnosis, treatment, and prevention of neurological and psychiatric disorders, such as the use of neuromodulation techniques. These techniques have shown promise in the treatment of various neurological and neuropsychiatric disorders, such as depression, anxiety, PD, and chronic pain [21,164–167]. The Special Issue on "Translational Laboratory and Experimental Medicine for Neurological Diseases and Mental Illnesses" is a testament to our unwavering dedication and innovation in the field of biomedicine as we celebrate a decade of pioneering exploration. The 21 papers presented herein demonstrate the diligent work of researchers and clinicians in understanding the intricacies of brain function and mental health. Their research provides encouraging perspectives on innovative therapeutic approaches and possible advancements. Our future goal is to connect the work carried out in laboratories with real-world applications, with the common objective of improving the lives of people affected by neurological disorders and mental illnesses.

Author Contributions: Conceptualization, M.T.; writing—original draft preparation, M.T.; writing—review and editing, M.T. and L.V. visualization, M.T.; supervision, M.T. and L.V.; project administration, M.T.; funding acquisition, M.T. and L.V. All authors have read and agreed to the published version of the manuscript.

Funding: This work was supported by the National Research, Development, and Innovation Office—NKFIH K138125, SZTE SZAOK-KKA No: 2022/5S729, and the HUN-REN Hungarian Research Network.

Institutional Review Board Statement: Not applicable.

Informed Consent Statement: Not applicable.

Data Availability Statement: Data sharing is not applicable to this article.

Conflicts of Interest: The authors declare no conflicts of interest.

Abbreviations

ALS	Amyotrophic lateral sclerosis
α-syn	Alpha-synuclein
BDNF	Brain-derived neurotrophic factor
CNS	Central nervous system
CR	Cognitive rehabilitation
DPSCs	Pulp stem cells
ENS	Enteric nervous system
GDF15	Growth/differentiation factor 15
HB-EGF	Heparin-binding epidermal growth factor-like growth factor
IF	Intentional forgetting
IL-6	Interleukin-6
KYN	Kynurenine
LAC	Light alcohol consumption
MDD	Major depressive disorder
MS	Multiple sclerosis
NLRP3	NOD-, LRR-, and pyrin domain-containing protein 3
PD	Parkinson's disease
SCN1A	Sodium voltage-gated channel alpha subunit 1
SOD1	Superoxide dismutase 1
SCI	Spinal cord injury
STS	Superior temporal sulcus
Trp	Tryptophan
UBL3	Ubiquitin-like 3

References

1. Campos, B.; Choi, H.; DeMarco, A.T.; Seydell-Greenwald, A.; Hussain, S.J.; Joy, M.T.; Turkeltaub, P.E.; Zeiger, W. Rethinking Remapping: Circuit Mechanisms of Recovery after Stroke. *J. Neurosci.* **2023**, *43*, 7489–7500. [CrossRef] [PubMed]
2. Gregorio, F.; Battaglia, S. The intricate brain-body interaction in psychiatric and neurological diseases. *Adv. Clin. Exp. Med.* **2024**, *33*, 321–326. [CrossRef] [PubMed]
3. Brewster, K.K.; Golub, J.S.; Rutherford, B.R. Neural circuits and behavioral pathways linking hearing loss to affective dysregulation in older adults. *Nat. Aging* **2021**, *1*, 422–429. [CrossRef] [PubMed]
4. Battaglia, S.; Avenanti, A.; Vécsei, L.; Tanaka, M. Neurodegeneration in Cognitive Impairment and Mood Disorders for Experimental, Clinical and Translational Neuropsychiatry. *Biomedicines* **2024**, *12*, 574. [CrossRef] [PubMed]
5. Luo, L. Architectures of neuronal circuits. *Science* **2021**, *373*, eabg7285. [CrossRef] [PubMed]
6. Battaglia, S.; Garofalo, S.; di Pellegrino, G. Context-dependent extinction of threat memories: Influences of healthy aging. *Sci. Rep.* **2018**, *8*, 12592. [CrossRef] [PubMed]
7. Andrade-Talavera, Y.; Fisahn, A.; Rodríguez-Moreno, A. Timing to be precise? An overview of spike timing-dependent plasticity, brain rhythmicity, and glial cells interplay within neuronal circuits. *Mol. Psychiatry* **2023**, *28*, 2177–2188. [CrossRef] [PubMed]
8. Tortora, F.; Hadipour, A.L.; Battaglia, S.; Falzone, A.; Avenanti, A.; Vicario, C.M. The role of serotonin in fear learning and memory: A systematic review of human studies. *Brain Sci.* **2023**, *13*, 1197. [CrossRef] [PubMed]
9. Cabrera, Y.; Koymans, K.J.; Poe, G.R.; Kessels, H.W.; Van Someren, E.J.W.; Wassing, R. Overnight neuronal plasticity and adaptation to emotional distress. *Nat. Rev. Neurosci.* **2024**, *25*, 253–271. [CrossRef]
10. Battaglia, S.; Avenanti, A.; Vécsei, L.; Tanaka, M. Neural Correlates and Molecular Mechanisms of Memory and Learning. *Int. J. Mol. Sci.* **2024**, *25*, 2724. [CrossRef]
11. Lawal, O.; Ulloa Severino, F.P.; Eroglu, C. The role of astrocyte structural plasticity in regulating neural circuit function and behavior. *Glia* **2022**, *70*, 1467–1483. [CrossRef] [PubMed]
12. Herzberg, M.P.; Gunnar, M.R. Early life stress and brain function: Activity and connectivity associated with processing emotion and reward. *Neuroimage* **2020**, *209*, 116493. [CrossRef] [PubMed]
13. Battaglia, S.; Di Fazio, C.; Mazzà, M.; Tamietto, M.; Avenanti, A. Targeting Human Glucocorticoid Receptors in Fear Learning: A Multiscale Integrated Approach to Study Functional Connectivity. *Int. J. Mol. Sci.* **2024**, *25*, 864. [CrossRef] [PubMed]
14. Di Gregorio, F.; Steinhauser, M.; Maier, M.E.; Thayer, J.F.; Battaglia, S. Error-related cardiac deceleration: Functional interplay between error-related brain activity and autonomic nervous system in performance monitoring. *Neurosci. Biobehav. Rev.* **2024**, *157*, 105542. [CrossRef] [PubMed]
15. Battaglia, S.; Nazzi, C.; Thayer, J.F. Genetic differences associated with dopamine and serotonin release mediate fear-induced bradycardia in the human brain. *Transl. Psychiatry* **2024**, *14*, 24. [CrossRef]
16. Tanaka, M.; Chen, C. Towards a mechanistic understanding of depression, anxiety, and their comorbidity: Perspectives from cognitive neuroscience. *Front. Behav. Neurosci.* **2023**, *17*, 1268156. [CrossRef]
17. Battaglia, S.; Schmidt, A.; Hassel, S.; Tanaka, M. Case reports in neuroimaging and stimulation. *Front. Psychiatry* **2023**, *14*, 1264669. [CrossRef] [PubMed]
18. Báez-Mendoza, R.; Vázquez, Y.; Mastrobattista, E.P.; Williams, Z.M. Neuronal Circuits for Social Decision-Making and Their Clinical Implications. *Front. Neurosci.* **2021**, *15*, 720294. [CrossRef] [PubMed]
19. Di Gregorio, F.; Battaglia, S. Advances in EEG-based functional connectivity approaches to the study of the central nervous system in health and disease. *Adv. Clin. Exp. Med.* **2023**, *32*, 607–612. [CrossRef]
20. Duerler, P.; Vollenweider, F.X.; Preller, K.H. A neurobiological perspective on social influence: Serotonin and social adaptation. *J. Neurochem.* **2022**, *162*, 60–79. [CrossRef]
21. Battaglia, S.; Nazzi, C.; Thayer, J. Heart's tale of trauma: Fear-conditioned heart rate changes in post-traumatic stress disorder. *Acta Psychiatr. Scand.* **2023**, *148*, 463–466. [CrossRef] [PubMed]
22. Valotto Neto, L.J.; Reverete de Araujo, M.; Moretti Junior, R.C.; Mendes Machado, N.; Joshi, R.K.; dos Santos Buglio, D.; Barbalho Lamas, C.; Direito, R.; Fornari Laurindo, L.; Tanaka, M. Investigating the Neuroprotective and Cognitive-Enhancing Effects of Bacopa monnieri: A Systematic Review Focused on Inflammation, Oxidative Stress, Mitochondrial Dysfunction, and Apoptosis. *Antioxidants* **2024**, *13*, 393. [CrossRef] [PubMed]
23. Martos, D.; Lőrinczi, B.; Szatmári, I.; Vécsei, L.; Tanaka, M. The Impact of C-3 Side Chain Modifications on Kynurenic Acid: A Behavioral Analysis of Its Analogs on Motor Domain. *Int. J. Mol. Sci.* **2024**, *25*, 3394. [CrossRef] [PubMed]
24. Jászberényi, M.; Thurzó, B.; Bagosi, Z.; Vécsei, L.; Tanaka, M. The Orexin/Hypocretin System, the Peptidergic Regulator of Vigilance, Orchestrates Adaptation to Stress. *Biomedicines* **2024**, *12*, 448. [CrossRef]
25. Tanaka, M.; Szabó, Á.; Körtési, T.; Szok, D.; Tajti, J.; Vécsei, L. From CGRP to PACAP, VIP, and Beyond: Unraveling the Next Chapters in Migraine Treatment. *Cells* **2023**, *12*, 2649. [CrossRef] [PubMed]
26. Tajti, J.; Szok, D.; Csáti, A.; Szabó, Á.; Tanaka, M.; Vécsei, L. Exploring novel therapeutic targets in the common pathogenic factors in migraine and neuropathic pain. *Int. J. Mol. Sci.* **2023**, *24*, 4114. [CrossRef] [PubMed]
27. Bássoli, R.; Audi, D.; Ramalho, B.; Audi, M.; Quesada, K.; Barbalho, S. The Effects of Curcumin on Neurodegenerative Diseases: A Systematic Review. *J. Herb. Med.* **2023**, *42*, 100771. [CrossRef]

28. Buglio, D.S.; Marton, L.T.; Laurindo, L.F.; Guiguer, E.L.; Araújo, A.C.; Buchaim, R.L.; Goulart, R.d.A.; Rubira, C.J.; Barbalho, S.M. The role of resveratrol in mild cognitive impairment and Alzheimer's disease: A systematic review. *J. Med. Food* **2022**, *25*, 797–806. [CrossRef]
29. Fraile-Ramos, J.; Garrit, A.; Reig-Vilallonga, J.; Giménez-Llort, L. Hepatic Oxi-Inflammation and Neophobia as Potential Liver–Brain Axis Targets for Alzheimer's Disease and Aging, with Strong Sensitivity to Sex, Isolation, and Obesity. *Cells* **2023**, *12*, 1517. [CrossRef]
30. Chen, J.; Huang, L.; Yang, Y.; Xu, W.; Qin, Q.; Qin, R.; Liang, X.; Lai, X.; Huang, X.; Xie, M. Somatic Cell Reprogramming for Nervous System Diseases: Techniques, Mechanisms, Potential Applications, and Challenges. *Brain Sci.* **2023**, *13*, 524. [CrossRef]
31. Skobeleva, K.; Shalygin, A.; Mikhaylova, E.; Guzhova, I.; Ryazantseva, M.; Kaznacheyeva, E. The STIM1/2-regulated calcium homeostasis is impaired in hippocampal neurons of the 5xFAD mouse model of Alzheimer's disease. *Int. J. Mol. Sci.* **2022**, *23*, 14810. [CrossRef]
32. Hong, F.; He, G.; Zhang, M.; Yu, B.; Chai, C. The establishment of a mouse model of recurrent primary dysmenorrhea. *Int. J. Mol. Sci.* **2022**, *23*, 6128. [CrossRef] [PubMed]
33. Garifulin, R.; Davleeva, M.; Izmailov, A.; Fadeev, F.; Markosyan, V.; Shevchenko, R.; Minyazeva, I.; Minekayev, T.; Lavrov, I.; Islamov, R. Evaluation of the autologous genetically enriched leucoconcentrate on the lumbar spinal cord morpho-functional recovery in a mini pig with thoracic spine contusion injury. *Biomedicines* **2023**, *11*, 1331. [CrossRef] [PubMed]
34. Bueno, C.R.d.S.; Tonin, M.C.C.; Buchaim, D.V.; Barraviera, B.; Junior, R.S.F.; Santos, P.S.d.S.; Reis, C.H.B.; Pastori, C.M.; Pereira, E.d.S.B.M.; Nogueira, D.M.B. Morphofunctional improvement of the facial nerve and muscles with repair using heterologous fibrin biopolymer and photobiomodulation. *Pharmaceuticals* **2023**, *16*, 653. [CrossRef]
35. Kalkman, H.O. Inhibition of microglial GSK3β activity is common to different kinds of antidepressants: A proposal for an in vitro screen to detect novel antidepressant principles. *Biomedicines* **2023**, *11*, 806. [CrossRef] [PubMed]
36. Zheng, Y.; Huo, J.; Yang, M.; Zhang, G.; Wan, S.; Chen, X.; Zhang, B.; Liu, H. ERK1/2 Signalling Pathway Regulates Tubulin-Binding Cofactor B Expression and Affects Astrocyte Process Formation after Acute Foetal Alcohol Exposure. *Brain Sci.* **2022**, *12*, 813. [CrossRef]
37. Li, T.; Xu, G.; Yi, J.; Huang, Y. Intraoperative Hypothermia Induces Vascular Dysfunction in the CA1 Region of Rat Hippocampus. *Brain Sci.* **2022**, *12*, 692. [CrossRef]
38. Martos, D.; Tuka, B.; Tanaka, M.; Vécsei, L.; Telegdy, G. Memory enhancement with kynurenic acid and its mechanisms in neurotransmission. *Biomedicines* **2022**, *10*, 849. [CrossRef]
39. Sivananthan, S.; Lee, L.; Anderson, G.; Csanyi, B.; Williams, R.; Gissen, P. Buffy coat score as a biomarker of treatment response in neuronal ceroid lipofuscinosis type 2. *Brain Sci.* **2023**, *13*, 209. [CrossRef]
40. Clement, A.; Wiborg, O.; Asuni, A.A. Steps towards developing effective treatments for neuropsychiatric disturbances in Alzheimer's disease: Insights from preclinical models, clinical data, and future directions. *Front. Aging Neurosci.* **2020**, *12*, 56. [CrossRef]
41. Socała, K.; Żmudzka, E.; Lustyk, K.; Zagaja, M.; Brighenti, V.; Costa, A.M.; Andres-Mach, M.; Pytka, K.; Martinelli, I.; Mandrioli, J. Therapeutic potential of stilbenes in neuropsychiatric and neurological disorders: A comprehensive review of preclinical and clinical evidence. *Phytother. Res.* **2024**, *38*, 1400–1461. [CrossRef]
42. Statsenko, Y.; Habuza, T.; Smetanina, D.; Simiyu, G.L.; Meribout, S.; King, F.C.; Gelovani, J.G.; Das, K.M.; Gorkom, K.N.-V.; Zaręba, K. Unraveling lifelong brain morphometric dynamics: A protocol for systematic review and meta-analysis in healthy neurodevelopment and ageing. *Biomedicines* **2023**, *11*, 1999. [CrossRef] [PubMed]
43. Fan, P.; Miranda, O.; Qi, X.; Kofler, J.; Sweet, R.A.; Wang, L. Unveiling the enigma: Exploring risk factors and mechanisms for psychotic symptoms in Alzheimer's disease through electronic medical records with deep learning models. *Pharmaceuticals* **2023**, *16*, 911. [CrossRef]
44. Di Gregorio, F.; La Porta, F.; Petrone, V.; Battaglia, S.; Orlandi, S.; Ippolito, G.; Romei, V.; Piperno, R.; Lullini, G. Accuracy of EEG biomarkers in the detection of clinical outcome in disorders of consciousness after severe acquired brain injury: Preliminary results of a pilot study using a machine learning approach. *Biomedicines* **2022**, *10*, 1897. [CrossRef]
45. Nani, A.; Manuello, J.; Mancuso, L.; Liloia, D.; Costa, T.; Vercelli, A.; Duca, S.; Cauda, F. The pathoconnectivity network analysis of the insular cortex: A morphometric fingerprinting. *NeuroImage* **2021**, *225*, 117481. [CrossRef]
46. Cauda, F.; Nani, A.; Liloia, D.; Manuello, J.; Premi, E.; Duca, S.; Fox, P.T.; Costa, T. Finding specificity in structural brain alterations through Bayesian reverse inference. *Hum. Brain Mapp.* **2020**, *41*, 4155–4172. [CrossRef] [PubMed]
47. Liloia, D.; Cauda, F.; Uddin, L.Q.; Manuello, J.; Mancuso, L.; Keller, R.; Nani, A.; Costa, T. Revealing the selectivity of neuroanatomical alteration in autism spectrum disorder via reverse inference. *Biol. Psychiatry Cogn. Neurosci. Neuroimaging* **2023**, *8*, 1075–1083. [CrossRef]
48. Liloia, D.; Crocetta, A.; Cauda, F.; Duca, S.; Costa, T.; Manuello, J. Seeking overlapping neuroanatomical alterations between dyslexia and attention-deficit/hyperactivity disorder: A meta-analytic replication study. *Brain Sci.* **2022**, *12*, 1367. [CrossRef] [PubMed]
49. Ippolito, G.; Bertaccini, R.; Tarasi, L.; Di Gregorio, F.; Trajkovic, J.; Battaglia, S.; Romei, V. The role of alpha oscillations among the main neuropsychiatric disorders in the adult and developing human brain: Evidence from the last 10 years of research. *Biomedicines* **2022**, *10*, 3189. [CrossRef]

50. Zhao, L.; Hou, B.; Ji, L.; Ren, D.; Yuan, F.; Liu, L.; Bi, Y.; Yang, F.; Yu, S.; Yi, Z. NGFR gene and single nucleotide polymorphisms, rs2072446 and rs11466162, playing roles in psychiatric disorders. *Brain Sci.* **2022**, *12*, 1372. [CrossRef]
51. Khan, S.R.; Al Rijjal, D.; Piro, A.; Wheeler, M.B. Integration of AI and traditional medicine in drug discovery. *Drug Discov. Today* **2021**, *26*, 982–992. [CrossRef] [PubMed]
52. Barbalho, S.M.; Direito, R.; Laurindo, L.F.; Marton, L.T.; Guiguer, E.L.; Goulart, R.d.A.; Tofano, R.J.; Carvalho, A.C.; Flato, U.A.P.; Capelluppi Tofano, V.A. Ginkgo biloba in the aging process: A narrative review. *Antioxidants* **2022**, *11*, 525. [CrossRef] [PubMed]
53. Senevirathne, D.K.L.; Mahboob, A.; Zhai, K.; Paul, P.; Kammen, A.; Lee, D.J.; Yousef, M.S.; Chaari, A. Deep brain stimulation beyond the clinic: Navigating the future of Parkinson's and Alzheimer's disease therapy. *Cells* **2023**, *12*, 1478. [CrossRef] [PubMed]
54. Vasiliu, O. Efficacy, tolerability, and safety of toludesvenlafaxine for the treatment of major depressive disorder—A narrative review. *Pharmaceuticals* **2023**, *16*, 411. [CrossRef] [PubMed]
55. Chu, P.-C.; Huang, C.-S.; Chang, P.-K.; Chen, R.-S.; Chen, K.-T.; Hsieh, T.-H.; Liu, H.-L. Weak ultrasound contributes to neuromodulatory effects in the rat motor cortex. *Int. J. Mol. Sci.* **2023**, *24*, 2578. [CrossRef] [PubMed]
56. Chojnowski, K.; Opiełka, M.; Gozdalski, J.; Radziwon, J.; Dańczyszyn, A.; Aitken, A.V.; Biancardi, V.C.; Winklewski, P.J. The role of arginine-vasopressin in stroke and the potential use of arginine-vasopressin type 1 receptor antagonists in stroke therapy: A narrative review. *Int. J. Mol. Sci.* **2023**, *24*, 2119. [CrossRef] [PubMed]
57. Adeel, M.; Chen, C.-C.; Lin, B.-S.; Chen, H.-C.; Liou, J.-C.; Li, Y.-T.; Peng, C.-W. Safety of Special Waveform of Transcranial Electrical Stimulation (TES): In Vivo Assessment. *Int. J. Mol. Sci.* **2022**, *23*, 6850. [CrossRef] [PubMed]
58. Tanaka, M.; Vécsei, L. From Lab to Life: Exploring Cutting-Edge Models for Neurological and Psychiatric Disorders. *Biomedicines* **2024**, *12*, 613. [CrossRef] [PubMed]
59. Tanaka, M.; Szabó, Á.; Vécsei, L.; Giménez-Llort, L. Emerging translational research in neurological and psychiatric diseases: From in vitro to in vivo models. *Int. J. Mol. Sci.* **2023**, *24*, 15739. [CrossRef]
60. Tanaka, M.; Szabó, Á.; Vécsei, L. Preclinical modeling in depression and anxiety: Current challenges and future research directions. *Adv. Clin. Exp. Med.* **2023**, *32*, 505–509. [CrossRef]
61. Cabral, D.F.; Fried, P.; Koch, S.; Rice, J.; Rundek, T.; Pascual-Leone, A.; Sacco, R.; Wright, C.B.; Gomes-Osman, J. Efficacy of mechanisms of neuroplasticity after a stroke. *Restor. Neurol. Neurosci.* **2022**, *40*, 73–84. [CrossRef] [PubMed]
62. Aderinto, N.; AbdulBasit, M.O.; Olatunji, G.; Adejumo, T. Exploring the transformative influence of neuroplasticity on stroke rehabilitation: A narrative review of current evidence. *Ann. Med. Surg.* **2023**, *85*, 4425–4432. [CrossRef] [PubMed]
63. Motolese, F.; Capone, F.; Di Lazzaro, V. New tools for shaping plasticity to enhance recovery after stroke. *Handb. Clin. Neurol.* **2022**, *184*, 299–315. [PubMed]
64. Lim, J.-S.; Lee, J.-J.; Woo, C.-W. Post-stroke cognitive impairment: Pathophysiological insights into brain disconnectome from advanced neuroimaging analysis techniques. *J. Stroke* **2021**, *23*, 297. [CrossRef] [PubMed]
65. Griffis, J.C.; Metcalf, N.V.; Corbetta, M.; Shulman, G.L. Damage to the shortest structural paths between brain regions is associated with disruptions of resting-state functional connectivity after stroke. *NeuroImage* **2020**, *210*, 116589. [CrossRef] [PubMed]
66. Rost, N.S.; Brodtmann, A.; Pase, M.P.; van Veluw, S.J.; Biffi, A.; Duering, M.; Hinman, J.D.; Dichgans, M. Post-Stroke Cognitive Impairment and Dementia. *Circ. Res.* **2022**, *130*, 1252–1271. [CrossRef]
67. Zotey, V.; Andhale, A.; Shegekar, T.; Juganavar, A. Adaptive Neuroplasticity in Brain Injury Recovery: Strategies and Insights. *Cureus* **2023**, *15*, e45873. [CrossRef] [PubMed]
68. De Luca, C.; Virtuoso, A.; Maggio, N.; Izzo, S.; Papa, M.; Colangelo, A.M. Roadmap for Stroke: Challenging the Role of the Neuronal Extracellular Matrix. *Int. J. Mol. Sci.* **2020**, *21*, 7554. [CrossRef] [PubMed]
69. Saceleanu, V.M.; Toader, C.; Ples, H.; Covache-Busuioc, R.A.; Costin, H.P.; Bratu, B.G.; Dumitrascu, D.I.; Bordeianu, A.; Corlatescu, A.D.; Ciurea, A.V. Integrative Approaches in Acute Ischemic Stroke: From Symptom Recognition to Future Innovations. *Biomedicines* **2023**, *11*, 2617. [CrossRef]
70. d'Annunzio, A.; Arboix, A.; García-Eroles, L.; Sánchez-López, M.-J. Vertigo in acute stroke is a predictor of brain location but is not related to early outcome: The experience of Sagrat Cor Hospital of Barcelona Stroke Registry. *Biomedicines* **2022**, *10*, 2830. [CrossRef]
71. Shimmyo, K.; Obayashi, S. Fronto–Cerebellar Diaschisis and Cognitive Dysfunction after Pontine Stroke: A Case Series and Systematic Review. *Biomedicines* **2024**, *12*, 623. [CrossRef] [PubMed]
72. Park, S.Y.; Lee, S.P.; Kim, D.; Kim, W.J. Gut Dysbiosis: A New Avenue for Stroke Prevention and Therapeutics. *Biomedicines* **2023**, *11*, 2352. [CrossRef]
73. Li, J.; Li, C.; Subedi, P.; Tian, X.; Lu, X.; Miriyala, S.; Panchatcharam, M.; Sun, H. Light Alcohol Consumption Promotes Early Neurogenesis Following Ischemic Stroke in Adult C57BL/6J Mice. *Biomedicines* **2023**, *11*, 1074. [CrossRef]
74. Gangemi, A.; De Luca, R.; Fabio, R.A.; Lauria, P.; Rifici, C.; Pollicino, P.; Marra, A.; Olivo, A.; Quartarone, A.; Calabrò, R.S. Effects of Virtual Reality Cognitive Training on Neuroplasticity: A Quasi-Randomized Clinical Trial in Patients with Stroke. *Biomedicines* **2023**, *11*, 3225. [CrossRef] [PubMed]
75. Chen, W.-C.; Wang, T.-S.; Chang, F.-Y.; Chen, P.-A.; Chen, Y.-C. Age, Dose, and Locomotion: Decoding Vulnerability to Ketamine in C57BL/6J and BALB/c Mice. *Biomedicines* **2023**, *11*, 1821. [CrossRef]
76. Nasini, S.; Tidei, S.; Shkodra, A.; De Gregorio, D.; Cambiaghi, M.; Comai, S. Age-Related Effects of Exogenous Melatonin on Anxiety-like Behavior in C57/B6J Mice. *Biomedicines* **2023**, *11*, 1705. [CrossRef] [PubMed]

77. Fišar, Z.; Hroudová, J.; Zvěřová, M.; Jirák, R.; Raboch, J.; Kitzlerová, E. Age-dependent alterations in platelet mitochondrial respiration. *Biomedicines* **2023**, *11*, 1564. [CrossRef]
78. Volnova, A.; Kurzina, N.; Belskaya, A.; Gromova, A.; Pelevin, A.; Ptukha, M.; Fesenko, Z.; Ignashchenkova, A.; Gainetdinov, R.R. Noradrenergic modulation of learned and innate behaviors in dopamine transporter knockout rats by guanfacine. *Biomedicines* **2023**, *11*, 222. [CrossRef]
79. Montanari, M.; Imbriani, P.; Bonsi, P.; Martella, G.; Peppe, A. Beyond the microbiota: Understanding the role of the enteric nervous system in Parkinson's disease from mice to human. *Biomedicines* **2023**, *11*, 1560. [CrossRef]
80. Chen, B.; Hasan, M.M.; Zhang, H.; Zhai, Q.; Waliullah, A.; Ping, Y.; Zhang, C.; Oyama, S.; Mimi, M.A.; Tomochika, Y. UBL3 Interacts with Alpha-synuclein in Cells and the Interaction is Downregulated by the EGFR Pathway Inhibitor Osimertinib. *Biomedicines* **2023**, *11*, 1685. [CrossRef]
81. Chiarini, A.; Gui, L.; Viviani, C.; Armato, U.; Dal Prà, I. NLRP3 Inflammasome's Activation in Acute and Chronic Brain Diseases—An Update on Pathogenetic Mechanisms and Therapeutic Perspectives with Respect to Other Inflammasomes. *Biomedicines* **2023**, *11*, 999. [CrossRef] [PubMed]
82. Polyák, H.; Galla, Z.; Nánási, N.; Cseh, E.K.; Rajda, C.; Veres, G.; Spekker, E.; Szabó, Á.; Klivényi, P.; Tanaka, M. The tryptophan-kynurenine metabolic system is suppressed in cuprizone-induced model of demyelination simulating progressive multiple sclerosis. *Biomedicines* **2023**, *11*, 945. [CrossRef] [PubMed]
83. Scalise, S.; Zannino, C.; Lucchino, V.; Lo Conte, M.; Scaramuzzino, L.; Cifelli, P.; D'Andrea, T.; Martinello, K.; Fucile, S.; Palma, E. Human iPSC modeling of genetic febrile seizure reveals aberrant molecular and physiological features underlying an impaired neuronal activity. *Biomedicines* **2022**, *10*, 1075. [CrossRef] [PubMed]
84. Younes, R.; Issa, Y.; Jdaa, N.; Chouaib, B.; Brugioti, V.; Challuau, D.; Raoul, C.; Scamps, F.; Cuisinier, F.; Hilaire, C. The Secretome of Human Dental Pulp Stem Cells and Its Components GDF15 and HB-EGF Protect Amyotrophic Lateral Sclerosis Motoneurons against Death. *Biomedicines* **2023**, *11*, 2152. [CrossRef]
85. Leone, G.E.; Shields, D.C.; Haque, A.; Banik, N.L. Rehabilitation: Neurogenic Bone Loss after Spinal Cord Injury. *Biomedicines* **2023**, *11*, 2581. [CrossRef] [PubMed]
86. Gamboa, O.L.; Chuan-Peng, H.; Salas, C.E.; Yuen, K.S. Obliviate! Reviewing Neural Fundamentals of Intentional Forgetting from a Meta-Analytic Perspective. *Biomedicines* **2022**, *10*, 1555. [CrossRef] [PubMed]
87. Battaglia, S.; Fabius, J.H.; Moravkova, K.; Fracasso, A.; Borgomaneri, S. The neurobiological correlates of gaze perception in healthy individuals and neurologic patients. *Biomedicines* **2022**, *10*, 627. [CrossRef] [PubMed]
88. Yoshimura, R.; Okamoto, N.; Chibaatar, E.; Natsuyama, T.; Ikenouchi, A. The serum brain-derived neurotrophic factor increases in serotonin reuptake inhibitor responders patients with first-episode, drug-naïve major depression. *Biomedicines* **2023**, *11*, 584. [CrossRef] [PubMed]
89. Cremone, I.M.; Nardi, B.; Amatori, G.; Palego, L.; Baroni, D.; Casagrande, D.; Massimetti, E.; Betti, L.; Giannaccini, G.; Dell'Osso, L. Unlocking the secrets: Exploring the biochemical correlates of suicidal thoughts and behaviors in adults with autism spectrum conditions. *Biomedicines* **2023**, *11*, 1600. [CrossRef]
90. Parolini, F.; Goethel, M.; Becker, K.; Fernandes, C.; Fernandes, R.J.; Ervilha, U.F.; Santos, R.; Vilas-Boas, J.P. Breaking Barriers: Artificial Intelligence Interpreting the Interplay between Mental Illness and Pain as Defined by the International Association for the Study of Pain. *Biomedicines* **2023**, *11*, 2042. [CrossRef]
91. Man Chan, Y.; Wong, Y.; Khalid, N.; Wastling, S.; Flores-Martin, A.; Frank, L.A.; Koohi, N.; Arshad, Q.; Davagnanam, I.; Kaski, D. Prevalence of acute dizziness and vertigo in cortical stroke. *Eur. J. Neurol.* **2021**, *28*, 3177–3181. [CrossRef] [PubMed]
92. Janacsek, K.; Evans, T.M.; Kiss, M.; Shah, L.; Blumenfeld, H.; Ullman, M.T. Subcortical Cognition: The Fruit Below the Rind. *Annu. Rev. Neurosci.* **2022**, *45*, 361–386. [CrossRef] [PubMed]
93. Cervellati, C.; Trentini, A.; Pecorelli, A.; Valacchi, G. Inflammation in Neurological Disorders: The Thin Boundary Between Brain and Periphery. *Antioxid. Redox Signal* **2020**, *33*, 191–210. [CrossRef]
94. Chen, Y.; Xu, J.; Chen, Y. Regulation of Neurotransmitters by the Gut Microbiota and Effects on Cognition in Neurological Disorders. *Nutrients* **2021**, *13*, 2099. [CrossRef] [PubMed]
95. Elkind, M.S.; Sciacca, R.; Boden-Albala, B.; Rundek, T.; Paik, M.C.; Sacco, R.L. Moderate alcohol consumption reduces risk of ischemic stroke: The Northern Manhattan Study. *Stroke* **2006**, *37*, 13–19. [CrossRef]
96. Shiotsuki, H.; Saijo, Y.; Ogushi, Y.; Kobayashi, S. Relationship between Alcohol Intake and Stroke Severity in Japanese Patients: A Sex- and Subtype-Stratified Analysis. *J. Stroke Cerebrovasc. Dis.* **2022**, *31*, 106513. [CrossRef]
97. Faria, A.L.; Pinho, M.S.; Bermúdez, I.B.S. A comparison of two personalization and adaptive cognitive rehabilitation approaches: A randomized controlled trial with chronic stroke patients. *J. Neuroeng. Rehabil.* **2020**, *17*, 78. [CrossRef]
98. Xuefang, L.; Guihua, W.; Fengru, M. The effect of early cognitive training and rehabilitation for patients with cognitive dysfunction in stroke. *Int. J. Methods Psychiatr. Res.* **2021**, *30*, e1882. [CrossRef]
99. VanGilder, J.L.; Hooyman, A.; Peterson, D.S.; Schaefer, S.Y. Post-stroke cognitive impairments and responsiveness to motor rehabilitation: A review. *Curr. Phys. Med. Rehabil. Rep.* **2020**, *8*, 461–468. [CrossRef]
100. van Balkom, T.D.; van den Heuvel, O.A.; Berendse, H.W.; van der Werf, Y.D.; Vriend, C. The Effects of Cognitive Training on Brain Network Activity and Connectivity in Aging and Neurodegenerative Diseases: A Systematic Review. *Neuropsychol. Rev.* **2020**, *30*, 267–286. [CrossRef]

101. Stumme, J.; Jockwitz, C.; Hoffstaedter, F.; Amunts, K.; Caspers, S. Functional network reorganization in older adults: Graph-theoretical analyses of age, cognition and sex. *Neuroimage* **2020**, *214*, 116756. [CrossRef] [PubMed]
102. Anatürk, M.; Kaufmann, T.; Cole, J.H.; Suri, S.; Griffanti, L.; Zsoldos, E.; Filippini, N.; Singh-Manoux, A.; Kivimäki, M.; Westlye, L.T.; et al. Prediction of brain age and cognitive age: Quantifying brain and cognitive maintenance in aging. *Hum. Brain Mapp.* **2021**, *42*, 1626–1640. [CrossRef] [PubMed]
103. Walsh, Z.; Mollaahmetoglu, O.M.; Rootman, J.; Golsof, S.; Keeler, J.; Marsh, B.; Nutt, D.J.; Morgan, C.J.A. Ketamine for the treatment of mental health and substance use disorders: Comprehensive systematic review. *BJPsych Open* **2021**, *8*, e19. [CrossRef] [PubMed]
104. Hartelius, G.; Muscat, S.A.; Bartova, L. Editorial: Bridging the gap: An interdisciplinary perspective on ketamine in psychiatric disorders. *Front. Psychiatry* **2023**, *14*, 1246891. [CrossRef] [PubMed]
105. Corkery, J.M.; Hung, W.C.; Claridge, H.; Goodair, C.; Copeland, C.S.; Schifano, F. Recreational ketamine-related deaths notified to the National Programme on Substance Abuse Deaths, England, 1997–2019. *J. Psychopharmacol.* **2021**, *35*, 1324–1348. [CrossRef] [PubMed]
106. Biggio, G.; Biggio, F.; Talani, G.; Mostallino, M.C.; Aguglia, A.; Aguglia, E.; Palagini, L. Melatonin: From Neurobiology to Treatment. *Brain Sci.* **2021**, *11*, 1121. [CrossRef] [PubMed]
107. Anghel, L.; Baroiu, L.; Popazu, C.R.; Pătraș, D.; Fotea, S.; Nechifor, A.; Ciubara, A.; Nechita, L.; Mușat, C.L.; Stefanopol, I.A.; et al. Benefits and adverse events of melatonin use in the elderly (Review). *Exp. Ther. Med.* **2022**, *23*, 219. [CrossRef] [PubMed]
108. Gunata, M.; Parlakpinar, H.; Acet, H.A. Melatonin: A review of its potential functions and effects on neurological diseases. *Rev. Neurol.* **2020**, *176*, 148–165. [CrossRef] [PubMed]
109. de Oliveira Zanuso, B.; Dos Santos, A.R.d.O.; Miola, V.F.B.; Campos, L.M.G.; Spilla, C.S.G.; Barbalho, S.M. Panax ginseng and aging related disorders: A systematic review. *Exp. Gerontol.* **2022**, *161*, 111731. [CrossRef]
110. de Souza, G.A.; de Marqui, S.V.; Matias, J.N.; Guiguer, E.L.; Barbalho, S.M. Effects of Ginkgo biloba on diseases related to oxidative stress. *Planta Medica* **2020**, *86*, 376–386.
111. Tripp, G.; Wickens, J. Using rodent data to elucidate dopaminergic mechanisms of ADHD: Implications for human personality. *Personal. Neurosci.* **2024**, *7*, e2. [CrossRef] [PubMed]
112. Kanarik, M.; Grimm, O.; Mota, N.R.; Reif, A.; Harro, J. ADHD co-morbidities: A review of implication of gene × environment effects with dopamine-related genes. *Neurosci. Biobehav. Rev.* **2022**, *139*, 104757. [CrossRef] [PubMed]
113. Cannon Homaei, S.; Barone, H.; Kleppe, R.; Betari, N.; Reif, A.; Haavik, J. ADHD symptoms in neurometabolic diseases: Underlying mechanisms and clinical implications. *Neurosci. Biobehav. Rev.* **2022**, *132*, 838–856. [CrossRef] [PubMed]
114. Feigin, V.L.; Vos, T.; Nichols, E.; Owolabi, M.O.; Carroll, W.M.; Dichgans, M.; Deuschl, G.; Parmar, P.; Brainin, M.; Murray, C. The global burden of neurological disorders: Translating evidence into policy. *Lancet Neurol.* **2020**, *19*, 255–265. [CrossRef] [PubMed]
115. Dumurgier, J.; Tzourio, C. Epidemiology of neurological diseases in older adults. *Rev. Neurol.* **2020**, *176*, 642–648. [CrossRef] [PubMed]
116. Popa-Wagner, A.; Dumitrascu, D.I.; Capitanescu, B.; Petcu, E.B.; Surugiu, R.; Fang, W.H.; Dumbrava, D.A. Dietary habits, lifestyle factors and neurodegenerative diseases. *Neural Regen. Res.* **2020**, *15*, 394–400. [CrossRef]
117. Mey, G.M.; Mahajan, K.R.; DeSilva, T.M. Neurodegeneration in multiple sclerosis. *WIREs Mech. Dis.* **2023**, *15*, e1583. [CrossRef] [PubMed]
118. Ratan, Y.; Rajput, A.; Maleysm, S.; Pareek, A.; Jain, V.; Pareek, A.; Kaur, R.; Singh, G. An Insight into Cellular and Molecular Mechanisms Underlying the Pathogenesis of Neurodegeneration in Alzheimer's Disease. *Biomedicines* **2023**, *11*, 1398. [CrossRef] [PubMed]
119. Mishra, A.K.; Dixit, A. Dopaminergic Axons: Key Recitalists in Parkinson's Disease. *Neurochem. Res.* **2022**, *47*, 234–248. [CrossRef]
120. Masrori, P.; Van Damme, P. Amyotrophic lateral sclerosis: A clinical review. *Eur. J. Neurol.* **2020**, *27*, 1918–1929. [CrossRef]
121. Direito, R.; Barbalho, S.M.; Sepodes, B.; Figueira, M.E. Plant-Derived Bioactive Compounds: Exploring Neuroprotective, Metabolic, and Hepatoprotective Effects for Health Promotion and Disease Prevention. *Pharmaceutics* **2024**, *16*, 577. [CrossRef]
122. Natale, G.; Ryskalin, L.; Morucci, G.; Lazzeri, G.; Frati, A.; Fornai, F. The Baseline Structure of the Enteric Nervous System and Its Role in Parkinson's Disease. *Life* **2021**, *11*, 732. [CrossRef]
123. Niesler, B.; Kuerten, S.; Demir, I.E.; Schäfer, K.H. Disorders of the enteric nervous system—A holistic view. *Nat. Rev. Gastroenterol. Hepatol.* **2021**, *18*, 393–410. [CrossRef] [PubMed]
124. Chanpong, A.; Borrelli, O.; Thapar, N. Recent advances in understanding the roles of the enteric nervous system. *Fac. Rev.* **2022**, *11*, 7. [CrossRef] [PubMed]
125. Hwang, J.T.; Lee, A.; Kho, C. Ubiquitin and Ubiquitin-like Proteins in Cancer, Neurodegenerative Disorders, and Heart Diseases. *Int. J. Mol. Sci.* **2022**, *23*, 5053. [CrossRef] [PubMed]
126. He, S.; Wang, F.; Yung, K.K.L.; Zhang, S.; Qu, S. Effects of α-Synuclein-Associated Post-Translational Modifications in Parkinson's Disease. *ACS Chem. Neurosci.* **2021**, *12*, 1061–1071. [CrossRef]
127. Sahoo, S.; Padhy, A.A.; Kumari, V.; Mishra, P. Role of Ubiquitin-Proteasome and Autophagy-Lysosome Pathways in α-Synuclein Aggregate Clearance. *Mol. Neurobiol.* **2022**, *59*, 5379–5407. [CrossRef] [PubMed]

128. Fornari Laurindo, L.; Aparecido Dias, J.; Cressoni Araújo, A.; Torres Pomini, K.; Machado Galhardi, C.; Rucco Penteado Detregiachi, C.; Santos de Argollo Haber, L.; Donizeti Roque, D.; Dib Bechara, M.; Vialogo Marques de Castro, M. Immunological dimensions of neuroinflammation and microglial activation: Exploring innovative immunomodulatory approaches to mitigate neuroinflammatory progression. *Front. Immunol.* **2024**, *14*, 1305933. [CrossRef]
129. Spiteri, A.G.; Wishart, C.L.; Pamphlett, R.; Locatelli, G.; King, N.J.C. Microglia and monocytes in inflammatory CNS disease: Integrating phenotype and function. *Acta Neuropathol.* **2022**, *143*, 179–224. [CrossRef]
130. Piancone, F.; La Rosa, F.; Marventano, I.; Saresella, M.; Clerici, M. The Role of the Inflammasome in Neurodegenerative Diseases. *Molecules* **2021**, *26*, 953. [CrossRef]
131. Fathi, M.; Vakili, K.; Yaghoobpoor, S.; Tavasol, A.; Jazi, K.; Mohamadkhani, A.; Klegeris, A.; McElhinney, A.; Mafi, Z.; Hajiesmaeili, M. Dynamic changes in kynurenine pathway metabolites in multiple sclerosis: A systematic review. *Front. Immunol.* **2022**, *13*, 1013784. [CrossRef] [PubMed]
132. Isık, S.M.T.; Onmaz, D.E.; Ekmekci, A.H.; Ozturk, S.; Unlu, A.; Abusoglu, S. Relationship of tryptophan metabolites with the type and severity of multiple sclerosis. *Mult. Scler. Relat. Disord.* **2023**, *77*, 104898. [CrossRef] [PubMed]
133. Török, N.; Tanaka, M.; Vécsei, L. Searching for peripheral biomarkers in neurodegenerative diseases: The tryptophan-kynurenine metabolic pathway. *Int. J. Mol. Sci.* **2020**, *21*, 9338. [CrossRef] [PubMed]
134. Ohno, Y.; Ishihara, S.; Mashimo, T.; Sofue, N.; Shimizu, S.; Imaoku, T.; Tsurumi, T.; Sasa, M.; Serikawa, T. Scn1a missense mutation causes limbic hyperexcitability and vulnerability to experimental febrile seizures. *Neurobiol. Dis.* **2011**, *41*, 261–269. [CrossRef]
135. Dutton, S.B.; Dutt, K.; Papale, L.A.; Helmers, S.; Goldin, A.L.; Escayg, A. Early-life febrile seizures worsen adult phenotypes in Scn1a mutants. *Exp. Neurol.* **2017**, *293*, 159–171. [CrossRef] [PubMed]
136. Yi, Y.; Zhong, C.; Wei-Wei, H. The long-term neurodevelopmental outcomes of febrile seizures and underlying mechanisms. *Front. Cell Dev. Biol.* **2023**, *11*, 1186050. [CrossRef] [PubMed]
137. Gugliandolo, A.; Mazzon, E. Dental mesenchymal stem cell secretome: An intriguing approach for neuroprotection and neuroregeneration. *Int. J. Mol. Sci.* **2021**, *23*, 456. [CrossRef] [PubMed]
138. Ueda, T.; Inden, M.; Ito, T.; Kurita, H.; Hozumi, I. Characteristics and therapeutic potential of dental pulp stem cells on neurodegenerative diseases. *Front. Neurosci.* **2020**, *14*, 407. [CrossRef]
139. Santilli, F.; Fabrizi, J.; Santacroce, C.; Caissutti, D.; Spinello, Z.; Candelise, N.; Lancia, L.; Pulcini, F.; Delle Monache, S.; Mattei, V. Analogies and Differences between Dental Stem Cells: Focus on Secretome in Combination with Scaffolds in Neurological Disorders. *Stem Cell Rev. Rep.* **2024**, *20*, 159–174. [CrossRef]
140. Invernizzi, M.; De Sire, A.; Renò, F.; Cisari, C.; Runza, L.; Baricich, A.; Carda, S.; Fusco, N. Spinal cord injury as a model of bone-muscle interactions: Therapeutic implications from in vitro and in vivo studies. *Front. Endocrinol.* **2020**, *11*, 204. [CrossRef]
141. Invernizzi, M.; de Sire, A.; Carda, S.; Venetis, K.; Renò, F.; Cisari, C.; Fusco, N. Bone muscle crosstalk in spinal cord injuries: Pathophysiology and implications for patients' quality of life. *Curr. Osteoporos. Rep.* **2020**, *18*, 422–431. [CrossRef]
142. Otzel, D.M.; Kok, H.J.; Graham, Z.A.; Barton, E.R.; Yarrow, J.F. Pharmacologic approaches to prevent skeletal muscle atrophy after spinal cord injury. *Curr. Opin. Pharmacol.* **2021**, *60*, 193–199. [CrossRef]
143. Cerna, C.; García, F.E.; Téllez, A. Brief mindfulness, mental health, and cognitive processes: A randomized controlled trial. *PsyCh J.* **2020**, *9*, 359–369. [CrossRef] [PubMed]
144. Denecke, K.; Vaaheesan, S.; Arulnathan, A. A mental health chatbot for regulating emotions (SERMO)-concept and usability test. *IEEE Trans. Emerg. Top. Comput.* **2020**, *9*, 1170–1182. [CrossRef]
145. Pachucki, M.C.; Ozer, E.J.; Barrat, A.; Cattuto, C. Mental health and social networks in early adolescence: A dynamic study of objectively-measured social interaction behaviors. *Soc. Sci. Med.* **2015**, *125*, 40–50. [CrossRef] [PubMed]
146. Matias, J.N.; Achete, G.; Campanari, G.S.d.S.; Guiguer, É.L.; Araújo, A.C.; Buglio, D.S.; Barbalho, S.M. A systematic review of the antidepressant effects of curcumin: Beyond monoamines theory. *Aust. N. Z. J. Psychiatry* **2021**, *55*, 451–462. [CrossRef]
147. Costanzi, M.; Cianfanelli, B.; Santirocchi, A.; Lasaponara, S.; Spataro, P.; Rossi-Arnaud, C.; Cestari, V. Forgetting Unwanted Memories: Active Forgetting and Implications for the Development of Psychological Disorders. *J. Pers. Med.* **2021**, *11*, 241. [CrossRef] [PubMed]
148. Anderson, M.C.; Hulbert, J.C. Active Forgetting: Adaptation of Memory by Prefrontal Control. *Annu. Rev. Psychol.* **2021**, *72*, 1–36. [CrossRef] [PubMed]
149. Luo, Y.; Wang, R.; Xie, H.; He, Z. The interplay between memory control and emotion regulation. *Ann. N. Y. Acad. Sci.* **2024**, *1533*, 73–80. [CrossRef]
150. McKay, K.T.; Grainger, S.A.; Coundouris, S.P.; Skorich, D.P.; Phillips, L.H.; Henry, J.D. Visual attentional orienting by eye gaze: A meta-analytic review of the gaze-cueing effect. *Psychol. Bull.* **2021**, *147*, 1269–1289. [CrossRef]
151. Kompatsiari, K.; Ciardo, F.; Wykowska, A. To follow or not to follow your gaze: The interplay between strategic control and the eye contact effect on gaze-induced attention orienting. *J. Exp. Psychol. Gen.* **2022**, *151*, 121–136. [CrossRef] [PubMed]
152. Hadders-Algra, M. Human face and gaze perception is highly context specific and involves bottom-up and top-down neural processing. *Neurosci. Biobehav. Rev.* **2022**, *132*, 304–323. [CrossRef]
153. Domínguez-Borràs, J.; Vuilleumier, P. Amygdala function in emotion, cognition, and behavior. *Handb. Clin. Neurol.* **2022**, *187*, 359–380. [CrossRef] [PubMed]
154. Barton, J.J.S. Face processing in the temporal lobe. *Handb. Clin. Neurol.* **2022**, *187*, 191–210. [CrossRef] [PubMed]

155. Rana, T.; Behl, T.; Sehgal, A.; Srivastava, P.; Bungau, S. Unfolding the Role of BDNF as a Biomarker for Treatment of Depression. *J. Mol. Neurosci.* **2021**, *71*, 2008–2021. [CrossRef]
156. Zelada, M.I.; Garrido, V.; Liberona, A.; Jones, N.; Zúñiga, K.; Silva, H.; Nieto, R.R. Brain-Derived Neurotrophic Factor (BDNF) as a Predictor of Treatment Response in Major Depressive Disorder (MDD): A Systematic Review. *Int. J. Mol. Sci.* **2023**, *24*, 14810. [CrossRef]
157. Nikolac Perkovic, M.; Gredicak, M.; Sagud, M.; Nedic Erjavec, G.; Uzun, S.; Pivac, N. The association of brain-derived neurotrophic factor with the diagnosis and treatment response in depression. *Expert. Rev. Mol. Diagn.* **2023**, *23*, 283–296. [CrossRef]
158. Abou Chahla, M.N.; Khalil, M.I.; Comai, S.; Brundin, L.; Erhardt, S.; Guillemin, G.J. Biological Factors Underpinning Suicidal Behaviour: An Update. *Brain Sci.* **2023**, *13*, 505. [CrossRef]
159. Berkelmans, G.; van der Mei, R.; Bhulai, S.; Gilissen, R. Identifying socio-demographic risk factors for suicide using data on an individual level. *BMC Public Health* **2021**, *21*, 1702. [CrossRef]
160. Roy, B.; Ochi, S.; Dwivedi, Y. Potential of Circulating miRNAs as Molecular Markers in Mood Disorders and Associated Suicidal Behavior. *Int. J. Mol. Sci.* **2023**, *24*, 4664. [CrossRef]
161. Lang, V.A.; Lundh, T.; Ortiz-Catalan, M. Mathematical and Computational Models for Pain: A Systematic Review. *Pain. Med.* **2021**, *22*, 2806–2817. [CrossRef]
162. Markfelder, T.; Pauli, P. Fear of pain and pain intensity: Meta-analysis and systematic review. *Psychol. Bull.* **2020**, *146*, 411–450. [CrossRef] [PubMed]
163. Puschmann, A.K.; Drießlein, D.; Beck, H.; Arampatzis, A.; Moreno Catalá, M.; Schiltenwolf, M.; Mayer, F.; Wippert, P.M. Stress and Self-Efficacy as Long-Term Predictors for Chronic Low Back Pain: A Prospective Longitudinal Study. *J. Pain. Res.* **2020**, *13*, 613–621. [CrossRef] [PubMed]
164. Tanaka, M.; Diano, M.; Battaglia, S. Insights into structural and functional organization of the brain: Evidence from neuroimaging and non-invasive brain stimulation techniques. *Front. Psychiatry* **2023**, *14*, 1225755. [CrossRef]
165. Candini, M.; Battaglia, S.; Benassi, M.; di Pellegrino, G.; Frassinetti, F. The physiological correlates of interpersonal space. *Sci. Rep.* **2021**, *11*, 2611. [CrossRef] [PubMed]
166. Ellena, G.; Battaglia, S.; Làdavas, E. The spatial effect of fearful faces in the autonomic response. *Exp. Brain Res.* **2020**, *238*, 2009–2018. [CrossRef]
167. Mendes, A.J.; de Souza Greco, A.I.; Pereira, R.S.; Malfará, W.R.; de Souza, M.d.S.S.; Barbalho, S.M.; Guiguer, E.L.; Araujo, A.C. Evaluation of the anxiolytic effects of acute administration of Passiflora alata extract in wistar rats submitted to swimming. *J. Med. Plants Res.* **2022**, *16*, 44–51.

biomedicines

MDPI

Article

Vertigo in Acute Stroke Is a Predictor of Brain Location but Is Not Related to Early Outcome: The Experience of Sagrat Cor Hospital of Barcelona Stroke Registry

Angela d'Annunzio [1], Adrià Arboix [1,*], Luís García-Eroles [1] and María-José Sánchez-López [2]

1 Department of Neurology, Hospital Universitari Sagrat Cor, Quirónsalud, Universitat de Barcelona, 08029 Barcelona, Catalunya, Spain
2 Medical Library, Hospital Universitari Sagrat Cor, Quirónsalud, Universitat de Barcelona, 08029 Barcelona, Catalunya, Spain
* Correspondence: aarboix@quironsalud.es; Tel.: +34–93–4948940

Abstract: Background: Vertigo is an uncommon symptom among acute stroke victims. Knowledge about the clinical profile, the brain location, and the early outcome in stroke patients with cerebrovascular diseases and vertigo remains limited. Objectives: In this study, the effects of vertigo on cerebral topography and early prognosis in cerebrovascular diseases were investigated. Methods: A comparative analysis in terms of demographics, risk factors, clinical characteristics, stroke subtypes, cerebral and vascular topography, and early outcome was performed between patients with presence or absence of vertigo on a sample of 3743 consecutive acute stroke patients available from a 24-year ongoing single-center hospital-based stroke registry. Results: Vertigo was present in 147 patients (3.9%). Multiple logistic regression analysis showed that variables independently associated with vertigo were: location in the cerebellum (OR 5.59, CI 95% 3.24–9.64), nausea or vomiting (OR 4.48, CI 95% 2.95–6.82), medulla (OR 2.87, CI 95% 1.31–6.30), pons (OR 2.39, CI 95% 1.26–4.51), basilar artery (OR 2.36, CI 95% 1.33–4.17), ataxia (OR 2.33, CI 95% 1.41–3.85), and headache (OR 2.31, CI 95% 1.53–3.49). Conclusion: The study confirmed that the presence of vertigo was not related with increased in-hospital mortality or poor prognosis at hospital discharge. Vertigo is mainly related to non-lacunar vertebrobasilar stroke with topographic localization in the cerebellum and/or brainstem.

Keywords: brainstem; cerebellum; cerebral infarction; cerebrovascular diseases; dizziness; imbalance; stroke; stroke registry; vertebrobasilar insufficiency; vertigo

Citation: d'Annunzio, A.; Arboix, A.; García-Eroles, L.; Sánchez-López, M.-J. Vertigo in Acute Stroke Is a Predictor of Brain Location but Is Not Related to Early Outcome: The Experience of Sagrat Cor Hospital of Barcelona Stroke Registry. *Biomedicines* **2022**, *10*, 2830. https://doi.org/10.3390/biomedicines10112830

Academic Editors: Masaru Tanaka and Eleonóra Spekker

Received: 7 October 2022
Accepted: 3 November 2022
Published: 6 November 2022

1. Introduction

According to WHO data, stroke is one of the predominant causes of mortality and disability worldwide, being the second leading cause of death in adult males and the first in adult females worldwide. Furthermore, stroke is one of the most important causes of disability and dementia in adults [1], leaving up to 50% of patients with chronic disability, which has a huge impact on health care and the economy. These impact data are expected to increase due to the aging of the population.

Age- and sex-standardized rates of stroke mortality have declined in recent decades; however, the absolute number of stroke sufferers, stroke survivors, and overall stroke conditions are large and increasing [2–6].

Hemiparesis and other types of motor weakness, sensory disturbances, visual symptoms, and aphasia or speech disturbances are the most common manifestations of cerebrovascular diseases. In contrast, vestibular syndromes and vertigo are uncommon clinical features in acute stroke patients [7–9].

Stroke has a complex pathophysiology. The most recent studies suggest that the brain is exquisitely sensitive to even short-duration ischemia, implicating multiple mechanisms in the tissue damage that results from cerebral ischemia [10–14]. Ischemic stroke initiates a cascade of events that generate ATP depletion, ionic dysregulation, increased glutamate release, and excessive free radical production, as well as edema and inflammation, all of which ultimately contribute to cell death [15–17]. In contrast, in intracerebral hemorrhage, the primary cause of injury is oppression and destruction of brain tissue by the hematoma, although the inflammation, coagulation response, and toxicity of released hemoglobin also play a key role [18–20].

Vertigo is an unpleasant distortion of static gravitational orientation and is defined as a specific type of dizziness consisting of a perception of spinning or tilting with nausea/vomiting and gait unsteadiness. It is the third most commonly presenting major symptom in general medicine clinics and accounts for 3%–5% of all visits [21].

Vertigo is usually due to acute peripheral vestibulopathy. However, the prevalence of central vertigo, especially due to acute stroke, is not negligible, accounting for 3–4% of vertigo cases in a clinical study [22].

It should be noted that dizziness/vertigo has also been recognized as a manifestation of epileptic seizures, migraine, or demyelinating disease [23]. Unpleasant vegetative effects, such as nausea and vomiting, associated with vertigo are related to clinical activation of the medullary vomiting center. Making this discrimination is considered a clinical challenge since, for example, acute vertebrobasilar strokes may present with analogous symptoms which mimic an acute vestibular syndrome [24].

The course and prognosis of vertigo syndrome in acute stroke patients are variable. The diagnosis of vertigo in patients presenting with posterior circulation acute stroke has recently increased markedly [25,26].

However, little is known about the cerebral location, clinical profile, prevalence in the different stroke subtypes, and early outcome in acute cerebrovascular patients with vertigo.

Most central ischemic vertigo syndromes are secondary to paramedian or lateral tegmental infratentorial lesions [27]. However, the specific sites of brain lesions and arterial vessel disruption that cause the characteristic vestibular syndromes or vertigo are less well known [24–27].

Furthermore, the clinical spectrum of acute stroke and vertigo includes brain-stem and cerebellar signs and symptoms, but vertigo has also long been recognized as an isolated manifestation of anterior inferior cerebellar artery ischemic stroke or in small insular acute strokes [27].

Previous studies have analyzed the presence of vertigo in a sample of acute strokes in general, as well as in ischemic or hemorrhagic strokes [22]. However, the study and data on the frequency of vestibular syndromes and vertigo, specifically in the different stroke subtypes (atherothrombotic infarcts, cardioembolic infarcts, lacunar stroke, infarcts of unusual etiology, and infarcts of unknown etiology or intracerebral hemorrhage), are lacking. However, this clinical aspect is potentially relevant due to the fact that the pathophysiology, prognosis, and clinical features of ischemic small vessel strokes are different from all other cerebral infarcts In addition, small demyelinating plaques or small lacunar infarcts in the root entry zone and/or in the vestibular nuclei may mimic a vestibular neuritis [28–31], which combines rotational vertigo and spontaneous nystagmus with abolition of caloric responses on the affected side. These are combined, in turn, with a masseteric paresis manifested by the masseter reflex or by ocular motor abnormalities such as saccadic pursuit [32].

Likewise, it is important to know the early clinical outcome in acute stroke patients. In previous clinical studies, older age, atrial fibrillation, NIHSS scale, presence of previous stroke, altered consciousness, and cerebral hemorrhage, but not the presence of vertigo, were predictors associated with both in-hospital mortality and hospital stay >12 days [2,33]. However, in one study, vertigo was a clinical symptom of hemorrhagic stroke of infratentorial location with a poor outcome [34].

The main objective of this single-center comparative clinical study was to expand and update the knowledge on the poorly understood relationship between vertigo and acute stroke, mainly in relation to the neurological clinical profile, the specific cerebral and vascular location, its frequency in the different stroke subtypes, and the association between vertigo, early evolution, and prognosis.

2. Materials and Methods

Since 1986, the Hospital Universitari Sagrat Cor (a 350-bed teaching hospital in Barcelona, Catalonia, Spain, serving a population of more than 300,000 inhabitants) established a hospital-based stroke registry [35,36]. Data for all patients included in our stroke registry were entered following a standardized protocol with 186 items (demographic data, risk factors, clinical characteristics, laboratory and neuroimaging data, complications, and outcomes). The use of the same protocol for all patients ensured the integrity of the information in the database. Stroke subtypes were classified according to the criteria of the Cerebrovascular Study Group of the Spanish Society of Neurology [37], which are similar to the classification of the National Institute of Neurological Disorders and Stroke [38]. The study protocol was approved by the Clinical Research Ethics Committee of the Hospital.

A total of 4600 consecutive acute stroke patients were screened for the study. The frequency of the different stroke subtypes in the acute stroke registry was as follows: 957 cardioembolic infarcts (20.8%), 946 atherothrombotic infarcts (20.5%), 865 lacunar strokes (18.8%), 128 cerebral infarctions of unusual cause (2.8%), 374 cerebral infarctions of unknown cause (8.1%), 761 transient ischemic attacks (16.5%), 473 intracerebral hemorrhages (10.3%), 52 subarachnoid hemorrhages (1.1%), and 44 spontaneous subdural hemorrhage/spontaneous epidural hemorrhages (1%).

For the purposes of the study, only patients diagnosed with cardioembolic infarction, atherothrombotic infarct, lacunar stroke, cerebral infarction of unusual cause, cerebral infarction of unknown cause, and intracerebral hemorrhage were selected from the stroke registry database (Figure 1).

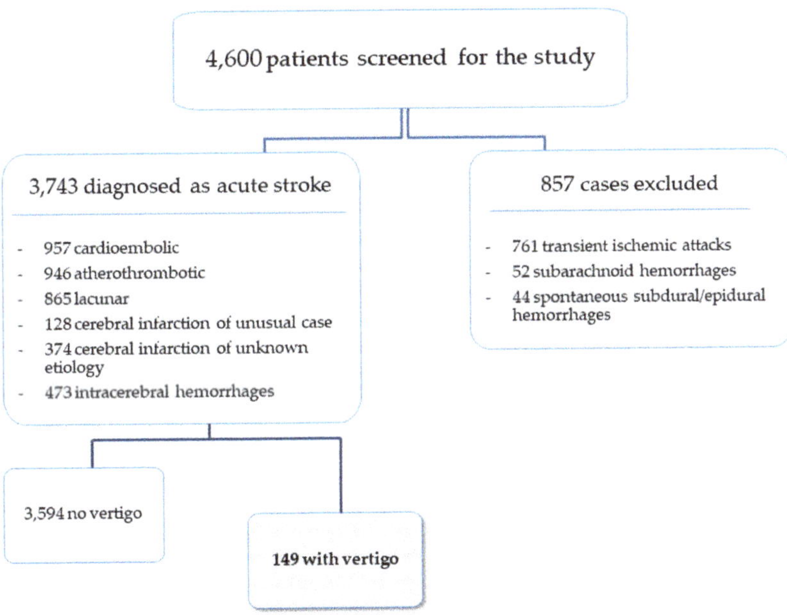

Figure 1. Flow-chart of patients included in the study.

The criterion for classifying patients with vertigo was the definition from the Bárány Society: "the sensation of self-motion when no self-motion is occurring or the sensation of distorted self-motion during an otherwise normal head movement" [39]. All patients were admitted within 48 hours of symptom onset. On admission, demographic data and the most salient clinical and neurological symptoms were recorded, as well as the results of the different laboratory tests (blood cell count, biochemical profile, serum electrolytes, and urinalysis), chest X-ray, 12-lead electrocardiogram, and brain CT (96.5%) and/or MRI (38.7%), together with the clinical investigations performed at the discretion of the responsible neurologist. The registry included medical complications—respiratory, cardiac, urinary, renal, and vascular—and mortality during the acute phase of the disease. The degree of clinical disability at hospital discharge was assessed according to the modified Rankin scale (mRS) [40].

Demographic characteristics of the patients included in the study are reported in Table 1.

Table 1. Demographics, data in acute stroke patients with vertigo versus non-vertigo.

Variable	Acute Stroke with Vertigo n = 149	Acute Stroke without Vertigo n = 3594	p Value
Age, years, mean (SD)	71.8 (12.9)	75.9 (11.6)	0.0001
Age $\geq$ 85 years old	21 (14.3)	827 (23)	0.013
	Sex		
Males	79 (52.6)	1757 (49.0)	0.676
Females	70 (47.6)	1837 (51.1)	

A comparative analysis was performed between patients with presence or absence of vertigo. Continuous data were summarized as mean and standard deviation (SD) and categorical data as frequencies and percentages. The distributions of variables in patients in both groups were compared with the chi-square (χ^2) test or the Fisher's exact test for categorical variables, and the Student's t test for quantitative variables. For all analyses, $p < 0.05$ was considered to indicate significance. Covariates with a p value < 0.20 in the univariate test were entered into a multivariate logistic regression model with a stepwise selection method, in which the presence of vertigo (versus absence of vertigo) was the dependent variable. The model was based on demographic data, cardiovascular risk factors, clinical characteristics, and cerebral and vascular topography.

To establish statistically significant criteria associated with both the presence and absence of vertigo, the odds ratio (OR), and the 95% confidence interval (CI) were calculated. The receiver operating characteristic (ROC) curve was used to assess the accuracy of the model for identifying vertigo in acute stroke; sensitivity, specificity, and positiveand negative predictive values were calculated. Statistics were analyzed using the package SPSS (Version 20 for Mac; SPSS Inc., Chicago, IL, USA).

3. Results

3.1. General Data

The study population included 3743 consecutive patients diagnosed with acute ischemic stroke or spontaneous intracerebral hemorrhage. Vertigo and laberynthine symptoms at stroke onset were present in 149 patients (3.9%) with a mean age of 71.8 $\pm$ 12.9 years. The remaining 3594 patients without vertigo were significantly older, with a mean age of 75.9 $\pm$ 11.6 years. The percentage of women was similar in the vertigo group (47.6%) compared to the non-vertigo group (51.1%) (Table 1).

3.2. Differences between the Vertigo and the Non-Vertigo Acute Stroke Groups

The presence of vertigo differed in the different stroke subtypes and was more frequent in atherothrombotic infarction (34.7%), intracerebral hemorrhage (19.7%), cardioembolic stroke (18.4%), and infarction of unknown cause (15%), and was less frequent in lacunar infarcts (9.5%) and infarction of unusual cause (2.7%).

The results of the differences between the vertigo and the non-vertigo groups by univariate analysis are presented in Table 2. Overall, hyperlipidemia, headache, nausea or vomiting, ataxia, and cranial nerve palsy were significantly more frequent in the vertigo group, whereas among those aged 85 years or older, atrial fibrillation (AF), nephropathy, limb weakness, and speech disturbances were significantly more frequent in the non-vertigo group.

Table 2. Cerebrovascular risk factors, neuroimaging, and outcomes in acute stroke patients with vertigo versus non-vertigo.

Variable	Acute Stroke with Vertigo $n = 149$	Acute Stroke without Vertigo $n = 3594$	p Value
	Risk factors		
Hypertension	96 (65.3)	2094 (58.3)	0.089
Atrial fibrillation	30 (20.4)	1024 (28.5)	0.033
Hyperlipidemia	38 (25.9)	664 (18.5)	0.025
Diabetes mellitus	34 (23.1)	826 (23.0)	0.517
Ischemic heart disease	22 (15.0)	537 (14.0)	0.534
Heavy smoking (>20 cigarettes/day)	22 (15.0)	372 (10.4)	0.074
Chronic obstructive pulmonary disease	7 (4.8)	320 (8.9)	0.081
Nephropathy	0 (0.0)	137 (3.8)	0.016
	Clinical findings		
Headache	58 (39.5)	442 (12.3)	0.0001
Nausea, vomiting	68 (46.3)	260 (7.2)	0.0001
Limb weakness	71 (48.3)	2728 (75.9)	0.0001
Speech disturbances (dysarthria, aphasia)	48 (32.7)	1821 (50.7)	0.0001
Ataxia	54 (36.7)	204 (5.7)	0.0001
Cranial nerve palsy	25 (17.0)	174 (4.8)	0.0001
	Neuroimaging findings topography		
Frontal lobe	9 (6.1)	505 (14.1)	0.006
Parietal lobe	14 (9.5)	873 (24.3)	0.0001
Temporal lobe	15 (10.2)	884 (24.6)	0.0001
Internal capsule	11 (7.5)	648 (18.0)	0.001
Basal ganglia	10 (6.8)	503 (14)	0.013
Midbrain	6 (4.1)	45 (1.3)	0.011
Pons	27 (18.4)	185 (5.1)	0.0001
Medulla	14 (9.5)	35 (1.0)	0.000
Cerebellum	46 (31.3)	83 (2.3)	0.0001
Middle cerebral artery	26 (17.7)	1921 (53.5)	0.0001
Vertebral artery	21 (14.3)	101 (2.8)	0.0001
Basilar artery	36 (24.5)	226 (6.3)	0.0001
Posteroinferior cerebellar artery	16 (10.9)	20 (0.6)	0.0001
Anteroinferior cerebellar artery	6 (4.1)	11 (0.3)	0.0001
Superior cerebellar artery	18 (12.2)	32 (0.9)	0.0001
	Stroke subtypes		0.0001
Atherothrombotic infarct	51 (34.7)	894 (24.9)	
Cardioembolic infarct	27 (18.4)	930 (25.9)	
Infarction of unknown cause	22 (15)	352 (9.8)	
Lacunar stroke	14 (9.5)	850 (23.7)	
Infarctions of unusual cause	4 (2.7)	124 (3.5)	
Intracerebral hemorrhage	29 (19.7)	444 (12.4)	
	Outcome		
Symptom-free at discharge	21 (14.3)	352 (15.4)	0.723
In-hospital death	17 (11.6)	523 (14.6)	0.312
Length of stay, days, median (interquartile range)	12 (8–20)	12 (8–20)	0.977
Prolonged hospital stay > 12 days	65 (44.2)	1695 (47.2)	0.269

Data expressed as numbers and percentages in parenthesis.

The distribution of lesions according to the cerebral location in the medulla, pons, and cerebellum, as well as vascular topography in the vertebral artery, basilar artery, posteroinferior cerebellar artery, anteroinferior cerebellar artery, and superior cerebellar artery were more frequent in the vertigo group.

Early outcome was similar in the vertigo group compared with the non-vertigo group, with a similar percentage of patients free of symptoms and with a mild neurological deficit at hospital discharge. In addition, the in-hospital mortality rate was not significantly higher in the vertigo group.

3.3. *Multivariate Analysis*

The results of the multivariate analysis are shown in Table 3. The logistic regression model based on demographics and cardiovascular risk factors, clinical characteristics, brain location, and vascular topography reported that cerebellar location (OR 5.59), presence of nausea or vomiting (OR 4.48), medulla location (OR 2.87), pons location (OR 2.39), basilar artery (OR 2.36), ataxia (OR 2.33), and headache (OR 2.31) were independently associated with the vertigo group, whereas speech disturbances (OR 0.63) and limb weakness (OR 0.47) were independent variables associated with patients without vertigo. According to this model, cases of acute stroke with vertigo versus non-vertigo were correctly classified in 84.4% of the cases.

Table 3. Results of multivariate analysis: variables independently associated with vertigo in acute stroke patients.

Regression Model	Coefficient (β)	Standard Error	Odds Ratio (95% Confidence Interval)	p Value
Model based on demographics, risk factors, clinical characteristics, and cerebral and vascular topography				
Cerebellum	1.721	0.278	5.59 (3.24–9.64)	0.0001
Nausea, vomiting	1.500	0.214	4.48 (2.95–6.82)	0.0001
Medulla involvement	1.055	0.401	2.87 (1.35–6.30)	0.009
Pons involvement	0.870	0.325	2.39 (1.26–4.51)	0.007
Basilar artery involvement	0.858	0.291	2.36 (1.33–4.17)	0.003
Ataxia	0.846	0.256	2.33 (1.41–3.85)	0.001
Headache	0.836	0.211	2.31 (1.53–3.49)	0.0001
Speech disturbances	−0.457	0.203	0.63 (0.42–0.94)	0.025
Limb weakness	−0.755	0.196	0.47 (0.32–0.69)	0.0001

Hosmer-Lemeshow goodness-of-fit test $\chi^2 = 1.070$, df = 4; $p = 0.0899$; vertigo versus non-vertigo acute stroke subjects were correctly classified in 84.4% of cases.

Figure 2 shows the ROC curve of the accuracy of the regression model. The area under the curve (AUC) was 0.857. The sensitivity was 73%, the specificity was 85%, the positive predictive value was 16%, and the negative predictive value was 99%.

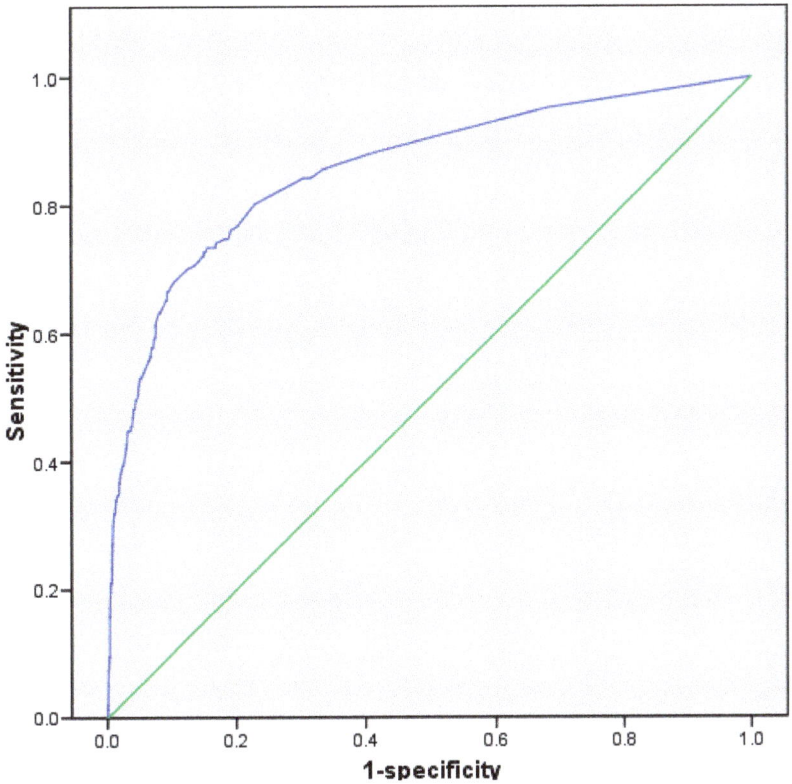

Figure 2. ROC curve for regression model including demographics, cardiovascular risk factors, clinical characteristics, and cerebral and vascular topography (AUC: 0.857). AUC: area under the curve; ROC: receiver operating characteristics.

4. Discussion

The main causes of vertigo are peripheral, such as Ménière's disease and vestibular migraine [27]. However, vertigo can be associated with acute stroke. In this study, 149 patients presented vertigo associated with acute cerebrovascular disease, with a prevalence of 3.9%: a percentage similar to that of other clinical studies [41–43].

The study of vertigo in acute stroke has usually been performed in patients selected by vascular topography, such as vertebrobasilar stroke, ischemic syndromes, or cerebral location such as brainstem and cerebellar strokes [44,45], whereas studies of acute strokes analyzed globally from stroke databanks, such as our study, are scarce. A study conducted at Hasan Sadikin General Hospital in Indonesia [43] also sought to determine the profile of stroke patients with vertigo in a retrospective clinical analysis and found that stroke patients experiencing vertigo were more likely to be women (59%), in contrast to our results in which no significant gender-related differences were obtained.

To our knowledge, our study is the only study conducted in a stroke database with the main objective of analyzing the clinical relevance of vertigo in acute stroke, and our sample is one of the largest to date with the aim of studying the clinical, topographic, and prognostic predictors of vertigo in acute stroke. In the present work, we have shown that the presence of vertigo is more frequent in non-lacunar acute stroke subtypes (atherothrombotic infarct, cardioembolic infarct, and intracerebral hemorrhage) compared to patients with lacunar infarcts, which are the subtype of ischemic stroke with a characteristic small lesional cerebral size (usually less than 15 mm maximum diameter) [46–50]. These results are similar to those of other studies in which an inverse relationship between volume and vertigo was demonstrated, with significantly less frequent presence of vertigo in small-sized acute strokes [51–56]. We speculate that larger cerebral infarcts or hemorrhages mediate the development of vascular vertigo through involvement of various posterior infratentorial brain structures and interconnections related to central vestibular connections.

As in previous studies, we found that vertebrobasilar stroke was significantly associated with the development of vertigo [44,57]. Tao et al. [58] reported a higher frequency of vertigo in vertebrobasilar strokes compared to anterior circulation strokes (18.9% vs. 1.7%). In relation to vascular topography, it should be noted that the basilar artery (24.5%), vertebral artery (14.3%), superior cerebellar artery (12.2%), posteroinferior cerebellar artery (10.9%), and anteroinferior cerebellar artery (4.1%) were the specific vascular arteries which were significantly more frequent in the group presenting with acute stroke and vertigo. This is in contrast to the results of other authors who differentiated only the vertebrobasilar arterial territory versus the carotid topography without analyzing—as we did in our study—the different cerebral arteries arising from the vertebrobasilar vascular territory in detail [22,25,41,59].

The study of Doijiri et al. [60] stated that the posteroinferior cerebellar artery was the most frequently associated with vertigo and stroke. Lee et al. [45] also found that the posteroinferior cerebellar artery was the most frequently disrupted, followed by the superior cerebellar artery with vertiginous symptomatology.

In our study, the presence of vertigo predicts the cerebral location of the acute ischemic or hemorrhagic stroke mainly in the cerebellum and brainstem. The development of vascular vertigo is associated with cerebrovascular lesions affecting the following central vestibular structures: the vestibular nuclei in the dorsolateral portion of the rostral medulla, the nucleus prepositus hypoglossi in the dorsal brainstem, and the dorsal insular cortex, as well as the cerebellar tonsil, the flocculus, the nodulus, and the inferior cerebellar peduncles [61]. The brainstem contains the neural structures involved in the integration and transmission of vestibular signals; therefore, lesions of the brainstem and cerebellum result in various vestibular symptoms and signs [62].

However, the clinical spectrum of cerebellar strokes rarely presents with isolated dizziness or without clear central neurological deficits (e.g., dysarthria or ataxia). Sometimes, isolated acute labyrinthine damage may herald impending pontocerebellar involvement in infarction in the territory of the anterior inferior cerebellar artery and may cause brainstem or cerebellar strokes; vestibular strokes are often overlooked because they mimic more benign hearing disorders [24,63]. Lee et al. [64] and Deng et al. [65] found that vestibular structures were more vulnerable to ischemia than any other structures in the brainstem and cerebellum, with the medial vestibular nucleus being the most vulnerable.

The unusual association of lacunar stroke with vertigo could be due to the usual lesional topography of lacunar strokes in the centrum semiovale, internal capsule, basis pontis, and ventroposterolateral thalamic nucleus, which are all cerebral structures that are remote from the central vestibular connecting pathways [66–68].

Furthermore, the odds ratio for the development of vertigo shows a clinical profile with three main associated clinical features: presence of nausea or vomiting, ataxia, and headache. These are all caused by the involvement of vascularized brain structures of the vertebrobasilar system. It should be noted that differentiation between brainstem and cerebellar lesions is, in most clinical cases of acute stroke, impossible because the major infratentorial arteries supply both the brainstem and the cerebellum [27].

Our results are consistent with those of Harriott et al. [69] and Levedova et al. [70], who found that headache at the onset of acute stroke is more frequent in posterior circulation stroke than in the carotid arterial territory. Additionally, it is noteworthy that the presence of speech disturbances and limb weakness is not associated with the vertigo group. These results agree with those of Elhfnawy et al. [51] who, in their study, found that there were fewer patients with associated neurological deficits in the vertigo group. We must differentiate our work from the interesting study published by Qiu et al. [71]. In their study, previous vertigo attacks were a risk factor or predictor of the presence of posterior acute ischemic stroke, whereas, in our study, vertigo and labyrinthine symptoms are clinical symptoms present at the onset of stroke in all patients analyzed.

Insular acute stroke is a rare and underreported pathology, and its clinical presentation is heterogeneous, although patients with acute insular stroke may also present a vestibular-like syndrome with isolated "vertigo" or "dizziness" with instability [72]. In our sample, we did not find any patient with this eventuality. However, it should be noted that, among anterior circulation strokes, isolated insular strokes are the most representative due to dysfunction of a relevant hub of the vestibular cortical network [72].

Little is known about the early outcome of patients with vestibular loss and vertigo associated with acute stroke. Importantly, acute cerebellar stroke may develop mass effect and intracranial hypertension. Although large cerebellar strokes can cause brainstem compression, resulting in hydrocephalus, cardiorespiratory complications, cerebral herniation, coma, and death [2,36,73,74], in our study, patients in the vertigo group had a similar percentage of patients free of symptoms and with mild neurological deficits at hospital discharge as the non-vertigo group. In addition, the in-hospital mortality rate was not significantly higher in either group. This negative association between vertigo and outcome is probably due in part to the fact that neurovegetative symptoms are very eloquent clinically, and, thus, we speculate that these patients are evaluated earlier in the emergency department and can receive the therapeutic regimens of choice in the most acute phase adequately.

From our results, we deduce that the presence of vertigo in patients with acute stroke is relevant to cerebral localization in cerebellum and brainstem (medulla and pons), as well as vertebrobasilar vascular topography, mainly in the territory of basilar artery; however, this does not translate into prognostic or early outcome significance. This study has several limitations. First, it is a retrospective, cross-sectional clinical study conducted at a single center. This design may have contributed to selection bias. In this regard, we recommend the performance of future large multicenter studies. Likewise, further validation of our results in a new prospective clinical study would be of interest. Secondly, the study did not consider the different vascular segments of the cerebral arteries (e.g., M1, M2, or M3 segments in MCA ischemic stroke), nor did it consider other specific partial brain locations such as insular topography. Another limitation of the study would be to not have analyzed the prognostic value of the presence of vertigo in acute stroke in the medium or long term.

In future studies, these aspects would be interesting lines of research. However, the assessment of the methods used in this study based on the results of a stroke data bank from a large sample of 4600 consecutive patients collected over a 24-year period is more or less objective. Additionally, a future line of investigation would be the use of high-resolution magnetic resonance imaging, which would allow for a better understanding between different vascular segments or small topographic brain locations and vertigo in acute stroke patients.

5. Conclusions

In conclusion, acute central vestibular syndrome is a strong predictor of cerebellar or dorsal brainstem acute stroke (pons or medulla involvement) and central vascular vertigo is more related to the larger volume cerebrovascular non-lacunar subtypes: atherothrombotic and cardioembolic infarcts, and spontaneous intracerebral hemorrhage. The presence of vertigo in acute stroke is associated with vertebrobasilar arterial location, mainly in basilar artery involvement.

Consequently, if a stroke occurs in the cerebellum or brainstem, areas that control balance in the brain, the patient may suffer vertigo. However, vertigo in the acute phase of stroke did not result in higher in-hospital mortality or worse early clinical outcomes. Possibly, since vertigo is a very eloquent clinical symptom, early treatment is more commonly carried out in acute stroke patients with vertigo, causing this early therapeutic assessment to be a probable clinical benefit.

Author Contributions: A.d. and A.A. designed the study and wrote the paper. A.d. and A.A. conducted the literature review and prepared the tables. L.G.-E. participated in the analysis of data and tables, and provided input in writing. M.-J.S.-L. contributed to write the paper, edited the manuscript, and provided editorial assistance. All authors have read and agreed to the published version of the manuscript.

Funding: This research received no external funding.

Institutional Review Board Statement: The study was conducted in accordance with the Declaration of Helsinki, and approved by the Clinical Research Ethics Committee of Sagrat Cor Hospital of Barcelona (protocol code: HSC-2008-05-NRL).

Informed Consent Statement: Informed consent was obtained from all subjects involved in the study.

Acknowledgments: We thank J. Massons, E. Verges, F. Gil, and L. Diez, for their help in this study.

Conflicts of Interest: The authors declare no conflict of interest.

References

1. Mao, R.; Zong, N.; Hu, Y.; Chen, Y.; Xu, Y. Neuronal Death Mechanisms and Therapeutic Strategy in Ischemic Stroke. *Neurosci. Bull.* **2022**, 1–19. [CrossRef] [PubMed]
2. Mencl, S.; Garz, C.; Niklass, S.; Braun, H.; Göb, E.; Homola, G.; Heinze, H.J.; Reymann, K.G.; Kleinschnitz, C.; Schreiber, S. Early Microvascular Dysfunction in Cerebral Small Vessel Disease Is Not Detectable on 3.0 Tesla Magnetic Resonance Imaging: A Longitudinal Study in Spontaneously Hypertensive Stroke-Prone Rats. *Exp. Transl. Stroke Med.* **2013**, *5*, 8. [CrossRef] [PubMed]
3. Ruchoux, M.M.; Domenga, V.; Brulin, P.; Maciazek, J.; Limol, S.; Tournier-Lasserve, E.; Joutel, A. Transgenic Mice Expressing Mutant Notch3 Develop Vascular Alterations Characteristic of Cerebral Autosomal Dominant Arteriopathy with Subcortical Infarcts and Leukoencephalopathy. *Am. J. Pathol.* **2003**, *162*, 329–342. [CrossRef]
4. Duering, M.; Csanadi, E.; Gesierich, B.; Jouvent, E.; Hervé, D.; Seiler, S.; Belaroussi, B.; Ropele, S.; Schmidt, R.; Chabriat, H.; et al. Incident Lacunes Preferentially Localize to the Edge of White Matter Hyperintensities: Insights into the Pathophysiology of Cerebral Small Vessel Disease. *Brain* **2013**, *136*, 2717–2726. [CrossRef]
5. Zonneveld, T.P.; Richard, E.; Vergouwen, M.D.I.; Nederkoorn, P.J.; de Haan, R.; Roos, Y.B.; Kruyt, N.D. Blood Pressure-Lowering Treatment for Preventing Recurrent Stroke, Major Vascular Events, and Dementia in Patients with a History of Stroke or Transient Ischaemic Attack. *Cochrane Database Syst. Rev.* **2018**, *7*, CD007858. [CrossRef]
6. Gasull, T.; Arboix, A. Molecular mechanisms and pathophysiology of acute stroke: Emphasis on biomarkers in the different stroke subtypes. *Int. J. Mol. Sci.* **2022**, *23*, 9476. [CrossRef]
7. Kim, K.Y.; Shin, K.Y.; Chang, K.A. Potential Biomarkers for Post-Stroke Cognitive Impairment: A Systematic Review and Meta-Analysis. *Int. J. Mol. Sci.* **2022**, *23*, 602. [CrossRef]
8. Cullell, N.; Gallego-Fábrega, C.; Cárcel-Márquez, J.; Muiño, E.; Llucià-Carol, L.; Lledós, M.; Martín-Campos, J.M.; Molina, J.; Casas, L.; Almería, M.; et al. ICA1L is associated with small vessel disease: A proteome-wide association study in small vessel stroke and intracerebral haemorrhage. *Int. J. Mol. Sci.* **2022**, *23*, 3161. [CrossRef]
9. Wang, W.X.; Springer, J.E.; Hatton, K.W. MicroRNAs as Biomarkers for Predicting Complications Following Aneurysmal Subarachnoid Hemorrhage. *Int. J. Mol. Sci.* **2021**, *22*, 9492. [CrossRef]
10. Giralt-Steinhauer, E.; Jiménez-Baladó, J.; Fernández-Pérez, I.; Rey, L.A.; Rodríguez-Campello, A.; Ois, A.; Cuadrado-Godia, E.; Jiménez-Conde, J.; Roquer, J. Genetics and epigenetics of spontaneous intracerebral hemorrhage. *Int. J. Mol. Sci.* **2022**, *23*, 6479. [CrossRef]
11. Kumar, S.; Selim, M.; Caplan, L. Medical complications after stroke. *Lancet Neurol.* **2010**, *9*, 105–118. [CrossRef]
12. Carrera, E.; Maeder-Ingvar, M.; Rossetti, A.O.; Devuyst, G.; Bogousslavsky, J. Trends in risk factors, paterns and causes in hospitalized strokes over 25 years: The Lausanne Stroke Registry. *Cerebrovasc. Dis.* **2007**, *24*, 97–103. [CrossRef] [PubMed]
13. WHO. The Top 10 Causes of Death 9 December 2020. Available online: https://www.who.int/news-room/fact-sheets/detail/the-top-10-causes-of-death (accessed on 30 March 2022).
14. Chung, J.; Marini, S.; Pera, J.; Norrving, B.; Jimenez-Conde, J.; Roquer, J.; Fernandez-Cadenas, I.; Tirschwell, D.L.; Selim, M.; Brown, D.L.; et al. Genome-Wide Association Study of Cerebral Small Vessel Disease Reveals Established and Novel Loci. *Brain* **2019**, *142*, 3176–3189. [CrossRef] [PubMed]
15. Luy, M.; Gast, K. Do women live longer or do men die earlier? Reflections on the causes of sex differences in life expectancy. *Gerontology* **2014**, *60*, 143–153. [CrossRef]
16. Kobayashi, L.C.; Beeken, R.J.; Meisel, S.F. Biopsychosocial predictors of perceived life expectancy in a national sample of older men and women. *Plos One* **2017**, *12*, e0189245.
17. Olindo, S.; Cabre, P.; Deschamps, R.; Chatot-Henry, C.; Rene-Corail, P.; Fournerie, P.; Saint-Vil, M.; May, F.; Smadja, D. Acute stroke in the very elderly. Epidemiological features, stroke subtypes, management, and outcome in Martinique, French West Indies. *Stroke* **2003**, *34*, 1593–1597. [CrossRef]

18. Montaner, J.; Perea-Gainza, M.; Delgado, P.; Ribó, M.; Chacón, P.; Rosell, A.; Quintana, M.; Palacios, M.E.; Molina, C.A.; Alvarez-Sabín, J. Etiologic Diagnosis of Ischemic Stroke Subtypes with Plasma Biomarkers. *Stroke* **2008**, *39*, 2280–2288. [CrossRef]
19. Powers, W.J.; Rabinstein, A.A.; Ackerson, T.; Adeoye, O.M.; Bambakidis, N.C.; Becker, K.; Biller, J.; Brown, M.; Demaerschalk, B.M.; Hoh, B.; et al. 2018 Guidelines for the Early Management of Patients with Acute Ischemic Stroke: A Guideline for Healthcare Professionals from the American Heart Association/American Stroke Association. *Stroke* **2018**, *49*, e46–e99. [CrossRef]
20. Palomeras Soler, E.; Fossas Felip, P.; Casado Ruiz, V.; Cano Orgaz, A.; Sanz Cartagena, P.; Muriana Batiste, D. The Mataró Stroke Registry: A 10-year registry in a community hospital. *Neurologia* **2015**, *30*, 283–289. [CrossRef]
21. Newman-Toker, D.E.; Della Santina, C.C.; Blitz, A. Vertigo and hearing loss. *Handb. Clin. Neurol.* **2016**, *136*, 905–921. [CrossRef]
22. Choi, K.D.; Kim, J.S. Vascular vertigo: Updates. *J. Neurol.* **2019**, *266*, 1835–1843. [CrossRef] [PubMed]
23. Phalgune, D.; Sankalia, D.; Kothari, S. Diagnosing Stroke in Acute Vertigo: Sensitivity and Specificity of HINTS Battery in Indian Population. *Neurol. India.* **2021**, *69*, 97–101. [CrossRef]
24. Tehrani, A.S.S.; Kattah, J.C.; Kerber, K.A.; Gold, D.R.; Zee, D.S.; Urrutia, V.C.; Newman-Toker, D.E. Diagnosing stroke in acute dizziness and vertigo. Pitfalls and pearls. *Stroke* **2018**, *49*, 788–795. [CrossRef]
25. Choi, J.; Lee, S.; Kim, J. Central vertigo. *Curr. Opin. Neurol.* **2018**, *31*, 81–89. [CrossRef] [PubMed]
26. Camps-Renom, P.; Delgado-Mederos, R.; Martínez-Domeño, A.; Prats-Sánchez, L.; Cortés-Vicente, E.; Simón-Talero, M.; Arboix, A.; Ois, Á.; Purroy, F.; Martí-Fàbregas, J. Clinical Characteristics and Outcome of the Capsular Warning Syndrome: A Multicenter Study. *Int. J. Stroke* **2015**, *10*, 571–575.
27. Dieterich, M.; Brandt, T.H. Vestibular syndromes and vertigo. In *Stroke Syndromes*; Bogousslavsky, J., Caplan, L., Eds.; Cambridge University Press: Cambridge, UK, 1995; pp. 80–90.
28. Jacova, C.; Pearce, L.A.; Costello, R.; McClure, L.A.; Holliday, S.L.; Hart, R.G.; Benavente, O.R. Cognitive impairment in lacunar strokes: The SPS3 trial. *Ann. Neurol.* **2012**, *72*, 351–362.
29. Liu, W.; Liu, R.; Sun, W.; Peng, Q.; Zhang, W.; Xu, E.; Cheng, Y.; Ding, M.; Li, Y.; Hong, Z.; et al. Different impacts of blood pressure variability on the progression of cerebral microbleeds and white matter lesions. *Stroke* **2012**, *43*, 2916–2922.
30. Arboix, A.; Roig, H.; Rossich, R.; Martínez, E.M.; García-Eroles, L. Differences between hypertensive and non-hypertensive ischemic stroke. *Eur. J. Neurol.* **2004**, *11*, 687–692.
31. Bejot, Y.; Catteau, A.; Caillier, M.; Rouaud, O.; Durier, J.; Marie, C.; Di Carlo, A.; Osseby, G.V.; Moreau, T.; Giroud, M. Trends in incidence, risk factors, and survival in symptomatic lacunar stroke in Dijon, France, from 1989 to 2006. A population-based study. *Stroke* **2008**, *39*, 1945–1951.
32. Hopf, H.C. Vertigo and masseter paresis. A new local brainstem syndrome probably of vascular origin. *J. Neurol.* **1987**, *235*, 42–45.
33. Brickman, A.M.; Reit, C.Z.; Luchsinger, J.A.; Manly, J.J.; Schupf, N.; Muraskin, J.; DeCarli, C.; Brown, T.R.; Mayeux, R. Long-term blood pressure fluctuation and cerebrovascular disease in an elderly cohort. *Arch. Neurol.* **2010**, *67*, 564–569. [CrossRef] [PubMed]
34. Arboix, A.; Rennie, M. Clinical study of 28 patients with cerebellar hemorrhage. *Med. Clin. (Barc.)* **2009**, *132*, 665–668. [CrossRef] [PubMed]
35. Arboix, A. Cardiovascular risk factors for acute stroke: Risk profiles in the different subtypes of ischemic stroke. *World. J. Clin. Cases.* **2015**, *3*, 418–429. [CrossRef]
36. Arboix, A.; Massons, J.; Oliveres, M.; García, L.; Titus, F. An analysis of 1000 consecutive patients with acute cerebrovascular disease. The registry of cerebrovascular disease of La Alianza-Hospital Central of Barcelona. *Med. Clin. (Barc.)* **1993**, *101*, 281–285. [PubMed]
37. Arboix, A.; Alvarez-Sabin, J.; Soler, L. Stroke. Classification and diagnostic criteria. Ad hoc Editorial Committee of the Task Force on Cerebrovascular Diseases of SEN. *Neurologia* **1998**, *13* (Suppl. S3), S3–S10.
38. Special Report from the National Institute of Neurological Disorders and Stroke: Classification of cerebrovascular diseases III. *Stroke* **1990**, *21*, 637–676. [CrossRef]
39. Bisdorff, A.; Von Brevern, M.; Lempert, T.; Newmn-Toker, D.E. Classification of vestibular symptoms: Towards an International classification of vestibular disorders. *J. Vestib. Res.* **2009**, *19*, 1–13. [CrossRef] [PubMed]
40. Bamford, J.M.; Sandercock, P.A.; Warlow, C.P.; Slattery, J. Interobserver agreement for the assessment of handicap in stroke patients. *Stroke* **1989**, *20*, 828. [CrossRef]
41. Choi, J.; Lee, S.; Kim, J. Ischemic syndromes causing dizziness and vertigo. *Handb. Clin. Neurol.* **2016**, *137*, 317–340. [CrossRef]
42. Von Brevern, M.; Süßmilch, S.; Zeise, D. Acute vertigo due to hemispheric stroke. *J. Neurol. Sci.* **2014**, *339*, 153–156. [CrossRef]
43. Khansa, A.; Cahyani, A.; Amalia, L. Clinical profile of stroke patients with vertigo in Hasan Sadikin General Hospital Bandung Neurology Ward. *J. Med. Health.* **2019**, *2*, 856–866. [CrossRef]
44. Kim, J.; Lee, H. Vertigo Due to Posterior Circulation Stroke. *Semin. Neurol.* **2013**, *33*, 179–184. [CrossRef] [PubMed]
45. Lee, H.; Sohn, S.I.; Cho, Y.W.; Lee, S.R.; Ahn, B.H.; Park, B.R.; Baloh, R.W. Cerebellar infarction presenting isolated vertigo: Frequency and vascular topographical patterns. *Neurology* **2006**, *67*, 1178–1183. [CrossRef]
46. Jiang, S.; Yan, Y.; Yang, T.; Zhu, Q.; Wang, C.; Bai, X.; Hao, Z.; Zhang, S.; Yang, Q.; Fan, Z.; et al. Plaque Distribution Correlates with Morphology of Lenticulostriate Arteries in Single Subcortical Infarctions. *Stroke* **2020**, *51*, 2801–2809. [CrossRef]
47. Boehme, A.K.; McClure, L.A.; Zhang, Y.; Luna, J.M.; Del Brutto, O.H.; Benavente, O.R.; Elkind, M.S.V. Inflammatory Markers and Outcomes after Lacunar Stroke: Levels of Inflammatory Markers in Treatment of Stroke Study. *Stroke* **2016**, *47*, 659–667. [CrossRef] [PubMed]

48. Lavallée, P.C.; Labreuche, J.; Faille, D.; Huisse, M.G.; Nicaise-Roland, P.; Dehoux, M.; Gongora-Rivera, F.; Jaramillo, A.; Brenner, D.; Deplanque, D.; et al. Circulating Markers of Endothelial Dysfunction and Platelet Activation in Patients with Severe Symptomatic Cerebral Small Vessel Disease on Behalf of the Lacunar-B.I.C.H.A.T. Investigators. *Cerebrovasc. Dis.* **2013**, *36*, 131–138. [CrossRef] [PubMed]
49. Zhang, Z.; Fan, Z.; Kong, Q.; Xiao, J.; Wu, F.; An, J.; Yang, Q.; Li, D.; Zhuo, Y. Visualization of the Lenticulostriate Arteries at 3T Using Black-Blood T1-Weighted Intracranial Vessel Wall Imaging: Comparison with 7T TOF-MRA. *Eur. Radiol.* **2019**, *29*, 1452–1459. [CrossRef] [PubMed]
50. Iadecola, C.; Parikh, N.S. Blood pressure ups and downs foreshadow cerebral microangiopathy. *J. Am. Coll. Cardiol.* **2020**, *75*, 2400–2402. [CrossRef]
51. Elhfnawy, A.M.; El-Raouf, M.A.; Volkmann, J.; Fluri, F.; Elsalamawy, D. Relation of infarction location and volume to vertigo in vertebrobasilar stroke. *Brain Behav.* **2020**, *10*, e01564. [CrossRef]
52. Maida, C.D.; Norrito, R.L.; Daidone, M.; Tuttolomondo, A.; Pinto, A. Neuroinflammatory mechanisms in ischemic stroke: Focus on cardioembolic stroke, background, and therapeutic approaches. *Int. J. Mol. Sci.* **2020**, *21*, 6454. [CrossRef]
53. Chen, Y.; Pu, J.; Liu, Y.; Tian, L.; Chen, X.; Gui, S.; Xu, S.; Song, X.; Xie, P. Pro-inflammatory cytokines are associated with the development of post-stroke depression in the acute stage of stroke: A meta-analysis. *Top Stroke Rehabil.* **2020**, *27*, 620–629. [CrossRef] [PubMed]
54. Mustanoja, S.; Putaala, J.; Koivunen, R.J.; Surakka, I.; Tatlisumak, T. Blood pressure levels in the acute phase after intracerebral hemorrhage are associated with mortality in young adults. *Eur. J. Neurol.* **2018**, *8*, 1034–1040. [CrossRef] [PubMed]
55. Arboix, A.; García-Plata, C.; García-Eroles, L.; Massons, J.; Comes, E. Clinical study of 99 patients with pure sensory stroke. *J. Neurol.* **2005**, *252*, 156–162. [CrossRef] [PubMed]
56. Rudilosso, S.; Rodríguez-Vázquez, A.; Urra, X.; Arboix, A. The Potential Impact of Neuroimaging and Translational Research on the Clinical Management of Lacunar Stroke. *Int. J. Mol. Sci.* **2022**, *23*, 1497. [CrossRef]
57. Purroy, F.; Montaner, J.; Molina, C.A.; Delgado, P.; Ribo, M.; Alvarez-Sabín, J. Patterns and predictors of early risk of recurrence after transient ischemic attack with respect to etiologic subtypes. *Stroke* **2007**, *38*, 3225–3229. [CrossRef]
58. Tao, W.D.; Liu, M.; Fisher, M.; Wang, D.R.; Li, J.; Furie, K.L.; Wu, B.O. Posterior versus anterior circulation infarction: How different are the neurological deficits? *Stroke* **2012**, *43*, 2060–2065. [CrossRef]
59. Kim, S.H.; Kim, H.J.; Kim, J.S. Isolated vestibular syndrome due to brainstem and cerebellar lesions. *J. Neurol.* **2017**, *264*, 63–69. [CrossRef]
60. Doijiri, R.; Uno, H.; Miyashita, K.; Ihara, M.; Nagatsuka, K. How commonly is stroke found in patients with isolated vertigo or dizziness attack? *J. Stroke Cerebrovasc. Dis.* **2016**, *25*, 2549–2552. [CrossRef]
61. Lee, J.O.; Park, S.H.; Kim, H.J.; Kim, M.S.; Park, B.R.; Kim, J.S. Vulnerability of the vestibular organs to transient ischemia: Implications for isolated vascular vertigo. *Neurosci. Lett.* **2014**, *558*, 180–185. [CrossRef]
62. Choi, K.D.; Lee, H.; Kim, J.S. Vertigo in brainstem and cerebellar strokes. *Curr. Opin. Neurol.* **2013**, *28*, 90–95. [CrossRef]
63. Zhang, D.P.; Li, H.R.; Ma, Q.K.; Yin, S.; Peng, Y.F.; Zhang, H.L.; Zhao, M.; Zhang, S.L. Prevalence of stroke and hypoperfusion in patients with isolated vertigo and vascular risk factors. *Front. Neurol.* **2018**, *9*, 974. [CrossRef] [PubMed]
64. Lee, S.H.; Kim, J.S. Acute diagnosis and management of stroke presenting dizziness or vertigo. *Neurol. Clin.* **2015**, *33*, 687–698. [CrossRef] [PubMed]
65. Deng, Y.; Zhang, L.; Zhang, R.; Duan, J.; Huang, J.; Qiu, D. Clinical features differ between patients with vertigo attack only and weakness attack accompanying vertigo before vertebrobasilar stroke: A retrospective study. *Front. Neurol.* **2022**, *13*, 928902. [CrossRef] [PubMed]
66. Smirnov, M.; Destrieux, C.; Maldonado, I.L. Cerebral White Matter Vasculature: Still Uncharted? *Brain* **2021**, *144*, 3561–3575. [CrossRef] [PubMed]
67. Muiño, E.; Fernández-Cadenas, I.; Arboix, A. Contribution of "Omic" Studies to the Understanding of CADASIL. A Systematic Review. *Int. J. Mol. Sci.* **2021**, *22*, 7357. [CrossRef] [PubMed]
68. Meissner, A. Hypertension and the brain: A risk factor for more than heart disease. *Cerebrovasc. Dis.* **2016**, *42*, 255–262. [CrossRef]
69. Harriott, A.M.; Karakaya, F.; Ayata, C. Headache after ischemic stroke: A systematic review and meta-analysis. *Neurology* **2020**, *94*, e75–e86. [CrossRef]
70. Lebedeva, E.R.; Ushenin, A.V.; Gurary, N.M.; Tsypushkina, T.S.; Gilev, D.V.; Kislyak, N.V.; Olesen, J. Headache at onset of first-ever ischemic stroke: Clinical characteristics and predictors. *Eur. J. Neurol.* **2021**, *28*, 852–860. [CrossRef]
71. Qiu, D.; Zhang, L.; Deng, J.; Xia, Z.; Duan, J.; Wang, J.; Zhang, R. New Insights Into Vertigo Attack Frequency as a Predictor of Ischemic Stroke. *Front. Neurol.* **2020**, *11*, 593524. [CrossRef]
72. Di Stefano, V.; De Angelis, M.V.; Montemitro, C.; Russo, M.; Carrini, C.; di Giannantonio, M.; Brighina, F.; Onofrj, M.; Werring, D.J.; Simister, R. Clinical presentation of strokes confined to the insula: A systematic review of literature. *Neurol. Sci.* **2021**, *42*, 1697–1704. [CrossRef]
73. Yokokawa, H.; Goto, A.; Terui, K.; Funami, Y.; Watanabe, K.; Yasumura, S. Prevalence of Metabolic Syndrome and Serum Marker Levels in Patients with Four Subtypes of Cerebral Infarction in Japan. *J. Clin. Neurosci.* **2008**, *15*, 769–773. [CrossRef]
74. Gąsecki, D.; Kwarciany, M.; Nyka, W.; Narkiewicz, K. Hypertension, brain damage and cognitive decline. *Curr. Hypertens. Rep.* **2013**, *15*, 547–558. [CrossRef]

biomedicines

Review

Fronto–Cerebellar Diaschisis and Cognitive Dysfunction after Pontine Stroke: A Case Series and Systematic Review

Kei Shimmyo [†] and Shigeru Obayashi [*,†]

Department of Rehabilitation Medicine, Saitama Medical Center, Saitama Medical University, 1981 Kamoda, Kawagoe 350-8550, Japan; kshimmyo@saitama-med.ac.jp
* Correspondence: ohbayash@saitama-med.ac.jp; Tel.: +81-49-228-3259
† These authors contributed equally to this work.

Abstract: It is well known that cortical damage may affect cognitive functions, whereas subcortical damage, especially brainstem stroke, would be far less likely to cause cognitive decline, resulting in this condition being overlooked. Few studies have focused on cognitive dysfunction after a pontine stroke. Here, we begin with describing our nine new case reports of in-depth neuropsychological findings from patients with pontine stroke. The dominant domain of cognitive dysfunction was commonly characterized by executive dysfunction, almost in line with previous studies. The severity was relatively mild. We give an overview of the available literature on cognitive decline following a pontine stroke. This is followed by discussions regarding the prognosis of the cognitive disabilities. Based on previous neuroimaging findings, we would like to get to the core of the neuropathology underlying the cognitive declines in the context of "diaschisis", a phenomenon of a broad range of brain dysfunctions remote from the local lesions. Specifically, our unique paper, with two modalities of neuroimaging techniques, may help us better understand the pathology. SPECT scans yield evidence of frontal and thalamic hyper-perfusion and cerebellar hypo-perfusion in patients with pontine stroke. Functional near-infrared spectroscopy, when focusing on the supplementary motor area (SMA) as one of the hyper-perfusion areas, exhibits that SMA responses may be subject to the severity of cognitive decline due to a pontine stroke and would also be related to the recovery. Finally, we posit that cognitive decline due to pontine stroke could be explained by the failure of hierarchical cognitive processing in the fronto–ponto–cerebellar–thalamic loop.

Keywords: brainstem; cerebellum; cognition; diaschisis; executive function; functional NIRS; near-infrared spectroscopy; SPECT; supplementary motor area; thalamus

Citation: Shimmyo, K.; Obayashi, S. Fronto–Cerebellar Diaschisis and Cognitive Dysfunction after Pontine Stroke: A Case Series and Systematic Review. *Biomedicines* **2024**, *12*, 623. https://doi.org/10.3390/biomedicines12030623

Academic Editors: Masaru Tanaka, Kuen-Jer Tsai and Eleonóra Spekker

Received: 13 December 2023
Revised: 4 March 2024
Accepted: 8 March 2024
Published: 11 March 2024

1. Introduction

Stroke not only causes functional disabilities, such as motor paresis and dependency in activities of daily living (ADL), but also affects cognition as an invisible disability [1–6]. Due to this invisibility, the cognitive impairment inherent in stroke survivors has often been overlooked during follow-ups. As clinical determinants, the prevalence of cognitive impairment differs according to the lesion location, such as cortical, subcortical, and infratentorial lesions (cerebellum and brainstem) [7–13]. A previous study demonstrated that cognitive impairment was present in 74% of acute-phase patients with a cortical stroke, 46% with a subcortical one, and 43% with an infratentorial one [14]. Insights into the modulatory role of the cerebellum in cognition are well documented [15–18]. On the other hand, conventional tenets posit that damage to the brainstem might not affect cognition. To date, the contribution of the brainstem in cognition is still underexplored [19–27]. Also, the prevalence, severity, and long-term trajectories of cognitive dysfunctions due to a brainstem stroke remain unknown. The literature is somewhat limited and lacks considerable data because of relatively small sample sizes in the relevant studies [28–32]. To the best of our knowledge, we only found a review article of this topic [19]. In addition, most of the cases in the literature have not always achieved sufficient neuropsychological outcome measures to document the broad range of cognitive impairments possible after a brainstem stroke. Pontine infarction accounts for about seven percent of all ischemic strokes [33,34]. Since most cases have not been limited to an isolated pontine stroke, we could not distinguish whether the type and severity of the cognitive symptom depends on the location of the injury. Accordingly, we focused on the pons to explore the cognitive function. An in-depth neuropsychological evaluation of the case series of pontine strokes could disclose the characteristics of the cognitive disabilities, which will help physicians and healthcare professionals better understand the deficits.

How do cognitive dysfunctions arise from subcortical or infratentorial lesions? The possible mechanism may be explained by the theory of "diaschisis", where the impact of focal lesions can expand into a widespread and diffuse brain network organization, remote from the lesion location [35–38]. Neuroimaging techniques would be suitable tools for visualizing the mechanisms underlying invisible disabilities in the form of diaschisis. In favor of this view, SPECT scans detected cerebral perfusion abnormalities or decreases in the regional cerebral blood flow in remote brain regions after a brainstem stroke [20,22,23,39,40]. Intriguingly, the injury is followed by morphological and degenerative changes in the brain, such as frontal and thalamic volume expansions or cerebellar atrophy [41–44], anterograde and retrograde degeneration in the corticospinal tracts [45–47], and aberrant functional connectivity [48–50].

The merit of the present case report lies in the fact that it helps the physicians become aware of cognitive decline after a pontine stroke and that further understanding of the neuropathology may eventually lead to the conquest of cognitive dysfunction. Now, we begin with the case series description by sharing the in-depth neuropsychological findings for nine patients with pontine stroke and provide clear definitions of the neuropsychological profiles. And we give an overview of the available literature on cognitive decline following a pontine stroke. The severity and prognosis of the cognitive decline is also discussed. Then, we will get to the main subject: all-encompassing deliberations about the neuropathology of the cognitive deficits based on neuroimaging findings and the theory of "diaschisis". While reviewing the literature dealing with the possible pathology of cognitive dysfunction, especially involving the frontal lobe and the cerebellum, we focus on the significant role of the fronto–ponto–cerebellar–thalamic loop in the neuropathology based on our recent data from two modalities of neuroimaging techniques, SPECT, and functional near-infrared spectroscopy (f-NIRS) [23]. Finally, we propose a specific function for pons in the hierarchical information processing system of this loop.

2. Case Series Description

The inclusion criteria were as follows: (1) under the age of 90; (2) first-ever isolated pontine infarct; (3) adequate mental state to participate (clear consciousness); (4) medically stable condition; (5) within two weeks of stroke onset at first time of the evaluation; and (6) independent ADL before admission. The exclusion criteria were as follows: (1) history of damage from a stroke (cerebral infarct, cerebral hemorrhage, subarachnoid hemorrhage, and lacunar infarct), brain injury, or brain tumor; (2) neurodegenerative disease; (3) mental illness; (4) dementia; (5) epilepsy; and (6) severe or moderate hemiparesis. Among a total of 163 cases, 9 cases of pontine stroke were selected, according to the inclusion and exclusion criteria. These cases consisted of seven men and two women with a mean age of 76.44 years and a range between 63 and 86 years (Table 1). In terms of the stroke type, the case series consisted of seven patients with pontine branch atheromatous disease (BAD), one patient with a paramedian lacunar infarct, and a one with a pontine hemorrhage. The stroke location was determined using an MRI scan or CT scan of the brain (Figure 1). Apart from the stroke in the pons, no additional lesions or atrophy were detected. Additionally, our case series was narratively described. All patients were assessed by well-trained neuropsychologists using the standardized Japanese translation of the Mini-Mental State Examination (MMSE) for the general intellectual ability [51] and a set of neuropsychological test batteries for attention, memory, and executive function. The batteries consisted of the following tests: the Trail Making Test [52,53], Japanese version-A (TMT-J part A), for assessing attention and processing speed; the TMT-B (Japanese version, where Kana letters replaced the Roman alphabet) and the ΔTMT, the temporal gap between B and A and the Frontal Assessment Battery (FAB) [54], for assessing the executive function; and the Standard Verbal Paired-Associate Learning Test (S-PA) [55] and Rey–Osterrieth Complex Figure Test (ROCFT) [56], for assessing the verbal memory and visual memory, respectively. Also, the reproduction of the ROCFT requires some strategy in term of its accuracy, involving executive function and visuospatial cognition. Abnormalities of both the TMT-J and S-PA were determined based on a database of normal healthy volunteers (mean and SD for people in their 60s and 70s). The Brunnstrom recovery stage (BRS) is designed to describe the motor recovery process of a sequence of limbs as well as the severity of hemiparesis, containing three items for the arm (shoulder/elbow/forearm: BRS-A), the hand/finger (BRS-H), and the leg (BRS-L), all of which are rated on a six-level scale (level 1 to 6) [57]. Subjects provided written informed consent after receiving a detailed explanation of the procedures. This study was reviewed and approved by the Ethics Committee of Saitama Medical Center (Approval number: 2021-093).

Table 1. Baseline characteristics of case series.

Case	Age	Gender	Type of Stroke	Laterality	Volume (mm^3)	BRS on Admission	BRS at Discharge
Case 1	63	M	BAD	Left	273.5	6,5,6	6,6,6
Case 2	72	M	BAD	Left	1328.5	6,6,6	6,6,6
Case 3	75	F	BAD	Left	259	3,3,4	3,3,4
Case 4	76	M	Lacunar infarct	Median	325	No paresis	
Case 5	77	M	BAD	Right	873	4,4,5	6,6,6
Case 6	77	F	Hemorrhage	Right	870.5	6,6,6	6,6,6
Case 7	79	M	BAD	Left	372.5	No paresis	
Case 8	82	M	BAD	Right	1147.5	5,5,5	5,5,5
Case 9	86	F	BAD	Left	1126	2,2,4	3,4,3

M: male; F: female; BAD: branch atheromatous disease; BRS: Brunnstrom recovery stage.

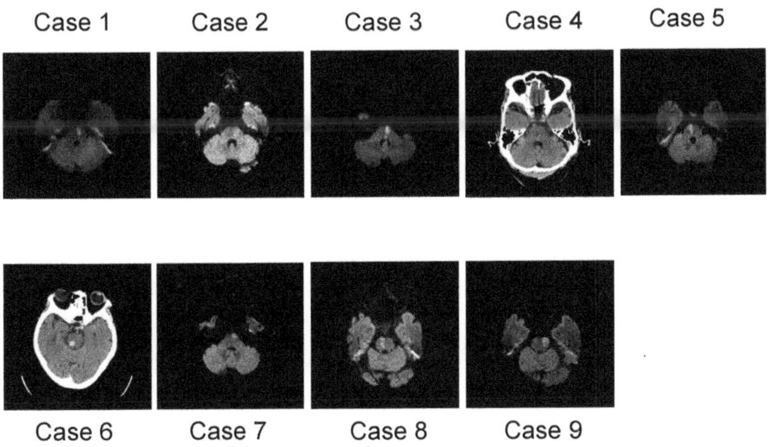

Figure 1. Distribution of strokes, which were detected using diffusion-weighted MRI and CT scans.

2.1. Case 1

A 63-year-old male, with a history of hypertension, developed right mild hemiparesis (BRS-A 6, BRS-H 5, and BRS-L 6) and dysarthria. Three days after onset, he walked into our hospital with his feet dragging. MRI imaging of his brain revealed a left ventral pontine infarction, which was considered to be BAD (branch atheromatous disease). He received conservative treatment with dual antiplatelet therapy (DAPT). Three days after admission, a neuropsychological investigation was started. The MMSE produced a normal score of 29/30, guaranteeing his preserved orientation. His executive function was mildly impaired, as he made a mistake on the similarities and lexical fluency, while he scored 16 on the FAB. The TMT-J -A response (43 s) was within the normal range, whereas the TMT-B (106 s) produced delayed responses, and the ΔTMT was gapped by 63 s, revealing his dysexecutive function syndrome. His verbal memory disturbance was delineated by the S-PA. A follow-up of the TMT, conducted five days after the initial test, showed an improvement in the ΔTMT to a normal range. Seven days after admission, he regained his hemiparesis fully and was discharged to his home.

2.2. Case 2

A 72-year-old male developed right mild hemiparesis (BRS-A 6, BRS-H 6, and BRS-L 6), numbness, and facial paralysis. The risk factors for stroke were hypertension and dyslipidemia. On the day of onset, he was admitted to the hospital, and MRI imaging of the brain revealed left pontine BAD. He received conservative treatment with DAPT, an anticoagulant and an antioxidant drug. On the day after admission, he complained of diplopia. At four days after admission, a neuropsychological investigation was started. He scored 26/30 on the MMSE, with points lost in Serial sevens test. Executive dysfunction was evident from the fact that he made a mistake in the Go/No-Go test on the FAB, and the reproduction of the ROCFT scored 31/36. The TMT-J was delayed in both A and B at 110 and 156 s, respectively. The S-PA was estimated to be within the normal range. Nine days after admission, he was transferred to the rehabilitation hospital.

2.3. Case 3

A 75-year-old female developed right hemiparesis (BRS-A 3, BRS-H 3, and BRS-L 4), diplopia, and dysarthria. One day after onset, she was admitted to our hospital. MRI imaging of the brain revealed isolated pontine BAD. She received conservative treatment with DAPT, an anticoagulant and an antioxidant drug. Two days after admission, she scored 29 on the MMSE. Her TMT-A response was at a normal level, but she had delayed TMT-B responses, showing a significant gap (ΔTMT 74 s), and low scores for the similarities and lexical fluency in the FAB revealed her dysexecutive function syndrome. A mild memory dysfunction was proven by the S-PA, with a delayed recall on the MMSE and ROCFT. At 17 days after the first assessment, a follow-up of the TMT-J recovered to normal range (A-39 s/B-69 s). At 27 days after admission, she transferred to the rehabilitation hospital, because her hemiparesis (BRS-A 3, BRS-H 3, and BRS-L 4) and dysarthria persisted.

2.4. Case 4

A 76-year-old male lost consciousness during work and was taken to the emergency room of our hospital. He regained consciousness and showed dysarthria and right medial longitudinal fasciculus syndrome. He had a history of myocardial infarction, and he carried an implanted cardioverter defibrillator in his body. A CT scan showed a high-density area at the top of the basilar artery. After he underwent tissue plasminogen activator (t-PA) therapy and an endovascular therapy with mechanical thrombectomy, his eye movement symptom disappeared. One day after admission, the dysarthria persisted, and the CT scan displayed a tiny low-density area isolated at the paramedian pons. He received conservative treatment with a direct oral anticoagulant (DOAC). Two days after admission, he made a mistake on the recall and three-stage command in the MMSE (resulting in a total score of 27). Lower S-PA scores and a delayed recall in the ROCFT revealed his mild verbal and visual memory deficits, respectively. He had a response delay in the TMT-A (118 s), showing an attention deficit. His executive dysfunction was evident using the TMT-B (292 s), showing a gap (ΔTMT 174 s), and he had low scores for lexical fluency and Go/No-Go on the FAB. At 12 days after the first assessment, a follow-up of the MMSE and FAB revealed no score changes relative to the first assessment. He was discharged to his home 16 days after admission.

2.5. Case 5

A 77-year-old male developed left hemiparesis (BRS-A 4, BRS-H 4, and BRS-L 5), facial palsy, and dysarthria. The risk factors for stroke were hypertension and sleep apnea syndrome. An MRI showed right pontine BAD. He received conservative treatment with DAPT, an anticoagulant and an antioxidant drug. His MMSE score was 28, guaranteeing his preserved orientation. Three days after admission, although within the normal limit of the TMT-A (60 s), he exhibited a significant gap between TMT-A and B (ΔTMT 149 s) and had low scores for the similarities and lexical fluency on the FAB, suggesting executive dysfunction. Copy of the ROCFT presented with his low scores, suggesting executive dysfunction and visuospatial recognition difficulty, and the S-PA revealed a normal verbal memory. He recovered from the left hemiparesis fully and was discharged to his home 30 days after admission.

2.6. Case 6

A 77-year-old female developed left facial numbness, dysarthria, and right abducens nerve palsy. The risk factor was diabetes mellitus. A CT scan showed a high-density area isolated at the right dorsal pons, diagnosing as pontine hemorrhage. The day after admission, a mild right hemiparesis (BRS-A 6, BRS-H 6, and BRS-L 6) was found. She received conservative treatment with an antihypertensive drug. Her MMSE score was 28, guaranteeing her good orientation. Three days after admission, low scores for similarities, lexical fluency, and the Go/No-Go test on the FAB suggested executive dysfunction. She showed normal responses to the TMT (A-60 s/B-79 s). A low score (11/36) of delayed recall in the ROCFT showed her visual memory disorder, while the S-PA displayed normal scores. Her numbness persisted, but she recovered from hemiparesis, dysarthria, and abducens nerve palsy. She was discharged 17 days after admission.

2.7. Case 7

A 79-year-old male developed dysarthria. His past history revealed microscopic polyangiitis. An MRI of the brain indicated left pontine BAD. He received conservative treatment with DAPT, an anticoagulant and an antioxidant drug. His MMSE score was 25, which included mistakes in attention and calculation, delayed recall, and picture copy. Three days after onset, he exhibited difficulties in similarities, lexical fluency, and the Go/No-Go task on the FAB, showing executive dysfunction. He had delayed TMT-A responses (84 s), indicating his attention deficits. A low score (16/36) of delayed recall in the ROCFT proved his visual memory impairment. He recovered from dysarthria and was discharged 42 days after admission.

2.8. Case 8

An 82-year-old male developed left mild hemiparesis (BRS-A 5, BRS-H 5, and BRS-L 5) on the day of onset, and MRI imaging of the brain revealed right pontine BAD. After admission, dysphagia was also revealed. He received conservative treatment with DAPT, an anticoagulant and an antioxidant drug. At two days after admission, his MMSE score was 23/30 (a total of 7 points were lost in Serial sevens and delayed recall), although guaranteeing his preserved orientation. It suggested his inattention and memory disturbance. He made a mistake on the similarities, lexical fluency, and Go/No-Go test, and his FAB score summed up to 10/18, suggesting his executive dysfunction. The response for TMT-A was delayed (243 s), suggesting his attention deficit. Because of his persistent hemiparesis, he was transferred to the rehabilitation hospital at eight days after admission.

2.9. Case 9

An 86-year-old female with independent activities of living developed right hemiparesis (BRS-A 2, BRS-H 2, and BRS-L 4), left facial palsy, and dysarthria. A risk factor for stroke was hypertension, and her past history included cardiac angina and skin cancer. An MRI revealed left pontine BAD. She received conservative treatment with DAPT, an anticoagulant and an antioxidant drug. Two days after admission, her MMSE had a full score of 30. Her dysexecutive function syndrome was evident in the low scores of similarities and lexical fluency on the FAB. Examinations of other domains were not available. Because of her persistent hemiparesis (BRS-A 3, BRS-H 4, and BRS-L 3), she was transferred to the rehabilitation hospital one month after admission.

3. Summary of Clinical Data and Review of the Literature

3.1. Summary of Case Series

Six patients presented with mild hemiparesis and eight patients showed dysarthria. Nine new cases of cognitive declines after a pontine stroke were documented. The domain-specific frequencies of cognitive dysfunction after a pontine stroke were delineated as follows: executive dysfunction (9/9 cases), visual memory disturbance (6/8), verbal memory disorder (2/8), and attention deficits (4/8) (Table 2). The dominant domains were characterized by a dysexecutive function and visual memory impairment, almost in line with previous studies. The severity was relatively mild. Short-term follow-up investigations were achieved in three cases, two in which the patients recovered from executive dysfunction following pons injury and the other with persistent cognitive decline. A long-term follow-up was not available for any patients. An overview from previous case studies, commonly demonstrating executive dysfunction as the most frequent domain, may support the above view. However, this domain tendency may be slightly different from our previous study [23], mainly because the latter was relatively lacking in in-depth neuropsychological investigations. An initial screen identified abstracts or titles. The second screening was based on the full-text review. Two investigators (KS and SO) independently assessed the full text for eligibility; discrepancies were resolved via discussion.

Table 2. Cognitive domain profiles of our case series.

	Executive Dysfunction (FAB; TMT-B; ROCFT Copy)	Visual Memory Disturbance (ROCFT Recall)	Verbal Memory Disturbance (S-PA)	Inattention (TMT-A)
Case 1	+ (16/18; 106 s; 26/36)	+ (16/36)	+	− (43 s)
Case 2	+ 16/18; 156 s; 31/36	NA	NA	+ (110 s)
Case 3	+ (16/18; 116 s; 36/36)	+ (16/36)	+/−	− (42 s)
Case 4	+ (15/18; 292 s; 31/36)	+ (11/36)	+/−	+ (118 s)
Case 5	+ (14/18; 209 s; 20/36)	NA	−	− (60 s)
Case 6	+ (15/18; 79 s; 32/36)	+ (11/36)	NA	− (60 s)
Case 7	+ (14/18; 98 s; 34/36)	+ (16/36)	NA	+ (84 s)
Case 8	+ (10/18; NA; NA)	NA	NA	+ (243 s)
Case 9	+ (15/18; NA; NA)	NA	NA	NA

FAB: frontal assessment battery; TMT: trail making test; ROCFT: Ray–Osterrieth complex figure test; S-PA: standard verbal paired-associate learning test; +: present; −: absent; +/−: borderline; NA: not available.

3.2. Literature Search and Study Eligibility

Online article databases including PubMed, MEDLINE, and Scopus were searched to identify the relevant literature on cases describing cognitive disturbances following brainstem stroke lesions. Cases with brain traumas, infections, or primary neurodegenerative diseases were excluded in this study, because they are typically associated with more widespread brain damage than an isolated brain injury, and they also often present functional deficits not limited to the lesion itself. The following keywords were used to search the electronic databases: *cognitive dysfunction, cognitive impairment, brainstem, pons, and pontine.* Only articles that provided details regarding the cases describing neurological tests were included in the overview. The articles were dated from 1998 until March 2023.

The initial literature search of keywords produced 345 results. Due to the inclusion criteria, 264 studies were excluded. Among the 19 studies screened for eligibility, 9 met our criteria. A flow chart summarizing the selection process is depicted in Figure 2.

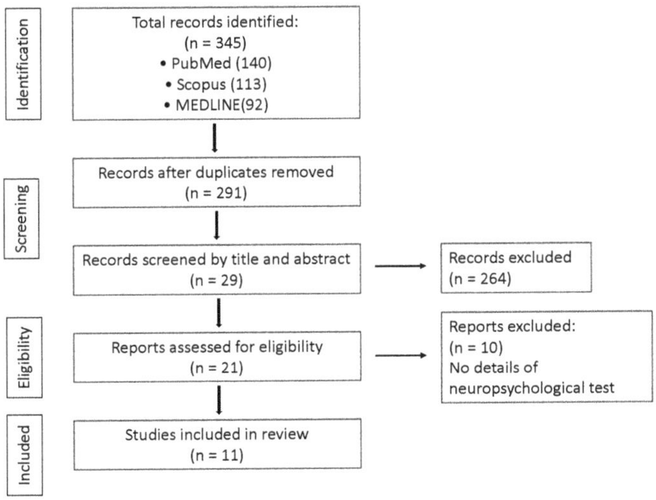

Figure 2. Flow diagram of included studies.

3.3. Methodological Quality and Risk of Bias

When undertaking a systematic review while considering the strengths and weaknesses of the research, the risk of bias in nonrandomized studies of interventions must be assessed. The Risk of Bias Assessment Tool for Nonrandomized Studies (RoBANS) version 2 is a comprehensive checklist instrument for assessing the risk of bias in cohort studies, case-control studies, and cross-sectional studies [58]. Case-control studies in the included papers were applied (Table 3). On the contrary, case reports and case series are uncontrolled study designs known for their increased risk of bias, but they have profoundly influenced the literature and continue to advance our knowledge. Murad presented a framework to evaluate the methodological qualities of case reports and case series based on the domains of selection, ascertainment, causality, and reporting [59]. The methodological quality of the included case reports was evaluated using this method (Table 4).

Table 3. Risk of bias in the case-control studies.

Author	Year	Comparability of the Target Group	Target GroupSelection	Confounders	Measurement of Interven-tion/Exposure	Blinding of Assessors	Outcome Assessment	Incomplete Outcome Data	Selective Outcome Reporting
Van Zandvoort [60]	2003	Low	Low	Low	High	Low	Low	Low	Low
Fu [61]	2017	Low	Low	Low	High	Low	Low	Low	Low
Wang [48]	2022	Low	Low	Low	High	Low	Low	Low	Low

Table 4. Risk of bias in case reports and case series.

Author	Year	Documentation	Uniqueness	Educational Value	Objectivity	Interpretation
Our case series	Present	2	1	2	2	2
Hoffman and Watts [58]	1998	2	2	2	2	2
Hoffman and Malek [26]	2005	2	2	2	2	2
Garrard et al. [21]	2002	2	1	2	2	2
Nishio et al. [62]	2007	2	1	2	2	2
Maeshima et al. [63]	2010	2	1	2	2	2
D'aes and Marien [19]	2014	2	1	2	2	2
Neki et al. [64]	2014	2	1	2	2	2
Obayashi [23]	2019	2	1	2	2	2

Questions 1–5 comprise the tool for the risk of bias assessment for case reports and case series: 1. Did the patient(s) represent the whole case(s) of the medical center? (The studies did not mention whether the reported patient(s) represented the whole case(s) of the medical center, and we assumed that the authors have reported all the cases in their center given the rarity of this association.). 2. Was the diagnosis correctly made? 3. Were other important diagnoses excluded? 4. Were all important data cited in the report? 5. Was the outcome correctly ascertained?

3.4. Synthesis

Owing to the study heterogeneity in terms of the quantitative designs and methods, a quantitative synthesis was not feasible in this systematic review. A narrative approach to synthesis was applied, including textual descriptions, tabulation, grouping, and content analysis for data translation [65]. One author (KS) carried out the initial synthesis, and iterative discussions with the other author (SO) were conducted to refine the essential elements of this review.

3.5. Characteristics of Cognitive Impairment after Pontine Stroke

There are some evidential items from group analyses that were determined by comparing the pontine stroke patients and healthy controls. Van Zandvoort et al. described the cognitive dysfunction of 17 patients with a brainstem stroke (including 13 with pons lesions) [60]. The brainstem group had significantly impaired language (naming), category fluencies, attention, executive functions, and visuospatial abilities compared with the age-matched control group. Wang et al. compared a set of neuropsychological results between 47 pontine stroke patients and 55 age-matched healthy control subjects. The pons lesion group had a significantly weakened executive function, working memory, and spatial memory relative to the healthy control group [48]. Fu et al. compared neuropsychological data between 34 brainstem stroke patients and a healthy control group. In regard to the distribution of the lesions, the mesencephalon was most common, followed by the pons. Significant differences in attention, visuospatial abilities, and language can be discerned in the brainstem lesion group [61].

So far, several individual details of the disabilities have been so far reported. D'aes and Mariën analyzed the cognitive characteristics of a brainstem stroke in 33 patients. This included 22 pontine stroke victims who exhibited executive dysfunctions, memory disturbances, and attention deficits [19]. They claimed that the most frequent domain was that of executive dysfunction, followed by inattention. For further details, Hoffmann and Watts reported that all four pontine stroke patients had executive dysfunction, such as disabilities in planning, initiating, and executing activities, as well as impaired self-monitoring [22]. Another case report demonstrated executive dysfunction [26]. Gerrard et al. reported that a pontine stroke generated executive dysfunction (3/4 patients), attention deficits (3/4), and memory disturbances (1/4) [21]. Four months of follow-up delineated persistent attention deficits and executive dysfunctions. All domain-specific outcomes were likely to be less severe and more transient than supratentorial cortical infarcts, which was also consistent with a previous study [37]. Obayashi described neuropsychological findings from 25 patients [23]. They presented attention deficits (20/25 patients), memory disturbances (15/25), executive dysfunctions (15/25), and social behavioral disturbances (1/25), with one cognitive domain for four patients, six for two domains, and seven for three domains [23]. A few months of follow-ups was achieved for two patients, who showed improvements in executive dysfunction. Nishio et al. [62] described a single case of a 74-year-old female with a pontine infarct who developed attention deficits and dysexecutive function syndrome while her memory, language, and visuospatial abilities remained intact. Maeshima et al. [63] reported a case of a 54 year-old male with a pontine hemorrhage presenting with executive dysfunction, attention deficits, and memory disturbance. Six months after onset, all domains were recovered. Neki et al. [64] demonstrated domain-specific impairments after a pontine hemorrhage. Five out of ten patients met the inclusion criteria, all of which affected the executive function. However, other domains were not investigated, except for the general intelligence ability.

The overall results of the included studies are summarized in Table 5.

Table 5. Summary of neuropsychological findings in the included studies.

Authors (Year) [Reference Number]	Classification	Executive Dysfunction	Inattention	Memory Disturbance	Linguistic Difficulty	Visuospatial Disability	General Intelligence
Hoffman and Watts (1998) [58]	5 cases	++	NA	±	NA	+	NA
Garrard et al. (2002) [21]	7 cases	++	+	+	−	−	±
Hoffman and Malek (2005) [26]	1 case	±	NA	−	+	+	NA
Nishio et al. (2007) [62]	1 case	++	+	±	−	−	±
Maeshima et al. (2010) [63]	1 case	+	+	+		−	−
D'aes and Marien (2014) [19]	3 cases	++	++	+	+	+	++
Neki et al. (2014) [64]	10 cases	+	NA	NA	NA	NA	+
Obayashi (2019) [23]	25 cases	+	++	+	−	NA	±
Van Zandvoort et al. (2003) [60]	Group comparison, 17 PS patients	+	+	NA	+	+	NA
Wang et al. (2022) [48]	Group comparison, 47 PS patients	++	−	+	NA	NA	NA
Fu et al. (2017) [61]	Group comparison, 34 PS patients	±	+	−	+	+	NA

PS: pontine stroke; ++: positive symptom; ±: suspected; −: negative.

4. Discussion

Conventionally, physicians and healthcare professionals posit that damage to the brainstem might not affect cognition. The aim of this paper is to make clear whether a pontine stroke could affect the cognition of patients and to disclose the domain characteristics while providing an overview of our case series and previous studies relevant to cognitive decline after a pontine stroke. Nine new cases of cognitive declines poststroke were commonly characterized by executive dysfunction, almost in line with previous studies. The severity was relatively mild. This trait is supported by a previous study exhibiting prefrontal dysfunction after brainstem damage [36] and is also based on anatomical evidence of the frontopontine pathways from the frontal association areas [66]. The ultimate goal of this paper is to disclose the presence and characteristics of cognitive impairments due to pontine stroke and to help the patients reintegrate into society. Awareness of the disabilities will enlighten physicians in clinical practice and likely improve clinical diagnostics and patient care. The notion that the damage to the pontine affects cognition may have a significant impact on rehabilitation interventions, usually assuming that all attention in the intervention should be directed toward pure sensorimotor recovery for patients with an isolated pontine stroke. It will be better if physicians and healthcare professionals decide more carefully whether the patient can return to work or drive a car. In the coming years, this research focus will be critical for further understanding the specific functions of the pons in cognition and the neuropathology of underlying cognitive deficits due to pontine injury. It will provide new insights into the neurobiology of cognition and develop new treatment and management strategies for cognitive impairment after a stroke. To achieve this goal, the use of non-invasive neuroimaging techniques would be very powerful and beneficial. Now, we highlight neuropsychological findings from our nine new case studies as well as the relevant studies, and as mentioned below, the literature review will get to the core of how the application of multi-modal neuroimaging techniques will expand our understanding of the neural mechanisms responsible for cognitive disabilities following a pontine stroke.

4.1. Neurobiology of Executive Function, Attention, and Memory

It is no wonder that the pathophysiology of cognitive impairment results from the disruptions of neural mechanisms underlying cognitive domains, such as executive function, attention, and memory. Basically, executive function is composed of self-control, self-monitoring, emotional control, flexibility, task initiation, organization, working memory, and planning and time management, so that the achievement of these functions may involve a widely distributed cortical network [67–69]. The function is responsible for the inferior frontal cortex (IFG), basal ganglia, and pre-supplementary motor area (pre-SMA) [70] or anterior cingulate cortex and parietal cortex [71], as well as the cerebellum [15]. Accordingly, these brain regions may be related to dysexecutive function after pontine injury. The molecular mechanism of executive function remains unknown, but an intriguing study has been reported. The paper stated the relevance of C-C chemokine receptor 5 (CCR5) expression in the cortex for cognitive recovery and motor recovery [72]. The authors claimed that CCR5 is upregulated in the cerebral cortex after a stroke and traumatic brain injury, and that spatial working memory disturbances due to traumatic brain injury were improved by the inhibition of CCR5, suggesting that cognitive recovery may be modulated by CCR5 expression.

On the other hand, models of attention have been postulated [73–76], but the definition of attention itself remains unclear. Posner proposed a hypothesis that attention may be composed of the alerting system, the orienting system, and the executive control of attention [77–79]. The neural mechanism for attention can be separated by three systems: (1) alerting that produces and maintains optimal vigilance, (2) orienting that prioritizes sensory input by selecting a modality or location, and (3) executive control that involves task switching, initiation, adjustments, and maintenance within trials in real time. The alerting system is attributed to the right frontal cortex and right parietal cortex. The orienting system involves the superior parietal and frontal lobes, and the executive control of attention is subserved by the frontoparietal system and cingulo–opercular system [80]. So, attention and executive function cannot be separated from each other. This suggests that attentional processing might be relatively fundamental and is involved in a broader range of brain areas than executive function. Moreover, it is possible that pontine damage is more likely to generate executive dysfunction and attentional deficits.

The mechanism of memory has been extensively studied in greater detail than other domains. A great deal about the cellular and molecular mechanisms of long-term memory storage has been learned at the level of the synapse [81–84], but the mechanism of consolidation at the level of neuronal systems has been relatively overlooked [85,86]. Episodic memory refers to a declarative memory that contains information specific to the time and place of acquisition [87]. Episodic memory retrieval is attributable to the frontal cortex, posterior parietal cortex, and medial temporal cortex [88], whereas semantic memory retrieval is responsible for the top–down signal from the prefrontal cortex and the subsequent memory processing of perirhinal circuits as a storage of memory in a hierarchical manner [89]. In contrast, procedural memory is achieved by the activation transition of the fronto–parieto–cerebellar circuit, such that the dorsolateral PFC and pre-SMA engage at an early stage of learning, the parietal IPS and precuneus work at an intermediate stage, and the cerebellum serves as storage at the final stage [90–93]. Also, the fronto–parieto–cerebellar circuit through the corpus callosum may contribute to the inter-manual transfer of procedural memory [92,93].

4.2. Short- and Long-Term Changes in Cognitive Decline Poststroke

The poststroke cognitive function changes temporally and dynamically over time. However, details of the longitudinal trajectory of domain-specific cognitive alterations after a stroke remain unknown [94–96]. Until now, few studies have dealt with the longitudinal effect of stroke on cognition, and they showed mixed results, either a trend toward deterioration [97], persistence [98,99], or improvements [100,101]. A recent study reported the probability of poststroke cognitive declines and follow-up alterations [101]. Cognitive impairment was present in 59% of survivors at three months poststroke, and 51% remained at eighteen months after onset. Some domains, such as executive function and language, improved during the follow-up period, but it is difficult to determine which of the cognitive domains was more inclined to recovery. Other reports suggested that executive and language functions as well as the visuospatial function may improve 3 months to 1.5 years after a stroke [101–103], while another report described that working memory may eventually recover years after a stroke [23]. In addition, the speed of recovery may differ depending on the lesion location. In the case of younger stroke survivors, domain-specific cognitive impairments improved or were stable 10 years after their stroke [102]. On the other hand, stroke is associated with an increased risk of dementia [104–107]. A previous report demonstrated about a twofold increase in cognitive decline after a stroke relative to before the stroke [94]. The executive performance has also been reported to be an excellent predictor of vascular dementia [108,109]. Most stroke survivors may fully recover from the decline between 3 and 15 months afterward [110,111], but others do not recover [112] and deteriorate to dementia. Likewise, a systematic review focusing on the natural history of cognitive impairment after a stroke replicated the mixed results [103]. As mentioned above, however, our observation and the relevant literature may seemingly imply the relatively early regaining of cognitive functions after pontine stroke. This view would be supported by the pathophysiology of cognitive decline after a pontine stroke in terms of "diaschisis".

4.3. Profile of "Diaschisis"

"Diaschisis" is well known as a phenomenon consisting of a broad range of depressed brain functions, which are remote from local lesions of the central nervous system. Age negatively influences the severity of diaschisis and determines the extent to which the patient recovers. The older the patient after a stroke, the more severe the neurological deficit and the less complete the neurological recovery. In addition, the corpus callosum plays a crucial role in the remote effects of diaschisis [113]. A previous study suggested that the deficits arising from infratentorial infarcts tended to be less severe and more transient than those from supratentorial cortical infarcts [37]. At the very least, given that cognitive decline after pontine injury represents cerebro-cerebellar diaschisis, it is expected that the decline would be less disabling and would eventually recover.

4.4. Neuroimaging of "Diaschisis": Single Photon Emission Tomography (SPECT)

A previous study reported that 58% of 55 patients with a supratentorial stroke presented with cerebellar hypoperfusion [114]. Some SPECT studies yielded evidence of cerebellar hypoperfusion after a pontine stroke [115] and lateral medullary infarcts [116]. Conversely, patients with a unilateral cerebellar stroke revealed contralateral cerebral hypoperfusion [117]. A previous SPECT study demonstrated that patients with a brainstem stroke presented an aberrant perfusion pattern in the ipsilateral frontoparietal lobes and the contralateral cerebellum, as evidence of diaschisis [20]. Another SPECT study reported frontal and parietal hypoperfusion in patients with brainstem infarcts [22]. The possible mechanism of the remote effects may be explained by the reciprocal connections of cortico-pontine-cerebellar fibers, which in turn project to the red nucleus and ventrolateral nucleus of the thalamus and to the frontoparietal cortex. Our recent SPECT study delineated that the patients in the acute phase of a pontine infarct showed hyper-perfusion in the bilateral frontal cortices, parietal cortices, and right thalamus and hypo-perfusion in bilateral cerebellum [23]. For more details regarding our earlier report [23], in our previously reported Case 3 (not the present case 3 mentioned above), SPECT revealed decreases in the right Brodmann area (BA) 39 and right putamen, and increases in the bilateral BAs 6 and 8, 44, bilateral 40, bilateral BA 24s and 32, and left putamen. In an earlier Case 4 from the report, SPECT revealed decreases in the left BAs 44 and 45, bilateral BAs 24 and 32, and bilateral putamen, and increases in the right BAs 6 and 8, BA 45, right BA 24, and bilateral BAs 39 and 40. In the earlier Case 5, SPECT showed decreases in the bilateral putamen, left BA 32, and right BA 39, and increases in the bilateral BAs 6 and 8, BAs 44 and 45, right BA 24, and bilateral BA 40. In the earlier Case 7, SPECT revealed decreases in the left BA 39, 40, and left putamen, and increases in the bilateral BA 6, 8, 44, right BA 39, 40, right BA 24, and bilateral putamen. It is very likely that depressed brain areas mainly represent vascular alterations of the diaschisis phenomenon in terms of neuro-vascular coupling, while hyper-perfusion brain areas largely reflect the compensating process for cognitive decline after a pontine stroke. Possible mechanisms for inter-subjective differences in perfusion abnormalities may be explained by the following: (1) differences in the cognitive domain, (2) severity of the deficits, (3) proportion of alterations to compensation process, and (4) inter-subject alterations of C-C chemokine receptor 5 (CCR5) expression after stroke [72,118]. Especially, poststroke CCR5 expression in the affected brain may be closely associated with the prognosis of motor and cognitive recovery after stroke. Pontine injury brings about secondary brain alterations remote from the damaged location during a term of diaschisis and is immediately followed by a compensation process for the diaschisis phenomenon. Very likely, after a stroke, diaschisis and compensation are mixed and, in some cases, competing with each other.

Another recent study investigated longitudinal regional cerebral blood flow (rCBF) changes in the acute phase, as well as follow-ups 1 week to 6 months after the pontine infarct (PI) [40]. There were significant rCBF decreases in the bilateral cerebellum and frontal (right supplementary motor area: SMA), parietal (right supramarginal gyrus), and occipital regions in the acute phase of the PI. The association of these alterations with the long-term cognitive outcome following a PI differed depending on the lesion location. In the left PI group, motor and memory recovery were associated with progressive increasing rCBF in the right supramarginal gyrus, whereas in the right PI group, memory and motor recovery were related to an increasing rCBF in the right SMA.

4.5. Morphological and Neurodegenerative Changes Induced by Pontine Stroke: MRI Studies

The brain may change after a pontine ictus. Voxel-based morphometry (VBM) can detect morphological brain changes after a pontine infarction [41,43,119,120]. A previous study revealed that a pontine infarction may reduce the gray matter volume (GMV) in the cerebellum and expand the ipsilateral GMV in the middle frontal gyrus, middle temporal gyrus, mediodorsal thalamus, superior frontal gyrus, and contralateral precuneus [41]. It was suggested that GMV expansion in the ipsilateral mediodorsal thalamus was associated with motor recovery after a pontine infarction, although the association of GMV with cognitive recovery was not addressed.

Diffusion tensor imaging (DTI) has been used to detect anterograde degeneration in the pyramidal tracts distal to a supratentorial lesion following an infarct [121]. Moreover, anterograde and retrograde degeneration remote from the primary lesion continuously deteriorates following a subcortical infarction, which interferes with poststroke functional recovery [45]. Likewise, DTI can detect the continuous deterioration of anterograde and retrograde degeneration in the pyramidal tract following a pontine infarct [45]. Therefore, it is plausible that similar progressive degeneration to that of the pyramidal tract following pontine injury may occur in any of the other tracts. In favor of this view, some functional connectivity MRI (fcMRI) studies have demonstrated alterations in the functional connectivity measures of the multiple pathways that are disrupted by focal damage to the pons [50,122,123]. The authors claimed that a pontine infarct may disrupt the prefrontal-cerebellar circuit. A decreased functional connectivity may be related to cognitive decline after a pontine stroke.

4.6. *Insights from Near-Infrared Spectroscopy*

As mentioned above, our recent SPECT study demonstrated frontal hyper-perfusion and cerebellar hypo-perfusion, which shared consistent results from four patients with a pontine infarct [23]. In particular, the hyper-perfusion areas commonly included the supplementary motor area (SMA: Brodmann areas 6 and 8). Although the function of the SMA is not yet fully understood, the SMA may contribute to speech production [124–126], word retrieval difficulty by aging [127], inhibitory control [128], and executive function [129–133]. Previously, Penfield postulated that the SMA might be a third speech area, based on the evidence of vocalization and speech arrest by direct electric stimulations of the SMA [134,135]. The SMA is functionally divided into at least two distinct areas: the SMA proper, posterior to the vertical anterior commissorial (VAC) line, and perpendicular to the anterior commissure (AC)—posterior commissure (PC) plane; and the pre-SMA, anterior to the VAC line [136–138]. In fact, the SMA proper and pre-SMA are anatomically different: the SMA proper receives information from all components of the motor system [139], whereas the pre-SMA is densely interconnected with the prefrontal cortex (PFC) and also receives input from basal ganglia and cerebellum, but has no connection with the motor system [140].

To clarify the neurophysiology of SMA hyper-perfusion, we measured dynamic changes in the SMA responses during the phonemic verbal fluency task (VFT) as an index of executive function using functional near-infrared spectroscopy (f-NIRS) [23]. The pontine infarct group had no significant difference in their fluency ability compared with the age-matched control group. Furthermore, no significant differences in SMA responses could be detected between the two groups, but the SMA responses had a moderate correlation with the executive function. On the other hand, the pontine infarct group had executive dysfunction, as proven by the delayed TMT, expectedly making this domain relatively mild and transient. In fact, we observed the recovery of this domain a few months later. Intriguingly, a follow-up f-NIRS demonstrated increased the SMA responses coupled with improving the TMT-B, suggesting the contribution of the SMA to cognitive recovery after pontine injury [23].

4.7. *Similarities and Differences of Functions among Pons, Cerebellum, and Thalamus*

D'aes and Mariën claimed that damage to the brainstem may affect cerebellar function, as cognitive decline due to a brainstem stroke seemed to share some common cognitive domains with the cerebellar cognitive affective syndrome (CCAS), comprising executive dysfunction, difficulties in spatial cognition, linguistic difficulties, and personality changes [15,19,141,142].

On the other hand, the function of the thalamus should be more complicated than those of the pons and cerebellum, and it would be different from those of pons or cerebellum in terms of cognition. The thalamus serves cognitive and language functions as the final hub of a sensory information relay to the neocortex, striatum, and hippocampus by divergent and convergent thalamocortical and corticothalamic pathways [143]. In a cognitive aspect, damage to the anterior portions of the thalamus generates memory loss [144,145], and damage to midline thalamic nuclei causes inattention and executive dysfunction [146,147]. In the linguistic aspect, it has been a matter of debate whether the thalamus plays a role in language [148,149]. According to a previous review [150], almost 90% of the left thalamic and bilateral thalamic patients 3 weeks to 4 months post-stroke presented with memory disturbances, inattention, executive dysfunctions, and behavioral and/or mood alterations. Linguistic difficulties, such as fluency (6.4%), repetition (15.1%), naming (72.2%), auditory comprehension (43.8%), reading (25%), and writing (65%) were found in patients with left thalamic lesions (n = 37), and comprehension (1/2), repetition (1/2), and naming (2/2) were found in those with bilateral thalamic lesions (n = 3). Our recent study reported that 25 of the 27 patients with acute thalamic stroke (92.6%) had cognition impairments, including inattention (18 patients), memory disturbances (15), executive dysfunctions (11), social behavioral disturbance (1), and aphasia (3) (Table 6). Also, we demonstrated that the thalamus plays a specific role in this loop, different from the pons or cerebellum, using two modalities of neuroimaging techniques such as SPECT and f-NIRS [151]. The SPECT results obtained from patients with thalamic lesions yielded evidence of common perfusion abnormalities in the fronto–parieto–cerebellar loop, including SMA, IFG, and surrounding language-relevant regions. In NIRS sessions during VFT, the thalamic stroke group encountered significant word retrieval difficulties relative to the age-matched healthy group. This implies that executive dysfunction due to a thalamic stroke may be more severe than a pontine ictus. Furthermore, a strong correlation between word retrieval and SMA responses has been demonstrated, and this suggests that there is a tight link between the thalamus and SMA. A follow-up NIRS revealed that increasing bilateral SMA responses may be associated with word retrieval improvements. The findings suggest that cognitive dysfunction after thalamic stroke may be related to the fronto–parieto–cerebellar loop, while language dysfunction is attributed to the SMA, inferior frontal gyrus (IFG), and language-related brain areas [151,152]. Together, the SMA may be responsible for the recovery of executive dysfunction after a thalamic stroke, as the SMA plays a role in cognitive recovery after a pontine stroke. These findings demonstrate that thalamic injury disrupts the SMA function more seriously than a pontine stroke, thus leading to cognitive impairments more directly than pons (Figure 3).

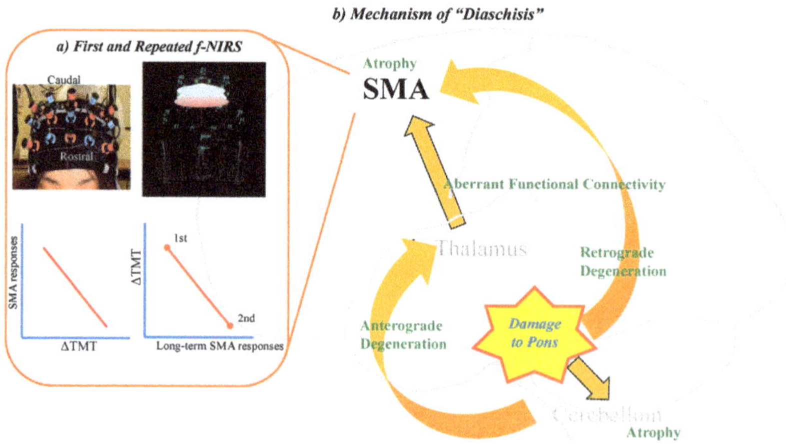

Figure 3. Neural mechanism underlying cognitive dysfunction after pontine stroke.

Table 6. Similarities and differences: pons, cerebellum, and thalamus.

	Executive Dysfunction	Inattention	Memory Disturbance	Linguistic Difficulties	Spatial Cognition Difficulties	Personality Changes
Pons						
Cerebellum						
Thalamus						

The color density of the rows in each lesion represents the occurrence frequency of each domain, i.e., the deep green scale as a higher frequency and the light green as lower. The color gradient may represent hierarchical cognitive processing in the fronto–ponto–cerebellar–thalamic loop.

4.8. Limitations and Future Direction

Our observations in the case series are derived from only a small sample of patients at a single institute, resulting in the prevention of generalized results. Larger sample sizes recruited from multicenter institutes are needed to confirm these findings. In Japanese medical circumstances, where most of the patients are transferred out for intensive rehabilitation within a few weeks of admission, we have difficulty in obtaining data from the long-term follow-up neuropsychological evaluation. Therefore, the prognoses and long-term trajectories of the cognitive disabilities remains largely unknown. We need to establish a criterion to decide exactly which measure would be more sensitive and more specific to each domain (executive function, attention, and memory) among the in-depth neuropsychological tests. To date, executive function is composed of multiple cognitive components. Further study is required to clarify which component is more vulnerable for pontine injury. The function of pons has been paid no attention in the basic neuroscience field. In the coming decade, an advance in translational research using an animal model with invasive techniques would provide us with great details on the molecular mechanisms of cognition. For example, the acquired genetic manipulation of a specific molecule at a specific legion by means of "optogenetics" technology might help us better understand how pons would be involved in cognitive processing at the molecular level. The application of new techniques and translational research would provide us with a new pharmacological intervention or other therapeutic approaches and management strategies which might facilitate the improvement of cognitive disabilities.

Pontine injury induces morphological and neurodegenerative changes in the fronto–ponto–cerebellar–thalamic loop, resulting in the failure of information processing and then leading to cognitive decline. f-NIRS could monitor the dynamic changes in the SMA associated with executive dysfunction due to pontine stroke and eventually, cognitive recovery by monitoring the SMA responses using f-NIRS and follow-ups.

5. Conclusions

Nine personal observations and a review of the literature showed that a range of cognitive symptoms may result from isolated pontine damage. In particular, executive dysfunction represents the most common cognitive symptom. In the vast majority of the previous neuroimaging literature dealing with cognitive deficits after a pontine stroke, a frontal perfusion abnormality was found. In addition, our unique data combining SPECT and f-NIRS show the involvement of fronto–cerebellar diaschisis, and also suggest that SMA responses might eventually reflect the severity of cognitive decline due to pontine stroke and may also be related to the recovery. We finally posit that cognitive decline after a pontine stroke may be attributable to fronto–ponto–cerebellar diaschisis. In other words, pons constitutes an intrinsic part of the fronto–cerebellar–thalamic loop, while each area performs its part in cognitive processing in a hierarchical manner, and that an isolated pontine injury can result in a variety of symptoms that are typically associated with a disrupted processing relay.

Author Contributions: Conceptualization, S.O.; writing—original draft preparation, K.S.; writing—review and editing, S.O.; project administration, S.O. All authors have read and agreed to the published version of the manuscript.

Funding: This research was partially supported by a grant from Grants-in-Aid for Scientific Research, Japan Society for the Promotion of Science (JSPS), awarded to S.O. (22K11425).

Institutional Review Board Statement: The study was conducted in accordance with the Declaration of Helsinki and was approved by the Ethics Committee of Saitama Medical Center (Approval number: 2021-093).

Informed Consent Statement: Informed consent was obtained from all subjects involved in the study.

Data Availability Statement: The data are available upon reasonable email request.

Acknowledgments: We thank the staff of our acute rehabilitation care unit for their support.

Conflicts of Interest: The authors declare no conflicts of interest.

References

1. Nys, G.M.S.; van Zandvoort, M.J.E.; de Kort, P.L.M.; van der Worp, H.B.; Jansen, B.P.W.; Algra, A.; de Haan, E.H.F.; Kappelle, L.J. The prognostic value of domain-specific cognitive abilities in acute first-ever stroke. *Neurology* **2005**, *64*, 821–827. [CrossRef] [PubMed]
2. Weimar, C.; Konig, I.R.; Kraywinkel, K.; Ziegler, A.; Diener, H.C.; German Stroke Study, C. Age and National Institutes of Health Stroke Scale Score within 6 hours after onset are accurate predictors of outcome after cerebral ischemia: Development and external validation of prognostic models. *Stroke* **2004**, *35*, 158–162. [CrossRef] [PubMed]
3. Jorgensen, H.S.; Reith, J.; Nakayama, H.; Kammersgaard, L.P.; Raaschou, H.O.; Olsen, T.S. What determines good recovery in patients with the most severe strokes? The Copenhagen Stroke Study. *Stroke* **1999**, *30*, 2008–2012. [CrossRef] [PubMed]
4. Hankey, G.J.; Jamrozik, K.; Broadhurst, R.J.; Forbes, S.; Anderson, C.S. Long-term disability after first-ever stroke and related prognostic factors in the Perth Community Stroke Study, 1989–1990. *Stroke* **2002**, *33*, 1034–1040. [CrossRef] [PubMed]
5. Miyai, I.; Suzuki, T.; Kang, J.; Volpe, B.T. Improved functional outcome in patients with hemorrhagic stroke in putamen and thalamus compared with those with stroke restricted to the putamen or thalamus. *Stroke* **2000**, *31*, 1365–1369. [CrossRef] [PubMed]
6. Thijs, V.N.; Lansberg, M.G.; Beaulieu, C.; Marks, M.P.; Moseley, M.E.; Albers, G.W. Is early ischemic lesion volume on diffusion-weighted imaging an independent predictor of stroke outcome? A multivariable analysis. *Stroke* **2000**, *31*, 2597–2602. [CrossRef] [PubMed]
7. Malm, J.; Kristensen, B.; Karlsson, T.; Carlberg, B.; Fagerlund, M.; Olsson, T. Cognitive impairment in young adults with infratentorial infarcts. *Neurology* **1998**, *51*, 433–440. [CrossRef]
8. Moliis, H.; Jokinen, H.; Parkkonen, E.; Kaste, M.; Erkinjuntti, T.; Melkas, S. Post-Stroke Cognitive Impairment is Frequent After Infra-Tentorial Infarct. *J. Stroke Cerebrovasc. Dis.* **2021**, *30*, 106108. [CrossRef]
9. Ferro, J.M. Hyperacute cognitive stroke syndromes. *J. Neurol.* **2001**, *248*, 841–849. [CrossRef]
10. van Zandvoort, M.J.; Kessels, R.P.; Nys, G.M.; de Haan, E.H.; Kappelle, L.J. Early neuropsychological evaluation in patients with ischaemic stroke provides valid information. *Clin. Neurol. Neurosurg.* **2005**, *107*, 385–392. [CrossRef]
11. Galski, T.; Bruno, R.L.; Zorowitz, R.; Walker, J. Predicting length of stay, functional outcome, and aftercare in the rehabilitation of stroke patients. The dominant role of higher-order cognition. *Stroke* **1993**, *24*, 1794–1800. [CrossRef] [PubMed]
12. Paolucci, S.; Antonucci, G.; Gialloreti, L.E.; Traballesi, M.; Lubich, S.; Pratesi, L.; Palombi, L. Predicting stroke inpatient rehabilitation outcome: The prominent role of neuropsychological disorders. *Eur. Neurol.* **1996**, *36*, 385–390. [CrossRef] [PubMed]
13. Tatemichi, T.K.; Desmond, D.W.; Stern, Y.; Paik, M.; Sano, M.; Bagiella, E. Cognitive impairment after stroke: Frequency, patterns, and relationship to functional abilities. *J. Neurol. Neurosurg. Psychiatry* **1994**, *57*, 202–207. [CrossRef] [PubMed]
14. Nys, G.M.S.; van Zandvoort, M.J.E.; de Kort, P.L.M.; Jansen, B.P.; de Haan, E.H.F.; Kappelle, L.J. Cognitive disorders in acute stroke: Prevalence and clinical determinants. *Cerebrovasc. Dis.* **2007**, *23*, 408–416. [CrossRef] [PubMed]
15. Schmahmann, J.D.; Sherman, J.C. Cerebellar cognitive affective syndrome. *Int. Rev. Neurobiol.* **1997**, *41*, 433–440. [PubMed]
16. De Smet, H.J.; Paquier, P.; Verhoeven, J.; Marien, P. The cerebellum: Its role in language and related cognitive and affective functions. *Brain Lang.* **2013**, *127*, 334–342. [CrossRef] [PubMed]
17. Baillieux, H.; De Smet, H.J.; Paquier, P.F.; De Deyn, P.P.; Marien, P. Cerebellar neurocognition: Insights into the bottom of the brain. *Clin. Neurol. Neurosurg.* **2008**, *110*, 763–773. [CrossRef] [PubMed]
18. Schmahmann, J.D.; Sherman, J.C. The cerebellar cognitive affective syndrome. *Brain* **1998**, *121 Pt 4*, 561–579. [CrossRef]
19. D'Aes, T.; Marien, P. Cognitive and affective disturbances following focal brainstem lesions: A review and report of three cases. *Cerebellum* **2015**, *14*, 317–340. [CrossRef]
20. Fazekas, F.; Payer, F.; Valetitsch, H.; Schmidt, R.; Flooh, E. Brain stem infarction and diaschisis. A SPECT cerebral perfusion study. *Stroke* **1993**, *24*, 1162–1166. [CrossRef]
21. Garrard, P.; Bradshaw, D.; Jäger, H.R.; Thompson, A.J.; Losseff, N.; Playford, D. Cognitive dysfunction after isolated brain stem insult. An underdiagnosed cause of long term morbidity. *J. Neurol. Neurosurg. Psychiatry* **2002**, *73*, 191–194. [CrossRef] [PubMed]

22. Hoffmann, M.; Watts, A. Cognitive dysfunction in isolated brainstem stroke: A neuropsychological and SPECT study. *J. Stroke Cerebrovasc. Dis.* **1998**, *7*, 24–31. [CrossRef] [PubMed]
23. Obayashi, S. Frontal dynamic activity as a predictor of cognitive dysfunction after pontine ischemia. *NeuroRehabilitation* **2019**, *44*, 251–261. [CrossRef] [PubMed]
24. Omar, R.; Warren, J.D.; Ron, M.A.; Lees, A.J.; Rossor, M.N.; Kartsounis, L.D. The neuro-behavioural syndrome of brainstem disease. *Neurocase* **2007**, *13*, 452–465. [CrossRef] [PubMed]
25. Evyapan, D.; Kumral, E. Pontine anosognosia for hemiplegia. *Neurology* **1999**, *53*, 647–649. [CrossRef] [PubMed]
26. Hoffmann, M.; Malek, A. Motor aprosodia due to isolated brainstem stroke in a young woman. *Acta Neurol. Scand.* **2005**, *112*, 197–198. [CrossRef] [PubMed]
27. Hurley, R.A.; Flashman, L.A.; Chow, T.W.; Taber, K.H. The brainstem: Anatomy, assessment, and clinical syndromes. *J. Neuropsychiatry Clin. Neurosci.* **2010**, *22*, 1–7. [CrossRef] [PubMed]
28. Trimble, M.R.; Cummings, J.L. Neuropsychiatric disturbances following brainstem lesions. *Br. J. Psychiatry* **1981**, *138*, 56–59. [CrossRef]
29. Arseni, C.; Goldenberg, M. Psychic disturbances in infiltrative gliomata of the brain stem. *Acta Neurochir.* **1959**, *7*, 292–300. [CrossRef]
30. Minabe, Y.; Kadono, Y.; Kurachi, M. A schizophrenic syndrome associated with a midbrain tegmental lesion. *Biol. Psychiatry* **1990**, *27*, 661–663. [CrossRef]
31. Greenberg, D.B.; Brown, G.L. Mania resulting from brain stem tumor. *J. Nerv. Ment. Dis.* **1985**, *173*, 434–436. [CrossRef] [PubMed]
32. Netsky, M.G.; Strobos, R.R. Neoplasms within the midbrain. *AMA Arch. Neurol. Psychiatry* **1952**, *68*, 116–129. [CrossRef] [PubMed]
33. Oh, S.; Bang, O.Y.; Chung, C.S.; Lee, K.H.; Chang, W.H.; Kim, G.M. Topographic location of acute pontine infarction is associated with the development of progressive motor deficits. *Stroke* **2012**, *43*, 708–713. [CrossRef] [PubMed]
34. Bassetti, C.; Bogousslavsky, J.; Barth, A.; Regli, F. Isolated infarcts of the pons. *Neurology* **1996**, *46*, 165–175. [CrossRef] [PubMed]
35. Carrera, E.; Tononi, G. Diaschisis: Past, present, future. *Brain* **2014**, *137*, 2408–2422. [CrossRef] [PubMed]
36. Salgado, J.V.; Costa-Silva, M.; Malloy-Diniz, L.F.; Siqueira, J.M.; Teixeira, A.L. Prefrontal cognitive dysfunction following brainstem lesion. *Clin. Neurol. Neurosurg.* **2007**, *109*, 379–382. [CrossRef] [PubMed]
37. Meyer, J.S.; Obara, K.; Muramatsu, K. Diaschisis. *Neurol. Res.* **1993**, *15*, 362–366. [CrossRef]
38. Feeney, D.M.; Baron, J.C. Diaschisis. *Stroke* **1986**, *17*, 817–830. [CrossRef]
39. Miyazawa, N.; Uchida, M.; Fukamachi, A.; Fukasawa, I.; Sasaki, H.; Nukui, H. Xenon contrast-enhanced CT imaging of supratentorial hypoperfusion in patients with brain stem infarction. *AJNR Am. J. Neuroradiol.* **1999**, *20*, 1858–1862.
40. Wei, Y.; Wu, L.; Wang, Y.; Liu, J.; Miao, P.; Wang, K.; Wang, C.; Cheng, J. Disrupted Regional Cerebral Blood Flow and Functional Connectivity in Pontine Infarction: A Longitudinal MRI Study. *Front. Aging Neurosci.* **2020**, *12*, 577899. [CrossRef]
41. Wang, P.; Jia, X.; Zhang, M.; Cao, Y.; Zhao, Z.; Shan, Y.; Ma, Q.; Qian, T.; Wang, J.; Lu, J.; et al. Correlation of Longitudinal Gray Matter Volume Changes and Motor Recovery in Patients After Pontine Infarction. *Front. Neurol.* **2018**, *9*, 312. [CrossRef] [PubMed]
42. Wang, Y.; Wang, C.; Miao, P.; Liu, J.; Wei, Y.; Wu, L.; Wang, K.; Cheng, J. An imbalance between functional segregation and integration in patients with pontine stroke: A dynamic functional network connectivity study. *Neuroimage Clin.* **2020**, *28*, 102507. [CrossRef]
43. Wang, C.; Zhao, L.; Luo, Y.; Liu, J.; Miao, P.; Wei, S.; Shi, L.; Cheng, J. Structural covariance in subcortical stroke patients measured by automated MRI-based volumetry. *Neuroimage Clin.* **2019**, *22*, 101682. [CrossRef] [PubMed]
44. Tien, R.D.; Ashdown, B.C. Crossed cerebellar diaschisis and crossed cerebellar atrophy: Correlation of MR findings, clinical symptoms, and supratentorial diseases in 26 patients. *Am. J. Roentgraphy* **1992**, *158*, 1155–1159. [CrossRef] [PubMed]
45. Liang, Z.; Zeng, J.; Zhang, C.; Liu, S.; Ling, X.; Xu, A.; Ling, L.; Wang, F.; Pei, Z. Longitudinal investigations on the anterograde and retrograde degeneration in the pyramidal tract following pontine infarction with diffusion tensor imaging. *Cerebrovasc. Dis.* **2008**, *25*, 209–216. [CrossRef] [PubMed]
46. Zhang, M.; Lin, Q.; Lu, J.; Rong, D.; Zhao, Z.; Ma, Q.; Liu, H.; Shu, N.; He, Y.; Li, K. Pontine infarction: Diffusion-tensor imaging of motor pathways-a longitudinal study. *Radiology* **2015**, *274*, 841–850. [CrossRef] [PubMed]
47. Liang, Z.; Zeng, J.; Zhang, C.; Liu, S.; Ling, X.; Wang, F.; Ling, L.; Hou, Q.; Xing, S.; Pei, Z. Progression of pathological changes in the middle cerebellar peduncle by diffusion tensor imaging correlates with lesser motor gains after pontine infarction. *Neurorehabil. Neural. Repair* **2009**, *23*, 692–698. [CrossRef] [PubMed]
48. Wang, Y.; Wang, C.; Wei, Y.; Miao, P.; Liu, J.; Wu, L.; Li, Z.; Li, X.; Wang, K.; Cheng, J. Abnormal functional connectivities patterns of multidomain cognitive impairments in pontine stroke patients. *Hum. Brain Mapp.* **2022**, *43*, 4676–4688. [CrossRef]
49. Wu, L.; Wang, C.; Liu, J.; Guo, J.; Wei, Y.; Wang, K.; Miao, P.; Wang, Y.; Cheng, J. Voxel-Mirrored Homotopic Connectivity Associated With Change of Cognitive Function in Chronic Pontine Stroke. *Front. Aging Neurosci.* **2021**, *13*, 621767. [CrossRef]
50. Jiang, L.; Geng, W.; Chen, H.; Zhang, H.; Bo, F.; Mao, C.N.; Chen, Y.C.; Yin, X. Decreased functional connectivity within the default-mode network in acute brainstem ischemic stroke. *Eur. J. Radiol.* **2018**, *105*, 221–226. [CrossRef]
51. Folstein, M.F.; Folstein, S.E.; McHugh, P.R. "Mini-mental state". A practical method for grading the cognitive state of patients for the clinician. *J. Psychiatr. Res.* **1975**, *12*, 189–198. [CrossRef] [PubMed]
52. Reitan, R.M. The relation of the trail making test to organic brain damage. *J. Consult. Psychol.* **1955**, *19*, 393–394. [CrossRef] [PubMed]

53. Sanchez-Cubillo, I.; Perianez, J.A.; Adrover-Roig, D.; Rodriguez-Sanchez, J.M.; Rios-Lago, M.; Tirapu, J.; Barcelo, F. Construct validity of the Trail Making Test: Role of task-switching, working memory, inhibition/interference control, and visuomotor abilities. *J. Int. Neuropsychol. Soc. JINS* **2009**, *15*, 438–450. [CrossRef] [PubMed]
54. Dubois, B.; Slachevsky, A.; Litvan, I.; Pillon, B. The FAB: A Frontal Assessment Battery at bedside. *Neurology* **2000**, *55*, 1621–1626. [CrossRef] [PubMed]
55. Adachi, K.; Ijuin, M.; Otsuki, M.; Koike, A.; Ishiai, S. Examining validity of the standardized verbal paired associates learning test. *High Brain Func. Res.* **2018**, *38*, 414–421. [CrossRef]
56. Shin, M.S.; Park, S.Y.; Park, S.R.; Seol, S.H.; Kwon, J.S. Clinical and empirical applications of the Rey-Osterrieth Complex Figure Test. *Nat. Protoc.* **2006**, *1*, 892–899. [CrossRef] [PubMed]
57. Brunnstrom, S. Motor testing procedures in hemiplegia: Based on sequential recovery stages. *Phys. Ther.* **1966**, *46*, 357–375. [CrossRef]
58. Seo, H.J.; Kim, S.Y.; Lee, Y.J.; Park, J.E. RoBANS 2: A Revised Risk of Bias Assessment Tool for Nonrandomized Studies of Interventions. *Korean J. Fam. Med.* **2023**, *44*, 249–260. [CrossRef]
59. Murad, M.H.; Sultan, S.; Haffar, S.; Bazerbachi, F. Methodological quality and synthesis of case series and case reports. *BMJ Evid. Based Med.* **2018**, *23*, 60–63. [CrossRef]
60. van Zandvoort, M.; de Haan, E.; van Gijn, J.; Kappelle, L.J. Cognitive functioning in patients with a small infarct in the brainstem. *J. Int. Neuropsychol. Soc. JINS* **2003**, *9*, 490–494. [CrossRef]
61. Fu, X.; Lu, Z.; Wang, Y.; Huang, L.; Wang, X.; Zhang, H.; Xiao, Z. A Clinical Research Study of Cognitive Dysfunction and Affective Impairment after Isolated Brainstem Stroke. *Front. Aging Neurosci.* **2017**, *9*, 400. [CrossRef] [PubMed]
62. Maeshima, S.; Osawa, A.; Kunishio, K. Cognitive dysfunction in a patient with brainstem hemorrhage. *Neurol. Sci.* **2010**, *31*, 495–499. [CrossRef]
63. Nishio, Y.; Ishii, K.; Kazui, H.; Hosokai, Y.; Mori, E. Frontal-lobe syndrome and psychosis after damage to the brainstem dopaminergic nuclei. *J. Neurol. Sci.* **2007**, *260*, 271–274. [CrossRef] [PubMed]
64. Neki, H.; Yamane, F.; Osawa, A.; Maeshima, S.; Ishihara, S. Cognitive Dysfunction in Patients with Pontine Hemorrhage. *No Shinkei Geka* **2014**, *42*, 109–113. [PubMed]
65. Popay, J.; Roberts, H.; Sowden, A.; Petticrew, M.; Arai, L.; Rodgers, M.; Britten, N.; Roen, K.; Duffy, S. *Guidance on the Conduct of Narrative Synthesis in Systematic Reviews: A Product from the ESRC Methods Programme Version 1*; University of Lancaster: Lancaster, UK, 2006.
66. Schmahmann, J.D.; Pandya, D.N. Anatomic organization of the basilar pontine projections from prefrontal cortices in rhesus monkey. *J. Neurosci.* **1997**, *17*, 438–458. [CrossRef] [PubMed]
67. Vataja, R.; Pohjasvaara, T.; Mantyla, R.; Ylikoski, R.; Leppavuori, A.; Leskela, M.; Kalska, H.; Hietanen, M.; Aronen, H.J.; Salonen, O.; et al. MRI correlates of executive dysfunction in patients with ischaemic stroke. *Eur. J. Neurol.* **2003**, *10*, 625–631. [CrossRef] [PubMed]
68. Fassbender, C.; Murphy, K.; Foxe, J.J.; Wylie, G.R.; Javitt, D.C.; Robertson, I.H.; Garavan, H. A topography of executive functions and their interactions revealed by functional magnetic resonance imaging. *Brain Res. Cogn. Brain Res.* **2004**, *20*, 132–143. [CrossRef]
69. Baker, S.C.; Rogers, R.D.; Owen, A.M.; Frith, C.D.; Dolan, R.J.; Frackowiak, R.S.; Robbins, T.W. Neural systems engaged by planning: A PET study of the Tower of London task. *Neuropsychologia* **1996**, *34*, 515–526. [CrossRef]
70. Aron, A.R.; Durston, S.; Eagle, D.M.; Logan, G.D.; Stinear, C.M.; Stuphorn, V. Converging evidence for a fronto-basal-ganglia network for inhibitory control of action and cognition. *J. Neurosci.* **2007**, *27*, 11860–11864. [CrossRef]
71. Uddin, L.Q. Cognitive and behavioural flexibility: Neural mechanisms and clinical considerations. *Nat. Rev. Neurosci.* **2021**, *22*, 167–179. [CrossRef]
72. Joy, M.T.; Ben Assayag, E.; Shabashov-Stone, D.; Liraz-Zaltsman, S.; Mazzitelli, J.; Arenas, M.; Abduljawad, N.; Kliper, E.; Korczyn, A.D.; Thareja, N.S.; et al. CCR5 Is a Therapeutic Target for Recovery after Stroke and Traumatic Brain Injury. *Cell* **2019**, *176*, 1143–1157. [CrossRef] [PubMed]
73. Desimone, R.; Duncan, J. Neural mechanisms of selective visual attention. *Annu. Rev. Neurosci.* **1995**, *18*, 193–222. [CrossRef] [PubMed]
74. Treisman, A.M.; Gelade, G. A feature-integration theory of attention. *Cogn. Psychol.* **1980**, *12*, 97–136. [CrossRef] [PubMed]
75. Chun, M.M.; Potter, M.C. A two-stage model for multiple target detection in rapid serial visual presentation. *J. Exp. Psychol. Hum. Percept Perform.* **1995**, *21*, 109–127. [CrossRef] [PubMed]
76. Walther, D.; Koch, C. Modeling attention to salient proto-objects. *Neural Netw.* **2006**, *19*, 1395–1407. [CrossRef] [PubMed]
77. Posner, M.I.; Petersen, S.E. The attention systems of the human brain. *Annu. Rev. Neurosci.* **1990**, *13*, 25–42. [CrossRef] [PubMed]
78. Posner, M.I. Orienting of attention. *Q. J. Exp. Psychol.* **1980**, *32*, 3–25. [CrossRef]
79. Petersen, S.E.; Posner, M.I. The attention system of the human brain: 20 years after. *Annu. Rev. Neurosci.* **2012**, *35*, 73–89. [CrossRef]
80. Fan, J.; McCandliss, B.D.; Sommer, T.; Raz, A.; Posner, M.I. Testing the efficiency and independence of attentional networks. *J. Cogn. Neurosci.* **2002**, *14*, 340–347. [CrossRef]
81. Moscovitch, M.; Cabeza, R.; Winocur, G.; Nadel, L. Episodic Memory and Beyond: The Hippocampus and Neocortex in Transformation. *Annu. Rev. Psychol.* **2016**, *67*, 105–134. [CrossRef]

82. Tonegawa, S.; Pignatelli, M.; Roy, D.S.; Ryan, T.J. Memory engram storage and retrieval. *Curr. Opin. Neurobiol.* **2015**, *35*, 101–109. [CrossRef] [PubMed]
83. Kandel, E.R. The molecular biology of memory storage—A dialogue between genes and synapses. *Science* **2001**, *294*, 1030–1038. [CrossRef]
84. Squire, L.R.; Wixted, J.T. The cognitive neuroscience of human memory since H.M. *Annu. Rev. Neurosci.* **2011**, *34*, 259–288. [CrossRef] [PubMed]
85. Asok, A.; Leroy, F.; Rayman, J.B.; Kandel, E.R. Molecular Mechanisms of the Memory Trace. *Trends Neurosci.* **2019**, *42*, 14–22. [CrossRef] [PubMed]
86. McGaugh, J.L. Memory—A Century of Consolidation. *Science* **2000**, *287*, 248–251. [CrossRef] [PubMed]
87. Tulving, E. Episodic memory: From mind to brain. *Annu. Rev. Psychol.* **2002**, *53*, 1–25. [CrossRef] [PubMed]
88. Miyamoto, K.; Osada, T.; Adachi, Y.; Matsui, T.; Kimura, H.M.; Miyashita, Y. Functional differentiation of memory retrieval network in macaque posterior parietal cortex. *Neuron* **2013**, *77*, 787–799. [CrossRef] [PubMed]
89. Miyashita, Y. Perirhinal circuits for memory processing. *Nat. Rev. Neurosci.* **2019**, *20*, 577–592. [CrossRef]
90. Sakai, K.; Hikosaka, O.; Miyauchi, S.; Takino, R.; Sasaki, Y.; Putz, B. Transition of brain activation from frontal to parietal areas in visuomotor sequence learning. *J. Neurosci.* **1998**, *18*, 1827–1840. [CrossRef]
91. Obayashi, S.; Suhara, T.; Kawabe, K.; Okauchi, T.; Maeda, J.; Akine, Y.; Onoe, H.; Iriki, A. Functional brain mapping of monkey tool use. *Neuroimage* **2001**, *14*, 853–861. [CrossRef]
92. Obayashi, S.; Suhara, T.; Kawabe, K.; Okauchi, T.; Maeda, J.; Nagai, Y.; Iriki, A. Fronto-parieto-cerebellar interaction associated with intermanual transfer of monkey tool-use learning. *Neurosci. Lett.* **2003**, *339*, 123–126. [CrossRef] [PubMed]
93. Obayashi, S. Possible mechanism for transfer of motor skill learning: Implication of the cerebellum. *Cerebellum* **2004**, *3*, 204–211. [CrossRef]
94. Rajan, K.B.; Aggarwal, N.T.; Wilson, R.S.; Everson-Rose, S.A.; Evans, D.A. Association of cognitive functioning, incident stroke, and mortality in older adults. *Stroke* **2014**, *45*, 2563–2567. [CrossRef] [PubMed]
95. Tham, W.; Auchus, A.P.; Thong, M.; Goh, M.L.; Chang, H.M.; Wong, M.C.; Chen, C.P. Progression of cognitive impairment after stroke: One year results from a longitudinal study of Singaporean stroke patients. *J. Neurol. Sci.* **2002**, *203–204*, 49–52. [CrossRef]
96. Wentzel, C.; Rockwood, K.; MacKnight, C.; Hachinski, V.; Hogan, D.B.; Feldman, H.; Ostbye, T.; Wolfson, C.; Gauthier, S.; Verreault, R.; et al. Progression of impairment in patients with vascular cognitive impairment without dementia. *Neurology* **2001**, *57*, 714–716. [CrossRef] [PubMed]
97. Levine, D.A.; Galecki, A.T.; Langa, K.M.; Unverzagt, F.W.; Kabeto, M.U.; Giordani, B.; Wadley, V.G. Trajectory of Cognitive Decline After Incident Stroke. *JAMA* **2015**, *314*, 41–51. [CrossRef] [PubMed]
98. Schaapsmeerders, P.; Maaijwee, N.A.; van Dijk, E.J.; Rutten-Jacobs, L.C.; Arntz, R.M.; Schoonderwaldt, H.C.; Dorresteijn, L.D.; Kessels, R.P.; de Leeuw, F.E. Long-term cognitive impairment after first-ever ischemic stroke in young adults. *Stroke* **2013**, *44*, 1621–1628. [CrossRef]
99. Delavaran, H.; Jonsson, A.C.; Lovkvist, H.; Iwarsson, S.; Elmstahl, S.; Norrving, B.; Lindgren, A. Cognitive function in stroke survivors: A 10-year follow-up study. *Acta Neurol. Scand.* **2017**, *136*, 187–194. [CrossRef]
100. Buvarp, D.; Rafsten, L.; Abzhandadze, T.; Sunnerhagen, K.S. A prospective cohort study on longitudinal trajectories of cognitive function after stroke. *Sci. Rep.* **2021**, *11*, 17271. [CrossRef]
101. Aam, S.; Einstad, M.S.; Munthe-Kaas, R.; Lydersen, S.; Ihle-Hansen, H.; Knapskog, A.B.; Ellekjaer, H.; Seljeseth, Y.; Saltvedt, I. Post-stroke Cognitive Impairment-Impact of Follow-Up Time and Stroke Subtype on Severity and Cognitive Profile: The Nor-COAST Study. *Front. Neurol.* **2020**, *11*, 699. [CrossRef]
102. Elgh, E.; Hu, X. Dynamic Trajectory of Long-Term Cognitive Improvement Up to 10 Years in Young Community-Dwelling Stroke Survivors: A Cohort Study. *Front. Neurol.* **2019**, *10*, 97. [CrossRef] [PubMed]
103. Tang, E.Y.; Amiesimaka, O.; Harrison, S.L.; Green, E.; Price, C.; Robinson, L.; Siervo, M.; Stephan, B.C. Longitudinal Effect of Stroke on Cognition: A Systematic Review. *J. Am. Heart Assoc.* **2018**, *7*, e006443. [CrossRef] [PubMed]
104. Katz, D.I.; Alexander, M.P.; Mandell, A.M. Dementia following strokes in the mesencephalon and diencephalon. *Arch. Neurol.* **1987**, *44*, 1127–1133. [CrossRef] [PubMed]
105. Meador, K.J.; Loring, D.W.; Sethi, K.D.; Yaghmai, F.; Styren, S.D.; DeKosky, S.T. Dementia associated with dorsal midbrain lesion. *J. Int. Neuropsychol. Soc. JINS* **1996**, *2*, 359–367. [CrossRef] [PubMed]
106. Rasquin, S.M.; Lodder, J.; Visser, P.J.; Lousberg, R.; Verhey, F.R. Predictive accuracy of MCI subtypes for Alzheimer's disease and vascular dementia in subjects with mild cognitive impairment: A 2-year follow-up study. *Dement. Geriatr. Cogn. Disord.* **2005**, *19*, 113–119. [CrossRef] [PubMed]
107. Savva, G.M.; Stephan, B.C.; Alzheimer's Society Vascular Dementia Systematic Review Group. Epidemiological studies of the effect of stroke on incident dementia: A systematic review. *Stroke* **2010**, *41*, e41–e46. [CrossRef] [PubMed]
108. Pohjasvaara, T.; Mantyla, R.; Ylikoski, R.; Kaste, M.; Erkinjuntti, T. Clinical features of MRI-defined subcortical vascular disease. *Alzheimer Dis. Assoc. Disord.* **2003**, *17*, 236–242. [CrossRef]
109. Roman, G.C.; Royall, D.R. Executive control function: A rational basis for the diagnosis of vascular dementia. *Alzheimer Dis. Assoc. Disord.* **1999**, *13* (Suppl. S3), S69–S80. [CrossRef]

110. Ballard, C.; Rowan, E.; Stephens, S.; Kalaria, R.; Kenny, R.A. Prospective follow-up study between 3 and 15 months after stroke: Improvements and decline in cognitive function among dementia-free stroke survivors >75 years of age. *Stroke* **2003**, *34*, 2440–2444. [CrossRef]
111. Desmond, D.W.; Moroney, J.T.; Sano, M.; Stern, Y. Recovery of cognitive function after stroke. *Stroke* **1996**, *27*, 1798–1803. [CrossRef]
112. Oh, H.; Park, J.; Seo, W. A 2-year prospective follow-up study of temporal changes associated with post-stroke cognitive impairment. *Int. J. Nurs. Pract.* **2018**, *24*, e12618. [CrossRef] [PubMed]
113. Kempinsky, W.H. Experimental study of distant effects of acute focal brain injury; a study of diaschisis. *AMA Arch. Neurol. Psychiatry* **1958**, *79*, 376–389. [CrossRef] [PubMed]
114. Pantano, P.; Baron, J.C.; Samson, Y.; Bousser, M.G.; Derouesne, C.; Comar, D. Crossed cerebellar diaschisis. Further studies. *Brain* **1986**, *109 Pt 4*, 677–694. [CrossRef] [PubMed]
115. Tsuda, Y.; Ayada, Y.; Izumi, Y.; Ichihara, S.; Hosomi, N.; Ohkawa, M.; Matsuo, H. Cerebellar diaschisis in pontine infarctions: A report of five cases. *Eur. J. Nucl. Med.* **1995**, *22*, 413–418. [CrossRef] [PubMed]
116. Rousseaux, M.; Steinling, M.; Mazingue, A.; Benaim, C.; Froger, J. Cerebral blood flow in lateral medullary infarcts. *Stroke* **1995**, *26*, 1404–1408. [CrossRef] [PubMed]
117. Botez, M.I.; Léveillé, J.; Lambert, R.; Botez, T. Single photon emission computed tomography (SPECT) in cerebellar disease: Cerebello-cerebral diaschisis. *Eur. Neurol.* **1991**, *31*, 405–412. [CrossRef] [PubMed]
118. Zhou, M.; Greenhill, S.; Huang, S.; Silva, T.K.; Sano, Y.; Wu, S.; Cai, Y.; Nagaoka, Y.; Sehgal, M.; Cai, D.J.; et al. CCR5 is a suppressor for cortical plasticity and hippocampal learning and memory. *eLife* **2016**, *5*, e20985. [CrossRef] [PubMed]
119. Wei, Y.; Wang, C.; Liu, J.; Miao, P.; Wu, L.; Wang, Y.; Wang, K.; Cheng, J. Progressive Gray Matter Atrophy and Abnormal Structural Covariance Network in Ischemic Pontine Stroke. *Neuroscience* **2020**, *448*, 255–265. [CrossRef]
120. Dang, C.; Liu, G.; Xing, S.; Xie, C.; Peng, K.; Li, C.; Li, J.; Zhang, J.; Chen, L.; Pei, Z.; et al. Longitudinal cortical volume changes correlate with motor recovery in patients after acute local subcortical infarction. *Stroke* **2013**, *44*, 2795–2801. [CrossRef]
121. Thomalla, G.; Glauche, V.; Weiller, C.; Rother, J. Time course of wallerian degeneration after ischaemic stroke revealed by diffusion tensor imaging. *J. Neurol. Neurosurg. Psychiatry* **2005**, *76*, 266–268. [CrossRef]
122. Lu, J.; Liu, H.; Zhang, M.; Wang, D.; Cao, Y.; Ma, Q.; Rong, D.; Wang, X.; Buckner, R.L.; Li, K. Focal pontine lesions provide evidence that intrinsic functional connectivity reflects polysynaptic anatomical pathways. *J. Neurosci.* **2011**, *31*, 15065–15071. [CrossRef] [PubMed]
123. Fox, M.D.; Raichle, M.E. Spontaneous fluctuations in brain activity observed with functional magnetic resonance imaging. *Nat. Rev. Neurosci.* **2007**, *8*, 700–711. [CrossRef] [PubMed]
124. Fontaine, D.; Capelle, L.; Duffau, H. Somatotopy of the supplementary motor area: Evidence from correlation of the extent of surgical resection with the clinical patterns of deficit. *Neurosurgery* **2002**, *50*, 297–303; discussion 303–305. [PubMed]
125. Tremblay, P.; Gracco, V.L. Contribution of the pre-SMA to the production of words and non-speech oral motor gestures, as revealed by repetitive transcranial magnetic stimulation (rTMS). *Brain Res.* **2009**, *1268*, 112–124. [CrossRef] [PubMed]
126. Fried, I.; Katz, A.; McCarthy, G.; Sass, K.J.; Williamson, P.; Spencer, S.S.; Spencer, D.D. Functional organization of human supplementary motor cortex studied by electrical stimulation. *J. Neurosci.* **1991**, *11*, 3656–3666. [CrossRef] [PubMed]
127. Obayashi, S.; Hara, Y. Hypofrontal activity during word retrieval in older adults: A near-infrared spectroscopy study. *Neuropsychologia* **2013**, *51*, 418–424. [CrossRef]
128. Kwon, Y.H.; Kwon, J.W. Response Inhibition Induced in the Stop-signal Task by Transcranial Direct Current Stimulation of the Pre-supplementary Motor Area and Primary Sensoriomotor Cortex. *J. Phys. Ther. Sci.* **2013**, *25*, 1083–1086. [CrossRef]
129. Alario, F.X.; Chainay, H.; Lehericy, S.; Cohen, L. The role of the supplementary motor area (SMA) in word production. *Brain Res.* **2006**, *1076*, 129–143. [CrossRef]
130. Shima, K.; Mushiake, H.; Saito, N.; Tanji, J. Role for cells in the presupplementary motor area in updating motor plans. *Proc. Natl. Acad. Sci. USA* **1996**, *93*, 8694–8698. [CrossRef]
131. Tanji, J.; Shima, K. Supplementary motor cortex in organization of movement. *Eur. Neurol.* **1996**, *36* (Suppl. 1), 13–19. [CrossRef]
132. Shima, K.; Tanji, J. Neuronal activity in the supplementary and presupplementary motor areas for temporal organization of multiple movements. *J. Neurophysiol.* **2000**, *84*, 2148–2160. [CrossRef] [PubMed]
133. Shima, K.; Tanji, J. Both supplementary and presupplementary motor areas are crucial for the temporal organization of multiple movements. *J. Neurophysiol.* **1998**, *80*, 3247–3260. [CrossRef] [PubMed]
134. Penfield, W. The supplementary motor area in the cerebral cortex of man. *Arch. Psychiatr. Nervenkr. Z. Gesamte Neurol. Psychiatr.* **1950**, *185*, 670–674. [CrossRef] [PubMed]
135. Penfield, W.; Welch, K. The supplementary motor area of the cerebral cortex; a clinical and experimental study. *AMA Arch. Neurol. Psychiatry* **1951**, *66*, 289–317. [CrossRef] [PubMed]
136. Matsuzaka, Y.; Aizawa, H.; Tanji, J. A motor area rostral to the supplementary motor area (presupplementary motor area) in the monkey: Neuronal activity during a learned motor task. *J. Neurophysiol.* **1992**, *68*, 653–662. [CrossRef] [PubMed]
137. Tanji, J. New concepts of the supplementary motor area. *Curr. Opin. Neurobiol.* **1996**, *6*, 782–787. [CrossRef] [PubMed]
138. Tanji, J.; Shima, K. Role for supplementary motor area cells in planning several movements ahead. *Nature* **1994**, *371*, 413–416. [CrossRef] [PubMed]
139. Jurgens, U. The efferent and afferent connections of the supplementary motor area. *Brain Res.* **1984**, *300*, 63–81. [CrossRef]

140. Akkal, D.; Dum, R.P.; Strick, P.L. Supplementary motor area and presupplementary motor area: Targets of basal ganglia and cerebellar output. *J. Neurosci.* **2007**, *27*, 10659–10673. [CrossRef]
141. Schmahmann, J.D.; Weilburg, J.B.; Sherman, J.C. The neuropsychiatry of the cerebellum—Insights from the clinic. *Cerebellum* **2007**, *6*, 254–267. [CrossRef]
142. Schmahmann, J.D.; Guell, X.; Stoodley, C.J.; Halko, M.A. The Theory and Neuroscience of Cerebellar Cognition. *Annu. Rev. Neurosci.* **2019**, *42*, 337–364. [CrossRef] [PubMed]
143. Wolff, M.; Vann, S.D. The Cognitive Thalamus as a Gateway to Mental Representations. *J. Neurosci.* **2019**, *39*, 3–14. [CrossRef] [PubMed]
144. Valenstein, E.; Bowers, D.; Verfaellie, M.; Heilman, K.M.; Day, A.; Watson, R.T. Retrosplenial amnesia. *Brain* **1987**, *110 Pt 6*, 1631–1646. [CrossRef] [PubMed]
145. Aggleton, J.P.; O'Mara, S.M.; Vann, S.D.; Wright, N.F.; Tsanov, M.; Erichsen, J.T. Hippocampal-anterior thalamic pathways for memory: Uncovering a network of direct and indirect actions. *Eur. J. Neurosci.* **2010**, *31*, 2292–2307. [CrossRef]
146. Van der Werf, Y.D.; Scheltens, P.; Lindeboom, J.; Witter, M.P.; Uylings, H.B.; Jolles, J. Deficits of memory, executive functioning and attention following infarction in the thalamus; a study of 22 cases with localised lesions. *Neuropsychologia* **2003**, *41*, 1330–1344. [CrossRef] [PubMed]
147. Carrera, E.; Bogousslavsky, J. The thalamus and behavior: Effects of anatomically distinct strokes. *Neurology* **2006**, *66*, 1817–1823. [CrossRef] [PubMed]
148. Johnson, M.D.; Ojemann, G.A. The role of the human thalamus in language and memory: Evidence from electrophysiological studies. *Brain Cogn.* **2000**, *42*, 218–230. [CrossRef]
149. Crosson, B. Thalamic mechanisms in language: A reconsideration based on recent findings and concepts. *Brain Lang.* **2013**, *126*, 73–88. [CrossRef]
150. De Witte, L.; Brouns, R.; Kavadias, D.; Engelborghs, S.; De Deyn, P.P.; Marien, P. Cognitive, affective and behavioural disturbances following vascular thalamic lesions: A review. *Cortex* **2011**, *47*, 273–319. [CrossRef]
151. Obayashi, S. The Supplementary Motor Area Responsible for Word Retrieval Decline After Acute Thalamic Stroke Revealed by Coupled SPECT and Near-Infrared Spectroscopy. *Brain Sci.* **2020**, *10*, 247. [CrossRef]
152. Obayashi, S. Cognitive and linguistic dysfunction after thalamic stroke and recovery process: Possible mechanism. *AIMS Neurosci.* **2022**, *9*, 1–11. [CrossRef]

biomedicines

MDPI

Article

Gut Dysbiosis: A New Avenue for Stroke Prevention and Therapeutics

Shin Young Park [1], Sang Pyung Lee [2], Dongin Kim [3] and Woo Jin Kim [3,*]

1 Department of Clinical Laboratory Science, Cheju Halla University, 38 Halladaehak-ro, Jeju-si 63092, Republic of Korea; kj901217@gmail.com
2 Department of Neurosurgery, Brain-Neuro Center, Cheju Halla General Hospital, 65 Doryeong-ro, Jeju-si 63127, Republic of Korea; nsdr745@gmail.com
3 Department of Laboratory Medicine, EONE Laboratories, 291 Harmony-ro, Incheon 22014, Republic of Korea; gtphrase@eonelab.co.kr
* Correspondence: oojinkim@hanmail.net; Tel.: +82-32-210-2108

Abstract: A stroke is a serious life-threatening condition and a leading cause of death and disability that happens when the blood vessels to part of the brain are blocked or burst. While major advances in the understanding of the ischemic cascade in stroke was made over several decades, limited therapeutic options and high mortality and disability have caused researchers to extend the focus toward peripheral changes beyond brain. The largest proportion of microbes in human body reside in the gut and the interaction between host and microbiota in health and disease is well known. Our study aimed to explore the gut microbiota in patients with stroke with comparison to control group. Fecal samples were obtained from 51 subjects: 25 stroke patients (18 hemorrhagic, 7 ischemic) and 26 healthy control subjects. The variable region V3–V4 of the 16S rRNA gene was sequenced using the Illumina MiSeq platform. PICRUSt2 was used for prediction of metagenomics functions. Our results show taxonomic dysbiosis in stroke patients in parallel with functional dysbiosis. Here, we show that stroke patients have (1) increased *Parabacteroides* and *Escherichia_Shigella*, but decreased *Prevotella* and *Fecalibacterium;* (2) higher transposase and peptide/nickel transport system substrate-binding protein, but lower RNA polymerase sigma-70 factor and methyl-accepting chemotaxis protein, which are suggestive of malnutrition. Nutrients are essential regulators of both host and microbial physiology and function as key coordinators of host–microbe interactions. Manipulation of nutrition is expected to alleviate gut dysbiosis and prognosis and improve disability and mortality in the management of stroke.

Keywords: stroke; gut; dysbiosis; nutrition; inflammation; microbiota; mortality; therapeutics; blood; brain

Citation: Park, S.Y.; Lee, S.P.; Kim, D.; Kim, W.J. Gut Dysbiosis: A New Avenue for Stroke Prevention and Therapeutics. *Biomedicines* **2023**, *11*, 2352. https://doi.org/10.3390/biomedicines11092352

Academic Editors: Masaru Tanaka and Eleonóra Spekker

Received: 10 July 2023
Revised: 20 August 2023
Accepted: 21 August 2023
Published: 23 August 2023

1. Introduction

A stroke is a serious life-threatening condition and a leading cause of death and disability that happens when the blood vessels to part of the brain are blocked or burst. Stroke has a long history but the research of its pathobiology has only been carried out recently. Stroke is a kind of inflammation, and upon neurological injury, inflammatory cascades are initiated and subsequent recovery follows. Initial ischemic event results in excitotoxicity and oxidative stress which leads to microglial and astrocyte activation to secrete cytokines, glial fibrillary acidic protein (GFAP), and matrix metalloproteinases (MMPs). These factors are proinflammatory and lead to upregulated expression of cell adhesion molecules such as selectins and intracellular adhesion molecule 1 (ICAM-1) on endothelial cells, and recruit blood-derived inflammatory cells including neutrophils, macrophages, and lymphocytes to the ischemic site. In addition, dying neuronal cells release danger-associated molecular patterns (DAMPs) which in turn stimulate microglia and peripheral immune cells such as neutrophils, macrophages, and lymphocytes, causing production of proinflammatory factors, resulting in further activation of microglia and astrocyte [1].

While major advances in the understanding of the ischemic cascade in stroke were made over several decades, limited therapeutic options and high mortality and disability have caused researchers to extend their focus toward peripheral changes beyond the brain [2].

Various research, including animal model studies and clinical trials, highlighted the bidirectional connection between the brain and the gut in the pathogenesis of neurological and neuropsychiatric diseases such as depressive diseases, anxiety, bipolar disorder, autism, schizophrenia, Parkinson's disease, Alzheimer's disease, dementia, multiple sclerosis, and epilepsy [3–12]. The brain–gut axis comprises brain, spinal cord, autonomic nervous system (ANS), enteric nervous system (ENS), and hypothalamic–pituitary–adrenal (HPA) axis [13]. The vagus nerve (tenth cranial nerve) represents the principal pathway from the gut lumen to the brain, which acts collaboratively with several neurotransmitters released from ENS and has immunomodulatory properties [14]. The integrated activity of HPA and the vagus nerve permits high-level communication between the brain and gastrointestinal tissues, in more detail enterochromaffin cells, interstitial cells of Cajal, smooth muscle cells, enteric neurons, epithelial cells, and immune cells, that in practice are all regulated by the gut microbiota [15]. The largest proportion of microbes in the human body reside in the gut, and balanced intestinal microbiota stimulate a regulatory environment in the gut-associated lymphoid tissue (GALT) through the production and release of various immunomodulatory compounds, like short-chain fatty acids (SCFAs) [16], but any alteration in gut microbiota and host communication can be considered as a triggering element in the pathogenesis of diseases [17].

A symbiotic relationship with gut microbiota is an essential component to maintain host heath both metabolically and immunologically. The host utilizes the gut microbiota to digest food to obtain the nutrients and energy and participates in host metabolism and organism immunity [18]. One-third of the small molecules in the blood derived from gut microbiota and the trillions of commensal or mutualistic bacteria and archaea are a huge chemical factory that can produce some essential amino acids and vitamins and many compounds that affect host energy homeostasis, body adiposity, glucose tolerance, insulin sensitivity, inflammation, and hormone secretion needed for their own existence and survival [19,20]. Fiber fermentation by fecal microbial flora produces the short-chain fatty acids (SCFAs) such as butyrate, propionate, and acetate, which affect host metabolism in various ways by acting on G protein-coupled receptors (GPCRs) expressed by enteroendocrine cells. Butyrate and acetate activate glucagon-like peptide 1 (GLP-1) and peptide YY (PYY) secretion with influence on GLP-1-induced insulin biosynthesis in the pancreas and on PYY-induced satiety in the brain, and acetate may upregulate fat storage by inducing release of ghrelin [21]. Microbiota-derived succinate stimulates expression of uncoupling protein 1 (UCP1), leading to upregulate thermogenesis in adipose tissue [22]. Primary bile acids are transformed by gut microbiota to secondary bile acids which act through the Takeda-G-protein-receptor-5 (TGR5) receptor to upregulate GLP-1 secretion causing thermogenesis in adipose tissue [23]. Indole and its derivatives are the ligand to the aryl hydrocarbon receptor (AhR). Indole-3-propionic acid is associated with enhanced insulin secretion and insulin sensitivity and thereby results in a reduced risk of type 2 diabetes [24]. Commensal bacteria produce N-acyl amide, a mimic of human signaling molecules, which controls host glucose metabolism by binding to the G protein-coupled receptor 119 (GPR119) [25]. Proteins released by gut symbionts also regulate paracrine or endocrine action. Caseinolytic peptidase B (ClpB), a protein released by *Escherichia coli*, is implicated in the control of appetite [26]. Melanocortin-like peptide of *E. coli* (MECO-1), which is a structural mimic to α-melanocyte stimulating hormone and adrenocorticotropin, functions through the mammalian melanocortin-1 receptor (MC1R), suppressing cytokine secretion in response to proinflammatory stimulation [27]. Amuc_1100, expressed on the outer membrane of *Akkermansia muciniphila*, ameliorates gut barrier function with elevated goblet cell population via Toll-like receptor 2 (TLR2) and shows the beneficial effect on insulin sensitivity and energy metabolism [28]. Host metabolism is also affected by synthesized neurotransmitters by gut commensals such as catecholamine, histamine, γ-aminobutyric acid and serotonin or gaseous neurotransmitters, including nitric oxide (NO) and hydrogen sulfide (H2S) [29].

The immune system orchestrates the maintenance of key features of host–microbe symbiosis while the microbiome plays critical roles in the training and development of major components of the host's immune system [30]. Trace metals are essential micronutrients required for survival across all kingdoms of life, and hosts have numerous strategies of metal limitation and intoxication that prevent bacterial proliferation, a process termed nutritional immunity [31]. Fecal calprotectin allows the chelation of essential divalent metal ions (e.g., calcium, iron, or zinc), limiting growth of invasive and commensal gut bacteria, while it represents a well-studied inflammatory biomarker in inflammatory bowel diseases [32].

We have previously measured fecal calprotectin (FC), a gut nutritional immunity marker of host, in stroke and reported two meaningful findings: (1) there was a significant increase in FC levels in stroke patients compared to those in controls; (2) FC in stroke patients had negative correlation with levels of albumin and lymphocyte but positive correlation with C-reactive protein (CRP) [33]. This study aimed further to explore the gut microbiota, the other side of host and microbe interaction, in stroke patients with comparison to control group.

2. Materials and Methods

Fecal samples were obtained from 51 subjects: 25 stroke patients (18 hemorrhagic, 7 ischemic) and 26 healthy control subjects. The variable region V3–V4 of the 16S rRNA gene was sequenced using the Illumina MiSeq platform. PICRUSt2 was used for prediction of metagenomics functions.

2.1. Subjects

In the present study, 51 subjects provided a single fecal sample including 25 stroke patients (STR) and 26 healthy control subjects (CON). Stroke patients, including both ischemic and hemorrhagic types, were directly admitted from the emergency department (ER) to the intensive care unit (ICU). Fecal samples were obtained from stroke patients as early as possible during their stay in the ICU. The enrolled period of the subjects was from September 2018 to April 2019. Informed consent was written and obtained from the subjects or their guardians. Diagnosis of stroke was based on head computed tomography (CT) scans or magnetic resonance imaging (MRI) studies. The study subjects had no history of colorectal or systemic inflammatory conditions. The present study was approved by the ethical review board of Cheju Halla University (IRB approval number: 1044348-20180713-HR-007-01).

2.2. Demographic Characteristics

Baseline demographic and clinical characteristics of the stroke patients were collected (Table 1). These included patient sex, age, body mass index (BMI), stroke type, and comorbidities. The age of the stroke patients ranged from 39 to 86 and the age of the control subjects ranged from 41 to 85. The male-to-female ratio of the stroke patients was 14:11 and the male-to-female ratio of the control subjects was 16:10. The body mass index (BMI) of the stroke patients ranged from 17.3 to 31.4 and the BMI of the control subjects ranged from 18.8 to 33.2. Comorbidities in stroke patients include diabetes mellitus (16%), hypertension (52%), and coronary artery disease (8%), while those in control subjects include diabetes mellitus (15.3%) and hypertension (30.7%). Regarding medication, intravenous antibiotics were taken along the progress in all stroke patients. While stroke was reported to be associated with obesity, hypertension, or diabetes mellitus, all of which had major effect on gut dysbiosis [34], the subject difference between stroke patients and control group was not as significant for those comorbidities.

Table 1. Demographic characteristics of the study population.

	CON (n = 26)	STR (n = 25)	p
Age (year)	58.6 ± 14.1	60.8 ± 15.2	0.579
Sex (M/F)	16/10	14/11	0.663
BMI (kg/m^2)	23.4 ± 3.5	23.7 ± 3.4	0.808
Comorbidities			
DM (number)	4	4	1
Hypertension (number)	8	13	0.159
CAD (number)	0	2	0.235
Medication			
PPI (number)	0	0	NA
NSAID (number)	0	0	NA
ABX (number)	0	25	<0.0001

Abbreviations: CON, healthy controls; STR, stroke patients; BMI, body mass index; DM, diabetes mellitus; CAD, coronary heart disease; PPI, proton pump inhibitor; NSAID, nonsteroidal anti-inflammatory drug; ABX, antibiotics; NA, not applicable. Values are expressed as mean ± SD.

2.3. DNA Extraction

Subjects provided a single fecal sample for measurement as early as possible during their ICU stay. Aliquots of fecal samples were kept frozen on receipt at −80 °C prior to DNA extraction [35]. DNA was extracted from 200 mg of fecal aliquot using a QIAamp PowerFecal Pro DNA Kit (Qiagen, Hilden, Germany) according to the manufacturer's instructions without modification. Briefly, the fecal sample and eight hundred microliters of lysis buffer (solution CD1) were applied to the PowerBead Pro tube. The mixture was horizontally agitated at maximum speed for 10 min until the fecal sample became homogeneous. The homogenate was then centrifuged at 15,000× g for one minute. The supernatant was removed and mixed with two hundred microliters of solution CD2, which contained inhibitor removal reagents. After the centrifuge stopped, the supernatant was removed and mixed with six hundred microliters of Solution CD3, which customized the DNA solution salt concentration and allowed more specific deoxynucleic acid binding to the column. The mixture was then carefully added to the QIAamp spin column and centrifuged at maximum speed for one minute. The column was cleaned two times with washing buffers (solutions EA and C5, respectively). Then, a total of one hundred microliters of elution buffer (Solution C6) was applied and centrifuged at maximum speed for one minute to elute the deoxynucleic acid [36].

2.4. Library Construction and Sequencing

The sequencing libraries were prepared according to the Illumina 16S Metagenomic Sequencing Library protocols to amplify the V3 and V4 regions. The 4 ng of input gDNA was polymerase chain reaction (PCR) amplified with 2× KAPA HiFi HotStart ReadyMix buffer (KAPA Biosystems, Wilmington, MA, USA), one micromole each of universal F/R PCR primers. The protocols for PCR were as follows: (1) heat activation for three minutes at 95 °C; (2) twenty five cycles of thirty seconds at 95 °C, thirty seconds at 55 °C, and thirty seconds at 72 °C; (3) five minutes final extension at 72 °C. The universal primer pairs with Illumina adapter overhang sequences were: V3-F: 5′-TCG TCG GCA GCG TCA GAT GTG TAT AAG AGA CAG CCT ACG GGN GGC WGC AG-3′, V4-R: 5′- GTC TCG TGG GCT CGG AGA TGT GTA TAA GAG ACA GGA CTA CHV GGG TAT CTA ATC C-3′. The PCR amplicon was cleaned with AMPure beads (Agencourt Bioscience, Beverly, MA, USA). After purification, five microliters of the PCR product was PCR amplified the second time for the last library construction holding the index using NexteraXT Indexed Primer. The protocol condition for the second PCR was equal to that for the first PCR except for eight cycles. The PCR amplicon was cleaned with AMPure beads. The last purified product was then quantitatively measured using a Qubit (Life Technologies, Carlsbad, CA, USA) 2.0 fluorometer, following the manufacturer's instructions, and qualified using the TapeStation D1000 ScreenTape (Agilent Technologies, Waldbronn, Germany). The paired-end (2 × 300 bp) sequencing was performed using the MiSeq™ platform (Illumina, San Diego, CA, USA) [37].

2.5. Sequencing Data Analysis

Forward and reverse paired-end 16S rRNA sequences were combined by Quantitative Insights Into Microbial Ecology (QIIME) 2 (version: 2021.2) pipeline. After merging, the sequences were demultiplexed and separated into samples using the index sequence of each sample. With the use of QIIME 2 plugin DADA2, a quality check was carried out, the noise was removed, and denoised and filtered amplicon sequencing variants were rarefied to a depth of 4000 sequences per sample. Analysis of alpha diversity was carried out by the diversity plugin of QIIME 2. Nonmetric multidimensional scaling (NMDS) diagrams were produced by R packages "phyloseq" and "ggplot2". Taxonomy was designated to the amplicon sequence variants (ASVs) by Vsearch pretrained on Silva reference database (silva-138-99-seqs) and the feature-classifier plugin of QIIME 2 [38]. The compositional microbiome data and their relative abundances were calculated on seven levels of taxonomy including species, genus, family, order, class, phylum, and kingdom. The linear discriminant analysis (LDA) effect size (LEfSe) method was carried out to reveal the differences in taxon abundance between groups at different taxonomic levels using default parameters except otherwise specified [39]. Two groups were regarded as significantly different at a p value less than 0.05 and a |log10(LDA score)| more than 2. The function prediction analysis was carried out using the Phylogenetic Investigation of Communities by Reconstruction of Unobserved States (PICRUSt) software based on Kyoto Encyclopedia of Genes and Genomes (KEGG) database. PICRUSt2 is a software for predicting functional abundances based only on marker gene sequences such as 16S rRNA gene sequencing data, and function refers to gene families such as KEGG orthologs [40].

2.6. Statistical Analysis

For each group, twenty-five participants were necessary to have a power of ninety percent with two-sided five percent significance and 0.2 for an effective size, according to Whitehead et al. [41]. The Student's t-test or Fisher's exact test were utilized according to variables to compare demographic characteristics of the study subjects. All tests were two-sided, and $p < 0.05$ was considered statistically significant. The statistical software package for data was the GraphPad Prism version 10.0.2 (GraphPad Software Inc., La Jolla, CA, USA). Beta diversity was measured from the pairwise PERMANOVA (permutational multivariate analysis of variance) test using Bray–Curtis and Jaccard distance metrices [42].

3. Results

3.1. Alpha Diversity

The Student's t-test indicated differences in the richness and evenness of the bacteria between two groups (Figure 1). The bacteria were less enriched in the STR group (p = 0.01) (Figure 2A). There was also a difference in the evenness between two groups (p = 0.001) (Figure 2B). To sum up, the gut microbiome was less abundant in the STR group.

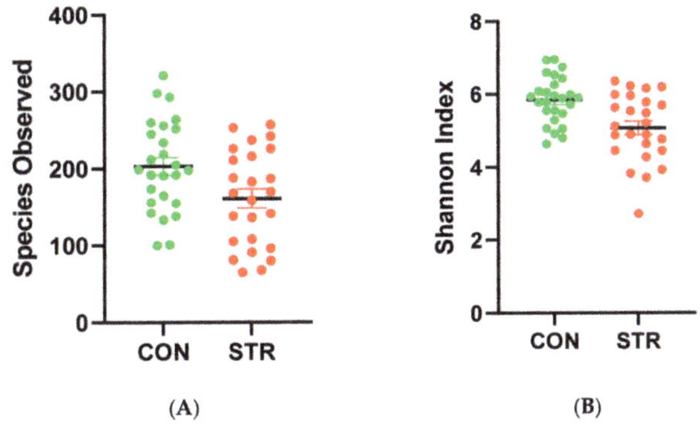

Figure 1. (**A**) Observed amplicon sequence variants (ASVs); $p = 0.01$. (**B**) Shannon index; $p = 0.001$; CON: control; STR: stroke; CON (green): control; STR (red): stroke.

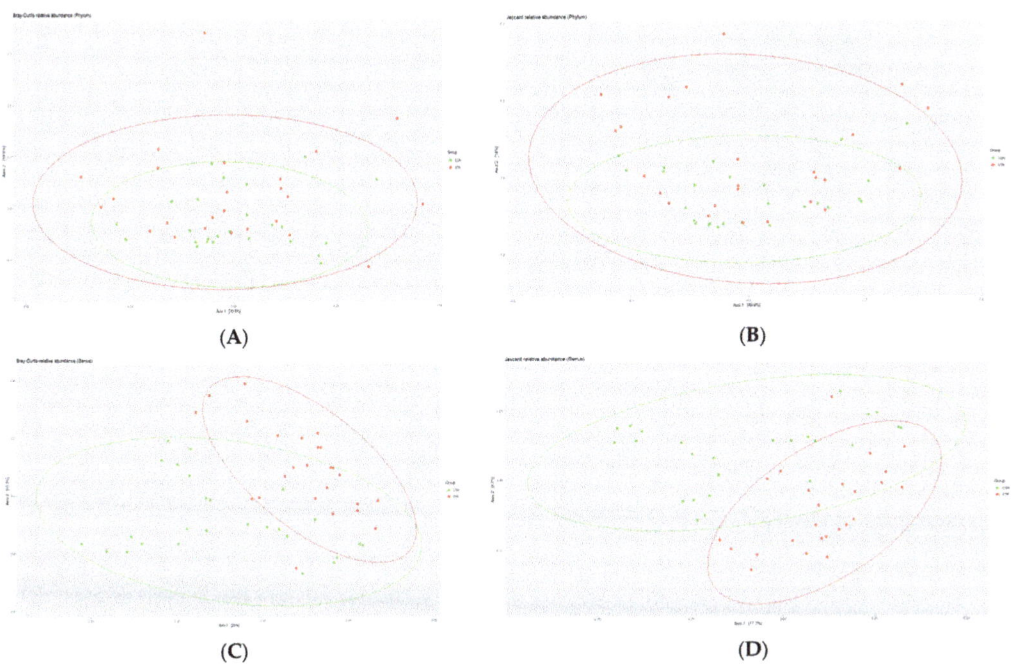

Figure 2. Beta diversity observation in two groups. (**A**) Bray–Curtis plot of control and stroke group microbiome communities at phylum level; $p = 0.354$. (**B**) Jaccard plot at phylum level; $p = 0.165$. (**C**) Bray–Curtis plot at genus level; $p = 0.001$. (**D**) Jaccard plot at genus level; $p = 0.001$; CON (green): control; STR (red): stroke.

3.2. Beta Diversity

When observed at the genus level, control and stroke groups differed significantly in bacterial diversity, while there was no difference at phylum level (Figure 2). In Figure 2, Bray–Curtis (A,C) and Jaccard (B,D) distance metrics were used. Consistent results were obtained in both metrics.

3.3. Taxonomy Changes

We performed LEfSe to identify bacterial taxa that could best explain the differences between CON and STR with alpha value for the factorial Kruskal–Wallis test among classes = 0.05, alpha value for the pairwise Wilcoxon test between subclasses = 0.05, and threshold on the logarithmic LDA score for discriminative features > 3.6. The clades that decreased in stroke patients included Prevotella, Fecalibacterium, Roseburia, Lachnospiraceae_NK4A136_group, Eubacterium, and Dialister while the clades that increased in stroke patients included Parabacteroides, Lachnoclostridium, Escherichia_Shigella, Fusobacterium, Lactobacillales, and Enterococcus (Figure 3).

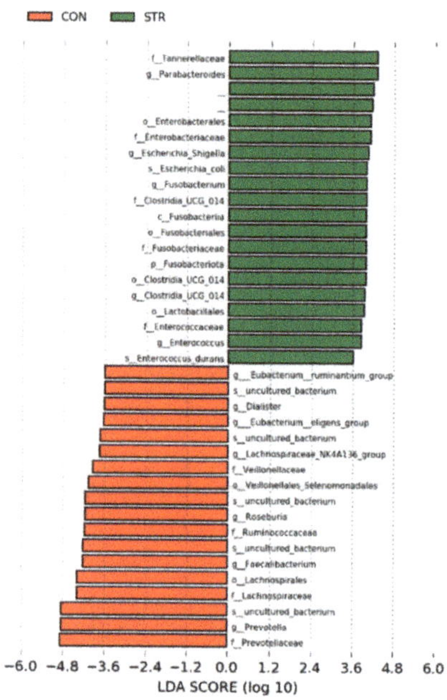

(A)

Taxon	Group	LDA (log 10)
d__Bacteria.p__Bacteroidota.c__Bacteroidia.o__Bacteroidales.f__Tannerellaceae.g__Parabacteroides	STR	4.306151865
d__Bacteria.p__Bacteroidota.c__Bacteroidia.o__Bacteroidales.f__Tannerellaceae	STR	4.306151865
d__Bacteria.p__Bacteroidota.c__Bacteroidia.o__Bacteroidales.f__Tannerellaceae.g__Parabacteroides.__	STR	4.187043005
d__Bacteria.p__Firmicutes.c__Clostridia.o__Lachnospirales.f__Lachnospiraceae.g__Lachnoclostridium.__	STR	4.151359968
d__Bacteria.p__Proteobacteria.c__Gammaproteobacteria.o__Enterobacterales	STR	4.111616596
d__Bacteria.p__Proteobacteria.c__Gammaproteobacteria.o__Enterobacterales.f__Enterobacteriaceae	STR	4.108084605
d__Bacteria.p__Proteobacteria.c__Gammaproteobacteria.o__Enterobacterales.f__Enterobacteriaceae.g__Escherichia_Shigella	STR	4.045739222
d__Bacteria.p__Proteobacteria.c__Gammaproteobacteria.o__Enterobacterales.f__Enterobacteriaceae.g__Escherichia_Shigella.s__Escherichia_coli	STR	3.989790143
d__Bacteria.p__Fusobacteriota.c__Fusobacteriia.o__Fusobacteriales.f__Fusobacteriaceae.g__Fusobacterium	STR	3.984260304
d__Bacteria.p__Firmicutes.c__Clostridia.o__Clostridia_UCG_014.f__Clostridia_UCG_014	STR	3.983844949
d__Bacteria.p__Fusobacteriota.c__Fusobacteriia	STR	3.982820676
d__Bacteria.p__Fusobacteriota.c__Fusobacteriia.o__Fusobacteriales	STR	3.982571044
d__Bacteria.p__Fusobacteriota.c__Fusobacteriia.o__Fusobacteriales.f__Fusobacteriaceae	STR	3.980900694
d__Bacteria.p__Fusobacteriota	STR	3.978338965
d__Bacteria.p__Firmicutes.c__Clostridia.o__Clostridia_UCG_014	STR	3.975436276
d__Bacteria.p__Firmicutes.c__Clostridia.o__Clostridia_UCG_014.f__Clostridia_UCG_014.g__Clostridia_UCG_014	STR	3.940907145
d__Bacteria.p__Firmicutes.c__Bacilli.o__Lactobacillales	STR	3.919167248
d__Bacteria.p__Firmicutes.c__Bacilli.o__Lactobacillales.f__Enterococcaceae	STR	3.844396716
d__Bacteria.p__Firmicutes.c__Bacilli.o__Lactobacillales.f__Enterococcaceae.g__Enterococcus	STR	3.844335133
d__Bacteria.p__Firmicutes.c__Bacilli.o__Lactobacillales.f__Enterococcaceae.g__Enterococcus.s__Enterococcus_durans	STR	3.616730215
d__Bacteria.p__Firmicutes.c__Clostridia.o__Lachnospirales.f__Lachnospiraceae.g__Eubacterium__ruminantium_group	CON	3.60361567
d__Bacteria.p__Firmicutes.c__Clostridia.o__Lachnospirales.f__Lachnospiraceae.g__Eubacterium__ruminantium_group.s__uncultured_bacterium	CON	3.60475793
d__Bacteria.p__Firmicutes.c__Negativicutes.o__Veillonellales_Selenomonadales.f__Veillonellaceae.g__Dialister	CON	3.619785108
d__Bacteria.p__Firmicutes.c__Clostridia.o__Lachnospirales.f__Lachnospiraceae.g__Eubacterium__eligens_group	CON	3.638031494
d__Bacteria.p__Firmicutes.c__Clostridia.o__Lachnospirales.f__Lachnospiraceae.g__Lachnospiraceae_NK4A136_group.s__uncultured_bacterium	CON	3.727208807
d__Bacteria.p__Firmicutes.c__Clostridia.o__Lachnospirales.f__Lachnospiraceae.g__Lachnospiraceae_NK4A136_group	CON	3.743330058
d__Bacteria.p__Firmicutes.c__Negativicutes.o__Veillonellales_Selenomonadales.f__Veillonellaceae	CON	3.937114284
d__Bacteria.p__Firmicutes.c__Negativicutes.o__Veillonellales_Selenomonadales	CON	4.059836163
d__Bacteria.p__Firmicutes.c__Clostridia.o__Lachnospirales.f__Lachnospiraceae.g__Roseburia.s__uncultured_bacterium	CON	4.159151509
d__Bacteria.p__Firmicutes.c__Clostridia.o__Lachnospirales.f__Lachnospiraceae.g__Roseburia	CON	4.160509029
d__Bacteria.p__Firmicutes.c__Clostridia.o__Oscillospirales.f__Ruminococcaceae	CON	4.171449528
d__Bacteria.p__Firmicutes.c__Clostridia.o__Oscillospirales.f__Ruminococcaceae.g__Faecalibacterium.s__uncultured_bacterium	CON	4.21574752
d__Bacteria.p__Firmicutes.c__Clostridia.o__Oscillospirales.f__Ruminococcaceae.g__Faecalibacterium	CON	4.215907664
d__Bacteria.p__Firmicutes.c__Clostridia.o__Lachnospirales	CON	4.390592848
d__Bacteria.p__Firmicutes.c__Clostridia.o__Lachnospirales.f__Lachnospiraceae	CON	4.391373187
d__Bacteria.p__Bacteroidota.c__Bacteroidia.o__Bacteroidales.f__Prevotellaceae.g__Prevotella.s__uncultured_bacterium	CON	4.834248183
d__Bacteria.p__Bacteroidota.c__Bacteroidia.o__Bacteroidales.f__Prevotellaceae.g__Prevotella	CON	4.83813673
d__Bacteria.p__Bacteroidota.c__Bacteroidia.o__Bacteroidales.f__Prevotellaceae	CON	4.865691256

(B)

Figure 3. (**A**) Linear discriminant analysis effect size (LEfse) result showing taxonomic difference between two groups. (**B**) Taxon table and LDA score of each taxon; CON (red): control; STR (green): stroke.

3.4. Biologic Function Changes

We performed LEfSe to identify biologic functions that could best explain the differences between CON and STR with alpha value for the factorial Kruskal–Wallis test among classes = 0.05, alpha value for the pairwise Wilcoxon test between subclasses = 0.05, and threshold on the logarithmic LDA score for discriminative features > 2.0 (Figure 4).

The functions that were decreased in stroke patients included RNA polymerase sigma-70 factor, putative ABC transporter system permease, methyl-accepting chemotaxis protein, glutamate synthase, thiamin-phosphate pyrophosphorylase, ATP-dependent DNA helicase RecG, putative hydrolase of the HAD superfamily, phosphoglycolate phosphatase, and GTP pyrophosphokinase.

The functions that were enriched included transposase, peptide/nickel transport system substrate-binding protein, cysteine-S-conjugate beta-lyase, peptide/nickel transport system ATP-binding protein, peptide/nickel transport system permease protein, ribose transport system ATP-binding protein, alpha-glucosidase, AraC family transcriptional regulator, iron (III) transport system permease protein, iron (III) transport system substrate binding protein, GntR family transcriptional regulator, iron complex transport system permease protein, putative tricarboxylic transport membrane, and dihydrolipoyl dehydrogenase [43]. Taken together, our results showed taxonomic dysbiosis in stroke patients in parallel with functional dysbiosis.

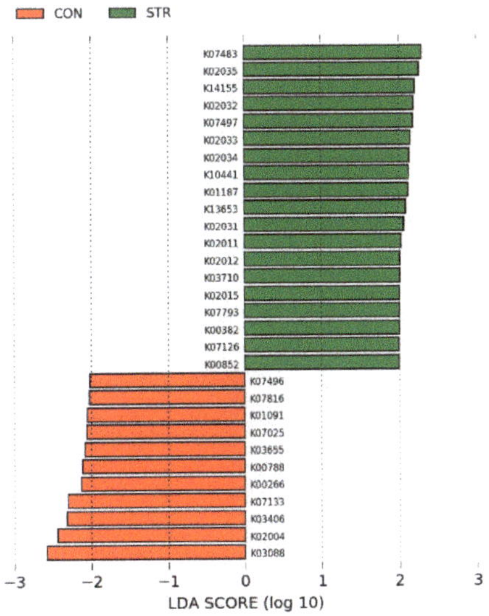

(A)

KEGG ortholog	Group	LDA (log 10)	Description
K07483	STR	2.302361	transposase
K02035	STR	2.272517	peptide/nickel transport system substrate-binding protein
K14155	STR	2.21992	cysteine-S-conjugate beta-lyase
K02032	STR	2.198913	peptide/nickel transport system ATP-binding protein
K07497	STR	2.192265	putative transposase
K02033	STR	2.158016	peptide/nickel transport system permease protein
K02034	STR	2.139416	peptide/nickel transport system permease protein
K10441	STR	2.135194	ribose transport system ATP-binding protein
K01187	STR	2.125917	alpha-glucosidase
K13653	STR	2.092386	AraC family transcriptional regulator
K02031	STR	2.072472	peptide/nickel transport system ATP-binding protein
K02011	STR	2.035077	iron(III) transport system permease protein
K02012	STR	2.019692	iron(III) transport system substrate-binding protein
K03710	STR	2.019187	GntR family transcriptional regulator
K02015	STR	2.012739	iron complex transport system permease protein
K07793	STR	2.012315	putative tricarboxylic transport membrane protein
K00382	STR	2.012189	dihydrolipoyl dehydrogenase
K07126	STR	2.001842	uncharacterized protein
K00852	STR	2.000867	Ribokinase
K07496	CON	2.014883	putative transposase
K07816	CON	2.016363	GTP pyrophosphokinase
K01091	CON	2.045491	phosphoglycolate phosphatase
K07025	CON	2.056074	putative hydrolase of the HAD superfamily
K03655	CON	2.078289	ATP-dependent DNA helicase RecG
K00788	CON	2.111514	thiamine-phosphate pyrophosphorylase
K00266	CON	2.129927	glutamate synthase (NADPH) small chain
K07133	CON	2.29469	uncharacterized protein
K03406	CON	2.31359	methyl-accepting chemotaxis protein
K02004	CON	2.433863	putative ABC transport system permease protein
K03088	CON	2.572468	RNA polymerase sigma-70 factor, ECF subfamily

(B)

Figure 4. (**A**) Linear discriminant analysis effect size (LEfse) result showing functional difference between two groups. (**B**) KEGG ortholog table and LDA score of each ortholog; CON (red): control; STR (green): stroke.

4. Discussion

While stroke is a vascular event and subsequent inflammation which occurs in the brain, the dysfunction of the gut–brain axis has been known to be a hopeful area of research for identifying preventive and treatment strategies against stroke [2]. Stroke is characterized by a disruption of blood supply to a specific region of the brain, leading to neuronal damage and death as well as disturbances in the blood–brain barrier and is complicated by functional deficits of motor, sensing, or cognition and seizures or depression [44]. Elaboration on the neural substrates affected by stroke can shed light on the potential mechanisms underlying the interaction between the gut and the brain, like in migraine or neuropathic pain, where such neural substrates or pathomechanisms which involve calcitonin gene-related peptide (CGRP), transient receptor potential channels (TRP channels), endocannabinoid system, glutamatergic system, tryptophan-kynurenin (Trp-KYN) metabolism, neuroinflammation, cytokines, and microglial activation were revealed [45]. In recent years, several efforts have been made and emerging evidence suggests that ischemic brain tissue and activated microglia release cytokines and damage-associated molecular patterns (DAMPs), which leads to triggering vascular endothelial cells to reveal adhesion molecules and to extravasate immune cells and inflammatory cells to the injury site of stroke from the blood circulation. In the meantime, production of cytokines and DAMPs in addition to stimulation of the vagus nerve, which acts as the primary communication pathway between the central nervous system (CNS) and enteric nervous system (ENS), lead to increase in intestinal permeability, gut dysbiosis, and gut dysmotility, resulting in bacterial translocation in intestine and migration of gut immune cells and inflammatory cells into the injury sites of stroke through circulating blood flow. Yet, the exact molecular landscape which underlies the alterations in the brain–gut axis is in its infancy [46,47].

We collected fecal samples and analyzed composition and KEGG function of gut microbiota both from patients with stroke and health controls. Our results showed (1) decreased alpha diversity, (2) different microbiomes from control subjects at genus-level beta-diversity, (3) dysbiotic change in bacterial abundances, and (4) dysbiotic change in biologic functions in stroke patients.

Peh et al. have recently reviewed 14 clinical human stroke studies globally and found that the main cohort background was Chinese (12/14). They revealed that alpha diversity in stroke was heterogeneous (no difference: 5, decreased: 4, increased: 3, N/A: 2) and 62 upregulated (e.g., *Streptococcus*, *Lactobacillus*, *Escherichia*) and 29 downregulated (e.g., *Eubacterium*, *Roseburia*) microbial taxa in stroke patients [48]. Our study showed that the patients group for the gut showed enrichment of *Parabacteroides*, Tannerellaceae, *Lachnoclostridium*, *Escherichia_Shigella*, *Fusobacterium*, *Clostridia_UCG_014*, and *Enterococcus* in contrast to depletion of *Prevotella*, *Faecalibacterium*, *Roseburia*, Selenomonadales, *Lachnospiraceae_NK4A136_group*, *Eubacterium*, and *Dialister*. Among the aforementioned 14 clinical studies, two studies had more than 300 in cohort size and our results are consistent with their results. Yin et al. studied 322 Chinese stroke patients and 231 controls and showed that Proteobacteria was enriched but *Bacteroides*, *Prevotella*, and *Faecalibacterium* were depleted [49]. Haak et al. studied 349 Dutch stroke patients and 51 controls and showed that Proteobacteria, *Escherichia/Shigella*, *Peptoniphilus*, *Ezakiella*, and *Enterococcus* were enriched while Firmicutes and Bacteroidetes were depleted [50].

There are few reports about the biologic function of gut microbiota in stroke. Sun et al. studied 132 Chinese stroke patients and divided them into good (n = 105) and poor (n = 27) outcome based on a 3-month modified Rankin scale. The functional potential was predicted using the Phylogenetic Investigation of Communities by Reconstruction of Unobserved States (PICRUSt), and showed that upregulated function in the poor outcome group included membrane transport, transcription, and metabolism while downregulated functions included amino acid metabolism, metabolism of cofactors and vitamins, and replication and repair [51]. These findings are consistent with our results. Our study showed that membrane transport (ribose transport system), transcription (AraC family transcription regulator), and metabolism (cysteine-S-conjugate beta-lyase) are enriched in stroke patients while amino acid metabolism (glutamate synthase), replication and repair (ATP-dependent DNA helicase RecG), and metabolism of cofactors and vitamins (thiamine-phosphate pyrophosphorylase) are depleted in stroke patients.

We have previously collected blood samples from stroke patients and analyzed various parameters; a notable finding is low levels of albumin and lymphocytes [33]. Albumin and lymphocytes serve as markers of prognostic nutritional index (PNI), and depletion of both markers is associated with poor functional outcome [52]. Stroke is regularly accompanied by dysphagia and other factors associated with decreased nutritional intake [53]. Nutrients are essential regulators of both host and microbial physiology and function as key coordinators of host–microbe interactions [54]. Trace metals are important nutrients for all forms of life. Murdoch et al. recently reviewed nutritional immunity for nutrient metals at the host–microbe interface [31], and oral cuprizone, copper chelator, was reported to induce a demyelination model simulating progressive multiple sclerosis and suppress the tryptophan-kynurenine metabolic system [55].

We have reported increase in stroke patients of fecal calprotectin [33], which binds to and sequesters Zn, Mn, Ni, Cu, and Fe in the extracellular milieu through the action of metal-binding sites [56], and our present study showed that gut microbiome in stroke patients became enriched in the nickel transport system and the iron transport system.

In this regard, malnutrition can increase bacterial infections and systemic inflammation that reversibly impact brain tissue degeneration. Recent preclinical studies have shown the protective effects of nutritional support. The administration of a high-protein (HP) diet in rats has been shown to reduce post-stroke neurological deficit [57], and a HP diet in radiation-induced acute-phase inflammation in rats led to increased percentage of lymphocytes and decreased percentage of neutrophils [58].

As the prognostic importance of nutrition led three German medical societies to publish the guidelines of clinical nutrition in patients with stroke [59] in 2013, the manipulation of diet contents for stroke management, understanding the specific roles and underlying mechanisms of nutrients in regulating the host–microbe interactions, and development of strategies for improving prognosis of stroke have been gaining attention.

The potential to modulate the activity of the immune system by interventions with specific nutrients is termed "immunonutrition", and this concept may be applied to stroke [60]. Immunonutrition has been reported to improve wound healing and reduce infectious complications and length of stay in hospital. Its formulation includes supplementation with arginine, glutamine, omega-3 fatty acids, vitamins, and trace minerals (zinc, selenium), some of which have commonly been classified as nonessential but have become essential in certain clinical situations, such as for trauma patient or patients at high risk for malnutrition [61]. Arginine has various effects on wound healing and immune function. Metabolically, arginine is a precursor for ornithine, which is essential for both polyamine synthesis and NO. It is also a precursor to proline, and is thus engaged specifically for collagen synthesis [62]. Recent studies in rodents and humans showed that supplemental arginine-induced gut microbiota remodeling with enrichment of *B. pseudolongum* boosts pulmonary immune defense against nontuberculous mycobacteria (NTM) infection by driving the protective gut–lung axis in vivo [63], and arginine treatment decreases neuronal death after rat cerebral ischemia/reperfusion (I/R) injury and improves functional recovery of stroke animals [64]. Glutamine is normally nonessential but has become "conditionally essential" during inflammatory conditions. Glutamine is important to cell proliferation in that it can act as a respiratory fuel and that it can enhance the function of stimulated immune cells [65]. Glutamine protects mouse brain from ischemic injury via upregulating heat shock protein 70 [66], and reduces the intestinal colonization and bacterial overgrowth or bacterial translocation [67]. Essential fatty acids play an important role in the immune system by regulating properties of cell membranes and controlling cell signaling, while Omega-3 fatty acids lessen inflammatory responses through their effects on production of specific chemokines and cytokines [68]. Omega-3 polyunsaturated fatty acids enhance cerebral angiogenesis and provide long-term protection after stroke [69], and they correlate with gut microbiome diversity [70]. Vitamin A is an essential micronutrient that comes in multiple forms, including retinols, retinals, and retinoic acids (RAs). It plays a role in the inflammatory phase of wound healing and has been demonstrated to enhance production of extracellular matrix components such as collagen type I and fibronectin [71]. Administration of a combination of vitamin A and D supplementation can significantly increase vitamin A and D serum levels, decrease IL-1β serum levels, and ultimately improve clinical outcome in ischemic stroke patients [72]. The gut flora was altered by a vitamin-A-deficient diet in rats and mice, and RA could restore *Lactobacilli* that were downregulated in a murine lupus model [73]. Vitamin C plays an essential role in collagen formation and post-translational modification, and its deficiency leads to scurvy with various cutaneous and wound manifestations. It acts as a cofactor in the hydroxylation of proline and lysine residues in procollagen, which are critical for the stability of collagen fibers. In addition, vitamin C enhances neutrophil motility [74]. Supplementation with vitamin C increased the abundance of bacteria of the genus *Bifidobacterium* [75], and post-stroke treatment with high-dose ascorbate protects the brain through epigenetic reprogramming and may function as a robust therapeutic against stroke injury [76]. Zinc is a cofactor in a number of intracellular enzymatic reactions pertaining to wound healing. It is also an antioxidant and confers resistance against epithelial apoptosis [77]. Zinc improves neurological recovery by promoting angiogenesis via the astrocyte-mediated HIF-1α/VEGF signaling pathway in experimental stroke [78]. Selenium has a strong antioxidant role, and organic selenium was associated with a higher concentration of total VFA, propionate, and butyrate, a higher number of DNA copies of *Lactobacillus*, and a trend to lower DNA copies of *Escherichia coli* [79]. Lower selenium levels were associated with worse stroke outcomes, and selenase improved the modified Rankin Scale and National Institute of Health Stroke Scale scores significantly [80].

Limitations of this study include relatively small sample size and predictive biologic functions based on PICRUSt2. Though our taxonomic result is consistent with a large cohort size of more than 300 Chinese and Dutch people and predictive biologic functions are consistent with a Chinese prognosis study, future directions include a larger sample size of Korean cohort and whole-genome sequencing-based biologic functions and metabolomics study.

5. Conclusions

Knowledge of the gut–brain axis and metabolic and immunological interaction between host and microbe is shedding light on the research of stroke with limited therapeutic options in spite of high mortality and disability. We previously reported that fecal calprotectin had an association with the Glasgow Coma Scale, which is suggestive of gut–brain axis and deficiency of blood albumin and lymphocytes in stroke patients. This time, we explored the gut microbiome, the other side of host and microbe interaction, through 16S rRNA sequencing and found taxonomic dysbiosis in stroke patients in parallel with functional dysbiosis, which is suggestive of malnutrition. Stroke is commonly accompanied by dysphagia and other gastrointestinal complications associated with decreased nutritional intake, and the nutritional status is correlated with prognosis of stroke patients. In contrast to symbiotic interaction, dysbiotic microbiome cannot provide beneficial metabolism to the host. Nutrients are common denominators and essential regulators of both host and microbial physiology and function as key coordinators of host–microbe interactions. The potential to modulate the activity of the immune system by interventions with specific nutrients is termed "immunonutrition", and this promising concept may be applied to enhanced management of stroke with regard to our current study.

Author Contributions: Conceptualization, S.Y.P., S.P.L. and W.J.K.; methodology, S.Y.P. and W.J.K.; software, W.J.K. and D.K.; validation, S.Y.P. and S.P.L.; investigation, W.J.K.; data curation, W.J.K. and S.P.L.; writing—original draft preparation, S.Y.P. and W.J.K.; writing—review and editing, S.Y.P., W.J.K. and S.P.L.; funding acquisition, S.Y.P. All authors have read and agreed to the published version of the manuscript.

Funding: This research was funded by the Basic Science Research Program through the National Research Foundation of Korea (NRF), funded by the Ministry of Science and ICT, Republic of Korea (grant number: NRF-2018R1D1A1B07041747).

Institutional Review Board Statement: The study was conducted according to the guidelines of the Declaration of Helsinki and approved by the ethical committee of Cheju Halla University (protocol code 1044348-20180713-HR-007-01, accessed on 13 July 2018).

Informed Consent Statement: Informed consent was obtained from all subjects or their guardians involved in the study.

Data Availability Statement: Data supporting the reported result can be accessed by corresponding with the authors.

Conflicts of Interest: The authors declare no conflict of interest.

References

1. Jayaraj, R.L.; Azimullah, S.; Beiram, R.; Jalal, F.Y.; Rosenberg, G.A. Neuroinflammation: Friend and foe for ischemic stroke. *J. Neuroinflamm.* **2019**, *16*, 142. [CrossRef] [PubMed]
2. Arya, A.K.; Hu, B. Brain-gut axis after stroke. *Brain Circ.* **2018**, *4*, 165–173. [CrossRef] [PubMed]

3. Mitrea, L.; Nemeş, S.A.; Szabo, K.; Teleky, B.E.; Vodnar, D.C. Guts Imbalance Imbalances the Brain: A Review of Gut Microbiota Association with Neurological and Psychiatric Disorders. *Front. Med.* **2022**, *9*, 813204. [CrossRef] [PubMed]
4. Montanari, M.; Imbriani, P.; Bonsi, P.; Martella, G.; Peppe, A. Beyond the Microbiota: Understanding the Role of the Enteric Nervous System in Parkinson's Disease from Mice to Human. *Biomedicines* **2023**, *11*, 1560. [CrossRef] [PubMed]
5. Kovtun, A.S.; Averina, O.V.; Angelova, I.Y.; Yunes, R.A.; Zorkina, Y.A.; Morozova, A.Y.; Pavlichenko, A.V.; Syunyakov, T.S.; Karpenko, O.A.; Kostyuk, G.P.; et al. Alterations of the Composition and Neurometabolic Profile of Human Gut Microbiota in Major Depressive Disorder. *Biomedicines* **2022**, *10*, 2162. [CrossRef]
6. González-Sanmiguel, J.; Schuh, C.M.A.P.; Muñoz-Montesino, C.; Contreras-Kallens, P.; Aguayo, L.G.; Aguayo, S. Complex Interaction between Resident Microbiota and Misfolded Proteins: Role in Neuroinflammation and Neurodegeneration. *Cells* **2020**, *9*, 2476. [CrossRef]
7. Sun, P.; Su, L.; Zhu, H.; Li, X.; Guo, Y.; Du, X.; Zhang, L.; Qin, C. Gut Microbiota Regulation and Their Implication in the Development of Neurodegenerative Disease. *Microorganisms* **2021**, *9*, 2281. [CrossRef]
8. Lee, G.A.; Lin, Y.K.; Lai, J.H.; Lo, Y.C.; Yang, Y.S.H.; Ye, S.Y.; Lee, C.J.; Wang, C.C.; Chiang, Y.H.; Tseng, S.H. Maternal Immune Activation Causes Social Behavior Deficits and Hypomyelination in Male Rat Offspring with an Autism-Like Microbiota Profile. *Brain Sci.* **2021**, *11*, 1085. [CrossRef]
9. Abuaish, S.; Al-Otaibi, N.M.; Abujamel, T.S.; Alzahrani, S.A.; Alotaibi, S.M.; AlShawakir, Y.A.; Aabed, K.; El-Ansary, A. Fecal Transplant and Bifidobacterium Treatments Modulate Gut Clostridium Bacteria and Rescue Social Impairment and Hippocampal BDNF Expression in a Rodent Model of Autism. *Brain Sci.* **2021**, *11*, 1038. [CrossRef]
10. Tanaka, M.; Vécsei, L. Editorial of Special Issue "Crosstalk between Depression, Anxiety, and Dementia: Comorbidity in Behavioral Neurology and Neuropsychiatry". *Biomedicines* **2021**, *9*, 517. [CrossRef]
11. Tanaka, M.; Toldi, J.; Vécsei, L. Exploring the Etiological Links behind Neurodegenerative Diseases: Inflammatory Cytokines and Bioactive Kynurenines. *Int. J. Mol. Sci.* **2020**, *21*, 2431. [CrossRef] [PubMed]
12. Tanaka, M.; Szabó, Á.; Spekker, E.; Polyák, H.; Tóth, F.; Vécsei, L. Mitochondrial Impairment: A Common Motif in Neuropsychiatric Presentation? The Link to the Tryptophan-Kynurenine Metabolic System. *Cells* **2022**, *11*, 2607. [CrossRef]
13. Carabotti, M.; Scirocco, A.; Maselli, M.A.; Severi, C. The gut-brain axis: Interactions between enteric microbiota, central and enteric nervous systems. *Ann. Gastroenterol.* **2015**, *28*, 203–209. [PubMed]
14. Thayer, J.F.; Sternberg, E.M. Neural aspects of immunomodulation: Focus on the vagus nerve. *Brain Behav. Immun.* **2010**, *24*, 1223–1228. [CrossRef]
15. Gwak, M.G.; Chang, S.Y. Gut-Brain Connection: Microbiome, Gut Barrier, and Environmental Sensors. *Immun. Netw.* **2021**, *21*, e20. [CrossRef] [PubMed]
16. Belkaid, Y.; Hand, T.W. Role of the microbiota in immunity and inflammation. *Cell* **2014**, *157*, 121–141. [CrossRef]
17. Busnelli, M.; Manzini, S.; Chiesa, G. The Gut Microbiota Affects Host Pathophysiology as an Endocrine Organ: A Focus on Cardiovascular Disease. *Nutrients* **2019**, *12*, 79. [CrossRef]
18. Wu, Z.A.; Wang, H.X. A Systematic Review of the Interaction Between Gut Microbiota and Host Health from a Symbiotic Perspective. *SN Compr. Clin. Med.* **2019**, *1*, 224–235. [CrossRef]
19. Fan, Y.; Pedersen, O. Gut microbiota in human metabolic health and disease. *Nat. Rev. Microbiol.* **2021**, *19*, 55–71. [CrossRef]
20. Owen, E.C. Nutrition and symbiosis. *Nature* **1947**, *159*, 78–81. [CrossRef]
21. Silva, Y.P.; Bernardi, A.; Frozza, R.L. The Role of Short-Chain Fatty Acids from Gut Microbiota in Gut-Brain Communication. *Front. Endocrinol.* **2020**, *11*, 25. [CrossRef] [PubMed]
22. Li, B.; Li, L.; Li, M.; Lam, S.M.; Wang, G.; Wu, Y.; Zhang, H.; Niu, C.; Zhang, X.; Liu, X.; et al. Microbiota Depletion Impairs Thermogenesis of Brown Adipose Tissue and Browning of White Adipose Tissue. *Cell Rep.* **2019**, *26*, 2720–2737.e5. [CrossRef] [PubMed]
23. Tian, F.; Huang, S.; Xu, W.; Chen, L.; Su, J.; Ni, H.; Feng, X.; Chen, J.; Wang, X.; Huang, Q. Compound K attenuates hyperglycemia by enhancing glucagon-like peptide-1 secretion through activating TGR5 via the remodeling of gut microbiota and bile acid metabolism. *J. Ginseng. Res.* **2022**, *46*, 780–789. [CrossRef] [PubMed]
24. Ye, X.; Li, H.; Anjum, K.; Zhong, X.; Miao, S.; Zheng, G.; Liu, W.; Li, L. Dual Role of Indoles Derived from Intestinal Microbiota on Human Health. *Front. Immunol.* **2022**, *13*, 903526. [CrossRef]
25. Cohen, L.J.; Esterhazy, D.; Kim, S.H.; Lemetre, C.; Aguilar, R.R.; Gordon, E.A.; Pickard, A.J.; Cross, J.R.; Emiliano, A.B.; Han, S.M.; et al. Commensal bacteria make GPCR ligands that mimic human signalling molecules. *Nature* **2017**, *549*, 48–53. [CrossRef] [PubMed]
26. Dominique, M.; Lucas, N.; Legrand, R.; Bouleté, I.M.; Bôle-Feysot, C.; Deroissart, C.; Léon, F.; Nobis, S.; do Rego, J.C.; Lambert, G.; et al. Effects of Bacterial CLPB Protein Fragments on Food Intake and PYY Secretion. *Nutrients* **2021**, *13*, 2223. [CrossRef]
27. Qiang, X.; Liotta, A.S.; Shiloach, J.; Gutierrez, J.C.; Wang, H.; Ochani, M.; Ochani, K.; Yang, H.; Rabin, A.; LeRoith, D.; et al. New melanocortin-like peptide of *E. coli* can suppress inflammation via the mammalian melanocortin-1 receptor (MC1R): Possible endocrine-like function for microbes of the gut. *NPJ Biofilms Microbiomes* **2017**, *3*, 31. [CrossRef] [PubMed]
28. Jian, H.; Liu, Y.; Wang, X.; Dong, X.; Zou, X. *Akkermansia muciniphila* as a Next-Generation Probiotic in Modulating Human Metabolic Homeostasis and Disease Progression: A Role Mediated by Gut-Liver-Brain Axes? *Int. J. Mol. Sci.* **2023**, *24*, 3900. [CrossRef]
29. Dicks, L.M.T. Gut Bacteria and Neurotransmitters. *Microorganisms* **2022**, *10*, 1838. [CrossRef]

30. Zheng, D.; Liwinski, T.; Elinav, E. Interaction between microbiota and immunity in health and disease. *Cell Res.* **2020**, *30*, 492–506. [CrossRef]
31. Murdoch, C.C.; Skaar, E.P. Nutritional immunity: The battle for nutrient metals at the host-pathogen interface. *Nat. Rev. Microbiol.* **2022**, *20*, 657–670. [CrossRef]
32. Jukic, A.; Bakiri, L.; Wagner, E.F.; Tilg, H.; Adolph, T.E. Calprotectin: From biomarker to biological function. *Gut* **2021**, *70*, 1978–1988. [CrossRef] [PubMed]
33. Park, S.Y.; Lee, S.P.; Kim, W.J. Fecal Calprotectin Is Increased in Stroke. *J. Clin. Med.* **2021**, *11*, 159. [CrossRef] [PubMed]
34. Koszewicz, M.; Jaroch, J.; Brzecka, A.; Ejma, M.; Budrewicz, S.; Mikhaleva, L.M.; Muresanu, C.; Schield, P.; Somasundaram, S.G.; Kirkland, C.E.; et al. Dysbiosis is one of the risk factor for stroke and cognitive impairment and potential target for treatment. *Pharmacol. Res.* **2021**, *164*, 105277. [CrossRef] [PubMed]
35. Blekhman, R.; Tang, K.; Archie, E.A.; Barreiro, L.B.; Johnson, Z.P.; Wilson, M.E.; Kohn, J.; Yuan, M.L.; Gesquiere, L.; Grieneisen, L.E.; et al. Common methods for fecal sample storage in field studies yield consistent signatures of individual identity in microbiome sequencing data. *Sci. Rep.* **2016**, *6*, 31519. [CrossRef] [PubMed]
36. Baek, C.; Kim, W.J.; Moon, J.; Moon, S.Y.; Kim, W.; Hu, H.J.; Min, J. Differences in the gut microbiome composition of Korean children and adult samples based on different DNA isolation kits. *PLoS ONE* **2022**, *17*, e0264291. [CrossRef]
37. Zheng, W.; Tsompana, M.; Ruscitto, A.; Sharma, A.; Genco, R.; Sun, Y.; Buck, M.J. An accurate and efficient experimental approach for characterization of the complex oral microbiota. *Microbiome* **2015**, *3*, 48. [CrossRef]
38. Ruff, W.E.; Dehner, C.; Kim, W.J.; Pagovich, O.; Aguiar, C.L.; Yu, A.T.; Roth, A.S.; Vieira, S.M.; Kriegel, C.; Adeniyi, O.; et al. Pathogenic Autoreactive T and B Cells Cross-React with Mimotopes Expressed by a Common Human Gut Commensal to Trigger Autoimmunity. *Cell Host Microbe* **2019**, *26*, 100–113.e8. [CrossRef]
39. Segata, N.; Izard, J.; Waldron, L.; Gevers, D.; Miropolsky, L.; Garrett, W.S.; Huttenhower, C. Metagenomic biomarker discovery and explanation. *Genome Biol.* **2011**, *12*, R60. [CrossRef]
40. Douglas, G.M.; Maffei, V.J.; Zaneveld, J.R.; Yurgel, S.N.; Brown, J.R.; Taylor, C.M.; Huttenhower, C.; Langille, M.G.I. PICRUSt2 for prediction of metagenome functions. *Nat. Biotechnol.* **2020**, *38*, 685–688. [CrossRef]
41. Whitehead, A.L.; Julious, S.A.; Cooper, C.L.; Campbell, M.J. Estimating the sample size for a pilot randomised trial to minimise the overall trial sample size for the external pilot and main trial for a continuous outcome variable. *Stat. Methods Med. Res.* **2016**, *25*, 1057–1073. [CrossRef]
42. Kers, J.G.; Saccenti, E. The Power of Microbiome Studies: Some Considerations on Which Alpha and Beta Metrics to Use and How to Report Results. *Front. Microbiol.* **2022**, *12*, 796025. [CrossRef] [PubMed]
43. Kanehisa, M.; Goto, S. KEGG: Kyoto encyclopedia of genes and genomes. *Nucleic Acids Res.* **2000**, *28*, 27–30. [CrossRef] [PubMed]
44. Bustamante, A.; García-Berrocoso, T.; Rodriguez, N.; Llombart, V.; Ribó, M.; Molina, C.; Montaner, J. Ischemic stroke outcome: A review of the influence of post-stroke complications within the different scenarios of stroke care. *Eur. J. Intern. Med.* **2016**, *29*, 9–21. [CrossRef] [PubMed]
45. Tajti, J.; Szok, D.; Csáti, A.; Szabó, Á.; Tanaka, M.; Vécsei, L. Exploring Novel Therapeutic Targets in the Common Pathogenic Factors in Migraine and Neuropathic Pain. *Int. J. Mol. Sci.* **2023**, *24*, 4114. [CrossRef]
46. Murthy, P.M.; Ca, J.; Kandi, V.; Reddy, M.K.; Harikrishna, G.V.; Reddy, K.; Jp, R.; Reddy, A.N.; Narang, J. Connecting the Dots: The Interplay Between Stroke and the Gut-Brain Axis. *Cureus* **2023**, *15*, e37324. [CrossRef]
47. Zhou, S.Y.; Guo, Z.N.; Yang, Y.; Qu, Y.; Jin, H. Gut-brain axis: Mechanisms and potential therapeutic strategies for ischemic stroke through immune functions. *Front. Neurosci.* **2023**, *17*, 1081347. [CrossRef]
48. Peh, A.; O'Donnell, J.A.; Broughton, B.R.S.; Marques, F.Z. Gut Microbiota and Their Metabolites in Stroke: A Double-Edged Sword. *Stroke* **2022**, *53*, 1788–1801. [CrossRef]
49. Yin, J.; Liao, S.X.; He, Y.; Wang, S.; Xia, G.H.; Liu, F.T.; Zhu, J.J.; You, C.; Chen, Q.; Zhou, L.; et al. Dysbiosis of Gut Microbiota with Reduced Trimethylamine-N-Oxide Level in Patients with Large-Artery Atherosclerotic Stroke or Transient Ischemic Attack. *J. Am. Heart Assoc.* **2015**, *4*, e002699. [CrossRef]
50. Haak, B.W.; Westendorp, W.F.; van Engelen, T.S.R.; Brands, X.; Brouwer, M.C.; Vermeij, J.D.; Hugenholtz, F.; Verhoeven, A.; Derks, R.J.; Giera, M.; et al. Disruptions of Anaerobic Gut Bacteria Are Associated with Stroke and Post-stroke Infection: A Prospective Case-Control Study. *Transl. Stroke Res.* **2021**, *12*, 581–592. [CrossRef]
51. Sun, H.; Gu, M.; Li, Z.; Chen, X.; Zhou, J. Gut Microbiota Dysbiosis in Acute Ischemic Stroke Associated With 3-Month Unfavorable Outcome. *Front. Neurol.* **2022**, *12*, 799222. [CrossRef] [PubMed]
52. Nergiz, S.; Ozturk, U. The Effect of Prognostic Nutritional Index on Infection in Acute Ischemic Stroke Patients. *Medicina* **2023**, *59*, 679. [CrossRef] [PubMed]
53. Tuz, A.A.; Hasenberg, A.; Hermann, D.M.; Gunzer, M.; Singh, V. Ischemic stroke and concomitant gastrointestinal complications- a fatal combination for patient recovery. *Front. Immunol.* **2022**, *13*, 1037330. [CrossRef] [PubMed]
54. Bang, Y.J. Vitamin A: A key coordinator of host-microbe interactions in the intestine. *BMB Rep.* **2023**, *56*, 133–139. [CrossRef]
55. Polyák, H.; Galla, Z.; Nánási, N.; Cseh, E.K.; Rajda, C.; Veres, G.; Spekker, E.; Szabó, Á.; Klivényi, P.; Tanaka, M.; et al. The Tryptophan-Kynurenine Metabolic System Is Suppressed in Cuprizone-Induced Model of Demyelination Simulating Progressive Multiple Sclerosis. *Biomedicines* **2023**, *11*, 945. [CrossRef]
56. Nakashige, T.G.; Zygielk, E.M.; Drennan, C.L.; Nolan, E.M. Nickel Sequestration by the Host-Defense Protein Human Calprotectin. *J. Am. Chem. Soc.* **2017**, *139*, 8828–8836. [CrossRef]

57. Ji, M.; Li, S.; Dong, Q.; Hu, W. Impact of Early High-protein Diet on Neurofunctional Recovery in Rats with Ischemic Stroke. *Med. Sci. Monit.* **2018**, *24*, 2235–2243. [CrossRef]
58. Kim, K.O.; Park, H.; Chun, M.; Kim, H.S. Immunomodulatory effects of high-protein diet with resveratrol supplementation on radiation-induced acute-phase inflammation in rats. *J. Med. Food* **2014**, *17*, 963–971. [CrossRef]
59. Wirth, R.; Smoliner, C.; Jäger, M.; Warnecke, T.; Leischker, A.H.; Dziewas, R.; DGEM Steering Committee. Guideline clinical nutrition in patients with stroke. *Exp. Transl. Stroke Med.* **2013**, *5*, 14. [CrossRef]
60. Calder, P.C. Immunonutrition. *BMJ* **2003**, *327*, 117–118. [CrossRef]
61. Chow, O.; Barbul, A. Immunonutrition: Role in Wound Healing and Tissue Regeneration. *Adv. Wound Care* **2014**, *3*, 46–53. [CrossRef] [PubMed]
62. Martí, I.; Líndez, A.A.; Reith, W. Arginine-dependent immune responses. *Cell. Mol. Life Sci.* **2021**, *78*, 5303–5324. [CrossRef]
63. Kim, Y.J.; Lee, J.Y.; Lee, J.J.; Jeon, S.M.; Silwal, P.; Kim, I.S.; Kim, H.J.; Park, C.R.; Chung, C.; Han, J.E.; et al. Arginine-mediated gut microbiome remodeling promotes host pulmonary immune defense against nontuberculous mycobacterial infection. *Gut Microbes* **2022**, *14*, 2073132. [CrossRef] [PubMed]
64. Chen, S.F.; Pan, M.X.; Tang, J.C.; Cheng, J.; Zhao, D.; Zhang, Y.; Liao, H.B.; Liu, R.; Zhuang, Y.; Zhang, Z.F.; et al. Arginine is neuroprotective through suppressing HIF-1α/LDHA-mediated inflammatory response after cerebral ischemia/reperfusion injury. *Mol. Brain* **2020**, *13*, 63. [CrossRef] [PubMed]
65. Newsholme, P. Why is L-glutamine metabolism important to cells of the immune system in health, postinjury, surgery or infection? *J. Nutr.* **2001**, *131*, 2515S–2522S. [CrossRef]
66. Luo, L.L.; Li, Y.F.; Shan, H.M.; Wang, L.P.; Yuan, F.; Ma, Y.Y.; Li, W.L.; He, T.T.; Wang, Y.Y.; Qu, M.J.; et al. L-glutamine protects mouse brain from ischemic injury via up-regulating heat shock protein 70. *CNS Neurosci. Ther.* **2019**, *25*, 1030–1041. [CrossRef]
67. Perna, S.; Alalwan, T.A.; Alaali, Z.; Alnashaba, T.; Gasparri, C.; Infantino, V.; Hammad, L.; Riva, A.; Petrangolini, G.; Allegrini, P.; et al. The Role of Glutamine in the Complex Interaction between Gut Microbiota and Health: A Narrative Review. *Int. J. Mol. Sci.* **2019**, *20*, 5232. [CrossRef]
68. Gutiérrez, S.; Svahn, S.L.; Johansson, M.E. Effects of Omega-3 Fatty Acids on Immune Cells. *Int. J. Mol. Sci.* **2019**, *20*, 5028. [CrossRef]
69. Wang, J.; Shi, Y.; Zhang, L.; Zhang, F.; Hu, X.; Zhang, W.; Leak, R.K.; Gao, Y.; Chen, L.; Chen, J. Omega-3 polyunsaturated fatty acids enhance cerebral angiogenesis and provide long-term protection after stroke. *Neurobiol. Dis.* **2014**, *68*, 91–103. [CrossRef]
70. Menni, C.; Zierer, J.; Pallister, T.; Jackson, M.A.; Long, T.; Mohney, R.P.; Steves, C.J.; Spector, T.D.; Valdes, A.M. Omega-3 fatty acids correlate with gut microbiome diversity and production of N-carbamylglutamate in middle aged and elderly women. *Sci. Rep.* **2017**, *7*, 11079. [CrossRef]
71. Polcz, M.E.; Barbul, A. The Role of Vitamin A in Wound Healing. *Nutr. Clin. Pract.* **2019**, *34*, 695–700. [CrossRef] [PubMed]
72. Kadri, A.; Sjahrir, H.; Juwita Sembiring, R.; Ichwan, M. Combination of vitamin A and D supplementation for ischemic stroke: Effects on interleukin-1ß and clinical outcome. *Med. Glas.* **2020**, *17*, 425–432. [CrossRef]
73. Liu, J.; Liu, X.; Xiong, X.Q.; Yang, T.; Cui, T.; Hou, N.L.; Lai, X.; Liu, S.; Guo, M.; Liang, X.H.; et al. Effect of vitamin A supplementation on gut microbiota in children with autism spectrum disorders—A pilot study. *BMC Microbiol.* **2017**, *17*, 204. [CrossRef]
74. Elste, V.; Troesch, B.; Eggersdorfer, M.; Weber, P. Emerging Evidence on Neutrophil Motility Supporting Its Usefulness to Define Vitamin C Intake Requirements. *Nutrients* **2017**, *9*, 503. [CrossRef] [PubMed]
75. Hazan, S.; Dave, S.; Papoutsis, A.J.; Deshpande, N.; Howell, M.C., Jr.; Martin, L.M. Vitamin C improves gut Bifidobacteria in humans. *Future Microbiol.* **2022**, *online ahead of print*. [CrossRef]
76. Morris-Blanco, K.C.; Chokkalla, A.K.; Kim, T.; Bhatula, S.; Bertogliat, M.J.; Gaillard, A.B.; Vemuganti, R. High-Dose Vitamin C Prevents Secondary Brain Damage After Stroke via Epigenetic Reprogramming of Neuroprotective Genes. *Transl. Stroke Res.* **2022**, *13*, 1017–1036. [CrossRef]
77. Lin, P.H.; Sermersheim, M.; Li, H.; Lee, P.H.U.; Steinberg, S.M.; Ma, J. Zinc in Wound Healing Modulation. *Nutrients* **2017**, *10*, 16. [CrossRef]
78. Li, Y.; Ma, T.; Zhu, X.; Zhang, M.; Zhao, L.; Wang, P.; Liang, J. Zinc improves neurological recovery by promoting angiogenesis via the astrocyte-mediated HIF-1α/VEGF signaling pathway in experimental stroke. *CNS Neurosci. Ther.* **2022**, *28*, 1790–1799. [CrossRef]
79. Pereira, A.M.; Pinna, C.; Biagi, G.; Stefanelli, C.; Maia, M.R.G.; Matos, E.; Segundo, M.A.; Fonseca, A.J.M.; Cabrita, A.R.J. Supplemental selenium source on gut health: Insights on fecal microbiome and fermentation products of growing puppies. *FEMS Microbiol. Ecol.* **2020**, *96*, fiaa212. [CrossRef]
80. Ramezani, M.; Simani, L.; Abedi, S.; Pakdaman, H. Is Selenium Supplementation Beneficial in Acute Ischemic Stroke? *Neurologist* **2021**, *27*, 51–55. [CrossRef]

biomedicines

Article

Light Alcohol Consumption Promotes Early Neurogenesis Following Ischemic Stroke in Adult C57BL/6J Mice

Jiyu Li [1], Chun Li [1], Pushpa Subedi [1], Xinli Tian [2], Xiaohong Lu [2], Sumitra Miriyala [1], Manikandan Panchatcharam [1] and Hong Sun [1,*]

[1] Department of Cellular Biology & Anatomy, LSUHSC-Shreveport, Shreveport, LA 71103, USA
[2] Department of Pharmacology, Toxicology & Neuroscience, LSUHSC-Shreveport, Shreveport, LA 71103, USA
* Correspondence: hong.sun@lsuhs.edu; Tel.: +1-(318)-675-4566; Fax: +1-(318)-675-5889

Abstract: Ischemic stroke is one of the leading causes of death and disability worldwide. Neurogenesis plays a crucial role in postischemic functional recovery. Alcohol dose-dependently affects the prognosis of ischemic stroke. We investigated the impact of light alcohol consumption (LAC) on neurogenesis under physiological conditions and following ischemic stroke. C57BL/6J mice (three months old) were fed with 0.7 g/kg/day ethanol (designed as LAC) or volume-matched water (designed as control) daily for eight weeks. To evaluate neurogenesis, the numbers of 5-bromo-2-deoxyuridine (BrdU)$^+$/doublecortin (DCX)$^+$ and BrdU$^+$/NeuN$^+$ neurons were assessed in the subventricular zone (SVZ), dentate gyrus (DG), ischemic cortex, and ischemic striatum. The locomotor activity was determined by the accelerating rotarod and open field tests. LAC significantly increased BrdU$^+$/DCX$^+$ and BrdU$^+$/NeuN$^+$ cells in the SVZ under physiological conditions. Ischemic stroke dramatically increased BrdU$^+$/DCX$^+$ and BrdU$^+$/NeuN$^+$ cells in the DG, SVZ, ischemic cortex, and ischemic striatum. The increase in BrdU$^+$/DCX$^+$ cells was significantly greater in LAC mice compared to the control mice. In addition, LAC significantly increased BrdU$^+$/NeuN$^+$ cells by about three folds in the DG, SVZ, and ischemic cortex. Furthermore, LAC reduced ischemic brain damage and improved locomotor activity. Therefore, LAC may protect the brain against ischemic stroke by promoting neurogenesis.

Keywords: alcohol; brain; neurogenesis; subventricular zone; dentate gyrus; ischemia; reperfusion; rotarod test; open field test; mice

Citation: Li, J.; Li, C.; Subedi, P.; Tian, X.; Lu, X.; Miriyala, S.; Panchatcharam, M.; Sun, H. Light Alcohol Consumption Promotes Early Neurogenesis Following Ischemic Stroke in Adult C57BL/6J Mice. *Biomedicines* **2023**, *11*, 1074. https://doi.org/10.3390/biomedicines11041074

Academic Editors: Masaru Tanaka and Zhenglin Gu

Received: 27 February 2023
Revised: 20 March 2023
Accepted: 31 March 2023
Published: 2 April 2023

1. Introduction

One of the leading global causes of death and permanent disability is stroke [1,2]. Ischemic strokes, which account for 87% of all strokes, result from an obstruction or narrowing in an artery that carries blood to the brain [3]. Two reperfusion therapies, pharmaceutical dissolution and mechanical blood clot removal, are the only ones currently approved for treating acute ischemic stroke [4,5]. Thus, transient focal cerebral ischemia has become one of the most common types of ischemic stroke. Unfortunately, although reperfusion/recanalization is critical for limiting ischemic brain damage, it may paradoxically worsen brain damage by inducing reperfusion injury [1,6]. Thus, most stroke survivors retain a variety of neurological deficits resulting from brain ischemia/reperfusion (I/R) damage. Therefore, it is imperative to develop novel therapeutic strategies to prevent and treat brain I/R damage [7–9].

After an ischemic stroke, endogenous regeneration occurs in the ischemic area [10,11]. Although the ability of intrinsic self-healing of the brain is limited, some neurological deficits, especially motor ones, show a spontaneous recovery during the chronic phase [12]. Thus, promoting endogenous regeneration may be an effective strategy to improve postischemic functional recovery. Evidence suggests that neurogenesis persists in the adult mammalian brain under physiological and pathological conditions [13–15]. The subventricular zone (SVZ) of the lateral ventricles and the subgranular zone (SGZ) of the dentate gyrus (DG) are the neurogenic regions in the adult brain [16,17]. Neuroblasts differentiated from the SVZ and SGZ migrate to the olfactory bulb, local parenchyma, and granule cell layer of the DG under physiological conditions [18]. After ischemic stroke, the proliferation and differentiation of neuronal progenitors dramatically increase. The neuroblasts formed before and after the stroke migrate to the lesion area and differentiate into functional neurons [19–21]. Postischemic neurogenesis starts as early as two days, peaks at two weeks, and continues for more than six weeks after the onset of ischemic stroke [15,22,23]. Although the precise mechanism driving postischemic neurogenesis is still unclear, many factors, including growth factors and inflammatory modulators, have been reported to be involved in the process [15,24,25].

Alcohol is one of the most frequently used chemicals. Its actions frequently target the brain [26]. Various epidemiological studies suggest that chronic alcohol consumption has dose-dependent effects on the incidence and prognosis of ischemic stroke. In contrast to heavy alcohol consumption (HAC), which increases the incidence and worsens the outcome of ischemic stroke, light-to-moderate alcohol consumption (LAC) lowers the incidence and improves the prognosis of ischemic stroke [27–35]. Recently, we discovered that LAC protected against brain I/R damage by promoting cerebral angiogenesis and reducing postischemic apoptosis, inflammation, and blood-brain barrier (BBB) disruption in rodents [36–38]. However, research studies that we are aware of have yet to investigate the impact of LAC on neurogenesis following ischemic stroke. In addition, the effect of chronic alcohol consumption on baseline neurogenesis has yet to be studied to a large extent. Previous studies focused on the impact of alcohol consumption on neurogenesis in the DG of the hippocampal formation [39–41]. However, the results lack consistency. Thus, our goal of the present study is to ascertain whether LAC impacts neurogenesis under physiological conditions and following transient focal cerebral ischemia.

$BrdU^+/DCX^+$ and $BrdU^+/NeuN^+$ cells, which represent newborn neurons, are commonly used to evaluate neurogenesis [42–45]. In the present study, we compared the number of $BrdU^+/DCX^+$ and $BrdU^+/NeuN^+$ neurons in the DG, SVZ, ischemic cortex, and ischemic striatum between LAC mice and control mice under physiological conditions and two weeks after transient focal cerebral ischemia. A previous study found that LAC reduced infarct size and improved motor function at an early reperfusion stage [37]. However, the impact of LAC on ischemic damage and locomotor activity has not been assessed at the late stage of reperfusion. Nissl staining is a valuable tool to determine postischemic brain atrophy [46,47]. On the other hand, the rotarod and open field tests are extensively used for analyzing locomotor activity [48–50]. Therefore, we further measured the effects of LAC on ischemic damage and locomotor activity at two weeks of reperfusion using the Nissl stain, rotarod test, and open field test.

2. Methods

2.1. Animal Models

The Louisiana State University Health Science Center (LSUHSC)-Shreveport Institutional Animal Care and Use Committee (IACUC) gave its approval to all the procedures and protocols in the present study, which were carried out following the ARRIVE (Animal Research: Reporting in Vivo Experiments) and National Institutes of Health (NIH) Guide for the Care and Use of Laboratory Animals. The protocol (P-21-038) was approved on 12 May 2021. Forty C57BL/6J mice (male, three months old, 25–30 g) were fed with 0.7 g/kg/day ethanol (designed as LAC, $n = 20$) or volume-matched water (designed as control, $n = 20$) through gavage once a day for eight weeks [36]. These mice were housed in Animal Resources at LSUHSC-Shreveport. The mice were housed in a room maintained at 20–22 °C and light cycle from 5:00 a.m. to 5:00 p.m. Body weight, blood pressure, heart rate, and fasting blood glucose level were measured at the end of the eight-week feeding period as previously described [36]. To measure fasting blood glucose level, mice fasted for 12 h during the daytime. In each group, five mice were used to assess neurogenesis under physiological conditions, five mice were used to determine the neurogenesis following transient focal cerebral ischemia, and the remaining ten mice were used to evaluate the locomotor activity under physiological conditions and following transient focal cerebral ischemia. To measure neurogenesis under physiological conditions, 50 mg/kg bromodeoxyuridine (BrdU) (B5002, Sigma-Aldrich, St. Louis, MI, USA) was intraperitoneally administered once a day for ten days from the fifth week. The mice were euthanized at the end of the eight-week feeding period.

2.2. *Transient Focal Cerebral Ischemia*

To determine postischemic neurogenesis and locomotor activity, transient focal cerebral ischemia was induced by unilateral middle cerebral artery occlusion (MCAO) for 60 min at the end of the eight-week feeding period as previously described [38]. Ethanol was not given on the day of or following the ischemic stroke to prevent the acute effect of alcohol and simulate in-hospital alcohol discontinuation. Isoflurane in a gas mixture that contained 30% oxygen and 70% nitrogen was used to induce (5%) and maintain (1.5%) the mouse's anesthesia. During the procedure, a temperature-controlled heating pad (Harvard Apparatus, March, Germany) was used to keep the body temperature stable, and a Laser-Doppler flow probe (PERIMED, PF 5010 LDPM Unit, Järfälla, Sweden) was used to monitor the regional cerebral blood flow (rCBF) of the right MCA territory. To occlude the right MCA, the right external carotid artery (ECA) and common carotid artery (CCA) were exposed and ligated. Then, a silicon rubber-coated monofilament was inserted into the right internal carotid artery (ICA) from the base of the right ECA cranially to the bifurcation, where the ICA splits into the MCA and the anterior cerebral artery (ACA). A sharp decline in rCBF in the MCA territory indicated the initiation of MCAO. Following 60 min of occlusion, the monofilament was withdrawn, and the CCA was reopened to allow for reperfusion. To measure postischemic neurogenesis, 75 mg/kg BrdU was intraperitoneally administered twice daily for five consecutive days from 48 h of reperfusion. Mice were euthanized at 14 days of reperfusion.

2.3. *Immunohistochemistry Staining*

To determine the impact of chronic alcohol consumption on neurogenesis, dual immunohistochemistry staining was performed. The anesthetized mice were transcardially perfused with phosphate-buffered saline (PBS) and 4% paraformaldehyde. The brains were removed, fixed for an overnight period in 4% paraformaldehyde, dehydrated for 72 h in a graded series of sugar solutions, embedded for 5 min in O.C.T. compound (23-730-571, Fisher Scientific, Waltham, MA, USA), and then immediately frozen in liquid nitrogen. Coronal sections of 14 μm thickness were cut from the frozen brains and placed on frost-free slides. Three sections (0.25 mm rostral and 0.47 mm and 2.15 mm caudal to bregma) from each mouse were incubated in 2 M HCl for 1 h and neutralized in 0.1 M sodium borate buffer (pH 8.5) for 10 min. After being washed with PBS, the sections were blocked with a mixture of 1% bovine serum albumin (BSA), 0.3% trypsin, and 5% goat serum for 1 h and incubated with 1:100 mouse anti-BrdU antibody (347580; BD Bioscience, La Jolla, CA, USA) at 4 °C overnight. Subsequently, the sections were washed with PBS, incubated with 1:200 biotinylated goat antimouse lgG antibody (BA-9200; Vector Labs, Newark, CA, USA) for 1 h, then 1:200 streptavidin Alexa Fluor™ 488 conjugate (s-32354; Thermo Fisher, Waltham, MA, USA) for 30 min. Following three washes, the sections were blocked with a mixture of 1% BSA, 0.3% trypsin, and 5% donkey serum for 1 h and then incubated with 1:100 rabbit antidoublecortin (DCX) (4604S; Cell Signaling, Danvers, MA, USA) or 1:100 rabbit anti-NeuN (MABN140; MilliporeSigma, Burlington, MA, USA) for 3 h. The sections were washed with PBS and incubated with 1:200 Alexa Fluor 546 conjugated donkey antirabbit IgG antibody (A10040; Thermo Fisher) for one hour. After three PBS washes, the sections were coated with a DAPI mounting medium (H-1800, VectorShield, Newark, CA, USA) and then observed under a fluorescence microscope (Nikon Eclipse Ts2). Five pictures from each region of interest in the SVZ, DG, ischemic cortex, and ischemic striatum were captured for quantitative analysis. BrdU- and DCX/NeuN-positive cells were counted, and their fold changes from the control were expressed.

2.4. Cresyl Violet Staining

The section at 0.47 mm caudal to bregma, which had the largest infarct at 24 h of reperfusion in the MCAO mouse model, was selected for Cresyl Violet staining to evaluate ischemic damage as described previously [37]. The sections were cleaned in xylene, dehydrated in ethanol, incubated in 0.01% Cresyl Violet acetate (C5042, Sigma-Aldrich) solution at 60 °C for 14 min, and mounted using xylene-a-based mounting media (8312-4, VWR, Radnor, PA, USA). The section was pictured under 1.0× magnification (Olympus) and analyzed using ImageJ. Instead of a complete lack of staining, which is defined as the infarct, a reduction in the volume of the ipsilateral hemisphere was observed. Thus, ischemic damage was represented by the ratio of the ipsilateral hemispheric volume to the contralateral hemispheric volume.

2.5. Locomotor Activity

The accelerating rotarod (LSI Letica Scientific Instruments, Barcelona, Spain) test was conducted as described previously to assess the induced motor activity [51]. Briefly, mice were trained to stay on the rotarod at a constant speed of 4 rpm for three consecutive days during the eighth week of gavage feeding and then underwent three test trials per day with the rotarod set at an acceleration rate of 4–40 rpm/10 min at the end of the eighth week of the feeding period and three days, seven days, and fourteen days of reperfusion following 60-min MCAO. The latency to fall from the accelerating rotarod was recorded for each trial. The average of the three trials on each day was used for statistical analysis.

To assess spontaneous motor activity, the open field test was performed. After habituation to the testing room, mice were placed into a square open field chamber (40 cm L × 40 cm W × 30 cm H) (AccuScan Instruments, Erie, PA, USA). Mice were allowed to explore freely for 30 min. The mice's movements within the chambers were recorded. The total distance traveled and the number of moves were analyzed using the Top Scan Lite-Top View Behavior Analyzing System (Noldus Information Technology, Wageningen, Gelderland, The Netherlands).

2.6. Statistical Analysis

The statistical analysis was done with Prism 9. The comparison of two independent groups was performed using an unpaired *t*-test. Means and the standard deviation (SD) are used to present the quantitative data. The differences are considered statistically significant when the *p*-value is less than 0.05.

3. Results

3.1. Control Conditions

Similar to the previously reported information, gavage feeding with 0.7 g/kg/day ethanol once a day for eight weeks did not significantly change body weight, mean arterial blood pressure (MABP), heart rate, or fasting glucose level (Table 1) [36,37].

Table 1. Effects of LAC on body weight, MABP, heart rate, and fasting blood glucose. Values are means ± SD for 6–20 mice in each group. Analyzed using an unpaired *t*-test.

	Control	0.7 g/kg/d EtOH	*p* Value
Body weight (g)	26.6 ± 2.0 (n = 20)	26.7 ± 1.4 (n = 20)	0.89
MABP (mmHg)	88.6 ± 13.3 (n = 10)	87.0 ± 14.2 (n = 10)	0.78
Heart rate (bpm)	653 ± 106 (n = 10)	598 ± 100 (n = 10)	0.23
Fasting blood glucose (mg/dL)	139.3 ± 19.7 (n = 6)	139.5 ± 26.5 (n = 6)	0.99

3.2. Effect of LAC on Neurogenesis under Physiological Conditions

$BrdU^+/DCX^+$ and $BrdU^+/NeuN^+$ cells were observed in the SVZ and DG under physiological conditions (Figures 1A,C and 2A,C) (Supplemental Figures S1 and S2 for higher magnification). Interestingly, an eight-week daily intake of 0.7 g/kg/day ethanol significantly increased the number of $BrdU^+/DCX^+$ and $BrdU^+/NeuN^+$ cells in the SVZ by two folds and six folds, respectively (Figure 1B,D). In addition, 0.7 g/kg/day of ethanol tended to increase the number of $BrdU^+/DCX^+$ and $BrdU^+/NeuN^+$ cells in the DG (Figure 2). However, the increase did not reach statistical significance compared to the control (Figure 2B,D).

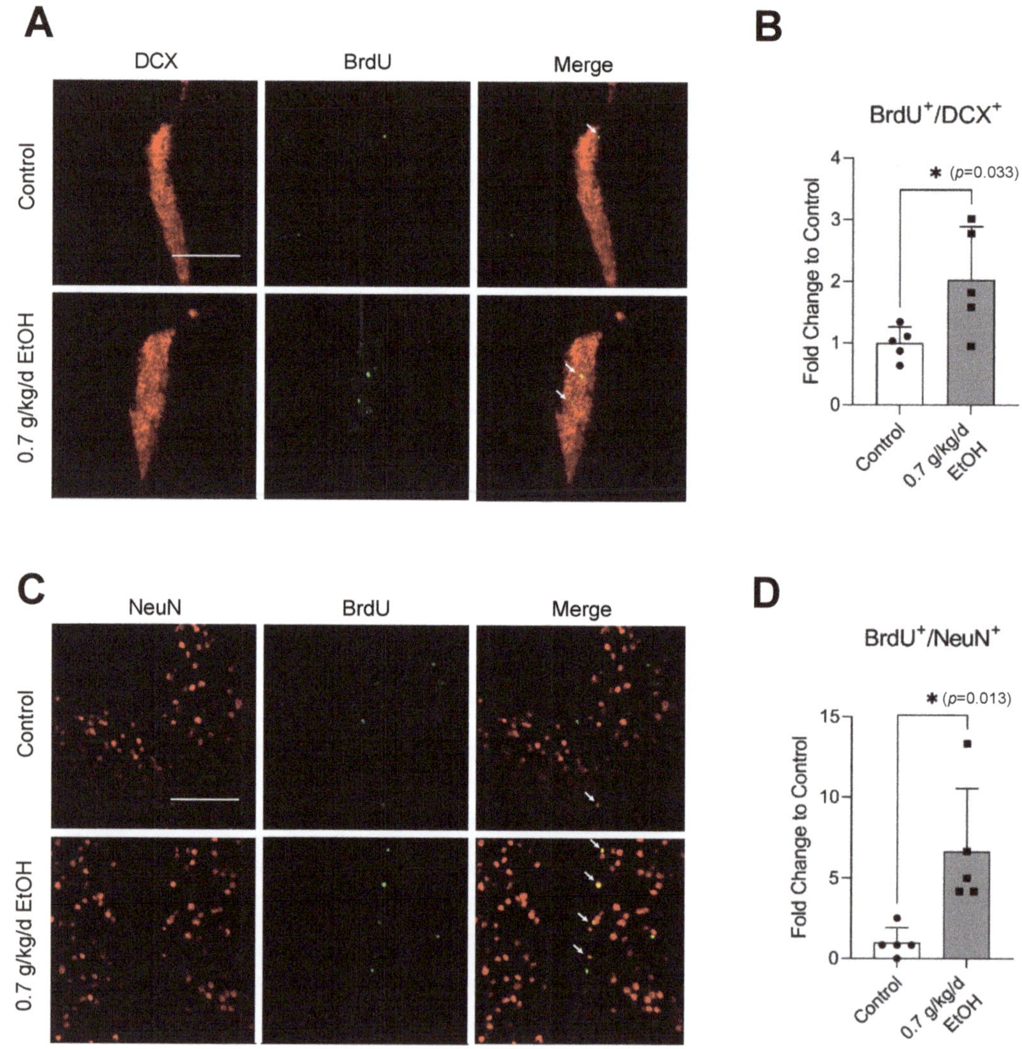

Figure 1. Effect of LAC on baseline neurogenesis in the SVZ. (**A**) Representative double staining of BrdU and DCX. Scale bar = 100 µm. (**B**) Values are means ± SD (n = 5). (**C**) Representative double staining of BrdU and NeuN. (**D**) Values are means ± SD (n = 5); * $p < 0.05$ vs. Control. Analyzed using an unpaired t-test.

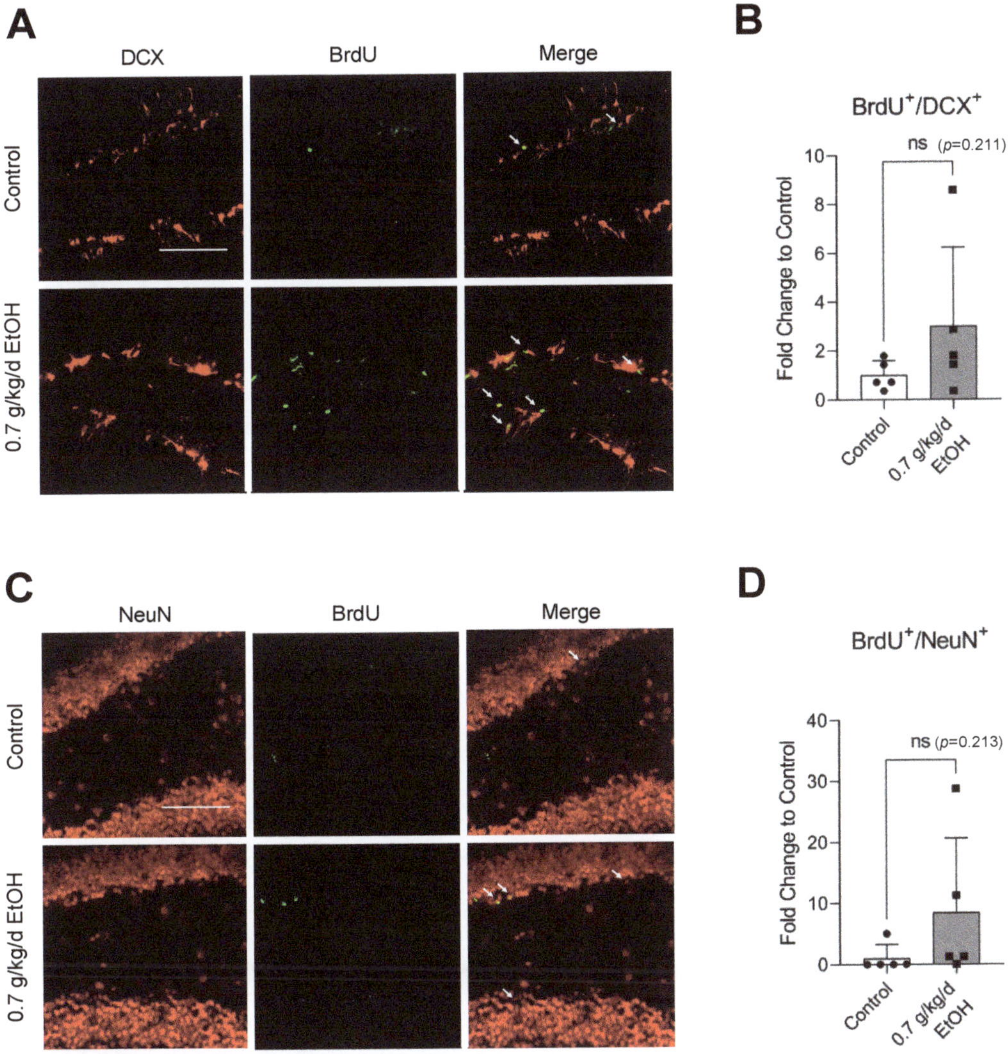

Figure 2. Effect of LAC on baseline neurogenesis in the DG. (**A**) Representative double staining of BrdU and DCX. Scale bar = 100 μm. (**B**) Values are means ± SD (n = 5). (**C**) Representative double staining of BrdU and NeuN. (**D**) Values are means ± SD (n = 5). Analyzed using an unpaired t-test.

3.3. *Effect of LAC on Brain I/R Damage*

Cresyl violet staining was performed to evaluate transient focal cerebral ischemia-induced brain damage. As shown in Figure 3A, a complete lack of staining, defined as the infarct lesion at 24 h of reperfusion [37,52], was not observed at 14 days of reperfusion following 60-min MCAO. Instead, a reduction in the volume of the ipsilateral (right) hemisphere was found. The ratio of the ipsilateral hemisphere to the contralateral hemisphere was significantly greater in 0.7 g/kg/day ethanol-fed mice compared to the control mice (Figure 3B).

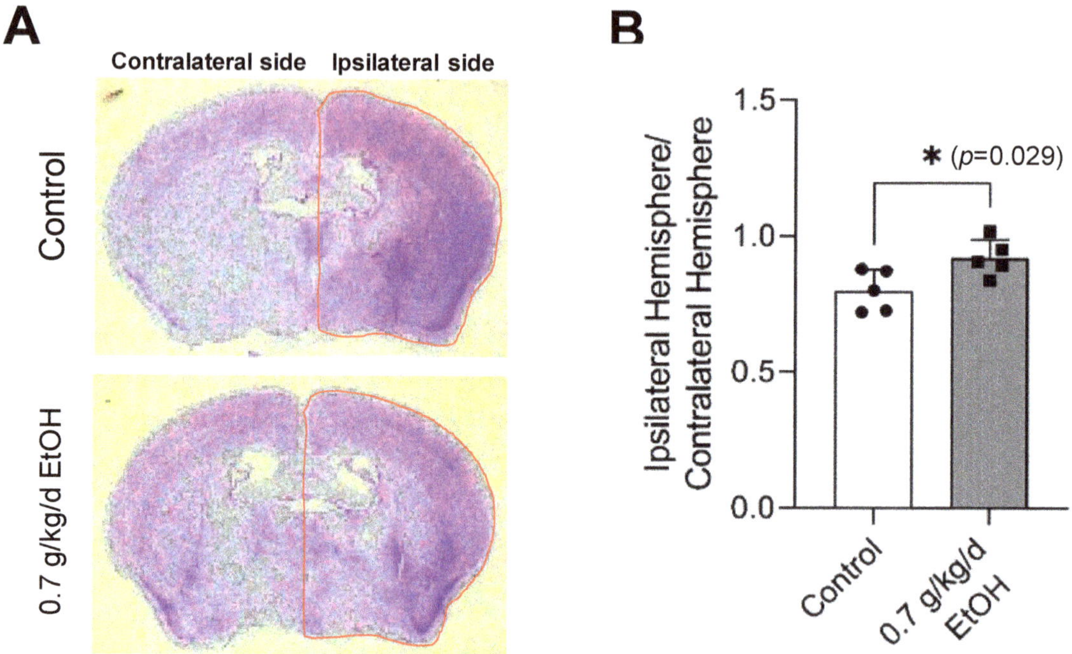

Figure 3. Effect of LAC on cerebral I/R damage at two weeks of reperfusion. (**A**) Representative brain sections stained with Cresyl violet. (**B**) The ratio of ipsilateral hemisphere/contralateral hemisphere. Values are means ± SD for five mice in each group; * $p < 0.05$ vs. Control. Analyzed using an unpaired t-test.

3.4. Effect of LAC on Neurogenesis Following Transient Focal Cerebral Ischemia

Sixty-minute MCAO significantly increased $BrdU^+/DCX^+$ and $BrdU^+/NeuN^+$ cells in the SVZ and DG at 14 days of reperfusion (Figures 4A,C and 5A,C) (Supplemental Figures S3 and S4 for higher magnification). In addition, $BrdU^+/DCX^+$ and $BrdU^+/NeuN^+$ cells can be seen in the ischemic cortex and ischemic striatum (Figures 6A,C and 7A,C). As shown in Figures 4B, 5B, 6B and 7B, the number of $BrdU^+/DCX^+$ cells in all observed areas was significantly greater in 0.7 g/kg/day ethanol-fed compared to the control mice. In addition, 0.7 g/kg/day of ethanol significantly augmented the postischemic increase in $BrdU^+/NeuN^+$ cells by about three folds in the SVZ and DG (Figures 4D and 5D). Furthermore, the number of $BrdU^+/NeuN^+$ cells in the ischemic cortex was significantly greater (by more than three folds) in 0.7 g/kg/day ethanol-fed mice compared to the control mice (Figure 6D). On the other hand, 0.7 g/kg/day of ethanol tended to increase the number of $BrdU^+/NeuN^+$ cells in the ischemic striatum. However, the increase did not reach statistical significance compared to the control (Figure 7D).

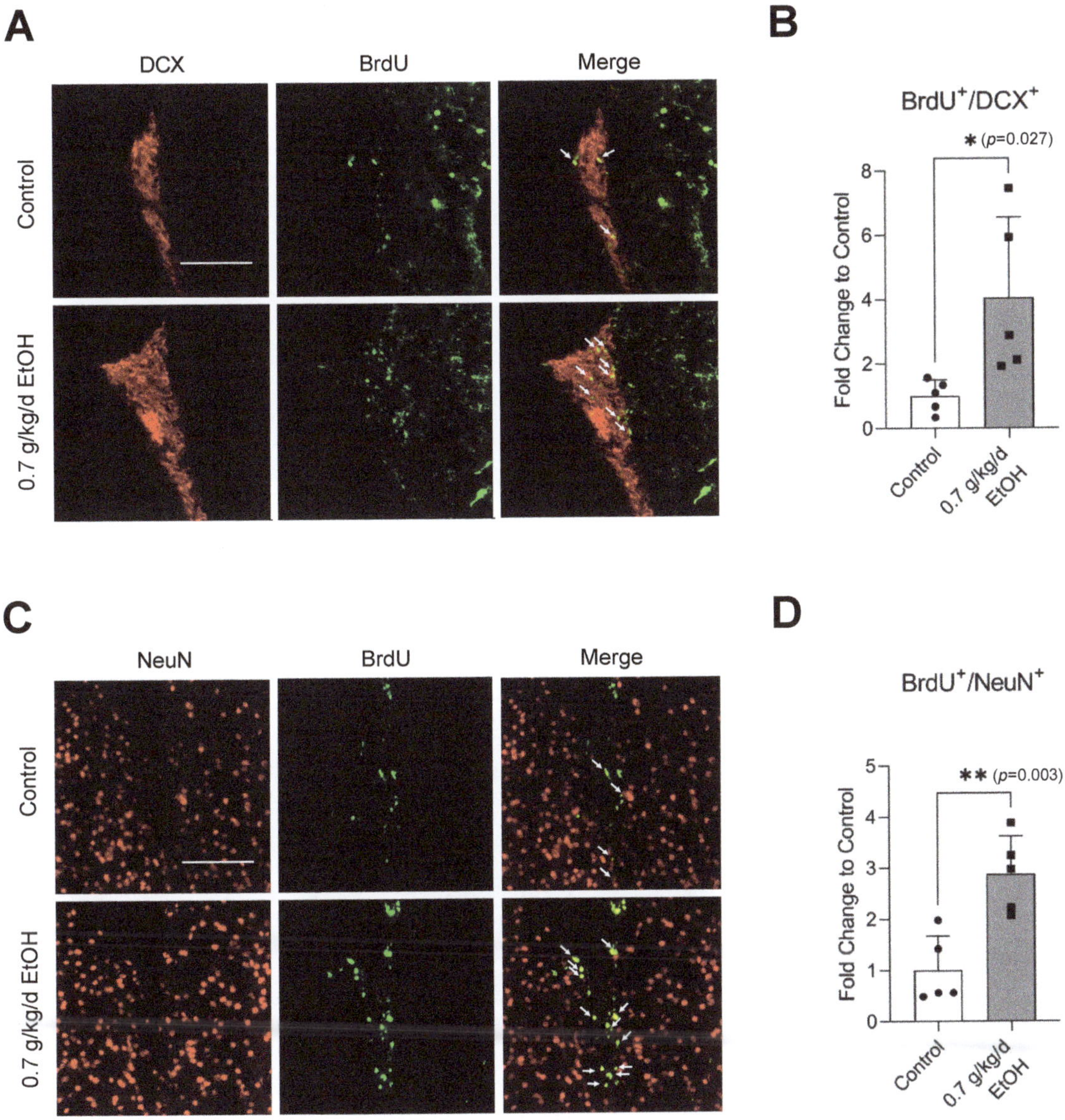

Figure 4. Effect of LAC on postischemic neurogenesis in the SVZ. (**A**) Representative double staining of BrdU and DCX. Scale bar = 100 μm. (**B**) Values are means ± SD (n = 5). (**C**) Representative double staining of BrdU and NeuN. (**D**) Values are means ± SD (n = 5); * $p < 0.05$; ** $p < 0.01$ vs. Control. Analyzed using an unpaired t-test.

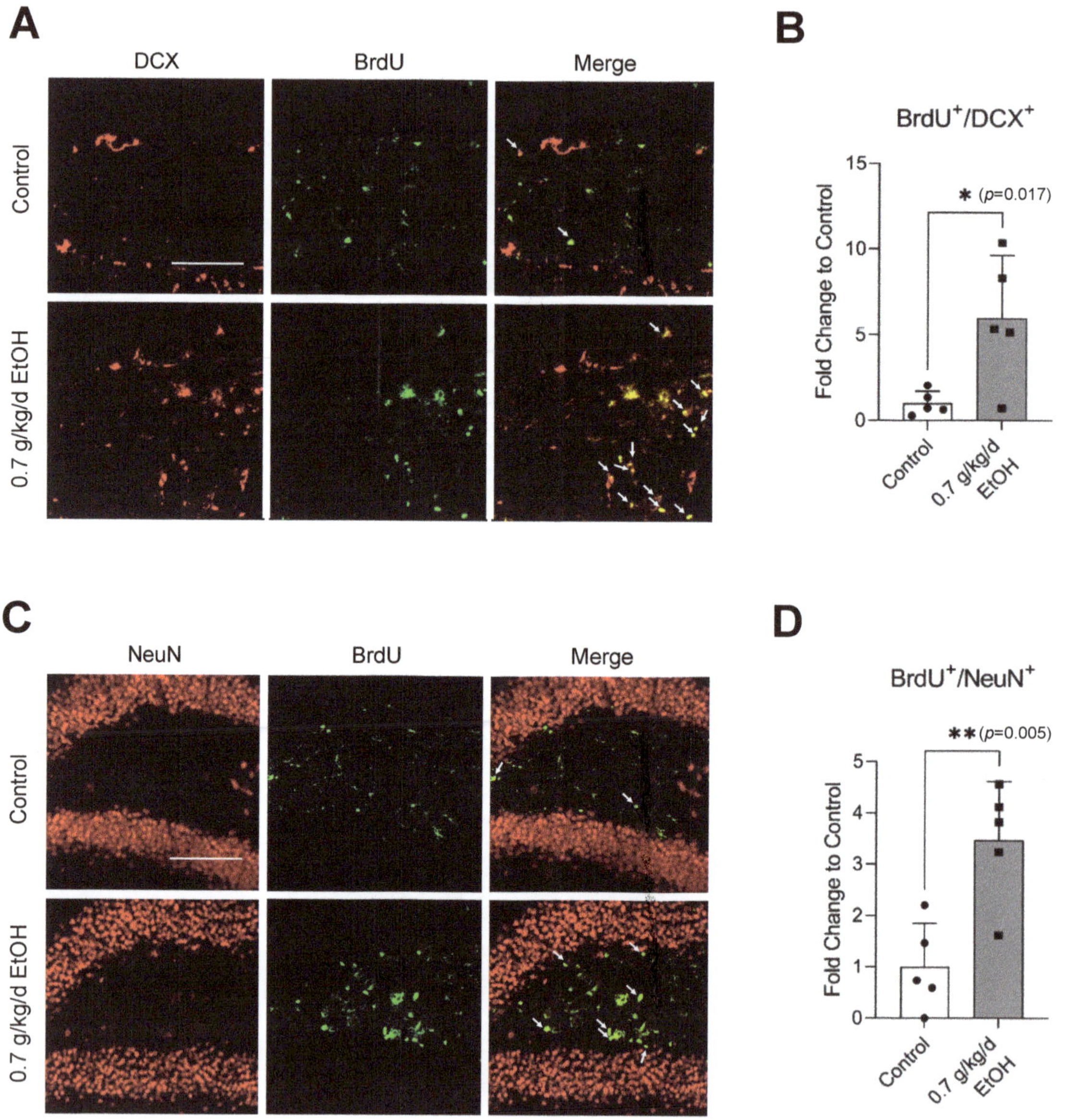

Figure 5. Effect of LAC on postischemic neurogenesis in the DG. (**A**) Representative double staining of BrdU and DCX. Scale bar = 100 μm. (**B**) Values are means ± SD (n = 5). (**C**) Representative double staining of BrdU and NeuN. (**D**) Values are means ± SD (n = 5); * $p < 0.05$; ** $p < 0.01$ vs. Control. Analyzed using an unpaired t-test.

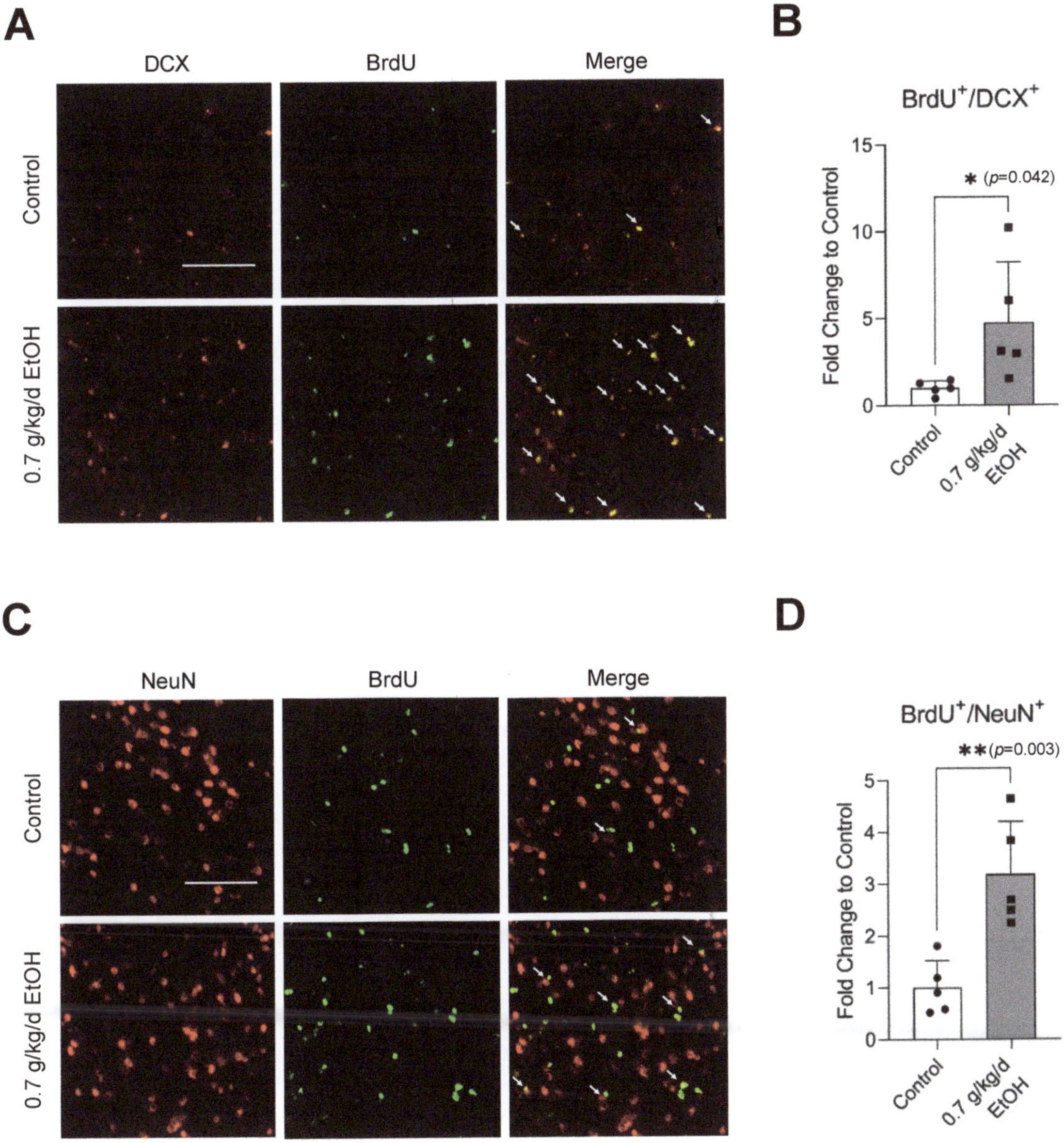

Figure 6. Effect of LAC on postischemic neurogenesis in the ischemic cortex. (**A**) Representative double staining of BrdU and DCX. Scale bar = 100 μm. (**B**) Values are means ± SD (n = 5). (**C**) Representative double staining of BrdU and NeuN. (**D**) Values are means ± SD (n = 5); * $p < 0.05$; ** $p < 0.01$ vs. Control. Analyzed using an unpaired t-test.

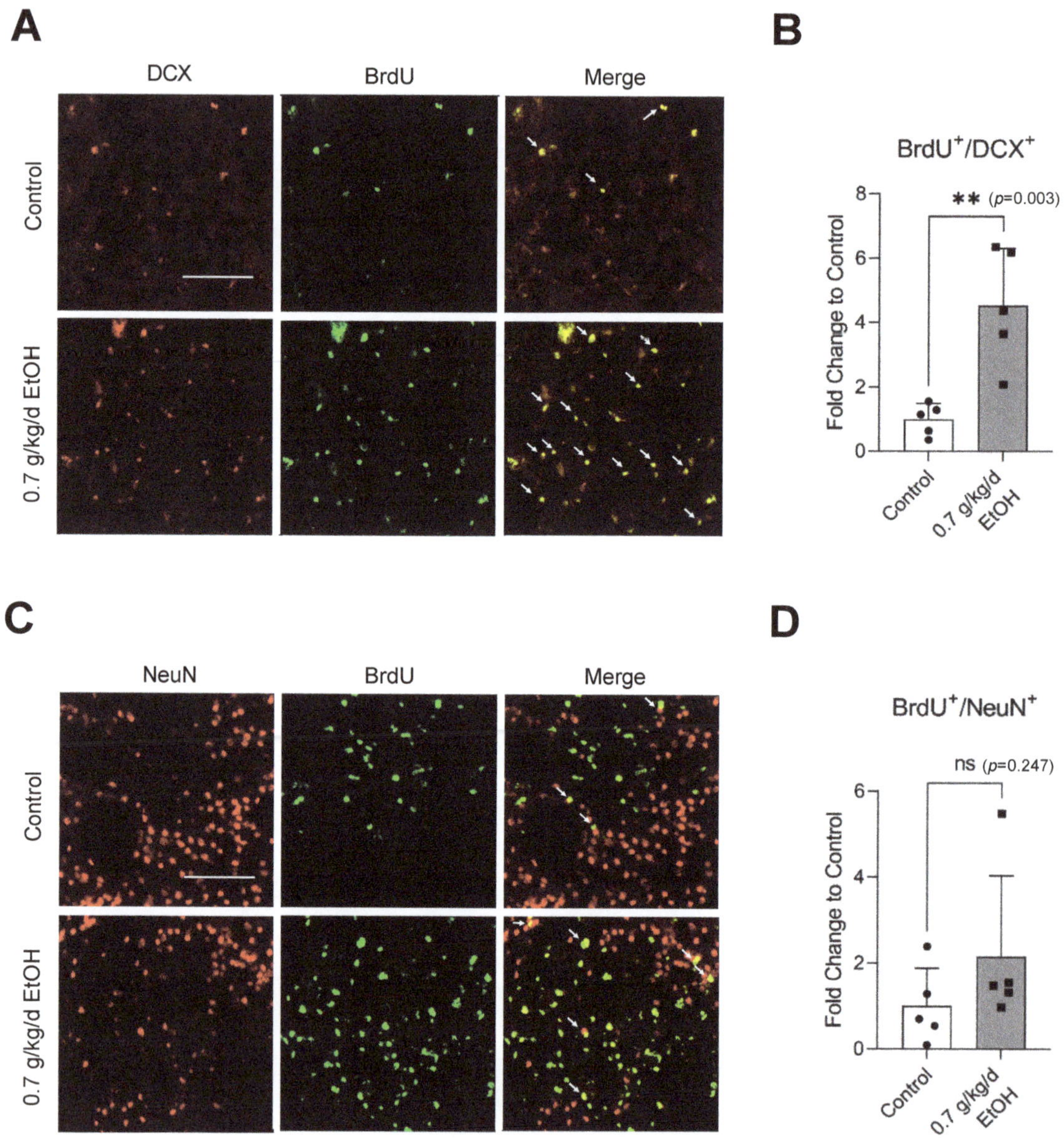

Figure 7. Effect of LAC on postischemic neurogenesis in the ischemic striatum. (**A**) Representative double staining of BrdU and DCX. Scale bar = 100 µm. (**B**) Values are means ± SD (n = 5). (**C**) Representative double staining of BrdU and NeuN. (**D**) Values are means ± SD (n = 5); ** $p < 0.01$ vs. Control. Analyzed using an unpaired t-test.

3.5. Effect of LAC on the Locomotor Activity

The accelerating rotarod and open field tests were conducted to assess the impact of LAC on locomotor activity. Interestingly, eight-week ingestion of 0.7 g/kg/day ethanol significantly strengthened the induced and spontaneous motor activities under physiological conditions (Figure 8). In order to allow the animals to recover from the surgical procedure of the MCAO for two days, both tests were not conducted for two days after ischemia. As shown in Figure 8A, the induced motor activity was significantly enhanced at 3 days, 7 days, and 14 days of reperfusion in 0.7 g/kg/day ethanol-fed mice compared to the control mice. In addition, 0.7 g/kg/day of ethanol significantly increased the total distance and number of movements in the open field test at seven days of reperfusion (Figure 8B,C).

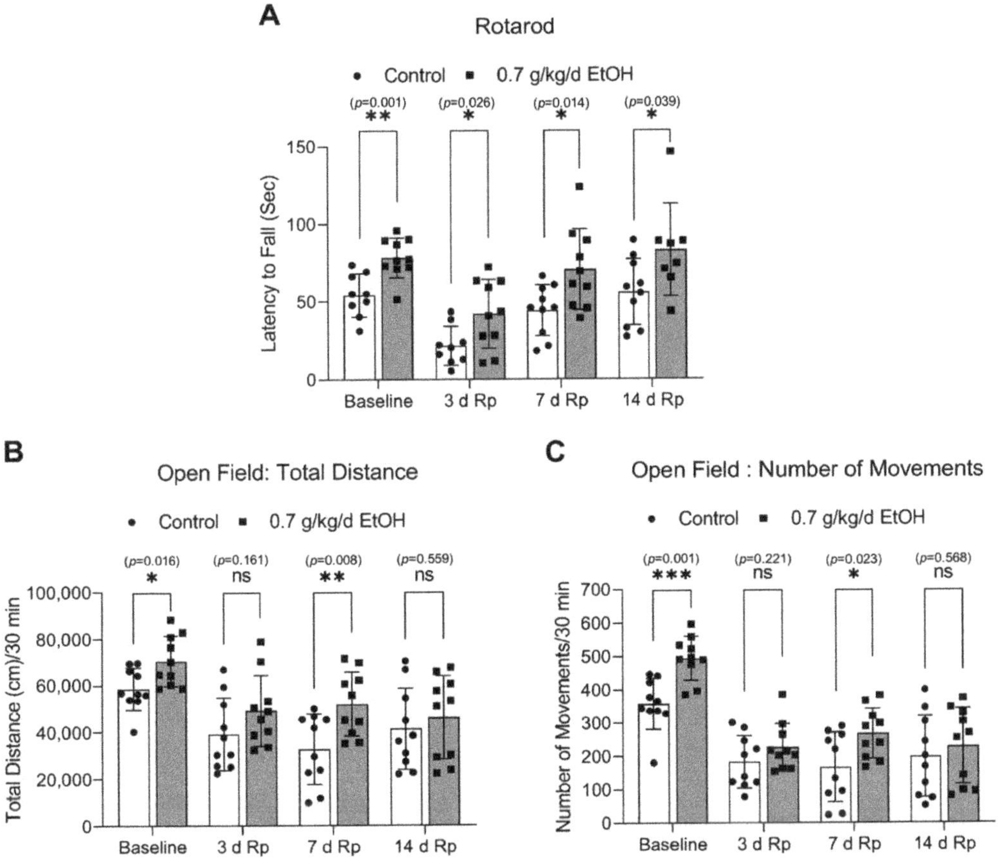

Figure 8. Effect of LAC on the locomotor activity under physiological conditions and following transient focal cerebral ischemia. (**A**) The latency to fall from the accelerating rotarod. (**B**) The total distance traveled during the 30-min trial in the open field test. (**C**) The number of movements during the 30-min trial in the open field test. Values are means ± SD (n = 10); * $p < 0.05$; ** $p < 0.01$; *** $p < 0.005$ vs. Control. Analyzed using an unpaired t-test.

In summary, LAC reduced postischemic brain atrophy at two weeks of reperfusion. Furthermore, LAC significantly increased newborn neurons in the SVZ, DG, and ischemic cortex and enhanced motor activity under physiological conditions and following ischemic stroke.

4. Discussion

The impact of LAC on neurogenesis under physiological circumstances and following transient focal cerebral ischemia was determined in the present study. There are several new findings. First, LAC increased neurogenesis in the SVZ but not DG under physiological conditions. Second, LAC alleviated transient focal cerebral ischemia-induced atrophy in the ipsilateral hemisphere at two weeks of reperfusion. Third, LAC augmented neurogenesis in the SVZ, DG, and ischemic cortex two weeks following transient focal cerebral ischemia. Fourth, LAC strengthened the locomotor activity under physiological conditions and following transient focal cerebral ischemia. We speculate that LAC may protect the brain against ischemic stroke by promoting neurogenesis. Thus, a future study appears necessary to elucidate the mechanism underlying LAC-induced neurogenesis, which may lead to new approaches for preventing and treating ischemic stroke and neurodegenerative diseases.

In our recent studies, eight-week daily consumption of 0.7 g/kg alcohol significantly diminished early ischemic brain damage in the mouse model of transient focal cerebral ischemia [36,37]. The peak blood alcohol concentration was about 9 mM [36], which can be seen following the consumption of 1.5 standard drinks (each containing 14 g of pure ethanol) in a male with average body weight [53]. Therefore, the dose of 0.7 g/kg/day ethanol in mice represents LAC in humans. BrdU, a thymidine analog readily incorporated into DNA during the S-phase of the cell cycle, is a reagent extensively used to label and quantify proliferating cells [54]. DCX is a nervous system-specific microtubule-associated protein (MAP) expressed in migrating neurons, and NeuN is a neuronal nuclear antigen expressed in mature neurons. Neural stem/progenitor cells (NSPCs) are present in the SVZ of the lateral ventricle and SGZ of the DG. After the ischemic process, NSPCs in these areas proliferate and migrate toward the lesion to participate in brain repair. Thus, the numbers of $DCX^+/BrdU^+$ and $NeuN^+/BrdU^+$ cells in the SVZ, DG, and ischemic areas of 0.7 g/kg/day ethanol-fed mice under physiological conditions and following ischemic stroke were counted to evaluate the impact of LAC on neurogenesis. In addition to mature neurons, NeuN is also expressed lightly in the maturing neurons [55,56]. Therefore, in the present study, both heavily and lightly NeuN-staining $BrdU^+$ cells were counted as newborn neurons to indicate neurogenesis.

The brain is one of the primary target organs of alcohol's effects. Several studies have examined the effects of acute and chronic alcohol exposure on hippocampal and subventricular neurogenesis under physiological conditions. However, the results appeared inconsistent. For example, an early study reported that acute alcohol intoxication dose-dependently inhibited NSPC proliferation in the DG and SVZ of male adolescent Sprague–Dawley rats [57]. Similarly, Taffe et al. found that long-term heavy alcohol consumption dramatically and persistently decreased hippocampal cell proliferation and neurogenesis in adolescent nonhuman primates [39]. Moreover, Liu and Crews found that adolescent intermittent ethanol exposure persistently decreased adult subventricular and hippocampal neurogenesis in Wistar rats [58]. Anderson et al. reported that chronic moderate alcohol consumption significantly decreased NSPC proliferation in the DG of either male or female adult Sprague-Dawley rats [40]. In contrast, Aberg et al. found that chronic moderate alcohol consumption improved hippocampal cell proliferation and neurogenesis in male adult C57BL/6 mice [41]. Xu et al. reported that two-month voluntary alcohol drinking stimulated neurogenesis in the DG and SVZ of male cHAP mice [59]. Recently, four-day binge drinking was shown to increase hippocampal neurogenesis in adult female Sprague-Dawley rats [60]. In the present study, although LAC tended to increase both $DCX^+/BrdU^+$ and $NeuN^+/BrdU^+$ cells in the DG, the increase did not reach statistically significant levels. However, LAC increased $DCX^+/BrdU^+$ cells by about two folds and $NeuN^+/BrdU^+$ cells by about six folds in the SVZ. The reason for the discrepancies between these studies regarding alcohol on hippocampal neurogenesis is unclear. It could be connected to the age, the length of alcohol exposure, the methods used to give alcohol, the timing of analysis, or the species differences in the effects of alcohol.

The present study is the first to examine the effects of LAC on postischemic neurogenesis. SVZ is the main source of neuroblasts generated following ischemic stroke [61–63]. It has been suggested that NSPCs in the SVZ proliferate, differentiate, and migrate to the infarct area, contributing to self-repair and repopulation of the injured area following ischemic stroke. Increasing evidence indicates that post-ischemic neurogenesis in the SVZ is involved in functional improvement [21,64]. On the other hand, the proliferation and differentiation of NSPCs in the DG are also remarkably stimulated following ischemic stroke. However, the production of new neurons in the DG influenced functional recovery negatively [15]. In the present study, LAC significantly increased both $DCX^+/BrdU^+$ and $NeuN^+/BrdU^+$ cells in the SVZ and DG. In addition, LAC significantly increased both $DCX^+/BrdU^+$ and $NeuN^+/BrdU^+$ cells in the ischemic cortex and $DCX^+/BrdU^+$ cells in the ischemic striatum. The ischemic injury's severity is a major factor affecting postischemic neurogenesis. However, our previous studies have shown that LAC reduces cerebral I/R damage at 24 h of reperfusion in this mouse model of transient focal cerebral ischemia [36,37,65]. In the present study, LAC significantly reduced postischemic atrophy of the cerebral hemisphere at two weeks of reperfusion. Thus, it is conceivable that LAC preconditioning may favor postischemic endogenous repair, but not due to the compensatory response.

In addition to the severity of the brain injury, inflammation, growth factors, and angiogenesis also have been demonstrated to affect neurogenesis [15,24,25,66]. However, mechanisms that account for the effects of alcohol on neurogenesis under physiological conditions remain unclear. Our recent studies found LAC did not significantly alter that baseline DNA fragmentation. In addition, LAC did not lead to neuronal apoptosis under physiological conditions in adult mice [37,65]. Thus, it does not appear that the proneurogenic effect of LAC under physiological conditions is a response to a brain injury. In contrast, LAC may stimulate neurogenesis through its anti-inflammatory effect. LAC significantly changed the inflammatory profile in the brain under physiological conditions and following ischemic stroke [36]. LAC tends to increase anti-inflammatory cytokines/chemokines. It reduced IL-1beta and increased IL-1ra following ischemic stroke. Moreover, early postischemic microglia activation, which is a proinflammatory phenotype of microglia, was significantly suppressed by LAC. A recent study found that the blockade of the IL-1 receptor promoted the proliferation of NSPCs in the SVZ, enhanced neuroblast migration, and increased the number of newly born neurons in the ischemic cortex [67]. On the other hand, the effect of microglia activation on neurogenesis seems dual. An early study reported that microglia activation impaired hippocampal neurogenesis [68]. Recently, the anti-inflammatory phenotype of microglia was found to enhance the proliferation and differentiation of neuronal progenitors in the SVZ after ischemic stroke [69]. Moreover, it is also possible that LAC stimulates neurogenesis by upregulating the vascular endothelial growth factor (VEGF) and VEGF receptor 2 (VEGFR2). We recently found that LAC upregulates VEGF and VEGFR2 and promotes cerebral angiogenesis [38]. It has been suggested that VEGF/VEGFR2 signaling can directly lead to neurogenesis by stimulating cell proliferation [70]. In addition, the increased angiogenesis may facilitate neuroblasts to reach the damaged area by migrating along the newly formed blood vessels [66,71]. In our recent study, LAC significantly increased postischemic angiogenesis in the ischemic cortex [38]. Both $DCX^+/BrdU^+$ and $NeuN^+/BrdU^+$ cells in the ischemic cortex were consistently significantly greater in LAC mice. Thus, the precise mechanism by which LAC induces neurogenesis must be elucidated in the future.

In the present study, we further evaluated the impact of LAC on locomotor activity. The accelerating rotarod has been reliably used to test motor coordination and motor learning. The open field test is generally used to assess spontaneous locomotion [51]. Interestingly, LAC strengthened the locomotor activity in both tests under physiological conditions. However, it is unknown whether the enhanced locomotor activity resulted from the increased neurogenesis in LAC mice. In addition, although LAC significantly increased $NeuN^+/BrdU^+$ cells in the DG, SVZ, and ischemic cortex at two weeks of reperfusion, the total distance and number of movements of the open field test only reached statistically significant at one week, but not two weeks of reperfusion. Thus, future studies are essential to investigate the impacts of LAC-induced neurogenesis under physiological conditions and whether postischemic (in-hospital) alcohol cessation compromises the neuroprotective effect of LAC following ischemic stroke.

The present study provided additional evidence that chronic alcohol consumption significantly alters the pathophysiology of ischemic stroke. Although LAC appears beneficial to the prognosis of ischemic stroke in many aspects, alcohol, especially heavy alcohol consumption, is associated with cancer and other diseases [72–74]. Therefore, drinking alcohol is not encouraged.

5. Conclusions

The present study determined the impact of LAC on neurogenesis. LAC promoted neurogenesis under physiological conditions following transient focal cerebral ischemia. Thus, the present study merits further investigation. A better understanding of how alcohol impacts neurogenesis will not only improve the clinical treatment of ischemic stroke in alcohol users, but will also result in new approaches for preventing and treating ischemic stroke and neurodegenerative diseases in nondrinkers.

Supplementary Materials: The following supporting information can be downloaded at: https://www.mdpi.com/article/10.3390/biomedicines11041074/s1, Figure S1. Effect of LAC on baseline neurogenesis in the SVZ; Figure S2. Effect of LAC on baseline neurogenesis in the DG; Figure S3. Effect of LAC on post-ischemic neurogenesis in the SVZ; Figure S4. Effect of LAC on post-ischemic neurogenesis in the DG.

Author Contributions: H.S. conceived the experiments. J.L., C.L., P.S., X.T., X.L., S.M., M.P. and H.S. conducted the experiments. J.L., C.L. and H.S. analyzed the results. J.L. wrote the main manuscript text and prepared all the figures. All authors have read and agreed to the published version of the manuscript.

Funding: H.S. received funding from Louisiana State University Health Sciences Center-Shreveport and an NIH grant (AA023610) to support this research. J.L. received a Malcolm Feist Predoctoral Fellowship from the Center of Excellence for Cardiovascular Diseases and Sciences at Louisiana State University Health Sciences Center-Shreveport.

Institutional Review Board Statement: The animal study protocol was approved by LSUHSC-Shreveport Institutional Animal Care and Use Committee (IACUC) (protocol code: P-21-038; date of approval: 12 May 2021).

Informed Consent Statement: Not applicable.

Data Availability Statement: The datasets generated and analyzed during the current study are available from the corresponding author on reasonable request.

Conflicts of Interest: The authors declare no conflict of interest.

References

1. Hernandez, I.H.; Villa-Gonzalez, M.; Martin, G.; Soto, M.; Perez-Alvarez, M.J. Glial Cells as Therapeutic Approaches in Brain Ischemia-Reperfusion Injury. *Cells* **2021**, *10*, 1639. [CrossRef] [PubMed]
2. Zhao, Y.; Zhang, X.; Chen, X.; Wei, Y. Neuronal injuries in cerebral infarction and ischemic stroke: From mechanisms to treatment (Review). *Int. J. Mol. Med.* **2022**, *49*, 15. [CrossRef] [PubMed]
3. Saini, V.; Guada, L.; Yavagal, D.R. Global Epidemiology of Stroke and Access to Acute Ischemic Stroke Interventions. *Neurology* **2021**, *97* (Suppl. S2), S6–S16. [CrossRef]
4. Politi, M.; Kastrup, A.; Marmagkiolis, K.; Grunwald, I.Q.; Papanagiotou, P. Endovascular Therapy for Acute Stroke. *Prog. Cardiovasc. Dis.* **2017**, *59*, 534–541. [CrossRef] [PubMed]
5. Hasan, T.F.; Hasan, H.; Kelley, R.E. Overview of Acute Ischemic Stroke Evaluation and Management. *Biomedicines* **2021**, *9*, 1486. [CrossRef]
6. Jean, W.C.; Spellman, S.R.; Nussbaum, E.S.; Low, W.C. Reperfusion injury after focal cerebral ischemia: The role of inflammation and the therapeutic horizon. *Neurosurgery* **1998**, *43*, 1382–1396. [CrossRef]
7. Zheng, X.; Haupt, M.; Bahr, M.; Tatenhorst, L.; Doeppner, T.R. Treating Cerebral Ischemia: Novel Therapeutic Strategies from Experimental Stroke Research. In *Cerebral Ischemia*; Pluta, R., Ed.; Exon Publications: Brisbane, Australia, 2021.
8. Lin, L.; Wang, X.; Yu, Z. Ischemia-reperfusion Injury in the Brain: Mechanisms and Potential Therapeutic Strategies. *Biochem. Pharmacol.* **2016**, *5*, 213.
9. Tuo, Q.Z.; Zhang, S.T.; Lei, P. Mechanisms of neuronal cell death in ischemic stroke and their therapeutic implications. *Med. Res. Rev.* **2022**, *42*, 259–305. [CrossRef]

10. Xing, Y.; Bai, Y. A Review of Exercise-Induced Neuroplasticity in Ischemic Stroke: Pathology and Mechanisms. *Mol. Neurobiol.* **2020**, *57*, 4218–4231. [CrossRef]
11. Tang, H.; Li, Y.; Tang, W.; Zhu, J.; Parker, G.C.; Zhang, J.H. Endogenous Neural Stem Cell-induced Neurogenesis after Ischemic Stroke: Processes for Brain Repair and Perspectives. *Transl. Stroke Res.* **2022**. *online ahead of print.* [CrossRef]
12. Cassidy, J.M.; Cramer, S.C. Spontaneous and Therapeutic-Induced Mechanisms of Functional Recovery After Stroke. *Transl. Stroke Res.* **2017**, *8*, 33–46. [CrossRef]
13. Ming, G.L.; Song, H.J. Adult Neurogenesis in the Mammalian Brain: Significant Answers and Significant Questions. *Neuron* **2011**, *70*, 687–702. [CrossRef] [PubMed]
14. Goncalves, J.T.; Schafer, S.T.; Gage, F.H. Adult Neurogenesis in the Hippocampus: From Stem Cells to Behavior. *Cell* **2016**, *167*, 897–914. [CrossRef] [PubMed]
15. Cuartero, M.I.; Garcia-Culebras, A.; Torres-Lopez, C.; Medina, V.; Fraga, E.; Vazquez-Reyes, S.; Jareno-Flores, T.; Garcia-Segura, J.M.; Lizasoain, I.; Moro, M.A. Post-stroke Neurogenesis: Friend or Foe? *Front. Cell Dev. Biol.* **2021**, *9*, 657846. [CrossRef] [PubMed]
16. Niklison-Chirou, M.V.; Agostini, M.; Amelio, I.; Melino, G. Regulation of Adult Neurogenesis in Mammalian Brain. *Int. J. Mol. Sci.* **2020**, *21*, 4869. [CrossRef]
17. Ribeiro, F.F.; Xapelli, S. An Overview of Adult Neurogenesis. *Adv. Exp. Med. Biol.* **2021**, *1331*, 77–94.
18. Jurkowski, M.P.; Bettio, L.; Woo, E.K.; Patten, A.; Yau, S.Y.; Gil-Mohapel, J. Beyond the Hippocampus and the SVZ: Adult Neurogenesis Throughout the Brain. *Front. Cell. Neurosci.* **2020**, *14*, 576444. [CrossRef]
19. Yamashita, T.; Ninomiya, M.; Hernandez Acosta, P.; Garcia-Verdugo, J.M.; Sunabori, T.; Sakaguchi, M.; Adachi, K.; Kojima, T.; Hirota, Y.; Kawase, T.; et al. Subventricular zone-derived neuroblasts migrate and differentiate into mature neurons in the post-stroke adult striatum. *J. Neurosci.* **2006**, *26*, 6627–6636. [CrossRef]
20. Lindvall, O.; Kokaia, Z. Stem cell research in stroke: How far from the clinic? *Stroke* **2011**, *42*, 2369–2375. [CrossRef]
21. Lindvall, O.; Kokaia, Z. Neurogenesis following Stroke Affecting the Adult Brain. *Cold Spring Harb. Perspect. Biol.* **2015**, *7*, a019034. [CrossRef]
22. Osman, A.M.; Porritt, M.J.; Nilsson, M.; Kuhn, H.G. Long-term stimulation of neural progenitor cell migration after cortical ischemia in mice. *Stroke* **2011**, *42*, 3559–3565. [CrossRef] [PubMed]
23. Ceanga, M.; Dahab, M.; Witte, O.W.; Keiner, S. Adult Neurogenesis and Stroke: A Tale of Two Neurogenic Niches. *Front. Neurosci.* **2021**, *15*, 700297. [CrossRef] [PubMed]
24. Moon, S.; Chang, M.S.; Koh, S.H.; Choi, Y.K. Repair Mechanisms of the Neurovascular Unit after Ischemic Stroke with a Focus on VEGF. *Int. J. Mol. Sci.* **2021**, *22*, 8543. [CrossRef] [PubMed]
25. Passarelli, J.P.; Nimjee, S.M.; Townsend, K.L. Stroke and Neurogenesis: Bridging Clinical Observations to New Mechanistic Insights from Animal Models. *Transl. Stroke Res.* 2022. *online ahead of print.* [CrossRef]
26. Abrahao, K.P.; Salinas, A.G.; Lovinger, D.M. Alcohol and the Brain: Neuronal Molecular Targets, Synapses, and Circuits. *Neuron* **2017**, *96*, 1223–1238. [CrossRef]
27. Hansagi, H.; Romelsjo, A.; de Verdier, M.G.; Andreasson, S.; Leifman, A. Alcohol consumption and stroke mortality. *Stroke* **1995**, *26*, 1768–1773. [CrossRef]
28. Berger, K.; Ajani, U.A.; Kase, C.S.; Gaziano, J.M.; Buring, J.E.; Glynn, R.J.; Hennekens, C.H. Light-to-moderate alcohol consumption and the risk of stroke among U.S. male physicians. *N. Engl. J. Med.* **1999**, *341*, 1557–1564. [CrossRef]
29. Sacco, R.L.; Elkind, M.; Boden-Albala, B.; Lin, I.F.; Kargman, D.E.; Hauser, W.A.; Shea, S.; Paik, M.C. The protective effect of moderate alcohol consumption on ischemic stroke. *JAMA* **1999**, *281*, 53–60. [CrossRef]
30. Reynolds, K.; Lewis, B.; Nolen, J.D.; Kinney, G.L.; Sathya, B.; He, J. Alcohol consumption and risk of stroke: A meta-analysis. *JAMA* **2003**, *289*, 579–588. [CrossRef]
31. Ikehara, S.; Iso, H.; Toyoshima, H.; Date, C.; Yamamoto, A.; Kikuchi, S.; Kondo, T.; Watanabe, Y.; Koizumi, A.; Wada, Y.; et al. Alcohol consumption and mortality from stroke and coronary heart disease among Japanese men and women: The Japan collaborative cohort study. *Stroke* **2008**, *39*, 2936–2942. [CrossRef]
32. Ronksley, P.E.; Brien, S.E.; Turner, B.J.; Mukamal, K.J.; Ghali, W.A. Association of alcohol consumption with selected cardiovascular disease outcomes: A systematic review and meta-analysis. *BMJ* **2011**, *342*, d671. [CrossRef] [PubMed]
33. Larsson, S.C.; Wallin, A.; Wolk, A.; Markus, H.S. Differing association of alcohol consumption with different stroke types: A systematic review and meta-analysis. *BMC Med.* **2016**, *14*, 178. [CrossRef]
34. Shiotsuki, H.; Saijo, Y.; Ogushi, Y.; Kobayashi, S.; Study, J.S.S.R. Relationships between Alcohol Intake and Ischemic Stroke Severity in Sex Stratified Analysis for Japanese Acute Stroke Patients. *J. Stroke Cerebrovasc.* **2019**, *28*, 1604–1617. [CrossRef] [PubMed]
35. Collaborators, G.B.D.A. Population-level risks of alcohol consumption by amount, geography, age, sex, and year: A systematic analysis for the Global Burden of Disease Study 2020. *Lancet* **2022**, *400*, 185–235.
36. Xu, G.; Li, C.; Parsiola, A.L.; Li, J.; McCarter, K.D.; Shi, R.; Mayhan, W.G.; Sun, H. Dose-Dependent Influences of Ethanol on Ischemic Stroke: Role of Inflammation. *Front. Cell. Neurosci.* **2019**, *13*, 6. [CrossRef] [PubMed]
37. Li, C.; Li, J.Y.; Xu, G.D.; Sun, H. Influence of Chronic Ethanol Consumption on Apoptosis and Autophagy Following Transient Focal Cerebral Ischemia in Male Mice. *Sci. Rep.* **2020**, *10*, 6164. [CrossRef]

38. Li, J.; Li, C.; Loreno, E.G.; Miriyala, S.; Panchatcharam, M.; Lu, X.; Sun, H. Chronic Low-Dose Alcohol Consumption Promotes Cerebral Angiogenesis in Mice. *Front. Cardiovasc. Med.* **2021**, *8*, 681627. [CrossRef] [PubMed]
39. Taffe, M.A.; Kotzebue, R.W.; Crean, R.D.; Crawford, E.F.; Edwards, S.; Mandyam, C.D. Long-lasting reduction in hippocampal neurogenesis by alcohol consumption in adolescent nonhuman primates. *Proc. Natl. Acad. Sci. USA* **2010**, *107*, 11104–11109. [CrossRef] [PubMed]
40. Anderson, M.L.; Nokia, M.S.; Govindaraju, K.P.; Shors, T.J. Moderate Drinking? Alcohol Consumption Significantly Decreases Neurogenesis in the Adult Hippocampus. *Neuroscience* **2012**, *224*, 202–209. [CrossRef]
41. Aberg, E.; Hofstetter, C.P.; Olson, L.; Brene, S. Moderate ethanol consumption increases hippocampal cell proliferation and neurogenesis in the adult mouse. *Int. J. Neuropsychoph.* **2005**, *8*, 557–567. [CrossRef]
42. Zhang, W.; Cheng, J.; Vagnerova, K.; Ivashkova, Y.; Young, J.; Cornea, A.; Grafe, M.R.; Murphy, S.J.; Hurn, P.D.; Brambrink, A.M. Effects of androgens on early post-ischemic neurogenesis in mice. *Transl. Stroke Res.* **2014**, *5*, 301–311. [CrossRef]
43. Kuhn, H.G.; Eisch, A.J.; Spalding, K.; Peterson, D.A. Detection and Phenotypic Characterization of Adult Neurogenesis. *Cold Spring Harb. Perspect. Biol.* **2016**, *8*, a025981. [CrossRef] [PubMed]
44. Zhao, X.; van Praag, H. Steps towards standardized quantification of adult neurogenesis. *Nat. Commun.* **2020**, *11*, 4275. [CrossRef]
45. Nawarawong, N.N.; Nickell, C.G.; Hopkins, D.M.; Pauly, J.R.; Nixon, K. Functional Activation of Newborn Neurons Following Alcohol-Induced Reactive Neurogenesis. *Brain Sci.* **2021**, *11*, 499. [CrossRef]
46. Teo, J.D.; Morris, M.J.; Jones, N.M. Hypoxic postconditioning reduces microglial activation, astrocyte and caspase activity, and inflammatory markers after hypoxia-ischemia in the neonatal rat brain. *Pediatr. Res.* **2015**, *77*, 757–764. [CrossRef] [PubMed]
47. Liu, D.; Bai, X.; Ma, W.; Xin, D.; Chu, X.; Yuan, H.; Qiu, J.; Ke, H.; Yin, S.; Chen, W.; et al. Purmorphamine Attenuates Neuro-Inflammation and Synaptic Impairments After Hypoxic-Ischemic Injury in Neonatal Mice via Shh Signaling. *Front. Pharmacol.* **2020**, *11*, 204. [CrossRef] [PubMed]
48. Yu, J.; Zhu, H.; Gattoni-Celli, S.; Taheri, S.; Kindy, M.S. Dietary supplementation of GrandFusion((R)) mitigates cerebral ischemia-induced neuronal damage and attenuates inflammation. *Nutr. Neurosci.* **2016**, *19*, 290–300. [CrossRef]
49. Manrique-Castano, D.; Sardari, M.; Silva de Carvalho, T.; Doeppner, T.R.; Popa-Wagner, A.; Kleinschnitz, C.; Chan, A.; Hermann, D.M. Deactivation of ATP-Binding Cassette Transporters ABCB1 and ABCC1 Does Not Influence Post-ischemic Neurological Deficits, Secondary Neurodegeneration and Neurogenesis, but Induces Subtle Microglial Morphological Changes. *Front. Cell. Neurosci.* **2019**, *13*, 412. [CrossRef]
50. Shvedova, M.; Islam, M.R.; Armoundas, A.A.; Anfinogenova, N.D.; Wrann, C.D.; Atochin, D.N. Modified middle cerebral artery occlusion model provides detailed intraoperative cerebral blood flow registration and improves neurobehavioral evaluation. *J. Neurosci. Methods* **2021**, *358*, 109179. [CrossRef]
51. Richard, A.D.; Tian, X.L.; El-Saadi, M.W.; Lu, X.H. Erasure of striatal chondroitin sulfate proteoglycan-associated extracellular matrix rescues aging-dependent decline of motor learning. *Neurobiol. Aging* **2018**, *71*, 61–71. [CrossRef]
52. Hase, Y.; Okamoto, Y.; Fujita, Y.; Kitamura, A.; Nakabayashi, H.; Ito, H.; Maki, T.; Washida, K.; Takahashi, R.; Ihara, M. Cilostazol, a phosphodiesterase inhibitor, prevents no-reflow and hemorrhage in mice with focal cerebral ischemia. *Exp. Neurol.* **2012**, *233*, 523–533. [CrossRef] [PubMed]
53. Fisher, H.R.; Simpson, R.I.; Kapur, B.M. Calculation of blood alcohol concentration (BAC) by sex, weight, number of drinks and time. *Can. J. Public Health* **1987**, *78*, 300–304.
54. Eminaga, S.; Teekakirikul, P.; Seidman, C.E.; Seidman, J.G. Detection of Cell Proliferation Markers by Immunofluorescence Staining and Microscopy Imaging in Paraffin-Embedded Tissue Sections. *Curr. Protoc. Mol. Biol.* **2016**, *115*, 14.25.1–14.25.14. [CrossRef] [PubMed]
55. Chareyron, L.J.; Amaral, D.G.; Lavenex, P. Selective lesion of the hippocampus increases the differentiation of immature neurons in the monkey amygdala. *Proc. Natl. Acad. Sci. USA* **2016**, *113*, 14420–14425. [CrossRef] [PubMed]
56. Chareyron, L.J.; Banta Lavenex, P.; Amaral, D.G.; Lavenex, P. Life and Death of Immature Neurons in the Juvenile and Adult Primate Amygdala. *Int. J. Mol. Sci.* **2021**, *22*, 6691. [CrossRef] [PubMed]
57. Crews, F.T.; Mdzinarishvili, A.; Kim, D.; He, J.; Nixon, K. Neurogenesis in adolescent brain is potently inhibited by ethanol. *Neuroscience* **2006**, *137*, 437–445. [CrossRef]
58. Liu, W.; Crews, F.T. Persistent Decreases in Adult Subventricular and Hippocampal Neurogenesis Following Adolescent Intermittent Ethanol Exposure. *Front. Behav. Neurosci.* **2017**, *11*, 151. [CrossRef]
59. Xu, H.; Liu, D.; Chen, J.; Li, H.; Xu, M.; Wen, W.; Frank, J.A.; Grahame, N.J.; Zhu, H.; Luo, J. Effects of Chronic Voluntary Alcohol Drinking on Thiamine Concentrations, Endoplasmic Reticulum Stress, and Oxidative Stress in the Brain of Crossed High Alcohol Preferring Mice. *Neurotox. Res.* **2019**, *36*, 777–787. [CrossRef]
60. Nawarawong, N.N.; Thompson, K.R.; Guerin, S.P.; Shaji, C.A.; Peng, H.; Nixon, K. Reactive, Adult Neurogenesis From Increased Neural Progenitor Cell Proliferation Following Alcohol Dependence in Female Rats. *Front. Neurosci.* **2021**, *15*, 689601. [CrossRef]
61. Zhang, R.L.; Zhang, Z.G.; Chopp, M. Ischemic stroke and neurogenesis in the subventricular zone. *Neuropharmacology* **2008**, *55*, 345–352. [CrossRef]
62. Palma-Tortosa, S.; Garcia-Culebras, A.; Moraga, A.; Hurtado, O.; Perez-Ruiz, A.; Duran-Laforet, V.; Parra, J.; Cuartero, M.I.; Pradillo, J.M.; Moro, M.A.; et al. Specific Features of SVZ Neurogenesis After Cortical Ischemia: A Longitudinal Study. *Sci. Rep.* **2017**, *7*, 16343. [CrossRef]

63. Dillen, Y.; Kemps, H.; Gervois, P.; Wolfs, E.; Bronckaers, A. Adult Neurogenesis in the Subventricular Zone and Its Regulation After Ischemic Stroke: Implications for Therapeutic Approaches. *Transl. Stroke Res.* **2020**, *11*, 60–79. [CrossRef]
64. Jin, K.; Wang, X.; Xie, L.; Mao, X.O.; Greenberg, D.A. Transgenic ablation of doublecortin-expressing cells suppresses adult neurogenesis and worsens stroke outcome in mice. *Proc. Natl. Acad. Sci. USA* **2010**, *107*, 7993–7998. [CrossRef] [PubMed]
65. Li, C.; Li, J.; Loreno, E.G.; Miriyala, S.; Panchatcharam, M.; Sun, H. Protective Effect of Low-Dose Alcohol Consumption against Post-Ischemic Neuronal Apoptosis: Role of L-PGDS. *Int. J. Mol. Sci.* **2021**, *23*, 133. [CrossRef]
66. Ruan, L.; Wang, B.; ZhuGe, Q.; Jin, K. Coupling of neurogenesis and angiogenesis after ischemic stroke. *Brain Res.* **2015**, *1623*, 166–173. [CrossRef] [PubMed]
67. Pradillo, J.M.; Murray, K.N.; Coutts, G.A.; Moraga, A.; Oroz-Gonjar, F.; Boutin, H.; Moro, M.A.; Lizasoain, I.; Rothwell, N.J.; Allan, S.M. Reparative effects of interleukin-1 receptor antagonist in young and aged/co-morbid rodents after cerebral ischemia. *Brain Behav. Immun.* **2017**, *61*, 117–126. [CrossRef]
68. Ekdahl, C.T.; Claasen, J.H.; Bonde, S.; Kokaia, Z.; Lindvall, O. Inflammation is detrimental for neurogenesis in adult brain. *Proc. Natl. Acad. Sci. USA* **2003**, *100*, 13632–13637. [CrossRef]
69. Choi, J.Y.; Kim, J.Y.; Kim, J.Y.; Park, J.; Lee, W.T.; Lee, J.E. M2 Phenotype Microglia-derived Cytokine Stimulates Proliferation and Neuronal Differentiation of Endogenous Stem Cells in Ischemic Brain. *Exp. Neurobiol.* **2017**, *26*, 33–41. [CrossRef]
70. Jin, K.; Zhu, Y.; Sun, Y.; Mao, X.O.; Xie, L.; Greenberg, D.A. Vascular endothelial growth factor (VEGF) stimulates neurogenesis in vitro and in vivo. *Proc. Natl. Acad. Sci. USA* **2002**, *99*, 11946–11950. [CrossRef]
71. Font, M.A.; Arboix, A.; Krupinski, J. Angiogenesis, neurogenesis and neuroplasticity in ischemic stroke. *Curr. Cardiol. Rev.* **2010**, *6*, 238–244. [CrossRef]
72. Rocco, A.; Compare, D.; Angrisani, D.; Sanduzzi Zamparelli, M.; Nardone, G. Alcoholic disease: Liver and beyond. *World J. Gastroenterol.* **2014**, *20*, 14652–14659. [CrossRef] [PubMed]
73. Rumgay, H.; Murphy, N.; Ferrari, P.; Soerjomataram, I. Alcohol and Cancer: Epidemiology and Biological Mechanisms. *Nutrients* **2021**, *13*, 3173. [CrossRef] [PubMed]
74. Barberia-Latasa, M.; Gea, A.; Martinez-Gonzalez, M.A. Alcohol, Drinking Pattern, and Chronic Disease. *Nutrients* **2022**, *14*, 1954. [CrossRef] [PubMed]

biomedicines

MDPI

Article

Effects of Virtual Reality Cognitive Training on Neuroplasticity: A Quasi-Randomized Clinical Trial in Patients with Stroke

Antonio Gangemi [1,†], Rosaria De Luca [1,†], Rosa Angela Fabio [2], Paola Lauria [1], Carmela Rifici [1], Patrizia Pollicino [1], Angela Marra [1], Antonella Olivo [1], Angelo Quartarone [1] and Rocco Salvatore Calabrò [1,*]

1 IRCCS Centro Neurolesi "Bonino-Pulejo", S.S. 113, Cda Casazza, 98124 Messina, Italy; antonio.gangemi@irccsme.it (A.G.); rosaria.deluca@irccsme.it (R.D.L.); paola.lauria@irccsme.it (P.L.); carmela.rifici@irccsme.it (C.R.); patrizia.pollicino@irccsme.it (P.P.); angela.marra@irccsme.it (A.M.); antonia.olivo@irccsme.it (A.O.); angelo.quartarone@irccsme.it (A.Q.)

2 Department of Economics, University of Messina, Via Consolare Valeria, 98125 Messina, Italy; rafabio@unime.it

* Correspondence: salbro77@tiscali.it or roccos.calabro@irccsme.it

† These authors contributed equally to this work.

Abstract: Cognitive Rehabilitation (CR) is a therapeutic approach designed to improve cognitive functioning after a brain injury, including stroke. Two major categories of techniques, namely traditional and advanced (including virtual reality—VR), are widely used in CR for patients with various neurological disorders. More objective outcome measures are needed to better investigate cognitive recovery after a stroke. In the last ten years, the application of electroencephalography (EEG) as a non-invasive and portable neuroimaging method has been explored to extract the hallmarks of neuroplasticity induced by VR rehabilitation approaches, particularly within the chronic stroke population. The aim of this study is to investigate the neurophysiological effects of CR conducted in a virtual environment using the VRRS device. Thirty patients with moderate-to-severe ischemic stroke in the chronic phase (at least 6 months after the event), with a mean age of 58.13 (±8.33) for the experimental group and 57.33 (±11.06) for the control group, were enrolled. They were divided into two groups: an experimental group and a control group, receiving neurocognitive stimulation using VR and the same amount of conventional neurorehabilitation, respectively. To study neuroplasticity changes after the training, we focused on the power band spectra of theta, alpha, and beta EEG rhythms in both groups. We observed that when VR technology was employed to amplify the effects of treatments on cognitive recovery, significant EEG-related neural improvements were detected in the primary motor circuit in terms of power spectral density and time-frequency domains. Indeed, EEG analysis suggested that VR resulted in a significant increase in both the alpha band power in the occipital areas and the beta band power in the frontal areas, while no significant variations were observed in the theta band power. Our data suggest the potential effectiveness of a VR-based rehabilitation approach in promoting neuroplastic changes even in the chronic phase of ischemic stroke.

Keywords: ischemic stroke; virtual reality; electroencephalography; neural plasticity; cognitive rehabilitation; brain injury; outcome measures; clinical trial; neurological disorders

Citation: Gangemi, A.; De Luca, R.; Fabio, R.A.; Lauria, P.; Rifici, C.; Pollicino, P.; Marra, A.; Olivo, A.; Quartarone, A.; Calabrò, R.S. Effects of Virtual Reality Cognitive Training on Neuroplasticity: A Quasi-Randomized Clinical Trial in Patients with Stroke. *Biomedicines* **2023**, *11*, 3225. https://doi.org/10.3390/biomedicines11123225

Academic Editors: Masaru Tanaka and Eleonóra Spekker

Received: 7 October 2023
Revised: 29 November 2023
Accepted: 4 December 2023
Published: 6 December 2023

1. Introduction

Global epidemiological data indicate that approximately 16.9 million people suffer a stroke each year, resulting in a global incidence rate of 258 per 100,000 individuals, with variations between high- and low-income countries [1]. Survivors of stroke face immediate challenges in coping with long-term sensory-motor and cognitive impairments [1,2]. Deficits in language and communication, attention, visuo-spatial processing, long-term and procedural memory, reasoning and problem solving, as well as executive functions, are often present following a stroke.

Motor and cognitive alterations are the primary targets of neurorehabilitation, as they hinder the patients' ability to perform activities of daily living (ADL) [3], and can lead to long-term medical issues (e.g., urinary incontinence), musculoskeletal problems (e.g., spasticity), and psychosocial complications (e.g., depression, emotional lability) [4]. Assessment of stroke impairment is fundamental to predict prognosis and functional recovery. To this aim, the NIH stroke scale [5,6], Barthel index [7], and FIM [8] are commonly used in clinical practice, although they lack a specific assessment of cognitive function. Moreover, evidence shows the validity and reliability of the modified Rankin scale (mRS) as a valuable instrument for assessing the impact of new stroke treatments [9]. Indeed, early, coordinated, and multidisciplinary rehabilitation plays a crucial role in promoting motor and cognitive recovery after stroke. Conventional stroke rehabilitation methods mainly involve physical therapy, occupational therapy, cognitive training, and speech therapy [10–12]. However, many stroke survivors still experience functional disabilities, with regard to cognitive deficits, which hinder their ability to perform daily activities.

Cognitive Rehabilitation (CR) is a therapeutic approach designed to improve cognitive functioning after a brain injury, including stroke, as well as in patients with neurodegenerative disorders. CR is a set of methods used to overcome cognitive deficits caused by stroke. Indeed, it includes different interventions aimed at improving the ability to perform cognitive tasks, achieved through the retraining of previously learned skills and/or the teaching of compensatory strategies. Two major categories of techniques, i.e., traditional and advanced, are widely used in the CR of patients with different neurological disorders [13]. Traditional techniques involve the use of cognitive strategies to retrain or alleviate deficits in the different cognitive domains by using a paper-and-pencil approach. On the other hand, advanced methods, including computer-assisted cognitive rehabilitation, use multimedia and informatics resources to potentiate neurocognitive performance [14].

In recent years, technology-based stroke rehabilitation interventions have shown promise in improving the motor and cognitive abilities and autonomy of stroke patients [15,16]. Advanced technology is increasingly being incorporated into stroke neurorehabilitation to enhance standard treatments, reduce neurological disability, and improve overall functioning. In fact, the integration of psychology, technology, and neuroscience allows for a better understanding of how virtual reality (VR) affects the cognition of the human brain [17]. VR is a commonly used advanced neurorehabilitation technology that aims to improve motor and cognitive abilities in stroke patients [18,19]. VR involves computer-based simulations that allow users to interact with multisensory environments, receiving real-time feedback on their performance [20,21]. This technology promotes repetitive and task-specific training, actively involving patients, providing constructive feedback, and accurately measuring functional improvement. Recent studies have shown the efficacy of cognitive and motor rehabilitation through the use of virtual environments, such as the VRRS Evo-4 machine, where patients interact with common images with physical properties simulated by the tool, allowing exercises designed to stimulate the main cognitive domains.

The assessment of cognitive outcomes is often performed using clinical tests (such as the Oxford Cognitive Screen), which are standardized and validated within this patient population [22]. However, they are user-dependent, and scores are influenced by the motivation and collaboration of the patient. Therefore, more objective tools to better investigate functional outcomes are needed for both research and clinical purposes.

Electrophysiological approaches, such as EEG, offer a high temporal resolution in the order of milliseconds and can serve as a set of biomarkers crucial for assessing cognitive changes observed during various activities and challenging conditions, including VR [23–26]. Studies on neurophysiological changes associated with VR neurorehabilitation are relatively new, with initial evidence from behavioral tasks in healthy individuals [27,28]. Currently, EEG is utilized in VR therapy to monitor and provide augmented feedback regarding cortical activation [29,30] during motor and cognitive tasks [31–33].

Therefore, our purpose is to investigate the neurophysiological effects of cognitive stimulation training conducted in a virtual environment using the VRRS device on patients in the chronic phase of ischemic stroke. We are particularly interested in evaluating how this advanced training impacts brain plasticity mechanisms. To this aim, we directly measured brain electrical activity through EEG recordings.

Through this investigation, we sought to gain valuable insights into the potential of VR cognitive stimulation as a neurorehabilitation approach for patients with neurological disorders and contribute to the understanding of the underlying neural mechanisms involved in the observed cognitive improvements. The findings from this study could have implications for the development of innovative and effective interventions to enhance cognitive recovery in individuals with stroke or other neurological events.

2. Materials and Methods

2.1. Study Setting and Participants

Thirty—moderate-to-severe patients with ischemic stroke in the chronic phase (at least 6 months after the event), with a mean age of 58.13 (±8.33) for the experimental group and 57.33 (±11.06) for the control group, were enrolled in this study. They attended the outpatient clinic of the Neurorehabilitation Unit of IRCCS Neurolesi "Bonino Pulejo" (Messina, Italy) from October 2022 to March 2023. A more detailed description of the two groups is provided in Table 1.

Table 1. Demographic and clinical description of both the experimental and control samples at the beginning of the study.

Subject	Age (Years)	Gender	Education (Years)	Barthel Index	Rankin Scale	Time Elapsed Since the Event
Experimental Group						
1	57	M	8	15	5	6
2	61	M	8	5	5	8
3	69	M	5	40	4	7
4	65	M	13	15	4	6
5	63	F	13	10	4	12
6	57	F	8	10	4	6
7	57	F	13	35	4	8
8	39	M	8	70	3	8
9	59	M	8	10	5	7
10	56	F	8	40	4	6
11	60	F	8	65	3	7
12	43	M	13	10	5	9
13	67	M	13	5	5	9
14	66	M	13	10	5	12
15	53	M	13	25	4	6
Control Group	Age (Years)	Gender	Education (Years)	Barthel Index	Rankin Scale	Time Elapsed Since the Event
1	38	M	8	10	5	6
2	73	M	13	5	5	6
3	59	M	8	35	4	7
4	73	M	5	20	4	12

Table 1. *Cont.*

Subject	Age (Years)	Gender	Education (Years)	Barthel Index	Rankin Scale	Time Elapsed Since the Event
5	68	M	5	15	4	8
6	69	F	8	15	4	8
7	65	F	8	30	4	6
8	64	M	13	70	3	6
9	55	M	5	40	5	6
10	54	F	8	40	4	7
11	48	M	13	55	4	9
12	55	F	10	5	5	6
13	50	M	13	20	4	7
14	45	F	8	15	4	7
15	44	F	8	10	4	8

The stroke patients and/or their family members were provided with adequate information about the study and offered the opportunity to participate with written consent. The study adhered to the principles set forth in the Declaration of Helsinki on Human Rights, and the local Ethics Committee approved the study (IRCCS-ME-CE 08/21). Inclusion criteria were as follows: (1) diagnosis of first right ischemic stroke in the chronic phase, i.e., $\geq$ 6 months after the event; (2) age range 18–75; and (3) absence of disabling sensory impairment (i.e., hearing and visual impairment); (4) Rankin Scale score $\geq$ 3; (5) Barthel index $\geq$ 5. Exclusion criteria were as follows: (1) intake of psychoactive drugs potentially interfering with the training; (2) presence of neurological disorders other than the first ever ischemic stroke; and (3) absence of the ability to understand verbal delivery of a simple order, Token Test $\leq$ 4; (4) presence of debilitating behavioral alterations and severe psychiatric symptoms.

2.2. Procedures

Thirty patients were randomly assigned to one of 2 groups, with fifteen allocated to the experimental group (EG) and the other fifteen to receive standard treatments, forming the control group (CG), based on the order of recruitment (in order to meet the criteria for a quasi-randomized study). We used the sample size calculator, a public service of Creative Research Systems survey software (https://www.surveysystem.com/sscalc.htm), to determine an adequate and minimal sample size to exclude systematic error, established with a confidence level of 95% and a confidence interval of 2%.

The experimental group received neurocognitive stimulation using virtual reality training (VRT) using the Virtual reality rehabilitation system (VRRS), while the control group received the same amount of standard neurorehabilitation (using a paper-and-pencil approach). To study neuroplasticity changes (that was the aim of the study), we focused on investigating the power band spectra of theta, alpha and beta EEG rhythms in both groups. Theta and alpha rhythms are specific frequency bands of brain electrical activity that have been associated with various cognitive processes and neural plasticity [34].

By comparing the changes in theta and alpha EEG rhythms between the experimental group (who received the VR cognitive stimulation) and the control group (who did not receive this advanced stimulation during the study period), we aimed to assess the specific effects of VR on brain plasticity in patients with chronic stroke.

EEG data were recorded by the neurophysiology technical staff under the supervision of the neurological physician and acquired using a gold-standard digital EEG amplifier (Micromed Medical System, Treviso, Italy). The system continuously recorded EEG signals on 19 channels. Electrodes were placed on the patient's scalp according to the International Measurement System 10/20 criteria, with the reference electrode positioned in the ear and the ground electrode placed posteriorly at Fz. The preprocessing of the EEG was performed in Matlab using EEGLAB [35].

First, the baseline was removed from each channel. Then, high EEG signals were filtered at 0.5 Hz to remove respiratory noises, while low signals were filtered at a cutoff frequency of 50 Hz to eliminate high-frequency noises. A notch filter was also applied at 50 Hz to remove power line interference. The signals were visually inspected to manually remove residual artifacts. The data were then segmented into epochs of 4 s free of artifacts.

For the analysis of brain rhythms, data were recorded from the occipital areas for alpha rhythm and from the fronto-temporal areas for beta and theta rhythms bilaterally, although the focus of our research was primarily aimed at investigating the neuroplastic effects in the right-lesioned hemisphere. Electrodes were placed across the entire scalp to ensure comprehensive coverage [36].

EEG data were recorded during a 20 min session where the patient was at psychosensory rest, with their eyes closed. For short periods, the patient was asked to open their eyes to assess the alpha rhythm reactivity recorded in the occipital cortex areas. Quantitative analysis was performed using custom algorithms developed in Matlab code. The power spectral density (PSD) was evaluated by transforming the signal from the time domain to the frequency domain using the Welch method [37]. PSDs were calculated for each epoch and then averaged. The absolute total power of the signal and absolute power of each band were computed for each electrode. The bands considered were theta (4–7 Hz), alpha (8–13 Hz), and beta (14–29 Hz).

2.3. Virtual Cognitive Task Using VRRS

VRRS is an advanced rehabilitation platform designed to facilitate the recovery process for patients with neurological conditions. It is one of the most comprehensive and clinically tested virtual reality systems for rehabilitation and tele-rehabilitation. The VRRS utilizes an exclusive magnetic kinematic acquisition system and offers a range of rehabilitative modules, including neurological, logopedic, and cognitive tasks, catering to a wide spectrum of neurological diseases. The system incorporates augmented feedback to enhance physiological learning, providing patients with specific information on their movements to improve the quality of their performance. Each exercise is accompanied by a preview that demonstrates to therapists and patients how the sensors should be positioned and the correct way to perform the exercise (Figure 1).

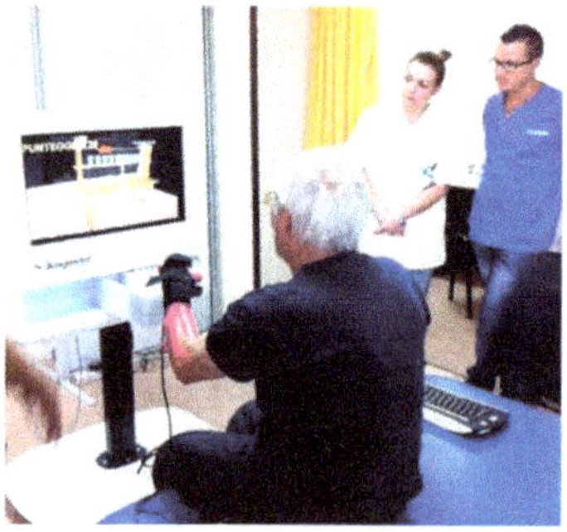

Figure 1. A patient supervised by two clinicians during the execution of a motor-cognitive exercise to improve sustained attention.

During the EEG execution, the enrolled patient sits in front of the VRRS device, actively engaging with it to perform the virtual cognitive task.

The VRRS cognitive module comprises a set of interactive activities designed for specific cognitive domains. These activities include tasks related to attention and memory, such as selective and sustained attention, as well as verbal or visual-spatial activities. The virtual tasks provided through the VRRS can be categorized into two main types based on the method of interaction with the virtual reality tool. The first category encompasses 2D exercises where the patient interacts with objects and scenarios through the touch screen or a specialized magnetic tracking sensor coupled with a squeezable object, effectively emulating mouse-like interaction capabilities. The second category involves 3D exercises, allowing patients to interact with virtual scenarios and objects using magnetic wearable sensors typically placed over the hand, enabling 3D position tracking of the end effector. This 3D modality allows for movements of the upper and lower limbs in three dimensions while interacting with the virtual environment (see Table 2). During the EEG examination, the virtual task is administered with three levels of difficulty for execution time, and the number of stimuli-targets and distractors presented to the patient is controlled (see Table 2) [38].

Table 2. Cognitive rehabilitation program: virtual reality task using VRRS—Evo and standard activities administered during EEG signal processing.

Domain	Sub-Domain	VRRS Task	Standard Activities
-Attention Processes	Selective	To administer the scanning exercise, the user must locate the target symbols in a grid and select the matching virtual symbols. To select and immediately recall feedback (audio and video) similar to various elements (colors, musical strings, geometric or abstract forms, animals, numbers) observed in the virtual environment, the patient touches the virtual target element within a specific time. This action causes a visual change with a specific audio feedback (positive reinforcement), using VVRS—interaction between the cognitive therapist and the patient. Otherwise, the element disappears (negative reinforcement).	To administer the attention exercise, the user must locate the target symbols while facing a paper-and-pencil grid and select the matching real symbols. To select and immediately recall feedback (audio and video) resembling various elements (colors, musical strings, geometric or abstract forms, animals, numbers) observed in the real environment, the patient touches the target element within a specific time, using a timer and the interaction between the cognitive therapist and the patient.
	Sustained	To stimulate sustained attention processes, the patient observes from 3 to 5 target stimuli for a variable and progressive time (10–15 min), with an attentional focus on the virtual tasks administered.	To stimulate sustained attention processes, the patient observes from 3 to 5 target stimuli for a variable and progressive time (10–15 min), with an attentional focus on the real activities administered.
Memory Abilities	Verbal	To work on recognition and remembrance in virtual tasks involving verbal material, reminiscence and validation therapy, mnemonic techniques, and strategic skills.	To work on recognition and remembrance in traditional tasks with paper-and-pencil verbal material, reminiscence and validation therapy, mnemonic techniques and strategic skills, face to face with a therapist, without a virtual tool.
	Visuo-Spatial	To work on recognition and remembrance virtual tasks with not verbal/visuo-spatial tasks (pictures; image; number; colors…) mnemonic techniques and strategic skills.	To work on recognition and remembrance using paper-and-pencil tasks without verbal/visuo-spatial tasks (pictures, images, numbers, colors), employing conventional mnemonic techniques and strategic skills, face to face with a therapist, without the use of virtual tools.

2.4. Standard Cognitive Training

The CG performed a conventional cognitive rehabilitation program with the same features and amount of time/intensity as the experimental one. However, cognitive domains were stimulated by using the classical paper-and-pencil approach instead of using VR (see Table 2). In fact, the cognitive-oriented intervention administered to CG included a series of standard face-to-face activities, organized for specific cognitive domains (attention processes and memory abilities) and relative cognitive sub-domains (selective and sustained attention processes, verbal and visuo-spatial memory), without the use of virtual systems.

2.5. Statistical Analysis

Data analysis was performed using IBM SPSS Statistics, Version 24 (IBM, Armonk, NY, USA) [39]. A mixed-model ANOVA for repeated measures was applied, with three repeated factors (bands: alpha, beta, and theta bands) and time (T0—pre-intervention baseline, T1—post-test) and one between subject factor (group: experimental group with VRT and control group without VRT). A Bonferroni correction was applied for multiple comparisons. The alpha level was set to $p < 0.05$ for all statistical tests. In the case of significant effects, the effect size of the test was reported. The effect sizes were computed and categorized according to eta squared (η^2).

3. Results

Regarding the demographic statistics, Table 3 shows the means and standard deviations for the provided demographics and indices. This table presents a comprehensive summary of the key characteristics of both the experimental and control groups, including age, years of education, Barthel index, Rankin Scale score, years from ischemic stroke, and sex distribution. To ensure the balance between the groups, we conducted a thorough analysis of the provided demographic and clinical characteristics. As shown in Table 3, we assessed the distribution of sex, age, years of education, Barthel index, Rankin Scale score, and years from ischemic stroke between the two groups. Independent samples *t*-tests for continuous variables and chi-squared tests for categorical variables were applied to compare the two groups. The statistical tests indicate that there are no significant differences between the groups for any of these variables. Therefore, we can confidently conclude that the groups are well-balanced in terms of these important baseline characteristics.

Table 3. Summary of the demographic and clinical description of both the experimental and control samples and statistical comparisons.

Socio-Demographic and Clinical Variables	Experimental Group	Control Group	Statistic	Pairwise Comparisons
Sex (male/female) [a]	M = 10	M = 10	0.00	($p = 1$)
	F = 5	F = 5		
Age (years) [b]	58.13 (8.33)	57.33	0.24	$p = 0.82$
Education level (years) [b]	10.13 (2.87)	8.96 (2.92)	1.19	$p = 0.24$
Barthel index (0–100) [b]	24.33 (21.20)	25.66 (19.07)	0.18	$p = 0.85$
Rankin Scale score (0–6) [b]	4.26 (0.70)	4.20 (0.56)	0.28	$p = 0.77$
Years from ischemic stroke [b]	7.8 (2.00)	7.26 (1.62)	0.80	$p = 0.43$

[a] Chi-square test (critical value). [b] Independent samples *t*-test: mean (standard deviation).

Concerning the right hemisphere, the mixed-model ANOVA for repeated measures revealed a significant effect of the factor "Group" (F (1, 28) = 4.11, $p < 0.05$, $\eta^2 = 0.09$), indicating a difference between the experimental and control groups.

The "Group × Time" interaction showed significant effects (F (1, 28) = 11.03, $p < 0.01$, $\eta^2 = 0.11$), suggesting that neurocognitive stimulation using VRT resulted in significant differences between the experimental and control groups in the alpha and beta bands. Furthermore, the "Group × Bands × Time" interaction exhibited significant effects (F (2, 56) = 78.45, $p < 0.01$, $\eta^2 = 0.12$), indicating that the alpha and beta bands showed an increase from pre-test to post-test for the experimental group, respectively, t (29) = 4.92, $p < 0.01$, d = 0.88 and t (29) = 6.01, $p < 0.01$, d = 0.91, while the theta band exhibited no increase, t (27) = 0.46, $p = 0.35$ (see Figures 1 and 2). With reference to the control group, no differences in the single bands were observed. To summarize, VR enhanced only the alpha and beta frequency bands (Table 4).

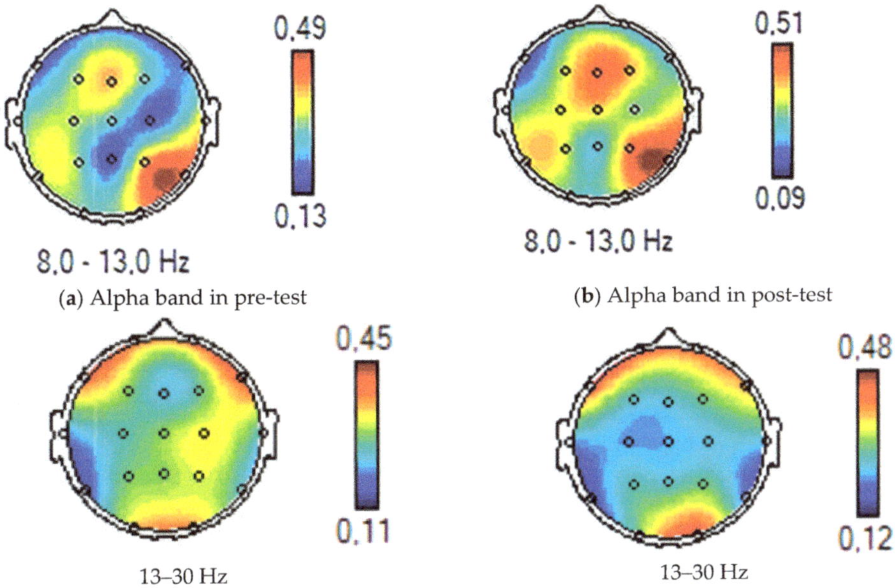

(**a**) Alpha band in pre-test (**b**) Alpha band in post-test (**c**) Beta band in pre-test (**d**) Beta band in post-test

Figure 2. The power spectral density in the alpha and beta bands of a subject randomly selected from a single case in the experimental group in both the pre-test and post-test phases is shown in the figure.

Regarding the left hemisphere, a significant effect of the factor "Group" (F (1, 28) = 5.93, $p < 0.05$, $\eta^2 = 0.11$) emerged, indicating a difference between the experimental and control groups within the left hemisphere.

The "Group × Time" interaction showed significant effects (F (1, 28) = 5.03, $p < 0.05$, $\eta^2 = 0.11$), suggesting that neurocognitive stimulation using VRT resulted in significant differences between the experimental and control groups in the alpha and beta bands. Furthermore, the "Group × Bands × Time" interaction exhibited significant effects (F (2, 56) = 8.19, $p < 0.01$, $\eta^2 = 0.09$), indicating that the alpha and beta bands showed an increase from pre-test to post-test for the experimental group, respectively, t (29) = 5.12, $p < 0.01$, d = 0.90; and t (29) = 5.87, $p < 0.01$, d = 0.91, while the theta band exhibited no increase, t (27) = 0.38, $p = 0.47$ (see picture n. 2 and picture n. 3). With reference to the control group, no differences in the single bands were observed. To summarize, VR enhanced only the alpha and beta frequency bands (Table 4).

In summary, the results suggest that neurocognitive stimulation through VRT had a significant impact on the alpha and beta frequency bands. These bands exhibited increased activity in the experimental group, while the control group showed no significant changes in any of the bands. This indicates that VRT enhances cognitive functioning primarily in the alpha and beta frequency bands.

Table 4. Means and standard deviations of theta, alpha, and beta bands (microvolts) in the right and left hemispheres, and *p*-values.

Right Hemisphere (Hz)	Pre-Test Experimental	*p* Control	Post-Test	*p* Experimental	Control	
Theta band	(M = 17.80; SD = 2.24)	(M = 18.30; SD = 1.73)	0.23	(M = 18.10; SD = 2.24)	(M = 18.02; SD = 1.76)	0.16
Alpha band	(M = 21.33; SD = 0.97)	(M = 21.41; SD = 1.02)	0.31	(M = 30.23; SD = 2.99)	(M = 21.8; SD = 1.02)	0.01
Beta band	(M = 23.13; SD = 2.74)	(M = 23.27; SD = 2.89)	0.37	(M = 28.27 SD = 2.37)	(M = 23.27; SD = 2.43)	0.01
Left Hemisphere *Theta band*	(M = 17.40; SD = 2.74)	(M = 18.03; SD = 1.77)	0.22	(M = 18.25; SD = 2.24)	(M = 18.72; SD = 1.76)	0.14
Alpha band	(M = 22.43; SD = 1.67)	(M = 21.32; SD = 1.32)	0.36	(M = 30.23; SD = 2.99)	(M = 21.8; SD = 1.02)	0.01
Beta band	(M = 23.53; SD = 3.15)	(M = 23.40; SD = 2.47)	0.49	(M = 26.97 SD = 3.81)	(M = 23.13; SD = 2.90)	0.05

4. Discussion

The study of EEG signals, correlated with their application in different technologies, holds great interest in neuroscience, particularly in the fields of assistive technology and neurorehabilitation, where the external stimulus can be provided through VR [40]. In a previous study, we investigated robotic-based rehabilitation (using the Lokomat, Hokoma, Zurigo) combined with VR in patients with chronic hemiparesis. We found that improvement in gait and balance was paralleled by important EEG signal modifications. In particular, the EEG data suggested that the use of VR may entrain several brain areas involved in motor planning and learning, thus leading to an enhanced motor performance [41].

In the current study, we confirmed that VR training could be effective in improving neuroplasticity (and potentially cognitive functioning) in patients in the chronic phase of ischemic stroke, as also demonstrated by other authors in different types of neurological disorders [42–46].

However, as far as we know, this is the first study focusing on the importance of assessment of electrophysiological changes after a brain injury involving only the right hemisphere. Indeed, our training and investigation mainly concerned visual working memory alterations, deficits of the speed of response and sustained attention, which are often underestimated signs of right hemisphere injury, despite their significant impact on rehabilitation outcomes [47,48]. This important role of the right hemisphere has also been demonstrated by neuroimaging studies [49–53].

The present study aimed to observe the effects of an innovative neurocognitive stimulation VR system, using training based on specific attention and memory abilities, on the EEG bands of two groups of subjects in the chronic phase of ischemic stroke. The findings revealed a significant increase in both alpha and beta bands in the experimental group following the intervention, suggesting that VRT had a positive impact on neural oscillatory activity, especially in the right hemisphere (which was the focus of the present study). The observed increase in alpha and beta bands aligns with previous research, demonstrating the neuroplasticity-inducing effects of VR-based interventions on brain activity [54–61]. The ability of VRT to engage and challenge the neural networks involved in cognitive processes may account for the observed changes in neural oscillations [62]. Specifically, the increase in alpha power could indicate enhanced attentional focus and cognitive resource allocation [29,56], while the rise in beta power may reflect improved cognitive control and motor planning [63]. The control group did not exhibit significant changes in alpha and beta bands over time. This suggests that the observed neurocognitive effects were specific to the VRT intervention and not simply a result of the passage of time or other non-specific factors.

These findings hold implications for the use of VRT as a potential tool in cognitive rehabilitation and neurocognitive enhancement. By understanding the neural mechanisms underlying the effects of VRT, future research can design targeted interventions to optimize cognitive training programs and potentially improve outcomes for individuals with cognitive deficits or neurodevelopmental disorders.

In addition to the implications for CR, the findings of this study could have broader implications for the field of neurorehabilitation and brain plasticity research. Virtual reality technology has shown promise as a versatile and effective tool for promoting neural plasticity in various clinical populations. By providing an immersive and engaging environment, VRT can target specific cognitive domains and facilitate neurocognitive, as well as motor reorganization. It has been shown that VR, due to the use of auditory and visual feedback, may affect different perceptual and experiential aspects. This complex sensory stimulation may increase the patient's awareness of his/her results (knowledge of results) as well as awareness of performance (knowledge of performance), inducing changes in neural plasticity processes with a consequent reinforcement of learning [20,45]. These positive effects on cortical plasticity could be due to the reactivation/amplification of brain neurotransmission within spared or unused circuits. It is also conceivably due to the involvement of mirror neurons, facilitated by the visual-motor information coming from the observation of the stimuli on the VR screen [24–45]. Moreover, by using VR environments, it is possible to perform tasks that may be too difficult, time-consuming, or impossible to perform in a natural world setting. It is noteworthy that VR enables healthcare professionals to provide standardized rehabilitation protocols, controlled stimulus presentations, and clinical progress and performance measures [45]. These outcome measures may become more objective if investigated (as in our study) with electrophysiological tools. Then, the advanced training can be tailored to the patient's clinical status and needs, also providing personalized feedback on performance.

Nonetheless, among the main concerns regarding the use of innovative technology, including VR, are system usability and the high costs. Most of the non-immersive VR devices, such as the VRRS, are easy to use and do not always require the presence of a caregiver to set the device if the patient does not have severe cognitive-behavioral problems [64]. On the contrary, semi-immersive and immersive devices (except for the Oculus) need a therapist to properly use the tool and supervise the training. As regards the costs, most VR devices have lower costs than other innovative rehabilitation treatments (e.g., robot-assisted training), and can be used at home if a telemedicine service is available.

5. Limitations of the Study and Future Perspectives

This pilot study has some limitations. First, the small sample size may prevent generalizing the results to the entire stroke population. However, the sample is homogeneous, as it is composed of patients with chronic right ischemic stroke. Moreover, to strengthen the generalizability of the results to neurorehabilitation, larger sample sizes and more diverse populations should be included.

Second, a quasi-randomization method may lead to selection bias. Therefore, randomized clinical trials are needed to confirm these promising results.

Third, we did not assess any behavioral tools after the intervention, so we cannot determine if and to what extent VR may have influenced our patients' clinical outcomes. Nonetheless, there is a lack of literature demonstrating that patients undergoing VR have superior (or at least the same) results than conventional training [65–69]. Also, this issue was outside the scope of the article, which focused on demonstrating the underpinning of VR-inducing neuroplasticity. The specific VRT protocols and tasks used in this study may have influenced the observed outcomes, and further research with a broader range of VR paradigms is warranted. Finally, although the present study focused on evaluating the power in the alpha, beta, and theta frequency bands in both hemispheres [70], we focused on the affected right hemisphere, and we did not calculate interhemispheric asymmetry. The latter would have definitely provided more detailed data on the functional connectivity (and therefore functional recovery) between the two cerebral hemispheres. However, the analysis of frequency band power is an approach also used in other studies to examine the intensity of brain waves in different regions or frequency bands [70]. This analysis can be conducted on data from individual hemispheres or specific regions without considering the symmetry parameter between the two hemispheres. In future research, we plan to integrate a comprehensive neurophysiological assessment of symmetry and coherence parameters along with hemodynamic studies using fNIRS to achieve improved spatiotemporal resolution.

As technology continues to advance, and VR platforms become more sophisticated, the potential of VRT for neurorehabilitation and cognitive enhancement will likely expand. Integrating VRT with other rehabilitative approaches, such as traditional therapy or brain stimulation techniques, could lead to synergistic effects and further optimize patient outcomes. Finally, by harnessing the potential of VRT as a tool for neurocognitive stimulation and brain plasticity, healthcare providers may be better equipped to design tailored and effective interventions for individuals with diverse cognitive challenges, including those with stroke, traumatic brain injury, and neurodevelopmental disorders.

6. Conclusions

In conclusion, this study may contribute valuable insights into the effects of neurocognitive stimulation using virtual reality training on brain oscillatory activity in the chronic phase of ischemic stroke. The significant increase in alpha and beta bands observed in the experimental group highlights the potential of VRT to induce neuroplastic changes in the brain. By providing an immersive and engaging environment, VRT can target specific cognitive domains and facilitate neurocognitive as well as motor reorganization. Through this investigation, we sought to contribute to the understanding of the underlying neural mechanisms involved in cognitive recovery after innovative approaches like virtual reality. In fact, the findings from this study could have implications for the development of innovative and effective interventions to enhance cognitive function in individuals with stroke, as well as other neurological disorders.

This research sets the stage for future investigations that can refine VRT protocols and pave the way for innovative and personalized cognitive rehabilitation interventions, ultimately benefiting individuals striving to overcome cognitive impairments and improve their attention processing and memory abilities. Indeed, larger multicenter RCTs are needed to confirm these promising findings and to investigate the role of other EEG bands as well as interhemispheric EEG coherence in the functional recovery following a brain injury.

Author Contributions: Conceptualization, A.G., R.D.L. and R.S.C.; methodology, A.G., R.A.F., A.O. and P.L.; software, A.G.; validation, all authors; formal analysis, A.G.; investigation, R.D.L., P.P., C.R., A.M. and P.L.; resources, A.Q., R.S.C. and R.D.L.; data curation, R.D.L. and A.G.; writing—original draft preparation, A.G., R.D.L. and R.S.C.; writing—review and editing, R.S.C. and A.Q.; visualization, all authors; supervision, R.S.C.; project administration, A.Q. and R.S.C.; funding acquisition, A.Q. All authors have read and agreed to the published version of the manuscript.

Funding: This work was supported by Current Research, Ministry of Health, 2023.

Institutional Review Board Statement: The study was conducted in accordance with the Declaration of Helsinki and approved by the Institutional Review Board (or Ethics Committee) of IRCCS Centro Neurolesi "Bonino-Pulejo" Code: IRCCS-ME-CE 08/21.

Informed Consent Statement: Informed consent was obtained from all subjects involved in the study.

Data Availability Statement: Data will be available on request to the corresponding author.

Conflicts of Interest: The authors declare no conflict of interest.

References

1. Chohan, S.A.; Venkatesh, P.K.; How, C.H. Long-term complications of stroke and secondary prevention: An overview for primary care physicians. *Singapore Med. J.* **2019**, *60*, 616–620. [CrossRef] [PubMed]
2. Katan, M.; Luft, A. Global Burden of Stroke. *Sem. Neurol.* **2018**, *38*, 208–211. [CrossRef]
3. Kim, K.; Kim, Y.M.; Kim, E.K. Correlation between the Activities of Daily Living of Stroke Patients in a Community Setting and Their Quality of Life. *J. Phys. Ther. Sci.* **2014**, *26*, 417–419. [CrossRef] [PubMed]
4. Lekoubou, A.; Nguyen, C.; Kwon, M.; Nyalundja, A.D.; Agrawal, A. Post-stroke Everything. *Curr. Neurol. Neurosci. Rep.* **2023**, *23*, 785–800. [CrossRef] [PubMed]
5. Béjot, Y.; Daubail, B.; Giroud, M. Epidemiology of stroke and transient ischemic attacks: Current knowledge and perspectives. *Revue Neurol.* **2016**, *172*, 59–68. [CrossRef]
6. Runde, D. Calculated decisions: NIH Stroke Scale/Score (NIHSS). *Emerg. Med. Pract.* **2021**, *23* (Suppl. 6), CD3–CD5.
7. Alijanpour, S.; Mostafazdeh-Bora, M.; Ahmadi Ahangar, A. Different Stroke Scales; Which Scale or Scales Should Be Used? *Caspian J. Int. Med.* **2021**, *4*, 1–21.
8. Sánchez-Herrera-Baeza, P.; Cano-de-la-Cuerda, R.; Serrada-Tejeda, S.; Fernández-Vázquez, D.; Navarro-López, V.; González-Alted, C.; Miangolarra-Page, J.C. Influence of Age, Gender and Education Level on Executive Functions and Functioning in People with Stroke. *Biomedicines* **2023**, *11*, 1603. [CrossRef]
9. Lai, S.M.; Duncan, P.W. Stroke recovery profile and the Modified Rankin assessment. *Neuroepidemiology* **2001**, *20*, 26–30. [CrossRef]
10. Rowland, T.J.; Cooke, D.M.; Gustafsson, L.A. Role of occupational therapy after stroke. *Ann. Indian Acad. Neurol.* **2008**, *11* (Suppl. 1), S99–S107. [CrossRef]
11. Gibson, E.; Koh, C.L.; Eames, S.; Bennett, S.; Scott, A.M.; Hoffmann, T.C. Occupational therapy for cognitive impairment in stroke patients. *Cochrane Database Syst. Rev.* **2022**, *3*, CD006430. [PubMed]
12. O'Donoghue, M.; Boland, P.; Leahy, S.; Galvin, R.; McManus, J.; Lisiecka, D.; Hayes, S. Exploring the perspectives of key stakeholders on the design and delivery of a cognitive rehabilitation intervention for people post-stroke. *PLoS ONE* **2022**, *17*, 0269961.
13. Torregrossa, W.; Torrisi, M.; De Luca, R.; Casella, C.; Rifici, C.; Bonanno, M.; Calabrò, R.S. Neuropsychological Assessment in Patients with Traumatic Brain Injury: A Comprehensive Review with Clinical Recommendations. *Biomedicines* **2023**, *11*, 1991. [CrossRef]
14. Parisi, A.; Bellinzona, F.; Di Lernia, D.; Repetto, C.; De Gaspari, S.; Brizzi, G.; Riva, G.; Tuena, C. Efficacy of Multisensory Technology in Post-Stroke Cognitive Rehabilitation: A Systematic Review. *J. Clin. Med.* **2022**, *11*, 6324. [CrossRef]
15. Malik, A.N.; Tariq, H.; Afridi, A.; Rathore, F.A. Technological advancements in stroke rehabilitation. *JPMA. J. Pak. Med. Assoc.* **2022**, *72*, 1672–1674. [PubMed]
16. Hatem, S.M.; Saussez, G.; Della Faille, M.; Prist, V.; Zhang, X.; Dispa, D.; Bleyenheuft, Y. Rehabilitation of Motor Function after Stroke: A Multiple Systematic Review Focused on Techniques to Stimulate Upper Extremity Recovery. *Front. Hum. Neurosci.* **2016**, *10*, 442. [CrossRef] [PubMed]
17. Laver, K.E.; Adey-Wakeling, Z.; Crotty, M.; Lannin, N.A.; George, S.; Sherrington, C. Telerehabilitation services for stroke. *Cochrane Database Syst.* **2020**, *1*, CD010255. [CrossRef]
18. Kleim, J.A.; Jones, T.A. Principles of experience-dependent neural plasticity: Implications for rehabilitation after brain damage. *J. Speech Hear. Res.* **2008**, *51*, S225–S239. [CrossRef]
19. Russo, M.; De Luca, R.; Naro, A.; Sciarrone, F.; Aragona, B.; Silvestri, G.; Manuli, A.; Bramanti, A.; Casella, C.; Bramanti, P.; et al. Does body shadow improve the efficacy of virtual reality-based training with BTS NIRVANA? A pilot study. *Medicine* **2017**, *96*, 8096. [CrossRef]
20. Maggio, M.G.; Latella, D.; Maresca, G.; Sciarrone, F.; Manuli, A.; Naro, A.; De Luca, R.; Calabrò, R.S. Virtual Reality and Cognitive Rehabilitation in People with Stroke: An Overview. *J. Neurosc. Nurs.* **2019**, *51*, 101–105. [CrossRef]
21. Proffitt, R.; Lange, B. Considerations in the efficacy and effectiveness of virtual reality interventions for stroke rehabilitation: Moving the field forward. *Phys. Ther.* **2015**, *95*, 441–448. [CrossRef]
22. Mancuso, M.; Varalta, V.; Sardella, L.; Capitani, D.; Zoccolotti, P.; Antonucci, G. Italian OCS Group. Italian normative data for a stroke specific cognitive screening tool: The Oxford Cognitive Screen. (OCS). *Neurol. Sci.* **2016**, *37*, 1713–1721. [CrossRef] [PubMed]
23. Laver, K.E.; Lange, B.; George, S.; Deutsch, J.E.; Saposnik, G.; Crotty, M. Virtual reality for stroke rehabilitation. *Cochrane Database Syst. Rev.* **2017**, *11*, CD008349. [CrossRef]
24. Calabrò, R.S.; Cerasa, A.; Ciancarelli, I.; Pignolo, L.; Tonin, P.; Iosa, M.; Morone, G. The Arrival of the Metaverse in Neurorehabilitation: Fact, Fake or Vision? *Biomedicines* **2022**, *10*, 2602. [CrossRef]
25. Maggio, M.G.; Naro, A.; Manuli, A.; Maresca, G.; Balletta, T.; Latella, D.; De Luca, R.; Calabrò, R.S. Effects of Robotic Neurorehabilitation on Body Representation in Individuals with Stroke: A Preliminary Study Focusing on an EEG-Based Approach. *Brain Topogr.* **2021**, *34*, 348–362. [CrossRef] [PubMed]
26. Steinisch, M.; Tana, M.G.; Comani, S. A post-stroke rehabilitation system integrating robotics, VR and high-resolution EEG imaging. *IEEE Trans. Neural Syst. Rehab. Eng.* **2013**, *21*, 849–859. [CrossRef] [PubMed]
27. Juhász, C.; Kamondi, A.; Szirmai, I. Spectral EEG analysis following hemispheric stroke: Evidences of transhemispheric diaschisis. *Acta Neurol. Scand.* **1997**, *96*, 397–400. [CrossRef] [PubMed]

28. Mingyu, L.; Jue, W.; Nan, Y.; Qin, Y. Development of EEG biofeedback system based on virtual reality environment. *IEEE Eng. Med. Biol. Soc.* **2005**, *3*, 5362–5364.
29. Pyasik, M.; Scandola, M.; Moro, V. Electrophysiological correlates of action monitoring in brain-damaged patients: A systematic review. *Neuropsychologia* **2022**, *174*, 108333. [CrossRef]
30. Mishra, S.; Kumar, A.; Padmanabhan, P.; Gulyás, B. Neurophysiological Correlates of Cognition as Revealed by Virtual Reality: Delving the Brain with a Synergistic Approach. *Brain Sci.* **2021**, *11*, 51. [CrossRef]
31. Arcuri, F.; Porcaro, C.; Ciancarelli, I.; Tonin, P.; Cerasa, A. Electrophysiological Correlates of Virtual-Reality Applications in the Rehabilitation Setting: New Perspectives for Stroke Patients. *Electronics* **2021**, *10*, 836. [CrossRef]
32. Gangemi, A.; Colombo, B.; Fabio, R.A. Effects of short- and long-term neurostimulation (tDCS) on Alzheimer's disease patients: Two randomized studies. *Aging Clin. Exp. Res.* **2021**, *33*, 383–390. [CrossRef]
33. Lin, C.T.; Chung, I.F.; Ko, L.W.; Chen, Y.C.; Liang, S.F.; Duann, J.R. EEG-based assessment of driver cognitive responses in a dynamic virtual-reality driving environment. *IEEE. Trans. Biomed. Eng.* **2007**, *54*, 1349–1352.
34. Beppi, C.; Ribeiro Violante, I.; Scott, G.; Sandrone, S. EEG, MEG and neuromodulatory approaches to explore cognition: Current status and future directions. *Brain Cogn.* **2021**, *148*, 105677. [CrossRef] [PubMed]
35. Delorme, A.; Makeig, S. EEGLAB: An open-source toolbox for analysis of single-trial EEG dynamics including independent component analysis. *J. Neurosci. Methods* **2004**, *134*, 9–21. [CrossRef] [PubMed]
36. Kamarajan, C.; Porjesz, B. Advances in electrophysiological research. *Alcohol. Res. Curr. Rev.* **2015**, *37*, 53–87.
37. Welch, P.D. The use of Fast Fourier Transform for the estimation of power spectra: A method based on time averaging over short, modified periodograms. *IEEE Trans. Audio Electroacoust.* **1967**, *AU-15*, 70–73. [CrossRef]
38. Georgiev, D.D.; Georgieva, I.; Gong, Z.; Nanjappan, V.; Georgiev, G.V. Virtual Reality for Neurorehabilitation and Cognitive Enhancement. *Brain Sci.* **2021**, *11*, 221. [CrossRef]
39. SPSS Statistics-IBM Data Science Community. Available online: https://www.ibm.com/products/spss-statistics (accessed on 30 June 2021).
40. Souza, R.H.C.E.; Naves, E.L.M. Attention Detection in Virtual Environments Using EEG Signals: A Scoping Review. *Front. Physiol.* **2021**, *12*, 727840. [CrossRef]
41. Calabrò, R.S.; Naro, A.; Russo, M.; Leo, A.; De Luca, R.; Balletta, T.; Buda, A.; La Rosa, G.; Bramanti, A.; Bramanti, P. The role of virtual reality in improving motor performance as revealed by EEG: A randomized clinical trial. *J. Neuroeng. Rehab* **2017**, *14*, 53. [CrossRef]
42. De Luca, R.; Bonanno, M.; Rifici, C.; Pollicino, P.; Caminiti, A.; Morone, G.; Calabrò, R.S. Does Non-Immersive Virtual Reality Improve Attention Processes in Severe Traumatic Brain Injury? Encouraging Data from a Pilot Study. *Brain Sci.* **2022**, *12*, 1211. [CrossRef]
43. Leonardi, S.; Maggio, M.G.; Russo, M.; Bramanti, A.; Arcadi, F.A.; Naro, A.; Calabrò, R.S.; De Luca, R. Cognitive recovery in people with relapsing/remitting multiple sclerosis: A randomized clinical trial on virtual reality-based neurorehabilitation. *Clin. Neurol. Neurosurg.* **2021**, *208*, 106828. [CrossRef] [PubMed]
44. De Luca, R.; Bonanno, M.; Marra, A.; Rifici, C.; Pollicino, P.; Caminiti, A.; Castorina, M.V.; Santamato, A.; Quartarone, A.; Calabrò, R.S. Can Virtual Reality Cognitive Rehabilitation Improve Executive Functioning and Coping Strategies in Traumatic Brain Injury? A Pilot Study. *Brain Sci.* **2023**, *13*, 578. [CrossRef]
45. Naro, A.; Calabrò, R.S. What Do We Know about The Use of Virtual Reality in the Rehabilitation Field? A Brief Overview. *Electronics* **2021**, *10*, 1042. [CrossRef]
46. Zak, M.; Wasik, M.; Sikorski, T.; Aleksandrowicz, K.; Miszczuk, R.; Courteix, D.; Dutheil, F.; Januszko-Szakiel, A.; Brola, W. Rehabilitation in Older Adults Affected by Immobility Syndrome, Aided by Virtual Reality Technology: A Narrative Review. *J. Clin. Med.* **2023**, *12*, 5675. [CrossRef] [PubMed]
47. Tompkins, C.A.; Lehman, M.T.; Wyatt, A.D.; Schulz, R. Functional outcome assessment of adults with right hemisphere brain damage. In *Seminars Speech Lang*; Thieme Medical Publishers, Inc.: New York, NY, USA, 1998; Volume 19, pp. 303–321.
48. Tompkins, C.A. Rehabilitation for cognitive-communication disorders in right hemisphere brain damage. *Arch. Phys. Med. Rehab.* **2012**, *93* (Suppl. 1), S61–S69. [CrossRef] [PubMed]
49. Sturm, W.; Willmes, K. On the functional neuroanatomy of intrinsic and phasic alertness. *NeuroImage* **2001**, *14*, S76–S84. [CrossRef]
50. Ferber, S.; Ruppel, J.; Danckert, J. Visual working memory deficits following right brain damage. *Brain Cogn.* **2020**, *142*, 105566. [CrossRef]
51. Audet, T.; Mercier, L.; Collard, S.; Rochette, A.; Hebert, R. Attention deficits: Is there a right hemisphere specialization for simple reaction time, sustained attention and phasic alertness? *Brain Cogn.* **2000**, *43*, 17–21. [PubMed]
52. Gillespie, D.C.; Bowen, A.; Chung, C.S.; Cockburn, J.; Knapp, P.; Pollock, A. Rehabilitation for post-stroke cognitive impairment: An overview of recommendations arising from systematic reviews of current evidence. *Clin. Rehabil.* **2015**, *29*, 120–128. [CrossRef]
53. Cicerone, K.D.; Langenbahn, D.M.; Braden, C.; Malec, J.F.; Kalmar, K.; Fraas, M.; Felicetti, T.; Laatsch, L.; Harley, J.P.; Bergquist, T.; et al. Evidence-based cognitive rehabilitation: Updated review of the literature from 2003 through 2008. *Arch. Phys. Med. Rehabil.* **2011**, *92*, 519–530. [CrossRef]
54. Spaccavento, S.; Marinelli, C.V.; Nardulli, R.; Macchitella, L.; Bivona, U.; Piccardi, L.; Zoccolotti, P.; Angelelli, P. Attention Deficits in Stroke Patients: The Role of Lesion Characteristics, Time from Stroke, and Concomitant Neuropsychological Deficits. *Behav. Neurol.* **2019**, *2019*, 7835710. [CrossRef] [PubMed]

55. Nieto-Escamez, F.; Cortés-Pérez, I.; Obrero-Gaitán, E.; Fusco, A. Virtual Reality Applications in Neurorehabilitation: Current Panorama and Challenges. *Brain Sci.* **2023**, *13*, 819. [CrossRef]
56. Feitosa, J.A.; Casseb, R.F.; Camargo, A.; Brandao, A.F.; Li, L.M.; Castellano, G. Graph analysis of cortical reorganization after virtual reality-based rehabilitation following stroke: A pilot randomized study. *Front. Neurol.* **2023**, *14*, 1241639. [CrossRef] [PubMed]
57. Alouani, A.T.; Elfouly, T. Traumatic Brain Injury (TBI) Detection: Past, Present, and Future. *Biomedicines* **2022**, *10*, 2472. [CrossRef]
58. Smith, E.E.; Reznik, S.J.; Stewart, J.L.; Allen, J.J. Assessing and conceptualizing frontal EEG asymmetry: An updated primer on recording, processing, analyzing, and interpreting frontal alpha asymmetry. *Int. J. Psychophysiol.* **2017**, *111*, 98–114. [CrossRef] [PubMed]
59. Johnson, D.A.; Wilson, G.S. Telemetry for Biosensor Systems. In *Electrochem. Methods for Neuroscience*; Michael, A.C., Ed.; CRC Press/Taylor & Francis: Boca Raton, FL, USA, 2007.
60. Lee, M.H.; Kwon, O.Y.; Kim, Y.J.; Kim, H.K.; Lee, Y.E.; Williamson, J.; Fazli, S.; Lee, S.W. EEG dataset and OpenBMI toolbox for three BCI paradigms: An investigation into BCI illiteracy. *GigaScience* **2019**, *8*, giz002. [CrossRef] [PubMed]
61. Klimesch, W. An algorithm for the EEG frequency architecture of consciousness and brain body coupling. *Front. Hum. Neurosci.* **2013**, *7*, 766. [CrossRef]
62. Park, W.; Kwon, G.H.; Kim, Y.H.; Lee, J.H.; Kim, L. EEG response varies with lesion location in patients with chronic stroke. *J. Neuroeng. Rehab.* **2016**, *13*, 21. [CrossRef]
63. Engel, A.K.; Fries, P.; Singer, W. Dynamic predictions: Oscillations and synchrony in top-down processing. *Nature Rev. Neurosci.* **2001**, *2*, 704–716. [CrossRef]
64. Torregrossa, W.; Raciti, L.; Rifici, C.; Rizzo, G.; Raciti, G.; Casella, C.; Naro, A.; Calabrò, R.S. Behavioral and Psychiatric Symptoms in Patients with Severe Traumatic Brain Injury: A Comprehensive Overview. *Biomedicines* **2023**, *11*, 1449. [CrossRef] [PubMed]
65. Maresca, G.; Maggio, M.G.; Latella, D.; Cannavò, A.; De Cola, M.C.; Portaro, S.; Stagnitti, M.C.; Silvestri, G.; Torrisi, M.; Bramanti, A.; et al. Toward Improving Poststroke Aphasia: A Pilot Study on the Growing Use of Telerehabilitation for the Continuity of Care. *J. Stroke Cerebrovasc. Dis.* **2019**, *28*, 104303. [CrossRef] [PubMed]
66. Calabrò, R.S.; Bonanno, M.; Torregrossa, W.; Cacciante, L.; Celesti, A.; Rifici, C.; Tonin, P.; De Luca, R.; Quartarone, A. Benefits of Telerehabilitation for Patients with Severe Acquired Brain Injury: Promising Results From a Multicenter Randomized Controlled Trial Using Nonimmersive Virtual Reality. *J. Med. Int. Res.* **2023**, *25*, e45458. [CrossRef] [PubMed]
67. Cha, K.; Wang, J.; Li, Y.; Shen, L.; Chen, Z.; Long, J. A novel upper-limb tracking system in a virtual environment for stroke rehabilitation. *J. Neuroeng. Rehab.* **2021**, *18*, 166. [CrossRef] [PubMed]
68. Dey, A.; Chatburn, A.; Billinghurst, M. Exploration of an EEG-Based Cognitively Adaptive Training System in Virtual Reality. In Proceedings of the IEEE Conference on Virtual Reality and 3D User Interfaces (VR), Osaka, Japan, 23–27 March 2019; pp. 220–226.
69. Magosso, E.; De Crescenzio, F.; Ricci, G.; Piastra, S.; Ursino, M. EEG Alpha Power Is Modulated by Attentional Changes during Cognitive Tasks and Virtual Reality Immersion. *Comput. Intell. Neurosci.* **2019**, *2019*, 7051079. [CrossRef] [PubMed]
70. Fabio, R.A.; Gangemi, A.; Capri, T.; Budden, S.; Falzone, A. Neurophysiological and cognitive effects of transcranial direct current stimulation in three girls with Rett Syndrome with chronic language impairments. *Res. Dev. Disabil.* **2018**, *76*, 76–87. [CrossRef]

biomedicines

MDPI

Article

Age, Dose, and Locomotion: Decoding Vulnerability to Ketamine in C57BL/6J and BALB/c Mice

Wen-Chien Chen [1], Tzong-Shi Wang [1], Fang-Yu Chang [1], Po-An Chen [2] and Yi-Chyan Chen [1,3,*]

[1] Department of Psychiatry, Taipei Tzu Chi Hospital, Buddhist Tzu Chi Medical Foundation, New Taipei City 231, Taiwan
[2] Department of Psychiatry, China Medical University Hsinchu Hospital, China Medical University, Hsinchu 302, Taiwan
[3] Department of Psychiatry, School of Medicine, Tzu Chi University, Hualien 970, Taiwan
* Correspondence: yichyanc@gmail.com or yichyanc@tzuchi.com.tw; Tel.: +886-2-6628-9779 (ext. 3710)

Abstract: Ketamine has been abused as a psychedelic agent and causes diverse neurobehavioral changes. Adolescence is a critical developmental stage but vulnerable to substances and environmental stimuli. Growing evidence shows that ketamine affects glutamatergic neurotransmission, which is important for memory storage, addiction, and psychosis. To explore diverse biological responses, this study was designed to assess ketamine sensitivity in mice of different ages and strains. Male C57BL/6J and BALB/c mice were studied in adolescence and adulthood separately. An open field test assessed motor behavioral changes. After a 30-min baseline habituation, mice were injected with ketamine (0, 25, and 50 mg/kg), and their locomotion was measured for 60 min. Following ketamine injection, the travelled distance and speed significantly increased in C57BL/6J mice between both age groups ($p < 0.01$), but not in BALB/c mice. The pattern of hyperlocomotion showed that mice were delayed at the higher dose (50 mg/kg) compared to the lower dose (25 mg/kg) of ketamine treatment. Ketamine accentuated locomotor activation in adolescent C57BL/6J mice compared to adults, but not in the BALB/c strain. Here, we show that ketamine-induced locomotor behavior is modulated by dose and age. The discrepancy of neurobehaviors in the two strains of mice indicates that sensitivity to ketamine is biologically determined. This study suggests that individual vulnerability to ketamine's pharmacological responses varies biologically.

Keywords: ketamine; NMDA; pharmacological response; locomotion; psychosis; learning; age; strain; dose effect; genetic diversity

Citation: Chen, W.-C.; Wang, T.-S.; Chang, F.-Y.; Chen, P.-A.; Chen, Y.-C. Age, Dose, and Locomotion: Decoding Vulnerability to Ketamine in C57BL/6J and BALB/c Mice. *Biomedicines* **2023**, *11*, 1821. https://doi.org/10.3390/biomedicines11071821

Academic Editors: Masaru Tanaka and Francesca Calabrese

Received: 28 February 2023
Revised: 15 June 2023
Accepted: 22 June 2023
Published: 25 June 2023

1. Introduction

Ketamine has been used as a dissociative anesthetic in human medicine since 1970. Growing research demonstrate that it has promising therapeutic effects for pain and treatment-resistant depression [1–3]. It can also be used recreationally as a party drug due to its hallucinogenic and dissociative effect for producing different states of mind, emotion, perception, and psychomotor behaviors. The acute effects of ketamine can induce trance-like states such as amnesia, visual distortion, hallucination, dissociative experiences, and psychotomimetic features [1,4]. Chronic abuse can influence the neuropsychological function and cognitive performance and induce urinary toxicity such as cystitis and lower urinary track syndrome [4,5]. An epidemiological survey showed an increasing prevalence of ketamine abuse in Asia, especially for teenagers [6]. Facing the public health and long-term impacts of ketamine abuse is a critical issue. Abuse behaviors not only potentiate the addiction tendency but also bring negative psychiatric consequences to the adolescent [4].

Ketamine principally acts as an *N*-methyl-D-aspartate (NMDA) receptor agonist and inhibits glutamatergic neurotransmission, which is responsible for anesthetic, analgesic, and psychotomimetic effects [7]. Moderations of the glutamate neuropathway can produce diverse neurophysiological and neurobehavioral changes, such as impairment of memory encoding and development of psychosis [8]. Growing evidence demonstrates the effectiveness for treatment-resistant depression through NMDA receptor inhibition and downstream α-amino-3-hydroxy-5-methyl-4-isoxazolepropionic acid (AMPA) receptor activation [9,10]. The prefrontal cortex, subgenual anterior cingulate cortex, and hippocampus are thought to be involved in this antidepressant effect [9,11]. Some studies have demonstrated overlapping neuronal circuits underlying depression and addiction, with a specific focus on the mesolimbic projection [10]. Recent studies have shown different pharmacodynamic properties between R- and S-ketamine regarding the risk of adverse effects [9,12]. Nonetheless, the long-term adverse effects of ketamine are worth noting in different populations. Based on the pharmacological properties, activation of the NMDA receptor can enhance long-term potentiation (LTP) and synaptic plasticity, which are important mechanisms for memory consolidation and recall. Compared to receptor activation, the inhibitory effects of ketamine on the NMDA receptor may deter LTP formation and cause negative impacts on memory and learning [13,14]. There are common symptoms of dissociation, hallucination, and delusion in acute ketamine users and cognitive deficits in chronic abusers. Many neuroscience studies have indicated that ketamine-induced psychotomimetic features are strongly correlated with NMDA hypofunction [15,16]. In a clinical observation survey, ketamine exacerbated psychotic symptoms in patients with schizophrenia [17]. Until now, ketamine-induced neurobehavioral change has seemed to be a powerful animal model for schizophrenia research [18,19].

Adolescence is a critical stage of brain development from the childhood to adulthood, possessing characteristics of rapid neuronal differentiation and neurocircuit remodeling [20]. During this stage, the brain is remarkably vulnerable to stressors and diverse substances, which can predispose the adolescent to many neuropsychiatric problems such as schizophrenia and bipolar disorder and may increase risk-taking behaviors, which can easily lead to substance addiction [20–22]. The putative mechanism of ketamine exerts negative impacts on information processing and episodic memory formation [23]. Moreover, ketamine abuse is often accompanied with adverse psychotomimetic effects in adolescent. On the other hand, growing evidence has demonstrated the efficacy of ketamine on treatment-resistant depression, which has a higher prevalent rate in adults and the elderly. In recent years, research into different treatments for major depressive disorders have led to a more comprehensive understanding of the mechanisms of depression beyond the monoamine hypothesis. For instance, studies involving ketamine have demonstrated the involvement of the glutamatergic system, while research on brain stimulation has revealed the significance of thalamocortical dysrhythmia [24,25]. Ketamine may prove to be a valuable tool, as it may help us to understand the pathophysiology of depression and identify possible biomarkers, such as genetic or electroencephalography biomarkers [26], in the future. Thus, the trade-off between the risks and benefits of ketamine treatment between the adolescent and the adult with refractory depressive features is a matter of dispute. To clarify, the age effect on the differences of ketamine sensitivity and pharmacological response is an important issue. In another aspect, geographical differences of ketamine abuse may also reflect the pharmacogenetic variation.

To explore the age effect and strain differences on the ketamine pharmacological response, an open field test was used as an animal model to evaluate neurobehavioral changes after acute ketamine injection. The mouse strains of BALB/c and C57BL/6J were used to compare the ketamine sensitivity between adolescence and adulthood. The study was designed (1) to investigate age-related behavioral changes following ketamine injection; (2) to compare the strain differences in terms of ketamine sensitivity; and (3) to compare the interactions of ketamine dose effects between strains and different age groups.

2. Materials and Methods

The study was designed to investigate the locomotor neurobehavioral changes following ketamine injection in mice with different ages and strains. An open field test with a computerized video-tracking system was adapted to measure the locomotor changes [27–29].

2.1. Subjects

Subjects were male C57BL/6J and BALB/c mice obtained from the National Laboratory Animal Center, Taiwan. C57BL/6J was chosen as a reference strain given its common use in models of explorative behavioral studies [27,28]. BALB/c mice, as genetic backgrounds for mutants, were chosen based on their frequent use in behavioral neuroscience, including sensitivity to stress and expectations of different reactions to ketamine [28,29]. Adolescent (4–6 week-old) and adult (9–15 week-old) mice were used. Mice were housed in groups of four in a temperature (22 ± 1 °C)- and humidity (50 ± 5%)-controlled vivarium under a 12-h light/dark cycle with ad libitum access to food and water. All experimental procedures were approved by the Institutional Animal Care and Use Committee, Buddhist Taipei General Hospital, and were conducted following the regulation of reduction and refinement. All efforts were made to minimize the number of animals used and their suffering.

2.2. Drug Preparation

Testing doses of ketamine were based on our pilot work [30,31] and previous studies of behavioral effects in mice [32–35]. In this experiment, the study dose was first assessed by the rotarod motor test to evaluate motor balance. To determine the doses without the effects of anesthesia and paralysis, testing doses were determined to the sub-anesthesia range from 25 to 50 mg/kg, with 0 mg/kg used as the negative control. The use of ketamine was approved by the Taiwan Food and Drug Administration, and ketamine was purchased from Pfizer (New York, NY, USA). Ketalar (ketamine hydrochloride, 50 mg/mL) was dissolved in 0.9% saline vehicle and intraperitoneally injected in volumes of 10 mL/kg.

2.3. Study Procedure

The novel open field apparatus was a 40 × 40 × 35 cm square arena with opaque Plexiglas walls and floor. The white composition was designed for C57BL/6J strain mice and the black for BALB/c mice. Before the behavioral study, mice were transferred to the testing room 1 h prior to testing for acclimation to the test environment. The open field apparatus was wiped with 70% ethanol prior to each trial and between trials. The study procedures (Figure 1) were performed in adolescent and adult stages of both strains with randomized assignment of different ketamine doses.

In the test, the open field apparatus was evenly illuminated to ~50 ± 5 lux [36]. A cross-over design was used for the mice to avoid environmental differences in the open field and to reduce the number of animals used. The apparatus was designed with black and white colors for C57BL/6J and BALB/c mice, respectively, to adjust the color differences for two strains under the monitoring system. The mice were initially placed in the apparatus for a 30-min habituation period. This design served as both an internal and systemic control. After 30 min of baseline locomotion habituation, mice were followed by intraperitoneal injection (i.p.) with ketamine (0, 25, and 50 mg/kg). The mice then were placed in the perimeter and allowed to explore the apparatus for 60 min. A computerized video tracking system (SINGA Trace MouseIIAnubis Track 1.6.4 pro) was adapted to measure the motor responses following ketamine injection.

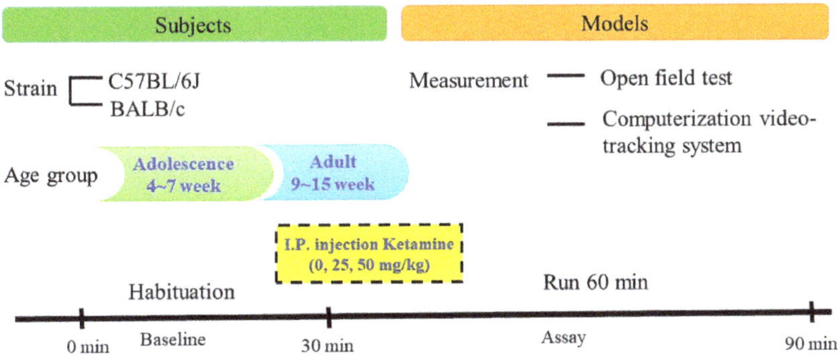

Figure 1. The study procedure: After 30 min baseline habituation, mice were followed by intraperitoneal injection with different ketamine dosages. The mice then were placed in the perimeter and allowed to explore the apparatus for 60 min. A video-tracking system was adapted to measure the motor response in the open field.

2.4. Statistical Analysis

All the statistical analyses were conducted using GraphPad Prism (GraphPad Software, Inc., San Diego, CA, USA) version 5.01.336 and Statistical Package for Social Science version 18 (SPSS Inc., Chicago, IL, USA). The moving paths and travelled distances in the open field were tracked before and after ketamine injection. The measurement parameters were analyzed by analysis of variance (ANOVA), followed by Bonferroni's post hoc tests. Probability (p) values under 0.05 were considered significant.

3. Results

Two strains of mice during adolescence and adulthood were randomized to different ketamine doses (0, 25, 50 mg/kg, separately) by i.p. injection. The number of experimental mice in each group was 11–12 (by strain × age × ketamine dose). Prior to the drug treatment, there were no differences in the baseline travelled distance within each strain × age group. The comparison between the two strains of mice at baseline, including the summation of 30 min of travelled distance and the total time spent in the central area of the apparatus, can be found in Supplementary Materials Figures S1 and S2.

The locomotor activity after ketamine injection was observed for 60 min. The travelled distance (cm) of each group is shown at the corresponding time points, by 2 min time intervals, in Figures 2A,B and 3A,B. The total distance travelled during 60 min after ketamine injection is demonstrated in Figures 2C,D and 3C,D. ANOVA tests and post hoc analysis were used to assess the differences of locomotor responses among each group. Table 1 summarizes the effect of different ketamine dosages on the total distance travelled during a 60 min period after ketamine injection.

Table 1. Comparisons of locomotor activity between C57BL/6J and BALB/c mice treated with different ketamine doses.

Ketamine	C57BL/6J		BALB/c	
	Adolescent	Adult	Adolescent	Adult
0 mg/kg	65.2 ± 7.0	66.8 ± 6.8	72.3 ± 13.8	82.3 ± 18.2
25 mg/kg	175.0 ± 13.8 [a]	130.9 ± 11.8 [b]	85.5 ± 12.8	103.4 ± 14.5
50 mg/kg	171.2 ± 15.7 [a]	136.9 ± 8.1 [b]	89.6 ± 17.5	52.8 ± 13.3 [c]

Values are means ± standard error of the mean (SEM); data represent the total travelled distance (meter) in 60 min following the ketamine injection. Differences among the study group were evaluated by multiple analysis of variance and post hoc tests. Subject numbers were 10–13 in each group. [a] $p < 0.001$ compared with 0 mg/kg ketamine group in adolescent C57BL/6J. [b] $p < 0.01$ compared with 0 mg/kg ketamine group in adult C57BL/6J, [c] $p < 0.05$ compared with 25 mg/kg ketamine group in adult BALB/c.

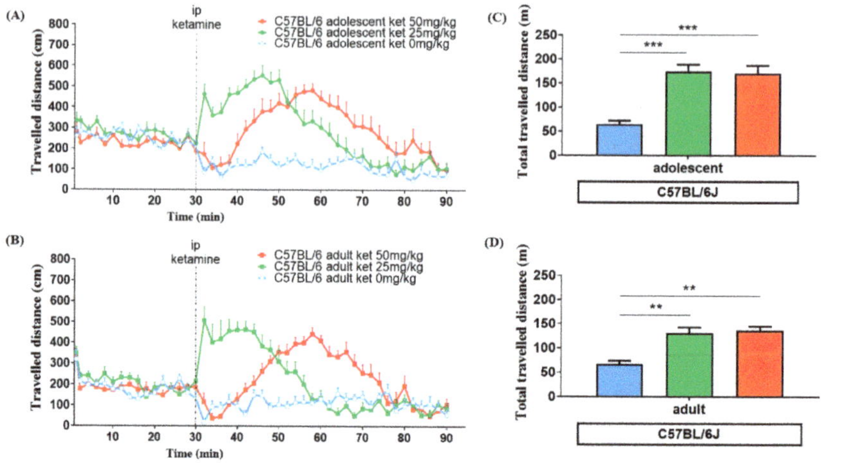

Figure 2. Travelled distance (cm) at the corresponding time points in the open field test for adolescent (**A**) and adult (**B**) C57BL/6J mice treated with different ketamine dose. The groups are denoted as adolescent (circle) and as adult (square), as well as different ketamine doses: 0 mg/kg (blue), 25 mg/kg (green), 50 mg/kg (red). Total travelled distance (m) represents the summation of moving distance following ketamine injection running for 60 min in both age groups: (**C**) for adolescent, and (**D**) for adulthood. Vertical bars represent standard error of the mean (SEM). Differences among the study groups were evaluated by analysis of variance (ANOVA) and post hoc tests. Statistically significant differences between groups: ** $p < 0.01$, *** $p < 0.001$ vs. vehicle.

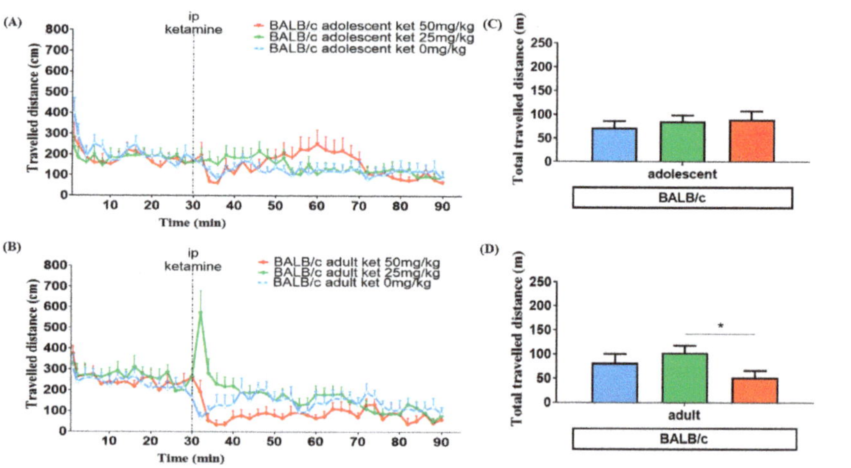

Figure 3. Travelled distance (cm) at the corresponding time points in the open field test for adolescent (**A**) and adult (**B**) BALB/c mice treated with different ketamine doses. The groups are denoted as adolescent (inverted triangle) and as adult (prism), as well as different ketamine doses: 0 mg/kg (blue), 25 mg/kg (green), 50 mg/kg (red). Total travelled distance (m) represents the summation of moving distance following ketamine injection running for 60 min in both age groups: (**C**) for adolescent, and (**D**) for adulthood. Vertical bars represent SEM. Differences among the study groups were evaluated by ANOVA and post hoc tests. Statistically significant differences between 25 mg and 50 mg dose groups: * $p < 0.05$.

3.1. Age Effect

ANOVA data showed a significant difference in the amount of distance traveled after receiving a ketamine injection between adolescent and adult C57BL/6G mice [F (1, 64) = 7.786, $p < 0.01$], but not in BALB/c [F (1, 62) = 0.057, $p = 0.81$]. In C57BL/6J strain mice treated with a low dose of ketamine (25 mg/kg), adolescents exhibited a more intense response of locomotor hyperactivity than adults. The summation of total travelled distance (in meters) following ketamine injection appeared to have a 3-fold increase in the adolescent (175.0 ± 13.8) and 2-fold increase in the adult (130.9 ± 11.8) group compared with the vehicle group (65.2 ± 7.0) ($p < 0.001$) according to Bonferroni's post hoc test. When treated with a higher dose of ketamine (50 mg/kg), delayed responses in moving behaviors were detected. Similarly, adult C57BL/6J mice (136.9 ± 8.1) showed less intensity in total distance traveled than adolescents (171.2 ± 15.7), as shown in Figure 2 C,D and Table 1. However, in BALB/c mice, locomotor activity transiently increased in adults with the 25 mg/kg ketamine injection. The total travelled distance was higher in adult mice with the 25 mg/kg dose than the adolescents, but there was no difference following the 50 mg/kg ketamine treatment (shown in Figure 3 and Table 1).

3.2. Ketamine Dose Effect

ANOVA data showed a significant difference in the amount of distance traveled among the three dosage groups after receiving a ketamine injection [F (2, 126) = 16.67, $p < 0.0001$]. In C57BL/6J mice following the 25 mg/kg injection, both adolescents and adults exhibited a strikingly faster increase in locomotor hyperactivity compared to those who received 50 mg/kg. In mice treated with a 50 mg/kg dose of ketamine, there was a notable delay in peak locomotor activation, which occurred 30 min after the injection. For BALB/c mice, the acute and transient locomotor activation was observed only in the adult group when treated with a low dose (25 mg/kg) of ketamine. Conversely, when adult BALB/c mice were treated with a high dose (50 mg/kg) of ketamine, they traveled a relatively shorter distance in 60 min compared to the low dose (25 mg/kg) of ketamine ($p < 0.05$). This was determined by Bonferroni's post hoc test, as shown in Figure 4 and Table 1.

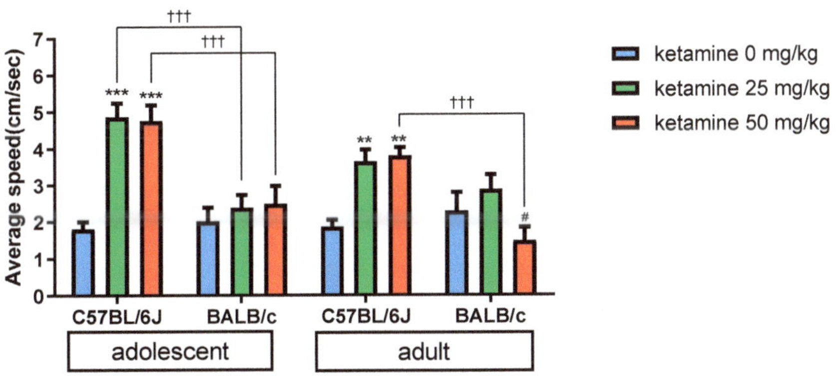

Figure 4. Comparisons of the average travelled speed (cm/sec) between C57BL/6J and BALB/c mice treated with 0, 25, and 50 mg/kg of ketamine. The differences of locomotor activity were measured following the ketamine injection running for 60 min in the open field. Vertical bars represent SEM. Differences among the study groups were evaluated by ANOVA and post hoc tests. Statistically significant differences between groups: # $p < 0.05$ vs. BALB/c treated with 25 mg/kg ketamine, ** $p < 0.01$, *** $p < 0.001$ vs. vehicle; ††† $p < 0.001$ vs. BALB/c.

3.3. Strain Variation

There were notable differences in locomotor behaviors between the C57BL/6J and BALB/c strains, as shown in Figures 2 and 3. ANOVA revealed a significant difference in total traveled distance [F (2, 126) = 32.17, $p < 0.0001$] and average traveled speed [F (1, 126) = 32.16, $p < 0.0001$] between the two strains following ketamine injection, as shown in Figure 4. In adolescent mice, the average travelled speed over 60 min after ketamine treatment was about 2-fold higher in C57BL/6J compared to BALB/c mice ($p < 0.001$), according to Bonferroni's post hoc test. For adult mice, there was no significant difference in average travelled speed between two strains after receiving a low dose (25 mg/kg) of ketamine. However, when given a high dose (50 mg/kg) of ketamine, adult BALB/c mice showed less travelled speed compared to adult C57BL/6J ($p < 0.001$).

4. Discussion

This integrative study aims to clarify the correlations of strain and age effects of ketamine by injecting low and high ketamine doses. The principal findings are as follows: (1) ketamine increased locomotor activity; (2) the pharmacological effects vary among different strains of mice, with hyperlocomotion consistently potentiated in the C57BL/6J group rather than in the BALB/c group; (3) adolescents demonstrated hypersensitive effects to ketamine in C57BL/6J mice; (4) both strains exhibited trends of dose-shift neurobehavioral changes, with acute onset following the low dose ketamine challenge in C57BL/6J mice.

4.1. Ketamine Induced Locomotor Change

The open field test is commonly used in rodent studies to evaluate anxiety and many neurobehaviors. In humans, ketamine can induce schizophrenic-like symptoms [37,38], and in animal studies, ketamine-induced locomotor hyperactivity may indicate excitatory behaviors and the psychotic domain for schizophrenia [35]. The plausible mechanism for how ketamine work is that it activates dopaminergic neurons via disinhibition of glutamatergic projections onto dopamine neurons.

Studies have also shown that acute or chronic ketamine use increases locomotor activity with rising dopamine levels in the cortex, striatum, and nucleus accumbens [16]. Additionally, this locomotor activation may also indicate drug reward [35] through the activation of dopaminergic circuits that contribute to the reinforcing effects. Past research has used the effects of ketamine on locomotion as a means of assessing acute drug sensitivity, as well as chronic exposure-related neuroplasticity and sensitization in addiction models [39–41].

4.2. Interactions between Strain, Age, and Dosage Effects

In open field tests, various strains of mice have exhibited distinct anxiety and locomotor responses [42,43]. For instance, BALB/c mice have been observed to display less exploratory behavior than C57BL/6J mice [43]. Our study aligns with this discovery, as C57BL6/J mice spent a greater amount of time in the central area during the 30-min habituation period in comparison to BALB/c mice (as evidenced in Supplementary Materials Figure S2).

Age differences also play an important role in our study. When treated with the same dosage of ketamine, adolescent C57BL/6J mice showed a higher peak response and longer duration of locomotor activation than adult C57BL/6J mice (Figure 2A,B). The literature suggests that a sub-anesthetic dosage of ketamine exceeding 10 mg/kg [28,35,44] could lead to an increase in ambulatory activities, particularly in adolescents when compared to adult animals [40,41,45]. Our findings are consistent with these reports, showing that adolescent mice are more sensitive to ketamine-related locomotor hyperactivity than adult mice. It is interesting to note that the locomotion of adult BALB/c mice was relatively low when treated with high dose (50 mg/kg) ketamine compared to the vehicle group. It is unlikely that this can be explained by the anesthetic effect of ketamine, which has been demonstrated to generally occur at doses above 100 mg/kg, resulting in initial immobilization followed by postanesthetic locomotion [32,46]. Instead, contributing factors may include the innate low sensitivity of BALB/c mice to ketamine-induced hyperlocomotion [43] and age difference. Many animal studies have demonstrated that various outcomes of ketamine seem to intensify during adolescence, such as emergent reactions of ketamine anesthesia, neurotoxic effects of high doses treatment, and antidepressant effects [41]. This study addresses the intriguing issue of discriminating age effects in the clinical application of ketamine for the treatment of refractory depression.

Different dosages of ketamine were found to produce different time course effects of locomotor activation in previous studies [32,46]. In our study of C57BL/6J mice, a lower dose (25 mg/kg) of ketamine produced the rapid onset of locomotor activation, while a higher dose (50 mg/kg) demonstrated a delayed effect of hyperlocomotion. The initial suppression effect may be due to high dose ketamine producing transient anesthetic effects, but not in BALB/c mice. These findings are consistent with previous reports [32,46]. The rapid pharmacological response not only caters to psychological desire but may also be prone to the development of ketamine addiction [47].

4.3. Adolescent Vulnerability to Ketamine-Related Psychotomimetic Effects

During adolescence, the brain is in a critical stage of development, making individuals more vulnerable to drug addiction [39]. The vulnerable brain in this stage and lack of self-control can lead to sensation-seeking and risky behaviors, which in turn can drive substance abuse. Furthermore, studies show that repeated use of ketamine produces sensitizing effects in rats, increasing the likelihood of addiction [40,41]. In view of psychotomimetic effects, the C57BL/6J mice in our study showed hypersensitivity to hyperlocomotion, especially adolescents. Ketamine is commonly used as the usual model of schizophrenia [18,19], a complex brain disorder that involves gene–gene and gene–environmental interactions. Substance exposure during adolescence has been proposed to deteriorate brain differentiation, remodeling, pruning, and maturation through disturbing dopaminergic projection, glutamate moderation, and myelination [22]. As such, exposure to ketamine during this period may increase the risk of developing schizophrenia or other psychotic disorders.

Some studies have demonstrated pharmacokinetic differences in drug biodistribution and metabolism depending on age. In particular, research suggests that adolescents have a shorter plasma half-life of many substances and psychotropics, both in human and in animal studies [40,48]. Furthermore, it has been observed that ketamine metabolism is slower in older rats than in adults [49]. These pharmacokinetic differences may influence the ketamine effects of anesthesia, pain reduction, and antidepressant effects [41,50–52]. Animal research has revealed changes in the densities of NMDA receptors over the lifespan. Specifically, the densities of NMDA receptors were found to increase from childhood to adolescence and then attenuate in certain brain areas during adulthood [53–56]. In this study, ketamine-induced hyperlocomotion may have originated from the interaction between ketamine and the NMDA system. A possibility is that the greater amount of NMDA receptors in the developmental shift predisposes the locomotor hypersensitivity of C57BL/6J adolescent mice [41].

The BALB/c strain has been observed to display anxious and aggressive behavior and differences in brain anatomy, such as hippocampal lamination defects and neuronal migration [57]. In this study, two strains of mice were used to symbolize how diverse human species may respond differently to ketamine exposure. Previous studies have also shown evidence of genetic factors in the differences of NMDA blockade responses between the two strains [28,58]. Recently, an altered endogenous tone of NMDA receptor-mediated neurotransmission was found in BALB/c mice, which was associated with a sociability deficit [27,59–61]. Pharmacodynamic diversity may be a factor to explain the discrepancy between the two mouse strains. Additionally, exploring the modulation of glutamatergic neurotransmission in BALB/c mice would be intriguing for psychiatry and neuropharmacological research.

4.4. Clinical Implications of the Study

Ketamine has been proven to exhibit a rapid antidepressant response and reduce suicidal ideation. However, the underlying neurobiological mechanism remains unclear, although numerous studies provide evidence that its treatment effects are achieved through modulating glutamate neurotransmission. The glutamate pathway also plays a role in memory consolidation by activating NMDA receptors and enhancing LTP formation. It also has diverse neuropathogenesis to modulate the development of neurocognitive disorders and provides novel opportunities for new drug discovery [54–56,62]. Our study demonstrates that adolescents may exhibit hypersensitivity to neurobehavioral changes, indicating vulnerability to memory formation, addiction tendencies, and psychotomimetic effects. Furthermore, experimental results reveal diverse pharmacological responses to ketamine in different strains, suggesting potential pharmacological variations in different populations. Based on these research findings, it is important to note the potential negative impacts of ketamine on learning, memory formation, and addiction in adolescents. Therefore, future research on ketamine or ketamine-like agents should consider genetic background and developmental stage differences to enable precise medicine approaches.

4.5. Limitation

The limitations of this study are as follows: (1) Acute injection could not reflect the real situation of ketamine abuse behaviors. (2) There was a lack of chronic ketamine effects to explore behavioral sensitization and neuroadaptation. (3) The open field test could only imply ketamine-related locomotion change; it will be important to investigate the other symptom domains by using more neurobehavioral models in the future. (4) If possible, it would be beneficial to have access to neuroimaging or neurochemistry data, as this could provide further insights into the mechanisms underlying neurobehavior. (5) Ketamine is a mixture of (R)-ketamine and (S)-ketamine, which have different affinities to the NMDA receptor and produce distinct neuropsychopharmacological effects. It would be helpful to clarify the correlations between strain and age effects on ketamine isomers.

5. Conclusions

The results of our study demonstrate that ketamine-induced locomotor behavior is influenced by both dose and age. The observed differences in neurobehavioral responses between the two mouse strains may suggest that sensitivity to ketamine is biologically determined. Furthermore, individual vulnerability to ketamine may indicate diverse pharmacological responses in different populations. It is noteworthy to consider the potential negative consequences of ketamine on learning, memory formation, addiction, and susceptibility to the development of psychosis, particularly in adolescents who are in a critical stage of brain development.

Supplementary Materials: The following supporting information can be downloaded at: https://www.mdpi.com/article/10.3390/biomedicines11071821/s1, Figure S1: Comparisons of total travelled distance (m) during the habituation period between C57BL/6J and BALB/c treated with 0, 25, and 50 mg/kg of ketamine; Figure S2: Comparisons of total central duration (sec) during the habituation period between C57BL/6J and BALB/c treated with 0, 25, and 50 mg/kg of ketamine.

Author Contributions: Conceptualization, Y.-C.C.; methodology, Y.-C.C. and W.-C.C.; software, Y.-C.C.; validation, Y.-C.C. and W.-C.C.; formal analysis, Y.-C.C. and W.-C.C.; investigation, Y.-C.C., W.-C.C., T.-S.W., F.-Y.C. and P.-A.C.; resources, Y.-C.C.; data curation, Y.-C.C.; writing—original draft preparation, W.-C.C., T.-S.W. and F.-Y.C.; writing—review and editing, Y.-C.C.; visualization, W.-C.C., T.-S.W., F.-Y.C. and P.-A.C.; supervision, Y.-C.C.; project administration, Y.-C.C.; funding acquisition, Y.-C.C. All authors have read and agreed to the published version of the manuscript.

Funding: This work was supported by the Ministry of Science and Technology (MOST 108-2314-B-303-016) and by the Taipei Tzu Chi Hospital (TCRD-TPE-106-47; TCRD-TPE-106-RT-10, TCRD-TPE-109-RT-4, TCRD-TPE-108-52 and TCRD-TPE-111-39), Taiwan.

Institutional Review Board Statement: In this animal study, all experimental procedures complied with the National Research Council's Guide for the Care and Use of Laboratory Animals and were approved by Institutional Animal Care and Use Committee, Buddhist Taipei General Hospital. This study was approved by the Laboratory Animal Care and Use Committee of Taipei Tzu Chi Hospital (Approval Number: 107-IAUCU-025).

Informed Consent Statement: This study is not involving humans.

Data Availability Statement: Data are available upon request to the corresponding author.

Conflicts of Interest: The authors declare no conflict of interest.

Abbreviations

NMDA, *N*-methyl-D-aspartate; LTP, long-term potentiation; i.p., intraperitoneal; ANOVA, analysis of variance; SEM, standard error of the mean.

References

1. Eldufani, J.; Nekoui, A.; Blaise, G. Nonanesthetic Effects of Ketamine: A Review Article. *Am. J. Med.* **2018**, *131*, 1418–1424. [CrossRef]
2. Zanos, P.; Gould, T.D. Mechanisms of ketamine action as an antidepressant. *Mol. Psychiatry* **2018**, *23*, 801–811. [CrossRef]
3. Ionescu, D.F.; Fu, D.J.; Qiu, X.; Lane, R.; Lim, P.; Kasper, S.; Hough, D.; Drevets, W.C.; Manji, H.; Canuso, C.M. Esketamine Nasal Spray for Rapid Reduction of Depressive Symptoms in Patients with Major Depressive Disorder Who Have Active Suicide Ideation with Intent: Results of a Phase 3, Double-Blind, Randomized Study (ASPIRE II). *Int. J. Neuropsychopharmacol.* **2021**, *24*, 22–31. [CrossRef]
4. Liu, Y.; Lin, D.; Wu, B.; Zhou, W. Ketamine abuse potential and use disorder. *Brain Res. Bull.* **2016**, *126*, 68–73. [CrossRef]
5. Middela, S.; Pearce, I. Ketamine-induced vesicopathy: A literature review. *Int. J. Clin. Pract.* **2011**, *65*, 27–30. [CrossRef]
6. Li, J.H.; Vicknasingam, B.; Cheung, Y.W.; Zhou, W.; Nurhidayat, A.W.; Jarlais, D.C.; Schottenfeld, R. To use or not to use: An update on licit and illicit ketamine use. *Subst. Abuse Rehabil.* **2011**, *2*, 11–20. [CrossRef]
7. Stahl, S.M. Mechanism of action of ketamine. *CNS Spectr.* **2013**, *18*, 171–174. [CrossRef]
8. Javitt, D.C. Glutamate as a therapeutic target in psychiatric disorders. *Mol. Psychiatry* **2004**, *9*, 984–997, 979. [CrossRef]
9. Hashimoto, K. Molecular mechanisms of the rapid-acting and long-lasting antidepressant actions of (R)-ketamine. *Biochem. Pharmacol.* **2020**, *177*, 113935. [CrossRef]
10. Kokane, S.S.; Armant, R.J.; Bolaños-Guzmán, C.A.; Perrotti, L.I. Overlap in the neural circuitry and molecular mechanisms underlying ketamine abuse and its use as an antidepressant. *Behav. Brain Res.* **2020**, *384*, 112548. [CrossRef]
11. Ionescu, D.F.; Felicione, J.M.; Gosai, A.; Cusin, C.; Shin, P.; Shapero, B.G.; Deckersbach, T. Ketamine-Associated Brain Changes: A Review of the Neuroimaging Literature. *Harv. Rev. Psychiatry* **2018**, *26*, 320–339. [CrossRef]
12. Bonaventura, J.; Lam, S.; Carlton, M.; Boehm, M.A.; Gomez, J.L.; Solís, O.; Sánchez-Soto, M.; Morris, P.J.; Fredriksson, I.; Thomas, C.J.; et al. Pharmacological and behavioral divergence of ketamine enantiomers: Implications for abuse liability. *Mol. Psychiatry* **2021**, *26*, 6704–6722. [CrossRef]
13. Yang, Y.; Ju, W.; Zhang, H.; Sun, L. Effect of Ketamine on LTP and NMDAR EPSC in Hippocampus of the Chronic Social Defeat Stress Mice Model of Depression. *Front. Behav. Neurosci.* **2018**, *12*, 229. [CrossRef]
14. Guo, D.; Gan, J.; Tan, T.; Tian, X.; Wang, G.; Ng, K.T. Neonatal exposure of ketamine inhibited the induction of hippocampal long-term potentiation without impairing the spatial memory of adult rats. *Cogn. Neurodyn.* **2018**, *12*, 377–383. [CrossRef]
15. Chatterjee, M.; Verma, R.; Ganguly, S.; Palit, G. Neurochemical and molecular characterization of ketamine-induced experimental psychosis model in mice. *Neuropharmacology* **2012**, *63*, 1161–1171. [CrossRef]
16. Kokkinou, M.; Ashok, A.H.; Howes, O.D. The effects of ketamine on dopaminergic function: Meta-analysis and review of the implications for neuropsychiatric disorders. *Mol. Psychiatry* **2018**, *23*, 59–69. [CrossRef]
17. Lahti, A.C.; Koffel, B.; LaPorte, D.; Tamminga, C.A. Subanesthetic doses of ketamine stimulate psychosis in schizophrenia. *Neuropsychopharmacology* **1995**, *13*, 9–19. [CrossRef]
18. Forrest, A.D.; Coto, C.A.; Siegel, S.J. Animal Models of Psychosis: Current State and Future Directions. *Curr. Behav. Neurosci. Rep.* **2014**, *1*, 100–116. [CrossRef]
19. Moghaddam, B.; Krystal, J.H. Capturing the angel in "angel dust": Twenty years of translational neuroscience studies of NMDA receptor antagonists in animals and humans. *Schizophr. Bull.* **2012**, *38*, 942–949. [CrossRef]
20. Crews, F.; He, J.; Hodge, C. Adolescent cortical development: A critical period of vulnerability for addiction. *Pharmacol. Biochem. Behav.* **2007**, *86*, 189–199. [CrossRef]
21. Casey, B.J.; Jones, R.M. Neurobiology of the adolescent brain and behavior: Implications for substance use disorders. *J. Am. Acad. Child Adolesc. Psychiatry* **2010**, *49*, 1189–1201. [CrossRef]
22. Owen, M.J.; Sawa, A.; Mortensen, P.B. Schizophrenia. *Lancet* **2016**, *388*, 86–97. [CrossRef]
23. Morgan, C.J.; Curran, H.V. Acute and chronic effects of ketamine upon human memory: A review. *Psychopharmacology* **2006**, *188*, 408–424. [CrossRef]

24. Leuchter, A.F.; Hunter, A.M.; Krantz, D.E.; Cook, I.A. Rhythms and blues: Modulation of oscillatory synchrony and the mechanism of action of antidepressant treatments. *Ann. N. Y. Acad. Sci.* **2015**, *1344*, 78–91. [CrossRef]
25. Ippolito, G.; Bertaccini, R.; Tarasi, L.; Di Gregorio, F.; Trajkovic, J.; Battaglia, S.; Romei, V. The Role of Alpha Oscillations among the Main Neuropsychiatric Disorders in the Adult and Developing Human Brain: Evidence from the Last 10 Years of Research. *Biomedicines* **2022**, *10*, 3189. [CrossRef]
26. Di Gregorio, F.; La Porta, F.; Petrone, V.; Battaglia, S.; Orlandi, S.; Ippolito, G.; Romei, V.; Piperno, R.; Lullini, G. Accuracy of EEG Biomarkers in the Detection of Clinical Outcome in Disorders of Consciousness after Severe Acquired Brain Injury: Preliminary Results of a Pilot Study Using a Machine Learning Approach. *Biomedicines* **2022**, *10*, 1897. [CrossRef]
27. Sankoorikal, G.M.; Kaercher, K.A.; Boon, C.J.; Lee, J.K.; Brodkin, E.S. A mouse model system for genetic analysis of sociability: C57BL/6J versus BALB/cJ inbred mouse strains. *Biol. Psychiatry* **2006**, *59*, 415–423. [CrossRef]
28. Akillioglu, K.; Melik, E.B.; Melik, E.; Boga, A. Effect of ketamine on exploratory behaviour in BALB/C and C57BL/6 mice. *Pharmacol. Biochem. Behav.* **2012**, *100*, 513–517. [CrossRef]
29. Brinks, V.; van der Mark, M.; de Kloet, R.; Oitzl, M. Emotion and cognition in high and low stress sensitive mouse strains: A combined neuroendocrine and behavioral study in BALB/c and C57BL/6J mice. *Front. Behav. Neurosci.* **2007**, *1*, 8. [CrossRef]
30. Lin, J.C.; Chan, M.H.; Lee, M.Y.; Chen, Y.C.; Chen, H.H. N,N-dimethylglycine differentially modulates psychotomimetic and antidepressant-like effects of ketamine in mice. *Prog. Neuropsychopharmacol. Biol. Psychiatry* **2016**, *71*, 7–13. [CrossRef]
31. Lin, J.C.; Lee, M.Y.; Chan, M.H.; Chen, Y.C.; Chen, H.H. Betaine enhances antidepressant-like, but blocks psychotomimetic effects of ketamine in mice. *Psychopharmacology* **2016**, *233*, 3223–3235. [CrossRef]
32. Irifune, M.; Shimizu, T.; Nomoto, M. Ketamine-induced hyperlocomotion associated with alteration of presynaptic components of dopamine neurons in the nucleus accumbens of mice. *Pharmacol. Biochem. Behav.* **1991**, *40*, 399–407. [CrossRef]
33. Karwacki, Z.; Kowiański, P.; Moryś, J. General anaesthesia in rats undergoing experiments on the central nervous system. *Folia Morphol.* **2001**, *60*, 235–242.
34. Becker, A.; Peters, B.; Schroeder, H.; Mann, T.; Huether, G.; Grecksch, G. Ketamine-induced changes in rat behaviour: A possible animal model of schizophrenia. *Prog. Neuropsychopharmacol. Biol. Psychiatry* **2003**, *27*, 687–700. [CrossRef]
35. Trujillo, K.A.; Smith, M.L.; Sullivan, B.; Heller, C.Y.; Garcia, C.; Bates, M. The neurobehavioral pharmacology of ketamine: Implications for drug abuse, addiction, and psychiatric disorders. *ILAR J.* **2011**, *52*, 366–378. [CrossRef]
36. Hefner, K.; Holmes, A. Ontogeny of fear-, anxiety- and depression-related behavior across adolescence in C57BL/6J mice. *Behav. Brain Res.* **2007**, *176*, 210–215. [CrossRef]
37. Krystal, J.H.; Karper, L.P.; Seibyl, J.P.; Freeman, G.K.; Delaney, R.; Bremner, J.D.; Heninger, G.R.; Bowers, M.B., Jr.; Charney, D.S. Subanesthetic effects of the noncompetitive NMDA antagonist, ketamine, in humans. Psychotomimetic, perceptual, cognitive, and neuroendocrine responses. *Arch. Gen. Psychiatry* **1994**, *51*, 199–214. [CrossRef]
38. Xu, K.; Krystal, J.H.; Ning, Y.; Chen, D.C.; He, H.; Wang, D.; Ke, X.; Zhang, X.; Ding, Y.; Liu, Y.; et al. Preliminary analysis of positive and negative syndrome scale in ketamine-associated psychosis in comparison with schizophrenia. *J. Psychiatr. Res.* **2015**, *61*, 64–72. [CrossRef]
39. Schramm-Sapyta, N.L.; Walker, Q.D.; Caster, J.M.; Levin, E.D.; Kuhn, C.M. Are adolescents more vulnerable to drug addiction than adults? Evidence from animal models. *Psychopharmacology* **2009**, *206*, 1–21. [CrossRef]
40. Rocha, A.; Hart, N.; Trujillo, K.A. Differences between adolescents and adults in the acute effects of PCP and ketamine and in sensitization following intermittent administration. *Pharmacol. Biochem. Behav.* **2017**, *157*, 24–34. [CrossRef]
41. Bates, M.L.S.; Trujillo, K.A. Long-lasting effects of repeated ketamine administration in adult and adolescent rats. *Behav. Brain Res.* **2019**, *369*, 111928. [CrossRef]
42. Carola, V.; D'Olimpio, F.; Brunamonti, E.; Mangia, F.; Renzi, P. Evaluation of the elevated plus-maze and open-field tests for the assessment of anxiety-related behaviour in inbred mice. *Behav. Brain Res.* **2002**, *134*, 49–57. [CrossRef]
43. Tang, X.; Orchard, S.M.; Sanford, L.D. Home cage activity and behavioral performance in inbred and hybrid mice. *Behav. Brain Res.* **2002**, *136*, 555–569. [CrossRef]
44. Wiley, J.L.; Evans, R.L.; Grainger, D.B.; Nicholson, K.L. Age-dependent differences in sensitivity and sensitization to cannabinoids and 'club drugs' in male adolescent and adult rats. *Addict. Biol.* **2008**, *13*, 277–286. [CrossRef]
45. Parise, E.M.; Alcantara, L.F.; Warren, B.L.; Wright, K.N.; Hadad, R.; Sial, O.K.; Kroeck, K.G.; Iñiguez, S.D.; Bolaños-Guzmán, C.A. Repeated ketamine exposure induces an enduring resilient phenotype in adolescent and adult rats. *Biol. Psychiatry* **2013**, *74*, 750–759. [CrossRef]
46. Hetzler, B.E.; Wautlet, B.S. Ketamine-induced locomotion in rats in an open-field. *Pharmacol. Biochem. Behav.* **1985**, *22*, 653–655. [CrossRef]
47. Allain, F.; Minogianis, E.A.; Roberts, D.C.; Samaha, A.N. How fast and how often: The pharmacokinetics of drug use are decisive in addiction. *Neurosci. Biobehav. Rev.* **2015**, *56*, 166–179. [CrossRef]
48. McCarthy, L.E.; Mannelli, P.; Niculescu, M.; Gingrich, K.; Unterwald, E.M.; Ehrlich, M.E. The distribution of cocaine in mice differs by age and strain. *Neurotoxicol. Teratol.* **2004**, *26*, 839–848. [CrossRef]
49. Veilleux-Lemieux, D.; Castel, A.; Carrier, D.; Beaudry, F.; Vachon, P. Pharmacokinetics of ketamine and xylazine in young and old Sprague-Dawley rats. *J. Am. Assoc. Lab. Anim. Sci.* **2013**, *52*, 567–570.
50. Olney, J.W.; Farber, N.B. NMDA antagonists as neurotherapeutic drugs, psychotogens, neurotoxins, and research tools for studying schizophrenia. *Neuropsychopharmacology* **1995**, *13*, 335–345. [CrossRef]

51. Farber, N.B.; Olney, J.W. Drugs of abuse that cause developing neurons to commit suicide. *Brain Res. Dev. Brain Res.* **2003**, *147*, 37–45. [CrossRef]
52. Cho, H.K.; Kim, K.W.; Jeong, Y.M.; Lee, H.S.; Lee, Y.J.; Hwang, S.H. Efficacy of ketamine in improving pain after tonsillectomy in children: Meta-analysis. *PLoS ONE* **2014**, *9*, e101259. [CrossRef]
53. Colwell, C.S.; Cepeda, C.; Crawford, C.; Levine, M.S. Postnatal development of glutamate receptor-mediated responses in the neostriatum. *Dev. Neurosci.* **1998**, *20*, 154–163. [CrossRef]
54. Henson, M.A.; Roberts, A.C.; Salimi, K.; Vadlamudi, S.; Hamer, R.M.; Gilmore, J.H.; Jarskog, L.F.; Philpot, B.D. Developmental regulation of the NMDA receptor subunits, NR3A and NR1, in human prefrontal cortex. *Cereb. Cortex* **2008**, *18*, 2560–2573. [CrossRef]
55. Insel, T.R.; Miller, L.P.; Gelhard, R.E. The ontogeny of excitatory amino acid receptors in rat forebrain--I. N-methyl-D-aspartate and quisqualate receptors. *Neuroscience* **1990**, *35*, 31–43. [CrossRef]
56. Luo, J.; Bosy, T.Z.; Wang, Y.; Yasuda, R.P.; Wolfe, B.B. Ontogeny of NMDA R1 subunit protein expression in five regions of rat brain. *Brain Res. Dev. Brain Res.* **1996**, *92*, 10–17. [CrossRef]
57. Nowakowski, R.S. The mode of inheritance of a defect in lamination in the hippocampus of BALB/c mice. *J. Neurogenet.* **1984**, *1*, 249–258. [CrossRef]
58. Akillioglu, K.; Binokay, S.; Kocahan, S. The effect of neonatal N-methyl-D-aspartate receptor blockade on exploratory and anxiety-like behaviors in adult BALB/c and C57BL/6 mice. *Behav. Brain Res.* **2012**, *233*, 157–161. [CrossRef]
59. Jacome, L.F.; Burket, J.A.; Herndon, A.L.; Deutsch, S.I. D-Cycloserine enhances social exploration in the Balb/c mouse. *Brain Res. Bull.* **2011**, *85*, 141–144. [CrossRef]
60. Jacome, L.F.; Burket, J.A.; Herndon, A.L.; Cannon, W.R.; Deutsch, S.I. D-serine improves dimensions of the sociability deficit of the genetically-inbred Balb/c mouse strain. *Brain Res. Bull.* **2011**, *84*, 12–16. [CrossRef]
61. Burket, J.A.; Herndon, A.L.; Deutsch, S.I. Locomotor activity of the genetically inbred Balb/c mouse strain is suppressed by a socially salient stimulus. *Brain Res. Bull.* **2010**, *83*, 255–256. [CrossRef]
62. Martos, D.; Tuka, B.; Tanaka, M.; Vécsei, L.; Telegdy, G. Memory Enhancement with Kynurenic Acid and Its Mechanisms in Neurotransmission. *Biomedicines* **2022**, *10*, 849. [CrossRef]

biomedicines

MDPI

Article

Age-Related Effects of Exogenous Melatonin on Anxiety-like Behavior in C57/B6J Mice

Sofia Nasini [1,†], Sara Tidei [1,†], Atea Shkodra [2,3], Danilo De Gregorio [2,3], Marco Cambiaghi [4,‡] and Stefano Comai [1,4,5,6,*,‡]

1 Department of Pharmaceutical and Pharmacological Sciences, University of Padova, 35131 Padova, Italy; sofia.nasini@phd.unipd.it (S.N.); saratidei@gmail.com (S.T.)
2 IRCCS San Raffaele Scientific Institute, 20132 Milan, Italy; shkodra.atea@hsr.it (A.S.); degregorio.danilo@hsr.it (D.D.G.)
3 School of Medicine, Vita Salute San Raffaele University, 20132 Milan, Italy
4 Department of Neurosciences, Biomedicine and Movement Sciences, University of Verona, 37134 Verona, Italy; marco.cambiaghi@univr.it
5 Department of Biomedical Sciences, University of Padova, 35131 Padova, Italy
6 Department of Psychiatry, McGill University, Montreal, QC H3A 1A1, Canada
* Correspondence: stefano.comai@unipd.it
† These authors contributed equally to this work and should be considered co-first authors.
‡ These authors contributed equally to this work and should be considered co-senior authorship.

Abstract: The synthesis of melatonin (MLT) physiologically decreases during aging. Treatment with MLT has shown anxiolytic, hypnotic, and analgesic effects, but little is known about possible age-dependent differences in its efficacy. Therefore, we studied the effects of MLT (20 mg/kg, intraperitoneal) on anxiety-like behavior (open field (OFT), elevated plus maze (EPMT), three-chamber sociability, and marble-burying (MBT) tests), and the medial prefrontal cortex (mPFC)-dorsal hippocampus (dHippo) circuit in adolescent (35–40 days old) and adult (three-five months old) C57BL/6 male mice. MLT did not show any effect in adolescents in the OFT and EPMT. In adults, compared to vehicles, it decreased locomotor activity and time spent in the center of the arena in the OFT and time spent in the open arms in the EPMT. In the MBT, no MLT effects were observed in both age groups. In the three-chamber sociability test, MLT decreased sociability and social novelty in adults, while it increased sociability in adolescents. Using local field potential recordings, we found higher mPFC-dHippo synchronization in the delta and low-theta frequency ranges in adults but not in adolescents after MLT treatment. Here, we show age-dependent differences in the effects of MLT in anxiety paradigms and in the modulation of the mPFC-dHippo circuit, indicating that when investigating the pharmacology of the MLT system, age can significantly impact the study outcomes.

Keywords: melatonin; anxiety; mice; adolescents; adults; local field potentials; hippocampus; medial prefrontal cortex; sociability; aging

Citation: Nasini, S.; Tidei, S.; Shkodra, A.; De Gregorio, D.; Cambiaghi, M.; Comai, S. Age-Related Effects of Exogenous Melatonin on Anxiety-like Behavior in C57/B6J Mice. *Biomedicines* **2023**, *11*, 1705. https://doi.org/10.3390/biomedicines11061705

Academic Editors: Masaru Tanaka and Zhenglin Gu

Received: 29 April 2023
Revised: 2 June 2023
Accepted: 9 June 2023
Published: 13 June 2023

1. Introduction

Anxiety disorders, including, among others, generalized anxiety disorder, social anxiety disorder, different phobias, obsessive-compulsive disorder, and panic disorder, are one of the major health problems of modern societies [1–3]. According to the World Health Organization, in 2019, more than 300 million people, of which 58 million were children and adolescents, were suffering from an anxiety disorder, and their prevalence increased by 26% in 2020 due to the COVID-19 pandemic [1]. The neurobiology of anxiety is still largely unknown [4,5], and the currently available medications, mostly benzodiazepines and antidepressants, induce several side effects and have many contraindications, especially when used in children, adolescents, and elderlies.

Melatonin (MLT) is a neuromodulator widely used in the world as an over-the-counter compound in both adolescents and adults for its anxiolytic and sedative/hypnotic properties despite the lack of clear evidence of activity [6,7]. MLT has shown anxiolytic-like effects in rodents in different behavioral paradigms of anxiety [8–13], as well as anxiolytic effects in humans, especially for reducing preoperative and postoperative anxiety in both children and adults [12,14,15]. The MLT system undergoes physiological changes from infants to adolescents, adults, and elderly. In particular, the synthesis of MLT in the pineal gland and, consequently, the peak of circulating levels occurring in the middle of the night is high between 5–10 years of age and then progressively declines with aging [16]. Although, as mentioned above, MLT is largely used in the population of all ages, and the MLT system undergoes changes according to aging, few preclinical and clinical studies have investigated possible age-dependent effects of MLT. MLT acts mainly by activating its two G-protein-coupled receptors named MT_1 and MT_2 [17], which display complementary or opposite effects in both the central nervous system and the periphery [18–20]. Concerning anxiety, preclinical data seem to indicate that the MLT receptor subtype most implicated is MT_2 [10,21–23]. Indeed, it has been found that (1) the selective MT_2 receptor's partial agonist UCM765 induces anxiolytic-like effects in rats in different preclinical paradigms of anxiety similar to those of MLT, which are blocked by the selective MT_2 antagonist 4P-PDOT [10]; (2) MT_2 receptors in knockout mice display altered levels of behaviors related to the anxiety spectrum [21,22,24]; and (3) activation of MT_2 receptors in the striatum produces anxiolytic-like effects in animal models of Parkinson's disease, which is characterized by high comorbidity with anxiety disorders (more than 50% of the affected individuals) [23]. It is important to mention that while changes in circulating levels of MLT according to aging have been shown, there is no information on whether the expression of MT_1 and MT_2 receptors in the different regions of the brain may also change during development and aging. If this occurs, it is plausible that the pharmacological effects induced by MLT could also vary depending on aging. For this reason, in this work, we investigated whether treatment with MLT induced different effects in adolescent (35–40 day-old) and adult (3–4-month-old) male mice when tested in behavioral paradigms covering the spectrum of anxiety disorders and in the oscillatory synchrony between the medial prefrontal cortex (mPFC) and the dorsal hippocampus (dHippo), two regions of the brain involved in anxiety and widely expressing both MT_1 and MT_2 receptors [25,26]. We decided to test MLT at the dose of 20 mg/kg since we previously found—in male adult rats—that it had anxiolytic-like activity in the elevated plus maze test (EPMT) and novelty-suppressed feeding test (NFST) without inducing any effect in the open field test (OFT) [10].

2. Materials and Methods

2.1. Experimental Design

Mice were habituated to the testing room by transferring them to the testing room 30 min prior to the beginning of the trials. All behavioral and in vivo electrophysiology tests were performed between 9:00 am and 4:00 pm. After testing, each mouse was removed from the apparatus and returned to its home cage, and all interior surfaces were thoroughly cleaned with 70% ethanol and then wiped dry to remove any trace of conspecific odor. Three groups of mice per treatment (vehicle or MLT) and age (adolescents or adults) were used. One group was tested in the open field test and then one week later in the elevated plus maze test, whereas the second group was tested in the three-chamber sociability test and then one week later in the marble burying test. We left one week between the two tests in each group to minimize the possible effects of one test over the other and to have the washout from MLT. The animals were randomized to each experimental session for their treatment (vehicle or MLT). The behavior was videotaped using an LCD camera connected to control and recording equipment. Automated tracking of the mice was achieved using ANY-maze software (Stoelting Europe, Dublin, Ireland). The third group was used for in vivo local field potential (LFP) recordings after implanting the electrodes into the two brain regions of interest, the medial prefrontal cortex (mPFC) and the dorsal hippocampus (dHippo). At the time of the behavioral and electrophysiology analyses, the experimenter was blind to the treatment received by each individual mouse.

2.2. Animals

Male C57BL/6J mice used for these experiments were reared in breeding colonies of the Department of Pharmaceutical and Pharmacological Sciences (University of Padua). The animals were kept in a temperature-controlled room (22 °C) on a 12:12 h light–dark cycle (light on at 7:00 AM) and fed a standard pellet diet and tap water ad libitum. Mature adult mice ranged in age from 3 to 5 months (n = 10–15), and adolescent mice between 35 and 40 days of age (n = 10–15) were used for the experimental procedures. All experimental protocols were performed after authorization from the Animal Care and Use Ethics Committee of the University of Padova and the Italian Ministry of Health and were in compliance with national and European guidelines for the handling and use of experimental animals.

2.3. Treatment

MLT (CAS Number: 73-31-4, Cayman Chemical Co., Ann Arbor, MI, USA) was used at the dose of 20 mg/kg based on previous research [10,27], and it was dissolved in a vehicle composed of 30% saline and 70% dimethyl sulfoxide (DMSO; Sigma–Aldrich, Steinheim, Germany). Each mouse received a single intraperitoneal (I.P.) injection (total volume 0.1 mL) of vehicle or MLT (20 mg/kg) 10 min before each behavioral or in vivo electrophysiology test.

2.4. Behavioral Testing

2.4.1. Open-Field Test (OFT)

The OFT to measure exploratory activity and locomotion was performed according to standardized protocols in the laboratory [28]. Briefly, mice were individually placed in the corner of a grey-painted open field arena (40 × 40 × 15 cm) and left to explore freely for 20 min. The experiment took place under standard room lighting (350 lx); a white lamp (100 W) was suspended 2 m above the arena. Anxiety-like behavior was measured by the frequency and total duration of visits to the central zone (20 × 20 cm) of the arena. Other ethological measures analyzed included grooming, rearing, and locomotor activity (total distance traveled).

2.4.2. Elevated Plus Maze Test (EPMT)

The EPM to assess anxiety-related behaviors relies on rodents' proclivity toward dark, enclosed spaces (approach) and an unconditioned fear of heights and open spaces (avoidance) [29]. It is plus-shaped, with two open arms (25 × 5 × 0.5 cm) and two enclosed arms (25 × 5 × 16 cm) with a central platform (5 × 5 × 0.5 cm). The closed arms are enclosed by two high walls (16 cm), whereas the open arms have no side wall. The apparatus was elevated to a height of 50 cm from the floor and weakly illuminated (350 lx). The walls of the enclosed arms were painted medium grey. Animals were placed in the center of the plus-maze and allowed to explore freely on the apparatus for 5 min. The time spent and number of entries into the open arm, as well as the time spent in the closed arm of the plus-maze, were measured according to our previous report [30].

2.4.3. Three-Chamber Sociability Test

Social anxiety was measured following our previous method in the three-chamber sociability test [28,31]. In this test, mice were first left free to explore the three-chamber apparatus for 10 min. Then, in each of the lateral chambers, an up-turned metal-grid pencil cup was placed: one remained empty as the novel object (O), while an age- and sex-matched WT stranger mouse (S1) was placed in the second up-turned cup. The stranger mouse was previously habituated to the cups for 3 × 10 min sessions. The testing mouse was left 10 min to explore the apparatus and to interact with either O or S1 (Sociability phase). Finally, to test for social preference, mice were presented for another 10 min with the choice of object (O), which now contained a second age- and sex-matched stranger mouse or the now familiar mouse (S1). Sociability and social novelty were determined manually by assessing the time spent by the tested mouse actively interacting with O or S1 in the sociability and social novelty phases, respectively.

2.4.4. Marble Burying Test

Marble burying test is used to depict anxiety or obsessive–compulsive disorder (OCD) behavior. It is based on the observation that mice will bury either harmful or harmless objects in their bedding [32]. This behavior is a correlational model for the detection of anxiolytics rather than an isomorphic model of anxiety [33]. Each mouse was placed in a cage filled approximately 5 cm deep with wood chip bedding, lightly tamped to make a flat, even surface, and left there for 30 min for habituation. Twenty glass marbles were then placed in a regular pattern, evenly spaced. The number of marbles buried (for at least 2/3 of the area) with bedding was counted to measure the obsessive–compulsive behavior.

2.5. *In Vivo LFP Recordings and Analysis*

Extracellular field potentials were recorded in freely moving mice in a 20 × 30 × 30 cm box 10 min after vehicle or 20 mg/kg MLT I.P. injection to examine the oscillatory synchrony between the mPFC and the dHippo in the two conditions. Following a standard procedure in the laboratory [34], stainless steel insulated wires (∅ 135 μm) were stereotaxically implanted unilaterally (right side) in the mPFC and the dHippo according to the following coordinates, in mm from bregma: mPFC, +1.8 AP, 0.3 ML, −2.4 DV and hippocampus, −2.1 AP, 1.5 ML, −1.4 DV. A screw over contralateral parietal areas served as a common reference and ground. All implants were secured using dental cement (Ketacem). After surgery, mice were allowed to recover for 5–6 days before testing [35]. LFPs were recorded and initially digitalized at 1 kHz and stored on a hard drive for offline analysis. LFP epochs were visually examined, and artifact-free segments were computed by analyzing 3 segments of 2 s each during the recording sessions after both vehicle and MLT 20 mg/kg injections. The coherence between LFP channels was measured by magnitude squared coherence (MSC), using the function mscohere in Matlab signal toolbox, which is a coherence estimate of the input signals x and y by using Welch's averaged, modified periodogram method. The MSC estimate is a function of frequency with values between 0 and 1 and indicates how well x corresponds to y at each frequency. The MSC estimate was calculated over the frequency range of 0.5–30 Hz for each mouse with a frequency resolution of 0.5 Hz. To test whether coherence values were significantly higher than expected by chance, we performed a permutation test in which coherence values were compared before inclusion in additional analyses with a shuffle procedure in which epochs were randomly shifted 5–10 s relative to each other. This process was repeated 1000 times to obtain the distribution of coherence expected by chance [34]. Differences in coherence were obtained by comparing coherence values (20 mg/kg MLT vs. vehicle), and statistics were performed on the normalized coherence within the frequency bands of interest: delta (1–4 Hz), low-theta (4–8 Hz), high-theta (8–12 Hz), beta (12–30 Hz) [34].

2.6. *Statistical Analysis*

Statistical analyses were conducted using GraphPad 8.0 (GraphPad Software, La Jolla, CA, USA) software. The normal distribution of data was verified with the Shapiro–Wilk test. Two-tailed unpaired Student's t-test was performed for the Open Field test, stereotypic behaviors, EPM, MBT, and coherence bands. Two-way ANOVA for repeated measures followed by Bonferroni post-hoc analysis was used for the three-chamber test. $p < 0.05$ was considered statistically significant. Data were presented as mean ± SEM. Tables 1 and 2 report statistical details for the different experiments in adolescent and adult mice, respectively.

Table 1. Statistical details for the behavioral and in vivo electrophysiology experimental comparisons between vehicle- and 20 mg/kg MLT-treated adolescent mice.

Figure	Panel		Test	Group-Size	Statistic	*p* Value	Pair-Wise Comparison	Statistic 2		
1	A	Open field test	Student's *t*-test	Vehicle = 11 mice	locomotor activity t = 0.9675; df = 21	*p* = 0.3443	N/A	N/A		
					time spent in centre t = 0.03147; df = 21	*p* = 0.9752				
				MLT 20 mg/kg = 10 mice	entries in centre t = 0.193; df = 21	*p* = 0.8488				
	B	Stereotypic behaviours	Student's *t*-test	Vehicle = 11 mice	grooming events t = 2.18; df = 20	*p* = 0.0414	N/A	N/A		
					time of grooming t = 1.167; df = 20	*p* = 0.257				
				MLT 20 mg/kg = 11 mice	rearing events t = 0.1577; df = 20	*p* = 0.8763				
					time of rearing t = 0.1773; df = 20	*p* = 0.8610				
	C	Elevated Plus Maze	Student's *t*-test	Vehicle = 13 mice	entries in open arms t = 0.3292; df = 23	*p* = 0.745	N/A	N/A		
				MLT 20 mg/kg = 12 mice	time in open arms t = 0.4702; df = 23	*p* = 0.6427				
					time in close arms t = 1.469; df = 23	*p* = 0.1553				
	D	3-chamber sociability test	Two-way ANOVA	Vehicle = 12 mice	Sociability:			**Test Details**	**t**	***p* Value**
					interaction F (1,17) = 8.171	*p* = 0.0109	Bonferroni post hoc comparison	empty cage vehicle vs. empty cage MLT 20 mg/kg	0.0695	*p* > 0.9999
					treatment F(1,17) = 2.886	*p* = 0.1076		mouse-1 vehicle vs. mouse-1 MLT 20 mg/kg	2.953	*p* = 0.0113
					sociability F (1,17) = 73.49	*p* < 0.0001		vehicle empty cage vs. vehicle mouse-1	4.707	*p* = 0.0004
								MLT 20 mg/kg empty cage vs. MLT 20 m/kg mouse-1	7.192	*p* < 0.0001
				MLT 20 mg/kg = 7 mice	Social novelty:		N/A	N/A		
					interaction F (1,17) = 0.1712	*p* = 0.6842				
					treatment F (1,17) = 4.570	*p* = 0.0474				
					social novelty F (1,17) = 22.93	*p* < 0.0001				

Table 1. *Cont.*

Figure	Panel		Test	Group-Size	Statistic	*p* Value	Pair-Wise Comparison	Statistic 2
	E	Marble burying test	Student's *t*-test	Vehicle = 15 mice; MLT 20 mg/kg = 10 mice	t = 1.454; df = 23	*p* = 0.1594	N/A	N/A
2	C	mPFC-dHippo coherence	Two-way ANOVA	Vehicle = 4 mice	interaction F (58,354) = 0.8933	*p* = 0.6937	N/A	N/A
					frequency F (58,354) = 3.410	*p* < 0.0001		
				MLT 20 mg/kg = 4 mice	treatment F (58,354) = 0.1881	*p* = 0.6648		
	D	mPFC-dHippo coherence	Student's *t*-test	Vehicle = 4 mice	1–4 Hz t = 0.4544; df = 6	*p* = 0.6655	N/A	N/A
					4–8 Hz t = 0.1669; df = 6	*p* = 0.8729		
				MLT 20 mg/kg = 4 mice	8–12 Hz t = 0.8479; df = 6	*p* = 0.4290		
					12–30 Hz t = 0.6128; df = 6	*p* = 0.5625		

Table 2. Statistical details for the behavioral and in vivo electrophysiology experimental comparisons between vehicle- and 20 mg/kg MLT-treated adult mice.

Figure	Panel		Test	Group-Size	Statistic	*p* Value	Pair-Wise Comparison	Statistic 2
3	A	Open field test	Student's *t*-test	Vehicle = 11 mice	locomotor activity t = 3.011; df = 19	*p* = 0.0072	N/A	N/A
					time spent in centre t = 3.096; df = 19	*p* = 0.006		
				MLT 20 mg/kg = 10 mice	entries in centre t = 3.319; df = 19	*p* = 0.0036		
	B	Stereotypic behaviours	Student's *t*-test	Vehicle = 10 mice	grooming events t = 3.138; df = 19	*p* = 0.0054	N/A	N/A
					time of grooming t = 2.336; df = 19	*p* = 0.0306		
				MLT 20 mg/kg = 11 mice	rearing events t = 1.102; df = 19	*p* = 0.2843		
					time of rearing t = 1.377; df = 19	*p* = 0.1844		

Table 2. *Cont.*

Figure	Panel		Test	Group-Size	Statistic	*p* Value	Pair-Wise Comparison	Statistic 2		
	C	Elevated Plus Maze	Student's *t*-test	Vehicle = 13 mice;	entries in open arms t = 1.32; df = 21	$p = 0.2011$	N/A	N/A		
				MLT 20 mg/kg = 10 mice	time in open arms t = 3.260; df = 21	$p = 0.0037$				
					time in close arms t = 0.177; df = 21	$p = 0.8612$				
	D	3-chamber sociability test	Two-way ANOVA	Vehicle = 11 mice	Sociability:			**Test Details**	**t**	***p* Value**
					interaction F(1,19) = 9.176	$p = 0.0069$	Bonferroni post hoc comparison	vehicle empty cage vs. vehicle mouse-1	7.172	$p < 0.0001$
					treatment F(1,19) = 26.29	$p < 0.0001$		MLT 20 mg/kg empty cage vs. MLT 20 mg/kg mouse-1	3.132	$p = 0.0110$
					sociability F(1,19) = 54.06	$p < 0.0001$		empty cage vehicle vs. empty cage MLT 20 mg/kg	2.53	$p = 0.0314$
								mouse-1 vehicle vs mouse-1 MLT 20 mg/kg	5.943	$p < 0.0001$
				MLT 20 mg/kg = 11 mice	Social novelty:			**Test Details**	**t**	***p* Value**
					interaction F(1,19) = 9.559	$p = 0.006$	Bonferroni post hoc comparison	vehicle mouse-1 vs. vehicle mouse-2	5.729	$p < 0.0001$
					treatment F(1,19) = 26.46	$p < 0.0001$		MLT 20 mg/kg mouse-1 vs. MLT 20 mg/kg mouse-2	1.529	$p = 0.2857$
					social novelty F(1,19) = 27.06	$p < 0.0001$		mouse-1 vehicle vs. mouse-1 MLT 20 mg/kg	3.11	$p = 0.0071$
								mouse-2 vehicle vs. mouse-2 MLT 20 mg/kg	6.246	$p < 0.0001$
	E	Marble burying test	Student's *t*-test	Vehicle = 12 mice; MLT 20 mg/kg = 12 mice	t = 0.6125; df = 22	$p = 0.5465$	N/A	N/A		
4	C	mPFC-dHippo coherence	Two-way ANOVA	Vehicle = 4 mice	interaction F (1,354)	$p = 0.3028$	N/A	N/A		
					frequency F (1,354)	$p = 0.0008$				
				MLT 20 mg/kg = 4 mice	treatment F (1,354)	$p = 0.0025$				
	D	mPFC-dHippo coherence	Student's *t*-test	Vehicle = 4 mice	1–4 Hz t = 2.613; df = 6	$p = 0.04$	N/A	N/A		
					4–8 Hz t = 2.553; df = 6	$p = 0.0433$				
				MLT 20 mg/kg = 4 mice	8–12 Hz t = 0.001256; df = 6	$p = 0.9990$				
					12–30 Hz t = 0.3003; df = 6	$p = 0.7741$				

3. Results

3.1. Adolescent Mice

3.1.1. Evaluation of the Effects of MLT on Anxiety-like Behaviors: OFT, Stereotypic Behaviors and EPMT

Evaluation of anxiety-like behavior was first conducted using the OFT, in which we also observed two stereotypic behaviors, grooming, and rearing. In adolescent mice (Figure 1A), the two-tailed unpaired Student's *t*-test showed no differences between mice treated with vehicle and 20 mg/kg MLT in the locomotor activity, in the time spent in the center of the arena and in the total number of entries in the center. The number of grooming events was higher in 20 mg/kg MLT-treated adolescents; instead, there was no difference in the number of rearing events. The total duration of both grooming and rearing events did not vary between adolescent mice treated with vehicle or 20 mg/kg MLT (Figure 1B).

The EPMT allows evaluation of the possible anxiolytic-like activity of a psychoactive compound by determining the time spent in the open arms of the apparatus and the number of entries in the open arms that represent a place where mice feel exposed to danger. No effects of 20 mg/kg MLT were observed concerning the time spent in open and closed arms and the number of entries in the open arms (Figure 1C).

3.1.2. Evaluation of the Effects of MLT on Sociability: Three-Chambers Sociability Test

In adolescents (Figure 1D), the two-way ANOVA for repeated measures analysis for the sociability stage resulted in a significant interaction between treatment and sociability and an effect of sociability, but no effect of treatment with both groups of animals spending more time interacting with the novel animal than the empty cage. Furthermore, MLT-treated mice interacted for a longer time with the novel animal than vehicle-treated mice. In the second stage of the test, there was no interaction between factors, but there was an effect of the social novelty, as both MLT-treated mice and vehicle-treated mice interacted more with the mouse-2 than mouse-1; moreover, there was an effect of treatment with animals treated with MLT interacting longer with both mouse 1 and mouse 2 than mice receiving vehicle.

3.1.3. Evaluation of the Effects of MLT on Obsessive–Compulsive Disorder (OCD) Behavior: Marble Burying Test

Obsessive–compulsive behavior, represented by repetitive actions, can be evaluated through the Marble Burying test by calculating the percentage of marbles buried. No statistically significant difference in the percent of marbles buried emerged from the two-tailed unpaired Student's *t*-test between vehicle-treated and 20 mg/kg MLT-treated mice (Figure 1E).

3.1.4. In Vivo Electrophysiology

A coherence analysis based on LFP recordings was used to measure the functional connectivity among different brain areas. We measured the effects of vehicle and MLT on the mPFC-dHippo synchrony in the adolescent mice (Figure 2A,B) and observed no overall changes in the spectrum due to MLT treatment but higher coherence values in the low-theta range in both vehicle and MLT treated mice (Figure 2C). Moreover, no difference between treatment with vehicle and MLT was found in the coherence levels for all the examined frequency bands (Figure 2D).

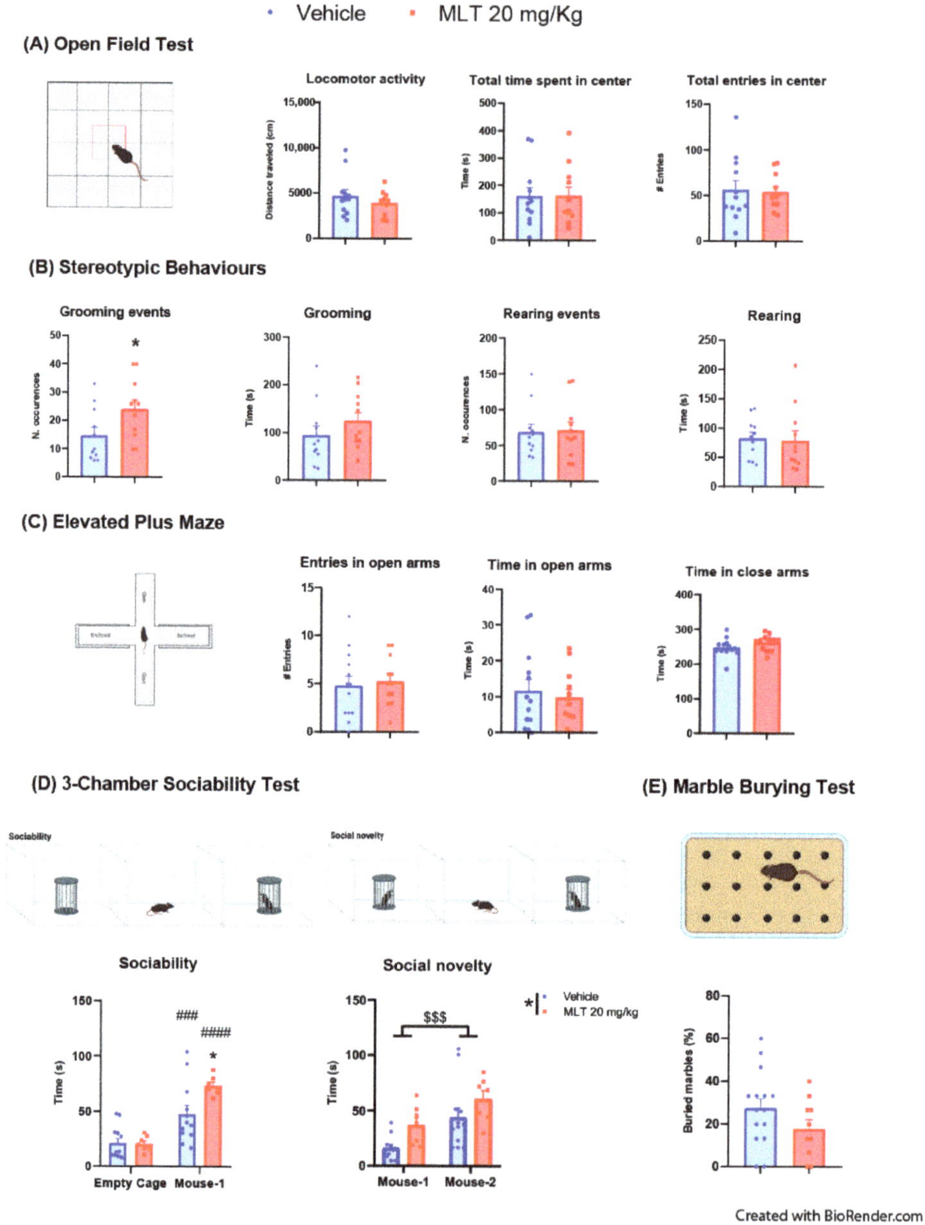

Figure 1. Behavioral testing in adolescent mice performed 10 min after the treatment with vehicle or 20 mg/kg melatonin (MLT). (**A**) Open Field test: the locomotor activity, the time spent in the central zone of the arena, and the number of entries in the center of the arena were evaluated during 20 min. (**B**) The number of occurrences and the duration of grooming and rearing (stereotypic behaviors) were measured during the 20 min Open Field test. Grooming events: statistical analyses were performed using the two-tailed unpaired Student's *t*-test; * $p < 0.05$ vehicle vs. 20 mg/kg. (**C**) Elevated Plus Maze test: effect of vehicle or MLT (20 mg/kg) on the number of entries in the open arms, the time spent in open arms, and the time spent in closed arms. (**D**) Three-Chamber Sociability test: the interaction time

of mice treated with vehicle or MLT (20 mg/kg) was measured during both the sociability (empty cage vs. cage with an unfamiliar conspecific mouse (mouse 1)) and social novelty (cage with the familiar conspecific mouse from the previous phase (mouse 1) vs. cage with a second unfamiliar conspecific mouse (mouse 2)). Statistical analyses were performed using the two-way ANOVA for repeated measures followed by Bonferroni's multiple comparison test; * $p < 0.05$ vehicle vs. MLT 20 mg/kg, ### $p < 0.001$ vehicle (empty cage vs. mouse-1), #### $p < 0.0001$ MLT 20 mg/kg (empty cage vs. mouse-1), $$$ $p < 0.001$ social novelty (mouse-1 vs. mouse-2). (**E**) Marble Burying test: the number of marbles buried during the 30 min test were measured, and the percentage of buried marbles was calculated. All datasets are represented as mean ± SEM; individual mice are represented as dots (vehicle) or squares (MLT).

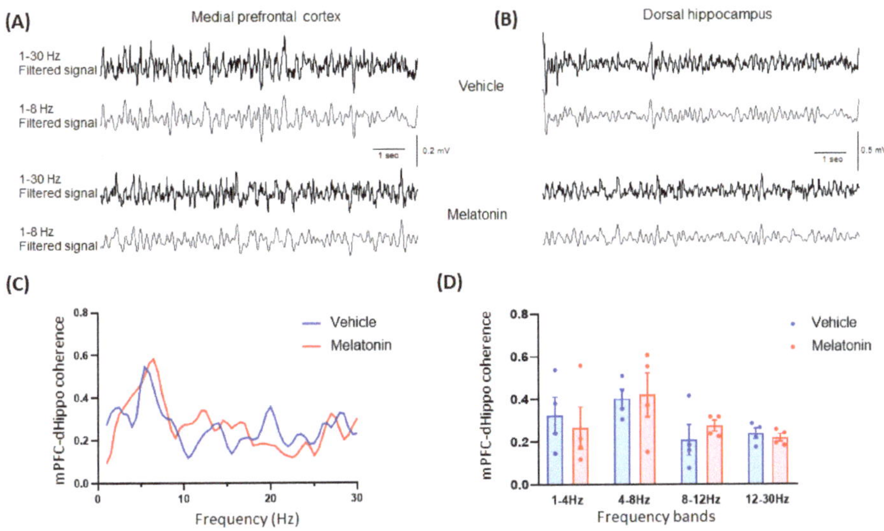

Figure 2. Medial prefrontal cortex (mPFC)-dorsal hippocampus (dHippo) synchrony in adolescent mice. Example of LFP traces (1–30 Hz filtered signal in black; 1–8 Hz filtered signal in grey) recorded in mPFC (**A**) and in dHippo (**B**) of an adolescent mouse 10 min after the treatment with vehicle (top) or 20 mg/kg melatonin (bottom). No effect of the treatment with MLT was observed in (**C**) the mPFC-dHippo coherence spectrum and in (**D**) all the analyzed frequency bands. Data are represented as mean ± SEM (n = 4).

3.2. Adult Mice

3.2.1. Evaluation of the Effects of MLT on Anxiety-like Behaviors: OFT, Stereotypic Behaviors and Epmt

In the OFT (Figure 3A), adult mice treated with 20 mg/kg MLT displayed lower locomotor activity, a lower time spent in the central zone of the arena, and a lower number of entries in the center compared to adult mice treated with vehicle. These results indicate that MLT at the dose of 20 mg/kg in our experimental conditions induced a sedative-like state in adult mice. Adult animals (Figure 3B) treated with 20 mg/kg MLT decreased the number (and the duration of grooming events) in keeping with a sedative-like activity of MLT at the tested dose. On the other hand, the number and duration of rearing events were not significantly affected by 20 mg/kg MLT.

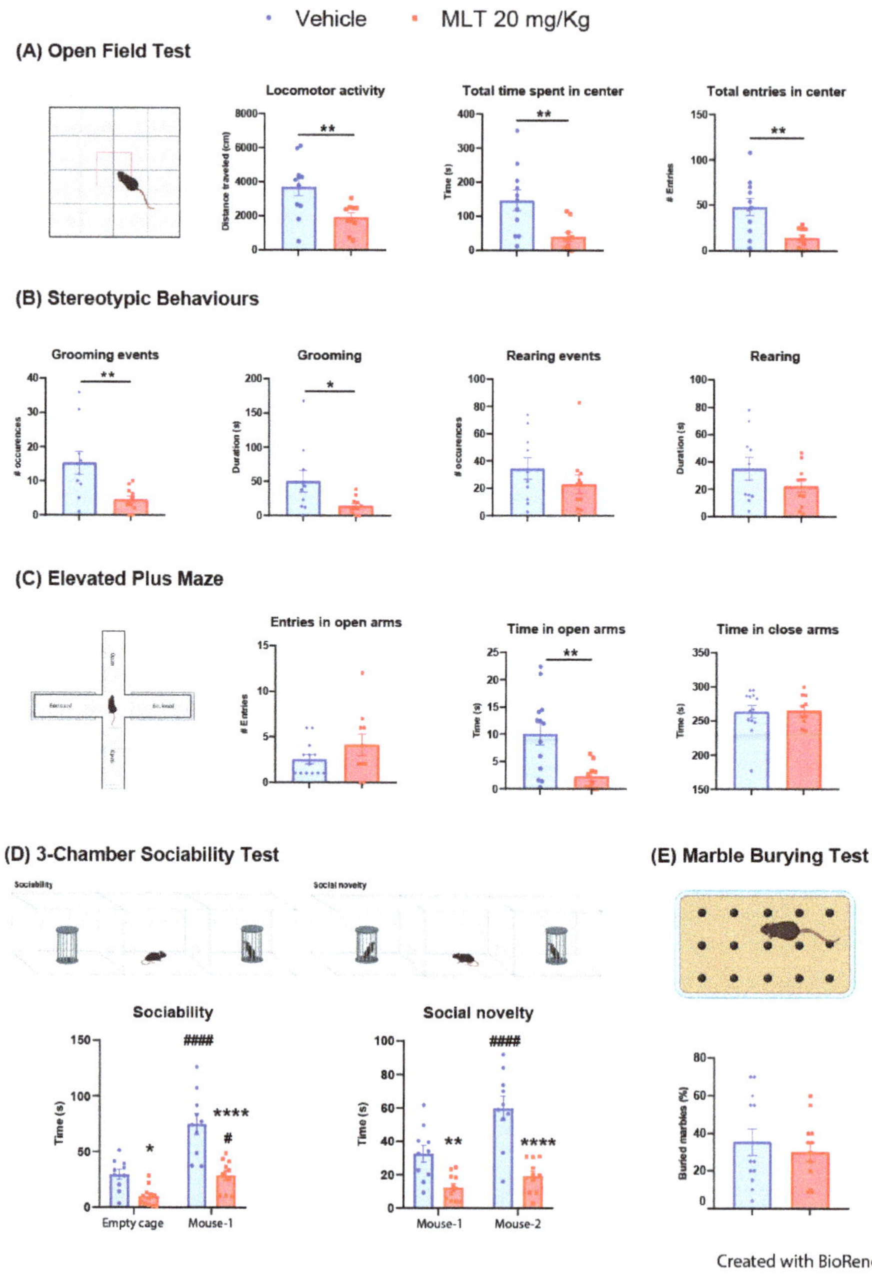

Figure 3. Behavioral testing in adult mice performed 10 min after the treatment with vehicle or 20 mg/kg melatonin (MLT). (**A**) Open Field test: the locomotor activity, the time spent in the central zone of the arena, and the number of entries in the center, were evaluated over 20 min. Statistical analyses were performed using the two-tailed unpaired Student's *t*-test; ** $p < 0.01$ vehicle vs. 20 mg/kg MLT. (**B**) The number of occurrences and the duration of grooming and rearing (stereotypic behaviors) were measured during the 20 min of the Open Field test. Statistical analyses were performed using

the two-tailed unpaired Student's *t*-test; * $p < 0.05$, ** $p < 0.01$ vehicle vs. 20 mg/kg MLT. (**C**) Elevated Plus Maze test: effect of vehicle or MLT (20 mg/kg) on the number of entries in the open arms, the time spent in open arms, and the time spent in closed arms. Statistical analyses were performed using the two-tailed unpaired Student's *t*-test; ** $p < 0.01$ vehicle vs. 20 mg/kg MLT. (**D**) Three-Chamber Sociability test: the interaction time of the mice treated with vehicle or MLT (20 mg/kg) was measured during both the sociability (empty cage vs. cage with an unfamiliar conspecific mouse (mouse 1)) and social novelty (cage with the familiar conspecific mouse from the previous phase (mouse 1) vs. cage with a second unfamiliar conspecific mouse (mouse 2)). Statistical analyses were performed using the two-way ANOVA for repeated measures followed by Bonferroni's multiple comparison test; * $p < 0.05$, ** $p < 0.01$, **** $p < 0.0001$ vehicle vs. 20 mg/kg MLT; # $p < 0.05$, #### $p < 0.0001$ empty cage vs. mouse 1, or mouse 1 vs. mouse 2. (**E**) Marble Burying test: the number of marbles buried during the 30 min of the test were measured, and the percentage of buried marbles was calculated. All datasets are represented as ± SEM, individual mice are represented as dots (vehicle) or squares (MLT).

In the EPMT, adult mice (Figure 3C) treated with 20 mg/kg MLT spent significantly less time in the open arms with respect to vehicle-treated mice. The number of entries in the open arms and the time spent in closed arms were not significant comparing adult mice treated with vehicle and MLT (20 mg/kg).

3.2.2. Evaluation of the Effects of MLT on Sociability: Three-Chambers Sociability Test

Two-way ANOVA for repeated measures analysis of sociability in adult mice (Figure 3D) showed an interaction between treatment and sociability and an effect of treatment and sociability. The 20 mg/kg MLT significantly reduced the time of interaction with the empty cage as well as the time of interaction with the familiar mouse compared to vehicle-treated mice. However, we found that both mice treated with vehicle and 20 mg/kg MLT interacted longer with mouse 1 than with the empty cage. In the social novelty stage, the two-way ANOVA for repeated measures analysis resulted in an interaction between treatment x social novelty and an effect of treatment and social novelty. In particular, we observed a reduction in the time of interaction of MLT-treated mice with familiar and unfamiliar mice compared to vehicle-treated mice. Finally, unlike mice treated with 20 mg/kg MLT, we found that mice treated with vehicle interacted significantly longer with mouse 2 than mouse 1.

3.2.3. Evaluation of the Effects of MLT on Obsessive–Compulsive Disorder (OCD) Behavior: Marble Burying Test

We did not find any effect of 20 mg/kg MLT on the percentage of marble buried by the mice during the 30 min test (Figure 3E).

3.2.4. In Vivo Electrophysiology

In adult mice, the effects of MLT on mPFC-dHippo synchronization led to an increased coherence at low frequencies below 10 Hz (Figure 4A–C). Indeed, compared with vehicle, the MLT treatment resulted in a more synchronized activity in the mPFC and dHippo at delta and lower theta frequencies (Figure 4D).

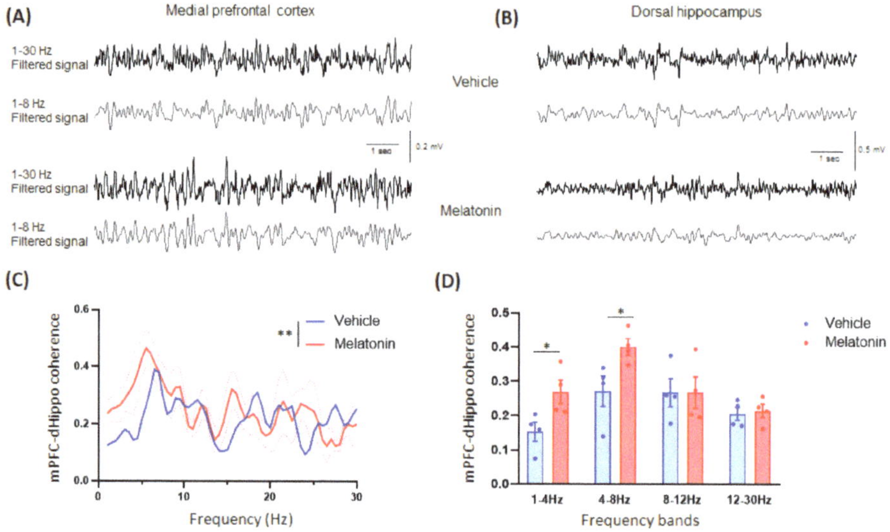

Figure 4. Medial prefrontal cortex (mPFC)–dorsal hippocampus (dHippo) synchrony in adult mice. Example of LFP traces (1–30 Hz filtered signal in black; 1–8 Hz filtered signal in grey) recorded in mPFC (**A**) and in dHippo (**B**) of an adult mouse 10 min after the treatment with vehicle (top) or 20 mg/kg melatonin (bottom). (**C**) A significant effect due to MLT treatment was observed in mPFC-dHippo coherence spectrum (two-way ANOVA, effect of treatment: $F(1,354) = 9.304$, $p = 0.0025$; ** $p < 0.01$ vehicle vs. 20 mg/kg MLT). (**D**) Melatonin induced a significant increase in coherence in the delta and low-theta frequency bands. Data are represented as mean ± SEM (n = 4). Statistical analyses were performed using the two-tailed unpaired Student's *t*-test; * $p < 0.05$ vehicle vs. 20 mg/kg MLT.

4. Discussion

In this study, we examined whether treatment with MLT could affect different aspects of anxiety-like phenotype according to age. Using adolescent (35–40 days of age) and adult (3–4 months of age) mice, we found that MLT, at a dose of 20 mg/kg, had different effects on behavioral paradigms of anxiety-like behaviors according to age. We found that in adolescent mice, 20 mg/kg MLT had no effect in the OFT, EPMT, and Marble Burying test and increased social behavior in the Three-Chamber Sociability test. On the contrary, in adult mice, 20 mg/kg MLT reduced the distance traveled in the OFT, an indication of a sedative effect induced by the drug at this dose. This sedative effect was also reflected in the reduced exploration of the center of the open field arena in the OFT, of the open arms of the EPMT, and in the overall social encounters in the Three-Chamber Sociability test. Similar to the adolescents, 20 mg/kg MLT in adults did not alter the number of marbles buried in the Marble Burying test. These distinct behavioral age-dependent effects induced by MLT were also paralleled by differences in activation of the mPFC-dHippo brain regions. Using LFP recordings in freely moving animals, we found that 20 mg/kg MLT induced a significant increase in mPFC-dHippo coherence in the low-frequency bands (delta and low theta) in adults but not in adolescents.

Previous research, in keeping with our findings, has shown different effects induced by exogenous administration of MLT according to age. Sharman et al. [36] found that treatment with MLT in aged mice was able to reverse the changes in the expression levels of various genes associated with inflammation and immune function, including lipocalin 2 mRNA, to the levels observed in younger mice. The same research group [37] also analyzed the expression levels of these genes following an inflammatory insult by lipopolysaccharide and found that MLT treatment was able to induce a pattern of response in the gene expression in the brain of aged mice that mirrored that of younger mice.

As already mentioned, we chose the dose of 20 mg/kg according to our previous study in adult rats, in which it induced anxiolytic-like effects in the EPMT and novelty-suppressed feeding test without affecting locomotor behavior [10]. Unlike our expectations, we found that in adult C57BL6/J mice, 20 mg/kg MLT induced a sedative-like effect, as measured by reduced locomotor activity in the OFT. It is noteworthy that C57 mice are characterized by a low synthesis of MLT in the pineal gland caused by a natural point mutation in the gene encoding for the enzyme aralkylamine N-acetyltransferase (AANAT), which transforms serotonin into N-acetyl serotonin (NAS) that is then converted into MLT by the enzyme hydroxy-o-methyltransferase [38]. However, they found that serotonin can be acetylated by arylamine transferase (NAT), an alternative to the AANAT enzyme [39,40], leading to the production of NAS in C57BL/6 mice despite genetic defects in AANAT. Of interest, redundancy in serotonin acetylation was also observed on the skin of hamsters [41], rats [42], and humans [43], and NAS can be detected in human serum [44]. Thus, we can hypothesize that the low circulating levels of MLT, along with a blunting of the daily circadian variation in its levels in C57 mice compared to rats, make the MLT receptors more sensitive to an exogenous injection of MLT in C57 mice than in rats. This translates to the fact that an anxiolytic dose in adult rats instead induced sedation in adult C57 mice, a condition characterized by a slightly more profound depressive state of the central nervous system. In keeping with a higher dose of MLT, 40 mg/kg induced sleep when administered to adult rats [45], while a lower dose (10 mg/kg) did not reduce the total distance traveled in adult C57 mice in the OFT (data not reported). However, future studies should confirm these hypotheses by showing dose-response curves after MLT treatment in the different behavioral tests and by investigating the age-dependent effects of MLT in other strains of mice, the so-called melatonin-proficient mice, including C3H and CBA, or the relatively newly developed Aanat+/+; Hiomt+/+ on the C57BL/6J genetic background model [46,47].

Although it has not been fully clarified yet, MT_1 and MT_2 receptors have been reported to desensitize according to circulating levels of MLT [48–50]. Therefore, it can be hypothesized that the effects of exogenous MLT may depend on the intrinsic status of MLT receptors. In line with this hypothesis, we and others have demonstrated in preclinical and clinical studies that the response to an exogenous injection of MLT or a melatonergic compound can vary according to the time of the day [19,45,51–53]. It is well known that the levels of circulating MLT change dramatically during development and aging, but whether also the expression of MLT receptors and their intrinsic functioning vary during development and with aging is yet to be elucidated. The fact that a dose of MLT inducing sedative-like effects in adults did not alter the explorative behavior of the adolescent mice, but enhanced social encounters with peers, can therefore be viewed in the different intrinsic functioning of the endogenous MLT system (MLT production, expression, and functioning of its receptors) between adolescents and adults. Future studies are needed to support this hypothesis and should investigate whether changes in the expression and function of MLT receptors in different brain regions can occur during development and aging. At the same time, pharmacokinetic aspects should also be considered, given that the absorption, distribution, metabolism, and excretion of drugs are known to undergo major changes during aging [54]. In our experimental conditions, given that we injected MLT intraperitoneally, it is unlikely that differences in the absorption of MLT occurred. In contrast, we cannot exclude the possibility that differences in the distribution, metabolism, and excretion of MLT could be present comparing adolescent and adult mice. Future studies measuring the levels of MLT, for example, in the brain, specifically in the mPFC and dHippo, may clarify these aspects.

Among the different symptoms of social anxiety disorder, there is an intense fear of interacting with strangers [55]. We assessed social anxiety using the Three-Chamber Sociability test, which is also used to study autism spectrum disorders (ASD). A high comorbidity between social anxiety disorder and ASD has been reported [56,57], and MLT is a compound largely used in individuals with ASD, as it seems to ameliorate their sleep dysfunctions, where improvement leads to better functioning during the day [58]. Our data seem to support this evidence mainly in adolescents since we found an increase in social encounters in the Three-Chamber Sociability test after treatment with MLT. In keeping, a recent study using the valproic acid mouse model of ASD showed that MLT was rescuing social deficits in the three-chamber test of 6-week-old valproic acid-exposed offspring [59], and another one using the same ASD model but in rats showed an improvement in social deficits given by agomelatine, a non-selective MT_1-MT_2 receptor agonist and a $5HT_{2C}$ receptor antagonist [60]. However, these studies did not investigate which one of the two MLT receptors could be involved in the prosocial effects of MLT. A study by Thomson et al. [22] examining the phenotype of MT_2 receptor knockout mice found that male mice had increased sociability compared to wild-type, highlighting a possible role for MT_2 receptors in the modulation of social behavior.

Collectively, this preclinical evidence, further supported by our findings, indicates that MLT may not only have efficacy in ASD due to its modulation of the sleep-wake cycle and circadian system [58], but also a direct effect on regulating social behavior cannot be excluded.

Age-dependent differences in the behavioral effects induced by MLT were paralleled by differences in the activation of the mPFC-dHippo crosstalk, as highlighted by the analysis of LFP synchrony of freely moving animals. While in adolescent mice, no significant differences were observed after MLT treatment, in the coherence between the mPFC and the dHippo, in adult mice, we found an increased synchronization in the lower frequencies, corresponding to delta and low-theta bands. Although mPFC is known to be highly correlated with the ventral hippocampus in anxiety [61], we opted to record the correlation between the mPFC and the dHippo mainly due to (1) the high presence of MLT receptors in both the mPFC and the rostral rather than the ventral region of the hippocampus [25,26], (2) the fact that preliminary work has suggested a possible role for the dHippo in the anxiolytic effects of MLT [62], (3) acting on the process of memory reconsolidation in which the dHippo plays a major role [63] and appears to be a novel promising strategy for treating anxiety disorders [64]. Remarkably, recent studies have demonstrated that direct functional projections from the dorsal hippocampus to the prelimbic cortex are necessary for strengthening aversive memory through different molecular mechanisms [65,66], while increased theta coherence is observed during spatial memory tasks and after the application of dopamine in mPFC [67]. In a mouse model of schizophrenia, dHippo-mPFC theta coherence was impaired during working memory performance [68], similar to the altered functional connectivity between the frontal and temporal lobes observed in patients with schizophrenia. Different findings suggest that a coupling within the low frequencies is observed between the medial frontal cortex and distant brain regions to guide behavioral performances. In particular, the mPFC seems to play a key role in regulating social and emotional-like behavior and also responses to stress and fear, recruiting cortical and subcortical areas within the low-frequency range [69,70].

Therefore, an important future step will be to examine—in a more comprehensive way—the effects of MLT during cognitive/behavioral tasks in association with a functional study of brain-related regions. In particular, these studies should be performed considering the factor of age, as multiple evidence has shown that the neural basis underlying the neurobiology of anxiety and emotional processing can be different when comparing adults and adolescents [71–74]. Accordingly, as we found, the modulatory effects of MLT on anxiety circuits may expect to vary during development and aging.

5. Conclusions

Here, we show that exogenous MLT may affect anxiety-like behaviors in a different manner according to age. In particular, it seems that adult mice may be more sensitive than adolescent mice to the pharmacological effects of MLT, as also evidenced by a different functional modulation of the prefrontal-hippocampal circuit. Given the significantly increased prevalence in the use of MLT supplements during the last two decades [75], and the lack of clinical studies in adolescents and analyzing the outcomes according to the age of the participants, our findings highlight the need to take into account the age factor when evaluating the therapeutics and the toxicity of MLT or MLT receptor ligands in both preclinical and clinical studies.

Author Contributions: Conceptualization, S.N., M.C. and S.C.; formal analysis, S.N., S.T., A.S., D.D.G. and M.C.; investigation, S.N., S.T., A.S., D.D.G., M.C. and S.C.; writing—original draft preparation, S.N., S.T., M.C. and S.C.; writing—review and editing, S.T., D.D.G., M.C. and S.C.; supervision, M.C. and S.C.; funding acquisition, S.C. All authors have read and agreed to the published version of the manuscript.

Funding: This research was supported by the Department of Pharmaceutical and Pharmacological Sciences, University of Padova (grant no. PRID-J 2021) to S.C. and M.C. was partially supported by Fondazione Umberto Veronesi (Post-doctoral Fellowships 2019).

Institutional Review Board Statement: This study was conducted in accordance with the guidelines established by the European Community Council (Directive 2010/63/EU, September 2010) and was approved by the Institutional Animal Care and Use Committee of the University of Padova and by the Italian Ministry of Health (IACUC permission number: A52).

Informed Consent Statement: Not applicable.

Data Availability Statement: The data presented in this study are available upon reasonable request from the corresponding author.

Acknowledgments: The authors would like to thank Francesca Niero, Andrea Giulia Barattin, Valentina Passuello, and Lisa Friso for their help in collecting preliminary behavioral data.

Conflicts of Interest: The authors declare no conflict of interest. The funders had no role in the design of the study; in the collection, analyses, or interpretation of data; in the writing of the manuscript; or in the decision to publish the results.

References

1. World Health Organization. *Mental Health and COVID-19: Early Evidence of the Pandemic's Impact: Scientific Brief, 2 March 2022 (No. WHO/2019-nCoV/Sci_Brief/Mental_health/2022.1)*; World Health Organization: Geneva, Switzerland, 2022.
2. Bandelow, B.; Michaelis, S. Epidemiology of Anxiety Disorders in the 21st Century. *Dialogues Clin. Neurosci.* **2015**, *17*, 327–335. [CrossRef] [PubMed]
3. Xiong, P.; Liu, M.; Liu, B.; Hall, B.J. Trends in the Incidence and DALYs of Anxiety Disorders at the Global, Regional, and National Levels: Estimates from the Global Burden of Disease Study 2019. *J. Affect. Disord.* **2022**, *297*, 83–93. [CrossRef] [PubMed]
4. Koskinen, M.K.; Hovatta, I. Genetic Insights into the Neurobiology of Anxiety. *Trends Neurosci.* **2023**, *46*, 318–331. [CrossRef]
5. The Psychobiology and Pathophysiology of Anxiety and Fear. In *Anxiety and the Anxiety Disorders*; Routledge: Londomn, UK, 2019; pp. 333–354. [CrossRef]
6. Moon, E.; Kim, K.; Partonen, T.; Linnaranta, O. Role of Melatonin in the Management of Sleep and Circadian Disorders in the Context of Psychiatric Illness. *Curr. Psychiatry Rep.* **2022**, *24*, 623–634. [CrossRef] [PubMed]
7. Marseglia, L.; D'Angelo, G.; Manti, S.; Aversa, S.; Arrigo, T.; Reiter, R.J.; Gitto, E. Analgesic, Anxiolytic and Anaesthetic Effects of Melatonin: New Potential Uses in Pediatrics. *Int. J. Mol. Sci.* **2015**, *16*, 1209–1220. [CrossRef]
8. Papp, M.; Litwa, E.; Gruca, P.; Mocaë, E. Anxiolytic-like Activity of Agomelatine and Melatonin in Three Animal Models of Anxiety. *Behav. Pharmacol.* **2006**, *17*, 9–18.
9. Golus, P.; King, M.G. The Effects of Melatonin on Open Field Behavior. *Pharmacol. Biochem. Behav.* **1981**, *15*, 883–885. [CrossRef]
10. Ochoa-Sanchez, R.; Rainer, Q.; Comai, S.; Spadoni, G.; Bedini, A.; Rivara, S.; Fraschini, F.; Mor, M.; Tarzia, G.; Gobbi, G. Anxiolytic Effects of the Melatonin MT2 Receptor Partial Agonist UCM765: Comparison with Melatonin and Diazepam. *Prog. Neuro-Psychopharmacol. Biol. Psychiatry* **2012**, *39*, 318–325. [CrossRef]
11. Golombek, D.A.; Martini, M.; Cardinali, D.P. Melatonin as an Anxiolytic in Rats: Time Dependence and Interaction with the Central GABAergic System. *Eur. J. Pharmacol.* **1993**, *237*, 231–236. [CrossRef]
12. Repova, K.; Baka, T.; Krajcirovicova, K.; Stanko, P.; Aziriova, S.; Reiter, R.J.; Simko, F. Melatonin as a Potential Approach to Anxiety Treatment. *Int. J. Mol. Sci.* **2022**, *23*, 16187. [CrossRef]
13. Crupi, R.; Mazzon, E.; Marino, A.; La Spada, G.; Bramanti, P.; Cuzzocrea, S.; Spina, E. Melatonin Treatment Mimics the Antidepressant Action in Chronic Corticosterone-Treated Mice. *J. Pineal Res.* **2010**, *49*, 123–129. [CrossRef] [PubMed]
14. Samarkandi, A.; Naguib, M.; Riad, W.; Thalaj, A.; Alotibi, W.; Aldammas, F.; Albassam, A. Melatonin vs. Midazolam Premedication in Children: A Double-Blind, Placebo-Controlled Study. *Eur. J. Anaesthesiol.* **2005**, *22*, 189–196. [CrossRef] [PubMed]

15. Madsen, B.K.; Zetner, D.; Møller, A.M.; Rosenberg, J. Melatonin for Preoperative and Postoperative Anxiety in Adults. *Cochrane Database Syst. Rev.* **2020**, *12*, 12. [CrossRef]
16. Karasek, M. Does Melatonin Play a Role in Aging Processes? *J. Physiol. Pharmacol.* **2007**, *58* (Suppl. S6), 105–113. [PubMed]
17. Jockers, R.; Delagrange, P.; Dubocovich, M.L.; Markus, R.P.; Renault, N.; Tosini, G.; Cecon, E.; Zlotos, D.P. Update on Melatonin Receptors: IUPHAR Review 20. *Br. J. Pharmacol.* **2016**, *173*, 2702–2725. [CrossRef]
18. Gobbi, G.; Comai, S. Sleep Well. Untangling the Role of Melatonin MT1 and MT2 Receptors in Sleep. *J. Pineal Res.* **2019**, *66*, e12544. [CrossRef]
19. López-Canul, M.; Hyun Min, S.; Posa, L.; De Gregorio, D.; Bedini, A.; Spadoni, G.; Gobbi, G.; Comai, S. Melatonin MT1 and MT2 Receptors Exhibit Distinct Effects in the Modulation of Body Temperature across the Light/Dark Cycle. *Int. J. Mol. Sci.* **2019**, *20*, 2452. [CrossRef]
20. Slominski, R.M.; Reiter, R.J.; Schlabritz-Loutsevitch, N.; Ostrom, R.S.; Slominski, A.T. Melatonin Membrane Receptors in Peripheral Tissues: Distribution and Functions. *Mol. Cell. Endocrinol.* **2012**, *351*, 152–166. [CrossRef]
21. Comai, S.; De Gregorio, D.; Posa, L.; Ochoa-Sanchez, R.; Bedini, A.; Gobbi, G. Dysfunction of Serotonergic Activity and Emotional Responses across the Light-Dark Cycle in Mice Lacking Melatonin MT2 Receptors. *J. Pineal Res.* **2020**, *69*, e12653. [CrossRef]
22. Thomson, D.M.; Mitchell, E.J.; Openshaw, R.L.; Pratt, J.A.; Morris, B.J. Mice Lacking Melatonin MT2 Receptors Exhibit Attentional Deficits, Anxiety and Enhanced Social Interaction. *J. Psychopharmacol.* **2021**, *35*, 1265–1276. [CrossRef]
23. Noseda, A.C.D.; Targa, A.D.S.; Rodrigues, L.S.; Aurich, M.F.; Lima, M.M.S. REM Sleep Deprivation Promotes a Dopaminergic Influence in the Striatal MT2 Anxiolytic-like Effects. *Sleep Sci.* **2016**, *9*, 47–54. [CrossRef] [PubMed]
24. Liu, J.; Clough, S.J.; Dubocovich, M.L. Role of the MT1 and MT2 Melatonin Receptors in Mediating Depressive- and Anxiety-like Behaviors in C3H/HeN Mice. *Genes Brain Behav.* **2017**, *16*, 546–553. [CrossRef] [PubMed]
25. Lacoste, B.; Angeloni, D.; Dominguez-Lopez, S.; Calderoni, S.; Mauro, A.; Fraschini, F.; Descarries, L.; Gobbi, G. Anatomical and Cellular Localization of Melatonin MT1 and MT2 Receptors in the Adult Rat Brain. *J. Pineal Res.* **2015**, *58*, 397–417. [CrossRef] [PubMed]
26. Ng, K.Y.; Leong, M.K.; Liang, H.; Paxinos, G. Melatonin Receptors: Distribution in Mammalian Brain and Their Respective Putative Functions. *Brain Struct. Funct.* **2017**, *222*, 2921–2939. [CrossRef]
27. Sugden, D. Psychopharmacological Effects of Melatonin in Mouse and Rat. *J. Pharmacol. Exp. Ther.* **1983**, *227*, 587–591.
28. Tassan Mazzocco, M.; Guarnieri, F.C.; Monzani, E.; Benfenati, F.; Valtorta, F.; Comai, S. Dysfunction of the Serotonergic System in the Brain of Synapsin Triple Knockout Mice Is Associated with Behavioral Abnormalities Resembling Synapsin-Related Human Pathologies. *Prog. Neuropsychopharmacol. Biol. Psychiatry* **2021**, *105*, 110135. [CrossRef]
29. Onaolapo, O.J.; Onaolapo, A.Y.; Akanmu, M.A.; Olayiwola, G.; Onaolapo, O.J.; Onaolapo, A.Y.; Akanmu, M.A.; Olayiwola, G. Changes in Spontaneous Working-Memory, Memory-Recall and Approach-Avoidance Following "Low Dose" Monosodium Glutamate in Mice. *AIMS Neurosci.* **2016**, *3*, 317–337. [CrossRef]
30. Comai, S.; Ochoa-Sanchez, R.; Dominguez-Lopez, S.; Bambico, F.R.; Gobbi, G. Melancholic-Like Behaviors and Circadian Neurobiological Abnormalities in Melatonin MT1 Receptor Knockout Mice. *Int. J. Neuropsychopharmacol.* **2015**, *18*, pyu075. [CrossRef]
31. Yang, M.; Silverman, J.L.; Crawley, J.N. Automated Three-Chambered Social Approach Task for Mice. *Curr. Protoc. Neurosci.* **2011**, *56*, 8–26. [CrossRef]
32. Deacon, R.M.J. Digging and Marble Burying in Mice: Simple Methods for in Vivo Identification of Biological Impacts. *Nat. Protoc.* **2006**, *1*, 122–124. [CrossRef]
33. Njung'e, K.; Handley, S.L. Evaluation of Marble-Burying Behavior as a Model of Anxiety. *Pharmacol. Biochem. Behav.* **1991**, *38*, 63–67. [CrossRef] [PubMed]
34. Cambiaghi, M.; Grosso, A.; Likhtik, E.; Mazziotti, X.R.; Concina, G.; Renna, A.; Sacco, T.; Gordon, J.A.; Benedetto Sacchetti, X.; Levi-Montalcini, R. Higher-Order Sensory Cortex Drives Basolateral Amygdala Activity during the Recall of Remote, but Not Recently Learned Fearful Memories. *J. Neurosci.* **2016**, *36*, 1647–1659. [CrossRef] [PubMed]
35. Concina, G.; Cambiaghi, X.M.; Renna, A.; Sacchetti, X.B. Behavioral/Cognitive Coherent Activity between the Prelimbic and Auditory Cortex in the Slow-Gamma Band Underlies Fear Discrimination. *J. Neurosci.* **2018**, *38*, 8313–8328. [CrossRef] [PubMed]
36. Sharman, E.H.; Bondy, S.C.; Sharman, K.G.; Lahiri, D.; Cotman, C.W.; Perreau, V.M. Effects of Melatonin and Age on Gene Expression in Mouse CNS Using Microarray Analysis. *Neurochem. Int.* **2007**, *50*, 336–344. [CrossRef] [PubMed]
37. Perreau, V.M.; Bondy, S.C.; Cotman, C.W.; Sharman, K.G.; Sharman, E.H. Melatonin Treatment in Old Mice Enables a More Youthful Response to LPS in the Brain. *J. Neuroimmunol.* **2007**, *182*, 22–31. [CrossRef]
38. Roseboom, P.H.; Namboodiri, M.A.A.; Zimonjic, D.B.; Popescu, N.C.; Rodriguez, I.R.; Gastel, J.A.; Klein, D.C. Natural Melatonin 'knockdown' in C57BL/6J Mice: Rare Mechanism Truncates Serotonin N-Acetyltransferase. *Mol. Brain Res.* **1998**, *63*, 189–197. [CrossRef]
39. Slominski, A.; Pisarchik, A.; Semak, I.; Sweatman, T.; Wortsman, J. Characterization of the Serotoninergic System in the C57BL/6 Mouse Skin. *Eur. J. Biochem.* **2003**, *270*, 3335–3344. [CrossRef]
40. Slominski, A.; Wortsman, J.; Tobin, D.J. The Cutaneous Serotoninergic/Melatoninergic System: Securing a Place under the Sun. *FASEB J.* **2005**, *19*, 176–194. [CrossRef]
41. Slominski, A.; Pisarchik, A.; Semak, I.; Sweatman, T.; Szczesniewski, A.; Wortsman, J. Serotoninergic System in Hamster Skin. *J. Investig. Dermatol.* **2002**, *119*, 934–942. [CrossRef]

42. Semak, I.; Korik, E.; Naumova, M.; Wortsman, J.; Slominski, A. Serotonin Metabolism in Rat Skin: Characterization by Liquid Chromatography-Mass Spectrometry. *Arch. Biochem. Biophys.* **2004**, *421*, 61–66. [CrossRef]
43. Slominski, A.; Pisarchik, A.; Semak, I.; Sweatman, T.; Wortsman, J.; Szczesniewski, A.; Slugocki, G.; McNulty, J.; Kauser, S.; Tobin, D.J.; et al. Serotoninergic and Melatoninergic Systems Are Fully Expressed in Human Skin. *FASEB J.* **2002**, *16*, 896–898. [CrossRef] [PubMed]
44. Slominski, A.T.; Kim, T.K.; Kleszczyński, K.; Semak, I.; Janjetovic, Z.; Sweatman, T.; Skobowiat, C.; Steketee, J.D.; Lin, Z.; Postlethwaite, A.; et al. Characterization of Serotonin and N-Acetylserotonin Systems in the Human Epidermis and Skin Cells. *J. Pineal Res.* **2020**, *68*, e12626. [CrossRef]
45. Ochoa-Sanchez, R.; Comai, S.; Spadoni, G.; Bedini, A.; Tarzia, G.; Gobbi, G. Melatonin, Selective and Non-Selective MT1/MT2 Receptors Agonists: Differential Effects on the 24-h Vigilance States. *Neurosci. Lett.* **2014**, *561*, 156–161. [CrossRef] [PubMed]
46. Zhang, C.; Clough, S.J.; Adamah-Biassi, E.B.; Sveinsson, M.H.; Hutchinson, A.J.; Miura, I.; Furuse, T.; Wakana, S.; Matsumoto, Y.K.; Okanoya, K.; et al. Impact of Endogenous Melatonin on Rhythmic Behaviors, Reproduction, and Survival Revealed in Melatonin-Proficient C57BL/6J Congenic Mice. *J. Pineal Res.* **2021**, *71*, e12748. [CrossRef] [PubMed]
47. Paulose, J.K.; Wang, C.; O'Hara, B.F.; Cassone, V.M. The Effects of Aging on Sleep Parameters in a Healthy, Melatonin-Competent Mouse Model. *Nat. Sci. Sleep* **2019**, *11*, 113–121. [CrossRef]
48. Witt-Enderby, P.A.; Bennett, J.; Jarzynka, M.J.; Firestine, S.; Melan, M.A. Melatonin Receptors and Their Regulation: Biochemical and Structural Mechanisms. *Life Sci.* **2003**, *72*, 2183–2198. [CrossRef]
49. Gobbi, G.; Comai, S. Differential Function of Melatonin MT1 and MT2 Receptors in REM and NREM Sleep. *Front. Endocrinol.* **2019**, *10*, 87. [CrossRef]
50. Pandi-Perumal, S.R.; Trakht, I.; Srinivasan, V.; Spence, D.W.; Maestroni, G.J.M.; Zisapel, N.; Cardinali, D.P. Physiological Effects of Melatonin: Role of Melatonin Receptors and Signal Transduction Pathways. *Prog. Neurobiol.* **2008**, *85*, 335–353. [CrossRef]
51. Ochoa-Sanchez, R.; Comai, S.; Lacoste, B.; Bambico, F.R.; Dominguez-Lopez, S.; Spadoni, G.; Rivara, S.; Bedini, A.; Angeloni, D.; Fraschini, F.; et al. Promotion of Non-Rapid Eye Movement Sleep and Activation of Reticular Thalamic Neurons by a Novel MT2 Melatonin Receptor Ligand. *J. Neurosci.* **2011**, *31*, 18439–18452. [CrossRef]
52. Harris, A.S.; Burgess, H.J.; Dawson, D. The Effects of Day-Time Exogenous Melatonin Administration on Cardiac Autonomic Activity. *J. Pineal Res.* **2001**, *31*, 199–205. [CrossRef]
53. Tzischinsky, O.; Lavie, P. Melatonin Possesses Time-Dependent Hypnotic Effects. *Sleep* **1994**, *17*, 638–645. [CrossRef] [PubMed]
54. Hutchison, L.C.; O'brien, C.E. Changes in Pharmacokinetics and Pharmacodynamics in the Elderly Patient. *J. Pharm. Pract.* **2007**, *20*, 4–12. [CrossRef]
55. Leary, M.R.; Kowalski, R.M. *Social Anxiety*; Guilford Press: New York, NY, USA, 1997.
56. Maddox, B.B.; White, S.W. Comorbid Social Anxiety Disorder in Adults with Autism Spectrum Disorder. *J. Autism Dev. Disord.* **2015**, *45*, 3949–3960. [CrossRef] [PubMed]
57. van Steensel, F.J.A.; Bögels, S.M.; Perrin, S. Anxiety Disorders in Children and Adolescents with Autistic Spectrum Disorders: A Meta-Analysis. *Clin. Child Fam. Psychol. Rev.* **2011**, *14*, 302–317. [CrossRef] [PubMed]
58. Rossignol, D.; Frye, R. Melatonin in Autism Spectrum Disorders. *Curr. Clin. Pharmacol.* **2014**, *9*, 326–334. [CrossRef]
59. Liu, X.; Cui, Y.; Zhang, Y.; Xiang, G.; Yu, M.; Wang, X.; Qiu, B.; Li, X.; Liu, W.; Zhang, D. Rescue of Social Deficits by Early-Life Melatonin Supplementation through Modulation of Gut Microbiota in a Murine Model of Autism. *Biomed. Pharmacother.* **2022**, *156*, 113949. [CrossRef]
60. Kumar, H.; Sharma, B.M.; Sharma, B. Benefits of Agomelatine in Behavioral, Neurochemical and Blood Brain Barrier Alterations in Prenatal Valproic Acid Induced Autism Spectrum Disorder. *Neurochem. Int.* **2015**, *91*, 34–45. [CrossRef]
61. Adhikari, A.; Topiwala, M.A.; Gordon, J.A. Synchronized Activity between the Ventral Hippocampus and the Medial Prefrontal Cortex during Anxiety. *Neuron* **2010**, *65*, 257–269. [CrossRef]
62. Arushanian, E.B.; Beier, E.V. The Participation of the Dorsal Hippocampus in the Antianxiety Action of Melatonin and Diazepam. *Eksperimental'naia Klin. Farmakol.* **1998**, *61*, 13–16.
63. Lee, J.L.C. Memory Reconsolidation Mediates the Strengthening of Memories by Additional Learning. *Nat. Neurosci.* **2008**, *11*, 1264–1266. [CrossRef]
64. Kindt, M.; Elsey, J.W.B. A Paradigm Shift in the Treatment of Emotional Memory Disorders: Lessons from Basic Science. *Brain Res. Bull.* **2023**, *192*, 168–174. [CrossRef] [PubMed]
65. Fukushima, H.; Zhang, Y.; Archbold, G.; Ishikawa, R.; Nader, K.; Kida, S. Enhancement of Fear Memory by Retrieval through Reconsolidation. *eLife* **2014**, *3*, e02736. [CrossRef] [PubMed]
66. Ye, X.; Kapeller-Libermann, D.; Travaglia, A.; Inda, M.C.; Alberini, C.M. Direct Dorsal Hippocampal-Prelimbic Cortex Connections Strengthen Fear Memories. *Nat. Neurosci.* **2017**, *20*, 52–61. [CrossRef]
67. Colgin, L.L. Oscillations and Hippocampal-Prefrontal Synchrony. *Curr. Opin. Neurobiol.* **2011**, *21*, 467–474. [CrossRef]
68. Sigurdsson, T.; Stark, K.L.; Karayiorgou, M.; Gogos, J.A.; Gordon, J.A. Impaired Hippocampal-Prefrontal Synchrony in a Genetic Mouse Model of Schizophrenia. *Nature* **2010**, *464*, 763–767. [CrossRef]
69. Kuga, N.; Abe, R.; Takano, K.; Ikegaya, Y.; Sasaki, T. Prefrontal-amygdalar oscillations related to social behavior in mice. *eLife* **2022**, *11*, e78428. [CrossRef] [PubMed]
70. Narayanan, N.S.; Cavanagh, J.F.; Frank, M.J.; Laubach, M. Common medial frontal mechanisms of adaptive control in humans and rodents. *Nat. Neurosci.* **2013**, *16*, 1888–1895. [CrossRef] [PubMed]

71. Britton, J.C.; Grillon, C.; Lissek, S.; Norcross, M.A.; Szuhany, K.L.; Chen, G.; Ernst, M.; Nelson, E.E.; Leibenluft, E.; Shechner, T.; et al. Response to Learned Threat: An FMRI Study in Adolescent and Adult Anxiety. *Am. J. Psychiatry* **2013**, *170*, 1195–1204. [CrossRef]
72. Jarcho, J.M.; Romer, A.L.; Shechner, T.; Galvan, A.; Guyer, A.E.; Leibenluft, E.; Pine, D.S.; Nelson, E.E. Forgetting the Best When Predicting the Worst: Preliminary Observations on Neural Circuit Function in Adolescent Social Anxiety. *Dev. Cogn. Neurosci.* **2015**, *13*, 21–31. [CrossRef]
73. Gunning-Dixon, F.M.; Gur, R.C.; Perkins, A.C.; Schroeder, L.; Turner, T.; Turetsky, B.I.; Chan, R.M.; Loughead, J.W.; Alsop, D.C.; Maldjian, J.; et al. Age-Related Differences in Brain Activation during Emotional Face Processing. *Neurobiol. Aging* **2003**, *24*, 285–295. [CrossRef]
74. Ganella, D.E.; Drummond, K.D.; Ganella, E.P.; Whittle, S.; Kim, J.H. Extinction of Conditioned Fear in Adolescents and Adults: A Human FMRI Study. *Front. Hum. Neurosci.* **2018**, *11*, 647. [CrossRef] [PubMed]
75. Li, J.; Somers, V.K.; Xu, H.; Lopez-Jimenez, F.; Covassin, N. Trends in Use of Melatonin Supplements Among US Adults, 1999–2018. *JAMA* **2022**, *327*, 483–485. [CrossRef] [PubMed]

biomedicines

MDPI

Article

Age-Dependent Alterations in Platelet Mitochondrial Respiration

Zdeněk Fišar *, Jana Hroudová, Martina Zvěřová, Roman Jirák, Jiří Raboch and Eva Kitzlerová

Department of Psychiatry, First Faculty of Medicine, Charles University and General University Hospital in Prague, Ke Karlovu 11, 120 00 Prague, Czech Republic; hroudova.jana@gmail.com (J.H.); zverova.martina@vfn.cz (M.Z.); roman.jirak@vfn.cz (R.J.); jiri.raboch@lf1.cuni.cz (J.R.); eva.kitzlerova@vfn.cz (E.K.)

* Correspondence: zfisar@lf1.cuni.cz; Tel.: +420-224-965-313

Abstract: Mitochondrial dysfunction is an important cellular hallmark of aging and neurodegeneration. Platelets are a useful model to study the systemic manifestations of mitochondrial dysfunction. To evaluate the age dependence of mitochondrial parameters, citrate synthase activity, respiratory chain complex activity, and oxygen consumption kinetics were assessed. The effect of cognitive impairment was examined by comparing the age dependence of mitochondrial parameters in healthy individuals and those with neuropsychiatric disease. The study found a significant negative slope of age-dependence for both the activity of individual mitochondrial enzymes (citrate synthase and complex II) and parameters of mitochondrial respiration in intact platelets (routine respiration, maximum capacity of electron transport system, and respiratory rate after complex I inhibition). However, there was no significant difference in the age-related changes of mitochondrial parameters between individuals with and without cognitive impairment. These findings highlight the potential of measuring mitochondrial respiration in intact platelets as a means to assess age-related mitochondrial dysfunction. The results indicate that drugs and interventions targeting mitochondrial respiration may have the potential to slow down or eliminate certain aging and neurodegenerative processes. Mitochondrial respiration in platelets holds promise as a biomarker of aging, irrespective of the degree of cognitive impairment.

Keywords: aging; platelet; mitochondria; respiratory chain complex; mitochondrial respiration; cognitive decline; neurodegenerative disease; neuroinflammation; neuroplasticity; oxidative stress

Citation: Fišar, Z.; Hroudová, J.; Zvěřová, M.; Jirák, R.; Raboch, J.; Kitzlerová, E. Age-Dependent Alterations in Platelet Mitochondrial Respiration. *Biomedicines* **2023**, *11*, 1564. https://doi.org/10.3390/biomedicines11061564

Academic Editor: Masaru Tanaka

Received: 22 April 2023
Revised: 25 May 2023
Accepted: 26 May 2023
Published: 28 May 2023

1. Introduction

The biology of aging is associated with metabolic and oxidative stress, inflammation, and DNA mutations, in which a number of cellular mechanisms are involved [1]. Twelve molecular cellular hallmarks for aging are proposed and discussed, which are interconnected among each other, and which include mitochondrial dysfunction [2]. Mitochondria are organelles that play a key role in bioenergetics and the maintenance and regulation of all brain functions, including neuroinflammation, neuroplasticity, oxidative stress, and apoptosis [3]. In aging, there is evidence of decreased mitochondrial function, increased oxidative stress, increased mitochondrial DNA (mtDNA) mutations, and miRNA dysregulation [4]. Due to the high energy demands of brain cells, mitochondrial dysfunction is associated not only with aging [5], but also with neurodegenerative diseases [6,7]. Age-related neurodegeneration involves complex interplay and synergy of various genetic and environmental factors [8], including mitochondrial impairment as a common motif in the pathophysiology of neuropsychiatric diseases [9]. Mitochondrial dysfunction is associated with decreasing neuroplasticity and weakening functional resilience [10], neuroinflammation [11], oxidative stress and apoptosis [12], excitotoxicity, neurotoxicity of protein agglomerates, and deficiencies in mitochondrial proteostasis and the protease-mediated quality control system [13].

Impaired mitochondrial function can be caused by genetic mutations and epigenetic modifications, environmental stressors, cellular senescence, and disturbed mitophagy. Mitochondrial dysfunction leading to reduced ATP production, increased ROS production, activation of the intrinsic apoptotic pathway, and impaired calcium buffering ultimately result in brain cell death, neurodegeneration, and cognitive decline [6,14].

The main current strategies for preventing or reversing processes associated with aging and neurodegeneration are aimed at mitochondrial quality control mechanisms to increase mitochondrial functions and include regulation of the OXPHOS system, generation of ATP through the electron transport system (ETS), reduction of oxidative stress with antioxidants, inhibition of apoptosis, autophagy enhancement, and stimulation of mitochondrial biogenesis by metabolic modulators, drugs, diet (caloric restriction), and exercise [15,16].

The mitochondrial hypothesis is based on the key role of mitochondria in the production of ATP through oxidative phosphorylation (OXPHOS) [17], the production of reactive oxygen species (ROS), the buffering of free calcium in the cytosol, the release of pro-apoptotic factors and the initiation of the intrinsic pathway of apoptosis, and the production of heat. According to the free radical theory of aging [18–20], aging and age-related diseases are associated with the generation of ROS, mainly from mitochondria, and subsequent damage to cellular proteins, lipids, and nucleic acids. Oxidative stress is accepted as a key modulator of the biological processes of aging and neurodegeneration [21], but useful endogenous mechanisms that can be initiated by ROS must also be taken into account in therapeutic interventions on cellular redox processes [22]. Attention is therefore paid to the role of other manifestations of mitochondrial dysfunction, such as inflammation, mtDNA, mitophagy, and retrograde signaling from the mitochondria to the nucleus [23–26].

The mitochondrial hypothesis of aging suggests that mitochondrial dysfunction over time leads to a decrease in cellular energy production, increased oxidative stress, calcium dysregulation, accumulation of mutations in mitochondrial DNA (mtDNA), apoptosis, alterations in mitochondrial dynamics (fusion and fission), accumulation of cellular waste products, disturbed mitophagy, and changes in cellular metabolism [27–30]. This, in turn, may be the biological basis for the development of age-related diseases, including neurodegenerative diseases, and the possibility of their treatment [31–34].

Mitochondrial dysfunctions are implicated in the pathophysiology of neurodegenerative diseases, such as Alzheimer's disease (AD), Parkinson's disease, and Huntington's disease [6,31,35]. Common features in their pathogenesis are mitochondrial dysfunction, progressive neurodegeneration, and cognitive decline leading to dementia. The main risk factors for the onset and development of AD are age, genetics and epigenetics, environmental factors, amyloidosis, tauopathy, and mitochondrial dysfunction. Neurodegenerative mechanisms in AD include early changes in mitochondria-associated endoplasmic reticulum membranes [36]. In the integrative hypothesis of Alzheimer's disease, interlinking and mutual synergistic connections between amyloid beta pathology, tau pathology, mitochondrial dysfunction, dysfunction of neurotransmitter systems, and disturbed neuroplasticity are assumed [37].

Mitochondrial dysfunction is also thought to contribute to the pathophysiology of psychiatric diseases such as major depressive disorder (MDD) [38] and bipolar disorder (BD) [39–42] through several pathways, including oxidative stress, neuroinflammation, genetics, and disturbed neuroplasticity. Affective disorders are a common comorbidity in neurodegenerative diseases [43]. Cognitive deficit is a principal component in MDD, especially in late-life depression [44,45]. Cognitive impairment in BD patients is associated with metabolic factors, gene polymorphisms, brain structural and functional changes, and neuroinflammation [46]. Emotions can improve or impair cognitive performance [47], and it is important to understand the functional interplay between the central and autonomic nervous systems and to elucidate how emotions are integrated into executive functions in neuropsychiatric diseases [48,49]. This is crucial for regulating physiological processes and maintaining homeostasis, with the prefrontal cortex playing a key role [50,51].

There are several assays to measure mitochondrial dysfunction in aging and neurodegeneration [7,52–54], and they use different biological models. It was proved that platelets are a suitable biological model for research on mitochondrial dysfunction [55–57]. Platelets are small blood components (nonnucleated cells) in mammals derived from the megakaryocytes. Average life span of circulating platelets is 8 to 9 days [58]; therefore, they reflect current systemic changes. Platelets contribute to hemostasis, innate immunity, and inflammatory response [59]. It was shown that platelet bioenergetics reflect muscle energetics and platelet mitochondrial function is altered in older adults [60]. Moreover, platelets are considered peripheral elements reflecting a variety of brain functions and neurochemical changes, including those leading to neurodegeneration [61–63]. Platelets can be easily separated from blood as platelet rich plasma (PRP) and mitochondrial function can be measured under physiological conditions [57]. Since platelets share many properties with brain cells [64], they appear to be a suitable biological model for monitoring processes of brain aging and neurodegeneration.

Age dependence of mitochondrial function in various organs and brain regions in rats has been described [65–67]; the results did not confirm the concept of a general pattern of age-dependent mitochondrial dysfunction. No such data are available in humans; measurements in humans are usually performed using skeletal muscle biopsies, fibroblasts, and circulating blood cells [68].

Due to intracellular homeostatic mechanisms, it is appropriate to monitor changes in both the activity of individual mitochondrial proteins and enzyme complexes, as well as complex mitochondrial functions characterizing a real physiological state. Frequently used methods for measuring mitochondrial function include measuring the rate of oxygen consumption, ATP production, hydrogen peroxide production, free calcium buffering in the cytosol, and the release of proapoptotic factors. In this study, the measurement of the activity of citrate synthase (CS), complexes I, II, III, and IV of the respiratory chain, and the parameters of mitochondrial respiration in the platelets of healthy individuals and individuals with neuropsychiatric disease is used to evaluate the age-dependent changes of mitochondrial dysfunction.

2. Materials and Methods

Commonly used mitochondrial function assays include oxygen consumption (complex I- and II-linked respiration, respiratory control ratio, uncoupling) [57,69], ATP production [70], hydrogen peroxide production [71,72], membrane potential [73,74], mitochondrial permeability transition, swelling, and calcium retention capacity [75,76], monoamine oxidase [77], release of pro-apoptotic factors (by ELISA and chromatography), membrane fluidity [78], mtDNA mutations [79], mitochondrial morphology and dynamics [80], mitochondrial biogenesis, and mitophagy [81].

Age-dependent changes in mitochondrial function were measured in people of different ages using spectrophotometric and high-resolution respirometry methods. Activity of CS, complex I, II, III, and IV and mitochondrial respiratory rate was measured in blood platelets of healthy subjects (CONTROL) and patients with Alzheimer's disease (AD), vascular dementia (VD), major depressive disorder (MDD), or bipolar disorder (BAD). For the analysis of the age dependence of measured mitochondrial parameters, data from control subjects included in our earlier analyses and publications on platelet mitochondrial parameters were used. Data from patients with AD, VD, MDD, or BAD were used to evaluate changes in age dependence of platelet mitochondrial parameters in people with cognitive impairment. The material and methods used have been published previously, so they are presented here only very briefly with the corresponding references.

2.1. Subjects and Their Clinical Evaluation

The subjects and their clinical evaluation have been described previously for AD patients [82–85], VD patients [83], MDD patients [86,87], and BAD patients [87,88]. Patients with AD, VD, MDD, or BAD were diagnosed and recruited from the Department of

Psychiatry of the First Faculty of Medicine, Charles University and General University Hospital in Prague, Czech Republic. In each individual study, the number of participants was determined to ensure adequate statistical power required to detect differences between groups of healthy subjects and neuropsychiatric patients.

Patients with probable AD or probable VD over 60 years of age were recruited. Criteria to diagnose AD and VD included the International Classification of Diseases, Tenth Edition (ICD-10), NINDS-AIREN VD criteria [89], the NINCDS-ADRDA Alzheimer's criteria [90,91], and the Hachinski Ischemic Score [92]. Depressive symptoms in AD patients were assessed by the Geriatric Depression Scale (GDS) [93]. Disease severity was assessed by the Addenbrooke's Cognitive Examination-Revised [94] inclusive of the Mini-Mental State Examination (MMSE) questionnaire. Other causes of dementia than AD or VD were excluded. Patients with a BAD diagnosis were clinically evaluated using diagnostic scales and questionnaires, including the Young Mania Rating Scale (YMRS), Clinical Global Impression—Severity Scale (CGI-01), and Brief Psychiatric Rating Scale (BPRS). The severity of depression in MDD patients was evaluated using the Hamilton Depressive Rating Scale, 21-item (HDRS-21) and the Clinical Global Impression—severity scale (CGI-01).

The controls included healthy volunteers, who underwent a psychiatric examination that was equivalent to that of neuropsychiatric patients, and they were without cognitive decline. The participants did not take mitochondria-targeting compounds.

Cognitive decline (decline in memory, attention, language, and executive function) is a common feature of aging and neuropsychiatric diseases. The Mini-Mental State Examination (MMSE) total score has been used to assess the progression of cognitive impairment as follows: no cognitive impairment with MMSE 24–30; mild cognitive impairment with MMSE 19–23; moderate cognitive impairment with MMSE 10–18; and severe cognitive impairment with MMSE ≤ 9 [95].

2.2. Chemicals and Solutions and Measurement Methods

The chemicals, solutions, and measurement methods are described in our earlier publications. Activities of CS and respiratory chain complexes were measured spectrophotometrically [84,96–98]; mitochondrial respiration in platelets was measured by high-resolution respirometry using the Oxygraph-2k (Oroboros Instruments Corp, Innsbruck, Austria) [57,99]. All chemicals were purchased from Sigma-Aldrich Co. (St. Louis, MO, USA)

Studies were conducted in accordance with the Code of Ethics of the World Medical Association (Declaration of Helsinki), and the study protocols were approved by the Ethical Review Board of the First Faculty of Medicine, Charles University and General University Hospital in Prague. Written informed consent was obtained from participants.

2.3. Data Analysis

Statistical analyses were performed using the STATISTICA data analysis software system (TIBCO Software Inc., Palo Alto, CA, USA). The relation between the measured platelet parameters and age was determined using the Pearson product-moment correlation coefficient. Simple regression was used to quantify the age dependence of all measured mitochondrial parameters.

DatLab software (Oroboros Instruments Corp, Innsbruck, Austria) was used for respirometry data acquisition and analysis. Oxygen consumption rates were normalized for platelet concentration (pmol O_2 per sec per 10^6 platelets). Activities of mitochondrial complexes I, II, III, and IV were normalized for CS activity.

3. Results

The age dependence of the mitochondrial parameters measured in platelets isolated from peripheral blood was evaluated. The following parameters were measured in platelets of healthy subjects (CONTROL) and patients with AD, VD, MDD, or BAD: (1) CS activity; (2) complex I activity normalized for CS (CI/CS); (3) complex II activity normalized for

CS (CII/CS); (4) complex III activity normalized for CS (CIII/CS); (5) complex IV activity normalized for CS (CIV/CS); (6) oxygen consumption rate in various respiratory states. Mitochondrial respiration in intact platelets was determined as (i) routine (basal) respiration (ROUT_i); (ii) oligomycin-induced respiration independent of ADP phosphorylation (LEAK_i); (iii) maximum capacity of electron transport system (ETSC_i); (iv) respiratory rate after complex I inhibition (ROT_i); and (v) using a derived parameter called respiration reserve capacity (RES_i) and calculated as 'ETSC_i-ROUT_i'.

A total of 637 blood samples from people aged 20 to 94 were included in the evaluation, including 162 samples of control subjects aged 20 to 80 years without serious somatic and neuropsychiatric diseases, 196 samples of AD patients aged 50 to 94 years, 120 samples of subjects with MDD aged 20 to 74 years, 127 subjects with BAD aged 20 to 73 years, and 32 people with VD aged 65 to 85 years. The age dependence of the mitochondrial parameters was calculated using correlation and regression analysis.

Correlation coefficients (Pearson r) between age and the mitochondrial parameters are summarized in Table 1. In the group of healthy controls, there is a significant negative correlation between age and CS, CII/CS, ROUT_i, ETSC_I, ROT_i, and RES_i; there is a significant positive correlation between age and CIV/CS.

Table 1. Correlation coefficients (Pearson r) between age and platelet mitochondrial variables.

Parameter	CONTROL	AD	MDD	BAD	VD	All
CS	**−0.2317** N = 100 $p = 0.020$	−0.0104 N = 86 $p = 0.924$	−0.1735 N = 111 $p = 0.069$	−0.0149 N = 98 $p = 0.884$	– N = 0 $p = -$	−0.0817 N = 395 $p = 0.105$
CI/CS	0.0435 N = 93 $p = 0.679$	0.1380 N = 81 $p = 0.219$	0.1283 N = 107 $p = 0.188$	−0.0321 N = 97 $p = 0.755$	–– N = 0 $p = -$	−0.0728 N = 378 $p = 0.158$
CII/CS	**−0.2529** N = 93 $p = 0.014$	−0.0625 N = 82 $p = 0.577$	−0.0211 N = 107 $p = 0.829$	0.0603 N = 96 $p = 0.559$	–– N = 0 $p = -$	0.0135 N = 378 $p = 0.794$
CIII/CS	0.2142 N = 31 $p = 0.247$	−0.2139 N = 22 $p = 0.339$	−0.0533 N = 15 $p = 0.850$	–– N = 0 $p = -$	–– N = 0 $p = -$	**0.3205** N = 68 $p = 0.008$
CIV/CS	**0.4072** N = 93 $p < 0.001$	−0.1393 N = 83 $p = 0.209$	0.0791 N = 107 $p = 0.418$	0.0023 N = 98 $p = 0.982$	–– N = 0 $p = -$	**0.1360** N = 381 $p = 0.008$
ROUT_i	**−0.2048** N = 101 $p = 0.040$	−0.0326 N = 132 $p = 0.710$	−0.0860 N = 42 $p = 0.588$	−0.0943 N = 113 $p = 0.320$	−0.1567 N = 29 $p = 0.417$	**−0.1617** N = 417 $p = 0.001$
LEAK_i	0.0919 N = 101 $p = 0.361$	−0.0224 N = 132 $p = 0.799$	**−0.3418** N = 42 $p = 0.027$	0.0091 N = 113 $p = 0.923$	−0.0806 N = 29 $p = 0.678$	**0.1150** N = 417 $p = 0.019$
ETSC_i	**−0.2812** N = 101 $p = 0.004$	0.0360 N = 132 $p = 0.682$	0.1013 N = 42 $p = 0.523$	−0.0602 N = 113 $p = 0.527$	−0.1964 N = 29 $p = 0.307$	**−0.2663** N = 417 $p < 0.001$
ROT_i	**−0.3688** N = 101 $p = 0.000$	0.0810 N = 132 $p = 0.356$	0.2249 N = 42 $p = 0.152$	0.0851 N = 113 $p = 0.370$	−0.0810 N = 29 $p = 0.676$	−0.0881 N = 417 $p = 0.072$
RES_i	**−0.2137** N = 101 $p = 0.032$	0.0967 N = 132 $p = 0.270$	0.2001 N = 42 $p = 0.204$	0.0151 N = 113 $p = 0.874$	−0.1811 N = 29 $p = 0.347$	**−0.2187** N = 417 $p = 0.000$

Mitochondrial parameters were measured in platelets of healthy subjects (CONTROL) and patients with Alzheimer's disease (AD), vascular dementia (VD), major depressive disorder (MDD), or bipolar disorder (BAD). Statistically significant correlation coefficients are in bold. CS, citrate synthase activity ($nmol \cdot min^{-1} \cdot mg^{-1}$); CI/CS, complex I activity normalized for CS; CII/CS, complex II activity normalized for CS; CIII/CS, complex III activity normalized for CS; CIV/CS, complex IV activity normalized for CS; ROUT_i, routine (basal) respiration; LEAK_i, oligomycin-induced respiration independent of ADP phosphorylation; ETSC_i, maximum capacity of electron transport system; ROT_i, respiratory rate after complex I inhibition by rotenone; and RES_i, respiration reserve capacity calculated as 'ETSC_i-ROUT_i' (all respiration parameters in pmol $O_2 \cdot 10^{-6}$ platelets$\cdot sec^{-1}$).

The age dependence of the measured parameters (ROUT_i, LEAL_i, ETSC_i, ROT_i, and RES_i) was quantified using the slopes of regression lines. For control group members, but not for AD, MDD, and BAD patients, a significant dependence on age was found for CS, CII/CS, and CIV/CS (Table 2, Figure 1), ROUT_i, ETSC_i, ROT_i, and RES_i (Tables 3–7). The greatest statistically significant decrease with age is shown in the mitochondrial respiratory parameter ETSC_i (Figure 2). Information was added to the results that the slope (linear regression result) for the age dependence of platelet count normalized for CS is not significantly different from zero.

Table 2. Age dependence of platelet concentration and mitochondrial enzyme activities.

Sample	Parameter	Slope	95% CI	p	Age range	N
CONTROL	PRP	−794	(−2400, 813)	0.331	20–80	162
	CS	* **−0.152**	(−0.280, −0.024)	**0.020**	20–77	100
	CI/CS	0.00055	(−0.00209, 0.00320)	0.679	20–77	93
	CII/CS	* **−0.00067**	(−0.00120, −0.00014)	**0.014**	20–77	93
	CIII/CS	0.00153	(−0.00112, 0.00418)	0.247	23–77	31
	CIV/CS	*** **0.00132**	(0.00070, 0.00194)	**0.000**	20–77	93
AD	PRP	−2009	(−4306, 288)	0.086	50–94	196
	CS	−0.020	(−0.426, 0.387)	0.924	56–91	86
	CI/CS	0.00482	(−0.00293, 0.01258)	0.219	56–91	81
	CII/CS	−0.00036	(−0.00166, 0.00093)	0.577	56–91	82
	CIII/CS	−0.00375	(−0.01175, 0.00424)	0.339	56–88	22
	CIV/CS	−0.00174	(−0.00448, 0.00100)	0.209	56–91	83
MDD	PRP	−728	(−2531, 1075)	0.426	20–74	120
	CS	−0.217	(−0.452, 0.017)	0.069	20–74	111
	CI/CS	0.00178	(−0.00088, 0.00445)	0.188	20–74	107
	CII/CS	−0.00006	(−0.00060, 0.00048)	0.829	20–74	107
	CIII/CS	−0.00038	(−0.00470, 0.00393)	0.850	29–74	15
	CIV/CS	0.00044	(−0.00063, 0.00152)	0.418	20–74	107
BAD	PRP	−686	(−2808, 1437)	0.524	20–73	127
	CS	−0.021	(−0.305, 0.263)	0.884	21–66	98
	CI/CS	−0.00180	(−0.01321, 0.00962)	0.755	21–66	97
	CII/CS	0.00044	(−0.00105, 0.00193)	0.559	21–66	96
	CIII/CS	ND	ND	ND	ND	0
	CIV/CS	0.00002	(−0.00179, 0.00183)	0.982	21–66	98
VD	PRP	5408	(−12379, 1561)	0.123	65–85	32
All	PRP	*** **−1606**	(−2205, −1006)	**<0.001**	20–94	637
	CS	−0.073	(−0.162, 0.015)	0.105	20–91	395
	CI/CS	−0.00172	(−0.00412, 0.00067)	0.158	20–91	378
	CII/CS	0.00004	(−0.00027, 0.00036)	0.794	20–91	378
	CIII/CS	** **0.00233**	(0.00064, 0.00402)	**0.008**	23–88	68
	CIV/CS	** **0.00064**	(0.00017, 0.00111)	**0.008**	20–91	381

Analyzed with simple regression. Statistically significant differences compared with controls is presented as * $p < 0.05$, ** $p < 0.01$, and *** $p < 0.001$. Statistically significant values are in bold. 95% CI, 95% confidence interval; PRP, platelet rich plasma. The meaning of the abbreviations is the same as in Table 1.

To assess the effect of depression on the age dependence of mitochondrial respiration, regression slopes were calculated for the subgroup of AD patients with depression (AD + DEP) and without depression (AD − DEP). Depression in AD was diagnosed for a Geriatric Depression Score > 6 [93]. A significant negative slope was found for the age dependence of ROUT_i and ROT_i in AD + DEP but not in AD − DEP (Tables 3 and 6).

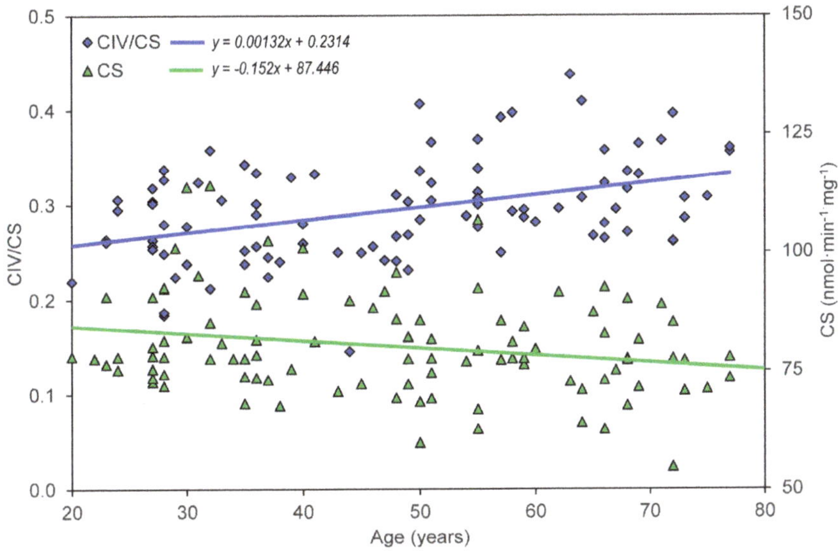

Figure 1. Age dependence of citrate synthase activity (CS) and complex IV activity normalized for CS (CIV/CS) in healthy controls.

Table 3. Age dependence of mitochondrial respiration parameter ROUT_i.

Sample	Slope (95% CI)		p	Age Range	Valid N
CONTROL	* **−0.309**	(−0.604, −0.014)	**0.040**	22–80	101
AD	−0.087	(−0.547, 0.374)	0.710	50–94	132
MDD	−0.013	(−0.308, 0.281)	0.927	20–74	42
BAD	−0.175	(−0.522, 0.172)	0.320	21–66	113
VD	−0.617	(−2.153, 0.919)	0.417	65–85	29
All	*** **−0.212**	(−0.328, −0.095)	**0.000**	20–94	417
AD + DEP	* **−1.067**	(−2.129, −0.005)	**0.049**	60–85	37
AD − DEP	0.023	(−0.474, 0.520)	0.927	50–94	95
MMSE > 23	* **−0.151**	(−0.295, −0.006)	**0.041**	20–91	328
MMSE <19,23>	−0.399	(−1.461, 0.663)	0.451	59–88	37
MMSE < 19	−0.100	(−1.049, 0.850)	0.834	67–94	52

Analyzed with simple regression. Statistically significant differences compared with controls is presented as * $p < 0.05$ and *** $p < 0.001$. Statistically significant values are in bold. DEP, depression; MMSE, Mini-Mental State Examination. The meaning of the other abbreviations is the same as in Tables 1 and 2. Slope is given in fmol $O_2 \cdot 10^{-6}$ platelets$\cdot$sec$^{-1}\cdot$year^{-1}.

Table 4. Age dependence of mitochondrial respiration parameter LEAK_i.

Sample	Slope (95% CI)		p	Age Range	Valid N
CONTROL	0.038	(−0.044, 0.121)	0.361	22–80	101
AD	−0.013	(−0.118, 0.091)	0.799	50–94	132
MDD	* **−0.117**	(−0.219, −0.014)	**0.027**	20–74	42
BAD	0.004	(−0.071, 0.078)	0.923	21–66	113
VD	−0.076	(−0.448, 0.296)	0.678	65–85	29
All	* **0.033**	(0.005, 0.060)	**0.019**	20–94	417
AD + DEP	−0.009	(−0.221, 0.203)	0.933	60–85	37
AD − DEP	−0.013	(−0.138, 0.113)	0.841	50–94	95
MMSE > 23	0.033	(−0.001, 0.067)	0.058	20–91	328
MMSE <19,23>	−0.047	(−0.298, 0.205)	0.707	59–88	37
MMSE < 19	0.008	(−0.214, 0.231)	0.941	67–94	52

Analyzed with simple regression. Statistically significant differences compared with controls is presented as * $p < 0.05$. Statistically significant values are in bold. The meaning of the abbreviations is the same as in Tables 1–3.

Table 5. Age dependence of mitochondrial respiration parameter ETSC_i.

Sample	Slope (95% CI)		p	Age Range	Valid N
CONTROL	** **−0.523**	(−0.879, −0.167)	**0.004**	22–80	101
AD	0.106	(−0.404, 0.616)	0.682	50–94	132
MDD	0.163	(−0.347, 0.673)	0.523	20–74	42
BAD	−0.155	(−0.640, 0.329)	0.527	21–66	113
VD	−1.057	(−3.139, 1.026)	0.307	65–85	29
All	*** **−0.412**	(−0.555, −0.268)	**0.000**	20–94	417
AD + DEP	−0.502	(−1.639, 0.636)	0.377	60–85	37
AD − DEP	0.041	(−0.503, 0.585)	0.881	50–94	95
MMSE > 23	** **−0.300**	(−0.483, −0.116)	**0.001**	20–91	328
MMSE <19,23>	−0.213	(−1.302, 0.876)	0.694	59–88	37
MMSE < 19	0.276	(−0.674, 1.226)	0.562	67–94	52

Analyzed with simple regression. Statistically significant differences compared with controls is presented as ** $p < 0.01$ and *** $p < 0.001$. Statistically significant values are in bold. The meaning of the abbreviations is the same as in Tables 1–3.

Table 6. Age dependence of mitochondrial respiration parameter ROT_i.

Sample	Slope (95% CI)		p	Age Range	Valid N
CONTROL	*** **−0.154**	(−0.231, −0.076)	**0.000**	22–80	101
AD	0.059	(−0.067, 0.185)	0.356	50–94	132
MDD	0.080	(−0.031, 0.191)	0.152	20–74	42
BAD	0.026	(−0.032, 0.085)	0.370	21–66	113
VD	−0.082	(−0.481, 0.317)	0.676	65–85	29
All	−0.028	(−0.058, 0.003)	0.072	20–94	417
AD + DEP	** **0.346**	(0.108, 0.584)	**0.006**	60–85	37
AD − DEP	−0.026	(−0.176, 0.123)	0.728	50–94	95
MMSE > 23	* **−0.042**	(−0.078, −0.006)	**0.023**	20–91	328
MMSE <19,23>	−0.145	(−0.487, 0.197)	0.394	59–88	37
MMSE < 19	0.032	(−0.260, 0.324)	0.827	67–94	52

Analyzed with simple regression. Statistically significant differences compared with controls is presented as * $p < 0.05$, ** $p < 0.01$, and *** $p < 0.001$. Statistically significant values are in bold. The meaning of the abbreviations is the same as in Tables 1–3.

Table 7. Age dependence of mitochondrial respiration parameter RES_i.

Sample	Slope (95% CI)		p	Age Range	Valid N
CONTROL	* **−0.214**	(−0.409, −0.019)	**0.032**	22–80	101
AD	0.193	(−0.151, 0.536)	0.270	50–94	132
MDD	0.197	(−0.127, 0.522)	0.226	20–74	42
BAD	0.020	(−0.224, 0.263)	0.874	21–66	113
VD	−0.440	(−1.382, 0.503)	0.347	65–85	29
All	*** **−0.201**	(−0.285, −0.117)	**0.000**	20–94	417
AD + DEP	0.565	(−0.192, 1.322)	0.139	60–85	37
AD − DEP	0.018	(−0.374, 0.410)	0.928	50–94	95
MMSE > 23	** **−0.148**	(−0.249, −0.046)	**0.004**	20–91	328
MMSE <19,23>	0.186	(−0.636, 1.008)	0.649	59–88	37
MMSE < 19	0.376	(−0.363, 1.115)	0.312	67–94	52

Analyzed with simple regression. Statistically significant differences compared with controls is presented as * $p < 0.05$, ** $p < 0.01$, and *** $p < 0.001$. Statistically significant values are in bold. The meaning of the abbreviations is the same as in Tables 1–3.

The effect of cognitive impairment on the age dependence of mitochondrial respiration was determined by dividing all included controls and neuropsychiatric patients into three groups according to their MMSE score: (1) no cognitive impairment with MMSE > 23 (N = 328), (2) mild cognitive impairment with MMSE in the interval <19,23> (N = 37), and (3) moderate or severe cognitive impairment with MMSE < 19 (N = 52). A significant negative slope was found for ROUT_i, ETSC_i, ROT_i, and RES_i in persons without significant cognitive impairment but not in persons with cognitive impairment (Tables 3–7, Figure 3).

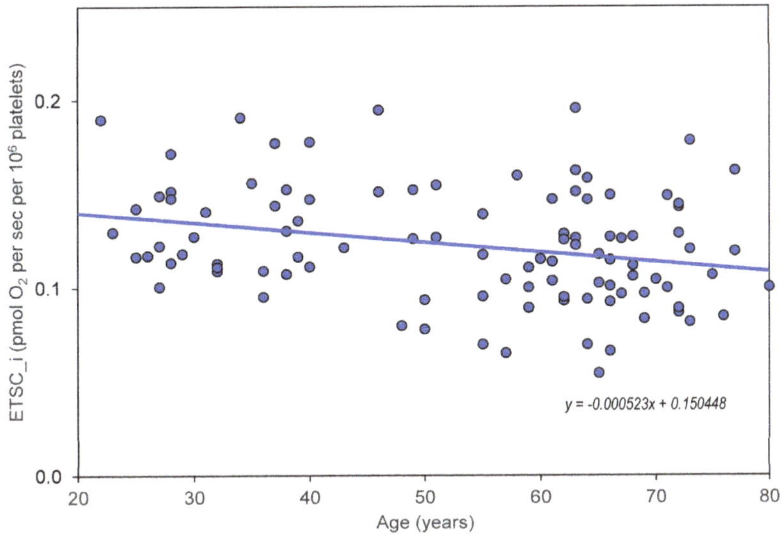

Figure 2. Age dependence of maximal capacity of electron transport system in intact platelets (ETSC_i) in healthy controls.

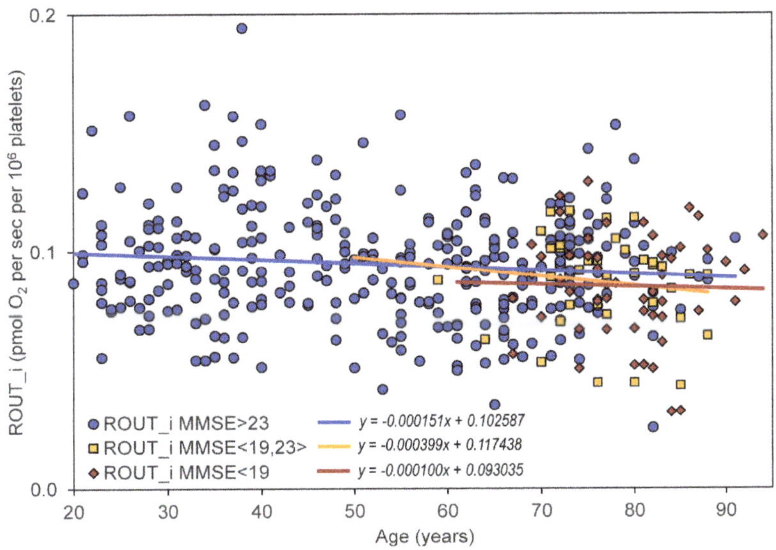

Figure 3. Age dependence of routine respiration state in intact platelets (ROUT_i) in subjects without cognitive impairment (MMSE > 23), with mild cognitive impairment (MMSE in the interval <19,23>), and with moderate or severe cognitive impairment (MMSE < 19).

4. Discussion

The age dependence of the mitochondrial parameters was assessed for citrate synthase activity, activity of respiratory chain complexes, and kinetics of mitochondrial oxygen consumption in platelets. The effect of cognitive impairment on the age dependence of the mitochondrial parameters was evaluated by comparing healthy individuals and individuals with neuropsychiatric disease (AD, VD, MDD, and BAD). Because it is not known whether the progression of mitochondrial dysfunction is different in neuropsychiatric diseases, we separately evaluated data for controls, AD, VD, MDD, and BAD, as well as pooled data. The progression of changes in mitochondrial respiration was evaluated using simple regression. We chose the significance and magnitude of the slope of the regression line as a criterion for evaluating the effect of age on the value of the mitochondrial parameter.

The significant correlation between age and measured mitochondrial parameters was negative in subjects without neuropsychiatric disease, except for complex IV activity normalized for CS (Table 1), whereas unnormalized complex IV activity did not change significantly with age. Because (i) CS activity normalized for platelet count does not show a significant correlation with age and (ii) CS and complex I activities are associated with mitochondrial content, while complex IV activity is associated with OXPHOS capacity [100], our data indicate that the reduction of mitochondrial content with age is due to a reduced number of platelets, rather than a reduced content of mitochondria in platelets. Regression analysis quantified the results by correlations (Tables 2–7). This indicates that the activity of individual mitochondrial enzymes and enzyme complexes involved in OXPHOS system, as well as the complex parameters of mitochondrial respiration, decrease with age. The increase in complex IV activity of the respiratory chain with age may be part of a compensatory mechanism to maintain cellular homeostasis. Non-significant correlations between age and measured mitochondrial parameters in persons with neuropsychiatric diseases (Table 1) indicate a small influence of these diseases on mitochondrial dysfunction associated with aging.

Our results indicate that the values of the mitochondrial respiration parameters ROUT_i, ETSC_i, LEAK_i, and RES_i decrease significantly with age, while the time course is not significantly different in controls and in neuropsychiatric diseases such as AD, VD, BAD, and MDD.

Depressive disorder is a frequent comorbidity in AD [101,102] and significantly reduces mitochondrial respiration in intact platelets. Therefore, age dependence was assessed separately in the subgroup of AD patients with and without depression. It can be speculated that the significant negative slope for ROUT_i, ETSC_i, ROT_i in AD patients with depression (Tables 5 and 6) is associated with comorbid depressive disorder.

The impairment of mitochondrial respiration in intact platelets was previously described in AD but was not associated with disease progression [82]. A significant negative slope for the age dependence of the respiratory parameters ROUT_i, ETSC_i, ROT_i, and RES_i in persons without cognitive impairment and statistically insignificant changes in slopes in persons with cognitive impairment (Tables 3–7) indicate that mitochondrial dysfunction (manifested by changes in mitochondrial oxygen consumption in intact platelets) is probably not a measure of cognitive impairment in AD. A certain limitation for this conclusion is the inclusion of a relatively low number of persons with severe cognitive impairment in the analysis.

The age dependence of mitochondrial respiration on cognitive impairment supports the hypothesis that progressive neurodegeneration in AD is associated with specific neurotoxicity of amyloid beta oligomers and tau protein, rather than direct consequences of mitochondrial dysfunction. However, age-related neuropsychiatric diseases can be initiated and regulated by mitochondrial dysfunction [37].

The results support a possible role of mitochondrial dysfunction in the regulation of aging processes. The progression of mitochondrial dysfunction (determined from measurements of mitochondrial respiration in intact platelets) does not appear to be significantly altered. Assuming that mitochondrial respiration in intact platelets reflects mitochondrial changes in the brain, the reduction of mitochondrial respiration does not appear to be the primary cause of neurodegeneration associated with aging but is only part of complex processes associated with impaired brain neuroplasticity.

In summary, a significant decrease with age was found for both the activity of mitochondrial enzymes (CS and CII) and the parameters of mitochondrial respiration measured in intact platelets (ROUT_i, ETSC_i, ROT_i, and RES_i). Compared to the large interindividual differences in these parameters, the rate of their reduction is not too large with age, and it can be expected that in healthy aging mitochondrial dysfunction can be regulated to a certain extent by lifestyle, primarily by exercise and diet [103–105].

Age-related neuropsychiatric diseases, such as AD, are associated with a decrease in mitochondrial respiration in intact platelets [82,84], but no significant change in age dependence of respiratory parameters was observed in platelets from AD patients compared to healthy controls (Figure 3). This supports the notion that the direct contribution of mitochondrial dysfunction to neurodegeneration is similar in aging and in AD, and that AD progression is driven more by amyloid beta and tau pathology. Mitochondrial dysfunction remains a promising candidate as an initial trigger and/or synergistic promoter of this specific pathophysiology [37]. For pathology of neurodegenerative diseases that have already begun, the therapeutic strategy should therefore include, in addition to dietary restriction and exercise (discussed under healthy aging), targeted pharmacological regulation of the causes of the development of specific disease pathology. To confirm the role of mitochondrial dysfunction in neurodegeneration associated with healthy aging and age-related neuropsychiatric diseases, repeated measurements of appropriate mitochondrial variables must be performed in the same persons over long periods of time. Suitable variables could be parameters of mitochondrial respiration in intact platelets.

4.1. Study Limitations

A limitation of mitochondrial dysfunction research in aging and neurodegeneration is the difficult identification of specific mitochondrial mechanisms and interactions with other cellular causes of aging and neurodegeneration. Mitochondrial biomarkers of aging and neurodegeneration are therefore still being sought. Some limitations in the study of age-dependent mitochondrial dysfunction include lack of standardized methods for measuring mitochondrial dysfunction. The development of standardized methods for measuring mitochondrial respiration is currently being intensively addressed. Most measurements have been conducted using animal models and cell lines. However, the use of intact platelets from peripheral blood [57] used in this study may solve the limited availability of human tissue samples to measure mitochondrial dysfunction. Although this is a good model for monitoring changes in a number of biochemical parameters in the brain that manifest themselves systemically, we must be aware that this is a model of brain changes. For comprehensive assessment of mitochondrial dysfunction during aging, a combination of commonly used methods should be used, including mitochondrial respiration, ROS production, mtDNA mutations, mitochondrial membrane potential, mitochondrial morphology and dynamics, mitochondrial biogenesis and mitophagy (see Introduction).

In this study, mitochondrial respiration was normalized for platelet concentration, and activities of mitochondrial complexes were normalized for CS activity. CS activity can be used as a marker for mitochondrial dysfunction, and a decline in CS activity has been linked to a decrease in the number of mitochondria and changes in mitochondrial morphology [100]. CS activity in the muscles may decrease with aging [106,107]. Thus, CS activity may be used to normalize other mitochondrial parameters, such as the activity of respiratory chain complexes [108] and mitochondrial respiration. However, using CS activity to normalize mitochondrial data means additional measurements and a possible source of error. Due to the good correlation between CS activity and platelet count in PRP, the normalization of mitochondrial respiration for platelet count appears to be suitable for measurements in intact platelets [82]. In the case of evaluating the overall activity of the OXPHOS system (measured by mitochondrial respiration), we consider normalization to the platelet count in the sample to be more accurate and simpler.

4.2. Clinical Significance

There are several treatment strategies to prevent or to slow down cognitive decline due to age or neurodegenerative disease. This is not only about medication (such as cholinesterase inhibitors, NMDA receptor antagonists, or targeting glutamatergic, noradrenergic, and endocannabinoid systems) [109], but also about exercise, diet, cognitive training, and social engagement.

Mitochondrial dysfunctions measured as disruption of mitochondrial respiration in intact platelets are common features of aging and neurodegenerative diseases, such as AD. Our study showed that the progression of mitochondrial respiration reduction with age does not appear to be significantly affected by neurodegenerative disease. The strategy of improving mitochondrial function in aging and neurodegeneration may include regulation of the OXPHOS system at the level of stimulation of mitochondrial biogenesis, activity of individual mitochondrial enzymes of the citrate cycle and respiratory chain complexes, and changing the availability of substrates of the electron transport system.

The research on age-dependent mitochondrial dysfunction has both theoretical implications and translational applications. Theoretical implications of age-dependent mitochondrial dysfunction research include understanding the mechanisms of aging and neurodegeneration. Translational applications of this research include finding new biomarkers for the diagnosis of age-related diseases and for finding molecular targets of new drugs effective in preventing or mitigating cognitive decline and neurodegeneration. Understanding the mechanism of onset and progression of mitochondrial dysfunction in aging and neurodegeneration can lead to the development of new therapies that target specific mitochondrial processes and are able to slow down the onset of age-related diseases.

4.3. Research Perspectives

Both aging and mitochondrial dysfunction are complex processes regulated by a variety of factors, including lifestyle and genetic, epigenetic, and internal and external environmental factors. The perspective in the research on mitochondrial dysfunction during aging and neurodegeneration therefore consists in a multidisciplinary approach combining genetic, biochemical, and physiological measurements with clinical evaluation of neurodegeneration and cognitive impairment. In addition, it is necessary to consider other biological processes that contribute to the process of aging and neurodegeneration, including neuroinflammation, apoptosis, and neurotoxicity. More accurate and detailed assessment of mitochondrial dysfunction in aging and neurodegeneration will be provided by advances in technologies for measuring mitochondrial function, further development of animal models of age-related diseases, development of methods for measuring mitochondrial biomarkers in peripheral blood, and application of omics technologies.

The regulation of mitochondrial function, neuroplasticity, and neurotransmission are means of prospective prevention and treatment of neurodegenerative diseases. However, careful monitoring of side and adverse effects of potential drugs is necessary, as stimulation of mitochondrial functions can also lead to negative neuroplasticity. Finally, there are large interindividual differences in mitochondrial dysfunction during aging and neurodegeneration, and treatment/regulation should therefore be based on a personalized approach.

5. Conclusions

Study of mitochondrial dysfunction during aging and neurodegeneration is an important area of research with the perspective of potential therapeutic use of new knowledge. The findings showed the potential of measuring mitochondrial respiration in intact platelets to assess age-related mitochondrial dysfunction and support the role of mitochondrial dysfunction in the aging process. Mitochondrial respiration in platelets may serve as a biomarker for aging and cognitive decline and it can be expected that interventions that improve mitochondrial function may prevent or slow cognitive decline in aging and age-related diseases. Mitochondrial respiration in platelets appears to be a potential biomarker of aging, regardless of the degree of cognitive impairment.

Author Contributions: Conceptualization, Z.F.; Methodology, Z.F., J.H., M.Z., R.J., E.K. and J.R.; Formal analysis, Z.F. and J.H.; Investigation, J.H.; Data curation, M.Z., R.J. and E.K.; Writing—original draft, Z.F.; Writing—review & editing, Z.F., J.H., M.Z., R.J., E.K. and J.R.; Supervision, Z.F. and J.R.; Funding acquisition, J.R. and Z.F. All authors have read and agreed to the published version of the manuscript.

Funding: This work was supported by Charles University, Prague, Czech Republic (project Cooperatio, research area Neurosciences) and by the Ministry of Health, Czech Republic (grant number MH CZ-DRO VFN64165).

Institutional Review Board Statement: The study was performed in line with the principles of the Declaration of Helsinki. Approval was granted by the Ethics Committee of the First Faculty of Medicine, Charles University and General University Hospital in Prague (protocol code 20/19 Grant AZV VES 2020 1.LF UK; date of approval: 29 May 2019).

Informed Consent Statement: Written informed consent was obtained from all subjects involved in the study. Each participant of the study signed an informed consent stating that all personal data will be anonymized during the evaluation of the study and that all medical data relevant to this study will be used for scientific research and may be published.

Data Availability Statement: The datasets generated and analyzed during the current study are not publicly available due to the fact that individual privacy could be compromised, but anonymized data are available from the corresponding author on reasonable request.

Acknowledgments: This work was supported by Charles University, Prague, Czech Republic (project Cooperatio, research area Neurosciences) and the Ministry of Health, Czech Republic (grant number MH CZ-DRO VFN64165).

Conflicts of Interest: The authors declare no conflict of interest.

References

1. Rysz, J.; Franczyk, B.; Rysz-Gorzynska, M.; Gluba-Brzozka, A. Ageing, Age-Related Cardiovascular Risk and the Beneficial Role of Natural Components Intake. *Int. J. Mol. Sci.* **2021**, *23*, 183. [CrossRef] [PubMed]
2. Lopez-Otin, C.; Blasco, M.A.; Partridge, L.; Serrano, M.; Kroemer, G. Hallmarks of aging: An expanding universe. *Cell* **2023**, *186*, 243–278. [CrossRef]
3. Zia, A.; Pourbagher-Shahri, A.M.; Farkhondeh, T.; Samarghandian, S. Molecular and cellular pathways contributing to brain aging. *Behav. Brain Funct.* **2021**, *17*, 6. [CrossRef] [PubMed]
4. Catanesi, M.; d'Angelo, M.; Tupone, M.G.; Benedetti, E.; Giordano, A.; Castelli, V.; Cimini, A. MicroRNAs Dysregulation and Mitochondrial Dysfunction in Neurodegenerative Diseases. *Int. J. Mol. Sci.* **2020**, *21*, 5986. [CrossRef] [PubMed]
5. Cui, H.; Kong, Y.; Zhang, H. Oxidative stress, mitochondrial dysfunction, and aging. *J. Signal Transduct.* **2012**, *2012*, 646354. [CrossRef] [PubMed]
6. Hroudova, J.; Singh, N.; Fisar, Z. Mitochondrial dysfunctions in neurodegenerative diseases: Relevance to Alzheimer's disease. *Biomed. Res. Int.* **2014**, *2014*, 175062. [CrossRef]
7. Grimm, A.; Eckert, A. Brain aging and neurodegeneration: From a mitochondrial point of view. *J. Neurochem.* **2017**, *143*, 418–431. [CrossRef]
8. Sini, P.; Dang, T.B.C.; Fais, M.; Galioto, M.; Padedda, B.M.; Luglie, A.; Iaccarino, C.; Crosio, C. Cyanobacteria, Cyanotoxins, and Neurodegenerative Diseases: Dangerous Liaisons. *Int. J. Mol. Sci.* **2021**, *22*, 8726. [CrossRef]
9. Tanaka, M.; Szabo, A.; Spekker, E.; Polyak, H.; Toth, F.; Vecsei, L. Mitochondrial Impairment: A Common Motif in Neuropsychiatric Presentation? The Link to the Tryptophan-Kynurenine Metabolic System. *Cells* **2022**, *11*, 2607. [CrossRef]
10. Tanaka, M.; Vecsei, L. Editorial of Special Issue "Dissecting Neurological and Neuropsychiatric Diseases: Neurodegeneration and Neuroprotection". *Int. J. Mol. Sci.* **2022**, *23*, 6991. [CrossRef]
11. Tanaka, M.; Toldi, J.; Vecsei, L. Exploring the Etiological Links behind Neurodegenerative Diseases: Inflammatory Cytokines and Bioactive Kynurenines. *Int. J. Mol. Sci.* **2020**, *21*, 2431. [CrossRef] [PubMed]
12. Picca, A.; Calvani, R.; Coelho-Junior, H.J.; Landi, F.; Bernabei, R.; Marzetti, E. Mitochondrial Dysfunction, Oxidative Stress, and Neuroinflammation: Intertwined Roads to Neurodegeneration. *Antioxidants* **2020**, *9*, 647. [CrossRef] [PubMed]
13. Brunetti, D.; Catania, A.; Viscomi, C.; Deleidi, M.; Bindoff, L.A.; Ghezzi, D.; Zeviani, M. Role of PITRM1 in Mitochondrial Dysfunction and Neurodegeneration. *Biomedicines* **2021**, *9*, 833. [CrossRef]
14. Wu, Y.; Chen, M.; Jiang, J. Mitochondrial dysfunction in neurodegenerative diseases and drug targets via apoptotic signaling. *Mitochondrion* **2019**, *49*, 35–45. [CrossRef] [PubMed]
15. Shen, X.; Sun, P.; Zhang, H.; Yang, H. Mitochondrial quality control in the brain: The physiological and pathological roles. *Front. Neurosci.* **2022**, *16*, 1075141. [CrossRef]

16. Lee, D.; Jo, M.G.; Kim, S.Y.; Chung, C.G.; Lee, S.B. Dietary Antioxidants and the Mitochondrial Quality Control: Their Potential Roles in Parkinson's Disease Treatment. *Antioxidants* **2020**, *9*, 1056. [CrossRef]
17. Mitchell, P.; Moyle, J. Chemiosmotic hypothesis of oxidative phosphorylation. *Nature* **1967**, *213*, 137–139. [CrossRef]
18. Harman, D. Aging: A theory based on free radical and radiation chemistry. *J. Gerontol.* **1956**, *11*, 298–300. [CrossRef]
19. Harman, D. The aging process. *Proc. Natl. Acad. Sci. USA* **1981**, *78*, 7124–7128. [CrossRef] [PubMed]
20. Harman, D. Origin and evolution of the free radical theory of aging: A brief personal history, 1954–2009. *Biogerontology* **2009**, *10*, 773–781. [CrossRef] [PubMed]
21. Singh, A.; Kukreti, R.; Saso, L.; Kukreti, S. Oxidative Stress: A Key Modulator in Neurodegenerative Diseases. *Molecules* **2019**, *24*, 1583. [CrossRef]
22. Ristow, M.; Schmeisser, S. Extending life span by increasing oxidative stress. *Free Radic. Biol. Med.* **2011**, *51*, 327–336. [CrossRef]
23. Jang, J.Y.; Blum, A.; Liu, J.; Finkel, T. The role of mitochondria in aging. *J. Clin. Invest.* **2018**, *128*, 3662–3670. [CrossRef]
24. Son, J.M.; Lee, C. Mitochondria: Multifaceted regulators of aging. *BMB Rep.* **2019**, *52*, 13–23. [CrossRef] [PubMed]
25. Son, J.M.; Lee, C. Aging: All roads lead to mitochondria. *Semin. Cell Dev. Biol.* **2021**, *116*, 160–168. [CrossRef]
26. Martini, H.; Passos, J.F. Cellular senescence: All roads lead to mitochondria. *FEBS J.* **2023**, *290*, 1186–1202. [CrossRef] [PubMed]
27. Chistiakov, D.A.; Sobenin, I.A.; Revin, V.V.; Orekhov, A.N.; Bobryshev, Y.V. Mitochondrial aging and age-related dysfunction of mitochondria. *Biomed. Res. Int.* **2014**, *2014*, 238463. [CrossRef]
28. Wei, Y.H.; Wu, S.B.; Ma, Y.S.; Lee, H.C. Respiratory function decline and DNA mutation in mitochondria, oxidative stress and altered gene expression during aging. *Chang Gung Med. J.* **2009**, *32*, 113–132.
29. Lee, H.C.; Wei, Y.H. Oxidative stress, mitochondrial DNA mutation, and apoptosis in aging. *Exp. Biol. Med.* **2007**, *232*, 592–606.
30. Jagtap, Y.A.; Kumar, P.; Kinger, S.; Dubey, A.R.; Choudhary, A.; Gutti, R.K.; Singh, S.; Jha, H.C.; Poluri, K.M.; Mishra, A. Disturb mitochondrial associated proteostasis: Neurodegeneration and imperfect ageing. *Front. Cell Dev. Biol.* **2023**, *11*, 1146564. [CrossRef] [PubMed]
31. Mangrulkar, S.V.; Wankhede, N.L.; Kale, M.B.; Upaganlawar, A.B.; Taksande, B.G.; Umekar, M.J.; Anwer, M.K.; Dailah, H.G.; Mohan, S.; Behl, T. Mitochondrial Dysfunction as a Signaling Target for Therapeutic Intervention in Major Neurodegenerative Disease. *Neurotox. Res.* **2023**, 1–22. [CrossRef]
32. Bagheri-Mohammadi, S.; Farjami, M.; Suha, A.J.; Zarch, S.M.A.; Najafi, S.; Esmaeili, A. The mitochondrial calcium signaling, regulation, and cellular functions: A novel target for therapeutic medicine in neurological disorders. *J. Cell. Biochem.* **2023**, *124*, 635–655. [CrossRef] [PubMed]
33. Rehman, M.U.; Sehar, N.; Dar, N.J.; Khan, A.; Arafah, A.; Rashid, S.; Rashid, S.M.; Ganaie, M.A. Mitochondrial dysfunctions, oxidative stress and neuroinflammation as therapeutic targets for neurodegenerative diseases: An update on current advances and impediments. *Neurosci. Biobehav. Rev.* **2023**, *144*, 104961. [CrossRef] [PubMed]
34. Li, R.L.; Wang, L.Y.; Duan, H.X.; Zhang, Q.; Guo, X.; Wu, C.; Peng, W. Regulation of mitochondrial dysfunction induced cell apoptosis is a potential therapeutic strategy for herbal medicine to treat neurodegenerative diseases. *Front. Pharmacol.* **2022**, *13*, 937289. [CrossRef] [PubMed]
35. Chaturvedi, R.K.; Flint Beal, M. Mitochondrial diseases of the brain. *Free Radic. Biol. Med.* **2013**, *63*, 1–29. [CrossRef]
36. Fernandes, T.; Resende, R.; Silva, D.F.; Marques, A.P.; Santos, A.E.; Cardoso, S.M.; Domingues, M.R.; Moreira, P.I.; Pereira, C.F. Structural and Functional Alterations in Mitochondria-Associated Membranes (MAMs) and in Mitochondria Activate Stress Response Mechanisms in an In Vitro Model of Alzheimer's Disease. *Biomedicines* **2021**, *9*, 881. [CrossRef]
37. Fišar, Z. Linking the Amyloid, Tau, and Mitochondrial Hypotheses of Alzheimer's Disease and Identifying Promising Drug Targets. *Biomolecules* **2022**, *12*, 1676. [CrossRef]
38. Klinedinst, N.J.; Regenold, W.T. A mitochondrial bioenergetic basis of depression. *J. Bioenerg. Biomembr.* **2015**, *47*, 155–171. [CrossRef]
39. Sigitova, E.; Fisar, Z.; Hroudova, J.; Cikankova, T.; Raboch, J. Biological hypotheses and biomarkers of bipolar disorder. *Psychiatry Clin. Neurosci.* **2017**, *71*, 77–103. [CrossRef]
40. Kato, T. The role of mitochondrial dysfunction in bipolar disorder. *Drug News Perspect.* **2006**, *19*, 597–602. [CrossRef]
41. Kato, T. Neurobiological basis of bipolar disorder: Mitochondrial dysfunction hypothesis and beyond. *Schizophr. Res.* **2017**, *187*, 62–66. [CrossRef]
42. Stork, C.; Renshaw, P.F. Mitochondrial dysfunction in bipolar disorder: Evidence from magnetic resonance spectroscopy research. *Mol. Psychiatry* **2005**, *10*, 900–919. [CrossRef] [PubMed]
43. Baquero, M.; Martin, N. Depressive symptoms in neurodegenerative diseases. *World J. Clin. Cases* **2015**, *3*, 682–693. [CrossRef]
44. Hammar, A.; Ronold, E.H.; Rekkedal, G.A. Cognitive Impairment and Neurocognitive Profiles in Major Depression—A Clinical Perspective. *Front. Psychiatry* **2022**, *13*, 764374. [CrossRef] [PubMed]
45. Masse, C.; Chopard, G.; Bennabi, D.; Haffen, E.; Vandel, P. Cognitive functions in late-life depression. *Geriatr. Psychol. Neuropsychiatr. Vieil.* **2022**, *19*, 202–210. [CrossRef]
46. Huang, Y.; Zhang, Z.; Lin, S.; Zhou, H.; Xu, G. Cognitive Impairment Mechanism in Patients with Bipolar Disorder. *Neuropsychiatr. Dis. Treat.* **2023**, *19*, 361–366. [CrossRef] [PubMed]
47. Pessoa, L.; Padmala, S.; Kenzer, A.; Bauer, A. Interactions between cognition and emotion during response inhibition. *Emotion* **2012**, *12*, 192–197. [CrossRef]

48. Battaglia, S.; Serio, G.; Scarpazza, C.; D'Ausilio, A.; Borgomaneri, S. Frozen in (e)motion: How reactive motor inhibition is influenced by the emotional content of stimuli in healthy and psychiatric populations. *Behav. Res. Ther.* **2021**, *146*, 103963. [CrossRef] [PubMed]
49. Battaglia, S.; Cardellicchio, P.; Di Fazio, C.; Nazzi, C.; Fracasso, A.; Borgomaneri, S. Stopping in (e)motion: Reactive action inhibition when facing valence-independent emotional stimuli. *Front. Behav. Neurosci.* **2022**, *16*, 998714. [CrossRef]
50. Roberts-Wolfe, D.; Sacchet, M.D.; Hastings, E.; Roth, H.; Britton, W. Mindfulness training alters emotional memory recall compared to active controls: Support for an emotional information processing model of mindfulness. *Front. Hum. Neurosci.* **2012**, *6*, 15. [CrossRef] [PubMed]
51. Salat, D.H.; Kaye, J.A.; Janowsky, J.S. Selective preservation and degeneration within the prefrontal cortex in aging and Alzheimer disease. *Arch. Neurol.* **2001**, *58*, 1403–1408. [CrossRef]
52. Lezi, E.; Swerdlow, R.H. Mitochondria in neurodegeneration. *Adv. Exp. Med. Biol.* **2012**, *942*, 269–286. [CrossRef]
53. Navarro, A.; Boveris, A. Brain mitochondrial dysfunction in aging, neurodegeneration, and Parkinson's disease. *Front. Aging Neurosci.* **2010**, *2*, 34. [CrossRef] [PubMed]
54. Brand, M.D.; Nicholls, D.G. Assessing mitochondrial dysfunction in cells. *Biochem. J.* **2011**, *435*, 297–312. [CrossRef] [PubMed]
55. Junker, A.; Wang, J.; Gouspillou, G.; Ehinger, J.K.; Elmer, E.; Sjovall, F.; Fisher-Wellman, K.H.; Neufer, P.D.; Molina, A.J.A.; Ferrucci, L.; et al. Human studies of mitochondrial biology demonstrate an overall lack of binary sex differences: A multivariate meta-analysis. *FASEB J.* **2022**, *36*, e22146. [CrossRef] [PubMed]
56. Jedlicka, J.; Kunc, R.; Kuncova, J. Mitochondrial respiration of human platelets in young adult and advanced age—Seahorse or O2k? *Physiol. Res.* **2021**, *70*, S369–S379. [CrossRef] [PubMed]
57. Fisar, Z.; Hroudova, J. Measurement of Mitochondrial Respiration in Platelets. *Methods Mol. Biol.* **2021**, *2277*, 269–276. [CrossRef]
58. Harker, L.A.; Roskos, L.K.; Marzec, U.M.; Carter, R.A.; Cherry, J.K.; Sundell, B.; Cheung, E.N.; Terry, D.; Sheridan, W. Effects of megakaryocyte growth and development factor on platelet production, platelet life span, and platelet function in healthy human volunteers. *Blood* **2000**, *95*, 2514–2522. [CrossRef] [PubMed]
59. Jenne, C.N.; Urrutia, R.; Kubes, P. Platelets: Bridging hemostasis, inflammation, and immunity. *Int. J. Lab. Hematol.* **2013**, *35*, 254–261. [CrossRef]
60. Braganza, A.; Corey, C.G.; Santanasto, A.J.; Distefano, G.; Coen, P.M.; Glynn, N.W.; Nouraie, S.M.; Goodpaster, B.H.; Newman, A.B.; Shiva, S. Platelet bioenergetics correlate with muscle energetics and are altered in older adults. *JCI Insight* **2019**, *5*, e128248. [CrossRef]
61. Leiter, O.; Walker, T.L. Platelets in Neurodegenerative Conditions-Friend or Foe? *Front. Immunol.* **2020**, *11*, 747. [CrossRef]
62. Canobbio, I. Blood platelets: Circulating mirrors of neurons? *Res. Pract. Thromb. Haemost.* **2019**, *3*, 564–565. [CrossRef] [PubMed]
63. Rivera, F.J.; Kazanis, I.; Ghevaert, C.; Aigner, L. Beyond Clotting: A Role of Platelets in CNS Repair? *Front. Cell Neurosci.* **2015**, *9*, 511. [CrossRef]
64. Burnouf, T.; Walker, T.L. The multifaceted role of platelets in mediating brain function. *Blood* **2022**, *140*, 815–827. [CrossRef] [PubMed]
65. Pandya, J.D.; Valdez, M.; Royland, J.E.; MacPhail, R.C.; Sullivan, P.G.; Kodavanti, P.R.S. Age-and Organ-Specific Differences in Mitochondrial Bioenergetics in Brown Norway Rats. *J. Aging Res.* **2020**, *2020*, 7232614. [CrossRef] [PubMed]
66. Pandya, J.D.; Royland, J.E.; MacPhail, R.C.; Sullivan, P.G.; Kodavanti, P.R. Age-and brain region-specific differences in mitochondrial bioenergetics in Brown Norway rats. *Neurobiol. Aging* **2016**, *42*, 25–34. [CrossRef]
67. Jedlicka, J.; Tuma, Z.; Razak, K.; Kunc, R.; Kala, A.; Proskauer Pena, S.; Lerchner, T.; Jezek, K.; Kuncova, J. Impact of aging on mitochondrial respiration in various organs. *Physiol. Res.* **2022**, *71*, S227–S236. [CrossRef]
68. Acin-Perez, R.; Beninca, C.; Shabane, B.; Shirihai, O.S.; Stiles, L. Utilization of Human Samples for Assessment of Mitochondrial Bioenergetics: Gold Standards, Limitations, and Future Perspectives. *Life* **2021**, *11*, 949. [CrossRef]
69. Doerrier, C.; Garcia-Souza, L.F.; Krumschnabel, G.; Wohlfarter, Y.; Meszaros, A.T.; Gnaiger, E. High-Resolution FluoRespirometry and OXPHOS Protocols for Human Cells, Permeabilized Fibers from Small Biopsies of Muscle, and Isolated Mitochondria. *Methods Mol. Biol.* **2018**, *1782*, 31–70. [CrossRef] [PubMed]
70. Chinopoulos, C.; Kiss, G.; Kawamata, H.; Starkov, A.A. Measurement of ADP-ATP exchange in relation to mitochondrial transmembrane potential and oxygen consumption. *Methods Enzymol.* **2014**, *542*, 333–348. [CrossRef]
71. Tretter, L.; Takacs, K.; Kover, K.; Adam-Vizi, V. Stimulation of H(2)O(2) generation by calcium in brain mitochondria respiring on alpha-glycerophosphate. *J. Neurosci. Res.* **2007**, *85*, 3471–3479. [CrossRef]
72. Makrecka-Kuka, M.; Krumschnabel, G.; Gnaiger, E. High-Resolution Respirometry for Simultaneous Measurement of Oxygen and Hydrogen Peroxide Fluxes in Permeabilized Cells, Tissue Homogenate and Isolated Mitochondria. *Biomolecules* **2015**, *5*, 1319–1338. [CrossRef]
73. Krumschnabel, G.; Eigentler, A.; Fasching, M.; Gnaiger, E. Use of safranin for the assessment of mitochondrial membrane potential by high-resolution respirometry and fluorometry. *Methods Enzymol.* **2014**, *542*, 163–181. [CrossRef]
74. Smiley, S.T.; Reers, M.; Mottola-Hartshorn, C.; Lin, M.; Chen, A.; Smith, T.W.; Steele, G.D., Jr.; Chen, L.B. Intracellular heterogeneity in mitochondrial membrane potentials revealed by a J-aggregate-forming lipophilic cation JC-1. *Proc. Natl. Acad. Sci. USA* **1991**, *88*, 3671–3675. [CrossRef]
75. Skrha, J., Jr.; Gall, J.; Buchal, R.; Sedlackova, E.; Platenik, J. Glucose and its metabolites have distinct effects on the calcium-induced mitochondrial permeability transition. *Folia Biol.* **2011**, *57*, 96–103.

76. Morkuniene, R.; Cizas, P.; Jankeviciute, S.; Petrolis, R.; Arandarcikaite, O.; Krisciukaitis, A.; Borutaite, V. Small A beta(1-42) Oligomer-Induced Membrane Depolarization of Neuronal and Microglial Cells: Role of N-Methyl-D-Aspartate Receptors. *J. Neurosci. Res.* **2015**, *93*, 475–486. [CrossRef] [PubMed]
77. Fisar, Z. Inhibition of monoamine oxidase activity by cannabinoids. *Naunyn-Schmiedeberg Arch. Pharmacol.* **2010**, *381*, 563–572. [CrossRef]
78. Aleardi, A.M.; Benard, G.; Augereau, O.; Malgat, M.; Talbot, J.C.; Mazat, J.P.; Letellier, T.; Dachary-Prigent, J.; Solaini, G.C.; Rossignol, R. Gradual alteration of mitochondrial structure and function by beta-amyloids: Importance of membrane viscosity changes, energy deprivation, reactive oxygen species production, and cytochrome c release. *J. Bioenerg. Biomembr.* **2005**, *37*, 207–225. [CrossRef] [PubMed]
79. Kato, T. DNA polymorphisms and bipolar disorder. *Am. J. Psychiatry* **2001**, *158*, 1169–1170. [CrossRef]
80. Song, W.; Bossy, B.; Martin, O.J.; Hicks, A.; Lubitz, S.; Knott, A.B.; Bossy-Wetzel, E. Assessing mitochondrial morphology and dynamics using fluorescence wide-field microscopy and 3D image processing. *Methods* **2008**, *46*, 295–303. [CrossRef]
81. Williams, J.A.; Zhao, K.; Jin, S.; Ding, W.X. New methods for monitoring mitochondrial biogenesis and mitophagy in vitro and in vivo. *Exp. Biol. Med.* **2017**, *242*, 781–787. [CrossRef] [PubMed]
82. Fisar, Z.; Hroudova, J.; Hansikova, H.; Spacilova, J.; Lelkova, P.; Wenchich, L.; Jirak, R.; Zverova, M.; Zeman, J.; Martasek, P.; et al. Mitochondrial Respiration in the Platelets of Patients with Alzheimer's Disease. *Curr. Alzheimer Res.* **2016**, *13*, 930–941. [CrossRef] [PubMed]
83. Fisar, Z.; Jirak, R.; Zverova, M.; Setnicka, V.; Habartova, L.; Hroudova, J.; Vanickova, Z.; Raboch, J. Plasma amyloid beta levels and platelet mitochondrial respiration in patients with Alzheimer's disease. *Clin. Biochem.* **2019**, *72*, 71–80. [CrossRef]
84. Fisar, Z.; Hansikova, H.; Krizova, J.; Jirak, R.; Kitzlerova, E.; Zverova, M.; Hroudova, J.; Wenchich, L.; Zeman, J.; Raboch, J. Activities of mitochondrial respiratory chain complexes in platelets of patients with Alzheimer's disease and depressive disorder. *Mitochondrion* **2019**, *48*, 67–77. [CrossRef]
85. Platenik, J.; Fisar, Z.; Buchal, R.; Jirak, R.; Kitzlerova, E.; Zverova, M.; Raboch, J. GSK3beta, CREB, and BDNF in peripheral blood of patients with Alzheimer's disease and depression. *Prog. Neuropsychopharmacol. Biol. Psychiatry* **2014**, *50*, 83–93. [CrossRef]
86. Hroudova, J.; Fisar, Z.; Kitzlerova, E.; Zverova, M.; Raboch, J. Mitochondrial respiration in blood platelets of depressive patients. *Mitochondrion* **2013**, *13*, 795–800. [CrossRef]
87. Zverova, M.; Hroudova, J.; Fisar, Z.; Hansikova, H.; Kalisova, L.; Kitzlerova, E.; Lambertova, A.; Raboch, J. Disturbances of mitochondrial parameters to distinguish patients with depressive episode of bipolar disorder and major depressive disorder. *Neuropsychiatr. Dis. Treat.* **2019**, *15*, 233–240. [CrossRef] [PubMed]
88. Hroudova, J.; Fisar, Z.; Hansikova, H.; Kalisova, L.; Kitzlerova, E.; Zverova, M.; Lambertova, A.; Raboch, J. Mitochondrial Dysfunction in Blood Platelets of Patients with Manic Episode of Bipolar Disorder. *CNS Neurol. Disord. Drug Targets* **2019**, *18*, 222–231. [CrossRef] [PubMed]
89. Roman, G.C.; Tatemichi, T.K.; Erkinjuntti, T.; Cummings, J.L.; Masdeu, J.C.; Garcia, J.H.; Amaducci, L.; Orgogozo, J.M.; Brun, A.; Hofman, A.; et al. Vascular dementia: Diagnostic criteria for research studies. Report of the NINDS-AIREN International Workshop. *Neurology* **1993**, *43*, 250–260. [CrossRef]
90. Dubois, B.; Feldman, H.H.; Jacova, C.; Dekosky, S.T.; Barberger-Gateau, P.; Cummings, J.; Delacourte, A.; Galasko, D.; Gauthier, S.; Jicha, G.; et al. Research criteria for the diagnosis of Alzheimer's disease: Revising the NINCDS-ADRDA criteria. *Lancet Neurol.* **2007**, *6*, 734–746. [CrossRef]
91. McKhann, G.; Drachman, D.; Folstein, M.; Katzman, R.; Price, D.; Stadlan, E.M. Clinical diagnosis of Alzheimer's disease: Report of the NINCDS-ADRDA Work Group under the auspices of Department of Health and Human Services Task Force on Alzheimer's Disease. *Neurology* **1984**, *34*, 939–944. [CrossRef] [PubMed]
92. Pantoni, L.; Inzitari, D. Hachinski's ischemic score and the diagnosis of vascular dementia: A review. *Ital. J. Neurol. Sci.* **1993**, *14*, 539–546. [CrossRef] [PubMed]
93. Yesavage, J.A. Geriatric Depression Scale. *Psychopharmacol. Bull.* **1988**, *24*, 709–711.
94. Mioshi, E.; Dawson, K.; Mitchell, J.; Arnold, R.; Hodges, J.R. The Addenbrooke's Cognitive Examination Revised (ACE-R): A brief cognitive test battery for dementia screening. *Int. J. Geriatr. Psychiatry* **2006**, *21*, 1078–1085. [CrossRef] [PubMed]
95. Tombaugh, T.N.; McIntyre, N.J. The mini-mental state examination: A comprehensive review. *J. Am. Geriatr. Soc.* **1992**, *40*, 922–935. [CrossRef]
96. Hroudova, J.; Fisar, Z. Activities of respiratory chain complexes and citrate synthase influenced by pharmacologically different antidepressants and mood stabilizers. *Neuro. Endocrinol. Lett.* **2010**, *31*, 336–342.
97. Singh, N.; Hroudova, J.; Fisar, Z. Cannabinoid-Induced Changes in the Activity of Electron Transport Chain Complexes of Brain Mitochondria. *J. Mol. Neurosci.* **2015**, *56*, 926–931. [CrossRef]
98. Luptak, M.; Fisar, Z.; Hroudova, J. Effect of Novel Antipsychotics on Energy Metabolism—In Vitro Study in Pig Brain Mitochondria. *Mol. Neurobiol.* **2021**, *58*, 5548–5563. [CrossRef]
99. Sjovall, F.; Morota, S.; Persson, J.; Hansson, M.J.; Elmer, E. Patients with sepsis exhibit increased mitochondrial respiratory capacity in peripheral blood immune cells. *Crit. Care* **2013**, *17*, R152. [CrossRef]
100. Larsen, S.; Nielsen, J.; Hansen, C.N.; Nielsen, L.B.; Wibrand, F.; Stride, N.; Schroder, H.D.; Boushel, R.; Helge, J.W.; Dela, F.; et al. Biomarkers of mitochondrial content in skeletal muscle of healthy young human subjects. *J. Physiol.* **2012**, *590*, 3349–3360. [CrossRef]

101. Heun, R.; Kockler, M.; Ptok, U. Depression in Alzheimer's disease: Is there a temporal relationship between the onset of depression and the onset of dementia? *Eur. Psychiatry* **2002**, *17*, 254–258. [CrossRef] [PubMed]
102. Lyketsos, C.G.; Carrillo, M.C.; Ryan, J.M.; Khachaturian, A.S.; Trzepacz, P.; Amatniek, J.; Cedarbaum, J.; Brashear, R.; Miller, D.S. Neuropsychiatric symptoms in Alzheimer's disease. *Alzheimers Dement.* **2011**, *7*, 532–539. [CrossRef] [PubMed]
103. Zoll, J.; Sanchez, H.; N'Guessan, B.; Ribera, F.; Lampert, E.; Bigard, X.; Serrurier, B.; Fortin, D.; Geny, B.; Veksler, V.; et al. Physical activity changes the regulation of mitochondrial respiration in human skeletal muscle. *J. Physiol.* **2002**, *543*, 191–200. [CrossRef]
104. Picard, M.; McEwen, B.S.; Epel, E.S.; Sandi, C. An energetic view of stress: Focus on mitochondria. *Front. Neuroendocrinol.* **2018**, *49*, 72–85. [CrossRef]
105. Hamrick, M.W.; Stranahan, A.M. Metabolic regulation of aging and age-related disease. *Ageing Res. Rev.* **2020**, *64*, 101175. [CrossRef]
106. Crane, J.D.; Devries, M.C.; Safdar, A.; Hamadeh, M.J.; Tarnopolsky, M.A. The effect of aging on human skeletal muscle mitochondrial and intramyocellular lipid ultrastructure. *J. Gerontol. A Biol. Sci. Med. Sci.* **2010**, *65*, 119–128. [CrossRef]
107. Houmard, J.A.; Weidner, M.L.; Gavigan, K.E.; Tyndall, G.L.; Hickey, M.S.; Alshami, A. Fiber type and citrate synthase activity in the human gastrocnemius and vastus lateralis with aging. *J. Appl. Physiol.* **1998**, *85*, 1337–1341. [CrossRef] [PubMed]
108. Fisar, Z.; Hroudova, J. Pig Brain Mitochondria as a Biological Model for Study of Mitochondrial Respiration. *Folia Biol.* **2016**, *62*, 15–25.
109. Battaglia, S.; Di Fazio, C.; Vicario, C.M.; Avenanti, A. Neuropharmacological Modulation of N-methyl-D-aspartate, Noradrenaline and Endocannabinoid Receptors in Fear Extinction Learning: Synaptic Transmission and Plasticity. *Int. J. Mol. Sci.* **2023**, *24*, 5926. [CrossRef]

biomedicines

MDPI

Article

Noradrenergic Modulation of Learned and Innate Behaviors in Dopamine Transporter Knockout Rats by Guanfacine

Anna Volnova [1,2,*], Natalia Kurzina [1], Anastasia Belskaya [1], Arina Gromova [2], Arseniy Pelevin [1,2], Maria Ptukha [1], Zoia Fesenko [1], Alla Ignashchenkova [3] and Raul R. Gainetdinov [1,4]

1 Institute of Translational Biomedicine, Saint Petersburg State University, Saint Petersburg 199034, Russia
2 Biological Faculty, Saint Petersburg State University, Saint Petersburg 199034, Russia
3 Nevsky Center of Scientific Collaboration, Saint Petersburg 192119, Russia
4 Saint Petersburg University Hospital, Saint Petersburg 199034, Russia
* Correspondence: a.volnova@spbu.ru

Abstract: Investigation of the precise mechanisms of attention deficit and hyperactivity disorder (ADHD) and other dopamine-associated conditions is crucial for the development of new treatment approaches. In this study, we assessed the effects of repeated and acute administration of α2A-adrenoceptor agonist guanfacine on innate and learned forms of behavior of dopamine transporter knockout (DAT-KO) rats to evaluate the possible noradrenergic modulation of behavioral deficits. DAT-KO and wild type rats were trained in the Hebb–Williams maze to perform spatial working memory tasks. Innate behavior was evaluated via pre pulse inhibition (PPI). Brain activity of the prefrontal cortex and the striatum was assessed. Repeated administration of GF improved the spatial working memory task fulfillment and PPI in DAT-KO rats, and led to specific changes in the power spectra and coherence of brain activity. Our data indicate that both repeated and acute treatment with a non-stimulant noradrenergic drug lead to improvements in the behavior of DAT-KO rats. This study further supports the role of the intricate balance of norepinephrine and dopamine in the regulation of attention. The observed compensatory effect of guanfacine on the behavior of hyperdopaminergic rats may be used in the development of combined treatments to support the dopamine–norepinephrine balance.

Keywords: dopamine (DA); norepinephrine (NE); dopamine transporter knockout (DAT-KO) rats; ADHD model; guanfacine (GF); spatial working memory; attention

Citation: Volnova, A.; Kurzina, N.; Belskaya, A.; Gromova, A.; Pelevin, A.; Ptukha, M.; Fesenko, Z.; Ignashchenkova, A.; Gainetdinov, R.R. Noradrenergic Modulation of Learned and Innate Behaviors in Dopamine Transporter Knockout Rats by Guanfacine. *Biomedicines* **2023**, *11*, 222. https://doi.org/10.3390/biomedicines11010222

Academic Editors: Masaru Tanaka and Eleonóra Spekker

Received: 30 November 2022
Revised: 11 January 2023
Accepted: 13 January 2023
Published: 15 January 2023

1. Introduction

Dysfunction of dopamine regulation leads to numerous neuropsychiatric disorders, such as Parkinson's disease, Huntington's disease, schizophrenia and depression, as well as attention deficit hyperactivity disorder (ADHD) [1–4]. ADHD is a neurodevelopmental disorder characterized by abnormalities in behavior, core features of which include hyperactivity, impulsivity and inattention [2,5–8]. Studies utilizing animal models are crucial for the understanding of the mechanisms underlying ADHD and for developing new approaches to the treatment of this disorder [9–15].

Dopamine transporter knockout (DAT-KO) rats, lacking the dopamine transporter gene, demonstrate elevated extracellular dopamine levels in the basal ganglia, a pronounced level of spontaneous hyperactivity and remarkable stereotypical patterns of locomotor activity. It is known that dopamine is involved in various cognitive processes, such as learning, memory and attention, as well as social interactions, goal-directed behaviors and motivation [16–21]. DAT-KO rats have impaired working memory, which affects object recognition tasks and conditioned–unconditioned stimulus association learning tasks [9,11,22–24]. DAT-KO rats are considered to be a valuable model for investigating putative neurophysiological mechanisms of ADHD, as well as other dopamine-associated pathologies. Due to pronounced hyperdopaminergia, the DAT-KO rat model is particularly promising to study the interactions of the dopaminergic and noradrenergic systems in the regulation of cognitive behaviors and other parameters [24].

It is known that patients with prefrontal DA hypofunction and striatal DA hyperfunction have difficulty differentiating the significance of the stimulus presented [25]. Dopamine and norepinephrine terminals coexist in the prefrontal cortex (PFC) [26] and their interactions seem to be crucial for controlling complex behaviors and, consequently, ADHD symptoms [27–33].

Innate forms of behavior also depend on the DA and NE balance. Pre-pulse inhibition (PPI), which is a measure of sensorimotor gating, is reduced in patients with such neuropsychiatric disorders as schizophrenia and autism, as well as ADHD, and may serve as a biomarker of these diseases. PPI, therefore, is invaluable for translational research in neuropsychiatry, since it can also easily be assessed in rodents [34–40].

In this study, we compare the effects of acute and repeated administration of α2A-adrenoceptor agonist guanfacine (GF) on learned and innate behaviors, as well as on the electrophysiological activity of the prefrontal cortex (PFC) and the striatum in DAT-KO rats. By means of adrenergic modulation, GF may have a compensatory effect on the DA imbalance seen in DAT-KO, since DA and NE balance in the PFC provides for its optimal functioning. We hypothesize that repeated GF administration may further improve spatial working memory task fulfillment in the Hebb–Williams maze in DAT-KO rats, because in our previous research on the acute GF administration led to an improvement in that aspect [41]. It is well established that GF is important for improving impairments of memory and attention processes through the modulation of PFC activity [27,42]. To evaluate possible biomarkers and correlates of behavioral changes, we also assessed the plausible effect of GF on brain activity and involuntary attention. Since guanfacine is widely used in long-term pharmacotherapy of ADHD, we aimed to evaluate the possible effects of acute and repeated GF on behavioral and neurophysiological deficits seen in DAT-KO rats.

2. Materials and Methods

2.1. Animals and Their Maintenance

30 DAT-KO and 30 wild-type (WT) littermate rats, males of the same age (3–4 months), were used in the experiments. During the experiments, the requirements of FELASA in RusLASA for studies using laboratory animals were implemented. The animal study protocol was approved by the Ethics Committee for Animal Research of Saint Petersburg State University, St. Petersburg, Russia No. 131-03-10 of 22 November 2021. Prior to the experiments, the rats were maintained in IVC cages (RAIR IsoSystem World Cage 500; Lab Products, Inc., Seaford, DE, USA) with free access to food and water, at a temperature of 22 ± 1 °C, 50–70% relative humidity and a 12 h light/dark cycle (light from 9 am.). Experiments were carried out between 2 pm. and 6 pm.

2.2. Hebb–Williams Maze

2.2.1. Apparatus

We used the Hebb–Williams maze to study animal spatial working memory [43,44]. The behavioral task entailed finding the path from start to finish to obtain food reinforcement (Figure 1). The detailed description of the maze and experimental procedure can be found in our previous work [41]. We compared data after the acute GF administration with the results obtained in this study after repeated GF administration. The data obtained after saline, acute (aGF) and repeated (rGF) guanfacine administration were compared.

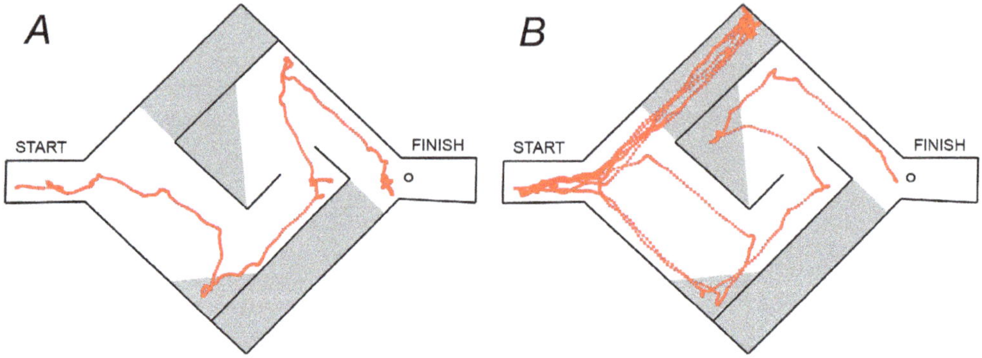

Figure 1. Samples of the visual tracks in the WT rats (**A**) and DAT-KO rats (**B**) in the Hebb–Williams maze. Error zones are indicated in grey color, the circle—the place of food reward.

2.2.2. Experimental Setup

For five days before training, the rats received food at a ratio of 90% of their regular diet (BioPro, Novosibirsk, Russia). Each animal was weighed daily prior to the experiment and throughout all of the experiments' duration. At the beginning of the experiment, all rats were familiarized with a wall-less maze for 2 days. Then, the rats were trained in the unchanging maze arena for 3 days (three trials for each animal per day) with a reward only in the goal box food well for habituation to the maze and task rules acquisition (for details, see [41]). Then, all animals were divided into groups and then trained for two consecutive days in a new maze arena configuration. Ten WT and 10 DAT-KO rats received repeated GF (rGF, Sigma, St. Louis, MO, USA; daily administration 0.25 mg/kg, i.p.) administration for two weeks leading up to the start of the experiments, and before all experimental sessions, 60 min before testing. Ten other WT and 10 DAT-KO rats were trained in the same way, but they received only the saline administration (0.3 mL, i.p.). The results of the experiments were compared with acute administration of GF (aGF, Sigma, USA; 0.25 mg/kg, i.p., 60 min before testing).

During the experiment, the following behavioral variables were recorded: distance traveled; time to reach the goal; time spent in error zones; the number of return runs. The analysis was performed using a video camera mounted above the maze and video tracking software EthoVision XT11.5 (Noldus Information Technology, Leesburg, VA, USA) (Figure 1).

2.3. Acoustic Startle Reaction (ASR) and the Pre-Pulse Inhibition (PPI)

Twenty DAT-KO and 20 wild-type (WT) rats were used in the experiments: 10 WT and 10 DAT-KO rats received saline administration and then (after 4–5 days) acute GF administration (0.25 mg/kg i.p., 60 min before testing), 10 other WT and 10 DAT-KO rats received repeated GF administration (daily, 0.25 mg/kg i.p., 60 min before testing, two weeks prior to the start of the experiments and during all the experimental sessions. The results after saline, acute and repeated administration of GF were compared.

2.3.1. Apparatus

A sensitive platform was used to measure the rat's startle response; the movement of the animal was detected via four floor-mounted vibration sensors. Force detected by the sensors was converted to voltage analog signals that were digitized and stored. Two high-frequency loudspeakers on the opposite sides of the experimental chamber, mounted above the platform, generated broadband background noise and acoustic stimuli, which were controlled by the data acquisition interface (CED Power1401-3A, Cambridge Electronic Design, Cambridge, UK) and Spike2 software. Sound and accelerometer sensitivity was routinely calibrated (DT-8820, CEM, Shenzhen, China).

2.3.2. Experimental Setup

Prior to the start of a series of experiments, each rat was placed on the platform and exposed to "white noise" with an intensity of 74 dB for 20 min for habituation to the experimental conditions. On the day of the experiment, each animal was presented with "white noise" with an intensity of 74 dB for 10 min, followed by 10 sound stimuli with an intensity of 78 dB and a duration of 50 ms (pre-pulse), which did not result in any motor reactions. Then, 20 acoustic stimuli with an intensity of 100 dB and a duration of 50 ms (pulse) followed, causing a pronounced ASR. Then, 20 combinations of acoustic stimuli (pre-pulse + pulse) followed to register the changes in the ASR amplitude and calculate the pre-pulse inhibition. The interval between pre-pulse and pulse stimuli was 100 ms. The interval between presentations of stimuli or paired stimuli varied from 10 to 14 s to avoid the animal's habituation to sounds. The trial presentation was controlled by a custom script for Spike2 software (CED, UK). PPI was calculated as a percentage score: ((startle response for pulse alone—startle response for pulse with pre-pulse)/startle response for pulse alone) × 100. The ASR amplitude was recorded by the Spike2 program using a data acquisition interface (CED Power1401, Cambridge Electronic Design, Cambridge, UK) synchronized with the sound stimulus delivery system. During the experiments, we encountered difficulties recording the DAT-KO rats' reactions, which are characterized by hyperactivity and an increased level of motor activity. In order to distinguish between the ASR and background motor reactions typical of DAT-KO rats, a video camera was synchronized with the recording of the motor reactions and the onset of the acoustic signal presentations. The synchronization allowed to avoid the overlapping of recorded motor responses with traveling around the platform, grooming, and rearings.

2.4. LFP Power Spectra and Coherence

The local field potential (LFP) recordings were carried out in 25 adult male rats: WT (n = 12) and DAT-KO (n = 13). Six WT and six DAT-KO rats received saline administration. Then, after 4–5 days, these animals received an acute administration of GF (0.25 mg/kg, i.p., 60 min before testing). Six WT and seven DAT-KO rats received repeated GF (daily administration, 0.25 mg/kg i.p., 60 min before testing) for two weeks leading up to the start of the experiments.

2.4.1. Surgical Procedures

Electrodes for brain activity recordings were mounted on the skull under isofluranum anesthesia with the use of a micromanipulator in a stereotaxic frame. Three electrodes were implanted into each animal. An epidural screw (1 mm in diameter; 1 mm in length; steel) was used as a reference electrode (AP= −7 mm; L= 3 mm, coordinates relative to bregma). Two intracerebral electrodes (50 μm in diameter; 2.5 mm/5 mm in length; tungsten wire in perfluoroalkoxy polymer isolation) were used for the LFP recordings. We used the following coordinates relative to bregma: prefrontal cortex (PFC) intracerebral electrode AP = +2 mm; L = 2.5 mm; 2.5 mm in length; and striatal (Str) intracerebral electrode AP = 0 mm; L = 3 mm; 5 mm in length.

2.4.2. EEG and LFP Recordings

The experiments were carried out 2–3 days after surgery. The experimental setting for LFP recordings consisted of an amplifier (×1000 gain), Cambridge Electronic Design (CED) Power1401-3A data acquisition interface, and Spike2 software (CED), sampling rate 25,000 Hz. During the recording process, the animals were placed in a 25 cm × 25 cm × 25 cm Plexiglas box located within a Faraday cage.

Brain activity was recorded in six WT and six DAT-KO rats, on two subsequent days: for one hour (60 min) after a saline injection (0.9% NaCl i.p., 30 min before the recording), and for one hour after an acute GF injection (0.25 mg/kg, i.p., 60 min before the recording). Brain activity after a repeated GF injection was recorded in six WT and seven DAT-KO rats (daily administration, 0.25 mg/kg i.p. for two weeks prior to the start of recording). For behavior monitoring, video was recorded simultaneously with brain activity. Only parts of the recordings where the animals were awake were used in the subsequent analysis.

2.4.3. Analysis of EEG and LFP Activity

For each recording, 200 s without artifacts were selected for analysis. The sampling rate of the recordings was 1000 Hz. The data were analyzed with a script (COHER.s2s, CED official website), which calculates the power spectra of a signal through fast Fourier transform and the coherence of two signals. A power spectrum is the ratio of frequency to power. Coherence values are between 0 and 1, and it is equal to 1 for a particular frequency if the phase shift between the waveforms is constant, and the amplitudes of the waves have a constant ratio. Data in the 0–0.8 range were excluded due to the abundance of artifacts. The following ranges for the electroencephalographic rhythms were used for analysis and interpretation (in Hz): delta (0.9–3), theta (4–8), alpha (9–11), lower beta (12–19), higher beta (20–29), lower gamma (30–48), higher gamma (52–74).

2.5. Statistical Analysis

The data were presented as the mean ± SEM; $p < 0.05$ was considered as statistically significant for all tests. A preliminary assessment of the normality of the data distribution was performed using the Kolmogorov–Smirnov test. To analyze the parameters of the rats' Hebb–Williams maze behavior and pre-pulse inhibition, we used a two-way analysis of variance (two-way ANOVA, genotype factor: WT or DAT-KO; treatment factor: saline, acute and repeated guanfacine administration) with a post-hoc Sidak multiple comparison test (WT vs. DAT-KO; saline vs. aGF; saline vs. rGF; saline vs. aGF). To analyze the power spectra density of the local field potential (LFP) from the striatum and prefrontal cortex, and Str-PFC coherence, we used an ordinary one-way ANOVA with post-test for linear trend or Sidak's multiple comparisons post-hoc test (for parametric data) and Kruskal–Wallis test with Dunn's multiple comparisons test (for nonparametric data).

3. Results

3.1. Hebb–Williams Maze

The analysis of the motor behavior of DAT-KO rats in the Hebb–Williams maze showed pronounced hyperactivity while performing behavioral tasks, in comparison to the WT rats [41]. We observed that the DAT-KO rats traveled significantly longer distances ($p < 0.01$) and required more time to reach the goal box ($p < 0.001$), in comparison with the WT rats (Figure 2A,B). Acute GF administration did not induce any significant changes in these behavioral parameters in WT rats. A comparative analysis of the behavioral parameters of WT and DAT-KO rats showed that following acute GF administration, the level of all behavioral parameters was also higher in DAT-KO, in comparison to WT rats after saline ($p < 0.05$ for the distance traveled and for the time to reach the finish zone). However, repeated GF administration resulted in a decrease in the distance traveled by the DAT-KO rats down to levels observed in the WT counterparts (Figure 2A). The time that DAT-KO rats spent in the arena before the rats reached the goal box, in comparison to that of the WT controls (Figure 2B), was longer after saline and acute GF administration ($p < 0.05$), but not after repeated GF administration.

The two-way ANOVA analysis revealed a significant difference in the distance traveled (genotype factor, $p < 0.05$, treatment factor $p < 0.0001$) and in the time of reaching the goal box (treatment factor $p < 0.001$) between the two groups of rats (Table S1 of Supplementary Materials and Table S2 of Supplementary Materials)

The amount of time the animal spent in the error zones (Figure 3A) was an indicator of the test performance: according to this indicator, the spatial navigation in the DAT-KO rats was worse than in the WT rats (## $p < 0.01$). In contrast to saline administration, acute and repeated GF administration decreased the time spent in the error zones approaching the parameters of the WT rats, resulting in no statistical differences from the WT rats after saline administration (Figure 3A, Table S3).

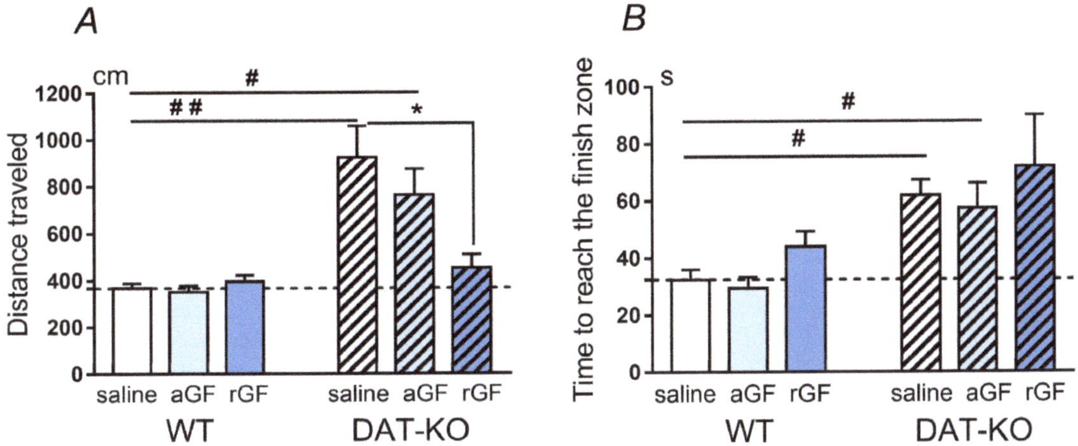

Figure 2. Comparison of the distance traveled (in cm), (**A**) and the time for reaching the goal box (in s), (**B**) of the Hebb–Williams maze in DAT-KO and WT rats after saline, acute (aGF) and repeated (rGF) guanfacine administration. Results are presented as the mean ± SEM, # $p < 0.05$; ## $p < 0.01$, Wilcoxon signed rank test; * $p < 0.05$; two-way ANOVA test combined with Sidak's multiple comparisons post-hoc test.

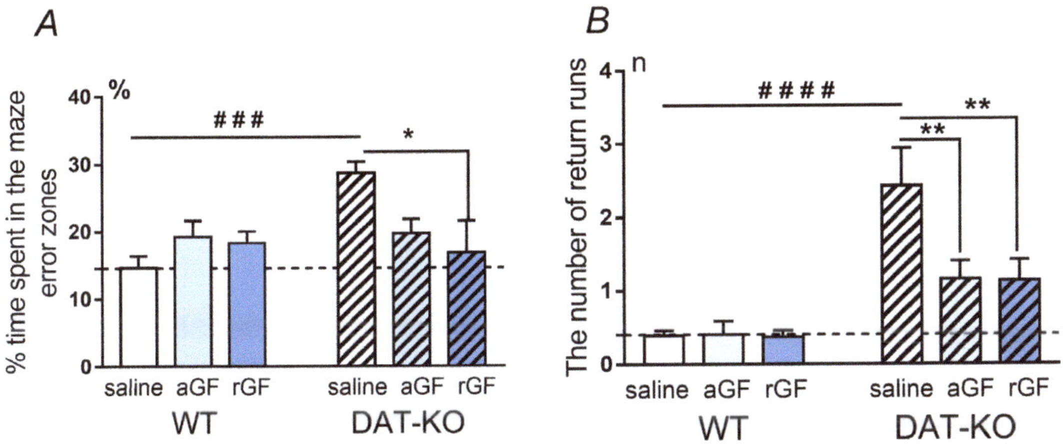

Figure 3. Comparison of the percentage of time spent in the Hebb–Williams maze error zones (in %), (**A**) and the number of return runs (**B**) by the DAT-KO and WT rats after saline, acute (aGF), and repeated (rGF) guanfacine administration. Results are presented as the mean ± SEM; ### $p < 0.001$, #### $p < 0.0001$; * $p < 0.05$, ** $p < 0.01$, two-way ANOVA test combined with Sidak's multiple comparisons post-hoc test.

The number of stereotypic (perseverative) reactions was recorded on the basis of the number of returns to the start zone of the maze (Figure 3B). Analysis of this parameter showed that the perseverative activity after saline administration in DAT-KO rats was significantly higher than in WT rats ($p < 0.001$, two-way ANOVA with Sidak's multiple comparisons post-hoc test). The acute and repeated GF administration lead to a decrease in the number of returns, in comparison to saline administration ($p < 0.01$; two-way ANOVA with Sidak's multiple comparisons post hoc test; Table S4 of Supplementary Materials) down to levels not reaching a significant difference from the control values for the WT rats.

The data obtained indicate that the prolonged period of the GF administration may further improve the spatial task performance. It is apparent that a longer period of GF treatment makes the task performance more effective, due to a significant decrease in the time spent in the error zones and the distance traveled. These findings suggest that GF may improve attention processes. A decrease in the stereotypical (perseverative) activity patterns of mutants was also observed after acute and repeated GF administration.

3.2. Acoustic Startle Reaction (ASR) and the Pre-Pulse Inhibition (PPI)

The ASR was investigated in the WT and DAT-KO rats after saline, acute, and repeated GF administration, and then the PPI values were calculated. In the WT rats after saline administration, ASR on the paired (pulse and pre-pulse) presentation was lower than the responses evoked by the pulse alone (Figure 4A). DAT-KO rats showed a similar trend, but the ASR amplitudes were lower than in the WT rats. In contrast to saline, the acute and repeated GF administration induced a decreased ASR for double acoustic stimuli (pre-pulse + pulse) in both WT and DAT-KO rats (Figure 4A–C).

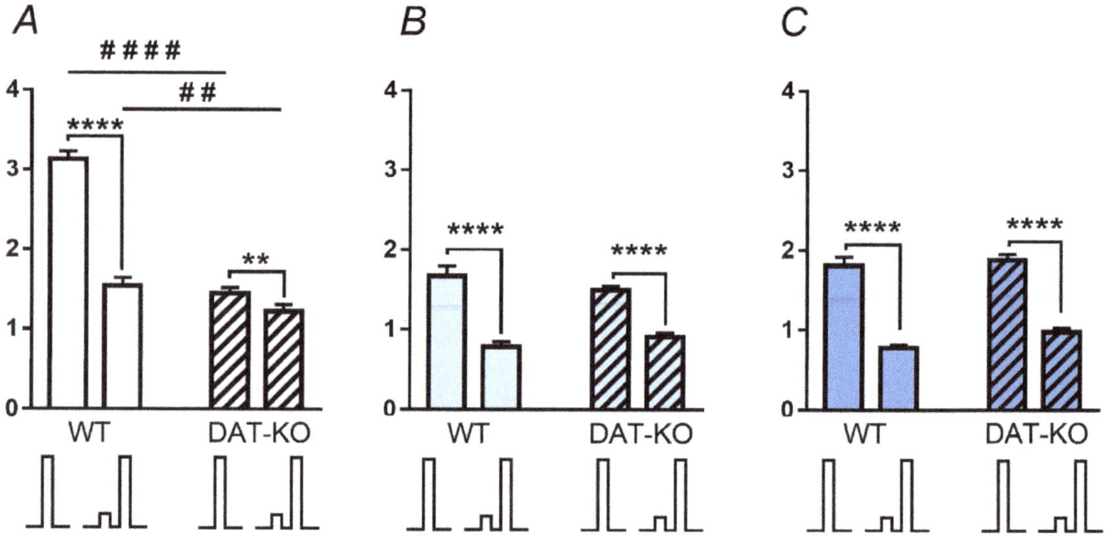

Figure 4. Comparison of the amplitude of the acoustic startle reaction (ASR, in mV) in DAT-KO and WT rats after saline (**A**), acute (**B**) and repeated (**C**) guanfacine administration. Results are presented as the mean ± SEM; ## $p < 0.01$, #### $p < 0.0001$, ** $p < 0.01$, **** $p < 0.0001$ according the Mann–Whitney test.

For detecting the differences between the amplitude of the responses elicited by the single (pulse) and double (pre-pulse + pulse) stimuli, we analyzed the value of the pre-pulse inhibition (PPI). Analysis of the data with a two-way ANOVA showed that for the factor "genotype", the differences are significant with $p < 0.0001$, while for the factor "treatment" —with $p < 0.01$, no interaction between the genotype and treatment was observed. Comparison of the results in DAT-KO and WT rats showed that after saline administration, the DAT-KO rats showed less a pronounced PPI than the WT rats (Figure 5A). Multiple comparisons between the mean values for the WT and DAT-KO (Fisher's least significant difference) showed significant differences between groups for each drug ($p < 0.01$) (Figure 5A, Table S5).

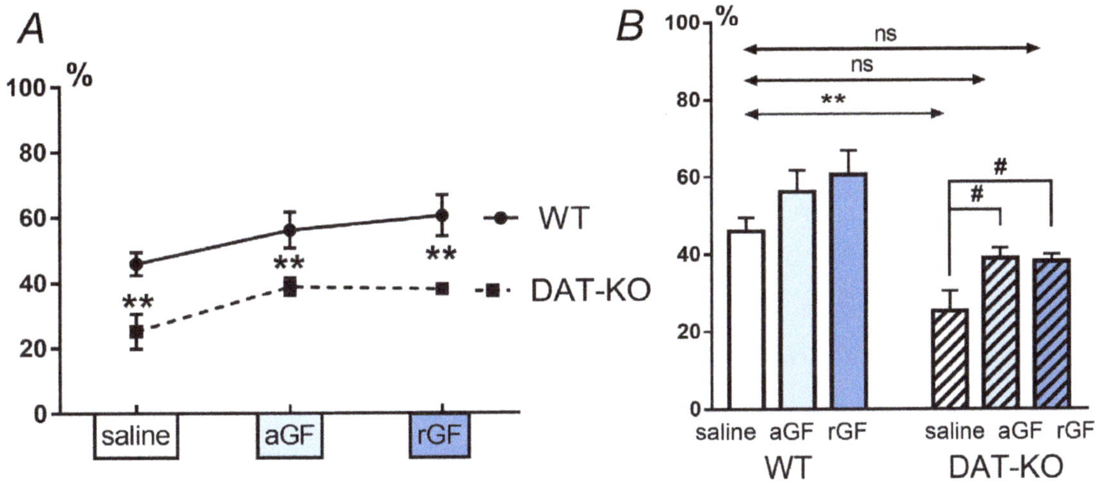

Figure 5. Pre-pulse inhibition (PPI) in DAT-KO and WT rats after saline, acute (aGF) and repeated (rGF) guanfacine administration; comparison of the groups of animals of different genotypes (**A**) and comparison of the effect of the administered drugs (**B**). Results are presented as the mean ± SEM; ** $p < 0.01$ according to Fisher's LSD post-test; # $p < 0.05$; ns–(not significant) two-way ANOVA and Dunn's multiple comparisons post-hoc test.

A comparison of the PPI index values for WT and DAT-KO rats after saline administration, showed significant differences (Figure 5B). These data show that the WT rats have normal processes of habitation and sensorimotor gating, while in the DAT-KO rats, the PPI index is reduced, indicating the impaired sensory information perception. In contrast to this, the GF administration (both acute and repeated) led to an increase in the PPI index in the knockout animals, and it became similar to that in the WT after saline (Figure 5B). An increase in the PPI index in the DAT-KO rats after acute and repeated administration of GF indicates an improvement in the sensorimotor gating.

3.3. Power Spectra and Coherence of the Brain Activity

To analyze the power spectra of the neural activity in the prefrontal cortex (PFC) and dorsal striatum (Str), we used a traditional LFP analysis, based on the signal decomposition into simpler sinusoidal harmonics. We compare electroencephalographic rhythms according to the following ranges (in Hz): delta (0.9–3), theta (4–8), alpha (9–11), lower beta (12–19), higher beta (20–29), lower gamma (30–48), higher gamma (52–74).

The differences between the power spectra and intrinsic coherence between PFC and Str in DAT-KO and WT rats after saline administration, were assessed in our previous article [45]. In the present research, we focused on the effects of acute and repeated guanfacine administration on these parameters.

Guanfacine administration to WT and DAT-KO rats led to an overall decrease in the power spectral density in the LFP activity of the dorsal striatum (Figure 6). This fact is confirmed by the presence of a linear trend, according to a 2-way ANOVA (indicated by the sloping line in the diagrams). Note the high reliability of the observed differences ($p < 0.0001$ and $p < 0.001$), as well as the unidirectional changes occurring in the power spectrum in WT and DAT-KO rats in all main bands (Table S6 of Supplementary Materials).

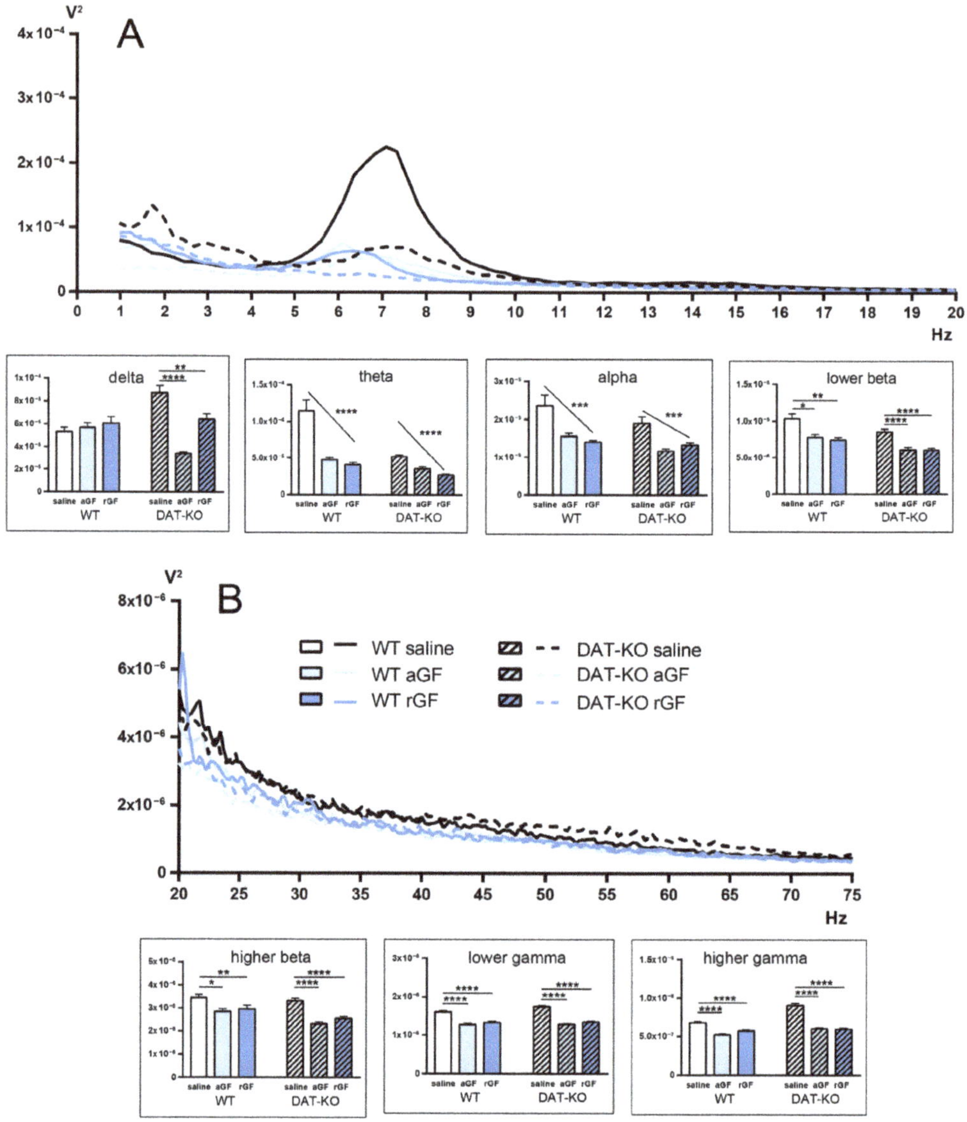

Figure 6. Power spectra of the brain activity in DAT-KO and WT rats recorded from the striatum (Str) after saline, acute (aGF) and repeated (rGF) guanfacine administration; (**A**) 1–20 Hz range, (**B**) 20–75 Hz range. Data are presented as the mean ± SEM. The diagrams represent the following electroencephalographic rhythms (in Hz): delta (0.9–3), theta (4–8), alpha (9–11), lower beta (12–19), higher beta (20–29), lower gamma (30–48), higher gamma (52–74); one-way ANOVA, * $p < 0.05$; *** $p < 0.001$; **** $p < 0.0001$ post test for the linear trend; ** $p < 0.01$ Dunnett's multiple comparisons test.

Analysis of the activity recorded in the PFC showed different tendencies (Figure 7; Table S7 of Supplementary Materials) in the two groups of rats. In WT rats, the aGF and rGF administration resulted in a significant decrease in the power spectra of the PFC activity in theta, alpha and lower beta ranges. On the contrary, in the high frequency areas, the power spectral density increased in the higher beta and gamma ranges. No changes were observed in the delta range in WT rats.

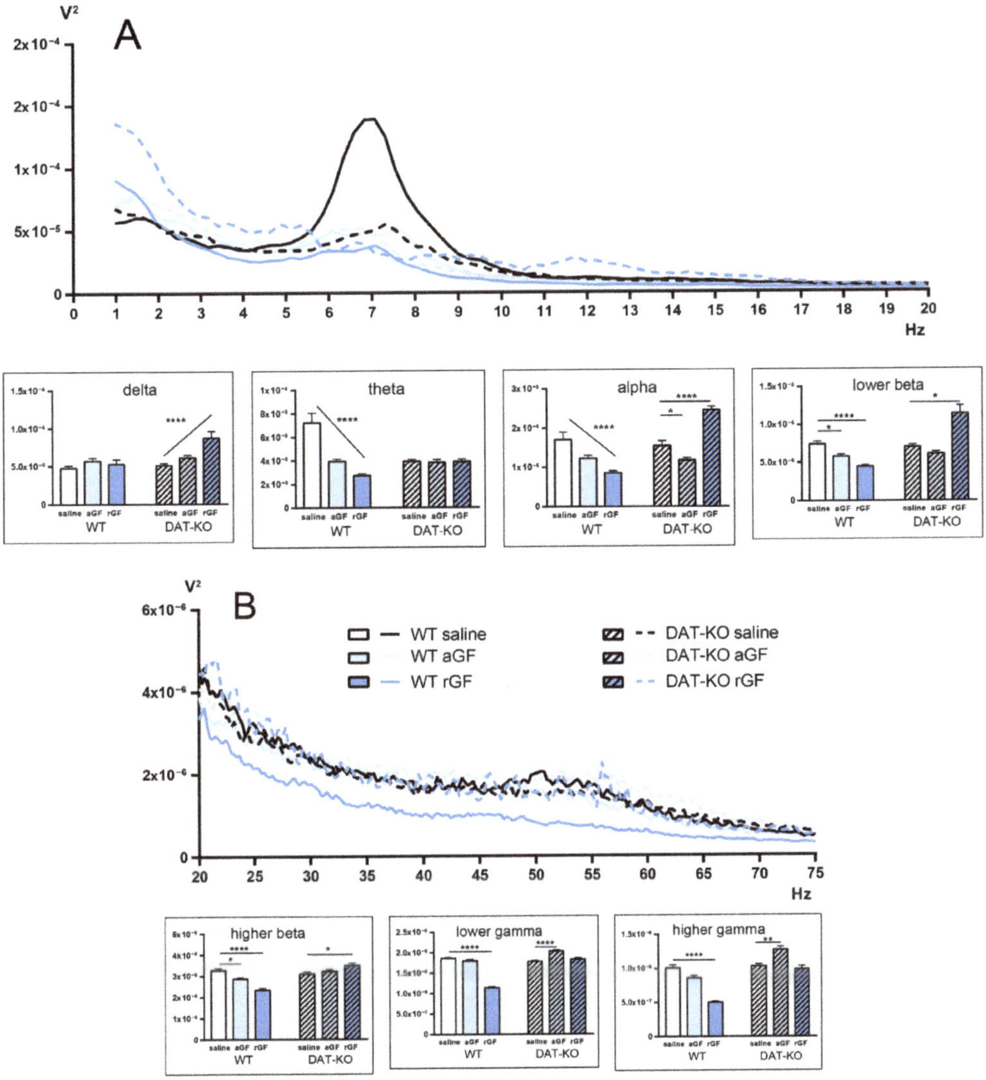

Figure 7. Power spectra of the brain activity of DAT-KO and WT rats recorded from the prefrontal cortex (PFC) after saline, acute (aGF) and repeated (rGF) guanfacine administration; (**A**) 1–20 Hz range, (**B**) 20–75 Hz range. Data are presented as mean ± SEM. The diagrams represent the following electroencephalographic rhythms (in Hz): delta (0.9–3), theta (4–8), alpha (9–11), lower beta (12–19), higher beta (20–29), lower gamma (30–48), higher gamma (52–74); one-way ANOVA, **** $p < 0.0001$ post test for the linear trend; * $p < 0.05$; ** $p < 0.01$ Dunnett's multiple comparisons test.

In DAT-KO rats, administration of aGF and rGF generally led to an increase in the power spectra of the PFC activity, except for the theta range. We found that this increase was unstable at different ranges. It can be concluded that aGF and rGF administration might lead to the changes in the frequency-time characteristics of LFP in DAT-KO rats, similarly to those in WT rats.

It was shown earlier that the DAT-KO rats are characterized by lower values of the coherence coefficient between the investigated brain regions (Str and PFC). In WT rats, we observed a decrease in coherence at lower frequencies and its increase in the gamma range (Figure 8; Table S8). In DAT-KO rats, a decrease in coherence was observed in all ranges except the lowest and highest. The delta range differed from the others in the character of the observed changes in both WT and DAT-KO rats. We assume that our findings might be associated with a greater variability of the values in this frequency range.

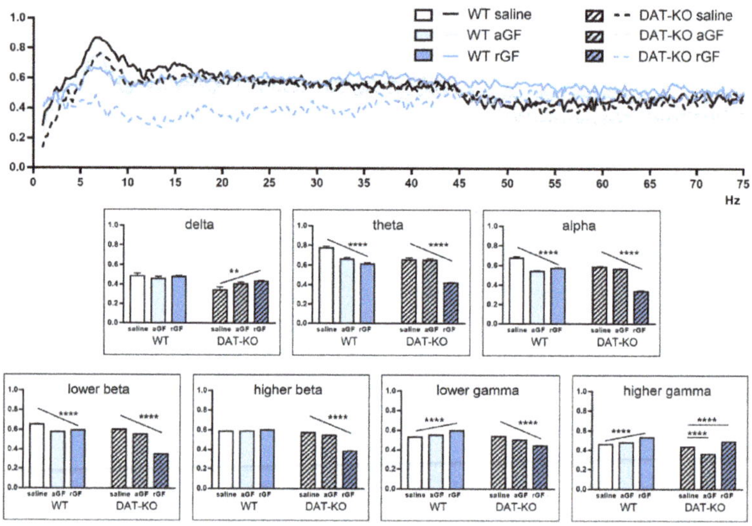

Figure 8. Coherence of the brain activity in PFC and Str in DAT-KO and WT rats after saline, acute (aGF) and repeated (rGF) guanfacine administration. Degree of coherence is expressed in fractions of one. Data are presented as the mean ± SEM. The diagrams represent the following electroencephalographic rhythms (in Hz): delta (0.9–3), theta (4–8), alpha (9–11), lower beta (12–19), higher beta (20–29), lower gamma (30–48), higher gamma (52–74); two-way ANOVA, ** $p < 0.01$ **** $p < 0.0001$ post test for the linear trend and Dunnett's multiple comparisons test.

4. Discussion

Attention deficit hyperactivity disorder is a common psychiatric disease in childhood. Over 50% of children with this disorder continue to exhibit its symptoms into adulthood [46,47]. ADHD is characterized by a high comorbidity with other developmental disorders, anxiety and mood disorders, tics, learning disabilities and sleep disorders [48–51]. Numerous studies underline the importance of normal brain development for the absence of neuropsychiatric symptoms [15]. Moreover, multiple factors contribute to the emergence of ADHD symptoms, including abnormalities in neurotransmission, structural changes in various brain structures, altered functional connectivity, stress, inflammation, etc. [48,49,52–54]. The precise mechanisms underlying ADHD symptoms are still poorly understood, thus investigation of pathophysiological states of the relevant brain networks in animal models becomes increasingly relevant. DAT-KO rats, lacking the dopamine transporter gene, mimic the main endophenotypes of ADHD patients, including hyperactivity, impulsivity and inattention. Behavioral changes seen in DAT-KO rats are, at least in part, due to extremely elevated levels of extracellular dopamine in the striatum [11], caused by the absence of DAT. Despite their pronounced hyperactivity, DAT-KO rats are able to learn a behavioral task in the 8-arm radial maze [22], and object recognition task in a RedBox paradigm [23], although with a poorer performance, in comparison to WT animals. It should be noted that, while learning cognitive tasks, hyperdopaminergic rats may use unique behavioral tactics [22]. In contrast to WT rats, DAT-KO rats show numerous perseverative reactions, which lead to an increased number of erroneous trials during the learned task performance [23].

In this study, we evaluated the changes in innate and learned behaviors of DAT-KO rats under the long-term treatment with guanfacine, a α2A-adrenoceptor agonist, which is widely used in ADHD treatment. In our experiments, repeated GF injections ameliorated the task fulfillment and PPI in DAT-KO rats. The distance traveled decreased, the time of reaching the goal box increased in mutants, while a decrease in the time spent in the erroneous zones was observed. The GF administration also reduced the number of perseverative reactions. Comparable tendencies were described in our previous work with acute administration of GF [41]. In contrast to WT rats, the DAT-KO rats demonstrated an improvement in the task fulfillment in the Hebb–Williams maze after repeated GF. These results might be indicative of improved attention processes in DAT-KO rats following GF administration.

There are numerous studies supporting the key role of DAT in the development of several dopamine related pathologies [55–59]. It is not surprising, since dopamine dynamics in dopaminergic terminals and synapses are, to a large extent, regulated by DAT. Since DA has a high affinity for NET, in NET-enriched areas, such as the PFC, NET regulates dopamine, as well as the norepinephrine levels. Thus, drugs affecting DAT or NET can regulate dopamine storage and release by noradrenergic neurons and are important for ADHD treatments [60].

It is well known that dopamine and norepinephrine terminals project to the prefrontal cortex (PFC) [61–63]. Dysfunction of the DA pathways leads to a lack of attention and to intolerance in waiting for a reward in patients with ADHD [64]. NE pathways from the locus coeruleus (LC) are involved in attention control. In ADHD patients, discharges from the LC are altered, thus complicating focused attention [65]. Dendritic spines on pyramidal neurons in the PFC have α2A adrenergic receptors and D1 dopamine receptors involved in PFC functions. NE has high affinity for the postsynaptic α2A adrenergic receptors and enhances cognitive functions. Normal PFC functioning requires optimal levels of catecholamines. A catecholamine deficiency leads to a reduced level of control of hyperactivity and attention deficit and the poor planning of goal-directed behavior [66]. Projections from the PFC to the Str constitute a part of the network that ensures the functioning of the working memory. The PFC-Str activities appear to contribute to a correct action after a period of working memory consolidation, resulting in the successful completion of the working memory task [67]. Moreover, it has been suggested that the PFC-Str projection pathways are selectively involved in inhibitory control [68]. It is also important to note that an individual's emotional state is strongly connected to a proper performance in a task. DA and NE are known to participate in many basic emotions. PFC is one of the main structures that controls cognitive functions and motivation-based behaviors. There is an opinion that DA induces happiness and pleasure, whereas NE is related to fear and anger states [69]. The control of emotional states may facilitate the interruption of ongoing behaviors, such as impulsive behavior, which is one of the core features of ADHD [70,71].

There are several approaches to ADHD treatment, including pharmacological interventions [72,73]. Psychostimulants, including amphetamine and methylphenidate, are most commonly used [74,75]. Non-psychostimulant drugs, such as guanfacine (α2A-adrenoceptor agonists) and atomoxetine (norepinephrine transporter inhibitor) are also used for counteracting ADHD symptoms as monotherapy and add-on therapy [73,76]. The action mechanism of NE-based treatments has been partially investigated in animal models. The acute and chronic administration of NE-based anti-ADHD drugs has been shown to selectively activate the prefrontal catecholamine systems in mice [77], while the systemic administration of the a2A-adrenoceptor agonist GF reduced the impulsive choice behavior in rats [78]. It has been shown that in juvenile SHR rats tested in a five-trial inhibitory avoidance task, the stimulation of postsynaptic α2A adrenergic receptors by GF leads to attention improvement [79]. There is an opinion that GF affects the dendritic branching of the pyramidal cells in the PFC, thus improving the cognitive performance [80]. There are also other data indicating that GF enhances the PFC regulation of attention and improves the performance in working memory tasks [14,78,81,82]. Thus, GF administration in DAT-KO rats induces neuropharmacological effects that may explain the results of the behavioral tests. The beneficial effects of GF are associated with an improved noradrenergic modulation in the neuronal circuits which involve the PFC [83]. It has been proposed that guanfacine is most specific for the treatment of the prefrontal attentional and working memory deficits [27,42].

Such innate behaviors as the acoustic startle reaction and pre-pulse inhibition (PPI) are linked to the habituation and ability to rapidly adjust to a changing environment. These innate forms of behavior may also depend on the DA-NE balance. It was described that the PPI, which is a measure of sensorimotor gating, is reduced in patients with several neuropsychiatric disorders, such as schizophrenia, autism and ADHD. PPI findings in humans and rodents show its importance for the translational investigation in neuropsychiatry [35–37,39,40]. Habituation to a new environment is deficient in DAT-KO rats [84,85]. DAT-KO mice and rats display a consistent PPI deficit [11,86]. The enhancement of cortical extracellular DA via the blockade of NET, leads to a reversal of this deficit in DAT-KO mice [87], as well as in rats [45]. It was also shown that the depletion of the prefrontal DA induces a PPI deficit in rats [88]. At the same time, it is known that dopaminergic agonists increase the ASR amplitude and shorten its latency [89]. Our findings showed that a PPI deficit in hyperdopaminergic rats is improved by both acute and repeated GF administration, thus validating an important role of the NE and DA interaction in the regulation of this process.

These observations show that activation of the alpha-2A-adrenoceptors ameliorate both learned and innate forms of behavior. In our previous work [45] we suggested that the possible NE-mediated activation of the PFC (produced by either adrenoceptor activation or NET-inhibition) seems to have a positive effect on attention and perseverative reactions, but not hyperactivity, which could be mainly regulated by the DA in the PFC. The results of the current study might challenge that assumption, since prolonged guanfacine administration significantly decreased the distance traveled by DAT-KO animals before reaching the goal box, making this parameter indistinguishable from that of the WT rats. This result may be interpreted as a correlate of reduced hyperactivity.

One of the significant neurophysiological correlates of ADHD is an increase in the power of the electrophysiological activity in the theta band [90]. However, as was described in detail in our previous study [45], a lower level of theta band power in knockout animals is observed both in the striatum and in the PFC. DAT-KO rats are also characterized by lower values of coherence in the investigated brain regions (Str and PFC). In this study, we attempted to evaluate the effects of acute and repeated administration of guanfacine on the electrophysiological characteristics of brain activity in DAT-KO rats to assess the possible biomarkers of behavioral changes. In WT rats, we observed a decrease in coherence at lower frequencies and its increase in the gamma range under GF. In DAT-KO rats under GF, a decrease in coherence was observed in almost all ranges, except the lowest and highest. The delta range differed from the others in the character of the observed changes in both WT and DAT-KO rats. We assume that these findings might be associated with a greater variability of values in this frequency range.

However, DAT-KO rats under GF showed a reduced activity in the alpha-beta bands consistent with the data reported for patients with ADHD [91,92]. An increase in power in the alpha range, which is ameliorated by GF, correlates with a slower reaction time, higher reaction time variability, and an overall tendency towards a lower performance in tasks measuring inhibition in children with ADHD [91].

It has been proposed that the GF action on α2A-adrenoceptors leads to a more effective regulation of attention and goal-directed behaviors by the PFC and to the strengthening of its functional connections with other cortical areas [93,94]. It was proposed that GF reduces the thalamo-cortical excitability and thereby effectively modulates cognitive processing [42]. In WT rats, GF induces a decrease in low-frequency bands, including the alpha band. Moreover, a decrease in the coherence index of the alpha band and lower frequencies was found. Phase synchronization of alpha bands in different areas of the brain provides effective network communication [95]. Therefore, changes in the coherence index may indicate that functional connectivity within the local brain areas becomes weaker, which is indicated by the decrease in the higher frequencies, while that of large scale areas becomes stronger. It is important to note that the increase in coherence in WT rats in the gamma band under GF indicates its influence on local neuronal ensembles involved in information processing. It means that despite the absence of external changes in the behavioral tests described above, GF stimulates a significant functional reorganization of the neuronal communication in WT rats.

The effect of GF injections on the time-frequency characteristics of neuronal activity in the brain of DAT-KO rats is not as unambiguous. Only acute administration of GF leads to a decrease in the alpha band in the PFC and striatum, whereas repeated administration has opposite effects eliciting a strong power spike. In contrast to acute administration, the cumulative effect of repeated administration of GF results in a stronger decline in the coherence index in the low-frequency range, including the alpha band. Our data are comparable to the findings on the changes in the task-related alpha (8–12 Hz) interhemispheric connectivity correlated with inattentive symptom severity [96].

To sum up, the effects of repeated GF administration on both DAT-KO and WT rats are mostly comparable to those seen after acute administration. In humans, however, both in childhood and adulthood, a gradual improvement in ADHD symptoms, has been reported under long-term GF, seen only after several weeks of treatment [42,97,98]. In our study, a similar tendency can be seen in locomotor activity in the Hebb–Williams maze, which improved more significantly under repeated GF than under acute GF and can be seen as a sign of reduced hyperactivity. Some differing changes can also be seen in brain activity, the interpretation of which requires further investigation, however, it may turn out to be correlates of certain behavioral changes.

5. Conclusions

In this study, acute and repeated guanfacine (GF), α2A-adrenoceptor agonist, was shown to influence innate and learned behaviors in DAT-KO (dopamine transporter knockout) rats, as well as the electrophysiological correlates of brain activity. The results obtained, combined with our previous studies, show that noradrenergic modulation improves different aspects of behavior in hyperdopaminergic knockout rats. GF administration improved the fulfillment of a learned spatial tasks and ameliorated the PPI in DAT-KO rats. Changes in the electrophysiological activity of the brain under GF proved to be similar to those observed in human patients. Overall, the impact of repeated GF was found to be mostly comparable to the acute administration, except for locomotor activity, which improved further with long-term administration, and some electrophysiological parameters in DAT-KO rats. The results obtained in DAT-KO and WT rats under acute and repeated GF, allow for the development of further hypotheses on differential effects of DA and NE networks on various forms of behavior and, possibly, in different cognitive and psychiatric disorders.

6. Limitations and Future Directions

Our study, as preclinical research on an animal model, has limitations, since there is no direct application of the findings to the clinical practice. Moreover, the DAT-KO rats, as an animal model, are supposedly not sufficient for assessing the full range of ADHD symptoms. The future directions of our studies involve the use of other non-psychostimulants in different cognitive tasks, a comparison of different doses of drugs, as well as other, more precise, methods of targeting specific neuronal networks. It is also important to conduct longitudinal experiments, in order to search for doses and combined drug applications for ameliorating ADHD symptoms long-term and understanding their mechanisms.

Supplementary Materials: The following supporting information can be downloaded at: https://www.mdpi.com/article/10.3390/biomedicines11010222/s1, Table S1. The statistical analyses of the distance traveled by DAT-KO and WT rats in Hebb-Williams maze after saline, acute (aGF), and repeated (rGF) guanfacine administration; Table S2. The statistical analyses of the time of reaching the goal box of Hebb-Williams maze by DAT-KO and WT rats after saline, acute (aGF), and repeated (rGF) guanfacine administration. Table S3. The statistical analyses of the percentage of time spent in the Hebb-Williams maze error zones by DAT-KO and WT rats after saline, acute (aGF), and repeated (rGF) guanfacine administration. Table S4. The statistical analyses of the number of return runs by DAT-KO and WT rats in the Hebb-Williams maze after saline, acute (aGF), and repeated (rGF) guanfacine administration. Table S5. The statistical analyses of pre-pulse inhibition index (PPI) for DAT-KO and WT rats after saline, acute (aGF), and repeated (rGF) guanfacine administration. Table S6. The statistical analyses of power spectral density of striatum LFP at various frequency bands in WT and DAT-KO rats. Table S7. The statistical analyses of power spectral density of prefrontal cortex LFP at various frequency bands in WT and DAT-KO rats. Table S8. The coherence between the activity of the prefrontal cortex (PFC) and Striatum (Str) at various frequency bands.

Author Contributions: Conceptualization, A.V., N.K. and R.R.G.; methodology, A.V., A.B., A.I. and A.G.; software, A.P.; validation, A.P., M.P. and Z.F.; formal analysis, A.B. and A.G.; data curation, A.B., A.G., M.P. and Z.F.; writing—original draft preparation, N.K. and Z.F.; writing—review and editing, A.V., N.K., M.P., Z.F., A.I. and R.R.G.; visualization, M.P. and Z.F.; supervision, A.V.; project administration, A.V. and R.R.G.; funding acquisition, A.V. and R.R.G. All authors have read and agreed to the published version of the manuscript.

Funding: This research was funded by the Russian Science Foundation grant number 21-75-20069. Z.F. and R.R.G. were supported by the project ID: 93018770 of the St. Petersburg State University, St. Petersburg, Russia.

Institutional Review Board Statement: The animal study protocol was approved by the Ethics Committee for Animal Research of Saint Petersburg State University, St. Petersburg, Russia No. 131-03-10 of 22 November 2021.

Informed Consent Statement: Not applicable.

Data Availability Statement: The raw data used in this study are available upon request from the corresponding author.

Conflicts of Interest: The authors declare no conflict of interest.

References

1. Nikolaus, S.; Antke, C.; Beu, M.; Müller, H.W.; Nikolaus, S. Cortical GABA, Striatal Dopamine and Midbrain Serotonin as the Key Players in Compulsiveand Anxiety Disorders—Results from in Vivo Imaging Studies. *Rev. Neurosci.* **2010**, *21*, 119–140. [CrossRef] [PubMed]
2. Grace, A.A. Dysregulation of the Dopamine System in the Pathophysiology of Schizophrenia and Depression. *Nat. Rev. Neurosci.* **2016**, *17*, 524–532. [CrossRef] [PubMed]
3. Klein, M.O.; Battagello, D.S.; Cardoso, A.R.; Hauser, D.N.; Bittencourt, J.C.; Correa, R.G. Dopamine: Functions, Signaling, and Association with Neurological Diseases. *Cell. Mol. Neurobiol.* **2018**, *39*, 31–59. [CrossRef]
4. Beaulieu, J.-M.; Gainetdinov, R.R. The Physiology, Signaling, and Pharmacology of Dopamine Receptors. *Pharmacol. Rev.* **2011**, *63*, 182–217. [CrossRef] [PubMed]
5. Galéra, C.; Cortese, S.; Orri, M.; Collet, O.; van der Waerden, J.; Melchior, M.; Boivin, M.; Tremblay, R.E.; Côté, S.M. Medical Conditions and Attention-Deficit/Hyperactivity Disorder Symptoms from Early Childhood to Adolescence. *Mol. Psychiatry* **2021**, *27*, 976–984. [CrossRef]
6. Bonvicini, C.; Faraone, S.V.; Scassellati, C. Attention-Deficit Hyperactivity Disorder in Adults: A Systematic Review and Meta-Analysis of Genetic, Pharmacogenetic and Biochemical Studies. *Mol. Psychiatry* **2016**, *21*, 872–884. [CrossRef]
7. Athanasiadou, A.; Buitelaar, J.K.; Brovedani, P.; Chorna, O.; Fulceri, F.; Guzzetta, A.; Scattoni, M.L. Early Motor Signs of Attention-Deficit Hyperactivity Disorder: A Systematic Review. *Eur. Child Adolesc. Psychiatry* **2020**, *29*, 903–916. [CrossRef]
8. Posner, J.; Polanczyk, G.V.; Sonuga-Barke, E. Attention-Deficit Hyperactivity Disorder. *Lancet* **2020**, *395*, 450–462. [CrossRef]

9. Cinque, S.; Zoratto, F.; Poleggi, A.; Leo, D.; Cerniglia, L.; Cimino, S.; Tambelli, R.; Alleva, E.; Gainetdinov, R.R.; Laviola, G.; et al. Behavioral Phenotyping of Dopamine Transporter Knockout Rats: Compulsive Traits, Motor Stereotypies, and Anhedonia. *Front. Psychiatry* **2018**, *9*, 43. [CrossRef] [PubMed]
10. Kocaturk, R.; Ozcan, O.; Karahan, M. Animal Models of Attention Deficit and Hyperactivity Disorder: A Critical Overview and Suggestions. *J. Neurobehav. Sci.* **2021**, *8*, 1. [CrossRef]
11. Leo, D.; Sukhanov, I.; Zoratto, F.; Illiano, P.; Caffino, L.; Sanna, F.; Messa, G.; Emanuele, M.; Esposito, A.; Dorofeikova, M.; et al. Pronounced Hyperactivity, Cognitive Dysfunctions, and BDNF Dysregulation in Dopamine Transporter Knock-out Rats. *J. Neurosci.* **2018**, *38*, 1959–1972. [CrossRef] [PubMed]
12. Natsheh, J.Y.; Shiflett, M.W. Dopaminergic Modulation of Goal-Directed Behavior in a Rodent Model of Attention-Deficit/Hyperactivity Disorder. *Front. Integr. Neurosci.* **2018**, *12*, 45. [CrossRef]
13. Russell, V.A.; Sagvolden, T.; Johansen, E.B. Animal Models of Attention-Deficit Hyperactivity Disorder. *Behav. Brain Funct.* **2005**, *1*, 9. [CrossRef]
14. Sagvolden, T.; Russell, V.A.; Aase, H.; Johansen, E.B.; Farshbaf, M. Rodent Models of Attention-Deficit/Hyperactivity Disorder. *Biol. Psychiatry* **2005**, *57*, 1239–1247. [CrossRef]
15. Tanaka, M.; Spekker, E.; Szabó, Á.; Polyák, H.; Vécsei, L. Modelling the Neurodevelopmental Pathogenesis in Neuropsychiatric Disorders. Bioactive Kynurenines and Their Analogues as Neuroprotective Agents—In Celebration of 80th Birthday of Professor Peter Riederer. *J. Neural Transm.* **2022**, *129*, 627–642. [CrossRef] [PubMed]
16. Bardgett, M.E.; Depenbrock, M.; Downs, N.; Points, M.; Green, L. Dopamine Modulates Effort-Based Decision Making in Rats. *Behav. Neurosci.* **2009**, *123*, 242–251. [CrossRef] [PubMed]
17. Cagniard, B.; Balsam, P.D.; Brunner, D.; Zhuang, X. Mice with Chronically Elevated Dopamine Exhibit Enhanced Motivation, but Not Learning, for a Food Reward. *Neuropsychopharmacology* **2006**, *31*, 1362–1370. [CrossRef]
18. Chudasama, Y.; Robbins, T.W. Dopaminergic Modulation of Visual Attention and Working Memory in the Rodent Prefrontal Cortex. *Neuropsychopharmacology* **2004**, *29*, 1628–1636. [CrossRef]
19. Tripp, G.; Wickens, J.R. Neurobiology of ADHD. *Neuropharmacology* **2009**, *57*, 579–589. [CrossRef] [PubMed]
20. Volkow, N.D.; Wang, G.J.; Newcorn, J.H.; Kollins, S.H.; Wigal, T.L.; Telang, F.; Fowler, J.S.; Goldstein, R.Z.; Klein, N.; Logan, J.; et al. Motivation Deficit in ADHD Is Associated with Dysfunction of the Dopamine Reward Pathway. *Mol. Psychiatry* **2011**, *16*, 1147–1154. [CrossRef]
21. Wise, R.A. Dopamine, Learning and Motivation. *Nat. Rev. Neurosci.* **2004**, *5*, 483–494. [CrossRef] [PubMed]
22. Kurzina, N.P.; Aristova, I.Y.; Volnova, A.B.; Gainetdinov, R.R. Deficit in Working Memory and Abnormal Behavioral Tactics in Dopamine Transporter Knockout Rats during Training in the 8-Arm Maze. *Behav. Brain Res.* **2020**, *390*, 112642. [CrossRef] [PubMed]
23. Kurzina, N.P.; Volnova, A.B.; Aristova, I.Y.; Gainetdinov, R.R. A New Paradigm for Training Hyperactive Dopamine Transporter Knockout Rats: Influence of Novel Stimuli on Object Recognition. *Front. Behav. Neurosci.* **2021**, *15*, 654469. [CrossRef] [PubMed]
24. Savchenko, A.; Müller, C.; Lubec, J.; Leo, D.; Korz, V.; Afjehi-Sadat, L.; Malikovic, J.; Sialana, F.J.; Lubec, G.; Sukhanov, I. The Lack of Dopamine Transporter Is Associated With Conditional Associative Learning Impairments and Striatal Proteomic Changes. *Front. Psychiatry* **2022**, *13*, 269. [CrossRef] [PubMed]
25. Jahn, C.I.; Varazzani, C.; Sallet, J.; Walton, M.E.; Bouret, S. Noradrenergic But Not Dopaminergic Neurons Signal Task State Changes and Predict Reengagement After a Failure. *Cereb. Cortex* **2020**, *30*, 4979–4994. [CrossRef]
26. Robbins, T.W.; Arnsten, A.F.T. The Neuropsychopharmacology of Fronto-Executive Function: Monoaminergic Modulation. *Annu. Rev. Neurosci.* **2009**, *32*, 267–287. [CrossRef]
27. Levy, F. Pharmacological and Therapeutic Directions in ADHD: Specificity in the PFC. *Behav. Brain Funct.* **2008**, *4*, 12. [CrossRef]
28. Gao, W.-J.; Wang, H.-X.; Snyder, M.A.; Li, Y.-C. The Unique Properties of the Prefrontal Cortex and Mental Illness. In *When Things Go Wrong: Diseases and Disorders of the Human Brain*; InTech: London, UK, 2012; pp. 1–26. [CrossRef]
29. Bokor, G.; Anderson, P.D. Attention-Deficit/Hyperactivity Disorder. *J. Pharm. Pract.* **2014**, *27*, 336–349. [CrossRef]
30. Zametkin, A.J.; Rapoport, J.L. Neurobiology of Attention Deficit Disorder with Hyperactivity: Where Have We Come in 50 Years? *J. Am. Acad. Child Adolesc. Psychiatry* **1987**, *26*, 676–686. [CrossRef]
31. Pliszka, S.R.; McCracken, J.T.; Maas, J.W. Catecholamines in Attention-Deficit Hyperactivity Disorder: Current Perspectives. *J. Am. Acad. Child Adolesc. Psychiatry* **1996**, *35*, 264–272. [CrossRef]
32. Ulke, C.; Rullmann, M.; Huang, J.; Luthardt, J.; Becker, G.A.; Patt, M.; Meyer, P.M.; Tiepolt, S.; Hesse, S.; Sabri, O.; et al. Adult Attention-Deficit/Hyperactivity Disorder Is Associated with Reduced Norepinephrine Transporter Availability in Right Attention Networks: A (S,S)-O-[^{11}C]Methylreboxetine Positron Emission Tomography Study. *Transl. Psychiatry* **2019**, *9*, 301. [CrossRef] [PubMed]
33. Biederman, J.; Spencer, T. Attention-Deficit/Hyperactivity Disorder (Adhd) as a Noradrenergic Disorder. *Biol. Psychiatry* **1999**, *46*, 1234–1242. [CrossRef]
34. Braff, D.L.; Geyer, M.A.; Swerdlow, N.R. Human Studies of Prepulse Inhibition of Startle: Normal Subjects, Patient Groups, and Pharmacological Studies. *Psychopharmacology* **2001**, *156*, 234–258. [CrossRef]
35. Valsamis, B.; Schmid, S. Habituation and Prepulse Inhibition of Acoustic Startle in Rodents. *J Vis Exp.* **2011**, *55*, e3446. [CrossRef]
36. Swerdlow, N.R.; Weber, M.; Qu, Y.; Light, G.A.; Braff, D.L. Realistic Expectations of Prepulse Inhibition in Translational Models for Schizophrenia Research. *Psychopharmacology* **2008**, *199*, 331–388. [CrossRef] [PubMed]

37. Swerdlow, N.R.; Geyer, M.A.; Braff, D.L. Neural Circuit Regulation of Prepulse Inhibition of Startle in the Rat: Current Knowledge and Future Challenges. *Psychopharmacology* **2001**, *156*, 194–215. [CrossRef]
38. Swerdlow, N.R.; Braff, D.L.; Geyer, M.A. Sensorimotor Gating of the Startle Reflex: What We Said 25 Years Ago, What Has Happened since Then, and What Comes Next. *J. Psychopharmacol.* **2016**, *30*, 1072–1081. [CrossRef] [PubMed]
39. Takahashi, H.; Kamio, Y. Acoustic Startle Response and Its Modulation in Schizophrenia and Autism Spectrum Disorder in Asian Subjects. *Schizophr. Res.* **2018**, *198*, 16–20. [CrossRef] [PubMed]
40. Swerdlow, N.R.; Light, G.A. Sensorimotor Gating Deficits in Schizophrenia: Advancing Our Understanding of the Phenotype, Its Neural Circuitry and Genetic Substrates. *Schizophr. Res.* **2018**, *198*, 1–5. [CrossRef]
41. Kurzina, N.; Belskaya, A.; Gromova, A.; Ignashchenkova, A.; Gainetdinov, R.R.; Volnova, A. Modulation of Spatial Memory Deficit and Hyperactivity in Dopamine Transporter Knockout Rats via A2A-Adrenoceptors. *Front. Psychiatry* **2022**, *13*, 503. [CrossRef]
42. Huss, M.; Chen, W.; Ludolph, A.G. Guanfacine Extended Release: A New Pharmacological Treatment Option in Europe. *Clin. Drug Investig.* **2016**, *36*, 1–25. [CrossRef]
43. Pritchett, K.; Mulder, G.B. Hebb-Williams Mazes. *Contemp. Top. Lab. Anim. Sci.* **2004**, *43*, 44–45.
44. Raut, S.B.; Jadhav, K.S.; Marathe, P.A. Role of Dopamine—D2 Receptor in Spatial Memory Retention and Retrieval Determined Using Hebb-Williams Complex Maze. *Indian J. Physiol. Pharmacol.* **2014**, *58*, 191–195.
45. Ptukha, M.; Fesenko, Z.; Belskaya, A.; Gromova, A.; Pelevin, A.; Kurzina, N.; Gainetdinov, R.R.; Volnova, A. Effects of Atomoxetine on Motor and Cognitive Behaviors and Brain Electrophysiological Activity of Dopamine Transporter Knockout Rats. *Biomolecules* **2022**, *12*, 1484. [CrossRef] [PubMed]
46. Leahy, L.G. Diagnosis and Treatment of ADHD in Children vs Adults: What Nurses Should Know. *Arch. Psychiatr. Nurs.* **2018**, *32*, 890–895. [CrossRef] [PubMed]
47. Leffa, D.T.; Caye, A.; Rohde, L.A. ADHD in Children and Adults: Diagnosis and Prognosis. *Curr. Top. Behav. Neurosci.* **2022**, *57*, 1–18. [CrossRef] [PubMed]
48. Saccaro, L.F.; Schilliger, Z.; Perroud, N.; Piguet, C. Inflammation, Anxiety, and Stress in Attention-Deficit/Hyperactivity Disorder. *Biomedicines* **2021**, *9*, 1313. [CrossRef] [PubMed]
49. Liloia, D.; Crocetta, A.; Cauda, F.; Duca, S.; Costa, T.; Manuello, J. Seeking Overlapping Neuroanatomical Alterations between Dyslexia and Attention-Deficit/Hyperactivity Disorder: A Meta-Analytic Replication Study. *Brain Sci.* **2022**, *12*, 1367. [CrossRef]
50. Reale, L.; Bartoli, B.; Cartabia, M.; Zanetti, M.; Costantino, M.A.; Canevini, M.P.; Termine, C.; Bonati, M.; Conte, S.; Renzetti, V.; et al. Comorbidity Prevalence and Treatment Outcome in Children and Adolescents with ADHD. *Eur. Child Adolesc. Psychiatry* **2017**, *26*, 1443–1457. [CrossRef]
51. Gnanavel, S.; Sharma, P.; Kaushal, P.; Hussain, S. Attention Deficit Hyperactivity Disorder and Comorbidity: A Review of Literature. *World J. Clin. Cases* **2019**, *7*, 2420–2426. [CrossRef]
52. Baboli, R.; Cao, M.; Halperin, J.M.; Li, X. Distinct Thalamic and Frontal Neuroanatomical Substrates in Children with Familial vs. Non-Familial Attention-Deficit/Hyperactivity Disorder (ADHD). *Brain Sci.* **2023**, *13*, 46. [CrossRef]
53. Nyatega, C.; Qiang, L.; Adamu, M.; Kawuwa, H.B. Atypical Functional Connectivity of Limbic Network in Attention Deficit/Hyperactivity Disorder. *Clin. Schizophr. Relat. Psychoses* **2022**, *16*, 2022. [CrossRef]
54. Arnsten, A.F. Catecholamine Modulation of Prefrontal Cortical Cognitive Function. *Trends Cogn. Sci.* **1998**, *2*, 436–447. [CrossRef] [PubMed]
55. Dresel, S.; Krause, J.; Krause, K.H.; LaFougere, C.; Brinkbäumer, K.; Kung, H.F.; Hahn, K.; Tatsch, K. Attention Deficit Hyperactivity Disorder: Binding of [99mTc]TRODAT-1 to the Dopamine Transporter before and after Methylphenidate Treatment. *Eur. J. Nucl. Med.* **2000**, *27*, 1518–1524. [CrossRef]
56. Purves-Tyson, T.D.; Owens, S.J.; Rothmond, D.A.; Halliday, G.M.; Double, K.L.; Stevens, J.; McCrossin, T.; Shannon Weickert, C. Putative Presynaptic Dopamine Dysregulation in Schizophrenia Is Supported by Molecular Evidence from Post-Mortem Human Midbrain. *Transl. Psychiatry* **2017**, *7*, e1003. [CrossRef] [PubMed]
57. Palermo, G.; Ceravolo, R. Molecular Imaging of the Dopamine Transporter. *Cells* **2019**, *8*, 872. [CrossRef]
58. Salatino-Oliveira, A.; Rohde, L.A.; Hutz, M.H. The Dopamine Transporter Role in Psychiatric Phenotypes. *Am. J. Med. Genet. Part B Neuropsychiatr. Genet.* **2018**, *177*, 211–231. [CrossRef]
59. Kourosh-Arami, M.; Komaki, A.; Zarrindast, M.-R. Dopamine as a Potential Target for Learning and Memory: Contribution to Related Neurological Disorders. *CNS Neurol. Disord.—Drug Targets* **2022**, *21*, 1. [CrossRef]
60. Madras, B.K.; Miller, G.M.; Fischman, A.J. The Dopamine Transporter and Attention-Deficit/Hyperactivity Disorder. *Biol. Psychiatry* **2005**, *57*, 1397–1409. [CrossRef]
61. Ramos, B.P.; Arnsten, A.F.T. Adrenergic Pharmacology and Cognition: Focus on the Prefrontal Cortex. *Pharmacol. Ther.* **2007**, *113*, 523–536. [CrossRef]
62. Ranjbar-Slamloo, Y.; Fazlali, Z. Dopamine and Noradrenaline in the Brain; Overlapping or Dissociate Functions? *Front. Mol. Neurosci.* **2020**, *12*, 334. [CrossRef]
63. Xing, B.; Li, Y.C.; Gao, W.J. Norepinephrine versus Dopamine and Their Interaction in Modulating Synaptic Function in the Prefrontal Cortex. *Brain Res.* **2016**, *1641*, 217–233. [CrossRef]

64. Plichta, M.M.; Vasic, N.; Wolf, R.C.; Lesch, K.-P.; Brummer, D.; Jacob, C.; Fallgatter, A.J.; Grön, G. Neural Hyporesponsiveness and Hyperresponsiveness during Immediate and Delayed Reward Processing in Adult Attention-Deficit/Hyperactivity Disorder. *Biol. Psychiatry* **2009**, *65*, 7–14. [CrossRef] [PubMed]
65. Devilbiss, D.M.; Berridge, C.W. Low-Dose Methylphenidate Actions on Tonic and Phasic Locus Coeruleus Discharge. *J. Pharmacol. Exp. Ther.* **2006**, *319*, 1327–1335. [CrossRef] [PubMed]
66. Arnsten, A.F.T.; Pliszka, S.R. Catecholamine Influences on Prefrontal Cortical Function: Relevance to Treatment of Attention Deficit/Hyperactivity Disorder and Related Disorders. *Pharmacol. Biochem. Behav.* **2011**, *99*, 211–216. [CrossRef] [PubMed]
67. Chernysheva, M.; Sych, Y.; Fomins, A.; Luis, J.; Warren, A.; Lewis, C.; Serratosa Capdevila, L.; Boehringer, R.; Amadei, E.A.; Grewe, B.F.; et al. Striatum-Projecting Prefrontal Cortex Neurons Support Working Memory Maintenance. *bioRxiv* **2021**. [CrossRef]
68. Terra, H.; Bruinsma, B.; de Kloet, S.F.; van der Roest, M.; Pattij, T.; Mansvelder, H.D. Prefrontal Cortical Projection Neurons Targeting Dorsomedial Striatum Control Behavioral Inhibition. *Curr. Biol.* **2020**, *30*, 4188–4200.e5. [CrossRef]
69. Wang, F.; Yang, J.; Pan, F.; Ho, R.C.; Huang, J.H. Editorial: Neurotransmitters and Emotions. *Front. Psychol.* **2020**, *11*, 21. [CrossRef]
70. Battaglia, S.; Cardellicchio, P.; Di Fazio, C.; Nazzi, C.; Fracasso, A.; Borgomaneri, S. The Influence of Vicarious Fear-Learning in "Infecting" Reactive Action Inhibition. *Front. Behav. Neurosci.* **2022**, *16*, 946263. [CrossRef]
71. Battaglia, S.; Cardellicchio, P.; Di Fazio, C.; Nazzi, C.; Fracasso, A.; Borgomaneri, S. Stopping in (e)Motion: Reactive Action Inhibition When Facing Valence-Independent Emotional Stimuli. *Front. Behav. Neurosci.* **2022**, *16*, 998714. [CrossRef]
72. Antshel, K.M.; Hargrave, T.M.; Simonescu, M.; Kaul, P.; Hendricks, K.; Faraone, S.V. Advances in Understanding and Treating ADHD. *BMC Med.* **2011**, *9*, 72. [CrossRef] [PubMed]
73. Mechler, K.; Banaschewski, T.; Hohmann, S.; Häge, A. Evidence-Based Pharmacological Treatment Options for ADHD in Children and Adolescents. *Pharmacol. Ther.* **2022**, *230*, 107940. [CrossRef] [PubMed]
74. Castells, X.; Blanco-Silvente, L.; Cunill, R. Amphetamines for Attention Deficit Hyperactivity Disorder (ADHD) in Adults. *Cochrane Database Syst. Rev.* **2018**, *8*, CD007813. [CrossRef] [PubMed]
75. Faraone, S.V. The Pharmacology of Amphetamine and Methylphenidate: Relevance to the Neurobiology of Attention-Deficit/Hyperactivity Disorder and Other Psychiatric Comorbidities. *Neurosci. Biobehav. Rev.* **2018**, *87*, 255–270. [CrossRef]
76. Budur, K.; Mathews, M.; Adetunji, B.; Mathews, M.; Mahmud, J. Non-Stimulant Treatment for Attention Deficit Hyperactivity Disorder. *Psychiatry* **2005**, *2*, 44. [PubMed]
77. Higgins, G.A.; Silenieks, L.B.; MacMillan, C.; Thevarkunnel, S.; Parachikova, A.I.; Mombereau, C.; Lindgren, H.; Bastlund, J.F. Characterization of Amphetamine, Methylphenidate, Nicotine, and Atomoxetine on Measures of Attention, Impulsive Action, and Motivation in the Rat: Implications for Translational Research. *Front. Pharmacol.* **2020**, *11*, 427. [CrossRef]
78. Nishitomi, K.; Yano, K.; Kobayashi, M.; Jino, K.; Kano, T.; Horiguchi, N.; Shinohara, S.; Hasegawa, M. Systemic Administration of Guanfacine Improves Food-Motivated Impulsive Choice Behavior Primarily via Direct Stimulation of Postsynaptic A2A-Adrenergic Receptors in Rats. *Behav. Brain Res.* **2018**, *345*, 21–29. [CrossRef]
79. Kawaura, K.; Karasawa, J.-I.; Chaki, S.; Hikichi, H. Stimulation of Postsynapse Adrenergic A2A Receptor Improves Attention/Cognition Performance in an Animal Model of Attention Deficit Hyperactivity Disorder. *Behav. Brain Res.* **2014**, *270*, 349–356. [CrossRef]
80. Álamo, C.; López-Muñoz, F.; Sánchez-García, J. Mechanism of Action of Guanfacine: A Postsynaptic Differential Approach to the Treatment of Attention Deficit Hyperactivity Disorder (Adhd). *Actas Esp. Psiquiatr.* **2016**, *44*, 107–112.
81. Ramos, B.P.; Stark, D.; Verduzco, L.; Van Dyck, C.H.; Arnsten, A.F.T. A2A-Adrenoceptor Stimulation Improves Prefrontal Cortical Regulation of Behavior through Inhibition of CAMP Signaling in Aging Animals. *Learn. Mem.* **2006**, *13*, 770–776. [CrossRef]
82. Arnsten, A.F.T. Guanfacine's Mechanism of Action in Treating Prefrontal Cortical Disorders: Successful Translation across Species. *Neurobiol. Learn. Mem.* **2020**, *176*, 107327. [CrossRef]
83. Sagvolden, T. The Alpha-2A Adrenoceptor Agonist Guanfacine Improves Sustained Attention and Reduces Overactivity and Impulsiveness in an Animal Model of Attention-Deficit/Hyperactivity Disorder (ADHD). *Behav. Brain Funct.* **2006**, *2*, 41. [CrossRef] [PubMed]
84. Adinolfi, A.; Zelli, S.; Leo, D.; Carbone, C.; Mus, L.; Illiano, P.; Alleva, E.; Gainetdinov, R.R.; Adriani, W. Behavioral Characterization of DAT-KO Rats and Evidence of Asocial-like Phenotypes in DAT-HET Rats: The Potential Involvement of Norepinephrine System. *Behav. Brain Res.* **2019**, *359*, 516–527. [CrossRef] [PubMed]
85. Mallien, A.S.; Becker, L.; Pfeiffer, N.; Terraneo, F.; Hahn, M.; Middelman, A.; Palme, R.; Creutzberg, K.C.; Begni, V.; Riva, M.A.; et al. Dopamine Transporter Knockout Rats Show Impaired Wellbeing in a Multimodal Severity Assessment Approach. *Front. Behav. Neurosci.* **2022**, *16*, 254. [CrossRef] [PubMed]
86. Ralph, R.J.; Paulus, M.P.; Fumagalli, F.; Caron, M.G.; Geyer, M.A. Prepulse Inhibition Deficits and Perseverative Motor Patterns in Dopamine Transporter Knock-Out Mice: Differential Effects of D1 and D2 Receptor Antagonists. *J. Neurosci.* **2001**, *21*, 305–313. [CrossRef]
87. Yamashita, M.; Fukushima, S.; Shen, H.; Hall, F.S.; Uhl, G.R.; Numachi, Y.; Kobayashi, H.; Sora, I. Norepinephrine Transporter Blockade Can Normalize the Prepulse Inhibition Deficits Found in Dopamine Transporter Knockout Mice. *Neuropsychopharmacology* **2006**, *31*, 2132–2139. [CrossRef] [PubMed]
88. Bubser, M.; Koch, M. Prepulse Inhibition of the Acoustic Startle Response of Rats Is Reduced by 6-Hydroxydopamine Lesions of the Medial Prefrontal Cortex. *Psychopharmacology* **1994**, *113*, 487–492. [CrossRef]

89. Gómez-Nieto, R.; Hormigo, S.; López, D.E. Prepulse Inhibition of the Auditory Startle Reflex Assessment as a Hallmark of Brainstem Sensorimotor Gating Mechanisms. *Brain Sci.* **2020**, *10*, 639. [CrossRef]
90. Snyder, S.M.; Hall, J.R. A Meta-Analysis of Quantitative EEG Power Associated with Attention-Deficit Hyperactivity Disorder. *J. Clin. Neurophysiol.* **2006**, *23*, 441–456. [CrossRef]
91. Loo, S.K.; Bilder, R.M.; Cho, A.L.; Sturm, A.; Cowen, J.; Walshaw, P.; Levitt, J.; Del'Homme, M.; Piacentini, J.; McGough, J.J.; et al. Effects of D-Methylphenidate, Guanfacine, and Their Combination on Electroencephalogram Resting State Spectral Power in Attention-Deficit/Hyperactivity Disorder. *J. Am. Acad. Child Adolesc. Psychiatry* **2016**, *55*, 674–682. [CrossRef]
92. Michelini, G.; Lenartowicz, A.; Vera, J.D.; Bilder, R.M.; McGough, J.J.; McCracken, J.T.; Loo, S.K. Electrophysiological and Clinical Predictors of Methylphenidate, Guanfacine, and Combined Treatment Outcomes in Children with Attention-Deficit/Hyperactivity Disorder. *J. Am. Acad. Child Adolesc. Psychiatry* 2022, *in press*. [CrossRef] [PubMed]
93. Arnsten, A.F.T.; Jin, L.E. Guanfacine for the Treatment of Cognitive Disorders: A Century of Discoveries at Yale. *Yale J. Biol. Med.* **2012**, *85*, 45. [PubMed]
94. Arnsten, A.F.T. Catecholamine Influences on Dorsolateral Prefrontal Cortical Networks. *Biol. Psychiatry* **2011**, *69*, e89–e99. [CrossRef] [PubMed]
95. Palva, S.; Palva, J.M. Functional Roles of Alpha-Band Phase Synchronization in Local and Large-Scale Cortical Networks. *Front. Psychol.* **2011**, *2*, 204. [CrossRef]
96. Luo, Y.; Adamek, J.H.; Crocetti, D.; Mostofsky, S.H.; Ewen, J.B. Dissociation in Neural Correlates of Hyperactive/Impulsive vs. Inattentive Symptoms in Attention-Deficit/Hyperactivity Disorder. *Front. Neurosci.* **2022**, *16*, 893239. [CrossRef]
97. Sallee, F.R.; Lyne, A.; Wigal, T.; McGough, J.J. Long-Term Safety and Efficacy of Guanfacine Extended Release in Children and Adolescents with Attention-Deficit/Hyperactivity Disorder. *J. Child Adolesc. Psychopharmacol.* **2009**, *19*, 215–226. [CrossRef]
98. Iwanami, A.; Saito, K.; Fujiwara, M.; Okutsu, D.; Ichikawa, H. Safety and Efficacy of Guanfacine Extended-Release in Adults with Attention-Deficit/Hyperactivity Disorder: An Open-Label, Long-Term, Phase 3 Extension Study. *BMC Psychiatry* **2020**, *20*, 485. [CrossRef]

biomedicines

Review

Beyond the Microbiota: Understanding the Role of the Enteric Nervous System in Parkinson's Disease from Mice to Human

Martina Montanari [1,2], Paola Imbriani [1,3], Paola Bonsi [1], Giuseppina Martella [1,*,†] and Antonella Peppe [3,*,†]

1 Laboratory of Neurophysiology and Plasticity, IRCCS Fondazione Santa Lucia, 00143 Rome, Italy; martina.montanari@students.uniroma2.eu (M.M.); p.imbriani@hsantalucia.it (P.I.); p.bonsi@hsantalucia.it (P.B.)
2 Department of Systems Neuroscience, University of Rome Tor Vergata, 00133 Rome, Italy
3 Clinical Neuroscience, IRCCS Fondazione Santa Lucia, 00179 Rome, Italy
* Correspondence: g.martella@hsantalucia.it (G.M.); a.peppe@hsantalucia.it (A.P.)
† These authors contributed equally to this work.

Abstract: The enteric nervous system (ENS) is a nerve network composed of neurons and glial cells that regulates the motor and secretory functions of the gastrointestinal (GI) tract. There is abundant evidence of mutual communication between the brain and the GI tract. Dysfunction of these connections appears to be involved in the pathophysiology of Parkinson's disease (PD). Alterations in the ENS have been shown to occur very early in PD, even before central nervous system (CNS) involvement. Post-mortem studies of PD patients have shown aggregation of α-synuclein (αS) in specific subtypes of neurons in the ENS. Subsequently, αS spreads retrogradely in the CNS through preganglionic vagal fibers to this nerve's dorsal motor nucleus (DMV) and other central nervous structures. Here, we highlight the role of the ENS in PD pathogenesis based on evidence observed in animal models and using a translational perspective. While acknowledging the putative role of the microbiome in the gut–brain axis (GBA), this review provides a comprehensive view of the ENS not only as a "second brain", but also as a window into the "first brain", a potentially crucial element in the search for new therapeutic approaches that can delay and even cure the disease.

Keywords: Parkinson's disease; gut–brain axis; enteric nervous system; central nervous system; neurons; glia cells; non-motor symptoms; gastrointestinal dysfunction; microbiota; rodent models; clinical evidence

Citation: Montanari, M.; Imbriani, P.; Bonsi, P.; Martella, G.; Peppe, A. Beyond the Microbiota: Understanding the Role of the Enteric Nervous System in Parkinson's Disease from Mice to Human. *Biomedicines* **2023**, *11*, 1560. https://doi.org/10.3390/biomedicines11061560

Academic Editors: Masaru Tanaka and Eleonóra Spekker

Received: 30 April 2023
Revised: 23 May 2023
Accepted: 25 May 2023
Published: 27 May 2023

1. Introduction

The gastrointestinal (GI) tract is a long tubular structure that harbors highly diverse and complex communities of microorganisms, including bacteria, archaea, microeukaryotes, and viruses that readily vary with diet, pharmacological intervention, and disease [1].

Numerous articles have described differences in the composition and function of the gut microbiome of healthy individuals and patients of metabolic, autoimmune, and neurodegenerative diseases [2–4]. In the pathogenesis of brain disorders, the possible involvement of peripheral organs has always been marginal. However, it is now well established that the environment of the GI tract and distant organs, such as the brain, is affected by the homeostasis of the gut microbiota and the host's health [4–6]. Early colonization of the gut microbiota is vital for brain function and behavior, considering that its absence results in impairment of the blood–brain barrier [7].

The enteric nervous system (ENS), the part of the nervous system closest to the microbiome, has recently become the subject of in-depth investigations [5,8,9]. It is now known that the microbiome affects the development and functioning of the ENS, modulating it throughout life [4]. Since a wide range of neuropathies are associated with ENS dysfunction, we believe it is worth taking a closer look at it [5,8,9].

The ENS is derived from pre-enteric (rhombencephalon) and sacral neural crest cells and includes efferent and afferent neurons, interneurons, and glial cells.

This well-organized and integrated network of plexuses is relatively independent because it can control gut function independently of CNS sympathetic and parasympathetic innervations [10,11]. However, the ENS is not autonomous, since several CNS structures monitor and regulate what is happening in the GI tract through biochemical signals [12,13]. The gut–brain axis (GBA) consists of a bidirectional communication between the CNS and ENS, linking the emotional and cognitive centers of the brain with peripheral gut functions [14–16]. The GBA connects CNS cognitive centers with gut centers, regulating immune activation, enteric reflex, entero-endocrine signaling, and intestinal permeability [5,17,18]. The bidirectional communication between the gut and the brain implies a vital role for the gut microbiome through regulating host metabolism and immune and vascular systems [19]. In addition, the gut microbiome can also influence the CNS through the vagus nerve by transmitting signals from the gut microbiome to the brain and vice versa in both health and disease through neuro-immuno-endocrine mediators [17,18,20,21].

Disruption of GBA results in alterations in intestinal motility and secretion causes visceral hypersensitivity and leads to cellular changes in the entero-endocrine and immune systems [20].

Considering this complexity, the GI tract can be affected by aging, irritable bowel syndrome, severe inflammatory conditions (Crohn's disease and ulcerative colitis), and even neurodegenerative diseases such as Parkinson's and Alzheimer's [19,21–24].

Parkinson's disease (PD) is a neurodegenerative disease characterized by the loss of dopaminergic cells in the Substantia Nigra pars compacta (SNpc) and brain accumulation of Lewy bodies (LB), which are abnormal aggregates of α-synuclein (αS) [25–27]. PD results from a synergistic interaction between genetic factors and environmental stressors in most patients, a condition termed "double-strike theory" [25,26,28,29].

Therefore, exploring the potential interaction between distinct genetic and environmental factors is essential to identify convergent pathways and potential molecular targets for neuroprotection [30]. PD patients are a heterogeneous group, varying in the age of disease onset, speed of progression, the severity of motor and non-motor symptoms, and the extent of central and peripheral inflammation [31–36]. Indeed, PD is characterized by motor features and numerous non-motor symptoms that include sensory abnormalities, fatigue, sleep disturbances, autonomic dysfunction, psychiatric disorders (depression, anxiety, and apathy), and others [32,36–38]. Orthostatic hypotension, urogenital system disorders, hypersalivation, swallowing impairment, delayed gastric emptying, and constipation are the common manifestations related to autonomic dysfunction in PD [39–45]. Constipation is one of the most frequent non-motor symptoms, affecting up to 80% of PD patients, and may precede the onset of motor symptoms by years [46–50]. In the premotor phase, idiopathic constipation is one of the most critical risk factors for the onset of PD and is associated with neurodegenerative changes in the ENS [12,51].

According to Braak's classic hypothesis [52], neurodegenerative diseases, particularly PD, may recognize a peripheral origin when putative pathogens enter the mucosa of the GI tract, inducing misfolding and aggregation of the hallmark αS in specific subtypes of CNS neurons, then spreading retrogradely to the CNS via preganglionic vagal fibers to the dorsal motor nucleus (DMV) and, finally, to other central nerve structures [12,51,53,54].

Recently, two categories of PD patients have been identified: a brain-first (top-down) type, in which the αS pathology arises initially in the CNS and then in the peripheral autonomic nervous system, and a body-first (bottom-up) type, in which the pathology originates in the ENS and then spreads to the CNS [55].

As pointed out earlier, PD is now considered a systemic disorder despite its typical neurological manifestations. Several autonomic changes in peripheral organs have been described as symptoms and prodromal markers [43,56–58]. The GI tract is primarily affected, hence the importance of assessing early changes occurring in the ENS and interpreting their role in the pathogenesis of PD [43,57,58]. This could help to understand the relationship between α-synucleinopathy, inflammation, neuroprotection, and neurotoxicity, which characterize patients with PD [4,6,7,59].

Data from patients and animal models suggest that PD affects distinct subsets of neurons and glia in the ENS and that the latter may participate in the pathogenesis of this disorder [10,12,56]. Moreover, numerous publications have pointed out the highly complex gut–brain link in PD, laying the foundation for developing new biomarkers and therapies [60]. However, the microbiome appears strongly influenced by environment and socioeconomic background, thus presenting extreme heterogeneity among individuals and little uniqueness [60,61].

In this context, the present article aims to go beyond the microbiome and focus on the involvement of the ENS in PD, elucidating the interactions with the GBA [43,62]. Our goal is to expand knowledge on the pathophysiology of PD by paying particular attention to peripheral biomarkers within the ENS to identify new therapeutic strategies. Furthermore, because the manifestations of neuropathologies are parallel in the ENS and CNS, we believe that the ENS may represent a more accessible target for studies of neural function, histopathology, and biochemistry in PD [56,62]. We envision the ENS not only as a "second brain", but also as a window into the "first brain".

2. Overview of the Enteric Nervous System: Anatomy and Function

The ENS, the intrinsic innervation of the GI tract, is the largest and most complex division of vertebrates' peripheral and autonomic nervous systems. In humans, the ENS contains 400–600 million neurons and an array of neurotransmitters and neuromodulators similar to those found in the CNS [11]. Unlike the CNS, in which efferent pathways are characterized by pre-ganglionic and post-ganglionic neurons [63], the axons of gut neurons in the ENS project to the sympathetic ganglia, brainstem, spinal cord, pancreas, gallbladder, and trachea [10]. The anatomy and physiology of the ENS have been studied since the 19th century, going so far as to demonstrate early in the last century how the peristaltic reflex (i.e., the pressure-induced propulsive activity of the intestines) is a local nervous mechanism that occurs in the absence of external nerve input [8]. Because of this autonomy and its complexity, Michael D. Gershon likened the ENS to a second brain [11]. Two-way communications between the ENS and the CNS are always active: the CNS can regulate or alter the normal functioning of the ENS and vice versa. For example, certain gut disorders impair the production of psychoactive substances such as serotonin (5-HT, 5-hydroxytryptamine), dopamine (DA), and opiates, which can affect mood [64]. Conversely, emotional states, such as intense anxiety, can cause colitis, constipation, irritable colon, or mucosal ulcers by stimulating peristalsis and hyperproduction of neurotransmitters [64]. The ENS originates around the eighth day of embryonic life from neural crest progenitor cells, endowed with stem-like properties, which migrate through the forming GI tract and colonize it within five days [11]. They subsequently differentiate into neurons and glia by integrating predetermined instructions with information from the microenvironment [9]. In humans, the ENS becomes functional in the last trimester of gestation and continues to develop after birth [9]. The ENS comprises small aggregations of nerve cells, the enteric ganglia, the neural connections between these ganglia, and the nerve fibers that supply effector tissues, including gut wall muscle, epithelial lining, intrinsic blood vessels, and gastroenteropancreatic endocrine cells [8,10,11,65]. Enteric neurons (NEs) are organized into ganglionic plexuses: the myenteric (Auerbach's) plexus and the submucosal (Meissner's) plexus. Ganglionic plexuses are enveloped by glial cells, such as CNS astrocytes, which form a proper blood–enteric barrier. Glial cells release enterocyte differentiation factors, participate in GI functions, and are involved in the pathogenesis of inflammatory disorders of the GI tract. Auerbach's myenteric plexus, located in the muscle tonaca between the layers of longitudinal and circular muscles, consists of linear chains of numerous interconnected neurons that span the length of the GI tract and regulate its movements. Meissner's submucosal plexus, located in the submucosa of the small and large intestines but absent in the esophagus and stomach, consists of ganglia stratified at different levels. It integrates sensory signals from the intestinal epithelium and contributes to the local control of secretion, intestinal absorption, blood flow, and submucosal muscle contraction [8,10,65] (Figure 1).

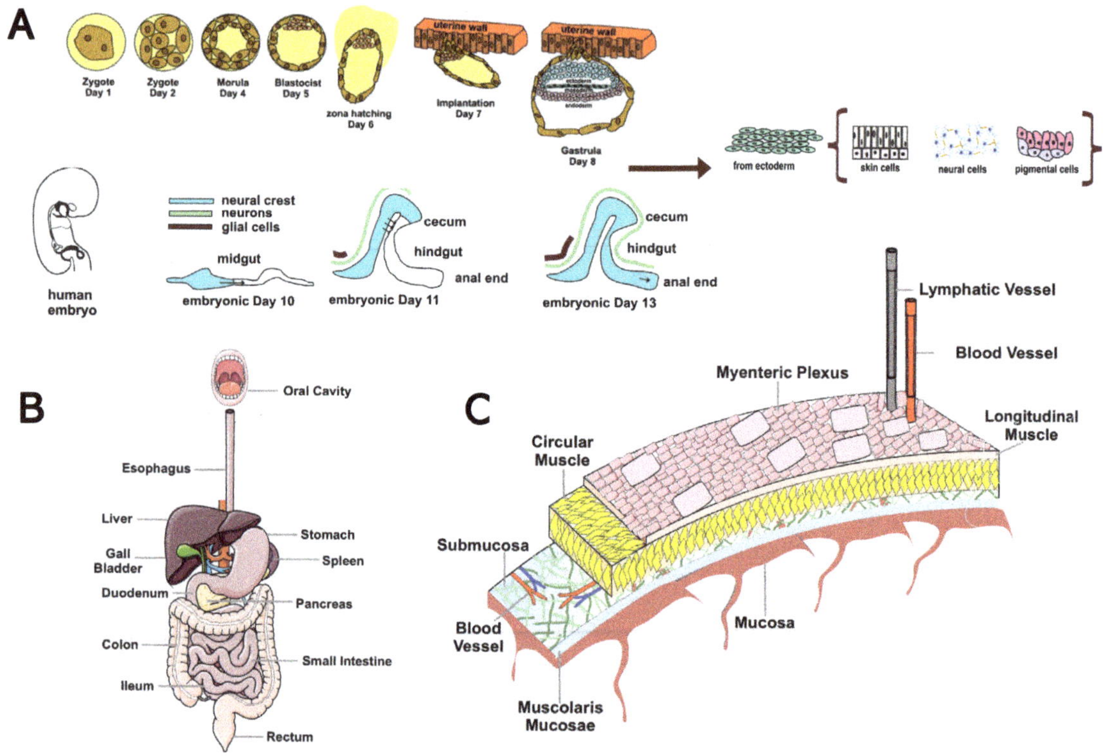

Figure 1. Overview of the anatomy and organization of the ENS. (**A**) Time course of ENS development. The ENS originates around the eighth day of embryonic life from neural crest progenitor cells (ENCDCs) with stem-like properties, which migrate through the GI tract and colonize it within five days. After invading the anterior intestine, these pre-ENCDCs migrate rostro-caudally, proliferating and differentiating into neurons and glia. During this process, the intestine elongates, changing shape from a straight line to a single curve, with the middle and small intestine closely adjacent. The cecal appendix grows and the entire intestine elongates further. At embryonic days 11 and 13, ENCDCs invade the colon by crossing the mesentery and transiting into the cecum. The cecal and trans mesenteric populations then fuse to form the ENS in the rostral colon. In humans, the ENS becomes functional in the last trimester of gestation and continues to develop after birth. (**B**) Schematic diagram of the human GI tract. (**C**) Organization of the ENS. NEs are organized into ganglionic plexuses: the myenteric plexus and the submucosal plexus. The ganglionic plexuses are enveloped by glial cells, such as CNS astrocytes, which form a proper blood–enteric barrier. The myenteric plexus is in the muscle tonaca between the layers of longitudinal and circular muscles. It consists of linear chains of numerous interconnected neurons that span the length of the GI and regulate its movements.

Twenty types of NEs characterized by different morphological, neurochemical, and electrophysiological aspects, connections, and functional roles have been identified [9,66,67]. Based on intracellular electrophysiological recordings, two types of NEs were detected: S and AH neurons. S neurons are characterized by high excitability and can exhibit rapid excitatory postsynaptic potentials, followed by a short-lived hyperpolarizing current (20–100 ms), rapidly restoring the membrane potential [66,68]. On the other hand, AH neurons exhibit large action potentials followed by a slow hyperpolarizing current (2–30 s) that makes them less excitable. NEs use more than 50 neurotransmitters in synaptic communications, from small neurotransmitters (e.g., ACh, acetylcholine, 5-HT) to neuropeptides (e.g., CGRP, calcitonin gene-related peptide, somatostatin, substance P, and VIP, vasoactive intestinal peptide) to gases (e.g., NO, nitric oxide) [67,68]. NEs are grouped into three functional classes: intrinsic sensory neurons called IPANs, muscle motor neurons, and interneurons. IPANs are large and equipped with numerous axons; they can sense mechanical, chemical, and thermal stimuli and transmit information about muscle tension state and endoluminal content to motor neurons [69], triggering reflexes that regulate motility, secretion, and blood flow. They make up about 10–30% of the neurons located in the submucosal and myenteric plexus of the small and large intestines; they are not present in the esophagus (whose motility is controlled by fibers originating from the CNS) and stomach (whose motility is under the control of vagal fibers) [69]. Motor neurons are divided into muscular and secretomotor-vasodilatory. The former (Dogiel's type I) innervate the circular and longitudinal musculature and the muscular mucosae, determining their contraction or relaxation; they have an elongated cell body, numerous dendrites, and a single slender axon; electrophysiologically, they correspond to type S. Neurons innervating circular and longitudinal musculature have their cell bodies in the myenteric plexus and are excitatory (using ACh and TK, tachykinin, and projecting orally) or inhibitory (using NO and VIP and projecting anally) [69]. Muscle motor neurons generate, following regional stimulation, coordinated and polarized muscle responses that allow the progression of intestinal contents, i.e., induce contraction in the oral direction and relaxation in the anal direction [69]. On the other hand, secretomotor-vasodilator neurons are located mainly in the submucosal ganglia, controlling both the secretion of ions and water via ACh and the vasodilation of submucosal arterioles via VIP [66,67]. Some influence glucose transport across the mucosa of the small intestine [70], a process also regulated by vagal-like reflexes; others modulate acid secretion in the stomach [70]. Interneurons integrate sensory afferents and organize effector responses [67,68]. In the myenteric plexus, they form chains that run in ascending and descending directions. They resemble type I neurons and are S-type [68]. In the course of life, the ENS undergoes plastic changes as a spatiotemporal adaptive response to external stimuli, which arrive through sensory afferents, and to internal stimuli that come from autonomic innervation [8]. In the complex microenvironment of the gut wall lodge, different types of cells (neurons, glia, Cajal cells, muscle cells, and immune cells) can communicate with each other in synaptic or paracrine ways. This interactive plurality modulates the functional state of NEs by influencing the digestive and secretory functions of the GI tract [71]. Changes in diet and perturbations in the gut microbiome, with its metabolites and neuroactive compounds, affect the functioning of the NE and its connections with the CNS, since they alter mucosal permeability and the secretion of hormones and immune cells. In addition, NEs are vulnerable to aging-related degeneration [71].

3. Evidence of the Role of the Enteric Nervous System in Animal Models of Parkinson's Disease

GI dysfunction is a common non-motor symptom of PD. While, in PD patients, it is present in 80–90% of cases and has been associated with αS aggregation and neuronal loss in the CNS, reports of GI symptoms in animal models of PD are known to vary, and the degree to which pathology in the CNS contributes to GI symptoms remains unclear [72].

PD benefits from a wide range of animal models whose diverse pharmacological, toxin, and genetic features are essential to study its etiology and neurobiology [73]. Animal models of PD rely on pharmacological or genetic approaches to simulate nigrostriatal neurodegeneration and disease pathogenesis [73]. However, much remains to be discovered and requires continuous questioning by the research community.

The most commonly used pharmacological models are based on neurotoxins administered to mice, rats, and nonhuman primates [74] (Figure 2).

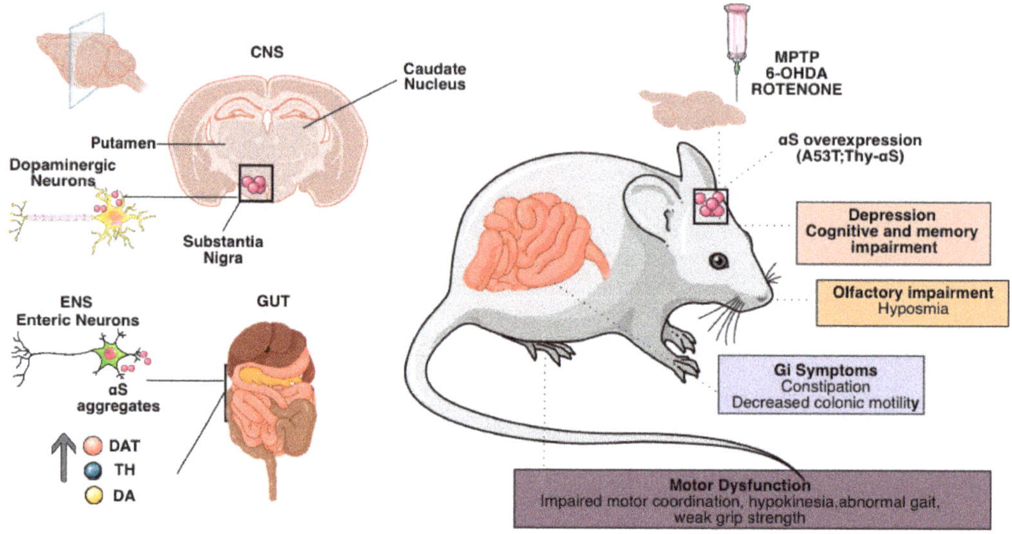

Figure 2. Schematic representation of the main physiological and behavioral changes in CNS and ENS of preclinical models of PD. PD is a heterogeneous disorder with varying ages of onset, symptoms, and progression rates. This heterogeneity requires the use of a variety of animal models to study different aspects of the disease. (Right) Neurotoxin-based approaches include exposure of rodents or nonhuman primates to 6-OHDA, MPTP, and agrochemicals such as the pesticide rotenone. Acute neurotoxin exposure induces motor deficits and rapid nigrostriatal dopaminergic cell death by disrupting mitochondrial function and increasing oxidative stress. Chronic neurotoxin administration induces progressive patterns that may include αS aggregates. Genetics-based approaches to modeling PD include transgenic and viral-vector-mediated models based on genes linked to monogenic PD. Among these, overexpression and introduction of preformed α-S fibrils induce toxic protein aggregates, nigrostriatal neurodegeneration, and variable motor deficits, depending on the specific model. (Left) GI dysfunction is the most common non-motor symptom of PD. Symptoms of GI dysmotility in PD include premature satiety and weight loss due to delayed gastric emptying and constipation due to altered colonic transit. We can find numerous alterations in the ENS in preclinical models of PD: neurodegeneration of NEs, which is the leading cause of behavioral and electrophysiological alterations in mouse models.

Both neurotoxins, 1-methyl-4-phenyl-1,2,3,6-tetrahydropyridine (MPTP) and 6-hydroxy dopamine (6-OHDA), consistently affect nigrostriatal dopaminergic pathways [74]. However, their impact on gut function and the CNS varies, depending on the agent, mode of administration, and assays used [75–78]. Systemic administration of MPTP in mice causes loss of dopaminergic neurons in the myenteric plexus but does not cause severe defects in GI motility [76,77]. Peripheral administration of MPTP in rats does not significantly affect the number of dopaminergic neurons and the expression of dopaminergic markers in the SNpc [79]. However, it significantly reduces tyrosine hydroxylase-immunoreactive (TH-IR) neurons in the GI tract, suggesting that the degeneration of dopaminergic neurons might start earlier than in the SNpc [48,76,79]. Parenteral administration of MPTP in mice simultaneously induces dopaminergic neurodegeneration in the ENS, which is associated with behavioral and electrophysiological alterations. Following MPTP intoxication, acceleration of motility (increased contraction) and decreased colonic relaxation are observed in response to electric field stimulation of the NE [80,81]. These complementary findings point to the altered function of enteric DA neurons. Several articles have shown that exogenous DA antagonizes colonic muscle contractility in a receptor-dependent manner [81].

Furthermore, confirming that MPTP is selectively toxic to dopaminergic neurons in the ENS, just as in the CNS, TH-positive neurons in the myenteric ganglia are reduced [82–84]. Most TH-positive neurons with cell bodies in the myenteric plexus can be considered dopaminergic, since adrenergic and noradrenergic inputs to the GI tract are mainly extrinsic [85].

Considering the neuropathological and electrophysiological findings, it is likely that dysfunction and death of dopaminergic neurons cause the transient increase in colonic motility observed after MPTP intoxication. Decreased dopaminergic inhibitory tone results in faster colonic transit due to the relative abundance of stimulatory neuronal input [80,81].

Neurotransmitters related to the GI dysfunction of PD could be involved in the intestinal dopaminergic, cholinergic, and oxidergic nitric systems [35]. To investigate the relationship between the GI dysfunction of PD and the alteration of GI neurotransmitters, 6-OHDA was microinjected into one side of the nigrostriatal system of the brain to generate an animal model of PD through the impairment of rat dopaminergic neurons, and the effect of neurotransmitter alterations in the CNS on GI function was observed [75].

GI dysfunction and changes in dopaminergic, nitric oxide synthase (NOS), and cholinergic neurons in the myenteric plexus were analyzed. Compared with control samples, 6-OHDA rats had delayed gastric emptying and constipation, which could be related to increased GI TH and decreased NOS. These symptoms were not associated with alterations in cholinergic transmitters [78].

Unfortunately, some of these studies did not analyze the submucosal plexus, making a direct comparison with more robust findings in complex PD patients [56]. Rats treated with 6-OHDA show elevated protein levels of TH and dopamine transporter (DAT) (dopaminergic markers) in both the epithelium and neurons of the GI tract, resulting in increased DA content in the gut and delayed gastric emptying [79]. In the epithelium and neurons of the GI tract, neurodegeneration of the SN by 6-OHDA increases the expression of TH and DAT proteins. It is hypothesized that the number of enteric dopaminergic neurons and cells may increase to compensate for the loss of DA in the SN in PD patients [79].

In contrast, the increased protein expression of TH and DAT in 6-OHDA-treated rats may increase the concentration of DA in the colon and the loss of DA in the SN, which may cause constipation [79].

Alterations in the monoaminergic system and decreased colonic motility were observed in rats microinjected with 6-OHDA in the bilateral SN [75].

DA, NE, and 5-HT play essential roles in regulating colonic motility: increased DA content, upregulation of β3-ARs, and decreased 5-HT4 receptors could contribute to the decreased spontaneous colonic contraction and constipation observed in rats with 6-OHDA [75].

Rats with lesions of SN dopaminergic neurons manifest GI dysmotility [86,87], including gastroparesis and constipation [87,88].

Animal models do not yet allow for an adequate study of how PD prodromal constipation occurs [89]. To date, there is a paucity of relevant experimental models of GI dysfunction associated with αS pathology; αS deposition in the ENS of PD patients has been reported in the myenteric and submucosal plexuses of GI tracts [90,91]. Transgenic mouse lines expressing a mutant form of human αS (A53T or A30P) under its promoter show colonic disorders similar to constipation and pathology characteristic of αS [92]. In a transgenic mouse model in which mutant human αS (A53T) was expressed under the control of the prion promoter [93], aggregates of αS were observed in the ENS prior to changes in the CNS [92]. This finding suggests that αS pathology may be initiated from the ENS and propagate to the CNS via the vagus nerve [52]. In support of this, in a transgenic mouse model, the accumulation of αS aggregates in the ENS precedes changes in the CNS [92].

Expression of human αS in the DMV, a region of the brain severely affected by PD, causes an age-related slowing in A53T mice of GI motility reminiscent of that observed in patients with PD [52,94]. The symptoms coincide with the disruption of efferent vagal processes that project from the DMV to the GI tract. This pattern parallels the pathology of postmortem specimens of PD patients and implicates the DMV as a possible mediator of GI neuropathology and symptomatology in PD [95].

However, αS mutations are only responsible for rare cases of PD [30]. Mice overexpressing wild-type human αS under the Thy-1 promoter (Thy1-αS) show increased transit time and colonic content compared with wild-type (WT) pups when tested at 12–14 months of age [96]. However, striatal dopamine loss occurs only after 14 months in Thy1-αS mice, manifesting motor and non-motor deficits, such as olfactory disturbances, as early as 2–3 months of age [97,98].

The mechanisms underlying colonic motor impairments may be related to αS overexpression in the colonic myenteric nervous system [96]. The reduced response to defecation stimuli in Thy1-αS could be related to the accumulation of αS in colonic myenteric plexuses [96].

The GI system is one of the most susceptible to environmental stressors, since it is in direct contact with environmental agents [99–101]. In a recent study, intra-gastric administration of rotenone in mice caused progressive αS deposition in both the ENS and CNS neurons affected by PD, such as neurons in the myenteric plexus, the vagus DMV, the spinal cord, and the sympathetic nervous system (SNS) [102]. These studies suggested that environmental stressors to the GI system could lead to αS pathology in the CNS.

Numerous preclinical pieces of evidence associate GI symptoms in toxic models of PD based on oral administration of rotenone [99]. Previous studies have shown that orally administered rotenone exposure induces PD-like changes in the ENS and triggers PD progression throughout the nervous system to the SN [100,102]. Interestingly, the latter changes appear as early as the first moments after rotenone administration (2 months) before the onset of motor symptoms (which occur after three months of exposure in this animal model), thus mimicking the pattern of progression observed in PD patients.

In two recent studies, rotenone exposure reduced sympathetic noradrenergic [103] and vagal cholinergic gut innervation [104].

The mechanism by which environmental agents induce αS aggregation is unknown. However, a recent study showed that αS expression in the ENS could be upregulated by agents that cause depolarization and increase cyclic AMP levels [105].

An emerging concept in gastroenterology is that a wide range of diseases, such as motility disorders, can be partially considered enteric neuropathies. In particular, aging is associated with various motility or gut disorders, including delayed gastric emptying and longer intestinal transit time [106]. Aged rats show neuronal loss and changes in neurochemical phenotype in the ENS, which may result in motility disorders [107]. Surprisingly, along with neuronal loss, these rats exhibit dystrophic NEs that contain αS aggregates reminiscent of Lewy pathology [108].

Braak et al. hypothesized that PD originated in the gut and subsequently progressed up, as if along a ladder, along the nerves connecting the gut to the brain [91].

Using double transgenic mice expressing mutant αS, it is possible to observe how early alterations in ENS can be identified as early disease markers. These animal models expressing mutant αS provide an opportunity to investigate the potential role of ENS as an early marker of disease [92]. Early ENS dysfunction would not only trigger disease but facilitate the entry of deleterious factors that cause progression and spread to the CNS [12,92]. A summary of animal models exhibiting each of these characteristics is provided in Table 1.

Table 1. Pathological features identified in animal models of PD. The table summarizes the significant alterations found in murine models of PD. The legend of the abbreviations is listed below.

PD Model	Affected Neuron Types	GI Symptoms	Alteration Biomarker	References
MPTP mice	Loss of dopaminergic neurons in the myenteric plexus.	Absence of severe defects in GI motility. Increased contraction and decreased relaxation of colon muscle in response to electric field stimulation of NEs.	Nd	[76,77]
MPTP rats (Peripheral administration)	Unaltered number of dopaminergic neurons in the SNpc. Presence of TH-IR neurons in the GI tract.	Nd	Unaltered expression of dopaminergic markers in the SNpc.	[79] [48,76,79]
6-OHDA rats	Alterations in the monoaminergic and cholinergic system.	Delayed gastric emptying and constipation, which could be related to increased GI TH and decreased NOS. Increased DA concentration in the colon, which is more likely to cause constipation. Decreased colonic motility.	Unaltered cholinergic transmitters. Elevated protein levels of TH and DAT both in the epithelium and neurons of the GI tract, resulting in increased DA content in the gut and delayed gastric emptying.	[75] [78,79] [86–88]
A53T mice (Expressing a mutant form of human αS)	Disruption of efferent vagal processes that project from the DMV to the GI tract.	Related slowing of GI motility caused by expression of human αS in the DMV.	Accumulation of αS aggregates in the ENS before changes in the CNS.	[92] [52,94] [95]
Thy1-αS mice	Nd	Striatal dopamine loss only after 14 months: manifesting motor and non-motor deficits, such as olfactory disturbances, as early as 2–3 months of age.	Increased transit time and colonic content. Overexpression of αS in the colonic myenteric nervous system. Reduced response to defecation stimuli.	[96] [97,98]
Rotenone mice model	Reduced sympathetic noradrenergic and vagal cholinergic gut innervation.	Aggregates of αS in both ENS neurons of the myenteric plexus and at the level of the DMV, spinal cord, and SNS.	Nd	[99] [102] [103] [104]
Fischer 344 rat	Neuronal loss and changes in neurochemical phenotype in the ENS.	Dystrophic enteric neurons that contain αS aggregates reminiscent of Lewy pathology.	Motility disorders	[108]

Parkinson's disease (PD); enteric neurons (NEs); 1-methyl-4-phenyl-1,2,3,6-tetrahydropyridine (MPTP); substantia nigra pars compacta (SNpc), tyrosine hydroxylase-immunoreactive (TH-IR); gastrointestinal tract (GI), tyrosine hydroxylase (TH); enteric nervous system (ENS); central nervous system (CNS); sympathetic nervous system (SNS); alpha-synuclein (αS); 6-hydroxydopamine (6-OHDA); dorsal motor nucleus of the vagus nerve (DMV); dopamine transporter (DAT); dopamine (DA); nitric oxide synthase (NOS); Not declared (Nd).

4. The Possible Role of the Enteric Nervous System in Parkinson's Disease: Clinical Evidence

In idiopathic PD, most patients show PD-related inclusions at CNS sites and in the ENS and sympathetic ganglia where LB and Lewy neurites (LN) can be found [91]. Based on postmortem studies performed on PD patients and healthy individuals, Braak et al. proposed a pathological disease staging [109] in which PD lesions follow a specific spatiotemporal pattern, as described. These lesions start in the olfactory bulb (OB) and/or at the intestinal level, maybe in the ENS, and progress to the CNS through synaptically connected structures. This pattern seems to correspond to early non-motor symptoms in Parkinsonian patients, such as hyposmia, GI manifestations, autonomic dysfunction, and pain [110] (Figure 3).

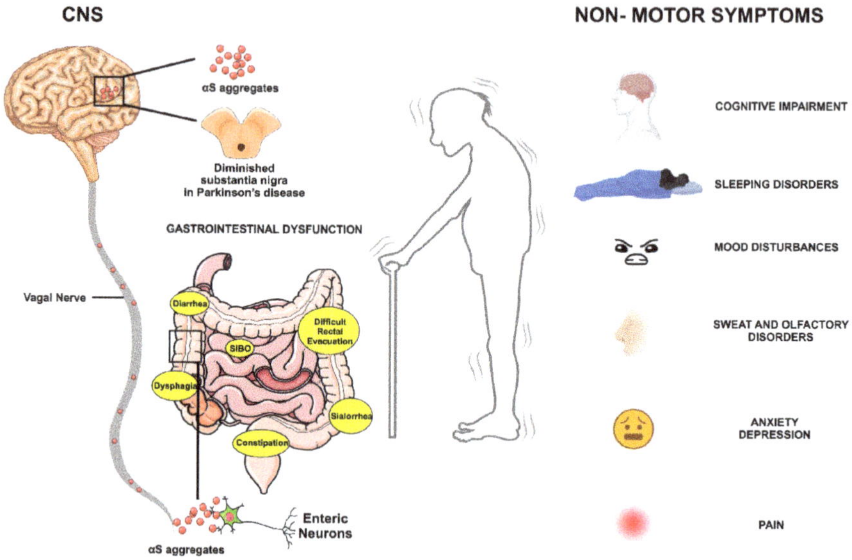

Figure 3. Non-motor features of Parkinson's disease, focus on gastrointestinal symptoms. Diagnosis of PD currently depends on motor deficits, including bradykinesia, rigidity, and tremor. The motor characteristics predominantly result from the loss of dopaminergic neurons in the SNpc. However, the non-motor symptoms of PD often begin before the more visible motor symptoms. These are called "pre-motor symptoms", such as loss of smell, depression, and constipation, which can appear years before diagnosis. The GI symptoms include excessive drooling, dysphagia, impaired gastric emptying, constipation, and impaired defecation. Moreover, alterations in the ENS levels have been reported in PD. It has been proposed that the appearance of αS aggregations beside the GI tract is an indicative tool that supports early diagnosis of PD before the onset of motor symptoms. αS has richly expressed throughout the ENS nerve plexus in healthy individuals and its growth rates with aging. Consequently, there is a need to evaluate the pathological relevance of αS carefully evaluated as a predictive biomarker of PD.

Little is known about the ENS degenerative process in PD patients. Although PD is mainly characterized by impaired extrapyramidal motor control, clinical studies have revealed delayed gastric emptying, external anal sphincter dystonia causing difficult rectal evacuation, and general slow-transit constipation caused by local loss of dopaminergic neurons [111–113]. Many researchers have recently studied the expression and modifications of enteric αS in PD patients, with controversial results. Although some have found increased inclusions of αS and phosphorylated αS in the above areas compared with control subjects [50], others point attention to high variability among patients [99].

All these hypotheses take advantage of clinical observation of both the prodromal symptoms and the non-motor ones [114–116], assuming that the involvement of the dopaminergic system in PD neurodegenerative process starts at the level of the DMV with a pattern of periphery–center (bottom–top) [115].

This process would initially occur in the enteric system (signs: constipation slow and transit alteration) and then progress to the brainstem with hypo/anosmia and sleep disturbances up to the mesencephalon (SNpc) with the appearance of the cardinal symptoms (rigidity, bradykinesia, tremor, and postural instability); finally, it would involve the cerebral cortex with the appearance of cognitive and behavioral disturbances [117]. Braak predicted six stages of the disease, of which only the third one involves the CNS with the appearance of motor symptoms [52,109], in a sort of dominos game [118]. An exciting and innovative hypothesis has shifted the interest of researchers in the last 20 years toward the discovery of early biomarkers [119]. The big challenge was to identify a potential pathogen capable of passing the mucosal barrier of the GI tract and, via postganglionic ENs, entering the CNS along unmyelinated preganglionic fibers generated from the viscero-motor projection cells of the vagus nerve [120]. There is also epidemiological evidence that complete but not partial vagotomy may protect against later PD [121,122].

Other studies suggested that not all cases of PD start in the ENS. Autopsy studies have shown that a minority of cases showing Lewy pathology do not present pathological inclusions in the DMV. Moreover, some cases display a distribution of αS inclusions that can be limbic-predominant, revealing less pathology in the brainstem [123–125]. Therefore, different subtypes of PD have been proposed according to these criteria: (i) a body-first (bottom-up) subtype, in which the disease starts in the enteric or peripheral autonomic nervous system and arises, via the sympathetic connectome of the vagus nerve, to the CNS [126]. This phenotype shows prolonged intestinal transit and constipation as prodromic symptoms; (ii) a brain-first (top-down) subtype in which the αS pathology originates in the brain or via the OB and descends to the peripheral autonomic nervous system. This phenotype shows hyposmia and sleep disturbances as prodromic signs [127]. These differences reflect the high variability between patients' phenotype and clinical signs in a "puzzle game" that PD seems to be.

The GBA comprises different functional, neuroendocrine, and neuroimmune systems, including the hypothalamic–pituitary–adrenal axis, the ENS through the sympathetic and parasympathetic systems, the vagus nerve, intestinal immune cells, and intestinal microbiota [128].

Approximately 100 trillion microbes in the human gut are involved in food fermentation, metabolic, and immune maturity. Many play a central role in developing the ENS and CNS and in the modulation of the pathogenesis of metabolic, neurodegenerative, and neurodevelopmental disorders [129].

The gut-hosted bacteria can impact brain function via different pathways. These bottom-up pathways include direct absorption through the gut–blood/lymphatic–brain pathways, as well as local signaling in the gut to prime immune cells, and the vagal retrograde transport pathways [130]. It was demonstrated that microbe-secreted products such as neurotransmitters, including catecholamines, GABA, 5-HT, and gut metabolites transit through the gut–blood and blood–brain barriers and elicit an immune response that changes the profiles of plasma proteomics and brain neurochemistry. In a different network, bacterial metabolites can activate immune cells [131,132]. A third pathway represented by the vagal route was identified too. Evidence shows that the vagus nerve can transport αS from the gut to the brain [133].

Studies performed by amplifying the rRNA gene or by "metagenomic sequencing" revealed changes in the intestinal microbiota of PD patients compared to healthy controls [134]. Modifications in the intestinal microbiome also correlate with PD progression. For example, a decrease in the microbiota producing short-chain fatty acids and an increase in proinflammatory bacteria seem to correlate with motor and cognitive severity in PD patients [135]. A longitudinal follow-up clinical study showed that a decreased amount of Roseburia intestinal bacteria is linked to a rapid progression of both motor and non-motor symptoms of PD. Moreover, a reduced amount of short-chain fatty-acid-producing bacteria, such as Fusicatenibacter and Faecalibacterium, is correlated to an increase in fecal inflammatory calprotectin levels in Parkinsonian patients [136].

Systemic and fecal inflammatory markers IFN-γ, TNF-α, and neutrophil gelatinase-associated lipocalin were also associated with an elevated expression of Bacteroides and Bifidobacterium in PD patients [41]. Thus, the intestinal microbiota composition in PD patients appears to influence pharmacological treatment responses [135]. A growing body of evidence supports the role of the microbiome in the pharmacokinetics of drugs used in PD treatment; at the same time, the drugs can alter the gut microbiome's composition [137]. This evidence highlights the concept that the intestinal microbiome may influence the treatment efficacy and the development of potential modified response to levodopa therapy [135].

In conclusion, in the very early stages of PD before CNS pathology, accumulation of enteric αS [138] may promote activation of immune/inflammatory signaling, including canonical caspase-1-dependent inflammasome pathways [139], resulting in a massive release of IL-1β, which, in turn, alters the intestinal epithelial barrier through the activation of IL-1 receptors on intestinal epithelial cells [139]. In this context, intestinal inflammation and altered intestinal epithelial barrier can induce changes in short-chain fatty acid levels, characterized by alterations in butyrate levels, which could contribute to the impairment of the intestinal epithelial barrier [140]. They can also cause an increase in the concentration of circulating lipopolysaccharide, contributing further to activating the intestinal immune/inflammatory pathways. This would induce a vicious circle that could bring the chronicization of inflammatory processes with the appearance of intestinal symptoms and brain pathology [141] (Table 2).

Table 2. Pathological features identified in PD patients.

PD Symptoms	Affected Neuron Types	GI Symptoms	Alteration Biomarker	References
Nd	Nd	Gastric emptying. Difficult rectal evacuation. Slow transit constipation.	Nd	[111–113]
Hypo/anosmia. Sleep disturbances. Rigidity, bradykinesia, tremor, and postural instability. Cognitive and behavioral disturbances.	Neurodegenerative process starting in the DMV with a pattern of periphery–center (bottom–top).	Nd	Increased inclusions of αS and phosphorylated αS.	[115] [117]
Nd	Nd	Prolonged intestinal transit and constipation.	Minority of cases with Lewy pathology without pathological inclusions in the DMV. Limbic-predominant distribution of αS inclusions with less pathology in the brainstem.	[123,124] [126]
Motor and cognitive symptoms.	Nd	Nd	Decrease in the short-chain fatty acids, including Fusicatenibacter and Faecalibacterium. Increase in proinflammatory bacteria.	[135] [136]
Nd	Nd	Nd	Systemic and fecal inflammatory markers IFN-γ, TNF-α, and neutrophil gelatinase-associated lipocalin, associated with an elevated expression of Bacteroides and Bifidobacterium.	[41]
Nd	Nd	Alteration in intestinal epithelial barrier.	Accumulation of enteric αS. Activation of immune/inflammatory signaling, including canonical caspase-1- dependent inflammasome pathways. Massive release of IL-1β.	[138] [139] [140]

Alpha-synuclein (αS); dorsal motor nucleus of the vagus nerve (DMV); tumor necrosis factor-alpha (TNF-α); interferon-gamma (IFN-γ); interleukin-1β (IL-1β); Not declared (Nd).

5. New Therapeutic Approach Targeting the Enteric Nervous System

The lessening of dopaminergic striatal and nigral innervation alters local microcircuits [142,143]. The emerging scenarios concerning enteric involvement in PD pathogenesis offer a new therapeutic approach.

The nonpharmacological approach based on the increase in enteric system motility has been well defined in the last few years. For example, a high fiber diet, appropriate fluid intake, and psyllium can represent an excellent approach to counteract the slowing of bowel pain in many PD patients, as well as exercise and physical activity directed to stimulate autonomic symptoms (impaired gastric motility, dysphagia, constipation, and bowel incontinence) [34]. These approaches are based on evidence that exercise may also change dopamine receptor availability in animal models of PD and patients [42,144].

Adjustment of anticholinergics and dopaminergic agents used for PD therapy can contribute to relieving intestinal and motor symptoms by demonstrating the connection between enteric and CNS [40]. Many other approaches were proposed to treat the co-morbidity of PD, considering that gut dysfunction may contribute to the symptomatic fluctuation in PD patients.

The microbiome is also well discussed in many other papers focusing on the enteric flora to explain different phenomena. Many different types of microbiomes have been found in different PD patients, which does not allow a unique key for reading, confirming the complexity of PD and the possibility that an inadequate therapeutic approach is used.

6. Discussion

PD is a frequent neurodegenerative disorder characterized by a constellation of clinical manifestations: apart from classic motor symptoms, patients also often experience non-motor manifestations, including hyposmia, sleep disturbances, depression, dementia, and GI dysfunction [144–148], some of them could appear even decades before the onset of motor signs [149–153].

GI dysfunction often occurs in the early stages of the disease [154]. This observation and the detection of misfolded αS protein in the ENS of PD patients [145,146] have directed interest toward the hypothesis that the disorder may originate in the gut. Indeed, it has become increasingly evident that, in PD, the neurodegenerative process involves several structures even distant from the CNS, such as the ENS, which is the dense neural network of neurons and glial cells regulating and co-ordinating gut function and motility, referred to as the "brain in the gut" or "second brain" [32,34–36,148]. Following this hypothesis, several studies have investigated the role of ENS in PD [91].

On the one hand, Lewy pathology could be induced in the ENS and transported to the CNS via the vagal nerve. The aggregation and propagation of enteric-derived αS probably indicate an early pathological stage that could subsequently initiate the motor and non-motor symptoms characteristic of PD.

On the other hand, the gut microbiota might also play a role through the effect of different molecules and proteins produced by gut bacteria that can act locally or be transported to the CNS through the vagus nerve fibers. This transport of substances between the gut and CNS has been verified in pathological conditions [5,20], finding implications in PD. Dysbiosis of the microbiota leads to an imbalance between beneficial and harmful microbial metabolites, causing increased intestinal permeability and inflammation and systemic inflammation. αS aggregates in the intestine likely induce enteric pathology and dysfunction, which can trigger enteric inflammation, dysbiosis, and intestinal hyperpermeability. The triggered inflammatory state impacts the CNS and promotes PD pathology.

In this review, we aim to discuss the ENS involvement in the pathophysiology of PD by providing evidence from preclinical and clinical studies.

Until April 2023, there have been 66,523 articles about the etiology of PD on the PubMed database. Of these papers, 13,060 are reviews. However, a substantial limitation is that only 0.99% of articles support the gut–brain axis theory, with only 546 works discussing the ENS and PD and 1119 articles focusing on PD and the microbiome. Current research focuses mainly on the microbiome rather than the relationships between the autonomic nervous system and the CNS, which probably underlie all etiological processes.

Therefore, our review attempted to go beyond the role of the microbiota, focusing mainly on other possible players in PD pathology, such as the ENS. Indeed, we believe that the field of microbiome research is complicated and highly heterogeneous. Moreover, the multitude of factors mediating the potential influence of the gut microbiome on PD is influenced by everyone's dietary and lifestyle habits, levels of inflammation, comorbidities, and use of supplements or medications [136]. Different compositions of the gut microbiomes in PD patients represent a limiting factor in interpreting results. In addition, ethnological, cultural, and lifestyle differences may cause heterogeneous results among studies, posing a limitation for further research. We believe more studies from different nations and regions are needed to explore the relationship between PD and GI diseases.

Although alterations in the GI tract have been highlighted in the pathogenesis of PD, the exact mechanism linking enteric inflammation and neurodegeneration remains to be elucidated. This would significantly aid in early diagnosis and intervention to slow or halt disease progression. Indeed, there is currently no therapy available to cure PD, and the early stages of the disease are probably best suited for personalized disease-modifying interventions. In this regard, animal models may represent an essential tool to study the pathogenesis of PD, as they offer the possibility of simultaneously observing behavioral abnormalities, in vivo imaging, and pathological assessments. However, as with clinical trials, preclinical research still has limitations, since animal models of PD do not accurately recapitulate human PD [155]. Moreover, animal models that recapitulate the prodromal stages of the disease still need to be developed [156] and few approaches are available to study alterations of the gut microbiome, inflammatory processes, and disease progression [157]. The limited availability of animal models that can recapitulate prodromal PD and reproduce both peripheral and central pathology has dramatically slowed the understanding of PD pathogenesis, including our comprehension of the precocious involvement of the autonomic nervous system, as well as of other non-motor symptoms that precede motor signs by several years. Nonetheless, different PD animal models may offer important insights into the role of the ENS and the gut microbiome.

Many approaches have been proposed to treat the comorbidity of PD, considering that bowel dysfunction may contribute to symptomatic fluctuation in PD patients. Based on this evidence, the ENS may be an excellent target to investigate some multifactorial aspects of PD and a potential biomarker for early diagnosis of PD. Identifying a reliable preclinical PD biomarker would be critical, enabling early intervention that could slow or prevent the disease [149,150]. A better understanding of the ENS and its relationships beyond the gut microbiome could represent a new scenario to better characterize a disease of which only the final stages are known but which probably represents the sum of numerous insults occurring over a lifetime of more than 20 years.

7. Conclusions

PD presents a significant challenge, unraveling with misrecognized symptoms that appear decades before motor features. PD studies have focused primarily on the CNS and associated motor dysfunction; however, the peripheral nervous system, including the ENS, is gaining prominence in the field of PD. Despite technological advances in neuroimaging, no fully validated biomarker is available for PD. There is an urgent need to identify biomarkers to differentiate PD from related disorders and assess disease severity and progression. From a clinical perspective, NEs and enteric glia could represent important new targets for the pharmacological treatment of neurodegenerative diseases.

Future studies should focus on this possibility, especially given the relative ease of studying these cells in humans. Considering the clinical and experimental evidence, the authors propose that Parkinsonism is only one aspect of a complex and multifaceted disorder, representing the last phase of a neuropathological process that begins at a young age in ENS.

Author Contributions: Conceptualization, M.M., A.P. and G.M.; resources, P.B.; writing—original draft preparation, M.M., A.P., P.I. and G.M.; writing—review and editing, P.I., M.M., P.B. and G.M.; visualization, P.I., M.M., P.B., A.P. and G.M.; supervision, P.B.; and G.M.; project administration, M.M., A.P. and G.M. All authors have read and agreed to the published version of the manuscript.

Funding: This work was partially supported by the Italian Ministry of Health "Ricerca Finalizzata" grants: RF-2019-12370182 to P.B. and RF-2021-12374979 to A.P. The funding source was not involved in the design or writing of the report or in the decision to submit the article for publication.

Data Availability Statement: All the data shown in this paper are available in PubMed Library. The authors created all representative draws appositely and are available on request.

Acknowledgments: All authors thank Massimo Tolu, and Massimiliano Di Virgilio for their excellent technical assistance.

Conflicts of Interest: The authors declare no conflict of interest. No sponsors participate in the choice of the items; the design of the paper; the collection of literature, the interpretation of analyzed papers; the writing of the manuscript; or in the decision to publish in the Biomedicine journal.

Abbreviations

(ach)	Acetylcholine
αS	α-synuclein
CNS	Central nervous system
DA	Dopamine
DAT	Dopamine transporter
DMV	Dorsal motor nucleus of the vagus nerve
ENS	Enteric nervous system
NEs	Enteric neurons
GI	Gastrointestinal tract
GBA	Gut–brain axis
GABA	G-aminobutyric acid
IFN-γ	Interferon-gamma
IL-1β	Interleukin-1β
Thy1-αS	Human αs under the thy-1 promoter
6-OHDA	6-hydroxydopamine
LB	Lewy bodies
MPTP	1-methyl-4-phenyl-1,2,3,6-tetrahydropyridine
NO	Nitric oxide
NOS	Nitric oxide synthase
OB	Olfactory bulb
PD	Parkinson's disease
SNS	Sympathetic Nervous System
SN	Substantia nigra
SNpc	Substantia nigra pars compacta
5-HT	5-hydroxytryptamine or serotonine
TNF-α	Tumor necrosis factor-alpha
TH	Tyrosine hydroxylase
TH-IR	Tyrosine Hydroxylase-Immunoreactive
VIP	Vasoactive intestinal peptide
WT	Wild-type

References

1. Endres, K.; Schäfer, K.-H. Influence of commensal microbiota on the enteric nervous system and its role in neurodegenerative diseases. *J. Innate Immun.* **2018**, *10*, 172–180. [CrossRef] [PubMed]
2. Clemente, J.C.; Ursell, L.K.; Parfrey, L.W.; Knight, R. The impact of the gut microbiota on human health: An integrative view. *Cell* **2012**, *148*, 1258–1270. [CrossRef] [PubMed]
3. Bercik, P.; Denou, E.; Collins, J.; Jackson, W.; Lu, J.; Jury, J.; Deng, Y.; Blennerhassett, P.; Macri, J.; McCoy, K.D.; et al. The intestinal microbiota affect central levels of brain-derived neurotropic factor and behavior in mice. *Gastroenterology* **2011**, *141*, 599–609.e3. [CrossRef] [PubMed]

4. Sampson, T.R.; Mazmanian, S.K. Control of brain development, function, and behavior by the microbiome. *Cell Host Microbe* **2015**, *17*, 565–576. [CrossRef]
5. Carabotti, M.; Scirocco, A.; Maselli, M.A.; Severi, C. The gut-brain axis: Interactions between enteric microbiota, central and enteric nervous systems. *Ann. Gastroenterol.* **2015**, *28*, 203–209.
6. Kabouridis, P.S.; Lasrado, R.; McCallum, S.; Chng, S.H.; Snippert, H.J.; Clevers, H.; Pettersson, S.; Pachnis, V. Microbiota controls the homeostasis of glial cells in the gut lamina propria. *Neuron* **2015**, *85*, 289–295. [CrossRef]
7. Belkaid, Y.; Hand, T.W. Role of the microbiota in immunity and inflammation. *Cell* **2014**, *157*, 121–141. [CrossRef]
8. Furness, J.B. *The Enteric Nervous System*; John Wiley & Sons: Hoboken, NJ, USA, 2008; ISBN 9781405173445.
9. Sasselli, V.; Pachnis, V.; Burns, A.J. The enteric nervous system. *Dev. Biol.* **2012**, *366*, 64–73. [CrossRef]
10. Furness, J.B. The enteric nervous system and neurogastroenterology. *Nat. Rev. Gastroenterol. Hepatol.* **2012**, *9*, 286–294. [CrossRef]
11. Gershon, M.D. The enteric nervous system: A second brain. *Hosp. Pract.* **1999**, *34*, 31–52. [CrossRef]
12. Natale, G.; Pasquali, L.; Paparelli, A.; Fornai, F. Parallel manifestations of neuropathologies in the enteric and central nervous systems. *Neurogastroenterol. Motil.* **2011**, *23*, 1056–1065. [CrossRef] [PubMed]
13. Furness, J.B. The organisation of the autonomic nervous system: Peripheral connections. *Auton. Neurosci.* **2006**, *130*, 1–5. [CrossRef] [PubMed]
14. Cryan, J.F.; O'Riordan, K.J.; Cowan, C.S.M.; Sandhu, K.V.; Bastiaanssen, T.F.S.; Boehme, M.; Codagnone, M.G.; Cussotto, S.; Fulling, C.; Golubeva, A.V.; et al. The Microbiota-Gut-Brain Axis. *Physiol. Rev.* **2019**, *99*, 1877–2013. [CrossRef] [PubMed]
15. Wang, H.X.; Wang, Y.P. Gut Microbiota-brain Axis. *Chin. Med. J.* **2016**, *129*, 2373–2380. [CrossRef]
16. Jaggar, M.; Rea, K.; Spichak, S.; Dinan, T.G.; Cryan, J.F. You've got male: Sex and the microbiota-gut-brain axis across the lifespan. *Front. Neuroendocrinol.* **2020**, *56*, 100815. [CrossRef]
17. Bauer, P.V.; Hamr, S.C.; Duca, F.A. Regulation of energy balance by a gut-brain axis and involvement of the gut microbiota. *Cell. Mol. Life Sci.* **2016**, *73*, 737–755. [CrossRef]
18. Margolis, K.G.; Cryan, J.F.; Mayer, E.A. The Microbiota-Gut-Brain Axis: From Motility to Mood. *Gastroenterology* **2021**, *160*, 1486–1501. [CrossRef]
19. Rhee, S.H.; Pothoulakis, C.; Mayer, E.A. Principles and clinical implications of the brain-gut-enteric microbiota axis. *Nat. Rev. Gastroenterol. Hepatol.* **2009**, *6*, 306–314. [CrossRef]
20. Kasarello, K.; Cudnoch-Jedrzejewska, A.; Czarzasta, K. Communication of gut microbiota and brain via immune and neuroendocrine signaling. *Front. Microbiol.* **2023**, *14*, 1118529. [CrossRef]
21. Mayer, E.A.; Savidge, T.; Shulman, R.J. Brain-gut microbiome interactions and functional bowel disorders. *Gastroenterology* **2014**, *146*, 1500–1512. [CrossRef]
22. Varesi, A.; Pierella, E.; Romeo, M.; Piccini, G.B.; Alfano, C.; Bjørklund, G.; Oppong, A.; Ricevuti, G.; Esposito, C.; Chirumbolo, S.; et al. The potential role of gut microbiota in alzheimer's disease: From diagnosis to treatment. *Nutrients* **2022**, *14*, 668. [CrossRef] [PubMed]
23. Quigley, E.M.M. Microbiota-Brain-Gut Axis and Neurodegenerative Diseases. *Curr. Neurol. Neurosci. Rep.* **2017**, *17*, 94. [CrossRef] [PubMed]
24. Sidransky, E.; Lopez, G. The link between the GBA gene and parkinsonism. *Lancet Neurol.* **2012**, *11*, 986–998. [CrossRef]
25. Schirinzi, T.; Martella, G.; Pisani, A. Double hit mouse model of Parkinson's disease. *Oncotarget* **2016**, *7*, 80109–80110. [CrossRef] [PubMed]
26. Martella, G.; Madeo, G.; Maltese, M.; Vanni, V.; Puglisi, F.; Ferraro, E.; Schirinzi, T.; Valente, E.M.; Bonanni, L.; Shen, J.; et al. Exposure to low-dose rotenone precipitates synaptic plasticity alterations in PINK1 heterozygous knockout mice. *Neurobiol. Dis.* **2016**, *91*, 21–36. [CrossRef]
27. Dickson, D.W. Parkinson's disease and parkinsonism: Neuropathology. *Cold Spring Harb. Perspect. Med.* **2012**, *2*, a009258. [CrossRef]
28. Hawkes, C.H.; Del Tredici, K.; Braak, H. Parkinson's disease: A dual-hit hypothesis. *Neuropathol. Appl. Neurobiol.* **2007**, *33*, 599–614. [CrossRef]
29. Tanner, C.M.; Goldman, S.M. Epidemiology of Parkinson's disease. *Neurol. Clin.* **1996**, *14*, 317–335. [CrossRef]
30. Vance, J.M.; Ali, S.; Bradley, W.G.; Singer, C.; Di Monte, D.A. Gene-environment interactions in Parkinson's disease and other forms of parkinsonism. *Neurotoxicology* **2010**, *31*, 598–602. [CrossRef]
31. Kline, E.M.; Houser, M.C.; Herrick, M.K.; Seibler, P.; Klein, C.; West, A.; Tansey, M.G. Genetic and environmental factors in parkinson's disease converge on immune function and inflammation. *Mov. Disord.* **2021**, *36*, 25–36. [CrossRef]
32. Poewe, W. Non-motor symptoms in Parkinson's disease. *Eur. J. Neurol.* **2008**, *15* (Suppl. S1), 14–20. [CrossRef] [PubMed]
33. Noyce, A.J.; Bestwick, J.P.; Silveira-Moriyama, L.; Hawkes, C.H.; Giovannoni, G.; Lees, A.J.; Schrag, A. Meta-analysis of early nonmotor features and risk factors for Parkinson disease. *Ann. Neurol.* **2012**, *72*, 893–901. [CrossRef] [PubMed]
34. Amara, A.W.; Memon, A.A. Effects of Exercise on Non-motor Symptoms in Parkinson's Disease. *Clin. Ther.* **2018**, *40*, 8–15. [CrossRef] [PubMed]
35. Postuma, R.B.; Aarsland, D.; Barone, P.; Burn, D.J.; Hawkes, C.H.; Oertel, W.; Ziemssen, T. Identifying prodromal Parkinson's disease: Pre-motor disorders in Parkinson's disease. *Mov. Disord.* **2012**, *27*, 617–626. [CrossRef]

36. Martinez-Martin, P.; Rodriguez-Blazquez, C.; Kurtis, M.M.; Chaudhuri, K.R. NMSS Validation Group The impact of non-motor symptoms on health-related quality of life of patients with Parkinson's disease. *Mov. Disord.* **2011**, *26*, 399–406. [CrossRef] [PubMed]
37. Battaglia, S.; Nazzi, C.; Thayer, J.F. Fear-induced bradycardia in mental disorders: Foundations, current advances, future perspectives. *Neurosci. Biobehav. Rev.* **2023**, *149*, 105163. [CrossRef] [PubMed]
38. Battaglia, S.; Di Fazio, C.; Vicario, C.M.; Avenanti, A. Neuropharmacological Modulation of N-methyl-D-aspartate, Noradrenaline and Endocannabinoid Receptors in Fear Extinction Learning: Synaptic Transmission and Plasticity. *Int. J. Mol. Sci.* **2023**, *24*, 5926. [CrossRef]
39. Tan, A.H.; Lim, S.Y.; Lang, A.E. The microbiome-gut-brain axis in Parkinson disease—From basic research to the clinic. *Nat. Rev. Neurol.* **2022**, *18*, 476–495. [CrossRef]
40. Mukherjee, A.; Biswas, A.; Das, S.K. Gut dysfunction in Parkinson's disease. *World J. Gastroenterol.* **2016**, *22*, 5742–5752. [CrossRef]
41. Zeng, J.; Wang, X.; Pan, F.; Mao, Z. The relationship between Parkinson's disease and gastrointestinal diseases. *Front. Aging Neurosci.* **2022**, *14*, 955919. [CrossRef]
42. Bhidayasiri, R.; Phuenpathom, W.; Tan, A.H.; Leta, V.; Phumphid, S.; Chaudhuri, K.R.; Pal, P.K. Management of dysphagia and gastroparesis in Parkinson's disease in real-world clinical practice—Balancing pharmacological and non-pharmacological approaches. *Front. Aging Neurosci.* **2022**, *14*, 979826. [CrossRef]
43. Chen, Z.; Li, G.; Liu, J. Autonomic dysfunction in Parkinson's disease: Implications for pathophysiology, diagnosis, and treatment. *Neurobiol. Dis.* **2020**, *134*, 104700. [CrossRef]
44. Chiang, H.-L.; Lin, C.-H. Altered gut microbiome and intestinal pathology in parkinson's disease. *J. Mov. Disord.* **2019**, *12*, 67–83. [CrossRef]
45. Devos, D.; Lebouvier, T.; Lardeux, B.; Biraud, M.; Rouaud, T.; Pouclet, H.; Coron, E.; Bruley des Varannes, S.; Naveilhan, P.; Nguyen, J.-M.; et al. Colonic inflammation in Parkinson's disease. *Neurobiol. Dis.* **2013**, *50*, 42–48. [CrossRef]
46. Cersosimo, M.G.; Benarroch, E.E. Pathological correlates of gastrointestinal dysfunction in Parkinson's disease. *Neurobiol. Dis.* **2012**, *46*, 559–564. [CrossRef]
47. Stocchi, F.; Torti, M. Constipation in parkinson's disease. *Int. Rev. Neurobiol.* **2017**, *134*, 811–826. [CrossRef]
48. Singaram, C.; Ashraf, W.; Gaumnitz, E.A.; Torbey, C.; Sengupta, A.; Pfeiffer, R.; Quigley, E.M. Dopaminergic defect of enteric nervous system in Parkinson's disease patients with chronic constipation. *Lancet* **1995**, *346*, 861–864. [CrossRef]
49. Pfeiffer, R.F.; Isaacson, S.H.; Pahwa, R. Clinical implications of gastric complications on levodopa treatment in Parkinson's disease. *Park. Relat. Disord.* **2020**, *76*, 63–71. [CrossRef]
50. Lebouvier, T.; Neunlist, M.; Bruley des Varannes, S.; Coron, E.; Drouard, A.; N'Guyen, J.-M.; Chaumette, T.; Tasselli, M.; Paillusson, S.; Flamand, M.; et al. Colonic biopsies to assess the neuropathology of Parkinson's disease and its relationship with symptoms. *PLoS ONE* **2010**, *5*, e12728. [CrossRef]
51. Zheng, H.; Shi, C.; Luo, H.; Fan, L.; Yang, Z.; Hu, X.; Zhang, Z.; Zhang, S.; Hu, Z.; Fan, Y.; et al. α-Synuclein in Parkinson's Disease: Does a Prion-like Mechanism of Propagation from Periphery to the Brain Play a Role? *Neuroscientist* **2021**, *27*, 367–387. [CrossRef]
52. Braak, H.; Del Tredici, K.; Rüb, U.; de Vos, R.A.I.; Jansen Steur, E.N.H.; Braak, E. Staging of brain pathology related to sporadic Parkinson's disease. *Neurobiol. Aging* **2003**, *24*, 197–211. [CrossRef] [PubMed]
53. Arotcarena, M.-L.; Dovero, S.; Prigent, A.; Bourdenx, M.; Camus, S.; Porras, G.; Thiolat, M.-L.; Tasselli, M.; Aubert, P.; Kruse, N.; et al. Bidirectional gut-to-brain and brain-to-gut propagation of synucleinopathy in non-human primates. *Brain* **2020**, *143*, 1462–1475. [CrossRef] [PubMed]
54. Natale, G.; Pasquali, L.; Ruggieri, S.; Paparelli, A.; Fornai, F. Parkinson's disease and the gut: A well known clinical association in need of an effective cure and explanation. *Neurogastroenterol. Motil.* **2008**, *20*, 741–749. [CrossRef]
55. Leclair-Visonneau, L.; Neunlist, M.; Derkinderen, P.; Lebouvier, T. The gut in Parkinson's disease: Bottom-up, top-down, or neither? *Neurogastroenterol. Motil.* **2020**, *32*, e13777. [CrossRef] [PubMed]
56. Chalazonitis, A.; Rao, M. Enteric nervous system manifestations of neurodegenerative disease. *Brain Res.* **2018**, *1693*, 207–213. [CrossRef]
57. Menozzi, E.; Macnaughtan, J.; Schapira, A.H.V. The gut-brain axis and Parkinson disease: Clinical and pathogenetic relevance. *Ann. Med.* **2021**, *53*, 611–625. [CrossRef]
58. Berg, D.; Borghammer, P.; Fereshtehnejad, S.-M.; Heinzel, S.; Horsager, J.; Schaeffer, E.; Postuma, R.B. Prodromal Parkinson disease subtypes—Key to understanding heterogeneity. *Nat. Rev. Neurol.* **2021**, *17*, 349–361. [CrossRef]
59. Elfil, M.; Kamel, S.; Kandil, M.; Koo, B.B.; Schaefer, S.M. Implications of the gut microbiome in parkinson's disease. *Mov. Disord.* **2020**, *35*, 921–933. [CrossRef]
60. Klann, E.M.; Dissanayake, U.; Gurrala, A.; Farrer, M.; Shukla, A.W.; Ramirez-Zamora, A.; Mai, V.; Vedam-Mai, V. The Gut-Brain Axis and Its Relation to Parkinson's Disease: A Review. *Front. Aging Neurosci.* **2021**, *13*, 782082. [CrossRef]
61. Ma, Z.S. Heterogeneity-disease relationship in the human microbiome-associated diseases. *FEMS Microbiol. Ecol.* **2020**, *96*, fiaa093. [CrossRef]
62. Natale, G.; Ryskalin, L.; Morucci, G.; Lazzeri, G.; Frati, A.; Fornai, F. The baseline structure of the enteric nervous system and its role in parkinson's disease. *Life* **2021**, *11*, 732. [CrossRef]
63. Brodal, P. *The Central Nervous System: Structure and Function*; Oxford University Press: Oxford, UK, 2004; ISBN 9780195165609.

64. Cussotto, S.; Strain, C.R.; Fouhy, F.; Strain, R.G.; Peterson, V.L.; Clarke, G.; Stanton, C.; Dinan, T.G.; Cryan, J.F. Differential effects of psychotropic drugs on microbiome composition and gastrointestinal function. *Psychopharmacology* **2019**, *236*, 1671–1685. [CrossRef]
65. The Enteric Nervous System and Regulation of Intestinal Motility—ProQuest. Available online: https://www.proquest.com/docview/222539969?pq-origsite=gscholar&fromopenview=true (accessed on 10 August 2022).
66. Brehmer, A. *Structure of Enteric Neurons*; Springer Science & Business Media: Berlin/Heidelberg, Germany, 2006; ISBN 9783540328742.
67. Costa, M.; Furness, J.B.; Gibbins, I.L. Chapter 15 Chemical coding of enteric neurons. In *Progress in Brain Research*; Elsevier: Amsterdam, The Netherlands, 1986; Volume 68, pp. 217–239. ISBN 9780444807625.
68. Furness, J.B.; Costa, M. Types of nerves in the enteric nervous system. In *Commentaries in the Neurosciences*; Elsevier: Amsterdam, The Netherlands, 1980; pp. 235–252. ISBN 9780080255019.
69. Furness, J.B.; Callaghan, B.P.; Rivera, L.R.; Cho, H.-J. The enteric nervous system and gastrointestinal innervation: Integrated local and central control. *Adv. Exp. Med. Biol.* **2014**, *817*, 39–71. [CrossRef]
70. Shirazi-Beechey, S.P.; Moran, A.W.; Batchelor, D.J.; Daly, K.; Al-Rammahi, M. Glucose sensing and signalling; regulation of intestinal glucose transport. *Proc. Nutr. Soc.* **2011**, *70*, 185–193. [CrossRef]
71. Saffrey, M.J. Cellular changes in the enteric nervous system during ageing. *Dev. Biol.* **2013**, *382*, 344–355. [CrossRef]
72. McQuade, R.M.; Singleton, L.M.; Wu, H.; Lee, S.; Constable, R.; Di Natale, M.; Ringuet, M.T.; Berger, J.P.; Kauhausen, J.; Parish, C.L.; et al. The association of enteric neuropathy with gut phenotypes in acute and progressive models of Parkinson's disease. *Sci. Rep.* **2021**, *11*, 7934. [CrossRef]
73. Lama, J.; Buhidma, Y.; Fletcher, E.J.R.; Duty, S. Animal models of Parkinson's disease: A guide to selecting the optimal model for your research. *Neuronal Signal.* **2021**, *5*, NS20210026. [CrossRef]
74. Tieu, K. A guide to neurotoxic animal models of Parkinson's disease. *Cold Spring Harb. Perspect. Med.* **2011**, *1*, a009316. [CrossRef]
75. Zhang, X.; Li, Y.; Liu, C.; Fan, R.; Wang, P.; Zheng, L.; Hong, F.; Feng, X.; Zhang, Y.; Li, L.; et al. Alteration of enteric monoamines with monoamine receptors and colonic dysmotility in 6-hydroxydopamine-induced Parkinson's disease rats. *Transl. Res.* **2015**, *166*, 152–162. [CrossRef]
76. Anderson, G.; Noorian, A.R.; Taylor, G.; Anitha, M.; Bernhard, D.; Srinivasan, S.; Greene, J.G. Loss of enteric dopaminergic neurons and associated changes in colon motility in an MPTP mouse model of Parkinson's disease. *Exp. Neurol.* **2007**, *207*, 4–12. [CrossRef]
77. Chaumette, T.; Lebouvier, T.; Aubert, P.; Lardeux, B.; Qin, C.; Li, Q.; Accary, D.; Bézard, E.; Bruley des Varannes, S.; Derkinderen, P.; et al. Neurochemical plasticity in the enteric nervous system of a primate animal model of experimental Parkinsonism. *Neurogastroenterol. Motil.* **2009**, *21*, 215–222. [CrossRef]
78. Zhu, H.C.; Zhao, J.; Luo, C.Y.; Li, Q.Q. Gastrointestinal dysfunction in a Parkinson's disease rat model and the changes of dopaminergic, nitric oxidergic, and cholinergic neurotransmitters in myenteric plexus. *J. Mol. Neurosci.* **2012**, *47*, 15–25. [CrossRef]
79. Tian, Y.M.; Chen, X.; Luo, D.Z.; Zhang, X.H.; Xue, H.; Zheng, L.F.; Yang, N.; Wang, X.M.; Zhu, J.X. Alteration of dopaminergic markers in gastrointestinal tract of different rodent models of Parkinson's disease. *Neuroscience* **2008**, *153*, 634–644. [CrossRef]
80. Li, Z.S.; Schmauss, C.; Cuenca, A.; Ratcliffe, E.; Gershon, M.D. Physiological modulation of intestinal motility by enteric dopaminergic neurons and the D2 receptor: Analysis of dopamine receptor expression, location, development, and function in wild-type and knock-out mice. *J. Neurosci.* **2006**, *26*, 2798–2807. [CrossRef]
81. Walker, J.K.; Gainetdinov, R.R.; Mangel, A.W.; Caron, M.G.; Shetzline, M.A. Mice lacking the dopamine transporter display altered regulation of distal colonic motility. *Am. J. Physiol. Gastrointest. Liver Physiol.* **2000**, *279*, G311–G318. [CrossRef]
82. Bové, J.; Prou, D.; Perier, C.; Przedborski, S. Toxin-induced models of Parkinson's disease. *NeuroRx* **2005**, *2*, 484–494. [CrossRef]
83. Jackson-Lewis, V.; Jakowec, M.; Burke, R.E.; Przedborski, S. Time course and morphology of dopaminergic neuronal death caused by the neurotoxin 1-methyl-4-phenyl-1,2,3,6-tetrahydropyridine. *Neurodegeneration* **1995**, *4*, 257–269. [CrossRef]
84. Heikkila, R.E.; Hess, A.; Duvoisin, R.C. Dopaminergic neurotoxicity of 1-methyl-4-phenyl-1,2,5,6-tetrahydropyridine in mice. *Science* **1984**, *224*, 1451–1453. [CrossRef]
85. Li, Z.S.; Pham, T.D.; Tamir, H.; Chen, J.J.; Gershon, M.D. Enteric dopaminergic neurons: Definition, developmental lineage, and effects of extrinsic denervation. *J. Neurosci.* **2004**, *24*, 1330–1339. [CrossRef]
86. Wakabayashi, K.; Takahashi, H.; Ohama, E.; Ikuta, F. Parkinson's disease: An immunohistochemical study of Lewy body-containing neurons in the enteric nervous system. *Acta Neuropathol.* **1990**, *79*, 581–583. [CrossRef]
87. Colucci, M.; Cervio, M.; Faniglione, M.; De Angelis, S.; Pajoro, M.; Levandis, G.; Tassorelli, C.; Blandini, F.; Feletti, F.; De Giorgio, R.; et al. Intestinal dysmotility and enteric neurochemical changes in a Parkinson's disease rat model. *Auton. Neurosci.* **2012**, *169*, 77–86. [CrossRef]
88. Zheng, L.F.; Song, J.; Fan, R.F.; Chen, C.L.; Ren, Q.Z.; Zhang, X.L.; Feng, X.Y.; Zhang, Y.; Li, L.S.; De Giorgio, R.; et al. The role of the vagal pathway and gastric dopamine in the gastroparesis of rats after a 6-hydroxydopamine microinjection in the substantia nigra. *Acta Physiol.* **2014**, *211*, 434–446. [CrossRef]
89. Rota, L.; Pellegrini, C.; Benvenuti, L.; Antonioli, L.; Fornai, M.; Blandizzi, C.; Cattaneo, A.; Colla, E. Constipation, deficit in colon contractions and alpha-synuclein inclusions within the colon precede motor abnormalities and neurodegeneration in the central nervous system in a mouse model of alpha-synucleinopathy. *Transl. Neurodegener.* **2019**, *8*, 5. [CrossRef]
90. Qualman, S.J.; Haupt, H.M.; Yang, P.; Hamilton, S.R. Esophageal Lewy bodies associated with ganglion cell loss in achalasia. *Gastroenterology* **1984**, *87*, 848–856. [CrossRef]

91. Braak, H.; de Vos, R.A.I.; Bohl, J.; Del Tredici, K. Gastric alpha-synuclein immunoreactive inclusions in Meissner's and Auerbach's plexuses in cases staged for Parkinson's disease-related brain pathology. *Neurosci. Lett.* **2006**, *396*, 67–72. [CrossRef]
92. Kuo, Y.-M.; Li, Z.; Jiao, Y.; Gaborit, N.; Pani, A.K.; Orrison, B.M.; Bruneau, B.G.; Giasson, B.I.; Smeyne, R.J.; Gershon, M.D.; et al. Extensive enteric nervous system abnormalities in mice transgenic for artificial chromosomes containing Parkinson disease-associated alpha-synuclein gene mutations precede central nervous system changes. *Hum. Mol. Genet.* **2010**, *19*, 1633–1650. [CrossRef]
93. Gispert, S.; Del Turco, D.; Garrett, L.; Chen, A.; Bernard, D.J.; Hamm-Clement, J.; Korf, H.-W.; Deller, T.; Braak, H.; Auburger, G.; et al. Transgenic mice expressing mutant A53T human alpha-synuclein show neuronal dysfunction in the absence of aggregate formation. *Mol. Cell. Neurosci.* **2003**, *24*, 419–429. [CrossRef]
94. Pfeiffer, R.F. Gastrointestinal dysfunction in Parkinson's disease. *Park. Relat. Disord.* **2011**, *17*, 10–15. [CrossRef]
95. Noorian, A.R.; Rha, J.; Annerino, D.M.; Bernhard, D.; Taylor, G.M.; Greene, J.G. Alpha-synuclein transgenic mice display age-related slowing of gastrointestinal motility associated with transgene expression in the vagal system. *Neurobiol. Dis.* **2012**, *48*, 9–19. [CrossRef]
96. Wang, L.; Fleming, S.M.; Chesselet, M.-F.; Taché, Y. Abnormal colonic motility in mice overexpressing human wild-type alpha-synuclein. *Neuroreport* **2008**, *19*, 873–876. [CrossRef]
97. Lam, H.A.; Wu, N.; Cely, I.; Kelly, R.L.; Hean, S.; Richter, F.; Magen, I.; Cepeda, C.; Ackerson, L.C.; Walwyn, W.; et al. Elevated tonic extracellular dopamine concentration and altered dopamine modulation of synaptic activity precede dopamine loss in the striatum of mice overexpressing human α-synuclein. *J. Neurosci. Res.* **2011**, *89*, 1091–1102. [CrossRef]
98. Chesselet, M.-F.; Richter, F. Modelling of Parkinson's disease in mice. *Lancet Neurol.* **2011**, *10*, 1108–1118. [CrossRef]
99. Schaffernicht, G.; Shang, Q.; Stievenard, A.; Bötzel, K.; Dening, Y.; Kempe, R.; Toussaint, M.; Gündel, D.; Kranz, M.; Reichmann, H.; et al. Pathophysiological Changes in the Enteric Nervous System of Rotenone-Exposed Mice as Early Radiological Markers for Parkinson's Disease. *Front. Neurol.* **2021**, *12*, 642604. [CrossRef]
100. Pan-Montojo, F.; Schwarz, M.; Winkler, C.; Arnhold, M.; O'Sullivan, G.A.; Pal, A.; Said, J.; Marsico, G.; Verbavatz, J.-M.; Rodrigo-Angulo, M.; et al. Environmental toxins trigger PD-like progression via increased alpha-synuclein release from enteric neurons in mice. *Sci. Rep.* **2012**, *2*, 898. [CrossRef]
101. Klingelhoefer, L.; Reichmann, H. Pathogenesis of Parkinson disease–the gut-brain axis and environmental factors. *Nat. Rev. Neurol.* **2015**, *11*, 625–636. [CrossRef]
102. Pan-Montojo, F.J.; Funk, R.H.W. Oral administration of rotenone using a gavage and image analysis of alpha-synuclein inclusions in the enteric nervous system. *J. Vis. Exp.* **2010**, *44*, e2123. [CrossRef]
103. Arnhold, M.; Dening, Y.; Chopin, M.; Arévalo, E.; Schwarz, M.; Reichmann, H.; Gille, G.; Funk, R.H.W.; Pan-Montojo, F. Changes in the sympathetic innervation of the gut in rotenone treated mice as possible early biomarker for Parkinson's disease. *Clin. Auton. Res.* **2016**, *26*, 211–222. [CrossRef]
104. Sharrad, D.F.; Chen, B.N.; Gai, W.P.; Vaikath, N.; El-Agnaf, O.M.; Brookes, S.J.H. Rotenone and elevated extracellular potassium concentration induce cell-specific fibrillation of α-synuclein in axons of cholinergic enteric neurons in the guinea-pig ileum. *Neurogastroenterol. Motil.* **2017**, *29*, e12985. [CrossRef]
105. Paillusson, S.; Tasselli, M.; Lebouvier, T.; Mahé, M.M.; Chevalier, J.; Biraud, M.; Cario-Toumaniantz, C.; Neunlist, M.; Derkinderen, P. α-Synuclein expression is induced by depolarization and cyclic AMP in enteric neurons. *J. Neurochem.* **2010**, *115*, 694–706. [CrossRef]
106. Camilleri, M.; Cowen, T.; Koch, T.R. Enteric neurodegeneration in ageing. *Neurogastroenterol. Motil.* **2008**, *20*, 185–196. [CrossRef]
107. Phillips, R.J.; Powley, T.L. Innervation of the gastrointestinal tract: Patterns of aging. *Auton. Neurosci.* **2007**, *136*, 1–19. [CrossRef]
108. Phillips, R.J.; Walter, G.C.; Ringer, B.E.; Higgs, K.M.; Powley, T.L. Alpha-synuclein immunopositive aggregates in the myenteric plexus of the aging Fischer 344 rat. *Exp. Neurol.* **2009**, *220*, 109–119. [CrossRef]
109. Braak, H.; Ghebremedhin, E.; Rüb, U.; Bratzke, H.; Del Tredici, K. Stages in the development of Parkinson's disease-related pathology. *Cell Tissue Res.* **2004**, *318*, 121–134. [CrossRef]
110. Wolters, E.C.; Braak, H. Parkinson's Disease: Premotor Clinico-Pathological Correlations. In *Parkinson's Disease and Related Disorders*; Journal of Neural Transmission. Supplementa; Springer: Vienna, Austria, 2006; pp. 309–319. [CrossRef]
111. Krogh, K.; Christensen, P. Neurogenic colorectal and pelvic floor dysfunction. *Best Pract. Res. Clin. Gastroenterol.* **2009**, *23*, 531–543. [CrossRef]
112. Marrinan, S.; Emmanuel, A.V.; Burn, D.J. Delayed gastric emptying in Parkinson's disease. *Mov. Disord.* **2014**, *29*, 23–32. [CrossRef]
113. Pfeiffer, R.F. Gastrointestinal dysfunction in Parkinson's disease. *Lancet Neurol.* **2003**, *2*, 107–116. [CrossRef]
114. Taguchi, T.; Ikuno, M.; Yamakado, H.; Takahashi, R. Animal model for prodromal parkinson's disease. *Int. J. Mol. Sci.* **2020**, *21*, 1961. [CrossRef]
115. Liepelt-Scarfone, I.; Ophey, A.; Kalbe, E. Cognition in prodromal Parkinson's disease. *Prog. Brain Res.* **2022**, *269*, 93–111. [CrossRef]
116. Solla, P.; Wang, Q.; Frau, C.; Floris, V.; Loy, F.; Sechi, L.A.; Masala, C. Olfactory impairment is the main predictor of higher scores at REM sleep behavior disorder (RBD) screening questionnaire in parkinson's disease patients. *Brain Sci.* **2023**, *13*, 599. [CrossRef]
117. Erkkinen, M.G.; Kim, M.-O.; Geschwind, M.D. Clinical neurology and epidemiology of the major neurodegenerative diseases. *Cold Spring Harb. Perspect. Biol.* **2018**, *10*, a033118. [CrossRef]

118. Braak, H.; Del Tredici, K. Neuropathological Staging of Brain Pathology in Sporadic Parkinson's disease: Separating the Wheat from the Chaff. *J. Park. Dis.* **2017**, *7*, S71–S85. [CrossRef]
119. Yilmaz, R.; Hopfner, F.; van Eimeren, T.; Berg, D. Biomarkers of Parkinson's disease: 20 years later. *J. Neural Transm.* **2019**, *126*, 803–813. [CrossRef]
120. Breit, S.; Kupferberg, A.; Rogler, G.; Hasler, G. Vagus Nerve as Modulator of the Brain-Gut Axis in Psychiatric and Inflammatory Disorders. *Front. Psychiatry* **2018**, *9*, 44. [CrossRef]
121. Liu, B.; Fang, F.; Pedersen, N.L.; Tillander, A.; Ludvigsson, J.F.; Ekbom, A.; Svenningsson, P.; Chen, H.; Wirdefeldt, K. Vagotomy and Parkinson disease: A Swedish register-based matched-cohort study. *Neurology* **2017**, *88*, 1996–2002. [CrossRef]
122. Kelly, M.J.; Breathnach, C.; Tracey, K.J.; Donnelly, S.C. Manipulation of the inflammatory reflex as a therapeutic strategy. *Cell Rep. Med.* **2022**, *3*, 100696. [CrossRef]
123. Parkkinen, L.; Pirttilä, T.; Alafuzoff, I. Applicability of current staging/categorization of alpha-synuclein pathology and their clinical relevance. *Acta Neuropathol.* **2008**, *115*, 399–407. [CrossRef]
124. Frigerio, R.; Fujishiro, H.; Ahn, T.-B.; Josephs, K.A.; Maraganore, D.M.; DelleDonne, A.; Parisi, J.E.; Klos, K.J.; Boeve, B.F.; Dickson, D.W.; et al. Incidental Lewy body disease: Do some cases represent a preclinical stage of dementia with Lewy bodies? *Neurobiol. Aging* **2011**, *32*, 857–863. [CrossRef]
125. Koga, S.; Sekiya, H.; Kondru, N.; Ross, O.A.; Dickson, D.W. Neuropathology and molecular diagnosis of Synucleinopathies. *Mol. Neurodegener.* **2021**, *16*, 83. [CrossRef]
126. Macefield, V.G.; Henderson, L.A. Identification of the human sympathetic connectome involved in blood pressure regulation. *NeuroImage* **2019**, *202*, 116119. [CrossRef]
127. Horsager, J.; Andersen, K.B.; Knudsen, K.; Skjærbæk, C.; Fedorova, T.D.; Okkels, N.; Schaeffer, E.; Bonkat, S.K.; Geday, J.; Otto, M.; et al. Brain-first versus body-first Parkinson's disease: A multimodal imaging case-control study. *Brain* **2020**, *143*, 3077–3088. [CrossRef]
128. Socała, K.; Doboszewska, U.; Szopa, A.; Serefko, A.; Włodarczyk, M.; Zielińska, A.; Poleszak, E.; Fichna, J.; Wlaź, P. The role of microbiota-gut-brain axis in neuropsychiatric and neurological disorders. *Pharmacol. Res.* **2021**, *172*, 105840. [CrossRef]
129. Dash, S.; Syed, Y.A.; Khan, M.R. Understanding the role of the gut microbiome in brain development and its association with neurodevelopmental psychiatric disorders. *Front. Cell Dev. Biol.* **2022**, *10*, 880544. [CrossRef]
130. Baj, A.; Moro, E.; Bistoletti, M.; Orlandi, V.; Crema, F.; Giaroni, C. Glutamatergic Signaling along the Microbiota-Gut-Brain Axis. *Int. J. Mol. Sci.* **2019**, *20*, 1482. [CrossRef]
131. Parker, A.; Fonseca, S.; Carding, S.R. Gut microbes and metabolites as modulators of blood-brain barrier integrity and brain health. *Gut Microbes* **2020**, *11*, 135–157. [CrossRef]
132. Caspani, G.; Kennedy, S.; Foster, J.A.; Swann, J. Gut microbial metabolites in depression: Understanding the biochemical mechanisms. *Microb. Cell* **2019**, *6*, 454–481. [CrossRef]
133. Kim, S.; Kwon, S.-H.; Kam, T.-I.; Panicker, N.; Karuppagounder, S.S.; Lee, S.; Lee, J.H.; Kim, W.R.; Kook, M.; Foss, C.A.; et al. Transneuronal Propagation of Pathologic α-Synuclein from the Gut to the Brain Models Parkinson's Disease. *Neuron* **2019**, *103*, 627–641. [CrossRef]
134. Wallen, Z.D.; Demirkan, A.; Twa, G.; Cohen, G.; Dean, M.N.; Standaert, D.G.; Sampson, T.R.; Payami, H. Metagenomics of Parkinson's disease implicates the gut microbiome in multiple disease mechanisms. *Nat. Commun.* **2022**, *13*, 6958. [CrossRef]
135. Zhu, M.; Liu, X.; Ye, Y.; Yan, X.; Cheng, Y.; Zhao, L.; Chen, F.; Ling, Z. Gut microbiota: A novel therapeutic target for parkinson's disease. *Front. Immunol.* **2022**, *13*, 937555. [CrossRef]
136. Chen, S.-J.; Lin, C.-H. Gut microenvironmental changes as a potential trigger in Parkinson's disease through the gut-brain axis. *J. Biomed. Sci.* **2022**, *29*, 54. [CrossRef]
137. Misera, A.; Łoniewski, I.; Palma, J.; Kulaszyńska, M.; Czarnecka, W.; Kaczmarczyk, M.; Liśkiewicz, P.; Samochowiec, J.; Skonieczna-Żydecka, K. Clinical significance of microbiota changes under the influence of psychotropic drugs. An updated narrative review. *Front. Microbiol.* **2023**, *14*, 1125022. [CrossRef]
138. Horsager, J.; Knudsen, K.; Sommerauer, M. Clinical and imaging evidence of brain-first and body-first Parkinson's disease. *Neurobiol. Dis.* **2022**, *164*, 105626. [CrossRef] [PubMed]
139. Molla, M.D.; Akalu, Y.; Geto, Z.; Dagnew, B.; Ayelign, B.; Shibabaw, T. Role of Caspase-1 in the Pathogenesis of Inflammatory-Associated Chronic Noncommunicable Diseases. *J. Inflamm. Res.* **2020**, *13*, 749–764. [CrossRef]
140. Parada Venegas, D.; De la Fuente, M.K.; Landskron, G.; González, M.J.; Quera, R.; Dijkstra, G.; Harmsen, H.J.M.; Faber, K.N.; Hermoso, M.A. Short Chain Fatty Acids (SCFAs)-Mediated Gut Epithelial and Immune Regulation and Its Relevance for Inflammatory Bowel Diseases. *Front. Immunol.* **2019**, *10*, 277. [CrossRef] [PubMed]
141. Pellegrini, C.; D'Antongiovanni, V.; Miraglia, F.; Rota, L.; Benvenuti, L.; Di Salvo, C.; Testa, G.; Capsoni, S.; Carta, G.; Antonioli, L.; et al. Enteric α-synuclein impairs intestinal epithelial barrier through caspase-1-inflammasome signaling in Parkinson's disease before brain pathology. *npj Park. Dis.* **2022**, *8*, 9. [CrossRef]
142. Muenter, M.D.; Tyce, G.M. L-dopa therapy of Parkinson's disease: Plasma L-dopa concentration, therapeutic response, and side effects. *Mayo Clin. Proc.* **1971**, *46*, 231–239.
143. Poewe, W.; Antonini, A. Novel formulations and modes of delivery of levodopa. *Mov. Disord.* **2015**, *30*, 114–120. [CrossRef]

144. Ouchi, Y.; Kanno, T.; Okada, H.; Yoshikawa, E.; Futatsubashi, M.; Nobezawa, S.; Torizuka, T.; Tanaka, K. Changes in dopamine availability in the nigrostriatal and mesocortical dopaminergic systems by gait in Parkinson's disease. *Brain* **2001**, *124*, 784–792. [CrossRef] [PubMed]
145. Kouli, A.; Torsney, K.M.; Kuan, W.-L. Parkinson's disease: Etiology, neuropathology, and pathogenesis. In *Parkinson's Disease: Pathogenesis and Clinical Aspects*; Stoker, T.B., Greenland, J.C., Eds.; Codon Publications: Brisbane, QLD, Australia, 2018; ISBN 9780994438164.
146. Chen, M.; Mor, D.E. Gut-to-Brain α-Synuclein Transmission in Parkinson's Disease: Evidence for Prion-like Mechanisms. *Int. J. Mol. Sci.* **2023**, *24*, 7205. [CrossRef]
147. Buhusi, M.; Olsen, K.; Yang, B.Z.; Buhusi, C.V. Stress-Induced Executive Dysfunction in GDNF-Deficient Mice, A Mouse Model of Parkinsonism. *Front. Behav. Neurosci.* **2016**, *10*, 114. [CrossRef]
148. Aarsland, D.; Andersen, K.; Larsen, J.P.; Perry, R.; Wentzel-Larsen, T.; Lolk, A.; Kragh-Sørensen, P. The rate of cognitive decline in Parkinson's disease. *Arch. Neurol.* **2004**, *61*, 1906–1911. [CrossRef]
149. Clairembault, T.; Leclair-Visonneau, L.; Neunlist, M.; Derkinderen, P. Enteric glial cells: New players in Parkinson's disease? *Mov. Disord.* **2015**, *30*, 494–498. [CrossRef] [PubMed]
150. Derkinderen, P.; Rouaud, T.; Lebouvier, T.; Bruley des Varannes, S.; Neunlist, M.; De Giorgio, R. Parkinson disease: The enteric nervous system spills its guts. *Neurology* **2011**, *77*, 1761–1767. [CrossRef] [PubMed]
151. Goldman, J.G.; Postuma, R. Premotor and nonmotor features of Parkinson's disease. *Curr. Opin. Neurol.* **2014**, *27*, 434–441. [CrossRef] [PubMed]
152. Del Rey, N.L.-G.; Quiroga-Varela, A.; Garbayo, E.; Carballo-Carbajal, I.; Fernández-Santiago, R.; Monje, M.H.G.; Trigo-Damas, I.; Blanco-Prieto, M.J.; Blesa, J. Advances in parkinson's disease: 200 years later. *Front. Neuroanat.* **2018**, *12*, 113. [CrossRef] [PubMed]
153. Simon, D.K.; Tanner, C.M.; Brundin, P. Parkinson's disease epidemiology, pathology, genetics, and pathophysiology. *Clin. Geriatr. Med.* **2020**, *36*, 1–12. [CrossRef]
154. Fasano, A.; Visanji, N.P.; Liu, L.W.C.; Lang, A.E.; Pfeiffer, R.F. Gastrointestinal dysfunction in Parkinson's disease. *Lancet Neurol.* **2015**, *14*, 625–639. [CrossRef]
155. Adler, C.H.; Beach, T.G. Neuropathological basis of nonmotor manifestations of Parkinson's disease. *Mov. Disord.* **2016**, *8*, 1114–1119. [CrossRef] [PubMed]
156. Alafuzoff, I.; Parkkinen, L. Staged pathology in Parkinson's disease. *Park. Relat. Disord.* **2014**, *20*, 57–61. [CrossRef]
157. Van Den Berge, N.; Ulusoy, A. Animal models of brain-first and body-first Parkinson's disease. *Neurobiol. Dis.* **2022**, *163*, 105599. [CrossRef]

biomedicines

MDPI

Article

UBL3 Interacts with Alpha-Synuclein in Cells and the Interaction Is Downregulated by the EGFR Pathway Inhibitor Osimertinib

Bin Chen [1], Md. Mahmudul Hasan [1], Hengsen Zhang [1], Qing Zhai [1], A. S. M. Waliullah [1], Yashuang Ping [1], Chi Zhang [1], Soho Oyama [1], Mst. Afsana Mimi [1], Yuna Tomochika [1], Yu Nagashima [2], Tomohiko Nakamura [3], Tomoaki Kahyo [1,4], Kenji Ogawa [5], Daita Kaneda [6], Minoru Yoshida [7,8,9] and Mitsutoshi Setou [1,4,10,*]

1 Department of Cellular and Molecular Anatomy, Hamamatsu University School of Medicine, 1-20-1 Handayama, Higashi-Ku, Hamamatsu 431-3192, Shizuoka, Japan; chenbin19101@gmail.com (B.C.); hasan.mahmudbio@gmail.com (M.M.H.); d19110@hama-med.ac.jp (H.Z.); zhaiqing199404@gmail.com (Q.Z.); wali.rubmb10@gmail.com (A.S.M.W.); pingyashuang1989@gmail.com (Y.P.); zhangchi07.pegasus@gmail.com (C.Z.); oyamasoho@gmail.com (S.O.); afsana.mimichem@gmail.com (M.A.M.); hhelibeyuna@gmail.com (Y.T.); kahyo@hama-med.ac.jp (T.K.)

2 Institute for Medical Photonics Research, Preeminent Medical Photonics Education and Research Center, Hamamatsu University School of Medicine, Hamamatsu 431-3192, Shizuoka, Japan; yunaga@hama-med.ac.jp

3 Department of Neurology, Hamamatsu University School of Medicine, 1-20-1 Handayama, Higashi-Ku, Hamamatsu 431-3192, Shizuoka, Japan; tomohiko@hama-med.ac.jp

4 International Mass Imaging Center, Hamamatsu University School of Medicine, 1-20-1 Handayama, Higashi-Ku, Hamamatsu 431-3192, Shizuoka, Japan

5 Laboratory of Veterinary Epizootiology, Department of Veterinary Medicine, Nihon University, Kameino 1866, Fujisawa 252-0880, Kanagawa, Japan; ogawa.kenji@nihon-u.ac.jp

6 Choju Medical Institute, Fukushimura Hospital, Yamanaka-19-14 Noyoricho, Toyohashi 441-8124, Aichi, Japan; kaneda@chojuken.net

7 Department of Biotechnology, Graduate School of Agricultural and Life Sciences, The University of Tokyo, Tokyo 113-8657, Japan; yoshidam@riken.jp

8 Collaborative Research Institute for Innovative Microbiology, The University of Tokyo, Tokyo 113-8657, Japan

9 RIKEN Center for Sustainable Resource Science, Wako 351-0198, Saitama, Japan

10 Department of Systems Molecular Anatomy, Institute for Medical Photonics Research, Preeminent Medical Photonics Education & Research Center, 1-20-1 Handayama, Higashi-Ku, Hamamatsu 431-3192, Shizuoka, Japan

* Correspondence: setou@hama-med.ac.jp; Tel.: +81-53-435-2086

Citation: Chen, B.; Hasan, M.M.; Zhang, H.; Zhai, Q.; Waliullah, A.S.M.; Ping, Y.; Zhang, C.; Oyama, S.; Mimi, M.A.; Tomochika, Y.; et al. UBL3 Interacts with Alpha-Synuclein in Cells and the Interaction Is Downregulated by the EGFR Pathway Inhibitor Osimertinib. *Biomedicines* **2023**, *11*, 1685. https://doi.org/10.3390/biomedicines11061685

Academic Editor: Ubaldo Armato

Received: 30 April 2023
Revised: 7 June 2023
Accepted: 8 June 2023
Published: 10 June 2023

Abstract: Ubiquitin-like 3 (UBL3) acts as a post-translational modification (PTM) factor and regulates protein sorting into small extracellular vesicles (sEVs). sEVs have been reported as vectors for the pathology propagation of neurodegenerative diseases, such as α-synucleinopathies. Alpha-synuclein (α-syn) has been widely studied for its involvement in α-synucleinopathies. However, it is still unknown whether UBL3 interacts with α-syn, and is influenced by drugs or compounds. In this study, we investigated the interaction between UBL3 and α-syn, and any ensuing possible functional and pathological implications. We found that UBL3 can interact with α-syn by the *Gaussia princeps* based split luciferase complementation assay in cells and immunoprecipitation, while cysteine residues at its C-terminal, which are considered important as PTM factors for UBL3, were not essential for the interaction. The interaction was upregulated by 1-methyl-4-phenylpyridinium exposure. In drug screen results, the interaction was significantly downregulated by the treatment of osimertinib. These results suggest that UBL3 interacts with α-syn in cells and is significantly downregulated by epidermal growth factor receptor (EGFR) pathway inhibitor osimertinib. Therefore, the UBL3 pathway may be a new therapeutic target for α-synucleinopathies in the future.

Keywords: UBL3; α-synuclein; interaction; drug screen; EGFR pathway inhibitor; osimertinib; downregulate; α-synucleinopathies

1. Introduction

Ubiquitin-like 3 (UBL3), a highly conserved ubiquitin-like protein first found in eukaryotes, is localized to the cell membrane by prenylation [1]. The UBL3 gene is widely expressed in human tissues, with the strongest expression in the testis, ovary, and brain tissues [2]. In our previous study, UBL3 was characterized as a post-translational modification (PTM) factor that regulates protein sorting to small extracellular vesicles (sEVs) [3]. Another recent study found that UBL3 is involved in adaptive immunity by regulating the transport of major histocompatibility complex II and CD86 through ubiquitination [4]. The downregulated expression of UBL3 has been reported to be associated with human diseases, such as cervical cancer [5], gastric cancer [6], esophageal cancer [7], and non-small-cell lung cancer [8]. Moreover, UBL3 can interact with more than 22 disease-related proteins, including neurodegenerative-disease-related molecules [3]. The previous proteomic analyses, however, may be insufficient since the results were only from MDA-MB-231 cells. Therefore, further exploration of the interactions between UBL3 and other proteins may facilitate the exploration of the potential effects of UBL3 and diseases.

Alpha-synuclein (α-syn), a highly conserved neuronal protein, is highly enriched in presynaptic nerve terminals. The physiological function of α-syn remains largely unclear; several biochemical activities have been proposed, including regulation of dopamine neurotransmission and synaptic function/plasticity [9]. α-syn knockout mice do not exhibit a distinct phenotype [10]. Misfolded α-syn is aggregated in α-synucleinopathies, such as Parkinson's disease (PD), dementia with Lewy bodies [11], and multiple system atrophy [12]. Various factors are involved in the transition of α-syn from a physiological state to pathological aggregation, including genetic mutation [13], protein–protein interactions [14], PTM [15], and oxidative stress [16]. The phosphorylation of α-syn at serine-129 may be important for the formation of inclusions in PD and related α-synucleinopathies [17]. The interactions between α-syn and its protein interactomes play key roles in the pathological accumulation of α-syn. For example, α-syn interacts with synphilin-1 promoting the inclusion formation of α-syn [14]. α-syn interacts with molecular chaperone proteins [18] and protein deglycase DJ-1 [19] preventing the formation of oligomerized α-syn. Exploring potential proteins that can interact with α-syn is an essential direction to elucidate the aggregation mechanism of α-syn and find new therapeutic targets for α-synucleinopathies. The pathological α-syn can be packaged in sEVs for cell-to-cell transport [20]. Until now, it is unknown whether α-syn interacts with UBL3.

Protein–protein interactions play essential roles in most biological processes [21] and these interactions provide a wide spectrum of therapeutic targets for the treatment of human diseases [22]. Many natural products and drugs have been reported to be excellent molecule candidates for stabilizing or inhibiting protein–protein interactions [23]. The interaction between α-syn and the target protein can be affected by different compounds or drugs. For example, the interaction between prolyl oligopeptidase and α-syn was downregulated by KYP-2047, a potent prolyl oligopeptidase inhibitor, reducing the accumulation of α-syn inclusion [24]. Furthermore, the protein–protein interaction of α-syn was downregulated by 03A10, a small molecule from the fruits of *Vernicia fordii* (Euphorbiaceae), inhibiting α-syn aggregation [25]. However, it remains unknown whether the interaction between UBL3 and α-syn will be affected by clinical drugs or chemical compounds.

We hypothesize that UBL3 may interact with α-syn in cells and mediate the loading of α-syn into sEVs. In this study, we aim to investigate whether UBL3 can interact with α-syn in cells, and explore whether the treatments of clinical drugs or chemical compounds affect the interaction between UBL3 and α-syn.

2. Materials and Methods

2.1. Animals

Age is an important risk factor for neurodegenerative diseases. People usually develop the disease around age 60 or older [26]. Therefore, we choose elder mice with an average age of 22 months to investigate the level of phosphorylated serine-129 (p-S-129). All of the mice in this study were on a C57BL/6J background. *Ubl3* knockout (*Ubl3*$^{-/-}$) mice were acquired from the previously established laboratory colony [3]. C57BL/6J strain wild type (WT) mice (SLC Inc., Hamamatsu, Shizuoka, Japan) were used as a control in this study. Three mice are included in each group. All mice were fed and bred at 12 h of light/dark cycles. The genotypes of mice were confirmed by polymerase chain reaction (PCR) according to our previous report [3].

2.2. Antibodies and Drugs

The antibodies used were: anti-p-S-129 α-syn antibody (Abcam, ab59264, 1:1000 dilution), anti-α-syn antibody (BioLegend, 834304, 1:1000 dilution), anti-UBL3 antibody (ABclonal, A4028, 1:1000 dilution), anti-MYC antibody (MBL, M1932, 1:1000 dilution), anti-Flag antibody (MERCK, F7425-.2MG, 1:1000 dilution), horseradish peroxidase (HRP)-conjugated anti-rabbit secondary antibody (Cell signaling, 7074, 1:5000 dilution), biotinylated anti-rabbit immunoglobulin G (Vector Laboratories, BA-1000, 1:1000 dilution).

Some representative drugs related to neurodegenerative diseases, that are available at our institution, were selected as our screening targets to explore whether they have the potential to influence the interaction of UBL3 with α-syn. Furthermore, some chemical compounds [27,28] and clinical drugs [29] that can affect the formation of α-syn aggregate in vitro were also selected as screening targets to explore whether they affect the interaction. Tyrosine kinase inhibition induces autophagy for neurodegenerative-disease-associated amyloid clearance and epidermal growth factor receptor tyrosine kinase inhibitor (EGFR-TKI) can reduce phosphorylated α-syn pathology [30]. Therefore, certain representative EGFR-TKI drugs that are available at our institution were also selected as our screening targets. All clinically approved drugs, natural products, and other bioactive components, a total of 32, were ordered from the corresponding suppliers (Supplementary Table S1) and used for drug screening. All drugs were dissolved in dimethyl sulfoxide (DMSO) (FUJIFILM Wako Pure Chemical Corporation, Osaka, Japan) to 10 mM stock solution.

2.3. Immunohistochemistry

Immunohistochemistry was conducted according to a previously published protocol [31]. Briefly, sections were incubated in 3% hydrogen peroxide (FUJIFILM Wako Pure Chemical Corporation, Osaka, Japan) in 1× phosphate-buffered saline (PBS, 0.1 mol/L, pH 7.4) for 15 min after serial deparaffinization, washed with 1× PBS three times, and treated with a solution containing 1% bovine serum albumin (Sigma-Aldrich, St. Louis, MI, USA) in 1× PBS for 1 h at room temperature. Then samples were incubated with primary antibody for 1 h at room temperature. After washing three times with 1× PBS, sections were treated for 1 h with secondary antibody washed three times in 1× PBS, and processed using the avidin–biotin complex (Vector Laboratories, Newark, NJ, USA) and prepared in 1× PBS for 30 min at room temperature. The reaction was visualized using DAB (FUJIFILM Wako Pure Chemical Corporation, Osaka, Japan). Finally, the sections were subsequently counterstained using hematoxylin (FUJIFILM Wako Pure Chemical Corporation, Osaka, Japan), dehydrated in graded alcohols (FUJIFILM Wako Pure Chemical Corporation, Osaka, Japan) (80%, 90%, 100%), transparentized with xylene (FUJIFILM Wako Pure Chemical Corporation, Osaka, Japan), and coverslipped with PathoMount (FUJIFILM Wako Pure Chemical Corporation, Osaka, Japan). Images of immunohistochemistry were acquired using a NanoZoomer 2.0 HT system (Hamamatsu Photonics, Hamamatsu, Shizuoka, Japan). To compare the differences in overall immunoreactive signal intensity between WT and *Ubl3*$^{-/-}$ mice, we randomly intercepted five areas with the same square from the substantia nigra of WT and *Ubl3*$^{-/-}$ mice, respectively. Quantification of the immunoreactivity signal intensity of each intercepted area was determined using the analysis software ImageJ 2.0 (National Institutes of Health, Bethesda, MD, USA).

2.4. Plasmids Construction

Following the previous paper [32], for the N-terminal region of *Gaussia princeps* luciferase (Gluc) sequence-tagged UBL3 (NGluc-UBL3) plasmid, the coding sequence of UBL3 (NM_007106) was inserted in the frame after the NGluc sequence in the pCI vector between the XhoI site and the MluI site. For the C-terminal region of Gluc sequence-tagged α-syn (α-syn-CGluc), the coding sequence of α-syn (NM_001146055.2) was inserted in the frame before the CGluc sequence in the pCI vector between the Xho I site and the Mlu I site. For the 3xFlag-UBL3 plasmid, the coding sequence of UBL3 was inserted in the frame after the 3xFlag sequence in the pcDNA3.1-3xFlag vector between the BamHI site and the EcoRI site. For the 6xMYC-α-syn plasmid, we amplified it by in vitro PCR using the corresponding primers. After digestion by XhoI and XbaI, the fragment of α-syn was inserted in the pcDNA3-6xMYC vector after the 6xMYC sequence. For the constructions of "CCVIL" amino acids in C-terminal region-deleted mutant of UBL3 (UBL3Δ5), including NGluc-UBL3Δ5 and 3xFlag-UBL3Δ5, we deleted the last five amino acids (5′-CCVIL-3′) of NGluc-UBL3 and 3xFlag-UBL3 by in vitro site-directed mutagenesis using the corresponding primers (Table 1). The fragments of NGluc-UBL3, NGluc-UBL3Δ5, and α-syn-CGluc were tagged with an immunoglobulin kappa secretory signal (IKSS) sequence after the start codon. Sequences of all constructed plasmids were verified by Sanger sequencing. All primers are listed in Table 1.

Table 1. Primer list.

Primer	Sequence
NGluc-UBL3Δ5	For 5′-TAAACGCGTGGTACCTCTAGAGTCG-3′ Rev 5′-ATTACTCTCTCCAGTCTTCTCACGATTCC-3′
3xFlag-UBL3Δ5	For 5′-TAAGAATTCTGCAGATATCCATCACAC-3′ Rev 5′-ATTACTCTCTCCAGTCTTCTCACGATTCC-3′
XhoI-α-syn-XbaI	For 5′-GCACTCGAGGCCACCATGGATG-3′ Rev 5′-GCCTCTAGATTA GGCTTCAGGTTCGTAGTC-3′

UBL3Δ5: CCVIL delete mutant of UBL3, For: Forward, Rev: Reverse.

2.5. Cell Culture and cDNA Transfection

Human embryonic kidney (HEK) 293 cells (RIKEN Cell Bank, Tsukuba, Ibaraki, Japan) were cultured in Dulbecco's modified Eagle's medium (DMEM, Thermo Fisher Scientific, Waltham, MA, USA) with 10% fetal bovine serum (FBS) (Sigma-Aldrich, St. Louis, MI, USA). Cell cultures were incubated at 37 °C in a 5% CO_2 humidified incubator. Cells were cultured in culture plates to 80–90% confluence, and transiently transfected with cDNA plasmids using Lipofectamine 2000 transfect reagent (Thermo Fisher Scientific, Waltham, MA, USA) diluted in Opti-MEM reduced serum medium (Thermo Fisher Scientific, Waltham, MA, USA) according to the reagent instructions.

2.6. Cell Treatment

For the detection of UBL3 and α-syn interaction, the cell culture medium (CM) of transfected HEK293 cells was changed to FBS (− Opti-MEM after being transfected with NGluc-UBL3, NGluc-UBL3Δ5, and α-syn-CGluc plasmids for 12 h. Then CM and cells were collected after further incubation for 3 days.

For the drug screening, HEK293 cells, after being transfected using NGluc-UBL3 and α-syn-CGluc plasmids for 18 h, were plated into 96-well cell culture plates in 100 μL/well of DMEM (10% FBS) and incubated overnight at 37 °C, in 5% CO_2 humidified incubator. The next day, all solutions of candidate drugs were diluted in prewarmed DMEM (10% FBS) to 1.5 μM and 15 μM, 1.5-fold of the final concentration. Then 50 μL/well of the old culture medium was replaced with 100 μL of pre-warmed DMEM (10% FBS) containing the different candidate drugs to the final concentrations of 1 μM and 10 μM. For the treatment of MPP^+, the solution of MPP^+ (Cayman Chemical, Ann Arbor, MI, USA) was diluted in pre-warmed DMEM (10% FBS) to 1.5-fold of final concentrations and 50 μL/well of the old culture medium was replaced with 100 μL of pre-warmed DMEM (10% FBS) containing MPP^+ to the final concentrations of 50 μM, 100 μM, 300 μM, 500 μM, and 600 μM, separately. The treatment of equivalent volume DMSO as the drug solution was set as a negative control. All cell culture plates were further incubated at 37 °C for another 48 h. Then the CM was collected for luminescence intensity assay.

2.7. Sample Preparation and Luciferase Assay

All CM were centrifuged at 1200 rpm for 5 min to remove the cell debris. The collected HEK293 cells were lysed using cell lysis buffer (1% [*v*/*v*] Triton X100 (Sigma-Aldrich, St. Louis, MI, USA) in 1× PBS) and then centrifuged at 15,000 rpm for 5 min to get supernatant. After adding 17 μg/mL coelenterazine (Cosmo Bio, Kyodo, Japan) diluted by Opti-MEM into CM and cell lysate (CL) supernatant, luminescence intensity was measured using a microplate reader (BioTek, Winooski, VT, USA) immediately. The luminescence intensity of untreated DMEM (10% FBS) was set as a background. For the drug screening, all luminescence intensities of CM were corrected by subtracting the luminescence intensity of the background. The background-corrected data were used to compute a ratio to the luminescence intensity compared to the background-corrected luminescence intensity of DMSO.

2.8. 3-(4,5-Dimethylthiazol-2-yl)-2,5-diphenyl-2H-tetrazolium Bromide (MTT) Assay

In live cells, MTT is a pale-yellow substrate that is cleaved by living cells to yield a dark blue formazan product. The quantity of formazan corresponds to the number of living cells. For transfected HEK293 cells, following the collection of CM for the measurement of luminescence intensity, we added fresh pre-warmed DMEM (10% FBS) medium to 100 μL/well, mixed with 10 μL of MTT reagent (Sigma-Aldrich, St. Louis, MI, USA) to each well, and incubated in the incubator at 37 °C for 4 h. After adding 100 μL/well of isopropanol (FUJIFILM Wako Pure Chemical Corporation, Osaka, Japan) with 0.04 N hydrogen chloride (KANTO Chemical Corporation, Tokyo, Japan) to dissolve the formazan, the absorbance (OD) of each well at a wavelength of 450 nm was detected using the microplate reader, cell viability $(OD_{\text{Intervention group}} - OD_{\text{Blank group}})/(OD_{\text{Control group}} - OD_{\text{Blank group}})$. The blank group had only medium without cells; the control group had medium and cells without intervention.

2.9. Bicinchoninic Acid (BCA) Assay

According to the instructions of the BCA test kit (Thermo Fisher Scientific, Waltham, MA, USA), we add 20 μL of each standard or 2 μL cell lysis sample diluted in 18 μL water replicate into a microplate well. Add 200 μL of the working solution to each well and mix the plate thoroughly on a plate shaker for 30 s. Incubate at 37 °C for 30 min. Then measure the absorbance at 562 nm on a microplate reader after cooling the plate to room temperature. Subtract the average absorbance measurement of the blank standard replicates from the measurements of all other individual standard and cell lysis sample replicates. Then calculate the protein concentration of each sample using the standard curve.

2.10. Co-Immunoprecipitation

For the validation of interaction between UBL3 and α-syn, HEK293 cells were transfected with 3xFlag-UBL3, 3xFlag-UBL3Δ5, and 6xMYC-α-syn plasmids in various combinations for 18 h. After replacing with new CM, the transfected HEK293 cells were continually incubated for 36 h, then washed and collected with ice-cold 1× PBS, pelleted by centrifugation at 1200 rpm for 5 min at 4 °C. Cell pellets were resuspended and lysed using 1% Triton lysate buffer (50 mM Tris-HCl [pH 7.4] (NAKALAI TESQUE, Kyoto, Japan), 100 mM NaCl (FUJIFILM Wako Pure Chemical Corporation, Osaka, Japan), and 1% [*v*/*v*] Triton X-100) for 30 min on ice. Cell debris and unbroken cells were removed by centrifugation at 15,000 rpm for 15 min at 4 °C. Protein content of the supernatant was measured using the BCA assay according to the manufacturer's instructions. The supernatants containing 500 µg of total protein were incubated with 50 µL of anti-Flag tag antibody magnetic beads (FUJIFILM Wako Pure Chemical Corporation, Osaka, Japan) with rotation for 10 h at 4 °C. An amount of 50 µL of 2-mercaptoethanol (−2× sodium dodecyl sulfate (SDS) sample loading buffer (100 mM Tris-HCl [pH 6.8], 4% SDS (NAKALAI TESQUE, Kyoto, Japan), 20% glycerol (FUJIFILM Wako Pure Chemical Corporation, Osaka, Japan), and 0.01% bromophenol blue (FUJIFILM Wako Pure Chemical Corporation, Osaka, Japan)) was added to the beads after the beads were washed three times using ice-cold wash buffer (50 mM Tris-HCl [pH 7.4], 100 mM NaCl), and boiled at 95 °C for 5 min. CL and precipitated proteins were separated by SDS-PAGE for Western blotting analysis.

2.11. Immunoblot

All samples, including CM and CL from transfected HEK293 cells and precipitated proteins, were loaded into 12% SDS-PAGE. Then the proteins were transferred to the polyvinylidene difluoride membrane (Cytiva, Tokyo, Japan). Membranes were blocked with shaking for 1 h at room temperature using 0.5% [*w*/*v*] skim milk (NAKALAI TESQUE, Kyoto, Japan) in Tween-20 (+) (FUJIFILM Wako Pure Chemical Corporation, Osaka, Japan) Tris Buffered Saline (TBS-T; 100 mM Tris-HCl [pH 8.0], 150 mM NaCl, 0.5% [*v*/*v*] Tween-20) and then incubated with shaking overnight at 4 °C using the appropriate primary antibodies. The membranes were incubated with HRP-conjugated anti-rabbit secondary antibody with shaking at room temperature for 1 h after three washes in TBS-T. The immunoreactive proteins were developed using the enhanced chemiluminescence kit (Thermo Fisher Scientific, Waltham, MA, USA) and detected on the FUSION FX imaging system (Vilber Lourmat, Collégien, Seine-et-Marne, France).

2.12. Statistical Analysis

Measurement data were analyzed using GraphPad Prism 7.0 (GraphPad Software, LaJolla, CA, USA) statistical software, and expressed as mean ± SD (standard deviation). The differences between groups of data were calculated using a Student's *t* test for unpaired data. Furthermore, the differences between groups of data were calculated using a one-way analysis of variance (ANOVA) and Dunnett's post hoc test for the comparison of multiple groups. $p < 0.05$ was considered statistically significant. All cell culture experiments were performed in triplicate.

3. Results

3.1. p-S-129 α-syn Was Upregulated in the Substantia Nigra of *Ubl3*$^{-/-}$ Mice

To explore whether UBL3 affects α-syn, the expression of p-S-129 α-syn was investigated in the brain tissues of the WT and *Ubl3*$^{-/-}$ mice using anti-p-S-129 α-syn antibody by immunohistochemistry. The immunoreactive signal intensity of p-S-129 α-syn was significantly increased in the substantia nigra of *Ubl3*$^{-/-}$ mice compared to WT mice ($p = 0.0005$) (Figure 1 and Figure S1A,B). The expression of p-S-129 α-syn was upregulated in the substantia nigra of *Ubl3*$^{-/-}$ mice, which suggested that UBL3 affects α-syn.

3.2. UBL3 Interacted with α-syn in Cells

Further, we examined whether UBL3 interacts with α-syn using a *Gaussia princeps* based split luciferase complementation assay (SLCA), a powerful approach taking advantage of the reconstruction of the N-terminal and C-terminal fragments of Gluc to detect protein–protein interactions in vitro (Figure 2A) [33]. In this study, we constructed the NGluc-UBL3, NGluc-UBL3Δ5, and α-syn-CGluc plasmids. All constructions contain an IKSS sequence after the start codon (Figure 2B), which allows successfully expressed fragments and their interacting complexes in cells to be secreted into the CM.

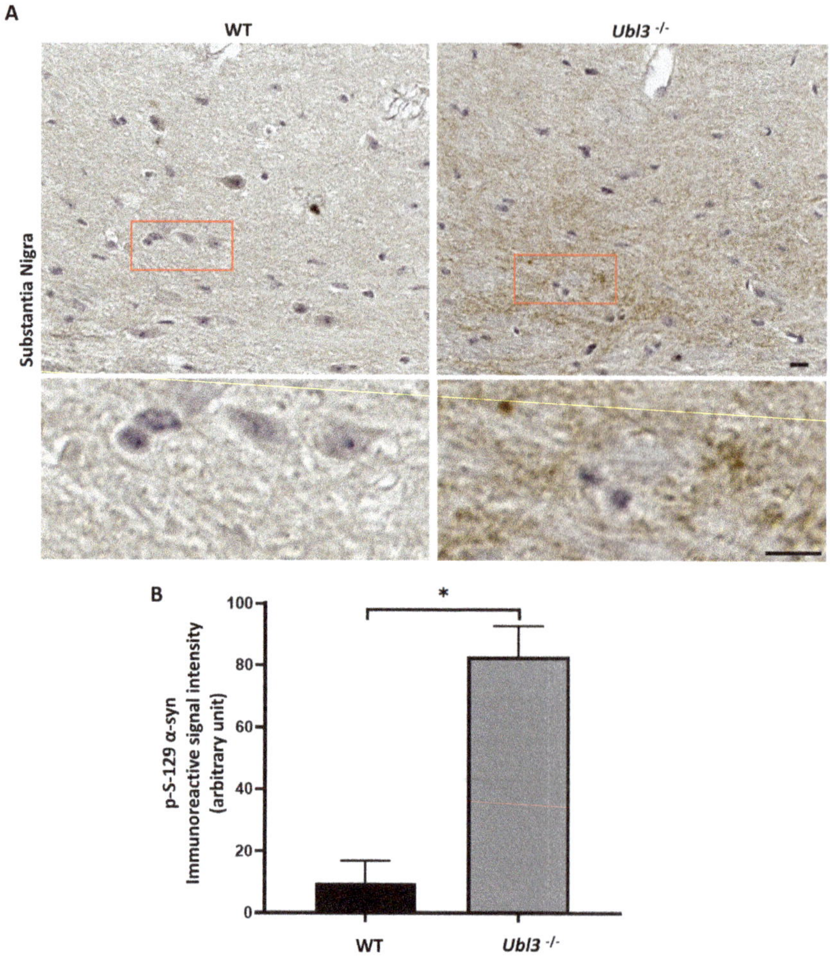

Figure 1. Expression of p-S-129 α-syn was upregulated in the substantia nigra of *Ubl3*$^{-/-}$ mice brain. (**A**) Representative images of immunocytochemistry staining of p-S-129 α-syn in the substantia nigra of WT and *Ubl3*$^{-/-}$ mice. Scale bars: 10 μm. The red boxes represent the selection zones of the magnified pictures below. (**B**) Quantification comparison of the immunohistochemical expression of p-S-129 α-syn in the substantia nigra between WT and *Ubl3*$^{-/-}$ mice. Histograms represent the mean + SD (the number of mice in each group was three). A Student's t test was performed. *: *p*-value < 0.05; WT: wild type; *Ubl3*$^{-/-}$: *Ubl3* knock out.

Then we co-expressed SLCA constructs in HEK293 cells in various combinations. Expressions of the SLCA constructs were confirmed by immunoblotting (Figure 2C). We measured the luminescence intensities of CM and CL from transfected HEK293 cells and found strong luminescence intensities from both fractions in the cells expressing NGluc-UBL3 + α-syn-CGluc, and NGluc-UBL3Δ5 + α-syn-CGluc. As a control, the CM and CL from Gluc over-expressing HEK293 cells showed markedly higher luminescence intensities compared to double-expression HEK293 cells. On the other hand, the luminescence intensities of CM and CL from HEK293 cells, expressing NGluc-UBL3, NGluc-UBL3Δ5, or α-syn-CGluc, were as low as the background level (Figure 2D).

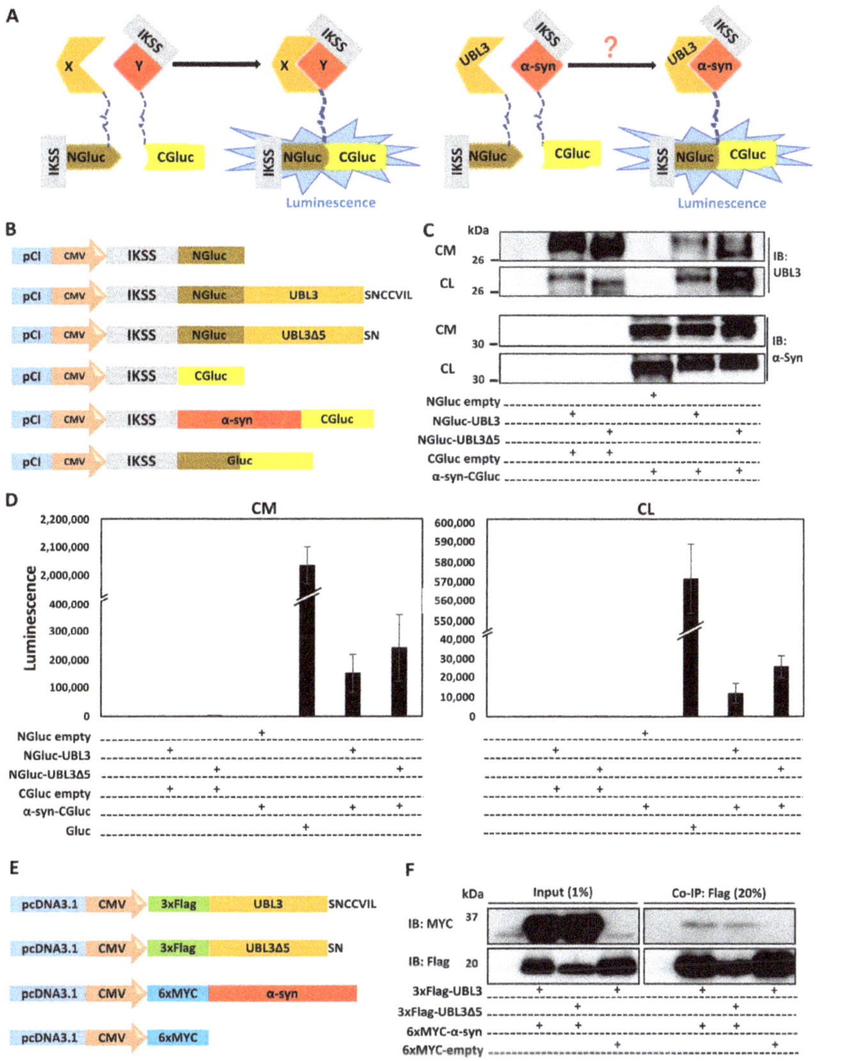

Figure 2. UBL3 interacted with α-syn in cells. (**A**) Schematic representation of the *Gaussia princeps* based split luciferase complementation assay system. Whether there is an interaction between UBL3 and α-syn is still unknown (Red question mark). (**B**) Schematic of split Gluc-tagged proteins, NGluc-UBL3, NGluc-UBL3Δ5, and α-syn-CGluc, both containing an IKSS sequence, which allows successfully expressed fragments and their interacting complexes to be secreted into the CM. (**C**) Immunoblotting of CM and CL from transfected HEK293 cells by anti-UBL3 polyclonal antibody and anti-α-syn antibody. (**D**) Luminescence of the CM (left) and CL (right) from transfected HEK293 cells. The luminescence ± SD in triplicate experiments is shown. (**E**) Schematic representation of 3xFlag-UBL3, 3xFlag-UBL3Δ5, and 6xMYC-α-syn for co-immunoprecipitation (Co-IP). (**F**) Co-immunoprecipitated 3xFlag-UBL3 and 3xFlag-UBL3Δ5 interact with 6xMYC-α-syn. The input lanes are 1% of the sample prior to Co-IP and the Co-IP lanes are 20% of the Co-IP products. IKSS: immunoglobulin kappa secretory signal; CM: cell culture medium; CL: cell lysate.

To validate the interaction between UBL3 and α-syn, we constructed and co-expressed 3xFlag-UBL3, 3xFlag-UBL3Δ5, and 6xMYC-α-syn in various combinations in HEK293 cells

(Figure 2E). The signal of 6xMYC-α-syn was detected from the co-immunoprecipitate of 3xFlag-UBL3 and 3xFlag-UBL3Δ5, while the signal of 6xMYC-α-syn in the co-immunoprecipitate of 3xFlag-UBL3Δ5 was less than that of 3xFlag-UBL3 (Figures 2F and S1C). These results showed that UBL3 interacts with α-syn in HEK293 cells.

3.3. The Interaction between UBL3 and α-syn in Cells Was Upregulated by the 1-Methyl-4-Phenylpyridinium (MPP+) Exposure

MPP^+, a bioactive derivative of 1-methyl-4-phenyl-1, 2, 3, 6-tetrahydropyridine (MPTP), has been reported to induce toxic aggregation of α-syn in cell models [34] and part of mouse models [35]. Furthermore MPTP has been reported to increase the α-syn immunoreactivity in the neurons of non-human primates [36]. To investigate whether the treatment of MPP^+ affects the interaction between UBL3 and α-syn, we treated HEK293 cells transfected by NGluc-UBL3 and α-syn-CGluc cDNA with 50 μM, 100 μM, 300 μM, 500 μM, and 600 μM of MPP^+. We collected CM and assayed the luminescence intensities after being treated with different concentrations of MPP^+ for 48 h and also assessed the cell viability using an MTT assay. Although the luminescence intensities of CM from transfected HEK293 cells treated with different concentrations of MPP^+ did not show significant differences (Figure 3A), the treatment with MPP^+ at concentrations between 100 μM and 600 μM significantly inhibited cell viability in a concentration-dependent manner (Figure 3B). Therefore, to exclude the effect of cell activity on the luminescence intensities of CM, we divided the luminescence intensities by the cell viability to calculate the ratio of luminescence relative to the cell number. The treatment of MPP^+ at concentrations between 300 μM and 600 μM significantly upregulated the interaction between UBL3 and α-syn in a dose-dependent manner (Figure 3C).

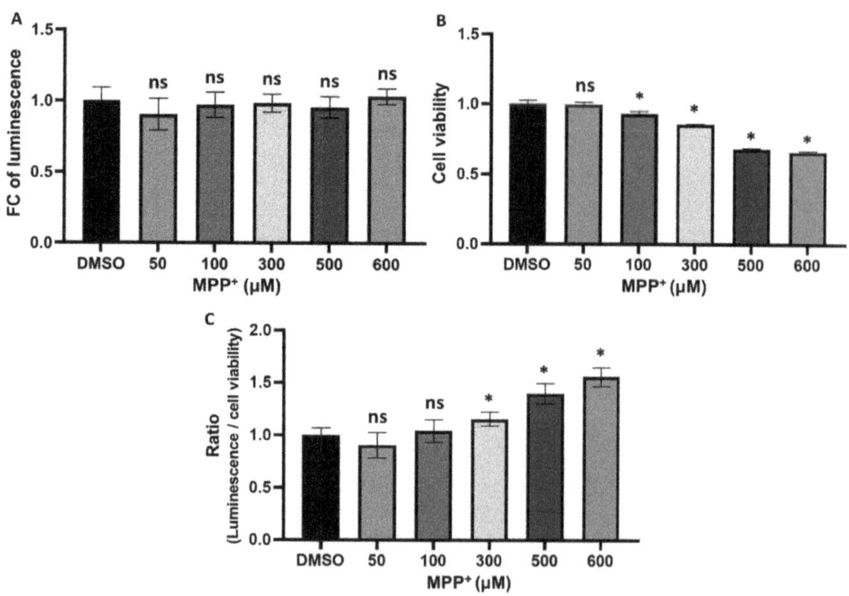

Figure 3. The interaction level in cells was upregulated by the treatment of MPP^+. (**A**) Luminescence of CM from transfected HEK293 cells, which were treated with 50, 100, 300, 500, and 600 μM of MPP^+ for 48 h. (**B**) The cell viability of transfected HEK293 cells was treated with different concentrations of MPP^+ for 48 h. (**C**) The ratios of luminescence divided by cell viability were calculated in triplicate. The luminescence ± SD, cell viability ± SD, and ratio ± SD in triplicate are shown. One-way ANOVA and Dunnett's post hoc test were performed. ns (non-significant) $p > 0.05$, * $p < 0.05$. FC: fold change; DMSO: dimethyl sulfoxide.

3.4. Interaction between UBL3 and α-syn in Cells Was Significantly Downregulated by Osimertinib

We used the HEK293 cells, transfected with NGluc-UBL3 and α-syn-CGluc cDNA, as a drug screening model to screen for drugs or compounds that can regulate the interaction between UBL3 and α-syn in cells. We assessed the luminescence intensities of CM from transfected HEK293 cells in the presence of 32 drugs (1 μM and 10 μM) at 48 h in triplicate. All luminescence intensities of CM under drug treatment were normalized to that of vehicle treatment. Under the treatment with a concentration of 1 μM (Figure 4A), one drug, sulfasalazine, upregulated the luminescence intensities of CM by more than 25%. In contrast, one drug, docetaxel, downregulated the luminescence intensities of CM by more than 25%. Under the treatment with a concentration of 10 μM (Figure 4B), three drugs upregulated the luminescence intensities of CM by more than 25%, sulfasalazine, pemetrexed, and gemcitabine. In contrast, four drugs downregulated the luminescence intensities of CM by more than 25%, methylcobalamin, erlotinib, docetaxel, and osimertinib.

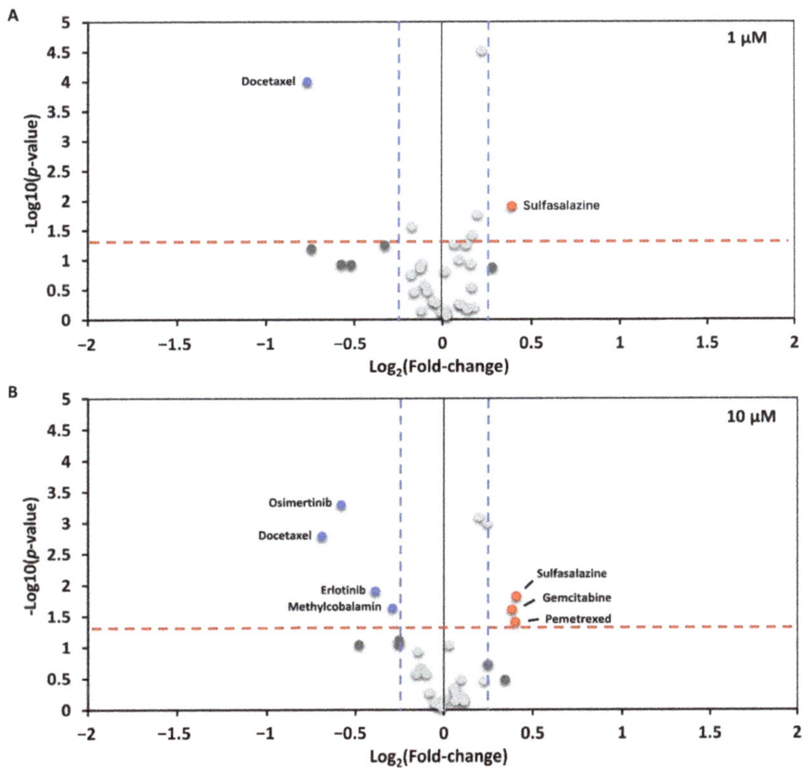

Figure 4. Drug screening results of 32 drugs at concentrations of 1 μM and 10 μM. (**A**) The screen result under a concentration of 1 μM. (**B**) The screen result under a concentration of 10 μM. The volcano plot showed the fold-change (x-axis) versus the significance (y-axis) of 32 drugs. The significance (non-adjusted p-value) and the fold-change are converted to $-$Log10(p-value) and Log2(fold-change), respectively. The vertical and horizontal dotted lines show the cut-off of fold-change $\pm$ 1.25, and p-value = 0.05, respectively. The luminescence of CM from transfected HEK293 cells was upregulated by >1.25-fold with a p-value < 0.05 (upper-right, dots colored red) and the luminescence of CM from transfected HEK293 cells was downregulated by < $-$1.25-fold with p-value < 0.05 (upper-left, dots colored blue). The luminescence $\pm$ SD in triplicate experiments is shown. One-way ANOVA and Dunnett's post hoc test were performed. $p < 0.05$ was considered statistically significant.

To exclude the effect of drug cytotoxicity on the drug screening results, we treated transfected HEK293 cells for 48 h using selected drugs that significantly affected the interaction between UBL3 and α-syn at concentrations of 1 μM and 10 μM, and assessed the cell viability with an MTT assay. As shown in Figure 5A, the cell viability was significantly decreased by the treatment of docetaxel, gemcitabine, osimertinib, and pemetrexed at concentrations of 10 μM. Thus, the ratios of luminescence intensities of CM to cell viability were calculated (Figure 5B). At the concentration of 1 μM, the interaction between UBL3 and α-syn was upregulated by the treatment of gemcitabine (ratio = 1.37, $p = 0.0008$), while it was significantly downregulated by the treatment of erlotinib (ratio = 0.73, $p < 0.0001$). At the concentration of 10 μM, the interaction was significantly upregulated by the treatment of gemcitabine (ratio = 1.53, $p < 0.0001$), while it was significantly downregulated by the treatment of erlotinib (ratio = 0.72, $p = 0.015$) and osimertinib (ratio = 0.55, $p < 0.0001$).

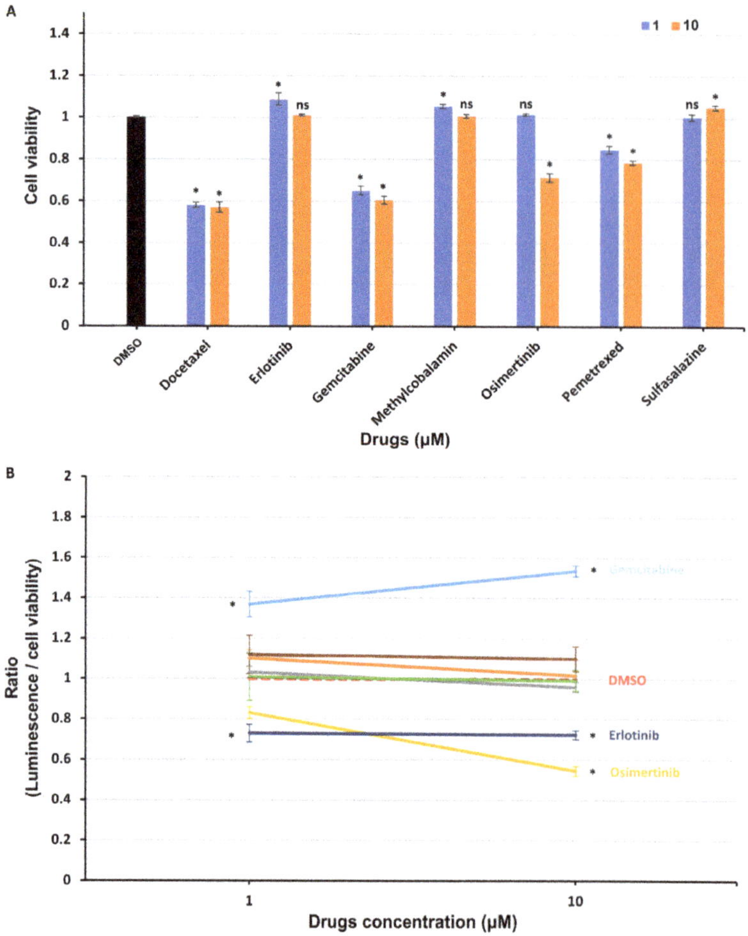

Figure 5. The interaction between UBL3 and α-syn in cells was regulated by the treatment of clinical drugs. (**A**) The cell viability of transfected HEK293 cells treated with selected drugs at concentrations of 1 and 10 μM for 48 h, including docetaxel, erlotinib, gemcitabine, methylcobalamin, osimertinib, pemetrexed, and sulfasalazine. (**B**) The ratio of luminescence divided by cell viability was calculated in triplicate. The cell viability ± SD and the ratio ± SD in triplicate experiments are shown. One-way ANOVA and Dunnett's post hoc test were performed. ns (non-significant) $p > 0.05$; * $p < 0.05$.

4. Discussion

This study first revealed that UBL3 could interact with α-syn in cells and was upregulated in response to the MPP^+ exposure. Furthermore, the interaction could also be regulated by the treatment of clinical drugs. These results provided the first evidence that UBL3 may be involved in α-synucleinopathies with the possibility of being a potential therapeutic target.

The expression level of p-S-129 α-syn was upregulated in the substantia nigra of *Ubl3*$^{-/-}$ mice. The phosphorylation of α-syn at S-129 is important for the formation of misfold α-syn in synucleinopathies [17]. Therefore, UBL3 may be related to the formation of misfold α-syn. On the other hand, α-syn can be secreted via sEVs in neurons for self-protection when they suffer cellular stress or pathological injury [37–39]. UBL3 plays a role in the sorting of proteins to sEVs by acting as a PTM factor [3]. Thus, these results suggested that the deletion of *Ubl3* may upregulate the formation of misfold α-syn and the UBL3 may play a role in the sorting of α-syn to sEVs.

Ageta et al. reported that UBL3 can modify its protein interactomes through disulfide binding depending on the cysteine residues at its C-terminal [3]. It is interesting to note that our results showed that the interaction between UBL3 and α-syn in HEK293 cells is not completely erased after the deleting mutation of the cysteine residues at its C-terminus. UBL3, as a member of the ubiquitin-like protein family, contains a ubiquitin-like domain. Ubiquitin and ubiquitin-like proteins, such as NEDD8, SUMO, FAT10, and ISG15, are covalently attached to lysine residues of their protein interactomes through the C-terminal glycine residues [40]. Taken together, it is possible that UBL3 interacts with α-syn in cells in another manner, rather than only dependent on cysteine residues at its C-terminal. In future studies, it is important to discover the interaction mechanism between UBL3 and α-syn.

Our results showed that the interaction between UBL3 and α-syn in cells was upregulated by the MPP^+ exposure. MPP^+, a key environmental risk factor of PD, has been widely used as a common neurotoxin for both in vivo and in vitro experiments [41]. MPP^+ exposure is known to disturb mitochondrial respiration by inhibiting the mitochondrial complex I, and this process plays a role in initiating mitochondrial dysfunction [42], which can induce and promote α-syn accumulation [43]. PD-like symptoms and aggregation of α-syn were observed in chronic MPP^+-exposed rodent models [35]. α-syn was involved in the process of induction of mitochondrial dysfunction by MPP^+ exposure [44]. These results suggested that the upregulation of interaction between UBL3 and α-syn induced by MPP^+ exposure might be a response to the mitochondrial dysfunction. However, MPP^+ exposure can also upregulate the expression of α-syn in SH-SY5Y cells [45]. Furthermore, in MPTP-induced non-human primates and partial rodent models it was reported that only α-syn immunoactivity was observed to be upregulated, without significant Lewy body or Lewy neurite formation [36]. Whether the upregulated interaction affects the accumulation of α-syn will need to be investigated in future studies.

The treatment of osimertinib significantly downregulated the interaction between UBL3 and α-syn in cells. Osimertinib, a third-generation EGFR-TKI, is widely used to treat non-small-cell lung cancer [46]. In recent years, it has been suggested that the EGFR signaling pathway and associated genes possibly play an essential role in dopamine neuron cell death [47]. An exogenous neurotrophic supply of EGFR ligands rescues dopaminergic neurons from cell death induced by neurotoxins, 6-hydroxydopamine, or MPTP [48]. An epidemiological study about the polymorphisms of the human EGFR gene found that rs730437 and rs11506105 polymorphisms of EGFR are possible in association with the susceptibility to PD [49]. The treatment of EGFR inhibitors can significantly reduce the p-S-129 α-syn pathology in mouse brain sections by reducing the level of seeding and propagation of pathological α-syn [30]. In our drug screening results, another EGFR inhibitor, erlotinib, also showed significant downregulation of interaction between UBL3 and α-syn. It is convincing from the view of the propagation pathway of α-syn pathology. α-syn can be secreted and transferred cell to cell via sEVs [20]. sEV-associated α-syn can facilitate the propagation of α-syn pathology through cell-to-cell transfer [32]. Furthermore, UBL3 interacts with its target proteins and regulates the sorting of them into sEVs [3]. Therefore, these results indicated that interaction between UBL3 and α-syn may be associated with the inhibition of the α-syn pathology propagation via sEVs by crosstalk with the EGFR pathway.

In addition, we found that the treatment of gemcitabine significantly upregulated the interaction between UBL3 and α-syn in cells. Gemcitabine, a nucleoside analog, has been widely used as an anticancer drug to treat a variety of cancers [50]. The activated gemcitabine triphosphate complex, formed by linking two phosphates, inhibits DNA synthesis by inhibiting ribonucleotide reductase [51]. The treatment of gemcitabine can induce the initiation of mitochondrial dysfunction [52]. This result is consistent with MPP^+ exposure, suggesting that the upregulation of interaction between UBL3 and α-syn might be a response to the mitochondrial dysfunction.

This study had some limitations. Firstly, the exact mechanisms by which drug treatment affects the interaction between NGluc-UBL3 and α-syn-CGluc, altering protein expression, degradation, or directly influencing the process of interactions, was not investigated. Secondly, whether the aggregation status of α-syn in HEK293 cells overexpressing α-syn affects the interaction between UBL3 and α-syn remains unstudied. Furthermore, due to technical constraints, it was not possible to determine whether the drug substance treatment would affect the activity of the luciferase itself. On the other hand, the impact of candidate drug treatments and UBL3 itself on the aggregation state of α-syn was also not investigated. In the future, we will further explore these limitations according to the methods summarized by the previous report [53]. In addition, the number and selection range of drugs used for drug screening in this study is limited and more drugs will need to be tested in future studies.

5. Conclusions

The results in this study showed that UBL3 interacts with α-syn in cells and the interaction between UBL3 and α-syn is upregulated in response to MPP^+ exposure. Moreover, it was downregulated by the treatment of EGFR inhibitor osimertinib. These findings provide the first evidence identifying UBL3 as an interacting protein of α-syn and UBL3 may be a new therapeutic option for α-synucleinopathies in the future. This study extends the horizon for further etiological and therapeutic studies of α-synucleinopathies.

Supplementary Materials: The following supporting information can be downloaded at: https://www.mdpi.com/article/10.3390/biomedicines11061685/s1, Figure S1: Reproduced IHC staining and co-immunoprecipitation (Co-IP results). Table S1: Drug list.

Author Contributions: Conceptualization, T.K. and M.S.; methodology, B.C., H.Z., Q.Z., Y.P., C.Z., S.O., Y.T., M.M.H., M.A.M., T.K. and M.S.; validation, B.C. and Q.Z.; formal analysis, B.C. and Q.Z.; investigation, B.C.; resources, B.C., M.Y., T.K. and A.S.M.W.; writing—original draft preparation, B.C.; writing—review and editing, Y.N., T.N., D.K., T.K., K.O. and M.S.; supervision, M.S.; project administration, B.C., T.K. and M.S. All authors have read and agreed to the published version of the manuscript.

Funding: This research was supported by Japan Society for the Promotion of Science Grant-in-Aid for Scientific Research under Grant Number JP22H02793, Japan Agency for Medical Research and Development program under Grant Number JP22nk0101641 and Grant Number JP20gm0910004, and HUSM Grant-in-Aid.

Institutional Review Board Statement: The animal study protocol was approved by the Institutional Animal Care and Use Committees of Hamamatsu University School of Medicine (protocol code: 2022024 and date of approval: May 2022).

Informed Consent Statement: Not applicable.

Data Availability Statement: All relevant data were reported within the article. Further supporting data will be provided upon a written request addressed to the corresponding author.

Acknowledgments: We thank all the members of Mitsutoshi Setou's lab for their support and help. We thank members of the Advanced Research Facilities & Services (ARFS), Hamamatsu University School of Medicine, for technical assistance.

Conflicts of Interest: The authors declare no conflict of interest.

References

1. Downes, B.P.; Saracco, S.A.; Lee, S.S.; Crowell, D.N.; Vierstra, R.D. MUBs, a family of ubiquitin-fold proteins that are plasma membrane-anchored by prenylation. *J. Biol. Chem.* **2006**, *281*, 27145–27157. [CrossRef] [PubMed]
2. Chadwick, B.P.; Kidd, T.; Sgouros, J.; Ish-Horowicz, D.; Frischauf, A.M. Cloning, mapping and expression of UBL3, a novel ubiquitin-like gene. *Gene* **1999**, *233*, 189–195. [CrossRef]
3. Ageta, H.; Ageta-Ishihara, N.; Hitachi, K.; Karayel, O.; Onouchi, T.; Yamaguchi, H.; Kahyo, T.; Hatanaka, K.; Ikegami, K.; Yoshioka, Y.; et al. UBL3 modification influences protein sorting to small extracellular vesicles. *Nat. Commun.* **2018**, *9*, 3936. [CrossRef]
4. Liu, H.; Wilson, K.R.; Firth, A.M.; Macri, C.; Schriek, P.; Blum, A.B.; Villar, J.; Wormald, S.; Shambrook, M.; Xu, B.; et al. Ubiquitin-like protein 3 (UBL3) is required for MARCH ubiquitination of major histocompatibility complex class II and CD86. *Nat. Commun.* **2022**, *13*, 1934. [CrossRef] [PubMed]
5. Huang, L.; Zheng, M.; Zhou, Q.M.; Zhang, M.Y.; Yu, Y.H.; Yun, J.P.; Wang, H.Y. Identification of a 7-gene signature that predicts relapse and survival for early stage patients with cervical carcinoma. *Med. Oncol. (Northwood Lon. Engl.)* **2012**, *29*, 2911–2918. [CrossRef]
6. Shi, Y.; Qi, L.; Chen, H.; Zhang, J.; Guan, Q.; He, J.; Li, M.; Guo, Z.; Yan, H.; Li, P. Identification of Genes Universally Differentially Expressed in Gastric Cancer. *BioMed Res. Int.* **2021**, *2021*, 7326853. [CrossRef]
7. Singh, V.; Singh, L.C.; Vasudevan, M.; Chattopadhyay, I.; Borthakar, B.B.; Rai, A.K.; Phukan, R.K.; Sharma, J.; Mahanta, J.; Kataki, A.C.; et al. Esophageal Cancer Epigenomics and Integrome Analysis of Genome-Wide Methylation and Expression in High Risk Northeast Indian Population. *Omics A J. Integr. Biol.* **2015**, *19*, 688–699. [CrossRef]
8. Zhao, X.; Yongchun, Z.; Qian, H.; Sanhui, G.; Jie, L.; Hong, Y.; Yanfei, Z.; Guizhen, W.; Yunchao, H.; Guangbiao, Z. Identification of a potential tumor suppressor gene, UBL3, in non-small cell lung cancer. *Cancer Biol. Med.* **2020**, *17*, 76–87. [CrossRef] [PubMed]
9. Winkler, S.; Hagenah, J.; Lincoln, S.; Heckman, M.; Haugarvoll, K.; Lohmann-Hedrich, K.; Kostic, V.; Farrer, M.; Klein, C. alpha-Synuclein and Parkinson disease susceptibility. *Neurology* **2007**, *69*, 1745–1750. [CrossRef]
10. Chandra, S.; Fornai, F.; Kwon, H.B.; Yazdani, U.; Atasoy, D.; Liu, X.; Hammer, R.E.; Battaglia, G.; German, D.C.; Castillo, P.E.; et al. Double-knockout mice for alpha- and beta-synucleins: Effect on synaptic functions. *Proc. Natl. Acad. Sci. USA* **2004**, *101*, 14966–14971. [CrossRef]
11. Spillantini, M.G.; Schmidt, M.L.; Lee, V.M.; Trojanowski, J.Q.; Jakes, R.; Goedert, M. Alpha-synuclein in Lewy bodies. *Nature* **1997**, *388*, 839–840. [CrossRef]
12. Mezey, E.; Dehejia, A.; Harta, G.; Papp, M.I.; Polymeropoulos, M.H.; Brownstein, M.J. Alpha synuclein in neurodegenerative disorders: Murderer or accomplice? *Nat. Med.* **1998**, *4*, 755–757. [CrossRef] [PubMed]
13. Dawson, T.M.; Dawson, V.L. Molecular pathways of neurodegeneration in Parkinson's disease. *Science (New York, N.Y.)* **2003**, *302*, 819–822. [CrossRef]
14. Xie, Y.Y.; Zhou, C.J.; Zhou, Z.R.; Hong, J.; Che, M.X.; Fu, Q.S.; Song, A.X.; Lin, D.H.; Hu, H.Y. Interaction with synphilin-1 promotes inclusion formation of alpha-synuclein: Mechanistic insights and pathological implication. *FASEB J. Off. Publ. Fed. Am. Soc. Exp. Biol.* **2010**, *24*, 196–205. [CrossRef]
15. Schmid, A.W.; Fauvet, B.; Moniatte, M.; Lashuel, H.A. Alpha-synuclein post-translational modifications as potential biomarkers for Parkinson disease and other synucleinopathies. *Mol. Cell. Proteom. MCP* **2013**, *12*, 3543–3558. [CrossRef]
16. Scudamore, O.; Ciossek, T. Increased Oxidative Stress Exacerbates α-Synuclein Aggregation In Vivo. *J. Neuropathol. Exp. Neurol.* **2018**, *77*, 443–453. [CrossRef]
17. Smith, W.W.; Margolis, R.L.; Li, X.; Troncoso, J.C.; Lee, M.K.; Dawson, V.L.; Dawson, T.M.; Iwatsubo, T.; Ross, C.A. Alpha-synuclein phosphorylation enhances eosinophilic cytoplasmic inclusion formation in SH-SY5Y cells. *J. Neurosci. Off. J. Soc. Neurosci.* **2005**, *25*, 5544–5552. [CrossRef] [PubMed]
18. Burmann, B.M.; Gerez, J.A.; Matečko-Burmann, I.; Campioni, S.; Kumari, P.; Ghosh, D.; Mazur, A.; Aspholm, E.E.; Šulskis, D.; Wawrzyniuk, M.; et al. Regulation of α-synuclein by chaperones in mammalian cells. *Nature* **2020**, *577*, 127–132. [CrossRef]
19. Shendelman, S.; Jonason, A.; Martinat, C.; Leete, T.; Abeliovich, A. DJ-1 is a redox-dependent molecular chaperone that inhibits alpha-synuclein aggregate formation. *PLoS Biol.* **2004**, *2*, e362. [CrossRef]
20. Danzer, K.M.; Kranich, L.R.; Ruf, W.P.; Cagsal-Getkin, O.; Winslow, A.R.; Zhu, L.; Vanderburg, C.R.; McLean, P.J. Exosomal cell-to-cell transmission of alpha synuclein oligomers. *Mol. Neurodegener.* **2012**, *7*, 42. [CrossRef] [PubMed]
21. Guo, W.; Wisniewski, J.A.; Ji, H. Hot spot-based design of small-molecule inhibitors for protein-protein interactions. *Bioorganic Med. Chem. Lett.* **2014**, *24*, 2546–2554. [CrossRef]
22. Arkin, M.R.; Tang, Y.; Wells, J.A. Small-molecule inhibitors of protein-protein interactions: Progressing toward the reality. *Chem. Biol.* **2014**, *21*, 1102–1114. [CrossRef]

23. Fischer, G.; Rossmann, M.; Hyvönen, M. Alternative modulation of protein-protein interactions by small molecules. *Curr. Opin. Biotechnol.* **2015**, *35*, 78–85. [CrossRef]
24. Savolainen, M.H.; Yan, X.; Myöhänen, T.T.; Huttunen, H.J. Prolyl oligopeptidase enhances α-synuclein dimerization via direct protein-protein interaction. *J. Biol. Chem.* **2015**, *290*, 5117–5126. [CrossRef] [PubMed]
25. Wang, Q.; Yao, S.; Yang, Z.X.; Zhou, C.; Zhang, Y.; Zhang, Y.; Zhang, L.; Li, J.T.; Xu, Z.J.; Zhu, W.L.; et al. Pharmacological characterization of the small molecule 03A10 as an inhibitor of α-synuclein aggregation for Parkinson's disease treatment. *Acta Pharmacol. Sin.* **2023**, *44*, 1122–1134. [CrossRef] [PubMed]
26. Rademacher, D.J. Potential for Therapeutic-Loaded Exosomes to Ameliorate the Pathogenic Effects of α-Synuclein in Parkinson's Disease. *Biomedicines* **2023**, *11*, 1187. [CrossRef] [PubMed]
27. Pandey, N.; Strider, J.; Nolan, W.C.; Yan, S.X.; Galvin, J.E. Curcumin inhibits aggregation of alpha-synuclein. *Acta Neuropathol.* **2008**, *115*, 479–489. [CrossRef]
28. Zaidi, F.K.; Deep, S. Scutellarin inhibits the uninduced and metal-induced aggregation of α-Synuclein and disaggregates preformed fibrils: Implications for Parkinson's disease. *Biochem. J.* **2020**, *477*, 645–670. [CrossRef] [PubMed]
29. Bar-On, P.; Crews, L.; Koob, A.O.; Mizuno, H.; Adame, A.; Spencer, B.; Masliah, E. Statins reduce neuronal α-synuclein aggregation in in vitro models of Parkinson's disease. *J. Neurochem.* **2008**, *105*, 1656–1667. [CrossRef]
30. Tavassoly, O.; Del Cid Pellitero, E.; Larroquette, F.; Cai, E.; Thomas, R.A.; Soubannier, V.; Luo, W.; Durcan, T.M.; Fon, E.A. Pharmacological Inhibition of Brain EGFR Activation By a BBB-penetrating Inhibitor, AZD3759, Attenuates α-synuclein Pathology in a Mouse Model of α-Synuclein Propagation. *Neurother. J. Am. Soc. Exp. NeuroTherapeut.* **2021**, *18*, 979–997. [CrossRef]
31. Guo, M.; Wang, J.; Zhao, Y.; Feng, Y.; Han, S.; Dong, Q.; Cui, M.; Tieu, K. Microglial exosomes facilitate α-synuclein transmission in Parkinson's disease. *Brain A J. Neurol.* **2020**, *143*, 1476–1497. [CrossRef]
32. Hashimoto, T.; Adams, K.W.; Fan, Z.; McLean, P.J.; Hyman, B.T. Characterization of oligomer formation of amyloid-β peptide using a split-luciferase complementation assay. *J. Biol. Chem.* **2011**, *286*, 27081–27091. [CrossRef]
33. Wille, T.; Blank, K.; Schmidt, C.; Vogt, V.; Gerlach, R.G. Gaussia princeps luciferase as a reporter for transcriptional activity, protein secretion, and protein-protein interactions in Salmonella enterica serovar typhimurium. *Appl. Environ. Microbiol.* **2012**, *78*, 250–257. [CrossRef] [PubMed]
34. Karmacharya, M.B.; Hada, B.; Park, S.R.; Choi, B.H. Low-Intensity Ultrasound Decreases α-Synuclein Aggregation via Attenuation of Mitochondrial Reactive Oxygen Species in MPP(+)-Treated PC12 Cells. *Mol. Neurobiol.* **2017**, *54*, 6235–6244. [CrossRef] [PubMed]
35. Sonsalla, P.K.; Zeevalk, G.D.; German, D.C. Chronic intraventricular administration of 1-methyl-4-phenylpyridinium as a progressive model of Parkinson's disease. *Park. Relat. Disord.* **2008**, *14* (Suppl. 2), S116–S118. [CrossRef]
36. Giráldez-Pérez, R.; Antolín-Vallespín, M.; Muñoz, M.; Sánchez-Capelo, A. Models of α-synuclein aggregation in Parkinson's disease. *Acta Neuropathol. Commun.* **2014**, *2*, 176. [CrossRef]
37. Jang, A.; Lee, H.J.; Suk, J.E.; Jung, J.W.; Kim, K.P.; Lee, S.J. Non-classical exocytosis of alpha-synuclein is sensitive to folding states and promoted under stress conditions. *J. Neurochem.* **2010**, *113*, 1263–1274. [CrossRef] [PubMed]
38. Fussi, N.; Höllerhage, M.; Chakroun, T.; Nykänen, N.P.; Rösler, T.W.; Koeglsperger, T.; Wurst, W.; Behrends, C.; Höglinger, G.U. Exosomal secretion of α-synuclein as protective mechanism after upstream blockage of macroautophagy. *Cell Death Dis.* **2018**, *9*, 757. [CrossRef]
39. Yamada, K.; Iwatsubo, T. Extracellular α-synuclein levels are regulated by neuronal activity. *Mol. Neurodegener.* **2018**, *13*, 9. [CrossRef]
40. Welchman, R.L.; Gordon, C.; Mayer, R.J. Ubiquitin and ubiquitin-like proteins as multifunctional signals. *Nat. Rev. Mol. Cell Biol.* **2005**, *6*, 599–609. [CrossRef]
41. Lotharius, J.; O'Malley, K.L. The parkinsonism-inducing drug 1-methyl-4-phenylpyridinium triggers intracellular dopamine oxidation. A novel mechanism of toxicity. *J. Biol. Chem.* **2000**, *275*, 38581–38588. [CrossRef] [PubMed]
42. Kim, H.Y.; Jeon, H.; Kim, H.; Koo, S.; Kim, S. Sophora flavescens Aiton Decreases MPP(+)-Induced Mitochondrial Dysfunction in SH-SY5Y Cells. *Front. Aging Neurosci.* **2018**, *10*, 119. [CrossRef] [PubMed]
43. Grünewald, A.; Kumar, K.R.; Sue, C.M. New insights into the complex role of mitochondria in Parkinson's disease. *Prog. Neurobiol.* **2019**, *177*, 73–93. [CrossRef]
44. Wu, F.; Poon, W.S.; Lu, G.; Wang, A.; Meng, H.; Feng, L.; Li, Z.; Liu, S. Alpha-synuclein knockdown attenuates MPP+ induced mitochondrial dysfunction of SH-SY5Y cells. *Brain Res.* **2009**, *1292*, 173–179. [CrossRef] [PubMed]
45. Lu, M.; Sun, W.L.; Shen, J.; Wei, M.; Chen, B.; Qi, Y.J.; Xu, C.S. LncRNA-UCA1 promotes PD development by upregulating SNCA. *Eur. Rev. Med. Pharmacol. Sci.* **2018**, *22*, 7908–7915. [CrossRef] [PubMed]
46. Tan, C.S.; Gilligan, D.; Pacey, S. Treatment approaches for EGFR-inhibitor-resistant patients with non-small-cell lung cancer. *Lancet Oncol.* **2015**, *16*, e447–e459. [CrossRef]
47. Kim, I.S.; Koppula, S.; Park, S.Y.; Choi, D.K. Analysis of Epidermal Growth Factor Receptor Related Gene Expression Changes in a Cellular and Animal Model of Parkinson's Disease. *Int. J. Mol. Sci.* **2017**, *18*, 430. [CrossRef] [PubMed]
48. Gill, S.S.; Patel, N.K.; Hotton, G.R.; O'Sullivan, K.; McCarter, R.; Bunnage, M.; Brooks, D.J.; Svendsen, C.N.; Heywood, P. Direct brain infusion of glial cell line-derived neurotrophic factor in Parkinson disease. *Nat. Med.* **2003**, *9*, 589–595. [CrossRef]
49. Jin, J.; Xue, L.; Bai, X.; Zhang, X.; Tian, Q.; Xie, A. Association between epidermal growth factor receptor gene polymorphisms and susceptibility to Parkinson's disease. *Neurosci. Lett.* **2020**, *736*, 135273. [CrossRef] [PubMed]

50. Pandit, B.; Royzen, M. Recent Development of Prodrugs of Gemcitabine. *Genes* **2022**, *13*, 466. [CrossRef]
51. Cerqueira, N.M.; Fernandes, P.A.; Ramos, M.J. Understanding ribonucleotide reductase inactivation by gemcitabine. *Chemistry (Weinheim an der Bergstrasse, Germany)* **2007**, *13*, 8507–8515. [CrossRef] [PubMed]
52. Inamura, A.; Muraoka-Hirayama, S.; Sakurai, K. Loss of Mitochondrial DNA by Gemcitabine Triggers Mitophagy and Cell Death. *Biol. Pharm. Bull.* **2019**, *42*, 1977–1987. [CrossRef] [PubMed]
53. Estaun-Panzano, J.; Arotcarena, M.L.; Bezard, E. Monitoring α-synuclein aggregation. *Neurobiol. Dis.* **2023**, *176*, 105966. [CrossRef] [PubMed]

biomedicines

Review

NLRP3 Inflammasome's Activation in Acute and Chronic Brain Diseases—An Update on Pathogenetic Mechanisms and Therapeutic Perspectives with Respect to Other Inflammasomes

Anna Chiarini [1,*], Li Gui [2], Chiara Viviani [1], Ubaldo Armato [1] and Ilaria Dal Prà [1,*]

1 Human Histology & Embryology Section, Department of Surgery, Dentistry, Pediatrics, and Gynecology, University of Verona, 37134 Verona, Italy; chiaraviviani.3@gmail.com (C.V.); uarmato@gmail.com (U.A.)
2 Department of Neurology, Southwest Hospital, Chongqing 400038, China; 2079244086@email.szu.edu.cn
* Correspondence: anna.chiarini@univr.it (A.C.); ilaria.dalpra@univr.it (I.D.P.)

Abstract: Increasingly prevalent acute and chronic human brain diseases are scourges for the elderly. Besides the lack of therapies, these ailments share a neuroinflammation that is triggered/sustained by different innate immunity-related protein oligomers called inflammasomes. Relevant neuroinflammation players such as microglia/monocytes typically exhibit a strong NLRP3 inflammasome activation. Hence the idea that NLRP3 suppression might solve neurodegenerative ailments. Here we review the recent Literature about this topic. First, we update conditions and mechanisms, including RNAs, extracellular vesicles/exosomes, endogenous compounds, and ethnic/pharmacological agents/extracts regulating NLRP3 function. Second, we pinpoint NLRP3-activating mechanisms and known NLRP3 inhibition effects in acute (ischemia, stroke, hemorrhage), chronic (Alzheimer's disease, Parkinson's disease, Huntington's disease, MS, ALS), and virus-induced (Zika, SARS-CoV-2, and others) human brain diseases. The available data show that (i) disease-specific divergent mechanisms activate the (mainly animal) brains NLRP3; (ii) no evidence proves that NLRP3 inhibition modifies human brain diseases (yet ad hoc trials are ongoing); and (iii) no findings exclude that concurrently activated other-than-NLRP3 inflammasomes might functionally replace the inhibited NLRP3. Finally, we highlight that among the causes of the persistent lack of therapies are the species difference problem in disease models and a preference for symptomatic over etiologic therapeutic approaches. Therefore, we posit that human neural cell-based disease models could drive etiological, pathogenetic, and therapeutic advances, including NLRP3's and other inflammasomes' regulation, while minimizing failure risks in candidate drug trials.

Keywords: neuroinflammation; inflammasomes; NLRP3; inhibitors; brain; neurodegenerative diseases; virus encephalitis; innate immunity

Citation: Chiarini, A.; Gui, L.; Viviani, C.; Armato, U.; Dal Prà, I. NLRP3 Inflammasome's Activation in Acute and Chronic Brain Diseases—An Update on Pathogenetic Mechanisms and Therapeutic Perspectives with Respect to Other Inflammasomes. *Biomedicines* **2023**, *11*, 999. https://doi.org/10.3390/biomedicines11040999

Academic Editors: Masaru Tanaka, Eleonóra Spekker and Massimo Grilli

Received: 7 February 2023
Revised: 16 March 2023
Accepted: 17 March 2023
Published: 23 March 2023

1. Introduction

1.1. An Overall Picture

Acute and chronic human brain diseases have been attracting the increased attention of scientists and the public. This has been due to the concurrence of several factors, i.e., brain illnesses' mounting prevalence, the persistent lack of effective therapies, increasingly huge healthcare and economic costs, hardships in assisting such patients particularly at home, marked psychopathological impacts on patients and relatives, a greater sensitivity to improper lifestyle consequences, and a common aspiration to long-lasting and healthy aging. To this must be added the growing concern about the serious risk that severe acute brain injuries surreptitiously evolve into chronic neuropathologies such as Alzheimer's disease (AD), Parkinson's disease (PD), and amyotrophic lateral sclerosis (ALS). Worldwide yearly estimates of acute brain injuries total about 42 million cases, while symptomatic AD by itself affects more than 50 million people. It is predicted that such figures will double or treble in twenty/thirty years unless effective therapies become available [1,2]. Yet, the latter quite understandable wish is hampered by ongoing controversies due to the still unclarified underlying pathogenetic mechanisms. A common feature in all brain diseases is ongoing neuroinflammation. From this observation, the hypothesis has been put forward that this inflammation is a main causative factor, whose mitigation or suppression would slow or stop the progression and/or improve the outcome [3,4].

"Inflammation" is a physiological defensive reaction of living tissues to harm, aiming at ridding the causative factor(s), disposing of cell debris, and restoring tissue integrity and homeostasis in the short term. In his treatise "*De Medicina*", Roman physician Aulus Cornelius Celsius (~14–37 AD; [5]) first described acute inflammation's five cardinal symptoms, i.e., "*rubor*" (Lat. reddening) and "*calor*" (Lat. heat), due to local increases in blood flow; "*tumor*" (Lat. swelling) caused by edema and leukocyte infiltration due to altered vessel permeability; "*dolor*" (Lat. pain), elicited by local acidosis overstimulating the nerves; and "*laesa functio*" (Lat. impaired function"), the injury's downstream upshot. Conversely, a persisting (chronic) inflammation is a pathological condition whose upshot can be severe.

Obviously, neuroinflammation has specific features, particularly in the various neurodegenerative diseases. In the latter, its onset can be early (familial cases) or surreptitious (sporadic cases). Its course is often quite slow, so that it can progress undetected for decades. However, while unnoticed, chronic neuroinflammation spreads from the site of origin (e.g., frontotemporal cerebral cortex, hippocampus, locus coeruleus, spinal cord) to other regions and in so doing progressively destroys the brain's neuronal functional reserve. When the reserve is depleted, the gray matter of the cerebral cortex, basal ganglia, thalamus, brain stem, cerebellum, spinal cord, and the white matter connectome (axons) are remarkably thinned. At this stage, the diseases become symptomatic. Progressive decreases in abilities, such as memory, cognition, emotions, psychic, and motor activities, render the patients unable to cope. Eventually, the neuropathology inexorably and more rapidly moves toward the *obitus* [6,7]. The etiologic factors also trigger various collateral cellular processes, such as the overproduction of hydroxyl radicals, superoxide anions (reactive oxygen species or ROS), nitric oxide (NO), peroxynitrite, ionic dyshomeostasis, mitochondrial, lysosomal, and autophagy disfunctions, and overproduction and accumulation of toxic protein species, which sustain the neuroinflammation. Other events concur, such as leukocyte infiltration and alterations in blood–brain barrier (BBB) function. Altogether, such *noxae* drive positive feedback loops, aggravating the neuropathology [8–13].

Since Celsius's time, and particularly in the last century, a huge amount of knowledge has been accumulating about the crucial relation between inflammation's drivers and immunity. Nowadays, we know that the innate immune system secures the first protection against harmful factors or "molecular patterns". The endogenous damage-associated molecular patterns (DAMPs) and homeostasis-altering molecular patterns (HAMPs) are sterile compounds (e.g., ATP, mitochondrial DNA), dysfunctional metabolism products, and cell debris. The exogenous pathogen-associated molecular patterns (PAMPs) are infectious (bacteria, fungi, viruses, prions) or toxic agents (chemicals, organic molecules). DAMPs/HAMPs/PAMPs form complexes with multiligand cellular "pattern recognition receptors" (PRRs). In turn, such complexes nucleate the assembly of multicomponent protein platforms, the "inflammasomes" [4,14], the activated signaling of which drives the tissue inflammation at the injury's site.

NLRs Assembly and Signaling Activation

The PRRs' group names are based upon shared structural domains. The most noted PRRs comprise the NLRs (NOD-like nucleotide-binding domain and leucine-rich-repeat (LRR) family of receptors); ALRs (absent in melanoma 2 receptors); and MEFV gene-encoded PYRIN receptors [15]. Currently, activated NLRs are the most intensely studied PRRs. In humans, NLRs having a PYRIN N-terminal homology domain (PYD) include 14 members, namely, NLRP1–NLR14. Physiologically, NLRs (excepting brain NLRs) keep an auto-inhibited conformation that winds up when they detect DAMPs/PAMPs/HAMPs. This drives the assembly and signaling activation of inflammasomes. NLRs' N-terminal

PYDs bind and nucleate the oligomerizing adaptor protein ASC (apoptosis-associated speck-like protein endowed with a caspase recruitment domain or CARD) [15,16]. Notably, the ASC gene encodes both a CARD and a PYD domain. Therefore, via CARD•CARD or PYD•PYD homotypic interactions, ASC proteins make complexes with the PYD or CARD domains of NLRs. PYDs and CARDs are conserved domains of 80–90 amino acids arranged in six anti-parallel α-helices forming an inner hydrophobic core with charged residues at the surface. Via CARD•CARD interactions, ASCs of canonical inflammasomes nucleate the inactive zymogens of caspase-1, a cysteine-type peptidase, causing their polymerization and proximity-mediated auto-catalytic self-cleavage, resulting in active caspase-1 duplets [16,17]. The latter produce mature interleukin (IL)-1β and IL-18 from their respective precursors and N-termini fragments of the gasdermin D protein (human, GSDMD; rodent, GsdmD), in addition to cleaving other proteins that share the YVHD/FESD consensus sequence [18]. Next, the GSDMD/GsdmD's N-fragments oligomerize, forming transmembrane pores that extracellularly release (i) mature proinflammatory IL-1β and IL-18; and (ii) K^+, causing an intracellular ion dyshomeostasis. Persistent K^+ losses lead to inflammatory death or pyroptosis of the involved cells. In turn, products released from pyroptotic cells (e.g., ATP, mitochondrial DNA) boost inflammation further [18]. NLRP oligomerization, ASC recruitment, and caspase-1 nucleated polymerization/activation are irreversible processes developing in a self-inducing prion-like fashion and promoting canonical inflammasome signaling [19].

Moreover, via CARD, domain-assembled NLRP1, NLRP2, NLRP3, and AIM2 inflammasomes activate the NF-κB signaling pathway, which transcriptionally regulates the genes encoding for the various inflammasomes' structural proteins [20]. Conversely, other NLRs, i.e., NLRC3, NLRP6, NLRP12, and NLRX1, impede the NF-κB pathway's activation, thereby mitigating or quelling inflammation [21]. Indeed, these *"anti-inflammasomes"* are crucially necessary, as they stop the onset of chronic inflammatory diseases. Moreover, CARD-only proteins (COPs) and PYD-only proteins (POPs) also regulate inflammasome activity [22]. Furthermore, epigenetic mechanisms, e.g., noncoding RNA expression, CpG island DNA methylation, and histone post-translational changes, modulate inflammasome function [23].

We recently reviewed the multiple roles of the NLRP1, NLRP2, AIM-2, and NLRC4 inflammasomes in human and rodent brain diseases [24]. Our work showed that several inflammasomes can partake in brain neuroinflammatory processes. This enticed us to review in this work the mounting literature specifically concerning the NLRP3 inflammasome, its modulation by endogenous and exogenous and pharmacological and ethnopharmacological agents/extracts, its pathogenetic implications in acute and chronic brain diseases, and the therapeutic potential of its inhibition. Based on the results we highlight that disease-specific divergent mechanisms activate the brain's NLRP3 in microglia/monocytes and other neural cell types. However, no proof is hitherto available that NLRP3 inhibition would be a human brain disease-modifying approach. Furthermore, no data have so far excluded the possible functional replacement of the inhibited NLRP3 by other concurrently activated inflammasomes. These facts led us to highlight that one of the causes of the persisting failures of human brain disease-related therapeutic attempts is the inadequate regard for its morpho-functional uniqueness based on the assumption that animal brain models are good enough. The consequent suggestion is to focus instead on human neural cell-based preclinical brain diseases models, which could drive etiological, pathogenetic, and therapeutic advances, including proper NLRP3 and other inflammasome regulation, and minimize failure risks concerning lead candidate drug testing in clinical trials.

The following paragraphs will delve into the main advances concerning the NLRP3 inflammasome, followed by specific paragraphs about its role in most relevant brain diseases, a discussion of the results, and a conclusion.

1.2. Brain NLRP3 Inflammasome

The inactive NLRP3 inflammasome (i.e., NLRP3-ASC or NOD-like receptor protein 3 (N-terminal PYD, central ATP-hydrolyzing NACHT (NAIP+CIITA+HET-E+TP1), and C-terminal LRR domains) molecules confine themselves to the endoplasmic reticulum (ER) membranes [25]. Upon activation, they bind adaptor ASC proteins by interacting with phosphatidylinositol-4-phosphate. ASC stabilizes the NLRP3•ASC complexes allowing their activation. Next, NLRP3•ASC complexes migrate to the perinuclear ER membranes and associated mitochondrial aggregates [9,26].

As monocytes/macrophages and microglia strongly express the NLRP3 inflammasome, the latter is involved in human brain diseases and is the most intensely studied and popular inflammasome. NLRP3 might be the "golden" therapeutic target of inflammatory morbidities, including neurodegenerative disorders (e.g., Alzheimer's disease [AD]) [27–29]. In advanced age, the NLRP3 inflammasome also partakes in low-grade sterile yet chronic inflammation called "*inflammaging*", driven by cell debris accumulating within tissues [30]. Moreover, NLRP3 gene mutations result in a spectrum of autoinflammatory diseases known as cryopyrin-associated periodic syndromes (CAPS) [31].

Table 1 lists the common brain NLRP3 inflammasome-activating diseases or agents.

Table 1. Main conditions and factors activating the brain NLRP3 inflammasome.

Condition/Factor	Mechanisms	References
Vascular ailments Stroke Intracerebral hemorrhage Hemorrhagic stroke	Mitochondrial disfunction after hypoxic ischemia/reperfusion (HI/R) Chronic hypoxia	[32–40]
Seizures Mesial lobe temporal epilepsy Soman or A255 (nerve agent) exposure	Acetyl- and butyryl-cholinesterase inhibition	[41,42]
Metal accumulation Manganese (Mn), Lead (Pb) Copper (Cu) Cadmium (Cd) Aluminium/alum	Metal-induced neurotoxicity ↑§ ROS & NF-κB-p65 pathway CaSR and GCP6RA signaling	[43–51] See also Box 1
Mechanical stresses and strains Skull trauma Optic nerve trauma Elevated intracranial pressure Glaucoma	Osteopontin NIMA-related kinase 7 (or NEK7) $P2X_7$ receptor activation HMGB1/caspase-8 pathway	[52–59] See also Box 2
Neurodegenerative diseases Alzheimer's disease (AD) Tauopathies Parkinson's disease (PD) Amyotrophic lateral sclerosis (ALS) Huntington's disease (HD) Prion disease (PrP^{Sc})	Aβs, autophagy block, NEK7 p-Taues paired helical filaments ER stress, ↑ ROS α-Synuclein aggregates Mutated *SOD1*, *TDP-43* Expanded CAG repeats in *HTT/OT15* gene Prion protein seeding	[60–71]
Environmental pollution PM2.5	Increased ROS production by microglia	[72,73]
Infectious diseases Sepsis (bacteria, fungi) West Nile Virus (WNV) HIV-1 Herpes Virus 1 Japanese Encephalitis Virus (JEV) Zika Virus (ZIKV) SARS-CoV-2 Encephalomyocarditis Virus (EMCV) Tuberculosis	Bacterial and fungal toxins Intensified IL-1β signaling Tat and gp120 proteins Gasdermin D-dependent pyroptosis ROS-dependent activation of Src/Ras/Raf/ERK/NF-κB signaling axis NS5 protein and ↑ ROS S1 spike glycoprotein, viroporin ORF3a/8 viroporin ORF2b Early secreted antigenic target protein of 6 kDa (ESAT-6)	[74–90]

Table 1. *Cont.*

Condition/Factor	Mechanisms	References
Metabolic disorders Atherosclerosis Gout Obesity/high-fat diet Nonalcoholic hepatic steatosis Type-2 diabetes mellitus (T2DM)	Hypercholesterolemia Urates→NEK7 Glucocorticoids and fatty acids surpluses→TNFR and Toll-like receptors ROS, NO, hydroperoxides, scavenger receptors, mTOR	[91–94]
Iatrogenic factors Postoperative cognitive dysfunction Cyclophosphamide cystitis $GdCl_3$, cinacalcet Glucocorticoids (elevated levels)	Drugs, infection, electrolyte imbalance TNF-α Calcimimetic•CaSR/ERK1/2/CaMKII NLRP1 and NLRP3 inflammasomes	[95–101] See also Box 1
Psychotropic drugs Cocaine Methamphetamine Scopolamine Ethanol Morphine Fentanyl	σ-1 receptor TLR-4 ↑ *Dhx58*, *S100a*, *Lrm4* genes TLR-4 μ-3 and κ opioid receptors μ-opioid receptor	[102–108]
Cellular stress and injury ATP Pore-inducing agents Phagocytosed protein polymers ROS Cardiolipin Raised IL-1β levels Reduced cyclic AMP (cAMP) levels Zn^{2+} deficiency K^+ efflux Ca^{2+} and Cl^- influx	Purinergic receptor signaling DDX3X protein/NLRP3 complexes Heat shock protein 60 (HSP60) and TLR-4-p38 MAPKs axis Oxidized mtDNA and proteins Lysosome-released cathepsin B Mitochondria-released hexokinase, ROS NLRP3 activation Ionic imbalances	[25,109–119]
Aging Inflammaging	↑ Membrane attack complexes (MAC) Reduced mitochondrial fission and fusion Declined mitophagy Mitochondrial damage Selective autophagy-mediated mitochondrial homeostasis (in microglia)	[33,112,120–122]

§ ↑ = increased.

1.2.1. NLRP3 Inflammasome Priming and Canonical Activation

Importantly, human, and rodent brain cells of all types preferentially express distinct inflammasomes, e.g., NLRP1 the neurons, NLRP2 the astrocytes, and NLRP3 the microglia [123–129]. However, under both normal and pathological conditions, all the neural cell types express the NLRP3 inflammasome, albeit with differing intensities and regulatory mechanisms [27,64]. Young mice brains physiologically express basal levels of NLRP3 inflammasome activity to upkeep conditioning-induced neuronal plasticity and memory consolidation in the ventral hippocampus and basolateral amygdala [130]. Discordant opinions exist about inflammasomes' roles in human brain diseases, as specific molecular lines of evidence are scanty [24,131,132].

Most studies have shown that NLRP3's canonical activation requires two initiating signals. The "Signal 1" or *"priming step"* is an endocytosed PAMP or an endogenous DAMP/HAMP evoking the signaling from Toll-like receptor 4 (TLR-4) or a NOD-like receptor (NLR) or the tumor necrosis factor receptor (TNFR). Furthermore, signaling from G-protein-coupled receptors (GPCRs) can affect NLRP3 activity (see Box 1 for further details and references).

Box 1. NLRP3 inflammasome regulation by G-protein coupled receptors (GPCRs).

The six GPCRs families (A–F) include eight hundred entities. The fact that 34% of FDA-approved drugs target GPCRs proves their clinical importance. For space reasons here, we discuss only a few GPCRs. For further information, see [133].

B1.1. *Calcium-Sensing Receptor (CaSR)*

The extracellular domain (i.e., venus flytrap) of the ubiquitously expressed CaSR of family C GPCRs binds not only Ca^{2+}, its orthosteric (type I) agonist, but also other mono-, bi-, and tri-valent cations, and various positively charged organic molecules, including polyamines, aminoglycoside antibiotics, and cationic peptides (e.g., amyloid-β [Aβ]) [134–136]. Moreover, CaSR's 7TM (seven-pass transmembrane domain) binds allosteric (type II) ligands (e.g., aromatic L-α-amino acids) and positive allosteric modulators (PAMs i.e., calcimimetics) and negative allosteric modulators (NAMs i.e., calcilytics). Ligand-activated CaSR signaling by its intracellular domains is mediated by various G-proteins and scaffold proteins (e.g., β-arrestin, homer-1) and turns on or off several pathways involving various enzymes, ion channels, and transcription factors [133]. Acting as a calciostat sensing changes in $[Ca^{2+}]_e$, the CaSR regulates systemic $[Ca^{2+}]_e$ homeostasis via parathormone secretion, modulating gut Ca^{2+} absorption, bone Ca^{2+} storage/release, and renal Ca^{2+} excretion [137]. All types of neural cells express the CaSR, and those in AD-relevant hippocampus very intensely [138]. Importantly, besides $[Ca^{2+}]_e$ homeostasis, the CaSR physiologically regulates neural cell growth, differentiation, migration, synaptic plasticity, and neurotransmission [133]. Moreover, the CaSR acts as a DAMP/HAMP/PAMP sensor, as inflammatory diseases affecting various organs, brain included, activate CaSR signaling [27]. In turn, CaSR signaling activates the NLRP3 inflammasome via a surge in phospholipase C-mediated $[Ca^{2+}]_i$ and a concurrent fall in the NLRP3-inhibiting cAMP [31], as well as a proteolytic cleavage of crucial NLRP3 regulators [139]. Moreover, increasing cAMP levels via an adenylate cyclase (AC) activator (e.g., PGE2) or a covalently changed (e.g., dibutyryl-) cAMP or a phosphodiesterase (PDE) inhibitor blocking cAMP catabolism to 5′-AMP (e.g., theophylline or milrinone) promotes cAMP binding to NLRP3, which hinders its activation [26,31,140]. CaSR PAM cinacalcet activates NLRP3 inflammasome via ERK1/2 signaling [98]. Wang et al. [99] showed that in subarachnoid hemorrhage-model mice, CaSR's expression surged in all CNS cell types. The CaSR agonist gadolinium trichloride ($GdCl_3$) upregulated the levels of phosphorylated CaMKII, NLRP3 inflammasome expression, active caspase-1, and mature IL-1β. Conversely, CaSR NAM NPS-2143 and CAMKII inhibitor KN-93 mitigated all CaSR signaling detrimental effects. Hence, CaSR signaling advanced the first stages of acute brain injury, and Aβ•CaSR signaling could drive human AD onset/progression [141].

B1.2. *G-Protein-Coupled Class C Group 6 Receptor A (GPC6RA)*

Alum has been and still is in use as an adjuvant in human vaccines. Alum's mechanism of action remained obscure until Quandt et al. [50] proved that in vitro and in vivo alum induced NLRP3 inflammasome activation via GPRC6A receptor signaling. GPC6RA, of the GPCR Family C Group 6, senses cations (e.g., Ca^{2+}), osteocalcin, L-α-amino acids, and testosterone. GPC6RA signaling partakes via MAPK and mTORC1 in prostatic carcinoma progression [51,142–145] and might contribute to the angiotensin II-driven hypertensive neuroinflammation promoted by 6β-hydroxytestosterone in male mice [146].

B1.3. *G protein-coupled estrogen receptors (GPERs)*

GPER1 and GPER30 are seven-pass transmembrane orphan receptors that rapidly mediate non-genomic estrogen-related kinase signaling. GPER signals prevented hippocampal neuron death due to transient global cerebral ischemia via a remarkable elevation of the endogenous interleukin-1 receptor antagonist (IL-1Ra), which suppresses the pro-inflammatory effects of IL-1β. GPER activation heightened the hippocampal levels of phosphorylated CREB (i.e., cAMP response element-binding) transcription factor, which promotes IL-1Ra expression. The G36 antagonist reversed GPER's neuroprotective effects, proving their specificity [147].

Clearly, CaSR, GPC6RA, and GPERs are PRRs whose roles in neuroinflammation are worthy of further investigation.

Signal 1 involves both translational and post-translational pathways linked to IFNR, PKA, MAPK, mTOR, complement proteins, AMPK/autophagy, IRAK1, TRIF (TIR[Toll/IL-1 receptor/resistance protein]-domain-containing adapter-inducing IFN-β), and NLRP3's de-ubiquination by BRCC3 (BRCA1/BRCA2-Containing Complex Subunit 3), a Lys^{63}-specific de-ubiquitinase. These pathways converge toward NF-κB pathway's activation, which mediates the genetic transcription of NLRP3, ASC, pro-caspase-1, pro-IL-1β, and pro-IL-18 [148–152]. The contours of "Signal 2" or the "*activation step*" of the NLRP3 inflammasome are less defined. A summary list of Signal 2 includes exogenous dead cell-released ATP, which is a ligand of purinergic receptors (see Box 2 for further details and references); cathepsin B released from destabilized lysosomes; phagocytosed protein polymers; reactive oxygen species (ROS); cardiolipin; oxidized mitochondrial DNA [112,114]; K^+ efflux or Ca^{2+} influx, independently of each other [153]; and cyclic AMP (cAMP) downregulation [154]. Importantly, also contact sites between mitochondria and ER membranes favor NLRP3 activation. ER-stress signal-released mitochondrial proteins, ER-released Ca^{2+} surges, lipid perturbations, and cholesterol trafficking critically partake in NLRP3 activation [155]. Moreover, a surge in extracellular Ca^{2+} ($[Ca^{2+}]_e$) triggers NLRP3 activation in monocytes [156]. Thus, $[Ca^{2+}]_i$ increases might be the signal shared by all the stimuli [155] and/or the final common NLRP3-activating pathway [157,158].

Box 2. Brain purinergic receptors.

CNS neural cells express diverse types of purinergic receptors, i.e., P1, for adenosine G protein-coupled receptors; P2X, for ATP-gated ion channels; and P2Y, for G protein-coupled receptors. Importantly, the intra-brain accumulation of Aβs induces the damaged neural cells to release ATP into the extracellular matrix (ECM). Exogenous ATP and the agonist 4-benzoyl-ATP (BzATP) activate the signaling from $P2X_7$ purinergic receptors expressed by neural cells. The upshots are an increased synthesis and release of pro-inflammatory cytokines and chemokines, and a decline in the α-secretase activity, causing a plunge in the extracellular shedding of the neurotrophic and neuroprotective soluble amyloid precursor protein (APP)-α. Yet, various (e.g., mechanical) stressing factors awaken the signaling of $P2X_7$ receptors, making the cells release their endogenous ATP through connexin 43 and pannexin hemichannels (i.e., "pathological pores") [159]. The results are the activation of the NF-κB axis and of the NLRP3•ASC•caspase-1 and IL-1β pathways in both the astrocytes and microglia, triggering the sterile neuroinflammation proper of AD within the brain and of glaucoma within the retina [57,160].

Moreover, the $P2X_7$ receptor agonist BzATP also elicits the release of various cytokines from the retinal ganglion neurons, i.e., IL-3 (in the presence of extracellular Ca^{2+}); IL-4; IL-10; IL-1Ra; TNF-α; MIG/CXCL9 (or monokine induced by IFN-γ/chemokine [C–X–C motif] ligand 9); VEGF; GM-CSF; MIP (macrophage inflammatory protein); CCL20 (or chemokine [C–C motif] ligand 20); and L-selectin, which altogether exert neuroprotective effects [161]. $P2X_7$ receptor stimulation also upregulates IL-6 release from the retinal astrocytes and neurons [162]. In microglial cells, $P2X_7$ receptors modulate the phagocytosis of exogenous debris in the absence of any ligand. However, signals from ligand-bound $P2X_7$ alter lysosome function, causing the cathepsin B-mediated NLRP3 inflammasome activation that a cathepsin B-blocker, CA-074, instead hinders [163].

$P2X_7\ ^{-/-}$ (KO), $P2X_7$ antagonists, such as Brilliant Blue G (BBG), A438079, A839977 and A740003, and the NF-κB inhibitor Bay 11-7082 blocked the effects elicited by purinergic receptors signaling. However, $P2X_7$-specific antagonists blocked only the purinergic receptor-dependent secretion of IL-6 and CCL2 but not TNF-α's release from microglia. These results revealed the differential regulation of the microglial secretion of such cytokines [164]. By contrast, the ATP-activated signaling from the $P2Y_2$ purinergic receptor exerted $P2X_7$-opposite, i.e., anti-inflammatory, and neuroprotective effects [165,166].

Nuclear receptors too control the NLRP3 inflammasome [167]. Thus, various positive and negative signaling pathways strictly regulate NLRP3's activation to prevent any harm while preserving the host tissues' homeostasis [168]. Various kinases, ubiquitin ligases, a de-ubiquitinase, and other enzymes crucially control both NLRP3's activation and function termination via ad hoc post-translational modifications of its protein components [169]. As an example, Bruton's tyrosine kinase (BTK) directly and positively regulates the NLRP3 inflammasome, which might have therapeutic implications [170]. Usually, sterile, and slow-acting DAMPs/HAMPs elicit weaker NLRP3 inflammasome responses than infectious PAMPS do [171]. Finally, inflammasome-interested scientists should note that species-related differences in animal models can crucially affect their results [172].

1.2.2. Noncanonical NLRP3 Activation

Hitherto, we have discussed NLRP3's "canonical activation", a concept valid also for NLRP1, NLRC4, and AIM2 inflammasomes. The more recently discovered "noncanonical activation" of inflammasomes is worth mentioning too. Concerning microglia's NLRP3, the noncanonical process involves the activation of caspase-11 and caspase-8 in mice and of caspase-4 and caspase-5 in humans [173–175]. These caspases behave as cytosolic sensors that directly bind and are activated by the lipopolysaccharide (LPS) of Gram-negative bacteria. This drives the secretion of mature IL-1β and IL-18. Additionally, the active caspases detach N-terminal fragments from the GSDMD/GsdmD proteins, which form transmembrane pores promoting K^+ efflux and thus causing both NLRP3's canonical activation and neurons' pyroptosis [176–179].

The HMGB1 (high mobility group box 1 protein)/caspase-8 pathway is an added mechanism of noncanonical NLRP3 activation proper of eye glaucoma. An acutely elevated intraocular pressure intensifies HMGB1's signaling, which activates the NLRP3 inflammasome by canonical and noncanonical (via caspase-8) mechanisms, producing higher amounts of mature IL-1β within the ischemic retinal tissue and thereby advancing neuroinflammation [59].

1.3. Brain NLRP3 Inflammasome's Modulation by RNAs

Cells express manifold kinds (ribosomal, messenger, and noncoding) of RNAs, which control most of their functions. Long noncoding (Lnc) RNAs have more than 200 base pairs but encode no or few proteins. However, LncRNAs importantly affect body development, cell differentiation, metabolism, autoimmunity, and immune function, and hence NLRP3 inflammasome activity [180,181]. MicroRNAs (or miRs) are ubiquitous 22-nucleotide-long single-stranded RNAs that post-transcriptionally control gene expression by silencing mRNAs via complementary base-pairing [182]. Notably, miRs abound (>2300 types) inside mammalian cells and are released via extracellular vesicles (EVs) or exosomes (Exos) into cerebrospinal fluid and blood. Circulating miRs are under investigation as biomarkers in various diseases and in the distinct stages of each illness. According to ongoing circumstances, distinct miRs promote or inhibit NLRP3 inflammasome activation.

Among noncoding RNAs, Alu-derived RNAs deserve a brief mention. They result from the transcription of primate-specific transposable "Alu elements" by small interspersed nuclear elements (SINEs). Alu-RNAs are plentiful, involving >10% of the human genome, with 102 to 103 copies released into the cytosol of each cell. Alu-RNAs regulate gene expression by binding and inhibiting RNA polymerase II (P2). Alu-RNAs accumulate in the brains of patients with dementia or sporadic Creutzfeldt–Jacob's disease (CJD), in which they drive neuroinflammation and neuron demise [183]. P3-transcribed Alu-RNAs (P3Alus) may advance NLRP3 inflammasome-driven neuroinflammation/neurodegeneration disorders, AD included [184]. Hence P3Alus may be therapeutic targets for such ailments. Later studies revealed that Alu-RNAs processing rates are elevated in mouse and human AD brains, tightly correlating with the up-regulated expression of HSF1 (heat shock transcription factor 1), a crucial stress response factor. The increased Alu-RNAs processing rates would fix into active mode the HSF1/Alu-RNA/stress response/cell death-promoting genes (e.g., p53) axis in AD patients [185,186].

This topic is bound to undergo further developments in regard not only to LncRNAs, miRs, and Alu-RNAs, but also to the recently discovered circular RNAs [187].

Table 2 reports details about LncRNAs/miRs and NLRP3 interactions.

Table 2. RNAs modulating brain NLRP3 inflammasome's function.

RNAs	Model	Mechanisms	References
(A) Activation.			
RNAs	**Model**	**Mechanisms**	**References**
LncRNA-Cox2	Murine microglia	↑ § Transcription of NLRP3 and ASC TLR-mediated signaling pathways Autophagy block Microglia activation	[180,181,188]
LncRNA-Meg3	Murine microglia	miR-7a-5 downregulation	[189]
miR-141	Brain tissue of diabetic mice	NF-κB-mediated NLRP3 expression	[190]
Exo-miR-124 Exo-miR-146a Exo-miR-155	LPS-primed N9 microglia cells	↑ TLR4/TLR2/NF-κB axis	[191]
miR-193	Murine brain cortex Murine microglia	↑ Expression of NLRP3, ASC, cleaved caspase-1 and mature IL-1β	[192]
miR-590-3	In silico AD patients' data	Promoted neurons' death via AMPK signaling	[193]
P3Alu-RNAs	Primary human retinal pigment cells	ERK1/2 and NLRP3 activation, neurons' death	[184]
(B) Inhibition.			
RNAs	**Model**	**Mechanisms**	**References**
circRNA_003564	Spinal cord injury (rat model)	↓ § NLRP3, caspase-1, mature IL-1β, Il-18, GsdmD ↓ Pyroptosis	[187]
LncRNA-Meg3	Rat hippocampal neuronal model of temporal epilepsy	PI3K/AKT/mTOR pathway activation	[194]
miR-7	Murine neural stem cells	NLRP3/caspase-1 suppressor	[195,196]
Exo-miR-21	APP/PS1 2xTg AD-model mouse	Improved memory	[197]
miR-22, Exo-miR-22	APP/PS1 2xTg AD-model mouse PC12 cells	Downregulated NLRP3	[198,199]
Exo-miR-23b	Rat model of intracerebral hemorrhage	Antioxidant effects via PTEN/NRF2 inhibition	[200]
miR-29c-3p Exo-miR-29c-3p	PC12 cells AD-model rat	Suppression of BACE1, p-Tau, and pyroptosis via Wnt/β-catenin pathway	[201,202]
miR-152	Microglial BV2 cell Hippocampal neuronal HT22 cell line Rat model of intracerebral hemorrhage	TXNIP-mediated block of NLRP3 activation	[203]
Exo-miR-188-3p	PD-model mouse MN9D dopaminergic neuronal cells	Suppression of NLR3/pyroptosis	[204]
miR-194-5p	Rat model of intra-cerebral hemorrhage	Blocked NLRP3/TRAF6 interaction	[205]
miR-223-3p	Serum samples from PD, AD, and MCI patients, and healthy controls	Negative NLRP3 regulation	[206]
miR-374a-5p	Rat model of hypoxic-ischemia encephalopathy	Suppressor of SMAD6/NLRP3 in microglia	[207]

§ ↑ = increased; ↓ = decreased.

1.4. Brain NLRP3 Inflammasome's Modulation by Extracellular Vesicles (EVs) and Exosomes (Exos)

EVs partake in neuroinflammation-promoting intercellular signaling. Exos are a class of EVs extruded by any cell type. Exos originate in multivesicular bodies, have sizes of 30–100 nm, and bear specific tetraspanin family markers on their membranes. Exos enclose and convey high numbers of functional proteins, lipids, and regulatory RNAs, which affect recipient cells' metabolic activities, proliferation, or death. Hence, nerve cell-released Exos can act as "either friends or foes" to neurons depending upon their cargoes (e.g., growth factors or Aβs or p-Taues) [60,208] (v. Table 2). In a model of microglial BV-2 cells, pyroptosis induced by O_2-glucose deprivation/reperfusion (OGD/R), human mesenchymal stem cells (MSC)-released Exos (huMSC-Exos) increased FOXO3a gene expression, thereby enhancing mitophagy while reducing the levels of NLRP3; cleaved caspase-1, IL-1β, IL-18; GsdmD-N fragments; and pyroptosis. Hence, huMSC-Exos might mitigate human neurons' OGD/R-induced pyroptosis [209]. Consistently, bone marrow MSC-derived Exos (BMMSC-Exos) intravenously injected 2 h after middle cerebral artery occlusion (MCAO) decreased brain infarct volume, NLRP3 protein expression, and neuron pyroptosis. Moreover, BMMSC-Exos administration shifted the ischemia-induced microglial proinflammatory M1 phenotype to the homeostatic M2 [210].

Cui et al. [197] reported that Exos released from hypoxia-preconditioned MSCs (MSC-Exos) downregulated TNF-α and IL-1β, hindered NF-κB and STAT3 (signal transducer and activator of transcription 3) activation, and decreased Aβ peptides levels and senile Aβ plaques, while upregulating anti-inflammatory IL-4 and IL-10, and exo-miR-21, which improved memory and learning in APP/PS1 AD-model mice. In another study, Cui et al. [211] used the CNS-specific rabies viral glycoprotein (RVG) to target intravenously infused Exos released from MSCs (MSC-RVG-Exos) to the cerebral cortex and hippocampi of transgenic APP/PS1 AD-model mice. MSC-RVG-Exos downregulated IL-1β, TNF-α, and IL-6, while upregulating anti-inflammatory IL-10, IL-4, and IL-13.

In summary, the available evidence about EVs' and Exos' beneficial or harmful roles in NLRP3-mediated neuroinflammation is still scanty. A further limitation is that most studies focused on the RNAs conveyed by EVs or Exos. However, EVs or Exos also transport high numbers of different proteins that either promote or hinder neuroinflammation. In fact, Exos from $A\beta_{25-35}$-exposed human cortical astrocytes conveyed significantly increased amounts of p-Taues [212], while Exos from human AD brains transported Aβ oligomers [213].

1.5. Other Brain NLRP3 Inflammasome Regulators

Under any situation, complex sets of endogenous factors control or restrain NLRP3 inflammasome assembly and/or function, trying to reestablish and/or upkeep tissue homeostasis. Zhang et al. [214] strengthened the relevance of the NLRP3 concept by proving that NLRP3 gene knockout or pharmacological blockage improved the course of various inflammatory diseases modeled in rodents. Hereafter we mention relevant NLRP3 regulators.

The zinc-finger protein A20, i.e., TNFAIP3 (TNF-α-induced protein 3), has two functions: it blocks apoptosis and crucially controls microglia function by inhibiting NF-κB activation in CNS physiological and pathological conditions. A20 knockout led to NLRP3 inflammasome's hyperactivation, increasing mature IL-1β secretion and neuroinflammation intensity [215].

Additionally, CD40 (i.e., cluster of differentiation 40) protein, a member of the TNFR superfamily, negatively affected the ATP•TLR4-signaling-mediated NLRP3 inflammasome's activation in microglia. Therefore, it regulated microglia's inflammation-initiating Th17 response triggered by DAMP-induced brain injuries [216].

Mitsugumin-53 (i.e., TRIM-72 or tripartite motif 72) protein partook in damaged plasma membranes repair and inhibited the NLRP3/caspase1/IL-1β pathway and TNF-α expression, thus mitigating neuroinflammation [217]. Conversely, the TRIM-21 protein promoted microglia's pro-inflammatory M1 phenotype polarization that TRIM-21's knockout reversed [218].

Osteopontin is a highly phosphorylated ECM sialoprotein expressed during the sub-acute phase following cerebral infarction. It stimulated microglia's chemotaxis while preventing NLRP3's activation and its sequels [52].

Worth mentioning here is PKR (i.e., protein kinase RNA-activated), a multirole serine–threonine kinase controlling mRNA transcription/translation, protein synthesis, cell proliferation, apoptosis, and brain function, in addition to shielding cells from viral infections. A dysfunctional PKR partook in cancer and neuroinflammation [219]. Moreover, by using wild-type and $PKR^{-/-}$ mouse macrophages, Lu et al. [220] showed that PKR needed to physically interact with NLRP3, NLRC4, and AIM-2 inflammasomes to activate them. However, using LPS-treated $PKR^{-/-}$ bone marrow-derived macrophages isolated from different mouse strains, He et al. [221] reported that following stimuli activating NLRP3, NLRC4, and AIM2 inflammasomes' PKR activity was critical for nitric oxide synthase-2 (NOS-2) induction, yet dispensable for pro-IL-1β and pro-IL-18 cleavage by caspase-1 [172]. Altogether the divergent results of Lu et al. [220] and Healy et al. [172] show that the animal species or strains investigated do significantly affect the kind of mechanisms activating or inactivating the NLRP3 and other inflammasomes. This adds a remarkable degree of complexity to the topic and stresses the importance of investigating corresponding mechanisms in human neural cells models.

1.6. Brain NLRP3 Inflammasome Inhibitors

Inhibiting the NLRP3 inflammasomes has been a tantalizing enterprise given its potential therapeutic applications in brain diseases. Table 3 lists the reported NLRP3 inhibitors, of which MCC950 is the most popular one in experimental works [222], although it failed in a clinical trial due to off-target toxic effects.

Table 3. Inhibitors of brain NLRP3 inflammasome.

Compound [References]	IUPAC Name	Main Molecular Activity	Main Biological Activity	Experimental Model
17β-Estradiol (E2) [223–225] **See also Box 1**	(8R,9S,13S,14S,17S)-13-methyl-6,7,8,9,11,12,14,15,16,17-decahydrocyclopenta[a]phenanthrene-3,17-diol	Ligand for estrogen receptor-α (ER-α) and -β (ER-β), and for G-protein coupled receptor 1 (GPER1)	↓§ NLRP3, ASC, cleaved caspase-1, IL-1β ↓§ M1 microglia ↑ M2 microglia	Male SOD1(G93A) ALS-model mice Global brain ischemia-model rodents
A43879 [226] **See also Box 2**	3-[[5-(2,3-dichlorophenyl)-tetrazol-1-yl]methyl]pyridine hydrochloride	$P2X_7$ purinergic receptor antagonist	↓ $P2X_7$ receptor signaling ↓ NLRP3	Spinal cord injury-model animal
Adiponectin [227]	Protein	Ligand for Adipo-R1 and Adipo-R2 receptors	↓ NLRP3, IL-1β, IL-18 ↑ Autophagy via AMPK pathway	Intracerebral hemorrhage-model rat
Amifostine [228]	2-(3-aminopropylamino)ethyl-sulfanylphosphonic acid	Protects against the DNA-damaging effects of ionizing radiations and chemotherapy drug-induced ROS	↓ ROS, pyroptosis	Experimental autoimmune encephalomyelitis (EAE)-model rat
α1-Antitrypsin (A1AT) [128]	Protein	Protease inhibitor	↓ $A\beta_{1-42}$-driven NLRP3 activation	Mouse primary cortical astrocytes
Anfibatide [229,230]	Dimeric protein	Antagonist of the glycoprotein Ib IX-V (GPIb) complex	↓ NLRP3/NF-κB axis, cleaved caspase-1 and -3, and Bax ↑ Bcl2	Cerebral HI/R injury-model rat
Atorvastatin [231]	(3*R*,5*R*)-7-[2-(4-fluorophenyl)-3-phenyl-4-(phenylcarbamoyl)-5-propan-2-ylpyrrol-1-yl]-3,5-dihydroxyheptanoic acid	Inhibitor of hydroxymethylglutaryl-coenzyme A (HMG-CoA) reductase	↓ NLRP3/NF-κB signaling axis	Surgery-induced BBB disruption in aged mice
Bay117082 [232]	(E)-3-(4-methylphenyl)-sulfonylprop-2-enenitrile	Calcium channel blocker	↓ ATPase activity of NLRP3	Spinal cord injury-model animal
BPBA [233]	(2-[2-(benzo[d]thiazol-2-yl) phenyl-amino] benzoic acid)	Inhibitor of self- and Cu^{2+}- or Zn^{2+}-induced Aβs aggregation	↓ Aβs aggregation and neurotoxicity ↓ NLRP3 and IL-1β	Aβ-induced paralysis in transgenic *Caenorabditis elegans*

Table 3. *Cont.*

Compound [References]	IUPAC Name	Main Molecular Activity	Main Biological Activity	Experimental Model
Caffeine [234]	1,3,7-trimethypurine-2,6-dione	Antagonist of all adenosine receptor subtypes (A1, A2a, A2b, A3) in the CNS PDE inhibitor	↓ Rapamycin (mTOR) axis and Bax ↑ Autophagy	EAE-model C57BL/6 mice Mouse microglia BV2 microglial cells
Calcitriol [235]	(1R,3S,5Z)-5-[(2E)-2-[(1R,3aS,7aR)-1-[(2R)-6-hydroxy-6-methylheptan-2-yl]-7a-methyl-2,3,3a,5,6,7-hexahydro-1H-inden-4-ylidene]ethylidene]-4-methylidenecyclohexane-1,3-diol	Ligand for vitamin D receptors	↓ ROS, NLRP3, caspase-1, IL-1β, CX3CR1, CCL17, Tbx21 ↓ Spinal cord demyelination	EAE-model C57BL/6 mice
Choline [236]	2-hydroxyethyl-(trimethyl)azanium	Methyl donor Ligand for choline transporters, CTL1 included	↓ NLRP3, Aβs deposition, and microgliosis	APP/PS1 AD-model mice
Dapansutrile (i.e., OLT1177) [237]	3-methylsulfonyl propanenitrile	Direct NLRP3 ATPase inhibitor	↓ Microglia activation and Aβs plaque numbers in the cerebral cortex ↓ IL-1β and IL-6 ↑ Dendritic spine density Successful Phase I clinical trial	APP/PS1 AD-model mice
Dexmedetomidine (Dexm) [96,238]	5-[(1S)-1-(2,3-dimethylphenyl)ethyl]-1H-imidazole	Specific and selective α-2 adrenoceptor agonist	↓ NF-κB and proinflammatory cytokines via miR-340 upregulation ↑ Autophagy	LPS-stimulated BV2 microglia cells
Dihydromyricetin [239]	(2*R*,3*R*)-3,5,7-trihydroxy-2-(3,4,5-trihydroxyphenyl)-2,3-dihydrochromen-4-one	Antioxidant, anti-binge hangover, and anti-cancer activity	↓ NLRP3 ↑ Aβ clearance ↑ Expression of neprilysin ↑ M2 microglial phenotype	APP/PS1 AD-model mice

Table 3. *Cont.*

Compound [References]	IUPAC Name	Main Molecular Activity	Main Biological Activity	Experimental Model
A-68930 [240]	1-(aminomethyl)-3-phenyl-3,4-dihydro-1*H*-isochromene-5,6-diol;hydrochloride	Potent and selective Dopamine D1-like receptor agonist	↓ NLRP3 activation	LPS-induced systemic inflammation mouse model
Bromocriptine	(6aR,9R)-5-bromo-N-[(1S,2S,4R,7S)-2-hydroxy-7-(2-methylpropyl)-5,8-dioxo-4-propan-2-yl-3-oxa-6,9-diazatricyclo[7.3.0.02,6]dodecan-4-yl]-7-methyl-6,6a,8,9-tetrahydro-4H-indolo[4,3-fg]quinoline-9-carboxamide	Dopamine D2 receptor agonist	↑ NLRP3 ubiquitination via cAMP	Neurotoxin MPTP-treated mice
Dopamine [226]	4-(2-aminoethyl)benzene-1,2-diol	Agonist for the five Dopamine receptor subtypes (D1, D2, D3, D4, D5)	↓ IL-1β and IL-18 secretion	Spinal cord injury-model rat
LY171555	(4aR,8aR)-5-propyl-1,4,4a,6,7,8,8a,9-octahydropyrazolo[3,4-g]quinoline;hydrochloride	Specific dopamine D2 receptor agonist		
Quinerolane	(5aR,9aR)-6-propyl-5a,7,8,9,9a,10-hexahydro-5H-pyrido[2,3-g]quinazolin-2-amine	Dopamine D2 and D3 receptors agonist		
EC144 [241]	5-[2-amino-4-chloro-7-[(4-methoxy-3,5-dimethylpyridin-2-yl)methyl]pyrrolo[2,3-d]pyrimidin-5-yl]-2-methylpent-4-yn-2-ol	Selective inhibitor of heat shock protein 90 (HSP90)	↓ IL-1β and IL-18	Peritonitis-model animal
Echinacoside [242]	[(2R,3R,4R,5R,6R)-6-[2-(3,4-dihydroxyphenyl)ethoxy]-5-hydroxy-2-[[(2R,3R,4S,5S,6R)-3,4,5-trihydroxy-6-(hydroxymethyl)oxan-2-yl]oxymethyl]-4-[(2S,3R,4R,5R,6S)-3,4,5-trihydroxy-6-methyloxan-2-yl]oxyoxan-3-yl] (E)-3-(3,4-dihydroxyphenyl)prop-2-enoate	Neuroprotective effects via undefined upstream mechanisms	↓ NLRP3, NF-κB-p65, and ROS	Spinal cord injury-model animal LPS-treated BV2 microglial cells

Table 3. *Cont.*

Compound [References]	IUPAC Name	Main Molecular Activity	Main Biological Activity	Experimental Model
Ellagic acid [243]	6,7,13,14-tetrahydroxy-2,9-dioxatetracyclo[6.6.2.04,16.011,15]hexadeca-1(15),4,6,8(16),11,13-hexaene-3,10-dione	ATP-competitive inhibitor of constitutively active CK2 Ser/Thr protein kinase	↓ caspase-1, IL-6, IL-10, IL-17A, TNF-α, GFAP, and Iba1	EAE-model mouse
Fimasartan [244]	2-[2-butyl-4-methyl-6-oxo-1-[[4-[2-(2H-tetrazol-5-yl) phenyl]phenyl]methyl]pyrimidin-5-yl]-N,N-dimethylethanethioamide	Angiotensin II receptor antagonist	↓ NLRP3/ASC/caspase-1 and NF-κB pathways	Intracerebral hemorrhage-model rat Hemolysate-treated BV2 microglia
Fluoxetine [245]	N-methyl-3-phenyl-3-[4-(trifluoromethyl)phenoxy]propan-1-amine	Serotonin reuptake inhibitor	↓ NF-κB, TLR-4, NLRP3, caspase-1, TNF-α, IL-1β ↓ AChE activity, Aβ, Tau protein, MDA	Depression- and AD-model animals
Ghrelin [246]	(4S)-4-[[(2S)-1-[(2S)-2-[[(2S)-2-[[(2S)-2-[[(2S)-2-[[(2S)-2-[(2-aminoacetyl)amino]-3-hydroxypropanoyl]amino]-3-hydroxypropanoyl]amino]-3-phenylpropanoyl]amino]-4-methylpentanoyl]amino]-1-oxopropan-2- . . . . yl]amino]-5-oxopentanoic acid	Ligand for GHS-R1a receptor	↓ NF-κB/NLRP3 axis, IL-6, COX2, TNF-α, NOS-2, and pyroptosis	EAE-model animal
Glibenclamide [247,248]	5-chloro-N-[2-[4-(cyclohexylcarbamoylsulfamoyl)phenyl]ethyl]-2-methoxybenzamide	Classic K_{ATP} channel blocker	ATP-sensitive K^+ channel inhibitor ↓ NLRP3 ↓ Release of HSP70 ↓ NLRP3, GsdmD-cleavage, ↓ Oxidative stress, demyelination, axon degeneration	Morphine-induced neuroinflammation animal and cellular models Hexanendione-induced neurotoxicity-model animal
HU-308 [249]	[(1R,4R,5R)-4-[2,6-dimethoxy-4-(2-methyloctan-2-yl)phenyl]-6,6-dimethyl-2-bicyclo[3.1.1]hept-2-enyl]methanol	Activator of cannabinoid receptor 2	↑ Autophagy	BV2 microglia cells EAE-model animals

Table 3. *Cont.*

Compound [References]	IUPAC Name	Main Molecular Activity	Main Biological Activity	Experimental Model
Indomethacin [250]	2-[1-(4-chlorobenzoyl)-5-methoxy-2-methylindol-3-yl]acetic acid	Prostaglandin G/H synthase 2 or cyclo-oxygenase (COX) enzyme inhibitor	↓ NLRC4 and NLRP3 genes ↓ IL-1β, caspase-1, and p-Taues	Streptozotocin (STZ)-induced AD-like model
Inzomelid [251]	1-(1,2,3,5,6,7-hexahydro-s-indacen-4-yl)-3-(1-propan-2-ylpyrazol-3-yl)sulfonylurea	Nonspecific and reversible inhibitor of the cyclo-oxygenase (COX) enzyme or prostaglandin G/H synthase	↓ NLRP3	ClinicalTrial.gov NCT04015076
JC124 [252]	5-chloro-2-methoxy-N-[2-[4-(methylsulfamoyl)phenyl]ethyl]benzamide)	Specific inhibitor of expression of NLRP3 and its adaptor protein ASC	↓ NLRP3, ASC, IL-1β, TNFα, NOS-2, caspase-1, and pyroptosis	Traumatic brain injury in male rats
Ketamine [253]	2-(2-chlorophenyl)-2-(methylamino)cyclohexan-1-one	NMDA receptors antagonist	↓ NF-κB, NLRP3, ASC, caspase-1, IL-1β ↑ Autophagy	Depressive-like-model rat
KPT-8602 [254]	(E)-3-[3-[3,5-bis(trifluoromethyl)phenyl]-1,2,4-triazol-1-yl]-2-pyrimidin-5-ylprop-2-enamide	Exportin 1 (XPO1) nuclear transport inhibitor	↓ Exportin 1 ↓ NLRP3/NF-κB signaling axis	LPS-treated macrophages LPS-induced inflammation mouse model MPTP mouse model of PD
Licochalcone B [255]	(E)-3-(3,4-dihydroxy-2-methoxyphenyl)-1-(4-hydroxyphenyl)prop-2-en-1-one	Specific inhibitor of NEK7-NLRP3 interaction	↓ Canonical and non-canonical NLRP3 inflammasome activation	Murine macrophages Mouse models of LPS-induced septic shock, peritonitis, and non-alcoholic steatohepatitis
Manoalide [256–259]	(2R)-2-hydroxy-3-[(2R,6R)-6-hydroxy-5-[(E)-4-methyl-6-(2,6,6-trimethylcyclohexen-1-yl)hex-3-enyl]-3,6-dihydro-2H-pyran-2-yl]-2H-furan-5-one	Inhibitor of NEK7-NLRP3 activating interaction	↓ Canonical and non-canonical NLRP3 inflammasome activation	EAE-model animal

Table 3. *Cont.*

Compound [References]	IUPAC Name	Main Molecular Activity	Main Biological Activity	Experimental Model
MCC950 (i.e., CRID3) [222,260]	1,2,3,5,6,7-hexahydro-s-indacen-4-ylcarbamoyl-[4-(2-hydroxypropan-2-yl)furan-2-yl]sulfonylazanide	Selectively and specifically binds NLRP3 NATCH domain hindering Walker B motif function thereby inhibiting NLRP3 conformational modifications and oligomerization	↓ NLRP3 ↑ Aβ-phagocytic capability of microglia ↓ IL-1β, IL-18, TNF-α, NLRP3, ASC, cleaved caspase-1, Iba1-, and GFAP-positive cells ↑ BDNF and PSD95 expression	APP/PS1 transgenic AD-model mouse LPS + ATP-induced microglia Perioperative neurocognitive disorders-model mice
Mefenamic, Tolfenamic, Flufenamic, Meclofenamic acids [261]	2-(2,3-dimethylanilino)benzoic acid 2-(3-chloro-2-methylanilino)benzoic acid 2-[3-(trifluoromethyl)anilino]benzoic acid 2-(2,6-dichloro-3-methylanilino)benzoic acid	Cyclooxygenase (COX) inhibitors Cl^- channel inhibitors	↓ NLRP3 and IL-1β processing and release	LPS-primed primary bone marrow-derived macrophages
Melatonin [262–266]	N-[2-(5-methoxy-1H-indol-3-yl)ethyl]acetamide	Natural hormone of the pineal gland acting through its receptors	↑ TFEB nuclear translocation ↑ mitophagy ↓ NLRP3, IL-18, IL-6, and IL-1β ↓ ROS ↑ Sirtuin 1 ↑ α7-nAChR-mediated "autophagic flux"	Aβ $_{25–35}$-treated SH-SY5Y cells APP/PS1 AD-model mice Chronic Gulf War syndrome
Metformin (MET) [267]	3-(diaminomethylidene)-1,1-dimethylguanidine	AMP-activated protein kinase (AMPK) agonist	↓ NF-κB signaling pathway ↑ Sirtuin 1 ↓ NLRP3-mediated ECs pyroptosis	LPS-stimulated lung tissues and pulmonary endothelial cells
Milrinone [268]	6-methyl-2-oxo-5-pyridin-4-yl-1H-pyridine-3-carbonitrile	Inhibitor of phosphodiesterase III	↑ cAMP ↓ TLR4/MyD88/NF-κB axis ↓ IL-1β, IL-6, TNF-α ↓ Aβ, p-Tau, ROS	LPS/Aβ-treated BV2 microglial cells APP/PS1 AD-model mouse
Minocycline [269,270]	(4*S*,4*aS*,5*aR*,12*aR*)-4,7-bis(dimethylamino)-1,10,11,12*a*-tetrahydroxy-3,12-dioxo-4*a*,5,5*a*,6-tetrahydro-4*H*-tetracene-2-carboxamide	Caspase-1 negative modulator	↓ TLR-2, MyD88, NLRP3/NF-κB axis, IL-1β	AD-like dementia-model mouse

Table 3. *Cont.*

Compound [References]	IUPAC Name	Main Molecular Activity	Main Biological Activity	Experimental Model
Mitoquinone (MitoQ) [271]	10-(4,5-dimethoxy-2-methyl-3,6-dioxocyclohexa-1,4-dien-1-yl)decyl-triphenylphosphanium	Selectively accumulates inside mitochondria with anti-oxidant action	↓ Mitochondrial ROS, NLRP3 activation, IL-1β, and IL-18 ↑ M2 phenotype microglia	Intracerebral hemorrhage-model mouse $FeCl_2$-treated microglia
N-acetylcysteine [272]	(2R)-2-acetamido-3-sulfanylpropanoic acid	Stimulator of glutathione synthetase	↓ ROS ↑ NRF2-induced NAD(P)H quinone dehydrogenase 1 (NQO1)	Ischemic stroke-model rat
Nafamostat mesylate [273]	(6-carbamimidoyl naphthalen-2-yl) 4-(diaminomethyl-ideneamino)benzoate	Synthetic inhibitor of serine proteases with a wide spectrum of activity	↓ NLRP3/NF-κB signaling ↓ TNF-α, IL-1β, NOS-2, COX-2, IL-18	Stroke-model animal
NT-0796 [274]	unknown	Orally available brain-penetrant NLRP3 inhibitor	↓ NLRP3	ANZCTR.org.au ACTRN126210010828-97
Phenyl vinyl sulfone [275]	ethenylsulfonylbenzene	Cysteine protease inhibitor	↓ NLRP3-mediated IL-1β release	LPS+ATP-treated J774A.1 cells LPS intraperitoneally injected C57BL/6 mouse
Phoenixin-14 [276] **See also Box 1**	protein	Ligand for the multiple function G protein-coupled receptor GPR173	↓ HMGB1-mediated NLRP3 activation ↓ IL-1β and IL-18	LPS-treated mouse primary astrocytes
Pramipexole [277]	6S)-6-N-propyl-4,5,6,7-tetrahydro-1,3-benzothiazole-2,6-diamine	Dopamine-D3 receptors agonist	↑ Autophagy ↓ NLRP3, ASC, cleaved caspase-1 IL-1β, IL-18	LPS+ATP-stimulated primary mouse astrocytes PD-model mouse
Prednisone (PDN) [278]	(8S,9S,10R,13S,14S,17R)-17-hydroxy-17-(2-hydroxyacetyl)-10,13-dimethyl-6,7,8,9,12,14,15,16-octahydrocyclopenta[a]phenanthrene-3,11-dione	Glucocorticoid receptor agonist	↓ NLRP3 activation ↓ TNF-α, CCL8, CXCL10, CXCL16 ↓ astrocytes and microglia activation	Cuprizone (CPZ)-induced demyelination-model mouse
Resolvin D1 [279] **See also Box 1**	(4Z,7S,8R,9E,11E,13Z,15E,17S,19Z)-7,8,17-trihydroxydocosa-4,9,11,13,15,19-hexaenoic acid	Ligand for *N*-formyl peptide receptor-2 and GPR-32	↑ A20 expression ↓ NLRP3/NF-κB axis	Subarachnoid hemorrhage-model rat

Table 3. *Cont.*

Compound [References]	IUPAC Name	Main Molecular Activity	Main Biological Activity	Experimental Model
Sildenafil [280]	5-[2-ethoxy-5-(4-methylpiperazin-1-yl)sulfonylphenyl]-1-methyl-3-propyl-6H-pyrazolo[4,3-d]pyrimidin-7-one	3′,5′-cyclic GMP (cGMP)-specific phosphodiesterase inhibitor	↓ NLRP3 ↓ Hippocampal Aβ_{1-40} and Aβ_{1-42} ↑ Brain cGMP levels	APP/PS1 AD-model mouse
TAK-242 (CLI-095) [103,281]	(R)-Ethyl 6-(N-(2-chloro-4-fluorophenyl)sulfamoyl)cyclohex-1-enecarboxylate	TLR-4 signal transduction inhibitors	↓ TLR-4-NF-κB-caspase-11 axis ↓ NLRP3, IL-1β, and IL-18	Methamphetamine-treated mouse and primary astrocytes Aβ_{1-42}-treated BV2 microglia and HT-22 neurons
1,2,4-TTB [282]	1,2,4-Trimethoxybenzene	Inhibitor of NLRP3 oligomer formation	↓ Nigericin- or ATP-mediated NLRP3 activation	Murine bone marrow-derived macrophages (BMDMs) Primary mouse microglia EAE-model mice
Urolithin A [283]	3,8-dihydroxybenzo[c]chromen-6-one	Gut microflora processed derivative of ellagic acid	↓ NLRP3 activation via mitophagy promotion in microglia	LPS- or MPTP-treated BV2 microglial cells MPTP PD-model mouse
VX-765 [284]	(2S)-1-[(2S)-2-[(4-amino-3-chlorobenzoyl)amino]-3,3-dimethylbutanoyl]-N-[(2R,3S)-2-ethoxy-5-oxooxolan-3-yl]pyrrolidine-2-carboxamide	Competitive inhibitor of ICE/caspase-1 (active metabolite: VRT-043198)	↓ NLRP3/caspase-1/GsdmD pathway	APP/PS1 AD-model mice BV2 microglial cells

§ ↑ = increased; ↓=decreased.

1.7. Brain NLRP3 Downregulation by Officinal Plant Agents/Herbal Extracts

Since time immemorial, plants were and still are the source of drugs helping human ailments. Although extracts of plant body portions are still in use in Traditional Chinese Medicine (TCM), the current more scientific attitude is to find the specific compound(s) of potential therapeutic use. Table 4 reports the most relevant agents and herbal extracts of interest regarding the brain NLRP3 inflammasome.

It is worth noting that save for ginsenoids, artemisinin, and artesunate, all the other hitherto-reported therapeutically promising plant agents/herbal extracts still need in-depth preclinical studies and well conducted clinical trials prior to becoming FDA-approved drugs. On the other hand, altogether the above-listed agents/extracts represent a treasure trove of future therapeutic assets.

Table 4. Brain NLRP3 inflammasome downregulation by officinal plants agents/extracts.

(**A**) Agents.

Natural Compounds and Sources	Chemical Class	Biological Activities	Experimental Models	References
Andrographolide from the roots and leaves of the plant Creat or Green chireta (*Andrographis paniculata* Wall. ex Nees)	labdane diterpenoid	↓§ $P2X_7$ receptor signaling ↓ HMGB1-induced TLR-4-NFκB signaling	LPS-activated mixed glial cells LPS-treated mouse	[285] see also Box 2
Artesunate/Artemisinin from *Artemisiae Iwayomogii* Herba	sesquiterpene lactone	↓ Inflammatory response and neuron death ↑§ Expression of BDNF, GDNF, and NT-3 neurotrophins	Traumatic brain injury-model mouse LPS-stimulated BV-2 microglial cells LPS-treated mouse	[286,287]
Astragaloside IV from *Astragalus membranaceus* (i.e., Huangqi)	pentacyclic triterpenoid	Antioxidant activity	Transient cerebral ischemia/reperfusion (I/R)-model mice	[288]
Baicalin from the root of *Scutellaria baicalensis* Georgi	flavonoid	↓ TLR-4/NF-κB/NLRP3 axis	APP/PS1 AD-model mice LPS/Aβ-stimulated BV2 microglial cells	[289]
Benzyl isothiocyanate from cruciferous vegetables	benzene	↓ NLRP3 activation via mitochondria-generated ROS inhibition ↓ NF-κB signaling	LPS-induced BV2 microglial cells	[290]
Bixin from the seeds of the Achiote tree (i.e., *Bixa orellana*)	apocarotenoid	Suppression of thioredoxin-interacting protein (TXNIP)-NLRP3 activity	EAE-model mouse	[291]
Carnosic acid (CA) **Carnosol (CS)** from *Rosmarinus officinalis*	abietane-type tricyclic diterpenes	↑ KEAP1 (Kelch-like ECH-associated protein 1)/NRF2 (erythroid 2–related factor 2) transcriptional pathway activation ↓ HSP 90 inhibition	APP/PS1 AD-model mice Primary mouse bone marrow-derived macrophages	[292,293]
Cucurbitacin B from *Cucurbitaceae*	tetracyclic triterpene	↓ NLRP3, caspase-1 self-activation, and IL-1β release	Ischemia/reperfusion injury-model rat	[294]
Dehydroisohispanolone diterpene (DT1) from *Ballota hispanica* (Labiatae)	labdane (bicyclic diterpene)	↓ NF-κB and NLRP3 signaling	Nigericin-activated murine bone marrow-derived macrophages	[295]

Table 4. *Cont.*

(A) Agents.

Natural Compounds and Sources	Chemical Class	Biological Activities	Experimental Models	References
Demethylene-tetrahydroberberine (DMTHB) from *Berberis vulgaris*, *Berberis aristata*	alkaloid	↓ NLRP3 inflammasome's activation ↓ IL-6 signaling	AD-model mice	[296]
Esculentoside A from the roots of Indian pokeweed (i.e., *Phytolacca esculenta* Van Houtte)	triterpene saponin	↓ NF-κB, MAPKs and NLRP3 pathways	LPS-activated murine primary microglia cells and BV2 microglia cells	[297]
Gastrodin from rhizome of *Gastrodia elata* Blume	phenolic glycoside	↓ TLR4-NF-κB-NLRP3 axis and microglia-mediated neuroinflammation	LPS-treated rats	[298]
Ginkgolide B (BN-52021) from *Ginkgo biloba* and *Machilus wangchiana*	diterpenoid esters	↓ NLRP3 and microglia-mediated neuroinflammation ↑ NLRP3 autophagic degradation	Aβ_{1-42}-induced BV2 cells LPS-primed BV2 cells senescence-accelerated male mouse prone 8 (SAMP8)	[299,300]
Ginsenosides (Rb1, Rg1, Rg3, Rg5, Rh1, Compound K, Chikusetsusaponin IVa, Gintonin, and **20(S)-Protopanaxatriol)** from *Panax ginseng* C.A. Meyer; *Panax quinquefolius* L. (i.e., American Ginseng); and *Panax japonicus* T. Nees	saponins	↓ NLRP3, NLRP1, AIM-2, and caspase-1 self-activation ↓ brain load of Aβs ↑ soluble (s)APP-α	AD in rodent models Depression-like behavior in rat model Post-traumatic stress disorder-like behavior in rodent model Stroke model High fat diet-model mouse	[301–305]
Isoformononetin from *Cicer arietinum* L. (chickpea)	methoxyisoflavone	↓ NLRP3, NLRP2, ASC, NFκB-p65, IL-1β, caspase-1 proteins, and ROS	Streptozotocin-treated rat	[306]
Isoliquiritigenin from the Chinese herbal medicine Glycyrrhiza (Guo Lao)	isoflavone	↓ NLRP3 ↑ NRF2-induced antioxidant activity	Hippocampal organotypic slice cultures after oxygen/glucose deprivation (OGD)	[307]
Isosibiricin from orange jasmine (i.e., *Murraya exotica* or *paniculata*)	coumarin	NLRP3-inhibition mediated by Dopamine D1/2 receptors	LPS-primed mouse BV-2 microglial cells	[308]

Table 4. *Cont.*

(**A**) Agents.				
Natural Compounds and Sources	**Chemical Class**	**Biological Activities**	**Experimental Models**	**References**
Kaempferol from several herbs in TCM	polyphenol flavonoid	↑ NLRP3 autophagic degradation	PD-model mouse LPS-primed BV-2 microglial cells	[309–311]
β-Lapachone from the Lapacho tree or Jacaranda (i.e., *Tabebuia Avellaneda* Lorentz)	benzochromenone	Antioxidant activity	Multiple sclerosis and AD-model animals	[312]
Lychee seed polyphenols (LSPs) from the *Litchi chinensis* tree	polyphenols	↑ Autophagy via the AMPK/mTOR/ULK1 axis ↑ Tight junctions' expression ↑ LRP1 (i.e., low-density lipoprotein receptor-related protein 1), Beclin 1, and LC-3II proteins	Aβ-induced BV2 microglia cells APP/PS1 AD-model mouse	[313,314]
Mangiferin from the rhizome of *Anemarrhena asphodeloides* Bunge	C-glucoside xanthone	↓ NF-κB and NLRP3 signaling ↓ Microglial M1 polarization	LPS-induced BV2 cells	[315]
Myricitrin from the root bark of the tallow shrub (i.e., *Myrica cerifera* L.)	polyphenol hydroxy flavonoid	↓ NLRP3/Bax/Bcl2 axis NF-κB inactivation Antioxidant activity	Rat model of sepsis-linked encephalopathy Brain HI-model rat	[316,317]
Neferine from the green seed embryos of the lotus plant (i.e., *Nelumbo nucifera* Gaertn)	bisbenzylisoquinoline alkaloid	↓ NLRP3-mediated neuronal pyroptosis	Neonatal HI brain damage model rat PC12 cells	[35]
Nobiletin from *Citrus* L. fruits	polymethoxylated flavonoid	↓ NLRP3 ↑ Autophagy via AMPK/mTOR/ULK1 axis	LPS-treated rat brain and BV2 cells	[318]
Oleocanthal from extra-virgin olive oil	phenylethanoid	↓ NLRP3 ↑ Autophagy via AMPK/mTOR/ULK1 axis	AD-model TgSwDI Mouse	[319]

Table 4. *Cont.*

(**A**) Agents.

Natural Compounds and Sources	Chemical Class	Biological Activities	Experimental Models	References
Oridonin from *Isodon Rubescens* (Hemsl.) H. Hara	(1S,2S,5S,8R,9S,10S,11R,15S,18R)-9,10,15,18-tetrahydroxy-12,12-dimethyl-6-methylidene-17-oxapentacyclo [7.6.2.15,8.01,11.02,8]octadecan-7-one	Binds NLRP3's NACHT domain blocking NEK-7-NLRP3 activating interaction ↓ NF-κB pathway, $A\beta_{1-42}$-elicited neuroinflammation, and pyroptosis	$A\beta_{1-42}$-induced AD mice	[320]
Osthole from the roots of various medicinal plants, including *Cnidium monnieri* L. and *Angelica pubescens* (Japan's Shishiudo).	7-methoxy-8-(3-methylpent-2-enyl) coumarin	↓ NLRP3 ↓ brain load of Aβs	Rat model of chronic cerebral ischemic hypoperfusion	[321]
Purpurin from *Rubia tinctorum* L. **Rhein** from *Rheum rhabarbarum*	anthraquinones	↓ NLRP3, caspase-1 self-activation, and IL-1β release	AD-model animals Perirhinal cortex high-fat-diet-induced animal model	[322]
Quercetin (plant pigment)	flavonoid	Antioxidant activity ↓ NLRP3-pyroptosis-mediated IL-1β release ↑ Sirtuin	LPS-induced primary microglial cells and BV2 cells LPS-induced PD model mouse Depression-model mouse SAMP8 mice	[323,324]
Sinomenine from the roots of the climbing plant *Sinomenium acutum* (Thumb.)	alkaloid	Antioxidant and anti-inflammatory activity	EAE-model mouse	[325]
Thonningianin A from *Penthorum chinense*	ellagitannin polyphenol	↑ NLRP3 autophagic degradation via AMPK/ULK1 and Raf/MEK/ERK axis	In vitro and in vivo AD models, including, *C. elegans*, APP/PS1 mice, BV-2 cells, and PC-12 cells	[119]
Withaferin from Indian ginseng (i.e., *Withania somnifera*)	steroidal lactone	↓ Gene expression of NF-κB and associated neuroinflammatory molecules	SH-SY5Y cells transfected with APP plasmid (SH-APP)	[326]

Table 4. *Cont.*

(**B**) Herbal Extracts.

Herbal/Fruit Extract	Source	Biological Activity	Experimental Model	References
Açaí extract	Berries of the *Euterpe oleracea* Mart. palm tree	Antioxidant activity	LPS- or nigericin-activated microglia (EOC 13.31) cells	[327]
***Crysanthemum indicum* extract (CIE)**	TCM (main components: chlorogenic acid, luteoloside, and 3,5-dicaffeoylquinic acid)	Antioxidant activity ↑ TrkB/Akt/CREB/BDNF and Akt/Nrf-2/ARE axes	H_2O_2-induced oxidative toxicity in hippocampal HT22 neuronal cell line	[328,329]
Glycyrrhiza (Guo Lao)	TCM (main components: licochalcone, isochalcone A, echinatin, isoliquiritigenin, and glycyrrhizin)	↓ NLRP3, TNF-α, IL-1β, and IL-18 ↑ AMPK/NRF2/antioxidant response element (ARE) signaling	LPS-induced chondrocyte pyroptosis LPS-induced macrophage cells Ischemic brain damage-model animal	[307,330]
Kutki	Ayurvedic medicine from rhizomes and roots of *Picrorhiza kurroa*	↓ NLRP3 and BACE-1 expression	5xFAD-model mice	[331]
Pien-Tze-Huang	TCM, including *Radix et Rhizoma Notoginseng*, *Moschus*, *Calculus Bovis*, and *Snake Gall*	↓ NLRP3 ↑ Autophagy via AMPK/mTOR/ULK1 axis	LPS-induced BV2 microglial cells cerebral ischemia/reperfusion impaired rats	[332]
Tojapride	TCM (main components: *Cyperus rotundus* L. (i.e., *Nagar motha* in India), *Perilla frutescens* L. (i.e., Basionym), and *Aurantii Fructus Immaturus* L., the natural flavanone glycosides *Naringin* and *Neohesperidin.*	↓ CaSR-mediated NLRP3 inflammasome's activation	Esophageal epithelial cells (reflux esophagitis)	[333] see also Box 1
Xingxiong	Extract from *Ginkgo biloba* L. or *Ginkgo folium* L. and tetramethylpyrazine sodium chloride	↓ NLRP3 ↑Akt/NRF2 axis	Focal cerebral I/R damage	[334]
Ze Lan	Rhizomes or rootstalks of *Lycopus lucidus*	↓ NLRP3	H_2O_2-induced oxidative injury in rat embryo cortical neurons	[335]

§ ↑ = increased; ↓ = decreased.

2. NLRP3 Inflammasome in Brain Acute Injuries

Glial NLRP3's role is controversial in HI/OGD (oxygen–glucose deprivation)-model animals. Denes et al. [336] reported that plasma IL-18 levels and brain infarction volume were alike in both wild-type and NLRP3-shRNA-silenced mice. Therefore, NLRP3's down-regulation was not as neuroprotective as expected because other inflammasomes took over and functioned in NLRP3's stead. In fact, after shRNA-induced NLRP3 depletion, OGD significantly increased AIM2 inflammasome's expression while NLRC4's expression did not change in BV-2 microglial cells.

Conversely, Yang et al. [337] showed that in newborn mouse astrocytes HI and OGD activated TRPV1 (transient receptor potential vanilloid 1), a non-selective cation channel of the TRP family. Next, the TRPV1 signaling drove the JAK2-STAT3 pathway, which mediated NLRP3 inflammasome's activation and increased IL-1β levels. Notably, in HI- and OGD-exposed TRPV1$^{-/-}$ mouse astrocytes, JAK2 and STAT3 activation and IL-1β upregulation were less intense. Interestingly, this study revealed different cell type-related timings of NLRP3 activation elicited by HI/OGD. In newborn mouse astrocytes of the hippocampus, striatum, and thalamic habenula, NLRP3's activity increased by 3 h, while in microglia it was insignificant at 3 h but increased remarkably by 72 h. Then again, Schölwer et al. [338] showed that OGD completely inactivated phagocytic activity in wild-type BV-2 cells, while HI restored phagocytic activity in NLRP3-shRNA-depleted BV-2 cells. Therefore, the authors posited that NLRP3 plays a minor replaceable role in the OGD-elicited neuroinflammation, at least in microglia. Conversely, an anti-inflammatory pleiotropic cytokine, IL-10, hindered NLRP3 activation in microglia by increasing STAT-3's function, which stifled the transcription/translation of pro-IL-1β and mature IL-1β production [339].

Relevant to this topic is IL-33, another IL-1 family member playing major pleiotropic roles in normal and pathological conditions [340]. In neonatal mouse astrocytes, IL-33 expression markedly increased by 24 h after a cerebral HI episode. Exogenously administered IL-33 did mitigate brain infarction volume by one week after the HI event. Astrocytes' basal expression of ST2 (or suppressor of tumorigenesis 2), the IL-33 receptor, was intense and after HI exposure increased further. Conversely, a ST2 shortfall worsened the HI-elicited brain infarction. The IL-33•ST2 signaling-activated pathways mitigated astrocytes' HI-elicited neuroinflammatory response and apoptosis. Moreover, in vitro IL-33-treated murine astrocytes released neurotrophic factors, which protected HI- and OGD-exposed neurons' viability [341]. Besides, administering IL-33 plus MCC950 and antimalarial drugs improved the outcome in a model of murine cerebral malaria [342] in which the *Plasmodium falciparum* overgrew inside the cortical capillaries, diffusely obstructing blood flow.

Franke et al. [36] showed that following stroke's onset, the early up-regulation of the NLRP3 inflammasome occurred in neurons, glia, and vascular endothelia, leading to blood–brain barrier (BBB) breakdown. Consistently, NLRP3 inhibition hindered endothelial pyroptosis induced by the thrombolytic agent rt-PA (or tissue plasminogen activator), thus preserving the BBB's integrity [11]. Similarly, NLRP3-inhibitor MCC950 protected brain endothelial cells from rt-PA's toxic effects in an in vitro HI-exposed BBB model [343]. Additionally, NLRP3's knockout alleviated the NF-κB pathway-mediated brain damage in a middle cerebral artery occlusion (MCAO)-induced focal ischemia mouse model [344]. Moreover, lithium (Li^+), the archetypal mood stabilizer, also impeded HI/R-induced NLRP3 inflammasome activation, and by stimulating STAT3's function improved motor behavior, cognition, and depression [345].

Figure 1 sums up the main signaling pathways involving NLRP3 in acute brain injuries.

Finally, electroacupuncture (EA) exerted analgesic effects by suppressing NLRP3 inflammasome function in the spinal dorsal horn of mice [346]. Moreover, EA at the skull's *Shenting* (DU24) and *Baihui* (DU20) acupoints attenuated cognitive impairment in rats with brain HI/R injury by regulating endogenous melatonin secretion through alkylamine N-acetyltransferase synthesis in the epiphysis. Next, melatonin acted neuroprotectively by blocking NLRP3 activation via upregulating mitophagy-associated proteins [347].

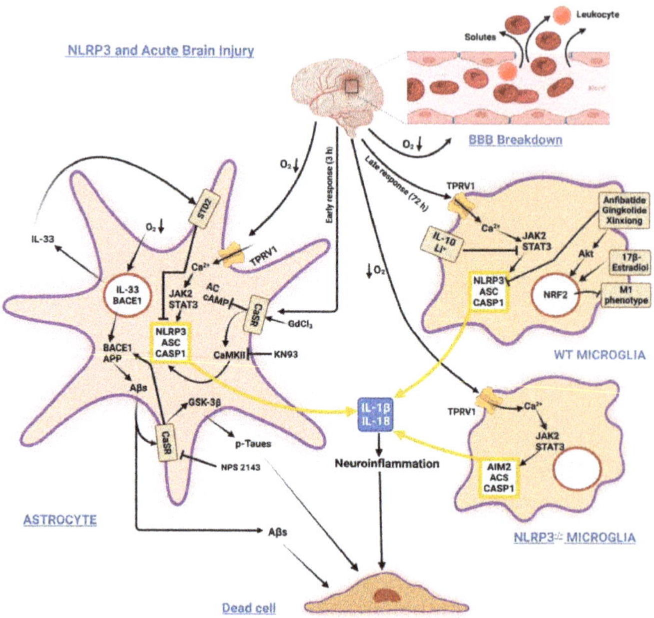

Figure 1. Schematic illustration of stressors and factors inducing/modulating NLRP3 inflammasome's activation and its sequels in astrocytes and microglia under acute injuries due to hypoxic ischemia, stroke, and hemorrhage. Left: Astrocyte's prompt response. Acute O_2 tension fall activates Ca^{2+} influx through TPRV1 channels, triggering the JAK2/STAT3 axis and NLRP3 inflammasome activation. It also increases BACE1 and IL-33 gene expression. Over-released IL-33 binds its STD2 receptor, whose signaling mitigates NLRP3 activity. Later, BACE1 increased activity overproduces Aβs. Extracellularly released excess Aβs bind and activate CaSR signaling, which contributes to NLRP3 inflammasome activation by reducing cAMP levels and activating CaMKII. Aβ•CaSR signaling also increases BACE1 and GSK-3β activities, driving the over production of Aβs from APP and p-Taues, which are both intracellularly accumulated and extracellularly released. CaSR NAM (Calcilytic) NPS2143 and CaMKII inhibitor KN93 suppress Aβs•CaSR signaling noxious effects (see for more details Box 1). Top right: Late wild-type microglia response. The NLRP3 activation is blocked by various agents, which activate via Akt the expression of NRF2 transcription factor. NRF2 activity reduces the M1 (proinflammatory) fraction of microglia. Bottom right: In a model of NLRP3 full-knockout microglia Ca^{2+} influx activates in NLRP3 stead the AIM2 inflammasome's signaling, the upshot being the same, i.e., the overproduction/release of IL-1β and IL-18 [336]. A yellow frame encloses the assembled inflammasomes, while nuclear envelopes are orange colored. Abbreviations: Aβs = amyloid-β peptides; AC = adenylyl cyclase; AIM2 = absent in melanoma 2 inflammasome; Akt = protein kinase B; APP = amyloid precursor protein; ASC = apoptosis-associated speck-like protein endowed with a caspase recruitment domain or CARD; BACE1 = β-secretase; BBB = blood-brain barrier; cAMP = 3′,5′-cyclic adenosine monophosphate; CASP1 = caspase-1; CaMKII = Ca^{2+}/calmodulin-dependent protein kinase II; CaSR, calcium-sensing receptor; $GdCl_3$ = gadolinium chloride; GSK-3β = glycogen synthase kinase-3β; JAK2 = Janus kinase 2; KN93 = N-[2-[[[(E)-3-(4-chlorophenyl)prop-2-enyl]-methylamino]methyl]phenyl]-N-(2-hydroxyethyl)-4-methoxybenzenesulfon-amide; NPS-2143 = 2-chloro-6-[(2R)-2-hydroxy-3-[(2-methyl-1-naphthalen-2-ylpropan-2-yl)amino]-propoxy]-benzonitrile; p-Taues = hyperphosphorylated Tau proteins; STAT3 = signal transducer and activator of transcription 3); STD2 = suppression of tumorigenicity 2 (receptor); TPRV1 = vanilloid type 1 receptor/channel; WT = wild-type. ↓O_2 = decrease in oxygen tension. The other arrows show the sequences of molecular events induced by stressors and factors. ⊥ = inhibition.

In conclusion, given the consistent risk that an acute brain injury triggers a chronic neurodegenerative disease entailing a lethal outcome, the therapeutic mitigation or better suppression of neuroinflammation within a brief time lag following the harmful event constitutes a quite valid target to be pursued.

3. NLRP3 Inflammasome in Chronic Neurodegenerative Disease

3.1. Alzheimer's Disease (AD)

AD is the most prevalent human dementia. Under healthy conditions, the NLRP3 inflammasome is inactive in microglia and astrocytes. Halle et al. [348] first showed that Aβ fibrils—AD's main drivers together with p-Taues and neuroinflammation—activate microglia's NLRP3 inflammasome in APP/PS1 AD-model mice. After phagocytosis by primary mouse microglia, $A\beta_{1-42}$ fibrils damaged the lysosomes, which released cathepsin B, activating the NLRP3 (previously named NALP3) inflammasome and IL-1β, TNF-α, and nitric oxide (NO) overproduction. In turn, the activated NLRP3 inflammasome intensified AD neuropathology in vivo well before Aβs senile plaques appeared [348–350]. Heneka et al. [349] also showed that NLRP3 inflammasome's downregulation shifted microglia's polarization toward the homeostatic M2 phenotype, concurrently depleting the brain's Aβs load. Hence, they posited that NLRP3 inflammasome activation remarkably partook in the microglia-mediated persistent neuroinflammation observed in AD-model mice. Consequently, NLRP3's inhibition would be a novel anti-AD therapeutic approach. Consistently, NLRP3-blocking dihydromyricetin [239] or MCC950 [222] promoted the brain's Aβs clearance, increased hippocampal and cortical M2 microglia fractions, and improved memory and cognition in APP/PS1 mice.

Astrocytes are by far the most abundant cell type populating the brain. Hence, any astrocytes' contributions to neuroinflammation are quite relevant to the progression/outcomes of neurodegenerative diseases. ASC is an adaptor protein forming stable NLRP3•ASC complexes acting as inflammasomal activation hubs. Studies using $ASC^{+/-}$ or $ASC^{-/-}$ 5xFAD newborn mice proved that Aβs do activate astrocytes' inflammasome(s). In $ASC^{+/-}$ mice, NLRP3 inflammasome activity was downregulated; concurrently, an upregulated MIP-1α/CCL3 release increased Aβs phagocytosis by lipopolysaccharide (LPS)-primed primary newborn 5xFAD mouse astrocytes. Moreover, in 7–8-month-old $ASC^{+/-}$ 5xFAD mice, Aβs' brain load downfall correlated with upregulated CCL3 gene expression and improved spatial reference memory [351,352]. Furthermore, ASC moieties released from pyroptotic neurons bound extracellular Aβs and cross-seeded Aβs' increase, promoting NLRP3 inflammasome's activation, neuronal pyroptosis, and neuroinflammation. In turn, these effects increased ASC's available moieties, triggering a self-sustaining feedforward vicious loop while undermining microglial Aβs clearance [353].

Murphy et al. [354] showed that exposure to Aβs increased cytosolic cathepsin B's protease activity, which drove NLRP3 inflammasome's activation and IL-1β over release from wild-type rat primary glial cultures. Consistently, the endogenous protease inhibitor α1-antitrypsin (A1AT) reduced $A\beta_{1-42}$-elicited NLRP3's activation and its sequels in primary cortical astrocytes from BALB/c mice [128,222].

More recent investigations using rodent astrocytes confirmed that exposure to $A\beta_{1-42}$ or LPS inhibited the autophagy/lysosome function while activating the NLRP3/ASC/caspase-1/IL-1β pathway. However, the administration of rapamycin or 17β-estradiol (E2) or progesterone rescued autophagic activity while curbing the $A\beta_{1-42}$- and LPS-activated NLRP3/caspase-/IL-1β pathway in the astrocytes. By contrast, 3-methyladenine, a specific autophagy inhibitor, blocked progesterone's neuroprotective effects and drove astrocytes' NLRP3 inflammasome activation and neuroinflammation [355,356].

Here, a mention is in order about the inducible thioredoxin-interacting protein (TXNIP), which partakes in oxidative stress and regulates thioredoxin (TRX), another redox controller. Both the unfolded protein response (UPR) and ER stress also activate TXNIP. Concurrently, UPR activates the IRE-1α (or inositol requiring enzyme-1α) stress sensor pathway, which in turn further increases TXNIP's amounts susceptible of activation [357].

Importantly, TXNIP's function is essential for the increased expression and activation of NLRP3's inflammatory cascade, both in the aging-associated chronic inflammaging, which goes along with senile cognitive decline, and in the hippocampal neurons and microglia of AD brains [66,67,358]. In rodent models of AD, $A\beta_{1-42}$ drove NLRP3 activation and oxidative damage via the formation of TXNIP•Keap1 (Kelch-like ECH-associated protein-1)•NRF2 (nuclear factor erythroid 2-related factor 2) complexes. Exposure to HJ105 or HJ22, both piperine derivatives, or 9-(NXPZ-2) or maxacalcitol, an active vitamin D analogue, directly inhibited the formation of Keap1•NRF2 complexes, upregulated NRF2's nuclear expression, hindered TXNIP-mediated NLRP3 inflammasome activation, and blocked $A\beta_{1-42}$ and oxidative stress noxious effects [359–362].

Figure 2 sums up the main signaling pathways involving NLRP3 in AD.

Notably, ER stress concurs with the depletion of the anti-aging and cognition-enhancing Klotho, FOXO-1, and mTOR proteins. Moreover, proteins partaking in ER stress development —such as BiP (binding immunoglobulin protein), eIF-2α (eukaryotic initiation factor-2α), and CHOP (C/EBP homology protein)—showed heightened levels of expression in the hippocampi of AD brains. Therefore, altogether TXNIP could link the chronic increases in glucocorticoids elicited by a persistent ER stress with AD's enduring NLRP3 activation and neuroinflammation [67,363].

A newly identified gene associated with the risk of AD is *TREML2* (triggering receptor expressed on myeloid cell-like 2), a protein expressed by microglia [364,365]. TREML2 protein expression levels rise along with AD progression in vivo [366] and after LPS stimulation in primary microglia in vitro, both proving TREML2 involvement in microglia-induced neuroinflammation [367]. Then again, Wang et al. [368] showed that LPS stimulation or lentivirus-mediated TREML2 overexpression remarkably upregulated NLRP3 inflammasome activation; IL-1β, IL-6, and TNF-α secretion; and proinflammatory M1-type polarization in microglia of APP/PS1 AD-model mice. Therefore, TREML2 inhibition would be a novel anti-AD therapeutic approach.

Two studies showed that caspase-1-mediated overproduction of IL-1β occurred in brain samples from mild cognitive impairment (MCI) and fully symptomatic AD patients. Hence, in both groups, microglial NLRP3 inflammasome activation advanced AD's persistent neuroinflammation [140,348]. Sokolowska et al. [140] also showed that phagocytosed $A\beta_{1-42}$ fibrils damaged human macrophages' lysosomes, which released cathepsin B into the cytosol, triggering the NLRP3•ASC•caspase-1 inflammasome's oligomerization and activation. Moreover, studies conducted on brain tissue samples from AD patients that had died because of intercurrent systemic infections and APP/PS1 AD-model mice revealed that any added proinflammatory insults intensified NLRP3 inflammasome's assembly/activation and IL-1β, IL-6, and various chemokines release from microglia, astrocytes, and neurons while increasing the brain's Aβs and p-Taues load. Hence, any concurring etiologic factor could worsen neuroinflammation and hasten AD progression in humans [71,369,370].

Saresella et al. [371] reported the occurrence of a significantly upregulated expression of mRNAs encoding for NLRP1; NLRP3; ASC/PYCARD; caspase-1, -5, and -8; pro-IL-1β; and pro-IL-18 in monocytes isolated from MCI or late-stage AD patients. However, both NLRP1 and NLRP3 inflammasomes functioned only in late-stage AD monocytes. Conversely, ASC/PYCARD and caspase-1 expression was normal in early MCI monocytes in which assembled/functional inflammasomes were missing. Hence, concurrently activated NLRP1 and NLRP3 inflammasomes aggravated neuroinflammation only in late AD.

Interestingly, in subjects with autistic spectrum disorders (ASD), Saresella et al. [131] found that both AIM2 and NLRP3 inflammasomes were active, overproducing IL-1β and IL-18. Simultaneously, there occurred an upregulation of the innate immunity suppressor IL-37, a decline of anti-inflammatory IL-33, and a rise in IFABP (intestinal fatty acid-binding protein—an altered gut permeability index). Therefore, multiple inflammasomes are active in both AD and ASD.

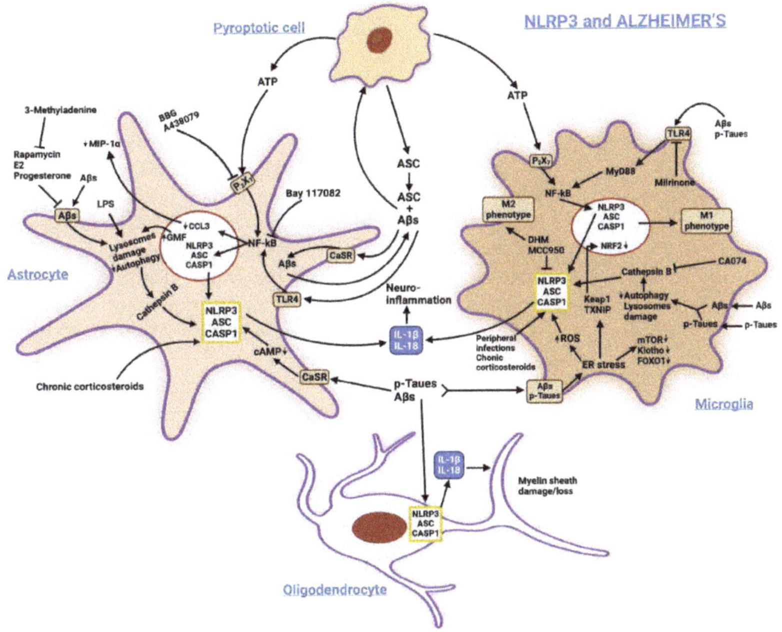

Figure 2. Schematic depiction of stressors and factors inducing/modulating glial cell NLRP3 inflammasome activation and its consequences in AD. Exogenous Aβs, p-Taues, ATP, ASC, IL-1β, and IL-18 interact with cell surface receptors, including CaSR (see Box 1), TL-4, and $P2X_7$ (see Box 2), or are endocytosed to activate NF-κB and NLRP3 inflammasome signaling. They also induce ER stress, release Cathepsin B from damaged lysosomes, and block autophagy, while over-releasing further amounts of Aβs, p-Taues, and inflammatory cytokines. Altogether, they damage myelin sheaths and cause M1 microglial phenotype polarization and neuron and oligodendrocyte pyroptotic death. NLRP3 and receptor inhibitors mitigate the just-mentioned noxious effects. Additionally, the CaSR NAM NPS-2143 blocks Aβs, p-Taues, and IL-6 over production and release and reactivates autophagy (not shown; [181,199,354]). Regarding the roles of other-than-NLRP3 inflammasomes, see [24]. A yellow frame encloses the assembled NLRP3 inflammasomes, while nuclear envelopes are orange-colored. Abbreviations: A438079 = 3-[[5-(2,3-dichlorophenyl)tetrazol-1-yl]methyl]pyridine; Aβs = amyloid-β peptides; ASC = apoptosis-associated speck-like protein endowed with a CARD; Bay117082 = (E)-3-(4-methylphenyl)sulfonylprop-2-enenitrile; BBB = blood-brain barrier; BBG = brilliant blue G; cAMP, 3′,5′-cyclic adenosine monophosphate; CA074 = CAS 134448-10-5; CASP1 = caspase-1; CaSR = calcium-sensing receptor; CCL3 = gene encoding MIP-1α chemokine; DHM = dihydromyricetin; E2 = estradiol; FOXO1 = forkhead box protein O1; GMF, glia maturation factor; JAK2 = Janus kinase 2; Keap1 = Kelch-like ECH-associated protein 1; KN93 = N-[2-[[[(E)-3-(4-chlorophenyl)prop-2-enyl]-methylamino]methyl]-phenyl]-N-(2-hydroxyethyl)-4-methoxybenzenesulfon-amide; LPS = bacterial lipopolysaccharide; MCC950, CAS 210826-40-7; MIP-1α = monocyte chemoattractant protein-1α; mTOR = mammalian target of rapamycin; MyD88 = myeloid differentiation primary response 88; NF-κB = nuclear factor κB; $P2X_7$ = purinergic receptor; p-Taues = hyperphosphorylated Tau proteins; STAT3 = signal transducer and activator of transcription 3; TLR-4 = Toll-like receptor 4; TPRV1, vanilloid type 1 receptor/channel; TXNIP = thioredoxin interacting protein. The small arrows close to a name indicate (↓) decrease, or (↑) increase in levels. ⊥ = inhibition. The other arrows show the sequences of molecular events induced by stressors and factors.

Immunohistochemical studies conducted on samples of temporal cerebral cortex of AD brains showed that the increased expression of NLRP3 inflammasome's constituents, including pro-caspase-1, and of IL-1β and IL-18, co-localized with glia maturation factor (GMF), APOE-ε4, sequestosome 1 (SQSTM1)/p62, LC3-positive autophagic vesicles, and LAMP1, a lysosomal marker. Notably, clusters of GMF overexpressing reactive astrocytes surrounded the amyloid senile plaques. GMF is a highly conserved proinflammatory protein that activates glial cells advancing human neurodegenerative processes. Conversely, in AD-model animals, GMF suppression mitigated the neurodegeneration. Altogether, these results showed that in humans, GMF could intensify NLRP3-driven neuroinflammation while concurrently hampering the autophagosomal pathway clearing Aβs aggregates [349]. Of note, Ahmed et al. [372] and Ramaswamy et al. [352] posited that GMF may advance neuroinflammation in all neurodegenerative diseases.

By sharp contrast, the results of another human postmortem study negated NLRP3 inflammasome function in the brains of advanced AD cases in which astrocyte activation was instead prominent [132].

In addition to Aβs and neuroinflammation, p-Taues are among AD's main drivers. Stancu et al. [373] and Ising et al. [71] proved that a causal link existed between p-Taues and inflammasomes' activation. They showed that following microglial endocytosis and lysosomal sorting, prion-like Tau seeds activated NLRP3 inflammasome signaling in the THY-Tau22 transgenic mouse line, a tauopathy-model animal. Moreover, the chronic intraventricular administration of NLRP3 inhibitor MCC950 significantly thwarted the neuropathology driven by the exogenous p-Tau seeds. Concurrently, NLRP3 suppression decreased the p-Taues levels and hindered their aggregation into neurofibrillary tangles by restraining Tau kinases' activity while increasing that of p-Tau phosphatases [71]. Then again, Jiang et al. [60] showed that p-Tau paired-helical filaments and p-Taues from human tauopathy brains primed and activated IL-1β production via MyD88 and NLRP3•ASC•caspase-1 pathways in primary human microglia. The authors also showed that p-Taues accumulation concurred with elevated ASC and IL-1β levels in postmortem brains of tauopathies patients.

Autophagy is a conserved process by which lysosomes remove dysfunctional cellular components and relevantly regulate NLRP3's role in inflammatory CNS diseases [10,374]. A reduced biogenesis and function of lysosomes/autophagosomes promotes the NLRP3's inflammasome activation driving the neuroinflammatory response in AD-model animals and cultured neural cells. In keeping with this, Zhou et al. [375] showed that overexpressing the transcription factor EB (TFEB), the primary regulator of lysosomal biogenesis, both improved the autophagosomes/lysosomes function and mitigated the neuroinflammation in AD-model cells.

Summing up, NLRP3 inflammasome targeting might hinder AD's etiopathogenetic tripod, i.e., Aβs, p-Taues, and neuroinflammation, and beneficially affect tauopathies too. This is indeed a sensible proposal, but hitherto its real effectiveness in stopping human AD's progression is unproven. Moreover, it does not consider inflammasomes' plurality, potential functional interchangeability, and their different expression levels in the distinct neural cell types.

3.2. Parkinson's Disease (PD)

PD is the second-most-common age-related human neurodegenerative disorder. The progressive spread of PD neuropathology causes motor disturbances and neuropsychiatric disorders (e.g., depression). PD's hallmarks are inclusions rich in misfolded α-synuclein (α-Syn) protein localized at the presynaptic terminals of melanin-rich dopaminergic neurons within the mesencephalic substantia nigra and subcortical corpus striatum. Zhang et al. [376] found the overexpression of IL-1β and IL-18 in cerebrospinal fluid samples from PD patients. Consistently, α-Syn mediated NLRP3 inflammasome activation in cultured human microglia [64]. In PD-model animals, β-hydroxybutyrate, a ketone body, did not inhibit NLRP3 [377] while blocking it in AD [378]. Therefore, α-Syn aggregates trigger chronic neuroinflammation sustained by mitochondrial dysfunction causing ROS overproduction and by unrestrained microglia activation advancing dopaminergic neurons' pyroptosis [379–381].

Figure 3 sums up the main signaling pathways involving NLRP3 in PD.

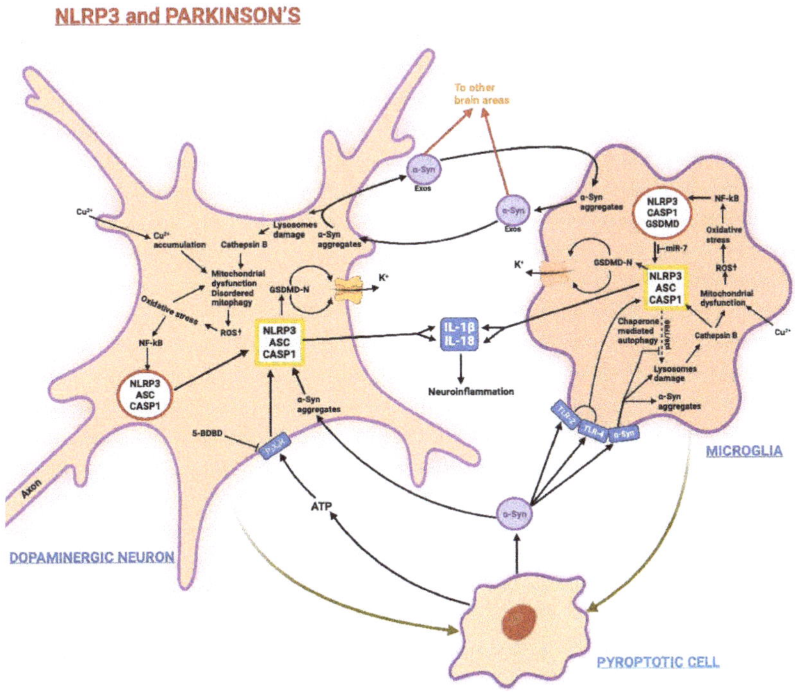

Figure 3. Summary illustration of stressors and factors inducing/modulating dopaminergic neurons' and microglia's NLRP3 inflammasome activation and its consequences in PD. Overproduced α-synuclein (α-Syn) forms cytosolic aggregates (named when massive Lewis bodies) that damage lysosomes releasing cathepsin B, a cysteine protease. The latter interferes with mitochondrial activities causing in sequence dysfunctional mitophagy, ROS surpluses, oxidative stress, NF-κB pathway signaling, and overexpression of NLRP3 inflammasome components, the latter's activation, and its downstream consequences. Exogenous ATP from pyroptotic cells helps activate NLRP3 inflammasome via the $P2X_7$ purinergic receptor signaling (see Box 2 for more details). The upshots are the release of IL-1β and IL-18 and K^+ efflux through pores made of GSDMD-N terminal fragments. α-Syn is also released extracellularly within exosomes that spread and are taken up by neighboring neural cells, expanding the neuropathology, or they circulate in the body fluids thus affecting peripheral tissues. Accumulated Cu^{2+} ions also harm mitochondria contributing to NLRP3 inflammasome's activation. The toxic α-Syn effects are similar in microglia, in which they are mediated by TLR-2 and TLR-4 receptors too. α-Syn also blocks the chaperone-mediated autophagy (CMA) pathway regulated by the p38 MAPK/TEFB axis. Eventually, both nigrostriatal dopaminergic neurons and microglia undergo pyroptotic death. A yellow frame encloses the assembled inflammasomes, while nuclear envelopes are orange-colored. Abbreviations: ASC = apoptosis-associated speck-like protein endowed with a CARD domain; 5-BDBD = 5-(3-Bromophenyl)-1,3-dihydro-2H-benzofuro[3,2-e]-1,4-diazepin-2-one; CASP1 = caspase-1; Exos = exosomes; GSDMD-N = gasdermin D N-terminal fragments; NF-κB = nuclear factor κB; $P2X_7$ = purinergic receptor; p38 MAPK = p38 mitogen activated protein kinase; ROS = reactive oxygen species; TFEB = transcription factor EB; TLR=Toll-like receptor. ↑ROS = increase in ROS levels. ⊥ = inhibition. The other arrows show the sequences of molecular events induced by stressors and factors.

Moreover, Scheiblich et al. [382] reported that the signaling triggered by the binding of α-Syn monomers or, to a lesser extent, α-Syn oligomers to TLR-2 and TLR-5 receptors activated the NLRP3 inflammasome in microglia with no priming needed. Using immunohistochemical and genetic approaches, von Herrmann et al. [383] supplied evidence that dopaminergic neurons are sites of NLRP3 activity in PD. Moreover, increases in NLRP3 inflammasome and NLRP3-dependent pro-inflammatory cytokines were detectable in the peripheral plasma of PD patients, proving NLRP3 inflammasome involvement in PD's pathogenesis [196,384]. The latter authors also showed that miR-7 inhibited NLRP3 gene expression in microglia, thereby reducing microglia activation, neuroinflammation, and nigrostriatal dopaminergic neuron pyroptosis. A patient-based study characterized NLRP3 in the first stages of midbrain nigral neurodegeneration and in the biofluids drawn from PD patients, suggesting that NLRP3 may be both a key inflammation mediator in the degenerating midbrain and a tractable therapeutic target [385]. Moreover, Wang et al. [386] showed that NLRP3 activation and IL-1β and IL-18 maturation occurred in the 6-OHDA (6-hydroxydopamine) neurotoxin-induced PD-model rat. The purinergic $P2X_4$-R siRNA-knockdown or block by the specific antagonist 5-BDBD (5-(3-bromophenyl)-1,3-dihydro-2H-benzofuro[3,2-e]-1,4-diazepin-2-one) counteracted NLRP3's effects, alleviated neuroinflammation, and reduced dopaminergic neuron pyroptosis. Therefore, the authors posited that the ATP•$P2X_4$-R signaling drives NLRP3 inflammasome's activation, which next regulates glial cell activation, nigrostriatal dopaminergic neurodegeneration, and dopamine levels (Figure 3; see also more details and the literature in Box 2) However, here one should be wary of extrapolating these data to PD patients. In PD-model rat brains, NLRP3 inflammasome's activation is not in fact equivalent to that proper of human PD brains. The present understanding of any beneficial effects of antagonizing ATP•$P2X_4$-R's signaling is too limited. Therefore, we need more studies to assess the pathophysiological relevance of nigrostriatal ATP•$P2X_4$R signaling in humans.

Consistently, inhibiting NLRP3 function with MCC950 evoked substantial neuroprotection in the 6-OHDA PD-model rats [387] and in MPTP (1-methyl-4-phenyl-1,2,3,6-tetrahydropyridine)-induced PD-model mice [384]. Moreover, NLRP3 inflammasome's activation in microglia promoted the extracellular release of α-Syn-conveying Exos, which could advance α-Syn spreading in PD brains [388]. Interestingly, copper (Cu^{2+}) accumulation also advanced PD's pathogenic mechanisms by inducing ROS-mediated oxidative stress, activating the NF-κB-p65 pathway in BV2 microglial cells [49]. A persistent intracellular Cu^{2+} buildup upregulated the NLRP3 pathway-related proteins, advancing proinflammatory cytokine secretion and a disordered mitochondrial autophagy (or mitophagy), altogether resulting in dopaminergic neuron pyroptosis. Of note, Cu^{2+} may drive AD neuropathology as well [48].

Finally, despite extensive investigations into the NLRP3 inflammasome-activating mechanisms in the diverse inflammatory brain diseases, their regulatory networks are still unclear in microglia and other neural cell types. Chen et al. [389] showed that NLRP3 is a substrate of chaperone-mediated autophagy (CMA). The p38/TFEB (transcription factor EB) axis regulated NLRP3 inflammasome degradation via CMA, inhibiting the overproduction of proinflammatory cytokines in microglial cells. Furthermore, both p38 and NLRP3 inhibitors could mitigate α-Syn aggregate-induced microglia activation and nigrostriatal dopaminergic neuron pyroptosis. Moreover, Panicker et al. [390] showed that the functional loss of Parkin, an E3 ubiquitin ligase, resulted in the priming and spontaneous activation of the NLRP3 inflammasome in mouse and human dopaminergic neurons, leading to their pyroptosis.

From a clinical standpoint, human PD is quite complex. Therefore, one may conclude that the roles of NLRP3 and other-than-NLRP3 inflammasomes in human PD require further investigations to be fully clarified and integrated to lead to effective therapeutic interventions.

3.3. Multiple Sclerosis (MS) and Experimental Autoimmune (or Allergic) Encephalomyelitis (EAE)

MS is a chronic autoimmune disease of unclear etiology affecting both the brain and spinal cord whose hallmarks include focal (plaque) demyelination and chronic neuroinflammation/neurodegeneration. One accredited theory posits that patients' T cells attack myelin sheath antigens, causing MS. The suggested relationship between MS and the NLRP3 inflammasome has linked autoimmunity with innate immunity and neuroinflammation [391–395]. Moreover, as gain-of-function genetic variants of the NLRP3 (e.g., Q705K) and NLRC4 inflammasomes associate with a more severe MS course, a constitutive NLRP3 inflammasome activation could be a risk factor for clinical MS presentation [396]. Moreover, Vidmar et al. [397] highlighted as pathogenetically important for MS patients the increased burden of rare variants in (i) *NLRP1* and *NLRP3* genes; (ii) genes partaking in inflammasome downregulation via autophagy and IFN-β; and (iii) genes involved in responses to type-1 IFNs (e.g., *PTPRC, TYK2*) and to DNA virus infections (e.g., *DHX58, POLR3A, IFIH1*).

Keane et al. [398] and Voet et al. [215] showed that following NLRP3 inflammasome activation, there occurred an increased IL-1β gene expression within MS demyelination plaques coupled with elevated levels of ASC, caspase-1, and IL-18 in the brains and cerebrospinal fluids of MS patients. Moreover, NLRP3 inflammasome pathway-related components were overexpressed in the blood monocytes isolated from the minor fraction of patients suffering from primary progressive (i.e., with no alternation of pauses and relapses) MS (PPMS), so entailing increased IL-1β production [393,394,399]. These results showed IL-1β as a prognostic factor in PPMS patients and the NLRP3 inflammasome as a prospective therapeutic target. Thus, a specific NLRP3 inhibitor may improve MS histopathology and reduce myelin sheath damage.

According to Farooqi et al. [400], EAE is a proper mouse model for pathogenetic and pharmacotherapeutic studies into human MS molecular mechanisms. In EAE-model mice, NLRP3 inflammasome's activation critically induced T-helper cell migration into the CNS. Next, the activated NLRP3 inflammasome of primed T cells (and microglia) drove the release of proinflammatory cytokines, thus partaking in MS pathogenesis [394,401]. In EAE-model mice the NLRP3 inhibitor MCC950 prevented the conversion of CNS astrocytes to the A1 neurotoxic reactive phenotype otherwise induced via the NF-κB pathway-mediated IL-18 production. Consistently, after the systemic delivery of NLRP3 inhibitor MCC950 axonal injury was mitigated within lysolecithin-induced demyelinated lesions in mice [402,403]. MCC950 also hindered complement C3 protein release from the astrocytes, which would have otherwise impaired hippocampal neuron viability [404]. IFN-β administration did improve this NLRP3-dependent EAE form. Conversely, when ad hoc experimental regimens brought about a NLRP3-independent, more aggressive EAE, the IFN-β treatment was ineffective. A similar NLRP3-independent mechanism might be at work in human MS cases not profiting from IFN-β therapy [405].

In conclusion, there is an intensely felt need to expand the study of NLRP3 and other-than-NLRP3 inflammasomes' role(s) in MS, using human neural cell-based experimental models to achieve a more detailed molecular picture and identify disease-modifying therapeutic targets.

3.4. Amyotrophic Lateral Sclerosis (ALS)

ALS is a devastatingly progressive multifactorial disorder characterized by the primary degeneration of the cerebral motor cortex, brain stem, and spinal cord motoneurons leading to skeletal muscle atrophy and paralysis. ALS patients may also develop cognitive and behavioral changes due to neurodegeneration-affected subcortical areas, e.g., diencephalon's dorsal thalamus. Typically, 90% of cases occur sporadically, and their etiological factors are poorly defined (smoking, violent sports, military service, exposure to insecticides and pesticides). About 10% of ALS cases are familiar due to heritable mutated genes. *SOD1* (superoxide dismutase 1) gene mutations occur in 20% of familiar cases [406].

The current belief is that *SOD1* mutations only trigger ALS onset within motoneurons but elicit only delayed and minor harm [407]. However, in astrocytes and/or microglia, *SOD1* mutations advance ALS progression [408]. TDP-43 (transactive response DNA binding 43 kDa) protein could be another ALS etiological agent as it accumulates in both sporadic and familial cases [409]. TDP-43 forms toxic ubiquitinated aggregates in the cytoplasm of neural cells of both ALS and frontotemporal lobar degeneration (FTLD) patients [410,411]. Neurons and astrocytes can secrete mutated or oxidized SOD1 and TDP-43 as misfolded proteins, which activate microglia by interacting with CD14, TLR-2, TLR-4, and scavenger receptors [412,413]. Thus, exogenous whole or fragmented, wild-type or mutated TDP-43 bound microglia's CD14 cell surface receptor activating AP1 and NF-κB pathways and upregulating NOX2 (SOD-generating NADPH oxidase 2), TNF-α, NLRP3•ASC•caspase-1, and IL-1β release. Importantly, TDP-43 was toxic to motoneurons only in the presence of microglia presence [414]. Using in situ hybridization and immunocytochemistry, Banerjee et al. [415] showed that an upregulated NLRP3 inflammasome occurred in neurons and glia of cognitively impaired ALS patients. Conversely, no differences were detectable between cognitively resilient ALS and healthy subjects.

Figure 4 sums up the main signaling pathways involving NLRP3 in ALS.

Johann et al. [127] showed that an activated NLRP3 inflammasome concurred with elevated levels of caspase-1, IL-1β, and IL-18, particularly in the spinal cord astrocytes of the SOD1G93A ALS-model mice and in the serum and spinal cord tissue of sporadic ALS patients—altogether findings confirming NLRP3 inflammasome's involvement in ALS. Moreover, Kadhim et al. [416] found that IL-18 was upregulated in the cerebral tissue of sporadic ALS patients vs. age-matched controls. Furthermore, Gugliandolo et al. [417] strengthened the concept that neuroinflammation plays a crucial role in ALS by confirming NLRP3 inflammasome activation and its sequels in SOD1G93A ALS-model rats. Immunofluorescent studies conducted on symptomatic SOD1G93A ALS-model mice revealed that NLRP3 and ASC expression intensity increased along with ALS progression, proving NLRP3's involvement in neuron death [418]. Moreover, Michaelson et al. [419] suggested a novel ALS pathogenetic mechanism mediated by the amino acid β-N-methylamino-l-alanine (BMAA), a *Cyanobacteria* product. BMAA is not a protein constituent, but a powerful neurotoxin inducing protein misfolding, NLRP3 inflammasome activation, and proinflammatory cytokine overexpression in spinal motoneurons.

In their work, Van Schoor et al. [420] observed increases in the NLRP3 inflammasome, GSDMD-N fragments, and IL-18 in the motor cortex and spinal cord microglia of human ALS patients, which suggested that an activated NLRP3 inflammasome had triggered the cells' pyroptosis. As compared to controls, in human ALS samples, a reduced array of neurons matched with an increased throng of cleaved-GSDMD-positive microglial cells in the underlying white matter of the premotor cortex. No alike findings were obtained in the human spinal cord. Similar findings were made in the cortex of TDP-43A315T transgenic mice in model ALS and FTLD [421]. In addition, these results stressed the relevance of ROS and ATP generation, both potential therapeutic targets, for microglial NLRP3 inflammasome activation and neuronal pyroptosis, which was confirmed in SOD1G93A-induced ALS-model mice. Importantly, both wild-type and mutant TDP-43 proteins activated the overexpressed NLRP3 and its downstream effects in the microglia of SOD1G93A mice. This proved that NLRP3 is the crucial microglial inflammasome mediating SOD1G93A-induced pyroptosis [65].

Lacking a suitable human microglia model, Quek et al. [422] characterized peripheral blood monocyte-derived microglia-like cells (ALS-MDMi) isolated from ALS patients at various stages. Importantly, ALS-MDMi recapitulated ALS neuropathology hallmarks, i.e., abnormal phosphorylated and non-phosphorylated TDP-43 cytoplasmic accumulation and phagocytosis impairment that paralleled ALS progression; altered neuroinflammatory cytology; DNA damage; NLRP3 inflammasome's activation; and microglia pyroptosis.

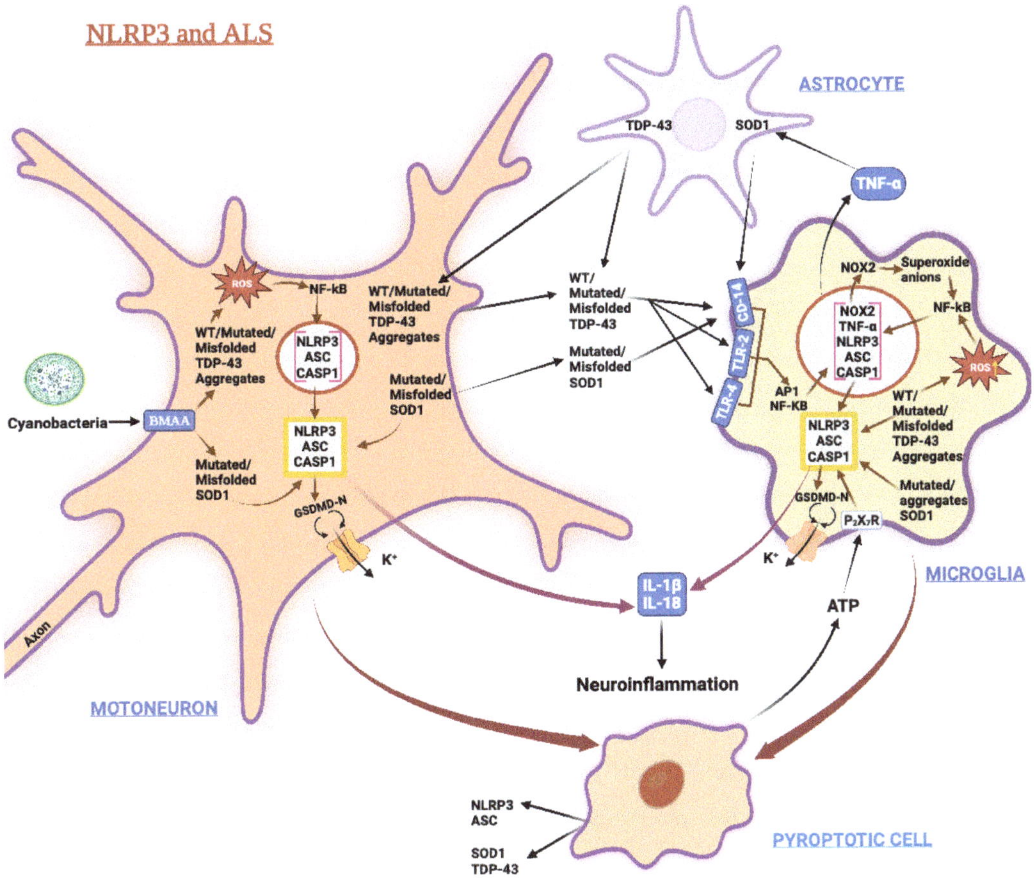

Figure 4. Summary depiction of stressors and factors inducing/modulating motoneurons' and microglia's NLRP3 inflammasome's activation and its consequences in ALS. Mutated/misfolded SOD1 and TDP-43 proteins as variously sized aggregates damage mitochondria, causing in sequence ROS surpluses release, oxidative stress, and NF-κB pathway signaling. These lead to NLRP3 inflammasome's component overexpression, NLRP3 inflammasome activation, over-release of IL-1β, IL-18, K^+ efflux, and eventually motoneurons' and microglia's pyroptosis. Exposure to the toxic BMMA amino-acid released by *Cyanobacteria* worsens the toxic effects of misfolded/mutated SOD1 and TDP-43. Toll-like receptors and CD-14 bind misfolded/mutated SOD1 and TDP-43 activating the AP1/NF-κB axis, and the expression and activation of NLRP3 inflammasome's components. ATP from pyroptotic cells partakes in NLRP3 activation via $P2X_7$ receptor signaling (see Box 2 for details). Astrocytes also release misfolded/mutated SOD1 and TDP3 that are engulfed by other neural cells, thus spreading the neuropathology. Besides ATP, pyroptotic cells also release NLRP3, SOD1, TDP-43, and ASC proteins that contribute to the neuroinflammation. A yellow frame encloses the assembled inflammasomes, while nuclear envelopes are orange-colored. Abbreviations: AP1 = activator protein 1; ASC = apoptosis-associated speck-like protein endowed with a CARD domain; BMAA = β-methylamino-L-alanine; CASP1 = caspase-1; CD14 = cluster of differentiation 14; GSDMD-N = gasdermin D N-terminal fragments; NF-κB = nuclear factor κB; NOX2 = NADPH oxidase 2; $P2X_7R$ = purinergic receptor; ROS = reactive oxygen species; SOD1 = superoxide dismutase 1; TDP-43 = TAR DNA-binding protein 43; TLR− =Toll-like receptor−; TNF-α = tumor necrosis factor-α; WT = wild-type. The arrows show the sequences of molecular events induced by stressors and factors.

It is seemly to consider the studies about NLRP3 and other-than-NLRP3 inflammasomes in human ALS are still in a preliminary phase even in the light of the groundbreaking results reported by Van Schoor et al. [420]. The latter should encourage scientists to delve deeper into the pathogenetic mechanisms of this devastating disease to find novel effective therapeutic approaches.

3.5. Huntington's Disease (HD)

HD is a rare autosomal dominant neurodegenerative disease caused by the unstable CAG repeat expansion in the Huntington (*HTT/IT15*) gene and presenting with motor, cognitive, and psychiatric symptoms [423] When the *HTT/IT15* gene holds 39 to 180 CAG repeats, the translated polyglutamine-containing mutant HTT protein (mHTT) complexes with and disrupts the normal function of several transcription factors, thereby altering the activities of neurons, astrocytes, and microglia. HD's harming mechanisms include mitochondrial dysfunction, excitotoxicity, CREB and BDNF downregulation, and microglia activation, altogether advancing neuronal death by apoptosis, necroptosis, ferroptosis, and NLRP3-linked pyroptosis [424,425].

Various HD-model animals were set up to clarify its molecular mechanisms and to try novel therapeutics for it. The transgenic R6/2 (B6CBA-Tg[HDexon1]62Gpb/1J) mouse line expressing the human *HTT* gene exon 1 carrying 120 $\pm$ 5 CAG repeats is the most popular HD animal model [426]. An upregulated NLRP3 inflammasome and caspase-1 expression already occurred in 13-week-old R6/2 HD-model mice, particularly in striatal parvalbumin interneurons and spiny GABAergic neurons, which preferentially undergo pyroptosis in HD [427]. Poly(ADP-ribose) polymerase-1 (PARP-1) is a nuclear enzyme whose activity is crucial for DNA repair in humans. Olaparib, a PARP-1 inhibitor presently sold as an anti-tumor drug, could also regulate NLRP3 inflammasome activation in the R6/2 HD-model mice. When given from the pre-symptomatic stage onwards, Olaparib mitigated neuronal pyroptosis, neurological symptoms, and neurobehavioral tests results, lengthening the survival of HD-model mice. Therefore, Olaparib could help human HD too [428]. Moreover, Chen et al. [429] showed that NLRP3 inhibitor MCC950 given to R6/2 HD-model mice suppressed IL-1β and ROS overproduction, mitigating neuroinflammation, motor dysfunction, and neuronal pyroptosis, while upregulating PSD-95 and NeuN proteins, and lengthening animals' lifespans. Therefore, inhibition of NLRP3's signaling, and its downstream effects would be therapeutically helpful in HD.

Interestingly, a role in HD etiopathogenesis may be played by galectins (i.e., *"S-type lectins"*)—soluble proteins specifically binding β-galactoside carbohydrates and playing multiple roles in autophagy, immune responses, and inflammation. Siew et al. [430] reported that galectin-3 (Gal-3) plasma levels increased well over healthy controls in HD patients and HD-model mice. In HD-mice, microglia Gal-3 levels increased prior to motor symptom presentation and stayed high while HD progressed. Gal-3 co-localized with microglial lysosomes, blocked the autophagic elimination of damaged endolysosomes, and partook in neuroinflammation via the NF-κB/NLRP3 axis. Gal-3 knockout improved HD-related neuropathology and survival in HD-model mice, showing Gal-3 as a potential therapeutic target. Conversely, Gal-1 and Gal-8 hindered neuroinflammation, promoting neuroprotective effects [431].

HD's rare occurrence is an adjunct hurdle to studies about the roles played in it by NLRP3 and other inflammasomes. However, this circumstance should not discourage attempts to increase our insights in this ailment, both in patients and animal models.

4. Brain NLRP3 and Neurotropic Viruses Infections

Both DNA and RNA neurotropic viruses activate the brain's NLRP3 inflammasome, causing neuroinflammation and sometimes triggering chronic neurodegenerative diseases [75]. Here, we review a few neurotropic viruses playing NLRP3-linked roles in human neuropathology.

4.1. *Zika Virus (ZIKV) Encephalitis*

The Zika Virus (ZIKV) is a single-stranded positive-sense RNA arbovirus of the *Flaviviridae* family (*Flavivirus* genus that also includes Dengue, West Nile, Yellow Fever, and Japanese Encephalitis viruses). ZIKV associates with congenital microcephaly in newborns and Guillain–Barré syndrome, myelopathy, and encephalitis in adults. Tricarico et al. [432] showed that ZIKV infected the U87-MG glioma cell line causing NLRP3 inflammasome activation and IL-1β oversecretion. Consistently, He et al. [82] made the same observations in the brains and sera of ZIKV-infected mice. ZIKV's NS5 protein drove ROS overproduction and NLRP3 inflammasome assembly, both needed for its activation. Conversely, in vitro and in vivo NLRP3 deficiency upregulated type-I IFN and strengthened the host's resistance to ZIKV, confirming NLRP3's role in ZIKV infection [433,434].

4.2. *West Nile Virus (WNV) Encephalitis*

Another *Flavivirus*, the West Nile Virus (WNV), causes an encephalitis entailing neurons' death and elevated IL-1β plasma levels. In a mouse model, WNV infection briskly induced IL-1β synthesis in cortical neurons. However, by cooperating with type-I IFN, the intensified IL-1β•IL-1β-R (receptor) signaling suppressed neuronal WNV replication, reducing the WNV brain load. Therefore, the NLRP3/IL-1β•IL-1β-R pathway regulated neuronal WNV infection and revealed a novel IL-1β antiviral action [435].

4.3. *Japanese Encephalitis Virus (JEV)*

By breaking the BBB, the Japanese Encephalitis Virus (JEV) enters the CNS where it induces a diffuse neuroinflammation. Thus, JEV infection activated (i) a ROS-dependent Src/Ras/Raf/ERK/NF-κB signaling axis in neurons/glia co-cultures [81]; (ii) a ROS/Src/PDGFR/PI3K/Akt/MAPK/AP-1 axis [436] and a PAK4/MAPK/NF-κB/AP-1 axis [437] in rat brain astrocytes; and (iii) via TLR-3 and RIG-I the ERK/MAPKp38/AP-1/NF-κB axis, ROS overproduction, and K^+ efflux in cultured mouse microglia. These effects both triggered NLRP3 inflammasome signaling and polarized microglia toward the proinflammatory/neurotoxic M1 phenotype. In all instances, JEV advanced cytokine overproduction and neural cell pyroptosis [438].

4.4. *Human Immunodeficiency Virus-1 (HIV-1) Encephalitis*

The immunosuppressive Lentiviruses efficiently infect macrophages and lymphoid cells. Human Immunodeficiency Virus-1 (HIV-1) belongs to the Retroviridae family (*Lentivirus* genus). Burdo et al. [439] showed that during the primary infection, HIV-1 productively infects brain macrophages and microglia. Studies using primary human microglia showed that IL-1β was released after HIV-1 infection. Walsh et al. [440] proved that HIV-1 infection induced an NLRP3 inflammasome-dependent ASC translocation, caspase-1 activation, and mature IL-1β release from cultured microglia. The authors highlighted the need to analyze the inflammasome inhibitors' effectiveness as novel therapeutics for HIV-1/AIDS.

4.5. *Viroporin Proteins*

Various RNA viruses, including *Coronaviridae*, express the viral-replication-indispensable small viroporin proteins. Being liposoluble, viroporins assemble hydrophilic transmembrane pores, allowing ions and/or small solutes to bidirectionally migrate along their electrochemical gradients. Viroporin activity could act as the "second signal" by increasing $[Ca^{2+}]_i$ or lowering the cytosolic pH due to H^+-releasing ion channel activity in the lysosomal acidic compartment [441].

4.6. *Encephalomyocarditis Virus (EMCV)*

The Encephalomyocarditis Virus (EMCV) of the *Cardiovirus* genus (Picornaviridae family) is a non-enveloped, positive single-stranded RNA virus. Via an unclear sensing mechanism, the NLRP3 inflammasome detects EMCVs [442,443]. In this regard, Ito et al. [89] reported that by releasing Ca^{2+} from intracellular stores into the cytosol, ECMV's viroporin ORF2b (or open reading frame 2b) triggered NLRP3 inflammasome activation.

4.7. SARS-CoV-2 Encephalitis

SARS-CoV-2 belongs to the β-*Coronavirus* genus (Coronaviridae family, also including 2003 SARS-CoV and 2012 MERS (Middle East Respiratory Syndrome)-CoV). SARS-CoV-2 is an enveloped single-stranded positive-sense RNA virus causing the COVID-19 (Coronavirus Disease 2019). The virus infects a wide spectrum of cell types. In the presence of Ca^{2+}, SARS-CoV-2's spike S1 glycoprotein binds ACE2 (angiotensin converting enzyme 2) and CD147 (cluster of differentiation 147) proteins, promoting virus endocytosis. Moreover, SARS-CoV-2's envelope (E) protein binds TLR-2, which also helps promote AD and PD [444]. Earlier epidemics proved Coronaviruses' neuroinvasive capability in humans [445,446]. SARS-CoV-2 infects neurons, astrocytes, microglia, and the BBB's endothelial cells [447,448]. Notably, microglia and astrocytes are major sources of proinflammatory cytokines. Moreover, Sepehrinezhad et al. [449] found SARS-CoV-2 virions in the cerebrospinal fluid of COVID-19 patients presenting severe neurological symptoms previously affected or unaffected by neuropathologies and in "long COVID" patients [450]. However, in the healthy CNS, ACE2 expression is weak, prevailing in the brainstem's respiratory centers—which explains the high prevalence of respiratory distress in COVID-19 patients [451]. However, uninfected AD patients showed upregulated ACE2 expression in the temporal and occipital neocortex and hippocampal CA1 subfield archicortex [452]. This ACE2 overexpression could advance SARS-CoV-2 infection in the same AD-hit inflamed areas, thus contributing to the high COVID19 mortality rates in aged AD patients [451].

Hitherto, SARS-CoV-2's priming triggers are uncertain. Theobald et al. [453] showed that S1 spike glycoprotein initiated NLRP3 inflammasome activation. Other SARS-CoV-2 proteins—i.e., S, N, E, and the pore-forming viroporins ORF3a and ORF8—are also NLRP3 activators by causing K^+ efflux and mitochondrial ROS over-release [83–88]. Moreover, Xu et al. [454] proved that viroporin ORF3a primed and activated the NLRP3 inflammasome through both ASC-dependent (canonical) and ASC-independent (noncanonical) pathways.

Notably, COVID-19 infection triggers a severe innate immune response producing elevated levels of multiple cytokines ("cytokines storm") and inflammatory mediators (e.g., IL-1β, IL-2, IL-2-R, IL-4, IL-10, IL-18, IFN-γ, C-reactive protein, GCSF (granulocyte colony-stimulating factor), IP10, MCP-1, MIP-1α, and TNF-α). BV2 microglial cells exposed to SARS-CoV-2's S1 spike glycoprotein expressed elevated levels of IL-1β, TNF-α, IL-6, NO, NLRP3, NF-κB signaling, and caspase-1 activity [88,455]. These cytokines cross the BBB inducing leukocyte infiltration, mitochondrial dysfunction, neuroinflammation, and neurons' pyroptosis [442]. Interestingly, a mix of melatonin, vitamin C, and Zn^{2+} inhibited SARS-CoV-2-driven inflammasome activation, hindering the cytokine storm in animals [456].

Additionally, Ding et al. [457] proved that hypercapnia enhanced NLRP3 inflammasome activation and IL-1β expression only in hypoxic BV-2 microglia cells. Therefore, the hypercapnia resulting from lung-protective ventilatory strategies used in acute respiratory distress syndrome (ARDS) patients may lead to neuroinflammation and cognitive impairment via a microglial NLRP3/IL-1β-dependent mechanism.

Based upon the above findings, Heneka et al. [458] posited that NLRP3 inflammasome activation during COVID-19 heightens the risk for the later development of chronic neurodegenerative diseases. Independent clinical and epidemiological investigations indicated that SARS-CoV-2 infection and the ensuing "long COVID" tightly relate to the onset of AD, PD, prion disease (PrD), and other ailments, particularly in patients in advanced age or suffering from intercurrent illnesses (CVD, T2DM, hypertension, other neurological disorders) or severe/fatal COVID-19 [459–461]. Even more alarming, the receptor-binding domain of SARS-CoV-2's S1 spike glycoprotein presents prion-like sequences. The latter diverge among viral variants, show a different affinity for ACE2, and promote immune-evasion, protein clustering, and protein aggregates' "seeding". The upshots would include prion-like proteins spreading, progressive dementia, or fast-evolving CJD [462–464].

Obviously, here we have considered only some of the known neurotropic viruses. The field of human brain-infecting viruses is more variegated and might also further expand in the future. Our knowledge about viral neuropathology is, we must admit, limited, particularly because viruses can target all stages of human life, from the uterus onward, with different age-related upshots. There is also a field that for the sake of brevity we omitted considering, i.e., the interactive relations between oncogenic viruses and inflammasomes, which deserves attention because of its potentially significant reflections on therapeutic outcomes.

5. Comments and Future Perspectives

An old dictum states that every disease starts with an inflammation. The prevalence of neuroinflammatory disease has been epidemically rising because of a lengthened lifespan and of little-appreciated toxic, environmental, and lifestyle-linked factors. To worsen this bleak situation, acute brain illnesses (e.g., stroke, hemorrhage, infection) too can trigger chronic neuroinflammation/neurodegeneration in a significant fraction of patients [465]. A steadily growing literature attests that NLRP3 inflammasome activation in CNS microglia and circulating monocytes plays a pivotal role in promoting the neuroinflammation driven by a host of etiologic factors (q.v. Table 1), potentially advancing the progression of neurodegenerative diseases [27,466,467]. Conversely, NLRP3's roles in the other neural cell types (i.e., neurons, astrocytes, and oligodendrocytes) [3,468–470] and in CNS pericytes and endothelial cells [126,471] have received less attention, probably because such cells preferentially express other types of inflammasomes. In fact, NLRP3 activity in such cells is modest and/or is the object of controversy, particularly in astrocytes, although NLRP3's inhibition still gives some therapeutic advantage. Moreover, these same neural cell types more intensely express various other-than-NLRP3 inflammasomes. The latter can also exert significant neuroinflammation-sustaining effects, as specific NLRP3 inhibitors do not hinder other-than-NLRP3 inflammasomes' activities [24]. We previously reviewed the known roles of various other-than-NLRP3 inflammasomes in human brain disease [24]. That work inspired us to delve deeply also into the role(s) of the brain's NLRP3 inflammasome. Indeed, the NLRP3-related extensive research works herein reviewed shows the high complexity of both the regulatory mechanisms involved and of the physiological, pathological, and ethnic/pharmacological factors that promote or hinder its activation. Particularly the abundance of blocking or preventative factors, many of them identified over millennia by TCM, bodes well for future therapeutic modulations of NLRP3 activity in various pathological settings. Various reports showed that particularly inhibiting microglial NLRP3 function exerted beneficial effects in rodent experimental models of human neurodegenerative illnesses. These favorable outcomes inspired and still inspire the opinion that therapeutically targeting the NLRP3 inflammasome will mitigate or stop both acute and progressive human neuroinflammatory diseases [472–474]. As just mentioned, despite or thanks to the intricacies of NLRP3 inflammasome's activating mechanisms, there are plenty of agents modulating its activity (Tables 2–4). At present, many small molecules are undergoing pharmaceutical research/development as novel candidate drugs targeting the NLRP3 inflammasome in various diseases [274]. At least five companies have started ad hoc clinical trials, of which Inflazome and NodThera have reported Phase I positive results of their brain-penetrating NLRP3 inflammasome inhibitors (Inzomelid [251] and NT-0796 [274], respectively), expecting to use them to treat central and peripheral nervous inflammatory diseases. These discoveries have even raised the possibility of a common cure for all or at least some human brain diseases. Moreover, Lupfer and Kanneganti [21] reported the existence of inflammasomes, such as NLRC3, NLRP6, NLRP12, and NLRX1, which hinder NF-κB pathway activation, thereby mitigating or switching off the incumbent or ongoing neuroinflammation. Such *"anti-inflammasomes"* deserve more consideration because in a hopefully not too far future, their pharmacological activation by proper means (yet to be established) could be a valuable therapeutic asset that will switch off neuroinflammation through physiological mechanisms.

Therefore, the intuitive conclusion is that reality is more intricate than it might appear at first sight. Furthermore, uncertainties and controversies about the etiological mechanisms driving human neurodegenerative diseases help confound the picture, as do other problems that we will briefly discuss below.

(*i*) *Are inflammasomes functionally interchangeable?* Hitherto the interplays that might occur between or among the distinct inflammasomes expressed by each human neural cell type remain mostly undefined. Yet, it is necessary to clarify them to better assess the therapeutic impact of NLRP3 inflammasome inhibitors. Denes et al.'s [336] study results in mice called for caution, as they showed that inflammasomes (e.g., AIM2) can functionally overtake a blocked NLRP3 (Figure 1). A (partial) solution to this problem might entail targeting the ASC protein, which would hinder the activation of all canonical inflammasomes instead of those of NLRP3s only [475]. The inflammasomes' noncanonical activation problem will persist but might be a minor one.

(*ii*) *The species difference problem.* Significant genomic differences apart, not all organs of humans and mammals are morpho-functionally alike. Acceptable similarities exist with liver, kidneys, and lungs. Yet, considering the CNS, while the human cerebral cortex consists mostly of a non-olfactory six-layered *neocortex*, the widely used rodent models have a less developed, structurally simpler, and mostly olfactory cortex. Moreover, fundamental cytological divergences in size, shape, connections, and functions distinguish the diverse types of neural cells of the human cortex from their rodent counterparts [476]. Human brain's molecular regulatory mechanisms, e.g., those involved in receptor signal transduction [133] and inflammasome regulation [24,27,477] (see also Boxes 1 and 2), also remarkably diverge from those of rodents. Moreover, human neurodegenerative diseases do not plague rodents in nature. Importantly, in rodent models of human neurodegenerative diseases, the astrocytes undergo an early death—which justifies the often-little attention paid to them—while neurons keep surviving. Conversely, human neurodegenerative diseases kill neurons first, while astrocytes survive and help advance the neuropathologies. Hence, a tight genomic, proteomic, and bio-pathological conformity between animal and human brains is lacking [478,479]. Although brilliant and highly praiseworthy, the manifold animal models of human neurodegenerative diseases in existence cannot surmount such inter-species differences [480]. A quite low animal-to-human translation rate of brain disease-targeting drugs has been persisting for decades, being ascribed to preclinical studies' faults in "internal consistency" (e.g., design flaws, uncontrolled bias) and/or "external consistency (i.e., animal models pre-testing). As a long trail of clinical trial failures shows, it is difficult to safely predict the effectiveness in humans of drugs pre-tested with favorable results in transgenic animal models [481]. Procedures involving animal models were necessary when nothing or truly little was known about human brain diseases. Now we know much more, albeit not yet enough. Moreover, in recent decades, the legislative/bureaucratic requirements to evaluate novel drugs have become increasingly burdensome to hinder the use of inadequately tested therapeutics. This trend has become stronger after rare events in which properly approved drugs unexpectedly elicited adverse reactions in the patients [482]. Moreover, the repurposing for neurodegenerative diseases of drugs previously evaluated for other ailments in clinical trials is not so easy to do, which precludes the faster testing of potentially useful drugs [483]. Hence, it would be wise to introduce some procedural changes. Animal and/or in silico studies should still help preselect lead drugs. Next, preclinical human untransformed neural cell models in vitro would allow for the assessment of the latter [24,141,212,484] prior to any clinical trial assessment. On rare occasions, animal studies might even be skipped in favor of preclinical human model studies [24,141,212,484]. Human neural cells models will help clarify specific etiopathogenetic mechanisms while supplying safer predicting information about effective drug benefits in clinical settings.

(*iii*) *Symptomatic and/or etiologic therapies?* Hitherto, no causal "brain disease modifying" therapies are available for human neurodegenerative diseases. An exception may be the just reported promising effects of Lecanemab, a humanized IgG1 monoclonal antibody binding soluble Aβ protofibrils. After 18 months, Lecanemab reduced brain amyloidosis and slowed cognition decline in early-stage AD patients vs. the placebo-given group. However, Lecanemab also caused collateral brain swelling and/or hemorrhage in some patients, particularly in case of APOE-ε4 homozygotes or anticoagulant therapy [485]. Hence, while Lecanemab's results confirm that Aβs play a key pathogenetic role in human AD, further studies will prove its etiologic or symptomatic value regarding Aβs/p-Taues' overproduction and accumulation and inflammasomes' activity.

6. Conclusions

In recent years, neuroinflammation has been attracting a lot of attention, particularly concerning one of its mediators, i.e., the NLRP3 inflammasome. In the present work, we systematically review the huge and still mounting evidence related to both NLRP3's involvement in human and animal models of acute and chronic brain diseases, and its many functional activators and inhibitors so far known. Unquestionably, no field expert should disregard the NLRP3 inflammasome, as it is intensely expressed by microglia and circulating monocytes. However, here we wish to stress the indisputable fact that human and animal neural cells of all types, whose morphologies and functions significantly diverge, also express many other inflammasomes and various "*anti-inflammasomes*"—the latter being tasked with mitigating neuroinflammation. Moreover, the so-called primary drivers of the distinct brain diseases should also be taken into due account because they can simultaneously trigger neurotoxicity and neuroinflammation. Hence, a more comprehensive view of the underlying molecular mechanisms of each brain disease would be beneficial. Importantly, the yet available data on the several inflammasomes' roles in *human* brain diseases are limited and controversial. Therefore, this is a field widely open to groundbreaking investigations. We are confident that choosing human untransformed neural cells as models for pathogenetic and pharmacological studies will advance our knowledge about each neuropathology and hasten the achievement of effective etiological therapies.

Author Contributions: Conceptualization, A.C. and U.A.; data curation and investigation, A.C., L.G., C.V., U.A. and I.D.P.; writing—original draft preparation, A.C., L.G., C.V., U.A. and I.D.P.; writing—review and editing, A.C., L.G., C.V., U.A. and I.D.P.; supervision, A.C., L.G., C.V., U.A. and I.D.P. All authors have read and agreed to the published version of the manuscript.

Funding: This work was supported by the FUR 2020 allotment from the Italian MUR (Ministry of University & Research) to A.C. and I.D.P. No funds were provided by private or commercial sources.

Institutional Review Board Statement: Not applicable.

Informed Consent Statement: Not applicable.

Data Availability Statement: Not applicable.

Acknowledgments: The authors want to thank Saqib Waheed for his expert help with the graphic materials.

Conflicts of Interest: The authors declare no conflict of interest.

References

1. World Health Organization endorses global action plan on rising incidence of dementia. *Nurs. Older People* **2017**, *29*, 7. [CrossRef]
2. Brett, B.L.; Gardner, R.C.; Godbout, J.; Dams-O'Connor, K.; Keene, C.D. Traumatic Brain Injury and Risk of Neurodegenerative Disorder. *Biol. Psychiatry* **2022**, *91*, 498–507. [CrossRef] [PubMed]
3. Walsh, J.G.; Muruve, D.A.; Power, C. Inflammasomes in the CNS. *Nat. Rev. Neurosci.* **2014**, *15*, 84–97. [CrossRef] [PubMed]
4. Guo, H.; Callaway, J.B.; Ting, J.P.Y. Inflammasomes: Mechanism of action, role in disease, and therapeutics. *Nat. Med.* **2015**, *21*, 677–687. [CrossRef] [PubMed]

5. Celsius, A.C. *De Medicina, Volume 3, Passim;* Spencer WG Loeb Classical Library, Translator; Harvard University Press: Cambridge, MA, USA, 1935; ISBN 978-067-499-370-9.
6. Dugger, B.N.; Dickson, D.W. Pathology of Neurodegenerative Diseases. *Cold Spring Harb. Perspect. Biol.* **2017**, *9*, a028035. [CrossRef]
7. Wilson, D.M., 3rd; Cookson, M.R.; Van Den Bosch, L.; Zetterberg, H.; Holtzman, D.M.; Dewachter, I. Hallmarks of neurodegenerative diseases. *Cell* **2023**, *186*, 693–714. [CrossRef]
8. Tanaka, M.; Toldi, J.; Vécsei, L. Exploring the Etiological Links behind Neurodegenerative Diseases: Inflammatory Cytokines and Bioactive Kynurenines. *Int. J. Mol. Sci.* **2020**, *21*, 2431. [CrossRef]
9. Zhou, R.; Yazdi, A.S.; Menu, P.; Tschopp, J. A role for mitochondria in NLRP3 inflammasome activation. *Nature* **2011**, *469*, 221–225. [CrossRef]
10. Cao, Z.; Wang, Y.; Long, Z.; He, G. Interaction between autophagy and the NLRP3 inflammasome. *Acta Biochim. Biophys. Sin.* **2019**, *51*, 1087–1095. [CrossRef]
11. Bellut, M.; Papp, L.; Bieber, M.; Kraft, P.; Stoll, G.; Schuhmann, M.K. NLPR3 Inflammasome Inhibition Alleviates Hypoxic Endothelial Cell Death in Vitro and Protects Blood–Brain Barrier Integrity in Murine Stroke. *Cell Death Dis.* **2021**, *13*, 20. [CrossRef]
12. Kreher, C.; Favret, J.; Maulik, M.; Shin, D. Lysosomal Functions in Glia Associated with Neurodegeneration. *Biomolecules* **2021**, *11*, 400. [CrossRef]
13. Chiarini, A.; Dal Pra, I.; Gottardo, R.; Bortolotti, F.; Whitfield, J.F.; Armato, U. BH(4) (tetrahydrobiopterin)-dependent activation, but not the expression, of inducible NOS (nitric oxide synthase)-2 in proinflammatory cytokine-stimulated, cultured normal human astrocytes is mediated by MEK-ERK kinases. *J. Cell. Biochem.* **2005**, *94*, 731–743. [CrossRef]
14. Martinon, F.; Burns, K.; Tschopp, J. The inflammasome: A molecular platform triggering activation of inflammatory caspases and processing of proIL-beta. *Mol. Cell* **2002**, *10*, 417–426. [CrossRef]
15. de Alba, E. Structure, interactions and self-assembly of ASC-dependent inflammasomes. *Arch. Biochem. Biophys.* **2019**, *670*, 15–31. [CrossRef]
16. Stehlik, C.; Lee, S.H.; Dorfleutner, A.; Stassinopoulos, A.; Sagara, J.; Reed, J.C. Apoptosis-associated speck-like protein containing a caspase recruitment domain is a regulator of procaspase-1 activation. *J. Immunol.* **2003**, *171*, 6154–6163. [CrossRef]
17. Julien, O.; Wells, J.A. Caspases and their substrates. *Cell Death Differ.* **2017**, *24*, 1380–1389. [CrossRef]
18. Ding, J.; Wang, K.; Liu, W.; She, Y.; Sun, Q.; Shi, J.; Sun, H.; Wang, D.C.; Shao, F. Pore-forming activity and structural autoinhibition of the gasdermin family. *Nature* **2016**, *535*, 111–116. [CrossRef]
19. Gambin, Y.; Giles, N.; O'Carroll, A.; Polinkovsky, M.; Hunter, D.; Sierecki, E. Single-molecule fluorescence reveals the oligomerization and folding steps driving the prion-like behavior of ASC. *J. Mol. Biol.* **2018**, *430*, 491–508. [CrossRef]
20. Kesavardhana, S.; Kanneganti, T.D. Mechanisms governing inflammasome activation, assembly and pyroptosis induction. *Int. Immunol.* **2017**, *29*, 201–210. [CrossRef]
21. Lupfer, C.; Kanneganti, T.D. Unsolved mysteries in NLR biology. *Front. Immunol.* **2013**, *4*, 285. [CrossRef]
22. Devi, S.; Stehlik, C.; Dorfleutner, A. An update on CARD only proteins (COPs) and PYD only proteins (POPs) as inflammasome regulators. *Int. J. Mol. Sci.* **2020**, *21*, 6901. [CrossRef] [PubMed]
23. Poli, G.; Fabi, C.; Bellet, M.M.; Costantini, C.; Nunziangeli, L.; Romani, L.; Brancorsini, S. Epigenetic mechanisms of inflammasome regulation. *Int. J. Mol. Sci.* **2020**, *21*, 5758. [CrossRef] [PubMed]
24. Chiarini, A.; Armato, U.; Gui, L.; Dal Prà, I. "Other Than NLRP3" Inflammasomes: Multiple Roles in Brain Disease. *Neuroscientist* **2022**, *11*, 10738584221106114. [CrossRef] [PubMed]
25. Mangan, M.S.J.; Olhava, E.J.; Roush, W.R.; Seidel, H.M.; Glick, G.D.; Latz, E. Targeting the NLRP3 inflammasome in inflammatory diseases. *Nat. Rev. Drug Discov.* **2018**, *17*, 588–606. [CrossRef]
26. Chen, J.; Chen, Z.J. PtdIns4P on dispersed trans-Golgi network mediates NLRP3 inflammasome activation. *Nature* **2018**, *564*, 71–76. [CrossRef]
27. Chiarini, A.; Armato, U.; Hu, P.; Dal Prà, I. Danger-sensing/pattern recognition receptors and neuroinflammation in Alzheimer's disease. *Int. J. Mol. Sci.* **2020**, *21*, 9036. [CrossRef]
28. Zhang, Y.; Zhao, Y.; Zhang, J.; Yang, G. Mechanisms of NLRP3 Inflammasome Activation: Its Role in the Treatment of Alzheimer's Disease. *Neurochem. Res.* **2020**, *45*, 2560–2572. [CrossRef]
29. Holbrook, J.A.; Jarosz-Griffiths, H.H.; Caseley, E.; Lara-Reyna, S.; Poulter, J.A.; Williams-Gray, C.H.; Peckham, D.; McDermott, M.F. Neurodegenerative Disease and the NLRP3 Inflammasome. *Front. Pharmacol.* **2021**, *12*, 643254. [CrossRef]
30. Mészáros, Á.; Molnár, K.; Nógrádi, B.; Hernádi, Z.; Nyúl-Tóth, Á.; Wilhelm, I.; Krizbai, I.A. Neurovascular Inflammaging in Health and Disease. *Cells* **2020**, *9*, 1614. [CrossRef]
31. Lee, G.S.; Subramanian, N.; Kim, A.I.; Aksentijevich, I.; Goldbach-Mansky, R.; Sacks, D.B.; Germain, R.N.; Kastner, D.L.; Chae, J.J. The calcium-sensing receptor regulates the NLRP3 inflammasome through Ca^{2+} and cAMP. *Nature* **2012**, *492*, 123–127. [CrossRef]
32. Gong, Z.; Pan, J.; Shen, Q.; Li, M.; Peng, Y. Mitochondrial dysfunction induces NLRP3 inflammasome activation during cerebral ischemia/reperfusion injury. *J. Neuroinflamm.* **2018**, *15*, 242. [CrossRef]
33. Su, S.H.; Wu, Y.F.; Wang, D.P.; Hai, J. Inhibition of excessive autophagy and mitophagy mediates neuroprotective effects of URB597 against chronic cerebral hypoperfusion. *Cell Death Dis.* **2018**, *9*, 733. [CrossRef]
34. Su, S.H.; Wu, Y.F.; Lin, Q.; Wang, D.P.; Hai, J. URB597 protects against NLRP3 inflammasome activation by inhibiting autophagy dysfunction in a rat model of chronic cerebral hypoperfusion. *J. Neuroinflamm.* **2019**, *16*, 260. [CrossRef]

35. Zhu, J.J.; Yu, B.Y.; Huang, X.K.; He, M.Z.; Chen, B.W.; Chen, T.T.; Fang, H.Y.; Chen, S.Q.; Fu, X.Q.; Li, P.J.; et al. Neferine Protects against Hypoxic-Ischemic Brain Damage in Neonatal Rats by Suppressing NLRP3-Mediated Inflammasome Activation. *Oxid. Med. Cell. Longev.* **2021**, *2021*, 6654954. [CrossRef]
36. Franke, M.; Bieber, M.; Kraft, P.; Weber, A.N.R.; Stoll, G.; Schuhmann, M.K. The NLRP3 inflammasome drives inflammation in ischemia/reperfusion injury after transient middle cerebral artery occlusion in mice. *Brain Behav. Immun.* **2021**, *92*, 223–233. [CrossRef]
37. Xu, Q.; Zhao, B.; Ye, Y.; Li, Y.; Zhang, Y.; Xiong, X.; Gu, L. Relevant mediators involved in and therapies targeting the inflammatory response induced by activation of the NLRP3 inflammasome in ischemic stroke. *J. Neuroinflamm.* **2021**, *18*, 123. [CrossRef]
38. Chen, S.H.; Scott, X.O.; Ferrer Marcelo, Y.; Almeida, V.W.; Blackwelder, P.L.; Yavagal, D.R.; Peterson, E.C.; Starke, R.M.; Dietrich, W.D.; Keane, R.W.; et al. Netosis and Inflammasomes in Large Vessel Occlusion Thrombi. *Front. Pharmacol.* **2021**, *11*, 607287. [CrossRef]
39. Xiao, L.; Zheng, H.; Li, J.; Wang, Q.; Sun, H. Neuroinflammation Mediated by NLRP3 Inflammasome after Intracerebral Hemorrhage and Potential Therapeutic Targets. *Mol. Neurobiol.* **2020**, *57*, 5130–5149. [CrossRef]
40. Yang, S.J.; Shao, G.F.; Chen, J.L.; Gong, J. The NLRP3 Inflammasome: An Important Driver of Neuroinflammation in Hemorrhagic Stroke. *Cell. Mol. Neurobiol.* **2018**, *38*, 595–603. [CrossRef]
41. Cristina de Brito Toscano, E.; Leandro Marciano Vieira, É.; Boni Rocha Dias, B.; Vidigal Caliari, M.; Paula Gonçalves, A.; Varela Giannetti, A.; Maurício Siqueira, J.; Kimie Suemoto, C.; Elaine Paraizo Leite, R.; Nitrini, R.; et al. NLRP3 and NLRP1 inflammasomes are up-regulated in patients with mesial temporal lobe epilepsy and may contribute to overexpression of caspase-1 and IL-β in sclerotic hippocampi. *Brain Res.* **2021**, *1752*, 147230. [CrossRef]
42. Wang, S.; He, H.; Long, J.; Sui, X.; Yang, J.; Lin, G.; Wang, Q.; Wang, Y.; Luo, Y. TRPV4 Regulates Soman-Induced Status Epilepticus and Secondary Brain Injury via NMDA Receptor and NLRP3 Inflammasome. *Neurosci. Bull.* **2021**, *37*, 905–920. [CrossRef] [PubMed]
43. Wang, D.; Zhang, J.; Jiang, W.; Cao, Z.; Zhao, F.; Cai, T.; Aschner, M.; Luo, W. The role of NLRP3-CASP1 in inflammasome-mediated neuroinflammation and autophagy dysfunction in manganese-induced, hippocampal-dependent impairment of learning and memory ability. *Autophagy* **2017**, *13*, 914–927. [CrossRef] [PubMed]
44. Sarkar, S.; Rokad, D.; Malovic, E.; Luo, J.; Harischandra, D.S.; Jin, H.; Anantharam, V.; Huang, X.; Lewis, M.; Kanthasamy, A.; et al. Manganese activates NLRP3 inflammasome signaling and propagates exosomal release of ASC in microglial cells. *Sci. Signal.* **2019**, *12*, eaat9900. [CrossRef] [PubMed]
45. Su, P.; Wang, D.; Cao, Z.; Chen, J.; Zhang, J. The role of NLRP3 in lead-induced neuroinflammation and possible underlying mechanism. *Environ. Pollut.* **2021**, *287*, 117520. [CrossRef] [PubMed]
46. Dong, J.; Wang, X.; Xu, C.; Gao, M.; Wang, S.; Zhang, J.; Tong, H.; Wang, L.; Han, Y.; Cheng, N.; et al. Inhibiting NLRP3 inflammasome activation prevents copper-induced neuropathology in a murine model of Wilson's disease. *Cell Death Dis.* **2021**, *12*, 87. [CrossRef]
47. Cai, J.; Guan, H.; Jiao, X.; Yang, J.; Chen, X.; Zhang, H.; Zheng, Y.; Zhu, Y.; Liu, Q.; Zhang, Z. NLRP3 inflammasome mediated pyroptosis is involved in cadmium exposure-induced neuroinflammation through the IL-1β/IkB-α-NF-κB-NLRP3 feedback loop in swine. *Toxicology* **2021**, *453*, 152720. [CrossRef]
48. Brewer, G.J. Divalent Copper as a Major Triggering Agent in Alzheimer's Disease. *J. Alzheimer's Dis.* **2015**, *46*, 593–604. [CrossRef]
49. Zhou, Q.; Zhang, Y.; Lu, L.; Zhang, H.; Zhao, C.; Pu, Y.; Yin, L. Copper induces microglia-mediated neuroinflammation through ROS/NF-κB pathway and mitophagy disorder. *Food Chem. Toxicol.* **2022**, *16*, 113369. [CrossRef]
50. Quandt, D.; Rothe, K.; Baerwald, C.; Rossol, M. GPRC6A mediates Alum-induced Nlrp3 inflammasome activation but limits Th2 type antibody responses. *Sci. Rep.* **2015**, *5*, 16719. [CrossRef]
51. Ye, R.; Pi, M.; Nooh, M.M.; Bahout, S.W.; Quarles, L.D. Human GPRC6A Mediates Testosterone-Induced Mitogen-Activated Protein Kinases and mTORC1 Signaling in Prostate Cancer Cells. *Mol. Pharmacol.* **2019**, *95*, 563–572. [CrossRef]
52. Zhang, X.; Shu, Q.; Liu, Z.; Gao, C.; Wang, Z.; Xing, Z.; Song, J. Recombinant osteopontin provides protection for cerebral infarction by inhibiting the NLRP3 inflammasome in microglia. *Brain Res.* **2021**, *1751*, 147170. [CrossRef]
53. Chen, Y.; Meng, J.; Bi, F.; Li, H.; Chang, C.; Ji, C.; Liu, W. NEK7 Regulates NLRP3 Inflammasome Activation and Neuroinflammation Post-traumatic Brain Injury. *Front. Mol. Neurosci.* **2019**, *12*, 202, Erratum in *Front. Mol. Neurosci.* **2019**, *12*, 247. [CrossRef]
54. Ji, X.; Song, Z.; He, J.; Guo, S.; Chen, Y.; Wang, H.; Zhang, J.; Xu, X.; Liu, J. NIMA-related kinase 7 amplifies NLRP3 inflammasome pro-inflammatory signaling in microglia/macrophages and mice models of spinal cord injury. *Exp. Cell Res.* **2021**, *398*, 112418. [CrossRef]
55. O'Brien, W.T.; Pham, L.; Symons, G.F.; Monif, M.; Shultz, S.R.; McDonald, S.J. The NLRP3 inflammasome in traumatic brain injury: Potential as a biomarker and therapeutic target. *J. Neuroinflamm.* **2020**, *17*, 104. [CrossRef]
56. Irrera, N.; Russo, M.; Pallio, G.; Bitto, A.; Mannino, F.; Minutoli, L.; Altavilla, D.; Squadrito, F. The Role of NLRP3 Inflammasome in the Pathogenesis of Traumatic Brain Injury. *Int. J. Mol. Sci.* **2020**, *21*, 6204. [CrossRef]
57. Albalawi, F.; Lu, W.; Beckel, J.M.; Lim, J.C.; McCaughey, S.A.; Mitchell, C.H. The P2X7 Receptor Primes IL-1β and the NLRP3 Inflammasome in Astrocytes Exposed to Mechanical Strain. *Front. Cell. Neurosci.* **2017**, *11*, 227. [CrossRef]
58. Ding, H.; Li, Y.; Wen, M.; Liu, X.; Han, Y.; Zeng, H. Elevated intracranial pressure induces IL1β and IL18 overproduction via activation of the NLRP3 inflammasome in microglia of ischemic adult rats. *Int. J. Mol. Med.* **2021**, *47*, 183–194. [CrossRef]

59. Chi, W.; Chen, H.; Li, F.; Zhu, Y.; Yin, W.; Zhuo, Y. HMGB1 promotes the activation of NLRP3 and caspase-8 inflammasomes via NF-κB pathway in acute glaucoma. *J. Neuroinflamm.* **2015**, *12*, 137. [CrossRef]
60. Jiang, S.; Maphis, N.M.; Binder, J.; Chisholm, D.; Weston, L.; Duran, W.; Peterson, C.; Zimmerman, A.; Mandell, M.A.; Jett, S.D.; et al. Proteopathic tau primes and activates interleukin-1β via myeloid-cell-specific MyD88- and NLRP3-ASC-inflammasome pathway. *Cell Rep.* **2021**, *36*, 109720. [CrossRef]
61. Shi, F.; Yang, L.; Kouadir, M.; Yang, Y.; Wang, J.; Zhou, X.; Yin, X.; Zhao, D. The NALP3 inflammasome engages in neurotoxic prion peptide-induced microglial activation. *J. Neuroinflamm.* **2012**, *9*, 73. [CrossRef]
62. Lai, M.; Yao, H.; Shah, S.Z.A.; Wu, W.; Wang, D.; Zhao, Y.; Wang, L.; Zhou, X.; Zhao, D.; Yang, L. The NLRP3-Caspase 1 Inflammasome Negatively Regulates Autophagy via TLR4-TRIF in Prion Peptide-Infected Microglia. *Front. Aging Neurosci.* **2018**, *10*, 116. [CrossRef] [PubMed]
63. Milner, M.T.; Maddugoda, M.; Götz, J.; Burgener, S.S.; Schroder, K. The NLRP3 inflammasome triggers sterile neuroinflammation and Alzheimer's disease. *Curr. Opin. Immunol.* **2021**, *68*, 116–124. [CrossRef] [PubMed]
64. Pike, A.F.; Varanita, T.; Herrebout, M.A.C.; Plug, B.C.; Kole, J.; Musters, R.J.P.; Teunissen, C.E.; Hoozemans, J.J.M.; Bubacco, L.; Veerhuis, R. α-Synuclein evokes NLRP3 inflammasome-mediated IL-1β secretion from primary human microglia. *Glia* **2021**, *69*, 1413–1428. [CrossRef] [PubMed]
65. Deora, V.; Lee, J.D.; Albornoz, E.A.; McAlary, L.; Jagaraj, C.J.; Robertson, A.A.B.; Atkin, J.D.; Cooper, M.A.; Schroder, K.; Yerbury, J.J.; et al. The microglial NLRP3 inflammasome is activated by amyotrophic lateral sclerosis proteins. *Glia* **2020**, *68*, 407–421, Erratum in *Glia* **2020**, *68*, 2167–2168. [CrossRef] [PubMed]
66. Ismael, S.; Nasoohi, S.; Li, L.; Aslam, K.S.; Khan, M.M.; El-Remessy, A.B.; McDonald, M.P.; Liao, F.F.; Ishrat, T. Thioredoxin interacting protein regulates age-associated neuroinflammation. *Neurobiol. Dis.* **2021**, *156*, 105399. [CrossRef]
67. Ismael, S.; Wajidunnisa; Sakata, K.; McDonald, M.P.; Liao, F.F.; Ishrat, T. ER stress associated TXNIP-NLRP3 inflammasome activation in hippocampus of human Alzheimer's disease. *Neurochem. Int.* **2021**, *148*, 105104. [CrossRef]
68. Shen, H.; Guan, Q.; Zhang, X.; Yuan, C.; Tan, Z.; Zhai, L.; Hao, Y.; Gu, Y.; Han, C. New mechanism of neuroinflammation in Alzheimer's disease: The activation of NLRP3 inflammasome mediated by gut microbiota. *Prog. Neuropsychopharmacol. Biol. Psychiatry* **2020**, *100*, 109884. [CrossRef]
69. Shukla, P.K.; Delotterie, D.F.; Xiao, J.; Pierre, J.F.; Rao, R.; McDonald, M.P.; Khan, M.M. Alterations in the Gut-Microbial-Inflammasome-Brain Axis in a Mouse Model of Alzheimer's Disease. *Cells* **2021**, *10*, 779. [CrossRef]
70. Yi, W.; Cheng, J.; Wei, Q.; Pan, R.; Song, S.; He, Y.; Tang, C.; Liu, X.; Zhou, Y.; Su, H. Effect of temperature stress on gut-brain axis in mice: Regulation of intestinal microbiome and central NLRP3 inflammasomes. *Sci. Total Environ.* **2021**, *772*, 144568. [CrossRef]
71. Ising, C.; Venegas, C.; Zhang, S.; Scheiblich, H.; Schmidt, S.V.; Vieira-Saecker, A.; Schwartz, S.; Albasset, S.; McManus, R.M.; Tejera, D.; et al. NLRP3 inflammasome activation drives tau pathology. *Nature* **2019**, *575*, 669–673. [CrossRef]
72. Wang, B.R.; Shi, J.Q.; Ge, N.N.; Ou, Z.; Tian, Y.Y.; Jiang, T.; Zhou, J.S.; Xu, J.; Zhang, Y.D. PM2.5 exposure aggravates oligomeric amyloid beta-induced neuronal injury and promotes NLRP3 inflammasome activation in an in vitro model of Alzheimer's disease. *J. Neuroinflamm.* **2018**, *15*, 132. [CrossRef]
73. Shi, J.Q.; Wang, B.R.; Jiang, T.; Gao, L.; Zhang, Y.D.; Xu, J. NLRP3 Inflammasome: A Potential Therapeutic Target in Fine Particulate Matter-Induced Neuroinflammation in Alzheimer's Disease. *J. Alzheimers Dis.* **2020**, *77*, 923–934. [CrossRef]
74. Yuan, L.; Zhu, Y.; Huang, S.; Lin, L.; Jiang, X.; Chen, S. NF-κB/ROS and ERK pathways regulate NLRP3 inflammasome activation in Listeria monocytogenes infected BV2 microglia cells. *J. Microbiol.* **2021**, *59*, 771–781. [CrossRef]
75. Zhao, Z.; Wang, Y.; Zhou, R.; Li, Y.; Gao, Y.; Tu, D.; Wilson, B.; Song, S.; Feng, J.; Hong, J.S.; et al. A novel role of NLRP3-generated IL-1β in the acute-chronic transition of peripheral lipopolysaccharide-elicited neuroinflammation: Implications for sepsis-associated neurodegeneration. *J. Neuroinflamm.* **2020**, *17*, 64. [CrossRef]
76. Danielski, L.G.; Giustina, A.D.; Bonfante, S.; de Souza Goldim, M.P.; Joaquim, L.; Metzker, K.L.; Biehl, E.B.; Vieira, T.; de Medeiros, F.D.; da Rosa, N.; et al. NLRP3 Activation Contributes to Acute Brain Damage Leading to Memory Impairment in Sepsis-Surviving Rats. *Mol. Neurobiol.* **2020**, *57*, 5247–5262. [CrossRef]
77. Chivero, E.T.; Guo, M.L.; Periyasamy, P.; Liao, K.; Callen, S.E.; Buch, S. HIV-1 Tat Primes and Activates Microglial NLRP3 Inflammasome-Mediated Neuroinflammation. *J. Neurosci.* **2017**, *37*, 3599–3609. [CrossRef]
78. Katuri, A.; Bryant, J.; Heredia, A.; Makar, T.K. Role of the inflammasomes in HIV-associated neuroinflammation and neurocognitive disorders. *Exp. Mol. Pathol.* **2019**, *108*, 64–72. [CrossRef]
79. He, X.; Yang, W.; Zeng, Z.; Wei, Y.; Gao, J.; Zhang, B.; Li, L.; Liu, L.; Wan, Y.; Zeng, Q.; et al. NLRP3-dependent pyroptosis is required for HIV-1 gp120-induced neuropathology. *Cell. Mol. Immunol.* **2020**, *17*, 283–299. [CrossRef]
80. Hu, X.; Zeng, Q.; Xiao, J.; Qin, S.; Wang, Y.; Shan, T.; Hu, D.; Zhu, Y.; Liu, K.; Zheng, K.; et al. Herpes Simplex Virus 1 Induces Microglia Gasdermin D-Dependent Pyroptosis through Activating the NLR Family Pyrin Domain Containing 3 Inflammasome. *Front. Microbiol.* **2022**, *13*, 838808. [CrossRef]
81. Chen, C.J.; Ou, Y.C.; Chang, C.Y.; Pan, H.C.; Lin, S.Y.; Liao, S.L.; Raung, S.L.; Chen, S.Y.; Chang, C.J. Src signaling involvement in Japanese encephalitis virus-induced cytokine production in microglia. *Neurochem. Int.* **2011**, *58*, 924–933. [CrossRef]
82. He, Z.; Chen, J.; Zhu, X.; An, S.; Dong, X.; Yu, J.; Zhang, S.; Wu, Y.; Li, G.; Zhang, Y.; et al. NLRP3 Inflammasome Activation Mediates Zika Virus-Associated Inflammation. *J. Infect. Dis.* **2018**, *217*, 1942–1951. [CrossRef] [PubMed]
83. Chen, I.Y.; Moriyama, M.; Chang, M.F.; Ichinohe, T. Severe Acute Respiratory Syndrome Coronavirus Viroporin 3a Activates the NLRP3 Inflammasome. *Front. Microbiol.* **2019**, *10*, 50. [CrossRef] [PubMed]

84. Siu, K.L.; Yuen, K.S.; Castano-Rodriguez, C.; Ye, Z.W.; Yeung, M.L.; Fung, S.Y.; Yuan, S.; Chan, C.P.; Yuen, K.Y.; Enjuanes, L.; et al. Severe acute respiratory syndrome coronavirus ORF3a protein activates the NLRP3 inflammasome by promoting TRAF3-dependent ubiquitination of ASC. *FASEB J.* **2019**, *33*, 8865–8877. [CrossRef] [PubMed]
85. de Rivero Vaccari, J.C.; Dietrich, W.D.; Keane, R.W.; de Rivero Vaccari, J.P. The inflammasome in times of COVID-19. *Front. Immunol.* **2020**, *11*, 583373. [CrossRef]
86. Pan, P.; Shen, M.; Yu, Z.; Ge, W.; Chen, K.; Tian, M.; Xiao, F.; Wang, Z.; Wang, J.; Jia, Y.; et al. SARS-CoV-2 N protein promotes NLRP3 inflammasome activation to induce hyperinflammation. *Nat. Commun.* **2021**, *12*, 4664, Erratum in *Nat. Commun.* **2021**, *12*, 5306. [CrossRef]
87. Yalcinkaya, M.; Liu, W.; Islam, M.N.; Kotini, A.G.; Gusarova, G.A.; Fidler, T.P.; Papapetrou, E.P.; Bhattacharya, J.; Wang, N.; Tall, A.R. Modulation of the NLRP3 inflammasome by SARS-CoV-2 Envelope protein. *Sci. Rep.* **2021**, *11*, 24432. [CrossRef]
88. Olajide, O.A.; Iwuanyanwu, V.U.; Adegbola, O.D.; Al-Hindawi, A.A. SARS-CoV-2 Spike Glycoprotein S1 Induces Neuroinflammation in BV-2 Microglia. *Mol. Neurobiol.* **2022**, *59*, 445–458. [CrossRef]
89. Ito, M.; Yanagi, Y.; Ichinohe, T. Encephalomyocarditis virus viroporin 2B activates NLRP3 inflammasome. *PLoS Pathog.* **2012**, *8*, e1002857. [CrossRef]
90. Moreira, J.D.; Iakhiaev, A.; Vankayalapati, R.; Jung, B.G.; Samten, B. Histone deacetylase-2 controls IL-1β production through the regulation of NLRP3 expression and activation in tuberculosis infection. *iScience* **2022**, *25*, 104799. [CrossRef]
91. Butterfield, D.A.; Halliwell, B. Oxidative stress, dysfunctional glucose metabolism and Alzheimer disease. *Nat. Rev. Neurosci.* **2019**, *20*, 148–160. [CrossRef]
92. Litwiniuk, A.; Bik, W.; Kalisz, M.; Baranowska-Bik, A. Inflammasome NLRP3 Potentially Links Obesity-Associated Low-Grade Systemic Inflammation and Insulin Resistance with Alzheimer's Disease. *Int. J. Mol. Sci.* **2021**, *22*, 5603. [CrossRef]
93. Sobesky, J.L.; D'Angelo, H.M.; Weber, M.D.; Anderson, N.D.; Frank, M.G.; Watkins, L.R.; Maier, S.F.; Barrientos, R.M. Glucocorticoids Mediate Short-Term High-Fat Diet Induction of Neuroinflammatory Priming, the NLRP3 Inflammasome, and the Danger Signal HMGB1. *eNeuro* **2016**, *3*, ENEURO.0113-16.2016. [CrossRef]
94. Keshk, W.A.; Ibrahim, M.A.; Shalaby, S.M.; Zalat, Z.A.; Elseady, W.S. Redox status, inflammation, necroptosis and inflammasome as indispensable contributors to high fat diet (HFD)-induced neurodegeneration; Effect of N-acetylcysteine (NAC). *Arch. Biochem. Biophys.* **2020**, *680*, 108227. [CrossRef]
95. Wei, P.; Yang, F.; Zheng, Q.; Tang, W.; Li, J. The Potential Role of the NLRP3 Inflammasome Activation as a Link between Mitochondria ROS Generation and Neuroinflammation in Postoperative Cognitive Dysfunction. *Front. Cell. Neurosci.* **2019**, *13*, 73. [CrossRef]
96. Zhang, L.; Xiao, F.; Zhang, J.; Wang, X.; Ying, J.; Wei, G.; Chen, S.; Huang, X.; Yu, W.; Liu, X.; et al. Dexmedetomidine Mitigated NLRP3-Mediated Neuroinflammation via the Ubiquitin-Autophagy Pathway to Improve Perioperative Neurocognitive Disorder in Mice. *Front. Pharmacol.* **2021**, *12*, 646265. [CrossRef]
97. Hirshman, N.A.; Hughes, F.M., Jr.; Jin, H.; Harrison, W.T.; White, S.W.; Doan, I.; Harper, S.N.; Leidig, P.D.; Purves, J.T. Cyclophosphamide-induced cystitis results in NLRP3-mediated inflammation in the hippocampus and symptoms of depression in rats. *Am. J. Physiol. Renal Physiol.* **2020**, *318*, F354–F362. [CrossRef]
98. D'Espessailles, A.; Mora, Y.A.; Fuentes, C.; Cifuentes, M. Calcium-sensing receptor activates the NLRP3 inflammasome in LS14 preadipocytes mediated by ERK1/2 signaling. *J. Cell. Physiol.* **2018**, *233*, 6232–6240. [CrossRef]
99. Wang, C.; Jia, Q.; Sun, C.; Jing, C. Calcium sensing receptor contribute to early brain injury through the CaMKII/NLRP3 pathway after subarachnoid hemorrhage in mice. *Biochem. Biophys. Res. Commun.* **2020**, *530*, 651–657. [CrossRef]
100. Hu, W.; Zhang, Y.; Wu, W.; Yin, Y.; Huang, D.; Wang, Y.; Li, W.; Li, W. Chronic glucocorticoids exposure enhances neurodegeneration in the frontal cortex and hippocampus via NLRP-1 inflammasome activation in male mice. *Brain Behav. Immun.* **2016**, *52*, 58–70. [CrossRef]
101. Maturana, C.J.; Aguirre, A.; Sáez, J.C. High glucocorticoid levels during gestation activate the inflammasome in hippocampal oligodendrocytes of the offspring. *Dev. Neurobiol.* **2017**, *77*, 625–642. [CrossRef]
102. Chivero, E.T.; Thangaraj, A.; Tripathi, A.; Periyasamy, P.; Guo, M.L.; Buch, S. NLRP3 Inflammasome Blockade Reduces Cocaine-Induced Microglial Activation and Neuroinflammation. *Mol. Neurobiol.* **2021**, *58*, 2215–2230. [CrossRef] [PubMed]
103. Du, S.H.; Qiao, D.F.; Chen, C.X.; Chen, S.; Liu, C.; Lin, Z.; Wang, H.; Xie, W.B. Toll-Like Receptor 4 Mediates Methamphetamine-Induced Neuroinflammation through Caspase-11 Signaling Pathway in Astrocytes. *Front. Mol. Neurosci.* **2017**, *10*, 409. [CrossRef] [PubMed]
104. Xu, E.; Liu, J.; Liu, H.; Wang, X.; Xiong, H. Inflammasome Activation by Methamphetamine Potentiates Lipopolysaccharide Stimulation of IL-1β Production in Microglia. *J. Neuroimmune Pharmacol.* **2018**, *13*, 237–253. [CrossRef] [PubMed]
105. Cheon, S.Y.; Koo, B.N.; Kim, S.Y.; Kam, E.H.; Nam, J.; Kim, E.J. Scopolamine promotes neuroinflammation and delirium-like neuropsychiatric disorder in mice. *Sci. Rep.* **2021**, *11*, 8376. [CrossRef]
106. Lippai, D.; Bala, S.; Petrasek, J.; Csak, T.; Levin, I.; Kurt-Jones, E.A.; Szabo, G. Alcohol-induced IL-1β in the brain is mediated by NLRP3/ASC inflammasome activation that amplifies neuroinflammation. *J. Leukoc. Biol.* **2013**, *94*, 171–182. [CrossRef]
107. Alfonso-Loeches, S.; Ureña-Peralta, J.; Morillo-Bargues, M.J.; Gómez-Pinedo, U.; Guerri, C. Ethanol-Induced TLR4/NLRP3 Neuroinflammatory Response in Microglial Cells Promotes Leukocyte Infiltration Across the BBB. *Neurochem. Res.* **2016**, *41*, 193–209. [CrossRef]

108. Carranza-Aguilar, C.J.; Hernández-Mendoza, A.; Mejias-Aponte, C.; Rice, K.C.; Morales, M.; González-Espinosa, C.; Cruz, S.L. Morphine and Fentanyl Repeated Administration Induces Different Levels of NLRP3-Dependent Pyroptosis in the Dorsal Raphe Nucleus of Male Rats via Cell-Specific Activation of TLR4 and Opioid Receptors. *Cell. Mol. Neurobiol.* **2020**, *42*, 677–694. [CrossRef]
109. Samir, P.; Kesavardhana, S.; Patmore, D.M.; Gingras, S.; Malireddi, R.K.S.; Karki, R.; Guy, C.S.; Briard, B.; Place, D.E.; Bhattacharya, A.; et al. DDX3X acts as a live-or-die checkpoint in stressed cells by regulating NLRP3 inflammasome. *Nature* **2019**, *573*, 590–594. [CrossRef]
110. Swaroop, S.; Mahadevan, A.; Shankar, S.K.; Adlakha, Y.K.; Basu, A. HSP60 critically regulates endogenous IL-1β production in activated microglia by stimulating NLRP3 inflammasome pathway. *J. Neuroinflamm.* **2018**, *15*, 177, Erratum in *J. Neuroinflamm.* **2018**, *15*, 317. [CrossRef]
111. Kim, M.J.; Yoon, J.H.; Ryu, J.H. Mitophagy: A balance regulator of NLRP3 inflammasome activation. *BMB Rep.* **2016**, *49*, 529–535. [CrossRef]
112. Mishra, S.R.; Mahapatra, K.K.; Behera, B.P.; Patra, S.; Bhol, C.S.; Panigrahi, D.P.; Praharaj, P.P.; Singh, A.; Patil, S.; Dhiman, R.; et al. Mitochondrial dysfunction as a driver of NLRP3 inflammasome activation and its modulation through mitophagy for potential therapeutics. *Int. J. Biochem. Cell Biol.* **2021**, *136*, 106013. [CrossRef]
113. Leemans, J.C.; Cassel, S.L.; Sutterwala, F.S. Sensing damage by the NLRP3 inflammasome. *Immunol. Rev.* **2011**, *243*, 152–162. [CrossRef]
114. Iyer, S.S.; He, Q.; Janczy, J.R.; Elliott, E.I.; Zhong, Z.; Olivier, A.K.; Sadler, J.J.; Knepper-Adrian, V.; Han, R.; Qiao, L.; et al. Mitochondrial cardiolipin is required for Nlrp3 inflammasome activation. *Immunity* **2013**, *39*, 311–323. [CrossRef]
115. Han, S.; He, Z.; Jacob, C.; Hu, X.; Liang, X.; Xiao, W.; Wan, L.; Xiao, P.; D'Ascenzo, N.; Ni, J.; et al. Effect of Increased IL-1β on Expression of HK in Alzheimer's Disease. *Int. J. Mol. Sci.* **2021**, *22*, 1306. [CrossRef]
116. Rivers-Auty, J.; Tapia, V.S.; White, C.S.; Daniels, M.J.D.; Drinkall, S.; Kennedy, P.T.; Spence, H.G.; Yu, S.; Green, J.P.; Hoyle, C.; et al. Zinc Status Alters Alzheimer's Disease Progression through NLRP3-Dependent Inflammation. *J. Neurosci.* **2021**, *41*, 3025–3038. [CrossRef]
117. Xu, Z.; Chen, Z.M.; Wu, X.; Zhang, L.; Cao, Y.; Zhou, P. Distinct Molecular Mechanisms Underlying Potassium Efflux for NLRP3 Inflammasome Activation. *Front. Immunol.* **2020**, *11*, 609441. [CrossRef]
118. Zhong, Z.; Liang, S.; Sanchez-Lopez, E.; He, F.; Shalapour, S.; Lin, X.J.; Wong, J.; Ding, S.; Seki, E.; Schnabl, B.; et al. New mitochondrial DNA synthesis enables NLRP3 inflammasome activation. *Nature* **2018**, *560*, 198–203. [CrossRef]
119. Zhou, X.G.; Qiu, W.Q.; Yu, L.; Pan, R.; Teng, J.F.; Sang, Z.P.; Law, B.Y.; Zhao, Y.; Zhang, L.; Yan, L.; et al. Targeting microglial autophagic degradation of the NLRP3 inflammasome for identification of thonningianin A in Alzheimer's disease. *Inflamm. Regen.* **2022**, *42*, 25. [CrossRef]
120. Zhao, T.; Gao, J.; Van, J.; To, E.; Wang, A.; Cao, S.; Cui, J.Z.; Guo, J.P.; Lee, M.; McGeer, P.L.; et al. Age-related increases in amyloid beta and membrane attack complex: Evidence of inflammasome activation in the rodent eye. *J. Neuroinflamm.* **2015**, *12*, 121. [CrossRef]
121. Reddy, P.H.; Oliver, D.M. Amyloid Beta and Phosphorylated Tau-Induced Defective Autophagy and Mitophagy in Alzheimer's Disease. *Cells* **2019**, *8*, 488. [CrossRef]
122. Eshraghi, M.; Adlimoghaddam, A.; Mahmoodzadeh, A.; Sharifzad, F.; Yasavoli-Sharahi, H.; Lorzadeh, S.; Albensi, B.C.; Ghavami, S. Alzheimer's Disease Pathogenesis: Role of Autophagy and Mitophagy Focusing in Microglia. *Int. J. Mol. Sci.* **2021**, *22*, 3330. [CrossRef] [PubMed]
123. Lech, M.; Avila-Ferrufino, A.; Skuginna, V.; Susanti, H.E.; Anders, H.J. Quantitative expression of RIG-like helicase, NOD-like receptor and inflammasome-related mRNAs in humans and mice. *Int. Immunol.* **2010**, *22*, 717–728. [CrossRef] [PubMed]
124. Minkiewicz, J.; de Rivero Vaccari, J.P.; Keane, R.W. Human astrocytes express a novel NLRP2 inflammasome. *Glia* **2013**, *61*, 1113–1121. [CrossRef]
125. de Rivero Vaccari, J.P.; Dietrich, W.D.; Keane, R.W. Activation and regulation of cellular inflammasomes: Gaps in our knowledge for central nervous system injury. *J. Cereb. Blood Flow Metab.* **2014**, *34*, 369–375. [CrossRef]
126. Nyúl-Tóth, Á.; Kozma, M.; Nagyőszi, P.; Nagy, K.; Fazakas, C.; Haskó, J.; Molnár, K.; Farkas, A.E.; Végh, A.G.; Váró, G.; et al. Expression of pattern recognition receptors and activation of the non-canonical inflammasome pathway in brain pericytes. *Brain Behav. Immun.* **2017**, *64*, 220–231. [CrossRef] [PubMed]
127. Johann, S.; Heitzer, M.; Kanagaratnam, M.; Goswami, A.; Rizo, T.; Weis, J.; Troost, D.; Beyer, C. NLRP3 inflammasome is expressed by astrocytes in the SOD1 mouse model of ALS and in human sporadic ALS patients. *Glia* **2015**, *63*, 2260–2273. [CrossRef]
128. Ebrahimi, T.; Rust, M.; Kaiser, S.N.; Slowik, A.; Beyer, C.; Koczulla, A.R.; Schulz, J.B.; Habib, P.; Bach, J.P. α1-antitrypsin mitigates NLRP3-inflammasome activation in amyloid β_{1-42}-stimulated murine astrocytes. *J. Neuroinflamm.* **2018**, *15*, 282. [CrossRef]
129. Sandhu, J.K.; Kulka, M. Decoding Mast Cell-Microglia Communication in Neurodegenerative Diseases. *Int. J. Mol. Sci.* **2021**, *22*, 1093. [CrossRef]
130. Komleva, Y.K.; Lopatina, O.L.; Gorina, Y.V.; Chernykh, A.I.; Trufanova, L.V.; Vais, E.F.; Kharitonova, E.V.; Zhukov, E.L.; Vahtina, L.Y.; Medvedeva, N.N.; et al. Expression of NLRP3 Inflammasomes in Neurogenic Niche Contributes to the Effect of Spatial Learning in Physiological Conditions but Not in Alzheimer's Type Neurodegeneration. *Cell. Mol. Neurobiol.* **2021**, *42*, 1355–1371. [CrossRef]

131. Saresella, M.; La Rosa, F.; Piancone, F.; Zoppis, M.; Marventano, I.; Calabrese, E.; Rainone, V.; Nemni, R.; Mancuso, R.; Clerici, M. The NLRP3 and NLRP1 inflammasomes are activated in Alzheimer's disease. *Mol. Neurodegener.* **2016**, *11*, 23. [CrossRef]
132. Tang, H.; Harte, M. Investigating markers of the NLRP3 inflammasome pathway in Alzheimer's disease: A human post-mortem study. *Genes* **2021**, *12*, 1753. [CrossRef]
133. Dal Prà, I.; Armato, U.; Chiarini, A. Family C G-Protein-Coupled Receptors in Alzheimer's Disease and Therapeutic Implications. *Front. Pharmacol.* **2019**, *10*, 1282. [CrossRef]
134. Dal Prà, I.; Armato, U.; Chioffi, F.; Pacchiana, R.; Whitfield, J.F.; Chakravarthy, B.; Gui, L.; Chiarini, A. The Aβ peptides-activated calcium-sensing receptor stimulates the production and secretion of vascular endothelial growth factor-A by normoxic adult human cortical astrocytes. *Neuromol. Med.* **2014**, *16*, 645–657. [CrossRef]
135. Dal Prà, I.; Chiarini, A.; Pacchiana, R.; Gardenal, E.; Chakravarthy, B.; Whitfield, J.F.; Armato, U. Calcium-sensing receptors of human Astrocyte-Neuron Teams: Amyloid-β-driven mediators and therapeutic targets of Alzheimer's Disease. *Curr. Neuropharmacol.* **2014**, *12*, 353–364. [CrossRef]
136. Dal Prà, I.; Armato, U.; Chiarini, A. Specific interactions of calcium-sensing receptors (CaSRs) with soluble amyloid-β peptides—A study using cultured normofunctioning adult human astrocytes. In Proceedings of the 2nd International Symposium on the Calcium-Sensing Receptor, San Diego, CA, USA, 3–4 March 2015; pp. 90–91.
137. Hofer, A.M.; Brown, E.M. Extracellular calcium sensing and signaling. *Nat. Rev. Mol. Cell Biol.* **2003**, *4*, 530–538. [CrossRef]
138. Gardenal, E.; Chiarini, A.; Armato, U.; Dal Prà, I.; Verkhratsky, A.; Rodríguez, J.J. Increased calcium-sensing receptor immunoreactivity in the hippocampus of a triple transgenic mouse model of Alzheimer's Disease. *Front. Neurosci.* **2017**, *11*, 81. [CrossRef]
139. Gutiérrez-López, T.Y.; Orduña-Castillo, L.B.; Hernández-Vásquez, M.N.; Vázquez-Prado, J.; Reyes-Cruz, G. Calcium sensing receptor activates the NLRP3 inflammasome via a chaperone-assisted degradative pathway involving Hsp70 and LC3-II. *Biochem. Biophys. Res. Commun.* **2018**, *505*, 1121–1127. [CrossRef]
140. Sokolowska, M.; Chen, L.Y.; Liu, Y.; Martinez-Anton, A.; Qi, H.Y.; Logun, C.; Alsaaty, S.; Park, Y.H.; Kastner, D.L.; Chae, J.J.; et al. Prostaglandin E2 Inhibits NLRP3 Inflammasome Activation through EP4 Receptor and Intracellular Cyclic AMP in Human Macrophages. *J. Immunol.* **2015**, *194*, 5472–5487. [CrossRef]
141. Armato, U.; Chiarini, A.; Chakravarthy, B.; Chioffi, F.; Pacchiana, R.; Colarusso, E.; Whitfield, J.F.; Dal Prà, I. Calcium-sensing receptor antagonist (calcilytic) NPS 2143 specifically blocks the increased secretion of endogenous Aβ42 prompted by exogenous fibrillary or soluble Aβ25-35 in human cortical astrocytes and neurons-therapeutic relevance to Alzheimer's disease. *Biochim. Biophys. Acta* **2013**, *1832*, 1634–1652. [CrossRef]
142. Pi, M.; Faber, P.; Ekema, G.; Jackson, P.D.; Ting, A.; Wang, N.; Fontilla-Poole, M.; Mays, R.W.; Brunden, K.R.; Harrington, J.J.; et al. Identification of a novel extracellular cation-sensing G-protein-coupled receptor. *J. Biol. Chem.* **2005**, *280*, 40201–40209. [CrossRef]
143. Pi, M.; Parrill, A.L.; Quarles, L.D. GPRC6A mediates the non-genomic effects of steroids. *J. Biol. Chem.* **2010**, *285*, 39953–39964. [CrossRef] [PubMed]
144. Pi, M.; Wu, Y.; Quarles, L.D. GPRC6A mediates responses to osteocalcin in β-cells in vitro and pancreas in vivo. *J. Bone Miner. Res.* **2011**, *26*, 1680–1683. [CrossRef] [PubMed]
145. Pi, M.; Quarles, L.D. GPRC6A regulates prostate cancer progression. *Prostate* **2011**, *72*, 399–409. [CrossRef] [PubMed]
146. Singh, P.; Dutta, S.R.; Song, C.Y.; Oh, S.; Gonzalez, F.J.; Malik, K.U. Brain Testosterone-CYP1B1 (Cytochrome P450 1B1) Generated Metabolite 6β-Hydroxytestosterone Promotes Neurogenic Hypertension and Inflammation. *Hypertension* **2020**, *76*, 1006–1018. [CrossRef] [PubMed]
147. Bai, N.; Zhang, Q.; Zhang, W.; Liu, B.; Yang, F.; Brann, D.; Wang, R. G-protein-coupled estrogen receptor activation upregulates interleukin-1 receptor antagonist in the hippocampus after global cerebral ischemia: Implications for neuronal self-defense. *J. Neuroinflamm.* **2020**, *17*, 45. [CrossRef]
148. Py, B.F.; Kim, M.S.; Vakifahmetoglu-Norberg, H.; Yuan, J. Deubiquitination of NLRP3 by BRCC3 critically regulates inflammasome activity. *Mol. Cell.* **2013**, *49*, 331–338. [CrossRef]
149. Yang, J.; Wise, L.; Fukuchi, K.I. TLR4 Cross-Talk with NLRP3 Inflammasome and Complement Signaling Pathways in Alzheimer's Disease. *Front. Immunol.* **2020**, *11*, 724. [CrossRef]
150. McKee, C.M.; Coll, R.C. NLRP3 inflammasome priming: A riddle wrapped in a mystery inside an enigma. *J. Leukoc. Biol.* **2020**, *108*, 937–952. [CrossRef]
151. Chen, M.-Y.; Ye, X.J.; He, X.H.; Ouyang, D.Y. The Signaling Pathways Regulating NLRP3 Inflammasome Activation. *Inflammation* **2021**, *44*, 1229–1245. [CrossRef]
152. Dierckx, T.; Haidar, M.; Grajchen, E.; Wouters, E.; Vanherle, S.; Loix, M.; Boeykens, A.; Bylemans, D.; Hardonnière, K.; Kerdine-Römer, S.; et al. Phloretin suppresses neuroinflammation by autophagy-mediated Nrf2 activation in macrophages. *J. Neuroinflamm.* **2021**, *18*, 148. [CrossRef]
153. Katsnelson, M.A.; Rucker, L.G.; Russo, H.M.; Dubyak, G.R. K+ efflux agonists induce NLRP3 inflammasome activation independently of Ca^{2+} signaling. *J. Immunol.* **2015**, *194*, 3937–3952. [CrossRef]
154. Elliott, E.I.; Sutterwala, F.S. Initiation and perpetuation of NLRP3 inflammasome activation and assembly. *Immunol. Rev.* **2015**, *265*, 35–52. [CrossRef]
155. Zhou, Y.; Tong, Z.; Jiang, S.; Zheng, W.; Zhao, J.; Zhou, X. The Roles of Endoplasmic Reticulum in NLRP3 Inflammasome Activation. *Cells* **2020**, *9*, 1219. [CrossRef]

156. Jäger, E.; Murthy, S.; Schmidt, C.; Hahn, M.; Strobel, S.; Peters, A.; Stäubert, C.; Sungur, P.; Venus, T.; Geisler, M.; et al. Calcium-sensing receptor-mediated NLRP3 inflammasome response to calciprotein particles drives inflammation in rheumatoid arthritis. *Nat. Commun.* **2020**, *11*, 4243. [CrossRef]
157. Murakami, T.; Ockinger, J.; Yu, J.; Byles, V.; McColl, A.; Hofer, A.M.; Horng, T. Critical role for calcium mobilization in activation of the NLRP3 inflammasome. *Proc. Natl. Acad. Sci. USA* **2012**, *109*, 11282–11287. [CrossRef]
158. Rossol, M.; Pierer, M.; Raulien, N.; Quandt, D.; Meusch, U.; Rothe, K.; Schubert, K.; Schöneberg, T.; Schaefer, M.; Krügel, U.; et al. Extracellular Ca2+ is a danger signal activating the NLRP3 inflammasome through G protein-coupled calcium sensing receptors. *Nat. Commun.* **2012**, *3*, 1329. [CrossRef]
159. Kim, K.; Kim, H.J.; Binas, B.; Kang, J.H.; Chung, I.Y. Inflammatory mediators ATP and S100A12 activate the NLRP3 inflammasome to induce MUC5AC production in airway epithelial cells. *Biochem. Biophys. Res. Commun.* **2018**, *503*, 657–664. [CrossRef]
160. Thawkar, B.S.; Kaur, G. Inhibitors of NF-κB and P2X7/NLRP3/Caspase 1 pathway in microglia: Novel therapeutic opportunities in neuroinflammation induced early-stage Alzheimer's disease. *J. Neuroimmunol.* **2019**, *326*, 62–74. [CrossRef]
161. Lim, J.C.; Lu, W.; Beckel, J.M.; Mitchell, C.H. Neuronal Release of Cytokine IL-3 Triggered by Mechanosensitive Autostimulation of the P2X7 Receptor Is Neuroprotective. *Front. Cell. Neurosci.* **2016**, *10*, 270. [CrossRef]
162. Lu, W.; Albalawi, F.; Beckel, J.M.; Lim, J.C.; Laties, A.M.; Mitchell, C.H. The P2X7 receptor links mechanical strain to cytokine IL-6 up-regulation and release in neurons and astrocytes. *J. Neurochem.* **2017**, *141*, 436–448. [CrossRef]
163. Campagno, K.E.; Mitchell, C.H. The $P2X_7$Receptor in Microglial Cells Modulates the Endolysosomal Axis, Autophagy, and Phagocytosis. *Front. Cell. Neurosci.* **2021**, *15*, 645244. [CrossRef] [PubMed]
164. Shieh, C.H.; Heinrich, A.; Serchov, T.; van Calker, D.; Biber, K. P2X7-dependent, but differentially regulated release of IL-6, CCL2, and TNF-α in cultured mouse microglia. *Glia* **2014**, *62*, 592–607. [CrossRef] [PubMed]
165. Cieślak, M.; Wojtczak, A. Role of purinergic receptors in the Alzheimer's disease. *Purinergic Signal.* **2018**, *14*, 331–344. [CrossRef] [PubMed]
166. Erb, L.; Woods, L.T.; Khalafalla, M.G.; Weisman, G.A. Purinergic signaling in Alzheimer's disease. *Brain Res. Bull.* **2019**, *151*, 25–37. [CrossRef]
167. Duez, H.; Pourcet, B. Nuclear Receptors in the Control of the NLRP3 Inflammasome Pathway. *Front. Endocrinol.* **2021**, *12*, 630536. [CrossRef]
168. Swanson, K.V.; Deng, M.; Ting, J.P. The NLRP3 inflammasome: Molecular activation and regulation to therapeutics. *Nat. Rev. Immunol.* **2019**, *19*, 477–489. [CrossRef]
169. Liang, Z.; Damianou, A.; Di Daniel, E.; Kessler, B.M. Inflammasome activation controlled by the interplay between post-translational modifications: Emerging drug target opportunities. *Cell Commun. Signal.* **2021**, *19*, 23. [CrossRef]
170. Weber, A.N.R. Targeting the NLRP3 Inflammasome via BTK. *Front. Cell Dev. Biol.* **2021**, *9*, 630479. [CrossRef]
171. Bezbradica, J.S.; Coll, R.C.; Schroder, K. Sterile signals generate weaker and delayed macrophage NLRP3 inflammasome responses relative to microbial signals. *Cell. Mol. Immunol.* **2017**, *14*, 118–126. [CrossRef]
172. Healy, L.M.; Yaqubi, M.; Ludwin, S.; Antel, J.P. Species differences in immune-mediated CNS tissue injury and repair: A (neuro)inflammatory topic. *Glia* **2020**, *68*, 811–829. [CrossRef]
173. Zhang, C.J.; Jiang, M.; Zhou, H.; Liu, W.; Wang, C.; Kang, Z.; Han, B.; Zhang, Q.; Chen, X.; Xiao, J.; et al. TLR-stimulated IRAKM activates caspase-8 inflammasome in microglia and promotes neuroinflammation. *J. Clin. Investing.* **2018**, *128*, 5399–5412. [CrossRef]
174. Kayagaki, N.; Warming, S.; Lamkanfi, M.; Vande Walle, L.; Louie, S.; Dong, J.; Newton, K.; Qu, Y.; Liu, J.; Heldens, S.; et al. Non-canonical inflammasome activation targets caspase-11. *Nature* **2011**, *479*, 117–121. [CrossRef] [PubMed]
175. Elizagaray, M.L.; Gomes, M.T.R.; Guimaraes, E.S.; Rumbo, M.; Hozbor, D.F.; Oliveira, S.C.; Moreno, G. Canonical and Non-canonical Inflammasome Activation by Outer Membrane Vesicles Derived from Bordetella pertussis. *Front. Immunol.* **2020**, *11*, 1879. [CrossRef]
176. Matikainen, S.; Nyman, T.A.; Cypryk, W. Function and Regulation of Noncanonical Caspase-4/5/11 Inflammasome. *J. Immunol.* **2020**, *204*, 3063–3069. [CrossRef]
177. Yi, Y.S. Caspase-11 Noncanonical Inflammasome: A Novel Key Player in Murine Models of Neuroinflammation and Multiple Sclerosis. *Neuroimmunomodulation* **2021**, *28*, 195–203. [CrossRef]
178. Zhang, D.; Qian, J.; Zhang, P.; Li, H.; Shen, H.; Li, X.; Chen, G. Gasdermin D serves as a key executioner of pyroptosis in experimental cerebral ischemia and reperfusion model both in vivo and in vitro. *J. Neurosci. Res.* **2019**, *97*, 645–660. [CrossRef]
179. Wang, K.; Sun, Z.; Ru, J.; Wang, S.; Huang, L.; Ruan, L.; Lin, X.; Jin, K.; Zhuge, Q.; Yang, S. Ablation of GSDMD Improves Outcome of Ischemic Stroke Through Blocking Canonical and Non-canonical Inflammasomes Dependent Pyroptosis in Microglia. *Front. Neurol.* **2020**, *11*, 577927. [CrossRef]
180. Carpenter, S.; Aiello, D.; Atianand, M.K.; Ricci, E.P.; Gandhi, P.; Hall, L.L.; Byron, M.; Monks, B.; Henry-Bezy, M.; Lawrence, J.B.; et al. A long noncoding RNA mediates both activation and repression of immune response genes. *Science* **2013**, *341*, 789–792. [CrossRef]
181. Heward, J.A.; Lindsay, M.A. Long non-coding RNAs in the regulation of the immune response. *Trends Immunol.* **2014**, *35*, 408–419. [CrossRef]
182. Bartel, D.P. Metazoan MicroRNAs. *Cell* **2018**, *173*, 20–51. [CrossRef]

183. Kiesel, P.; Gibson, T.J.; Ciesielczyk, B.; Bodemer, M.; Kaup, F.J.; Bodemer, W.; Zischler, H.; Zerr, I. Transcription of Alu DNA elements in blood cells of sporadic Creutzfeldt-Jakob disease (sCJD). *Prion* **2010**, *4*, 87–93. [CrossRef] [PubMed]
184. Polesskaya, O.; Kananykhina, E.; Roy-Engel, A.M.; Nazarenko, O.; Kulemzina, I.; Baranova, A.; Vassetsky, Y.; Myakishev-Rempel, M. The role of Alu-derived RNAs in Alzheimer's and other neurodegenerative conditions. *Med. Hypotheses* **2018**, *115*, 29–34. [CrossRef] [PubMed]
185. Cheng, Y.; Saville, L.; Gollen, B.; Isaac, C.; Belay, A.; Mehla, J.; Patel, K.; Thakor, N.; Mohajerani, M.H.; Zovoilis, A. Increased processing of SINE B2 ncRNAs unveils a novel type of transcriptome deregulation in amyloid beta neuropathology. *eLife* **2020**, *9*, e61265. [CrossRef]
186. Cheng, Y.; Saville, L.; Gollen, B.; Veronesi, A.A.; Mohajerani, M.; Joseph, J.T.; Zovoilis, A. Increased Alu RNA processing in Alzheimer brains is linked to gene expression changes. *EMBO Rep.* **2021**, *22*, e52255. [CrossRef]
187. Zhao, Y.; Chen, Y.; Wang, Z.; Xu, C.; Qiao, S.; Liu, T.; Qi, K.; Tong, D.; Li, C. Bone Marrow Mesenchymal Stem Cell Exosome Attenuates Inflammasome-Related Pyroptosis via Delivering circ_003564 to Improve the Recovery of Spinal Cord Injury. *Mol. Neurobiol.* **2022**, *59*, 6771–6789. [CrossRef]
188. Xue, Z.; Zhang, Z.; Liu, H.; Li, W.; Guo, X.; Zhang, Z.; Liu, Y.; Jia, L.; Li, Y.; Ren, Y.; et al. lincRNA-Cox2 regulates NLRP3 inflammasome and autophagy mediated neuroinflammation. *Cell Death Differ.* **2019**, *26*, 130–145. [CrossRef]
189. Meng, J.; Ding, T.; Chen, Y.; Long, T.; Xu, Q.; Lian, W.; Liu, W. LncRNA-Meg3 promotes Nlrp3-mediated microglial inflammation by targeting miR-7a-5p. *Int. Immunopharmacol.* **2021**, *90*, 107141. [CrossRef]
190. Docrat, T.F.; Nagiah, S.; Chuturgoon, A.A. Metformin protects against neuroinflammation through integrated mechanisms of miR-141 and the NF-κB-mediated inflammasome pathway in a diabetic mouse model. *Eur. J. Pharmacol.* **2021**, *903*, 174146. [CrossRef]
191. Cunha, C.; Gomes, C.; Vaz, A.R.; Brites, D. Exploring New Inflammatory Biomarkers and Pathways during LPS-Induced M1 Polarization. *Mediat. Inflamm.* **2016**, *2016*, 6986175. [CrossRef]
192. Si, L.; Wang, H.; Wang, L. Suppression of miR-193a alleviates neuroinflammation and improves neurological function recovery after traumatic brain injury (TBI) in mice. *Biochem. Biophys. Res. Commun.* **2020**, *523*, 527–534. [CrossRef]
193. Cao, Y.; Tan, X.; Lu, Q.; Huang, K.; Tang, X.; He, Z. miR-590-3 and SP1 Promote Neuronal Apoptosis in Patients with Alzheimer's Disease via AMPK Signaling Pathway. *Contrast Media Mol. Imaging* **2021**, *2021*, 6010362. [CrossRef] [PubMed]
194. Zhang, H.; Tao, J.; Zhang, S.; Lv, X. LncRNA MEG3 Reduces Hippocampal Neuron Apoptosis via the PI3K/AKT/mTOR Pathway in a Rat Model of Temporal Lobe Epilepsy. *Neuropsychiatr. Dis. Treat.* **2020**, *16*, 2519–2528. [CrossRef]
195. Fan, Z.; Lu, M.; Qiao, C.; Zhou, Y.; Ding, J.H.; Hu, G. MicroRNA-7 Enhances Subventricular Zone Neurogenesis by Inhibiting NLRP3/Caspase-1 Axis in Adult Neural Stem Cells. *Mol. Neurobiol.* **2016**, *53*, 7057–7069. [CrossRef]
196. Zhou, Y.; Lu, M.; Du, R.H.; Qiao, C.; Jiang, C.Y.; Zhang, K.Z.; Ding, J.H.; Hu, G. MicroRNA-7 targets Nod-like receptor protein 3 inflammasome to modulate neuroinflammation in the pathogenesis of Parkinson's disease. *Mol. Neurodegener.* **2016**, *11*, 28. [CrossRef] [PubMed]
197. Cui, G.H.; Wu, J.; Mou, F.F.; Xie, W.H.; Wang, F.B.; Wang, Q.L.; Fang, J.; Xu, Y.W.; Dong, Y.R.; Liu, J.R.; et al. Exosomes derived from hypoxia-preconditioned mesenchymal stromal cells ameliorate cognitive decline by rescuing synaptic dysfunction and regulating inflammatory responses in APP/PS1 mice. *FASEB J.* **2018**, *32*, 654–668. [CrossRef]
198. Han, C.; Guo, L.; Yang, Y.; Guan, Q.; Shen, H.; Sheng, Y.; Jiao, Q. Mechanism of microRNA-22 in regulating neuroinflammation in Alzheimer's disease. *Brain Behav.* **2020**, *10*, e01627. [CrossRef]
199. Zhai, L.; Shen, H.; Sheng, Y.; Guan, Q. ADMSC Exo-MicroRNA-22 improve neurological function and neuroinflammation in mice with Alzheimer's disease. *J. Cell. Mol. Med.* **2021**, *25*, 7513–7523, Erratum in *J. Cell. Mol. Med.* **2021**, *25*, 11037–11038. [CrossRef]
200. Hu, L.T.; Wang, B.Y.; Fan, Y.H.; He, Z.Y.; Zheng, W.X. Exosomal miR-23b from bone marrow mesenchymal stem cells alleviates oxidative stress and pyroptosis after intracerebral hemorrhage. *Neural Regen. Res.* **2023**, *18*, 560–567. [CrossRef]
201. Cao, Y.; Tan, X.; Lu, Q.; Huang, K.; Tang, X.; He, Z. MiR-29c-3p May Promote the Progression of Alzheimer's Disease through BACE1. *J. Healthc. Eng.* **2021**, *2021*, 2031407. [CrossRef]
202. Sha, S.; Shen, X.; Cao, Y.; Qu, L. Mesenchymal stem cells-derived extracellular vesicles ameliorate Alzheimer's disease in rat models via the microRNA-29c-3p/BACE1 axis and the Wnt/β-catenin pathway. *Aging* **2021**, *13*, 15285–15306. [CrossRef]
203. Hu, L.; Zhang, H.; Wang, B.; Ao, Q.; He, Z. MicroRNA-152 attenuates neuroinflammation in intracerebral hemorrhage by inhibiting thioredoxin interacting protein (TXNIP)-mediated NLRP3 inflammasome activation. *Int. Immunopharmacol.* **2020**, *80*, 106141. [CrossRef] [PubMed]
204. Li, Q.; Wang, Z.; Xing, H.; Wang, Y.; Guo, Y. Exosomes derived from miR-188-3p-modified adipose-derived mesenchymal stem cells protect Parkinson's disease. *Mol. Ther. Nucleic Acids* **2021**, *23*, 1334–1344. [CrossRef] [PubMed]
205. Wan, S.Y.; Li, G.S.; Tu, C.; Chen, W.L.; Wang, X.W.; Wang, Y.N.; Peng, L.B.; Tan, F. MicroNAR-194-5p hinders the activation of NLRP3 inflammasomes and alleviates neuroinflammation during intracerebral hemorrhage by blocking the interaction between TRAF6 and NLRP3. *Brain Res.* **2021**, *1752*, 147228. [CrossRef] [PubMed]
206. Mancuso, R.; Agostini, S.; Hernis, A.; Zanzottera, M.; Bianchi, A.; Clerici, M. Circulatory miR-223-3p Discriminates Between Parkinson's and Alzheimer's Patients. *Sci. Rep.* **2019**, *9*, 9393. [CrossRef] [PubMed]
207. Chen, Z.; Hu, Y.; Lu, R.; Ge, M.; Zhang, L. MicroRNA-374a-5p inhibits neuroinflammation in neonatal hypoxic-ischemic encephalopathy via regulating NLRP3 inflammasome targeted Smad6. *Life Sci.* **2020**, *252*, 117664. [CrossRef]

208. Kaur, S.; Verma, H.; Dhiman, M.; Tell, G.; Gigli, G.L.; Janes, F.; Mantha, A.K. Brain Exosomes: Friend or Foe in Alzheimer's Disease? *Mol. Neurobiol.* **2021**, *58*, 6610–6624. [CrossRef]
209. Hu, Z.; Yuan, Y.; Zhang, X.; Lu, Y.; Dong, N.; Jiang, X.; Xu, J.; Zheng, D. Human Umbilical Cord Mesenchymal Stem Cell-Derived Exosomes Attenuate Oxygen-Glucose Deprivation/Reperfusion-Induced Microglial Pyroptosis by Promoting FOXO3a-Dependent Mitophagy. *Oxid. Med. Cell. Longev.* **2021**, *2021*, 6219715. [CrossRef]
210. Liu, X.; Zhang, M.; Liu, H.; Zhu, R.; He, H.; Zhou, Y.; Zhang, Y.; Li, C.; Liang, D.; Zeng, Q.; et al. Bone marrow mesenchymal stem cell-derived exosomes attenuate cerebral ischemia-reperfusion injury-induced neuroinflammation and pyroptosis by modulating microglia M1/M2 phenotypes. *Exp. Neurol.* **2021**, *341*, 113700. [CrossRef]
211. Cui, G.H.; Guo, H.D.; Li, H.; Zhai, Y.; Gong, Z.B.; Wu, J.; Liu, J.S.; Dong, Y.R.; Hou, S.X.; Liu, J.R. RVG-modified exosomes derived from mesenchymal stem cells rescue memory deficits by regulating inflammatory responses in a mouse model of Alzheimer's disease. *Immun. Ageing* **2019**, *16*, 10. [CrossRef]
212. Chiarini, A.; Armato, U.; Gardenal, E.; Gui, L.; Dal Prà, I. Amyloid β-exposed human astrocytes overproduce phospho-Tau and overrelease it within exosomes, effects suppressed by calcilytic NPS 2143. Further implications for Alzheimer's therapy. *Front. Neurosci.* **2017**, *11*, 217. [CrossRef]
213. Sardar Sinha, M.; Ansell-Schultz, A.; Civitelli, L.; Hildesjö, C.; Larsson, M.; Lannfelt, L.; Ingelsson, M.; Hallbeck, M. Alzheimer's disease pathology propagation by exosomes containing toxic amyloid-beta oligomers. *Acta Neuropathol.* **2018**, *136*, 41–56. [CrossRef]
214. Zhang, Q.; Sun, Y.; He, Z.; Xu, Y.; Li, X.; Ding, J.; Lu, M.; Hu, G. Kynurenine regulates NLRP2 inflammasome in astrocytes and its implications in depression. *Brain Behav. Immun.* **2020**, *88*, 471–481. [CrossRef]
215. Voet, S.; Mc Guire, C.; Hagemeyer, N.; Martens, A.; Schroeder, A.; Wieghofer, P.; Daems, C.; Staszewski, O.; Vande Walle, L.; Jordao, M.J.C.; et al. A20 critically controls microglia activation and inhibits inflammasome-dependent neuroinflammation. *Nat. Commun.* **2018**, *9*, 2036. [CrossRef]
216. Gaikwad, S.; Patel, D.; Agrawal-Rajput, R. CD40 Negatively Regulates ATP-TLR4-Activated Inflammasome in Microglia. *Cell. Mol. Neurobiol.* **2017**, *37*, 351–359. [CrossRef]
217. Ma, S.; Wang, Y.; Zhou, X.; Li, Z.; Zhang, Z.; Wang, Y.; Huang, T.; Zhang, Y.; Shi, J.; Guan, F. MG53 Protects hUC-MSCs against Inflammatory Damage and Synergistically Enhances Their Efficacy in Neuroinflammation Injured Brain through Inhibiting NLRP3/Caspase-1/IL-1β Axis. *ACS Chem. Neurosci.* **2020**, *11*, 2590–2601. [CrossRef]
218. Xiao, T.; Wan, J.; Qu, H.; Li, Y. Tripartite-motif protein 21 knockdown extenuates LPS-triggered neurotoxicity by inhibiting microglial M1 polarization via suppressing NF-κB-mediated NLRP3 inflammasome activation. *Arch. Biochem. Biophys.* **2021**, *706*, 108918. [CrossRef]
219. Gal-Ben-Ari, S.; Barrera, I.; Ehrlich, M.; Rosenblum, K. PKR: A Kinase to Remember. *Front. Mol. Neurosci.* **2019**, *11*, 480. [CrossRef]
220. Lu, B.; Nakamura, T.; Inouye, K.; Li, J.; Tang, Y.; Lundbäck, P.; Valdes-Ferrer, S.I.; Olofsson, P.S.; Kalb, T.; Roth, J.; et al. Novel role of PKR in inflammasome activation and HMGB1 release. *Nature* **2012**, *488*, 670–674. [CrossRef]
221. He, Y.; Franchi, L.; Núñez, G. The protein kinase PKR is critical for LPS-induced iNOS production but dispensable for inflammasome activation in macrophages. *Eur. J. Immunol.* **2013**, *43*, 1147–1152. [CrossRef]
222. Dempsey, C.; Rubio Araiz, A.; Bryson, K.J.; Finucane, O.; Larkin, C.; Mills, E.L.; Robertson, A.; Cooper, M.A.; O'Neill, L.; Lynch, M.A. Inhibiting the NLRP3 inflammasome with MCC950 promotes non-phlogistic clearance of amyloid-β and cognitive function in APP/PS1 mice. *Brain Behav. Immun.* **2017**, *61*, 306–316. [CrossRef]
223. Heitzer, M.; Kaiser, S.; Kanagaratnam, M.; Zendedel, A.; Hartmann, P.; Beyer, C.; Johann, S. Administration of 17beta-Estradiol Improves Motoneuron Survival and Down-regulates Inflammasome Activation in Male SOD1 (G93A) ALS Mice. *Mol. Neurobiol.* **2017**, *54*, 8429–8443. [CrossRef] [PubMed]
224. Aryanpour, R.; Zibara, K.; Pasbakhsh, P.; Jame'ei, S.B.; Namjoo, Z.; Ghanbari, A.; Mahmoudi, R.; Amani, S.; Kashani, I.R. 17β-Estradiol Reduces Demyelination in Cuprizone-fed Mice by Promoting M2 Microglia Polarity and Regulating NLRP3 Inflammasome. *Neuroscience* **2021**, *463*, 116–127. [CrossRef] [PubMed]
225. Thakkar, R.; Wang, R.; Wang, J.; Vadlamudi, R.K.; Brann, D.W. 17β-Estradiol Regulates Microglia Activation and Polarization in the Hippocampus Following Global Cerebral Ischemia. *Oxid. Med. Cell. Longev.* **2018**, *2018*, 4248526. [CrossRef] [PubMed]
226. Jiang, W.; Huang, Y.; He, F.; Liu, J.; Li, M.; Sun, T.; Ren, W.; Hou, J.; Zhu, L. Dopamine D1 Receptor Agonist A-68930 Inhibits NLRP3 Inflammasome Activation, Controls Inflammation, and Alleviates Histopathology in a Rat Model of Spinal Cord Injury. *Spine (Phila Pa 1976)* **2016**, *41*, E330–E334. [CrossRef]
227. Wang, S.; Yao, Q.; Wan, Y.; Wang, J.; Huang, C.; Li, D.; Yang, B. Adiponectin reduces brain injury after intracerebral hemorrhage by reducing NLRP3 inflammasome expression. *Int. J. Neurosci.* **2020**, *130*, 301–308. [CrossRef]
228. Li, J.; Wu, D.M.; Yu, Y.; Deng, S.H.; Liu, T.; Zhang, T.; He, M.; Zhao, Y.Y.; Xu, Y. Amifostine ameliorates induction of experimental autoimmune encephalomyelitis: Effect on reactive oxygen species/NLRP3 pathway. *Int. Immunopharmacol.* **2020**, *88*, 106998. [CrossRef]
229. Li, B.X.; Dai, X.; Xu, X.R.; Adili, R.; Neves, M.A.D.; Lei, X.; Shen, C.; Zhu, G.; Wang, Y.; Zhou, H.; et al. In vitro assessment and phase I randomized clinical trial of anfibatide a snake venom derived anti-thrombotic agent targeting human platelet GPIbα. *Sci. Rep.* **2021**, *11*, 11663. [CrossRef]
230. Li, R.; Si, M.; Jia, H.Y.; Ma, Z.; Li, X.W.; Li, X.Y.; Dai, X.R.; Gong, P.; Luo, S.Y. Anfibatide alleviates inflammation and apoptosis via inhibiting NF-kappaB/NLRP3 axis in ischemic stroke. *Eur. J. Pharmacol.* **2022**, *926*, 175032. [CrossRef]

231. Liu, P.; Gao, Q.; Guan, L.; Hu, Y.; Jiang, J.; Gao, T.; Sheng, W.; Xue, X.; Qiao, H.; Li, T. Atorvastatin attenuates surgery-induced BBB disruption and cognitive impairment partly by suppressing NF-κB pathway and NLRP3 inflammasome activation in aged mice. *Acta Biochim. Biophys. Sin.* **2021**, *53*, 528–537. [CrossRef]
232. Jiang, W.; Li, M.; He, F.; Zhou, S.; Zhu, L. Targeting the NLRP3 inflammasome to attenuate spinal cord injury in mice. *J. Neuroinflamm.* **2017**, *14*, 207. [CrossRef]
233. Yang, T.; Zhang, L.; Shang, Y.; Zhu, Z.; Jin, S.; Guo, Z.; Wang, X. Concurrent suppression of Aβ aggregation and NLRP3 inflammasome activation for treating Alzheimer's disease. *Chem. Sci.* **2022**, *13*, 2971–2980. [CrossRef]
234. Wang, H.Q.; Song, K.Y.; Feng, J.Z.; Huang, S.Y.; Guo, X.M.; Zhang, L.; Zhang, G.; Huo, Y.C.; Zhang, R.R.; Ma, Y.; et al. Caffeine Inhibits Activation of the NLRP3 Inflammasome via Autophagy to Attenuate Microglia-Mediated Neuroinflammation in Experimental Autoimmune Encephalomyelitis. *J. Mol. Neurosci.* **2022**, *72*, 97–112. [CrossRef]
235. de Oliveira, L.R.C.; Mimura, L.A.N.; Fraga-Silva, T.F.C.; Ishikawa, L.L.W.; Fernandes, A.A.H.; Zorzella-Pezavento, S.F.G.; Sartori, A. Calcitriol Prevents Neuroinflammation and Reduces Blood-Brain Barrier Disruption and Local Macrophage/Microglia Activation. *Front. Pharmacol.* **2020**, *11*, 161. [CrossRef]
236. Wang, Y.; Guan, X.; Chen, X.; Cai, Y.; Ma, Y.; Ma, J.; Zhang, Q.; Dai, L.; Fan, X.; Bai, Y. Choline Supplementation Ameliorates Behavioral Deficits and Alzheimer's Disease-Like Pathology in Transgenic APP/PS1 Mice. *Mol. Nutr. Food Res.* **2019**, *63*, e1801407. [CrossRef]
237. Lonnemann, N.; Hosseini, S.; Marchetti, C.; Skouras, D.B.; Stefanoni, D.; D'Alessandro, A.; Dinarello, C.A.; Korte, M. The NLRP3 inflammasome inhibitor OLT1177 rescues cognitive impairment in a mouse model of Alzheimer's disease. *Proc. Natl. Acad. Sci. USA* **2020**, *117*, 32145–32154. [CrossRef]
238. Bao, Y.; Zhu, Y.; He, G.; Ni, H.; Liu, C.; Ma, L.; Zhang, L.; Shi, D. Dexmedetomidine Attenuates Neuroinflammation in LPS-Stimulated BV2 Microglia Cells Through Upregulation Of miR-340. *Drug Des. Devel. Ther.* **2019**, *13*, 3465–3475. [CrossRef]
239. Feng, J.; Wang, J.X.; Du, Y.H.; Liu, Y.; Zhang, W.; Chen, J.F.; Liu, Y.J.; Zheng, M.; Wang, K.J.; He, G.Q. Dihydromyricetin inhibits microglial activation and neuroinflammation by suppressing NLRP3 inflammasome activation in APP/PS1 transgenic mice. *CNS Neurosci. Ther.* **2018**, *24*, 1207–1218. [CrossRef]
240. Yan, Y.; Jiang, W.; Liu, L.; Wang, X.; Ding, C.; Tian, Z.; Zhou, R. Dopamine controls systemic inflammation through inhibition of NLRP3 inflammasome. *Cell* **2015**, *160*, 62–73. [CrossRef]
241. Nizami, S.; Arunasalam, K.; Green, J.; Cook, J.; Lawrence, C.B.; Zarganes-Tzitzikas, T.; Davis, J.B.; Di Daniel, E.; Brough, D. Inhibition of the NLRP3 inflammasome by HSP90 inhibitors. *Immunology* **2021**, *162*, 84–91. [CrossRef]
242. Gao, S.; Xu, T.; Guo, H.; Deng, Q.; Xun, C.; Liang, W.; Sheng, W. Ameliorative effects of echinacoside against spinal cord injury via inhibiting NLRP3 inflammasome signaling pathway. *Life Sci.* **2019**, *237*, 116978. [CrossRef]
243. Kiasalari, Z.; Afshin-Majd, S.; Baluchnejadmojarad, T.; Azadi-Ahmadabadi, E.; Esmaeil-Jamaat, E.; Fahanik-Babaei, J.; Fakour, M.; Fereidouni, F.; Ghasemi-Tarie, R.; Jalalzade-Ogvar, S.; et al. Ellagic acid ameliorates neuroinflammation and demyelination in experimental autoimmune encephalomyelitis: Involvement of NLRP3 and pyroptosis. *J. Chem. Neuroanat.* **2021**, *111*, 101891. [CrossRef] [PubMed]
244. Yang, X.; Sun, J.; Kim, T.J.; Kim, Y.J.; Ko, S.B.; Kim, C.K.; Jia, X.; Yoon, B.W. Pretreatment with low-dose fimasartan ameliorates NLRP3 inflammasome-mediated neuroinflammation and brain injury after intracerebral hemorrhage. *Exp. Neurol.* **2018**, *310*, 22–32. [CrossRef] [PubMed]
245. Abu-Elfotuh, K.; Al-Najjar, A.H.; Mohammed, A.A.; Aboutaleb, A.S.; Badawi, G.A. Fluoxetine ameliorates Alzheimer's disease progression and prevents the exacerbation of cardiovascular dysfunction of socially isolated depressed rats through activation of Nrf2/HO-1 and hindering TLR4/NLRP3 inflammasome signaling pathway. *Int. Immunopharmacol.* **2022**, *104*, 108488. [CrossRef] [PubMed]
246. Liu, F.; Li, Z.; He, X.; Yu, H.; Feng, J. Ghrelin Attenuates Neuroinflammation and Demyelination in Experimental Autoimmune Encephalomyelitis Involving NLRP3 Inflammasome Signaling Pathway and Pyroptosis. *Front. Pharmacol.* **2019**, *10*, 1320. [CrossRef] [PubMed]
247. Qu, J.; Tao, X.Y.; Teng, P.; Zhang, Y.; Guo, C.L.; Hu, L.; Qian, Y.N.; Jiang, C.Y.; Liu, W.T. Blocking ATP-sensitive potassium channel alleviates morphine tolerance by inhibiting HSP70-TLR4-NLRP3-mediated neuroinflammation. *J. Neuroinflamm.* **2017**, *14*, 228. [CrossRef]
248. Hou, L.; Yang, J.; Li, S.; Huang, R.; Zhang, D.; Zhao, J.; Wang, Q. Glibenclamide attenuates 2.5-hexanedione-induced neurotoxicity in the spinal cord of rats through mitigation of NLRP3 inflammasome activation, neuroinflammation and oxidative stress. *Toxicol. Lett.* **2020**, *331*, 152–158. [CrossRef]
249. Shao, B.Z.; Wei, W.; Ke, P.; Xu, Z.Q.; Zhou, J.X.; Liu, C. Activating cannabinoid receptor 2 alleviates pathogenesis of experimental autoimmune encephalomyelitis via activation of autophagy and inhibiting NLRP3 inflammasome. *CNS Neurosci. Ther.* **2014**, *20*, 1021–1028. [CrossRef]
250. Karkhah, A.; Saadi, M.; Pourabdolhossein, F.; Saleki, K.; Nouri, H.R. Indomethacin attenuates neuroinflammation and memory impairment in an STZ-induced model of Alzheimer's like disease. *Immunopharmacol. Immunotoxicol.* **2021**, *43*, 758–766. [CrossRef]
251. Cooper, M.A. Inzomelid is a CNS penetrant anti-inflammatory drug that blocks NLRP3 inflammasome activation targeted to prevent Synuclein Pathology and Dopaminergic Degeneration in Parkinson's disease. In Proceedings of the 7th International Conference on Parkinson's & Movement Disorders, London, UK, 11–12 November 2019.

252. Kuwar, R.; Rolfe, A.; Di, L.; Xu, H.; He, L.; Jiang, Y.; Zhang, S.; Sun, D. A novel small molecular NLRP3 inflammasome inhibitor alleviates neuroinflammatory response following traumatic brain injury. *J. Neuroinflamm.* **2019**, *16*, 81. [CrossRef]
253. Lyu, D.; Wang, F.; Zhang, M.; Yang, W.; Huang, H.; Huang, Q.; Wu, C.; Qian, N.; Wang, M.; Zhang, H.; et al. Ketamine induces rapid antidepressant effects via the autophagy-NLRP3 inflammasome pathway. *Psychopharmacology* **2022**, *239*, 3201–3212. [CrossRef]
254. Liu, S.; Wang, S.; Gu, R.; Che, N.; Wang, J.; Cheng, J.; Yuan, Z.; Cheng, Y.; Liao, Y. The XPO1 Inhibitor KPT-8602 Ameliorates Parkinson's Disease by Inhibiting the NF-κB/NLRP3 Pathway. *Front. Pharmacol.* **2022**, *13*, 847605. [CrossRef] [PubMed]
255. Li, Q.; Feng, H.; Wang, H.; Wang, Y.; Mou, W.; Xu, G.; Zhang, P.; Li, R.; Shi, W.; Wang, Z.; et al. Licochalcone B specifically inhibits the NLRP3 inflammasome by disrupting NEK7-NLRP3 interaction. *EMBO Rep.* **2022**, *23*, e53499. [CrossRef] [PubMed]
256. Wang, H.R.; Tang, J.Y.; Wang, Y.Y.; Farooqi, A.A.; Yen, C.Y.; Yuan, S.F.; Huang, H.W.; Chang, H.W. Manoalide Preferentially Provides Antiproliferation of Oral Cancer Cells by Oxidative Stress-Mediated Apoptosis and DNA Damage. *Cancers* **2019**, *11*, 1303. [CrossRef] [PubMed]
257. Folmer, F.; Jaspars, M.; Schumacher, M.; Dicato, M.; Diederich, M. Marine Natural Products Targeting Phospholipases A2. *Biochem. Pharmacol.* **2010**, *80*, 1793–1800. [CrossRef] [PubMed]
258. Salam, K.A.; Furuta, A.; Noda, N.; Tsuneda, S.; Sekiguchi, Y.; Yamashita, A.; Moriishi, K.; Nakakoshi, M.; Tsubuki, M.; Tani, H.; et al. Inhibition of Hepatitis C Virus NS3 Helicase by Manoalide. *J. Nat. Prod.* **2012**, *75*, 650–654. [CrossRef]
259. Li, C.; Lin, H.; He, H.; Ma, M.; Jiang, W.; Zhou, R. Inhibition of the NLRP3 Inflammasome Activation by Manoalide Ameliorates Experimental Autoimmune Encephalomyelitis Pathogenesis. *Front. Cell Dev. Biol.* **2022**, *10*, 822236. [CrossRef]
260. Fu, Q.; Li, J.; Qiu, L.; Ruan, J.; Mao, M.; Li, S.; Mao, Q. Inhibiting NLRP3 inflammasome with MCC950 ameliorates perioperative neurocognitive disorders, suppressing neuroinflammation in the hippocampus in aged mice. *Int. Immunopharmacol.* **2020**, *82*, 106317. [CrossRef]
261. Swanton, T.; Beswick, J.A.; Hammadi, H.; Morris, L.; Williams, D.; de Cesco, S.; El-Sharkawy, L.; Yu, S.; Green, J.; Davis, J.B.; et al. Selective inhibition of the K+ efflux sensitive NLRP3 pathway by Cl- channel modulation. *Chem. Sci.* **2020**, *11*, 11720–11728. [CrossRef]
262. Muñoz-Jurado, A.; Escribano, B.M.; Caballero-Villarraso, J.; Galván, A.; Agüera, E.; Santamaría, A.; Túnez, I. Melatonin and multiple sclerosis: Antioxidant, anti-inflammatory and immunomodulator mechanism of action. *Inflammopharmacology* **2022**, *5*, 1569–1596. [CrossRef]
263. Madhu, L.N.; Kodali, M.; Attaluri, S.; Shuai, B.; Melissari, L.; Rao, X.; Shetty, A.K. Melatonin improves brain function in a model of chronic Gulf War Illness with modulation of oxidative stress, NLRP3 inflammasomes, and BDNF-ERK-CREB pathway in the hippocampus. *Redox Biol.* **2021**, *43*, 101973. [CrossRef]
264. Fan, L.; Zhaohong, X.; Xiangxue, W.; Yingying, X.; Xiao, Z.; Xiaoyan, Z.; Jieke, Y.; Chao, L. Melatonin Ameliorates the Progression of Alzheimer's Disease by Inducing TFEB Nuclear Translocation, Promoting Mitophagy, and Regulating NLRP3 Inflammasome Activity. *Biomed. Res. Int.* **2022**, *2022*, 8099459. [CrossRef]
265. Farré-Alins, V.; Narros-Fernández, P.; Palomino-Antolín, A.; Decouty-Pérez, C.; Lopez-Rodriguez, A.B.; Parada, E.; Muñoz-Montero, A.; Gómez-Rangel, V.; López-Muñoz, F.; Ramos, E.; et al. Melatonin Reduces NLRP3 Inflammasome Activation by Increasing α7 nAChR-Mediated Autophagic Flux. *Antioxidants* **2020**, *9*, 1299. [CrossRef]
266. Zheng, R.; Ruan, Y.; Yan, Y.; Lin, Z.; Xue, N.; Yan, Y.; Tian, J.; Yin, X.; Pu, J.; Zhang, B. Melatonin Attenuates Neuroinflammation by Down-Regulating NLRP3 Inflammasome via a SIRT1-Dependent Pathway in MPTP-Induced Models of Parkinson's Disease. *J. Inflamm. Res.* **2021**, *14*, 3063–3075. [CrossRef]
267. Zhang, Y.; Zhang, H.; Li, S.; Huang, K.; Jiang, L.; Wang, Y. Metformin Alleviates LPS-Induced Acute Lung Injury by Regulating the SIRT1/NF-κB/NLRP3 Pathway and Inhibiting Endothelial Cell Pyroptosis. *Front. Pharmacol.* **2022**, *13*, 801337. [CrossRef]
268. Chen, Q.; Yin, Y.; Li, L.; Zhang, Y.; He, W.; Shi, Y. Milrinone Ameliorates the Neuroinflammation and Memory Function of Alzheimer's Disease in an APP/PS1 Mouse Model. *Neuropsychiatr. Dis. Treat.* **2021**, *17*, 2129–2139. [CrossRef]
269. Garcez, M.L.; Mina, F.; Bellettini-Santos, T.; da Luz, A.P.; Schiavo, G.L.; Macieski, J.M.C.; Medeiros, E.B.; Marques, A.O.; Magnus, N.Q.; Budni, J. The Involvement of NLRP3 on the Effects of Minocycline in an AD-Like Pathology Induced by β-Amyloid Oligomers Administered to Mice. *Mol. Neurobiol.* **2019**, *56*, 2606–2617. [CrossRef]
270. Cruz, S.L.; Armenta-Reséndiz, M.; Carranza-Aguilar, C.J.; Galván, E.J. Minocycline prevents neuronal hyperexcitability and neuroinflammation in medial prefrontal cortex, as well as memory impairment caused by repeated toluene inhalation in adolescent rats. *Toxicol. Appl. Pharmacol.* **2020**, *395*, 114980. [CrossRef]
271. Chen, W.; Guo, C.; Huang, S.; Jia, Z.; Wang, J.; Zhong, J.; Ge, H.; Yuan, J.; Chen, T.; Liu, X.; et al. MitoQ attenuates brain damage by polarizing microglia towards the M2 phenotype through inhibition of the NLRP3 inflammasome after ICH. *Pharmacol. Res.* **2020**, *161*, 105122. [CrossRef]
272. Chen, W.; Teng, X.; Ding, H.; Xie, Z.; Cheng, P.; Liu, Z.; Feng, T.; Zhang, X.; Huang, W.; Geng, D. Nrf2 protects against cerebral ischemia-reperfusion injury by suppressing programmed necrosis and inflammatory signaling pathways. *Ann. Transl. Med.* **2022**, *10*, 285. [CrossRef]
273. Li, C.; Wang, J.; Fang, Y.; Liu, Y.; Chen, T.; Sun, H.; Zhou, X.F.; Liao, H. Nafamostat mesilate improves function recovery after stroke by inhibiting neuroinflammation in rats. *Brain Behav. Immun.* **2016**, *56*, 230–245. [CrossRef]
274. Coll, R.C.; Schroder, K.; Pelegrín, P. NLRP3 and pyroptosis blockers for treating inflammatory diseases. *Trends Pharmacol. Sci.* **2022**, *43*, 653–668. [CrossRef] [PubMed]

275. Zhang, X.; Xu, A.; Ran, Y.; Wei, C.; Xie, F.; Wu, J. Design, synthesis, and biological evaluation of phenyl vinyl sulfone based NLRP3 inflammasome inhibitors. *Bioorg. Chem.* **2022**, *128*, 106010. [CrossRef] [PubMed]
276. Wang, J.; Zheng, B.; Yang, S.; Tang, X.; Wang, J.; Wei, D. The protective effects of phoenixin-14 against lipopolysaccharide-induced inflammation and inflammasome activation in astrocytes. *Inflamm. Res.* **2020**, *69*, 779–787. [CrossRef] [PubMed]
277. Dong, A.Q.; Yang, Y.P.; Jiang, S.M.; Yao, X.Y.; Qi, D.; Mao, C.J.; Cheng, X.Y.; Wang, F.; Hu, L.F.; Liu, C.F. Pramipexole inhibits astrocytic NLRP3 inflammasome activation via Drd3-dependent autophagy in a mouse model of Parkinson's disease. *Acta Pharmacol. Sin.* **2022**, *44*, 32–43. [CrossRef]
278. Yu, H.; Wu, M.; Lu, G.; Cao, T.; Chen, N.; Zhang, Y.; Jiang, Z.; Fan, H.; Yao, R. Prednisone alleviates demyelination through regulation of the NLRP3 inflammasome in a C57BL/6 mouse model of cuprizone-induced demyelination. *Brain Res.* **2018**, *1678*, 75–84. [CrossRef]
279. Wei, C.; Guo, S.; Liu, W.; Jin, F.; Wei, B.; Fan, H.; Su, H.; Liu, J.; Zhang, N.; Fang, D.; et al. Resolvin D1 ameliorates Inflammation-Mediated Blood-Brain Barrier Disruption After Subarachnoid Hemorrhage in rats by Modulating A20 and NLRP3 Inflammasome. *Front. Pharmacol.* **2021**, *11*, 610734. [CrossRef]
280. Zhang, J.; Guo, J.; Zhao, X.; Chen, Z.; Wang, G.; Liu, A.; Wang, Q.; Zhou, W.; Xu, Y.; Wang, C. Phosphodiesterase-5 inhibitor sildenafil prevents neuroinflammation, lowers beta-amyloid levels and improves cognitive performance in APP/PS1 transgenic mice. *Behav. Brain Res.* **2013**, *250*, 230–237. [CrossRef]
281. Liu, Y.; Dai, Y.; Li, Q.; Chen, C.; Chen, H.; Song, Y.; Hua, F.; Zhang, Z. Beta-amyloid activates NLRP3 inflammasome via TLR4 in mouse microglia. *Neurosci. Lett.* **2020**, *736*, 135279. [CrossRef]
282. Liao, Y.J.; Pan, R.Y.; Kong, X.X.; Cheng, Y.; Du, L.; Wang, Z.C.; Yuan, C.; Cheng, J.B.; Yuan, Z.Q.; Zhang, H.Y. Correction: 1,2,4-Trimethoxybenzene selectively inhibits NLRP3 inflammasome activation and attenuates experimental autoimmune encephalomyelitis. *Acta Pharmacol. Sin.* **2022**, *43*, 504, Erratum in *Acta Pharmacol. Sin.* **2020**, *42*, 1769–1779. [CrossRef]
283. Qiu, J.; Chen, Y.; Zhuo, J.; Zhang, L.; Liu, J.; Wang, B.; Sun, D.; Yu, S.; Lou, H. Urolithin A promotes mitophagy and suppresses NLRP3 inflammasome activation in lipopolysaccharide-induced BV2 microglial cells and MPTP-induced Parkinson's disease model. *Neuropharmacology* **2022**, *207*, 108963. [CrossRef]
284. Tian, D.; Xing, Y.; Gao, W.; Zhang, H.; Song, Y.; Tian, Y.; Dai, Z. Sevoflurane Aggravates the Progress of Alzheimer's Disease Through NLRP3/Caspase-1/Gasdermin D Pathway. *Front. Cell Dev. Biol.* **2022**, *9*, 801422. [CrossRef]
285. Das, S.; Mishra, K.P.; Ganju, L.; Singh, S.B. Andrographolide - A promising therapeutic agent, negatively regulates glial cell derived neurodegeneration of prefrontal cortex, hippocampus and working memory impairment. *J. Neuroimmunol.* **2017**, *313*, 161–175. [CrossRef]
286. Gugliandolo, E.; D'Amico, R.; Cordaro, M.; Fusco, R.; Siracusa, R.; Crupi, R.; Impellizzeri, D.; Cuzzocrea, S.; Di Paola, R. Neuroprotective Effect of Artesunate in Experimental Model of Traumatic Brain Injury. *Front. Neurol.* **2018**, *9*, 590. [CrossRef]
287. Ju, I.G.; Huh, E.; Kim, N.; Lee, S.; Choi, J.G.; Hong, J.; Oh, M.S. Artemisiae Iwayomogii Herba inhibits lipopolysaccharide-induced neuroinflammation by regulating NF-κB and MAPK signaling pathways. *Phytomedicine* **2021**, *84*, 153501. [CrossRef]
288. Li, M.; Li, H.; Fang, F.; Deng, X.; Ma, S. Astragaloside IV attenuates cognitive impairments induced by transient cerebral ischemia and reperfusion in mice via anti-inflammatory mechanisms. *Neurosci. Lett.* **2017**, *639*, 114–119. [CrossRef]
289. Jin, X.; Liu, M.Y.; Zhang, D.F.; Zhong, X.; Du, K.; Qian, P.; Yao, W.F.; Gao, H.; Wei, M.J. Baicalin mitigates cognitive impairment and protects neurons from microglia-mediated neuroinflammation via suppressing NLRP3 inflammasomes and TLR4/NF-κB signaling pathway. *CNS Neurosci. Ther.* **2019**, *25*, 575–590. [CrossRef]
290. Lee, C.M.; Lee, D.S.; Jung, W.K.; Yoo, J.S.; Yim, M.J.; Choi, Y.H.; Park, S.; Seo, S.K.; Choi, J.S.; Lee, Y.M.; et al. Benzyl isothiocyanate inhibits inflammasome activation in E. coli LPS-stimulated BV2 cells. *Int. J. Mol. Med.* **2016**, *38*, 912–918. [CrossRef]
291. Yu, Y.; Wu, D.M.; Li, J.; Deng, S.H.; Liu, T.; Zhang, T.; He, M.; Zhao, Y.Y.; Xu, Y. Bixin Attenuates Experimental Autoimmune Encephalomyelitis by Suppressing TXNIP/NLRP3 Inflammasome Activity and Activating NRF2 Signaling. *Front. Immunol.* **2020**, *11*, 593368. [CrossRef]
292. Satoh, T.; Trudler, D.; Oh, C.K.; Lipton, S.A. Potential Therapeutic Use of the Rosemary Diterpene Carnosic Acid for Alzheimer's Disease, Parkinson's Disease, and Long-COVID through NRF2 Activation to Counteract the NLRP3 Inflammasome. *Antioxidants* **2022**, *11*, 124. [CrossRef]
293. Shi, W.; Xu, G.; Zhan, X.; Gao, Y.; Wang, Z.; Fu, S.; Qin, N.; Hou, X.; Ai, Y.; Wang, C.; et al. Carnosol inhibits inflammasome activation by directly targeting HSP90 to treat inflammasome-mediated diseases. *Cell Death Dis.* **2020**, *11*, 252. [CrossRef]
294. Chu, X.; Zhang, L.; Zhou, Y.; Fang, Q. Cucurbitacin B alleviates cerebral ischemia/reperfusion injury by inhibiting NLRP3 inflammasome-mediated inflammation and reducing oxidative stress. *Biosci. Biotechnol. Biochem.* **2022**, *11*, zbac065. [CrossRef] [PubMed]
295. González-Cofrade, L.; Cuadrado, I.; Amesty, Á.; Estévez-Braun, A.; de Las Heras, B.; Hortelano, S. Dehydroisohispanolone as a Promising NLRP3 Inhibitor Agent: Bioevaluation and Molecular Docking. *Pharmaceuticals* **2022**, *15*, 825. [CrossRef] [PubMed]
296. Zhang, Y.; Liu, D.; Yao, X.; Wen, J.; Wang, Y.; Zhang, Y. DMTHB ameliorates memory impairment in Alzheimer's disease mice through regulation of neuroinflammation. *Neurosci. Lett.* **2022**, *785*, 136770. [CrossRef] [PubMed]
297. Yang, H.; Chen, Y.; Yu, L.; Xu, Y. Esculentoside A exerts anti-inflammatory activity in microglial cells. *Int. Immunopharmacol.* **2017**, *51*, 148–157. [CrossRef]
298. Zheng, X.; Gong, T.; Tang, C.; Zhong, Y.; Shi, L.; Fang, X.; Chen, D.; Zhu, Z. Gastrodin improves neuroinflammation-induced cognitive dysfunction in rats by regulating NLRP3 inflammasome. *BMC Anesthesiol.* **2022**, *22*, 371. [CrossRef]

299. Zhang, Y.; Zhao, Y.; Zhang, J.; Gao, Y.; Li, S.; Chang, C.; Yu, D.; Yang, G. Ginkgolide B inhibits NLRP3 inflammasome activation and promotes microglial M2 polarization in Aβ1-42-induced microglia cells. *Neurosci. Lett.* **2021**, *764*, 136206. [CrossRef]
300. Shao, L.; Dong, C.; Geng, D.; He, Q.; Shi, Y. Ginkgolide B inactivates the NLRP3 inflammasome by promoting autophagic degradation to improve learning and memory impairment in Alzheimer's disease. *Metab. Brain Dis.* **2022**, *37*, 329–341. [CrossRef]
301. Jiang, J.; Sun, X.; Akther, M.; Lian, M.L.; Quan, L.H.; Koppula, S.; Han, J.H.; Kopalli, S.R.; Kang, T.B.; Lee, K.H. Ginsenoside metabolite 20(S)-protopanaxatriol from Panax ginseng attenuates inflammation-mediated NLRP3 inflammasome activation. *J. Ethnopharmacol.* **2020**, *251*, 112564. [CrossRef]
302. Gao, Y.; Li, J.; Wang, J.; Li, X.; Li, J.; Chu, S.; Li, L.; Chen, N.; Zhang, L. Ginsenoside Rg1 prevent and treat inflammatory diseases: A review. *Int. Immunopharmacol.* **2020**, *87*, 106805. [CrossRef]
303. Wang, J.; Wang, D.; Zhou, Z.; Zhang, X.; Zhang, C.; He, Y.; Liu, C.; Yuan, C.; Yuan, D.; Wang, T. Saponins from Panax japonicus alleviate HFD-induced impaired behaviors through inhibiting NLRP3 inflammasome to upregulate AMPA receptors. *Neurochem. Int.* **2021**, *148*, 105098. [CrossRef]
304. Yi, Y.S. Roles of ginsenosides in inflammasome activation. *J. Ginseng Res.* **2019**, *43*, 172–178. [CrossRef]
305. Yi, Y.S. New mechanisms of ginseng saponin-mediated anti-inflammatory action via targeting canonical inflammasome signaling pathways. *J. Ethnopharmacol.* **2021**, *278*, 114292. [CrossRef]
306. Chaturvedi, S.; Tiwari, V.; Gangadhar, N.M.; Rashid, M.; Sultana, N.; Singh, S.K.; Shukla, S.; Wahajuddin, M. Isoformononetin, a dietary isoflavone protects against streptozotocin induced rat model of neuroinflammation through inhibition of NLRP3/ASC/IL-1 axis activation. *Life Sci.* **2021**, *286*, 119989. [CrossRef]
307. Zeng, J.; Chen, Y.; Ding, R. Isoliquiritigenin alleviates early brain injury after experimental intracerebral hemorrhage via suppressing ROS- and/or NF-κB-mediated NLRP3 inflammasome activation by promoting Nrf2 antioxidant pathway. *J. Neuroinflamm.* **2017**, *14*, 119. [CrossRef]
308. Wang, Y.H.; Lv, H.N.; Cui, Q.H.; Tu, P.F.; Jiang, Y.; Zeng, K.W. Isosibiricin inhibits microglial activation by targeting the dopamine D1/D2 receptor-dependent NLRP3/caspase-1 inflammasome pathway. *Acta Pharmacol. Sin.* **2020**, *41*, 173–180. [CrossRef]
309. Manach, C.; Scalbert, A.; Morand, C.; Rémésy, C.; Jiménez, L. Polyphenols: Food sources and bioavailability. *Am. J. Clin. Nutr.* **2004**, *79*, 727–747. [CrossRef]
310. Nabavi, S.F.; Sureda, A.; Dehpour, A.R.; Shirooie, S.; Silva, A.S.; Devi, K.P.; Ahmed, T.; Ishaq, N.; Hashim, R.; Sobarzo-Sánchez, E.; et al. Regulation of autophagy by polyphenols: Paving the road for treatment of neurodegeneration. *Biotechnol. Adv.* **2018**, *36*, 1768–1778. [CrossRef]
311. Han, X.; Sun, S.; Sun, Y.; Song, Q.; Zhu, J.; Song, N.; Chen, M.; Sun, T.; Xia, M.; Ding, J.; et al. Small molecule-driven NLRP3 inflammation inhibition via interplay between ubiquitination and autophagy: Implications for Parkinson disease. *Autophagy* **2019**, *15*, 1860–1881. [CrossRef]
312. Mokarizadeh, N.; Karimi, P.; Erfani, M.; Sadigh-Eteghad, S.; Fathi Maroufi, N.; Rashtchizadeh, N. β-Lapachone attenuates cognitive impairment and neuroinflammation in beta-amyloid induced mouse model of Alzheimer's disease. *Int. Immunopharmacol.* **2020**, *81*, 106300. [CrossRef]
313. Xiong, R.; Zhou, X.G.; Tang, Y.; Wu, J.M.; Sun, Y.S.; Teng, J.F.; Pan, R.; Law, B.Y.; Zhao, Y.; Qiu, W.Q.; et al. Lychee seed polyphenol protects the blood-brain barrier through inhibiting Aβ(25-35)-induced NLRP3 inflammasome activation via the AMPK/mTOR/ULK1-mediated autophagy in bEnd.3 cells and APP/PS1 mice. *Phytother. Res.* **2021**, *35*, 954–973. [CrossRef]
314. Qiu, W.Q.; Pan, R.; Tang, Y.; Zhou, X.G.; Wu, J.M.; Yu, L.; Law, B.Y.; Ai, W.; Yu, C.L.; Qin, D.L.; et al. Lychee seed polyphenol inhibits Aβ-induced activation of NLRP3 inflammasome via the LRP1/AMPK mediated autophagy induction. *Biomed. Pharmacother.* **2020**, *130*, 110575. [CrossRef] [PubMed]
315. Lei, L.Y.; Wang, R.C.; Pan, Y.L.; Yue, Z.G.; Zhou, R.; Xie, P.; Tang, Z.S. Mangiferin inhibited neuroinflammation through regulating microglial polarization and suppressing NF-κB, NLRP3 pathway. *Chin. J. Nat. Med.* **2021**, *19*, 112–119. [CrossRef] [PubMed]
316. Gong, J.; Luo, S.; Zhao, S.; Yin, S.; Li, X.; Mou, T. Myricitrin attenuates memory impairment in a rat model of sepsis-associated encephalopathy via the NLRP3/Bax/Bcl pathway. *Folia Neuropathol.* **2019**, *57*, 327–334. [CrossRef] [PubMed]
317. Yan, T.; Lu, M. Myricitrin attenuates hypoxic-ischemia-induced brain injury in neonatal rats by mitigating oxidative stress and nuclear factor erythroid 2-related factor 2/hemeoxygenase-1/antioxidant response element signaling pathway. *Phcog. Mag.* **2021**, *17*, 828–835. [CrossRef]
318. Wang, H.; Guo, Y.; Qiao, Y.; Zhang, J.; Jiang, P. Nobiletin Ameliorates NLRP3 Inflammasome-Mediated Inflammation Through Promoting Autophagy via the AMPK Pathway. *Mol. Neurobiol.* **2020**, *57*, 5056–5068. [CrossRef]
319. Al Rihani, S.B.; Darakjian, L.I.; Kaddoumi, A. Oleocanthal-Rich Extra-Virgin Olive Oil Restores the Blood-Brain Barrier Function through NLRP3 Inflammasome Inhibition Simultaneously with Autophagy Induction in TgSwDI Mice. *ACS Chem. Neurosci.* **2019**, *10*, 3543–3554. [CrossRef]
320. Wang, S.; Yang, H.; Yu, L.; Jin, J.; Qian, L.; Zhao, H.; Xu, Y.; Zhu, X. Oridonin attenuates Aβ1-42-induced neuroinflammation and inhibits NF-κB pathway. *PloS ONE* **2014**, *9*, e104745. [CrossRef]
321. Liu, Y.; Chen, X.; Gong, Q.; Shi, J.; Li, F. Osthole Improves Cognitive Function of Vascular Dementia Rats: Reducing Aβ Deposition via Inhibition NLRP3 Inflammasome. *Biol. Pharm. Bull.* **2020**, *43*, 1315–1323. [CrossRef]
322. Chen, D.B.; Gao, H.W.; Peng, C.; Pei, S.Q.; Dai, A.R.; Yu, X.T.; Zhou, P.; Wang, Y.; Cai, B. Quinones as preventive agents in Alzheimer's diseases: Focus on NLRP3 inflammasomes. *J. Pharm. Pharmacol.* **2020**, *72*, 1481–1490. [CrossRef]

323. Han, X.; Xu, T.; Fang, Q.; Zhang, H.; Yue, L.; Hu, G.; Sun, L. Quercetin hinders microglial activation to alleviate neurotoxicity via the interplay between NLRP3 inflammasome and mitophagy. *Redox Biol.* **2021**, *44*, 102010. [CrossRef]
324. Li, H.; Chen, F.J.; Yang, W.L.; Qiao, H.Z.; Zhang, S.J. Quercetin improves cognitive disorder in aging mice by inhibiting NLRP3 inflammasome activation. *Food Funct.* **2021**, *12*, 717–725. [CrossRef]
325. Kiasalari, Z.; Afshin-Majd, S.; Baluchnejadmojarad, T.; Azadi-Ahmadabadi, E.; Fakour, M.; Ghasemi-Tarie, R.; Jalalzade-Ogvar, S.; Khodashenas, V.; Tashakori-Miyanroudi, M.; Roghani, M. Sinomenine Alleviates Murine Experimental Autoimmune Encephalomyelitis Model of Multiple Sclerosis through Inhibiting NLRP3 Inflammasome. *J. Mol. Neurosci.* **2021**, *71*, 215–224. [CrossRef]
326. Atluri, V.S.R.; Tiwari, S.; Rodriguez, M.; Kaushik, A.; Yndart, A.; Kolishetti, N.; Yatham, M.; Nair, M. Inhibition of Amyloid-Beta Production, Associated Neuroinflammation, and Histone Deacetylase 2-Mediated Epigenetic Modifications Prevent Neuropathology in Alzheimer's Disease in vitro Model. *Front. Aging Neurosci.* **2020**, *11*, 342. [CrossRef]
327. Cadoná, F.C.; de Souza, D.V.; Fontana, T.; Bodenstein, D.F.; Ramos, A.P.; Sagrillo, M.R.; Salvador, M.; Mota, K.; Davidson, C.B.; Ribeiro, E.E.; et al. Açaí (*Euterpe oleracea* Mart.) as a Potential Anti-neuroinflammatory Agent: NLRP3 Priming and Activating Signal Pathway Modulation. *Mol. Neurobiol.* **2021**, *58*, 4460–4476. [CrossRef]
328. Yu, S.H.; Sun, X.; Kim, M.K.; Akther, M.; Han, J.H.; Kim, T.Y.; Jiang, J.; Kang, T.B.; Lee, K.H. Chrysanthemum indicum extract inhibits NLRP3 and AIM2 inflammasome activation via regulating ASC phosphorylation. *J. Ethnopharmacol.* **2019**, *239*, 111917. [CrossRef]
329. Jeong, Y.H.; Kim, T.I.; Oh, Y.C.; Ma, J.Y. Chrysanthemum indicum Prevents Hydrogen Peroxide-Induced Neurotoxicity by Activating the TrkB/Akt Signaling Pathway in Hippocampal Neuronal Cells. *Nutrients* **2021**, *13*, 3690. [CrossRef]
330. Wang, Z.; Xu, G.; Li, Z.; Xiao, X.; Tang, J.; Bai, Z. NLRP3 Inflammasome Pharmacological Inhibitors in Glycyrrhiza for NLRP3-Driven Diseases Treatment: Extinguishing the Fire of Inflammation. *J. Inflamm. Res.* **2022**, *15*, 409–422. [CrossRef]
331. Kim, N.; Do, J.; Ju, I.G.; Jeon, S.H.; Lee, J.K.; Oh, M.S. Picrorhiza kurroa Prevents Memory Deficits by Inhibiting NLRP3 Inflammasome Activation and BACE1 Expression in 5xFAD Mice. *Neurotherapeutics* **2020**, *17*, 189–199. [CrossRef]
332. Huang, Z.; Zhou, X.; Zhang, X.; Huang, L.; Sun, Y.; Cheng, Z.; Xu, W.; Li, C.G.; Zheng, Y.; Huang, M. Pien-Tze-Huang, a Chinese patent formula, attenuates NLRP3 inflammasome-related neuroinflammation by enhancing autophagy via the AMPK/mTOR/ULK1 signaling pathway. *Biomed. Pharmacother.* **2021**, *141*, 111814. [CrossRef]
333. Yin, X.L.; Wu, H.M.; Zhang, B.H.; Zhu, N.W.; Chen, T.; Ma, X.X.; Zhang, L.Y.; Lv, L.; Zhang, M.; Wang, F.Y.; et al. Tojapride prevents CaSR-mediated NLRP3 inflammasome activation in oesophageal epithelium irritated by acidic bile salts. *J. Cell. Mol. Med.* **2020**, *24*, 1208–1219. [CrossRef]
334. Zhu, T.; Fang, B.Y.; Meng, X.B.; Zhang, S.X.; Wang, H.; Gao, G.; Liu, F.; Wu, Y.; Hu, J.; Sun, G.B.; et al. Folium Ginkgo extract and tetramethylpyrazine sodium chloride injection (Xingxiong injection) protects against focal cerebral ischaemia/reperfusion injury via activating the Akt/Nrf2 pathway and inhibiting NLRP3 inflammasome activation. *Pharm. Biol.* **2022**, *60*, 195–205. [CrossRef] [PubMed]
335. Kim, H.; Hong, J.Y.; Jeon, W.J.; Lee, J.; Baek, S.H.; Ha, I.H. Lycopus lucidus Turcz Exerts Neuroprotective Effects Against H_2O_2-Induced Neuroinflammation by Inhibiting NLRP3 Inflammasome Activation in Cortical Neurons. *J. Inflamm. Res.* **2021**, *14*, 1759–1773. [CrossRef]
336. Denes, A.; Coutts, G.; Lénárt, N.; Cruickshank, S.M.; Pelegrin, P.; Skinner, J.; Rothwell, N.; Allan, S.M.; Brough, D. AIM2 and NLRC4 inflammasomes contribute with ASC to acute brain injury independently of NLRP3. *Proc. Natl. Acad. Sci. USA* **2015**, *112*, 4050–4055. [CrossRef] [PubMed]
337. Yang, X.L.; Wang, X.; Shao, L.; Jiang, G.T.; Min, J.W.; Mei, X.Y.; He, X.H.; Liu, W.H.; Huang, W.X.; Peng, B.W. TRPV1 mediates astrocyte activation and interleukin-1β release induced by hypoxic ischemia (HI). *J. Neuroinflamm.* **2019**, *16*, 114. [CrossRef] [PubMed]
338. Schölwer, I.; Habib, P.; Voelz, C.; Rolfes, L.; Beyer, C.; Slowik, A. NLRP3 Depletion Fails to Mitigate Inflammation but Restores Diminished Phagocytosis in BV-2 Cells After In Vitro Hypoxia. *Mol. Neurobiol.* **2020**, *57*, 2588–2599. [CrossRef]
339. Sun, Y.; Ma, J.; Li, D.; Li, P.; Zhou, X.; Li, Y.; He, Z.; Qin, L.; Liang, L.; Luo, X. Interleukin-10 inhibits interleukin-1β production and inflammasome activation of microglia in epileptic seizures. *J. Neuroinflamm.* **2019**, *16*, 66. [CrossRef]
340. Liew, F.; Girard, J.P.; Turnquist, H. Interleukin-33 in health and disease. *Nat. Rev. Immunol.* **2016**, *16*, 676–689. [CrossRef]
341. Jiao, M.; Li, X.; Chen, L.; Wang, X.; Yuan, B.; Liu, T.; Dong, Q.; Mei, H.; Yin, H. Neuroprotective effect of astrocyte-derived IL-33 in neonatal hypoxic-ischemic brain injury. *J. Neuroinflamm.* **2020**, *17*, 251. [CrossRef]
342. Strangward, P.; Haley, M.J.; Albornoz, M.G.; Barrington, J.; Shaw, T.; Dookie, R.; Zeef, L.; Baker, S.M.; Winter, E.; Tzeng, T.C.; et al. Targeting the IL33-NLRP3 axis improves therapy for experimental cerebral malaria. *Proc. Natl. Acad. Sci. USA* **2018**, *115*, 7404–7409. [CrossRef]
343. Bellut, M.; Raimondi, A.T.; Haarmann, A.; Zimmermann, L.; Stoll, G.; Schuhmann, M.K. NLRP3 Inhibition Reduces rt-PA Induced Endothelial Dysfunction under Ischemic Conditions. *Biomedicines* **2022**, *10*, 762. [CrossRef]
344. Xu, Q.; Ye, Y.; Wang, Z.; Zhu, H.; Li, Y.; Wang, J.; Gao, W.; Gu, L. NLRP3 Knockout Protects against Lung Injury Induced by Cerebral Ischemia-Reperfusion. *Oxid. Med. Cell. Longev.* **2022**, *2022*, 6260102. [CrossRef]
345. Chen, B.; Zhang, M.; Ji, M.; Zhang, D.; Chen, B.; Gong, W.; Li, X.; Zhou, Y.; Dong, C.; Wen, G.; et al. The neuroprotective mechanism of lithium after ischaemic stroke. *Commun. Biol.* **2022**, *5*, 105. [CrossRef]

346. Zhang, Y.; Wang, Y.; Zhao, W.; Li, L.; Li, L.; Sun, Y.; Shao, J.; Ren, X.; Zang, W.; Cao, J. Role of spinal RIP3 in inflammatory pain and electroacupuncture-mediated analgesic effect in mice. *Life Sci.* **2022**, *306*, 120839. [CrossRef]
347. Zhong, X.; Chen, B.; Li, Z.; Lin, R.; Ruan, S.; Wang, F.; Liang, H.; Tao, J. Electroacupuncture Ameliorates Cognitive Impairment Through the Inhibition of NLRP3 Inflammasome Activation by Regulating Melatonin-Mediated Mitophagy in Stroke Rats. *Neurochem. Res.* **2022**, *47*, 1917–1930, Erratum in *Neurochem. Res.* **2022**, *47*, 1931–1933. [CrossRef]
348. Halle, A.; Hornung, V.; Petzold, G.C.; Stewart, C.R.; Monks, B.G.; Reinheckel, T.; Fitzgerald, K.A.; Latz, E.; Moore, K.J.; Golenbock, D.T. The NALP3 inflammasome is involved in the innate immune response to amyloid-beta. *Nat. Immunol.* **2008**, *9*, 857–865. [CrossRef]
349. Heneka, M.T.; Kummer, M.P.; Stutz, A.; Delekate, A.; Schwartz, S.; Vieira-Saecker, A.; Griep, A.; Axt, D.; Remus, A.; Tzeng, T.C.; et al. NLRP3 is activated in Alzheimer's disease and contributes to pathology in APP/PS1 mice. *Nature* **2013**, *493*, 674–678. [CrossRef]
350. Lučiūnaitė, A.; McManus, R.M.; Jankunec, M.; Rácz, I.; Dansokho, C.; Dalgėdienė, I.; Schwartz, S.; Brosseron, F.; Heneka, M.T. Soluble Aβ oligomers and protofibrils induce NLRP3 inflammasome activation in microglia. *J. Neurochem.* **2020**, *155*, 650–661. [CrossRef]
351. Couturier, J.; Stancu, I.C.; Schakman, O.; Pierrot, N.; Huaux, F.; Kienlen-Campard, P.; Dewachter, I.; Octave, J.N. Activation of phagocytic activity in astrocytes by reduced expression of the inflammasome component ASC and its implication in a mouse model of Alzheimer disease. *J. Neuroinflamm.* **2016**, *13*, 20. [CrossRef]
352. Ramaswamy, S.B.; Bhagavan, S.M.; Kaur, H.; Giler, G.E.; Kempuraj, D.; Thangavel, R.; Ahmed, M.E.; Selvakumar, G.P.; Raikwar, S.P.; Zaheer, S.; et al. Glia Maturation Factor in the Pathogenesis of Alzheimer's disease. *Open Access J. Neurol. Neurosurg.* **2019**, *12*, 79–82.
353. Friker, L.L.; Scheiblich, H.; Hochheiser, I.V.; Brinkschulte, R.; Riedel, D.; Latz, E.; Geyer, M.; Heneka, M.T. β-Amyloid Clustering around ASC Fibrils Boosts Its Toxicity in Microglia. *Cell Rep.* **2020**, *30*, 3743–3754.e6. [CrossRef]
354. Murphy, N.; Grehan, B.; Lynch, M.A. Glial uptake of amyloid beta induces NLRP3 inflammasome formation via cathepsin-dependent degradation of NLRP10. *Neuromolecular Med.* **2014**, *16*, 205–215. [CrossRef] [PubMed]
355. Slowik, A.; Lammerding, L.; Zendedel, A.; Habib, P.; Beyer, C. Impact of steroid hormones E2 and P on the NLRP3/ASC/Casp1 axis in primary mouse astroglia and BV-2 cells after in vitro hypoxia. *J. Steroid Biochem. Mol. Biol.* **2018**, *183*, 18–26. [CrossRef] [PubMed]
356. Hong, Y.; Liu, Y.; Yu, D.; Wang, M.; Hou, Y. The neuroprotection of progesterone against Aβ-induced NLRP3-Caspase-1 inflammasome activation via enhancing autophagy in astrocytes. *Int. Immunopharmacol.* **2019**, *74*, 105669. [CrossRef] [PubMed]
357. Chen, D.; Dixon, B.J.; Doycheva, D.M.; Li, B.; Zhang, Y.; Hu, Q.; He, Y.; Guo, Z.; Nowrangi, D.; Flores, J.; et al. IRE1α inhibition decreased TXNIP/NLRP3 inflammasome activation through miR-17-5p after neonatal hypoxic-ischemic brain injury in rats. *J. Neuroinflamm.* **2018**, *15*, 32. [CrossRef]
358. Wang, C.Y.; Xu, Y.; Wang, X.; Guo, C.; Wang, T.; Wang, Z.Y. Dl-3-n-Butylphthalide Inhibits NLRP3 Inflammasome and Mitigates Alzheimer's-Like Pathology via Nrf2-TXNIP-TrX Axis. *Antioxid. Redox. Signal.* **2019**, *30*, 1411–1431. [CrossRef]
359. Sun, Y.; Huang, J.; Chen, Y.; Shang, H.; Zhang, W.; Yu, J.; He, L.; Xing, C.; Zhuang, C. Direct inhibition of Keap1-Nrf2 Protein-Protein interaction as a potential therapeutic strategy for Alzheimer's disease. *Bioorg. Chem.* **2020**, *103*, 104172. [CrossRef]
360. Saad El-Din, S.; Rashed, L.; Medhat, E.; Emad Aboulhoda, B.; Desoky Badawy, A.; Mohammed ShamsEldeen, A.; Abdelgwad, M. Active form of vitamin D analogue mitigates neurodegenerative changes in Alzheimer's disease in rats by targeting Keap1/Nrf2 and MAPK-38p/ERK signaling pathways. *Steroids* **2020**, *156*, 108586. [CrossRef]
361. Yang, X.; Ji, J.; Liu, C.; Zhou, M.; Li, H.; Ye, S.; Hu, Q. HJ22, a Novel derivative of piperine, attenuates ibotenic acid-induced cognitive impairment, oxidative stress, apoptosis and inflammation via inhibiting the protein-protein interaction of Keap1-Nrf2. *Int. Immunopharmacol.* **2020**, *83*, 106383. [CrossRef]
362. Yang, X.; Zhi, J.; Leng, H.; Chen, Y.; Gao, H.; Ma, J.; Ji, J.; Hu, Q. The piperine derivative HJ105 inhibits $A\beta_{1-42}$-induced neuroinflammation and oxidative damage via the Keap1-Nrf2-TXNIP axis. *Phytomedicine* **2021**, *87*, 153571. [CrossRef]
363. Bharti, V.; Tan, H.; Zhou, H.; Wang, J.F. Txnip mediates glucocorticoid-activated NLRP3 inflammatory signaling in mouse microglia. *Neurochem. Int.* **2019**, *131*, 104564. [CrossRef]
364. Gussago, C.; Casati, M.; Ferri, E.; Arosio, B. The Triggering Receptor Expressed on Myeloid Cells-2 (TREM-2) as Expression of the Relationship between Microglia and Alzheimer's Disease: A Novel Marker for a Promising Pathway to Explore. *J. Frailty Aging* **2019**, *8*, 54–56. [CrossRef]
365. Wang, S.Y.; Gong, P.Y.; Yan, E.; Zhang, Y.D.; Jiang, T. The role of TREML2 in Alzheimer's disease. *J. Alzheimers Dis.* **2020**, *76*, 799–806. [CrossRef]
366. Sierksma, A.; Lu, A.; Mancuso, R.; Fattorelli, N.; Thrupp, N.; Salta, E.; Zoco, J.; Blum, D.; Buee, L.; De Strooper, B.; et al. Novel Alzheimer risk genes determine the microglia response to amyloid-beta but not to TAU pathology. *EMBO Mol. Med.* **2020**, *12*, e10606. [CrossRef]
367. Zheng, H.; Liu, C.C.; Atagi, Y.; Chen, X.F.; Jia, L.; Yang, L.; He, W.; Zhang, X.; Kang, S.S.; Rosenberry, T.L.; et al. Opposing roles of the triggering receptor expressed on myeloid cells 2 and triggering receptor expressed on myeloid cells-like transcript 2 in microglia activation. *Neurobiol. Aging* **2016**, *42*, 132–141. [CrossRef]

368. Wang, S.Y.; Fu, X.X.; Duan, R.; Wei, B.; Cao, H.M.; Yan, E.; Chen, S.Y.; Zhang, Y.D.; Jiang, T. The Alzheimer's disease-associated gene TREML2 modulates inflammation by regulating microglia polarization and NLRP3 inflammasome activation. *Neural Regen. Res.* **2023**, *18*, 434–438. [CrossRef]
369. Tejera, D.; Mercan, D.; Sanchez-Caro, J.M.; Hanan, M.; Greenberg, D.; Soreq, H.; Latz, E.; Golenbock, D.; Heneka, M.T. Systemic inflammation impairs microglial Aβ clearance through NLRP3 inflammasome. *EMBO J.* **2019**, *38*, e101064. [CrossRef]
370. Lopez-Rodriguez, A.B.; Hennessy, E.; Murray, C.L.; Nazmi, A.; Delaney, H.J.; Healy, D.; Fagan, S.G.; Rooney, M.; Stewart, E.; Lewis, A.; et al. Acute systemic inflammation exacerbates neuroinflammation in Alzheimer's disease: IL-1β drives amplified responses in primed astrocytes and neuronal network dysfunction. *Alzheimers Dement.* **2021**, *17*, 1735–1755. [CrossRef]
371. Saresella, M.; Piancone, F.; Marventano, I.; Zoppis, M.; Hernis, A.; Zanette, M.; Trabattoni, D.; Chiappedi, M.; Ghezzo, A.; Canevini, M.P.; et al. Multiple inflammasome complexes are activated in autistic spectrum disorders. *Brain Behav. Immun.* **2016**, *57*, 125–133. [CrossRef]
372. Ahmed, M.E.; Iyer, S.; Thangavel, R.; Kempuraj, D.; Selvakumar, G.P.; Raikwar, S.P.; Zaheer, S.; Zaheer, A. Co-Localization of Glia Maturation Factor with NLRP3 Inflammasome and Autophagosome Markers in Human Alzheimer's Disease Brain. *J. Alzheimers Dis.* **2017**, *60*, 1143–1160. [CrossRef]
373. Stancu, I.C.; Cremers, N.; Vanrusselt, H.; Couturier, J.; Vanoosthuyse, A.; Kessels, S.; Lodder, C.; Brône, B.; Huaux, F.; Octave, J.N.; et al. Aggregated Tau activates NLRP3-ASC inflammasome exacerbating exogenously seeded and non-exogenously seeded pathology in vivo. *Acta Neuropathol.* **2019**, *137*, 599–617. [CrossRef]
374. Zhao, S.; Li, X.; Wang, J.; Wang, H. The Role of the Effects of Autophagy on NLRP3 Inflammasome in Inflammatory Nervous System Diseases. *Front. Cell Dev. Biol.* **2021**, *9*, 657478. [CrossRef] [PubMed]
375. Zhou, W.; Xiao, D.; Zhao, Y.; Tan, B.; Long, Z.; Yu, L.; He, G. Enhanced Autolysosomal Function Ameliorates the Inflammatory Response Mediated by the NLRP3 Inflammasome in Alzheimer's Disease. *Front. Aging Neurosci.* **2021**, *13*, 629891. [CrossRef] [PubMed]
376. Zhang, P.; Shao, X.Y.; Qi, G.J.; Chen, Q.; Bu, L.L.; Chen, L.J.; Shi, J.; Ming, J.; Tian, B. Cdk5-Dependent Activation of Neuronal Inflammasomes in Parkinson's Disease. *Mov. Disord.* **2016**, *31*, 366–376. [CrossRef] [PubMed]
377. Deora, V.; Albornoz, E.A.; Zhu, K.; Woodruff, T.M.; Gordon, R. The Ketone Body β-Hydroxybutyrate Does Not Inhibit Synuclein Mediated Inflammasome Activation in Microglia. *J. Neuroimmune Pharmacol.* **2017**, *12*, 568–574. [CrossRef]
378. Shippy, D.C.; Wilhelm, C.; Viharkumar, P.A.; Raife, T.J.; Ulland, T.K. β-Hydroxybutyrate inhibits inflammasome activation to attenuate Alzheimer's disease pathology. *J. Neuroinflamm.* **2020**, *17*, 280. [CrossRef]
379. Sarkar, S.; Malovic, E.; Harishchandra, D.S.; Ghaisas, S.; Panicker, N.; Charli, A.; Palanisamy, B.N.; Rokad, D.; Jin, H.; Anantharam, V.; et al. Mitochondrial impairment in microglia amplifies NLRP3 inflammasome proinflammatory signaling in cell culture and animal models of Parkinson's disease. *NPJ Park. Dis.* **2017**, *3*, 30. [CrossRef]
380. Gordon, R.; Albornoz, E.A.; Christie, D.C.; Langley, M.R.; Kumar, V.; Mantovani, S.; Robertson, A.A.B.; Butler, M.S.; Rowe, D.B.; O'Neill, L.A.; et al. Inflammasome Inhibition Prevents a-Synuclein Pathology and Dopaminergic Neurodegeneration in Mice. *Sci. Transl. Med.* **2018**, *10*, eaah4066. [CrossRef]
381. Li, Y.; Xia, Y.; Yin, S.; Wan, F.; Hu, J.; Kou, L.; Sun, Y.; Wu, J.; Zhou, Q.; Huang, J.; et al. Targeting Microglial α-Synuclein/TLRs/NF-kappaB/NLRP3 Inflammasome Axis in Parkinson's Disease. *Front. Immunol.* **2021**, *12*, 719807. [CrossRef]
382. Scheiblich, H.; Bousset, L.; Schwartz, S.; Griep, A.; Latz, E.; Melki, R.; Heneka, M.T. Microglial NLRP3 Inflammasome Activation upon TLR2 and TLR5 Ligation by Distinct α-Synuclein Assemblies. *J. Immunol.* **2021**, *207*, 2143–2154. [CrossRef]
383. von Herrmann, K.M.; Salas, L.A.; Martinez, E.M.; Young, A.L.; Howard, J.M.; Feldman, M.S.; Christensen, B.C.; Wilkins, O.M.; Lee, S.L.; Hickey, W.F.; et al. NLRP3 expression in mesencephalic neurons and characterization of a rare NLRP3 polymorphism associated with decreased risk of Parkinson's disease. *NPJ Park. Dis.* **2018**, *4*, 24. [CrossRef]
384. Fan, Z.; Pan, Y.T.; Zhang, Z.Y.; Yang, H.; Yu, S.Y.; Zheng, Y.; Ma, J.H.; Wang, X.M. Systemic Activation of NLRP3 Inflammasome and Plasma a-Synuclein Levels Are Correlated with Motor Severity and Progression in Parkinson's Disease. *J. Neuroinflamm.* **2020**, *17*, 11. [CrossRef]
385. Anderson, F.L.; von Herrmann, K.M.; Andrew, A.S.; Kuras, Y.I.; Young, A.L.; Scherzer, C.R.; Hickey, W.F.; Lee, S.L.; Havrda, M.C. Plasma-borne indicators of inflammasome activity in Parkinson's disease patients. *NPJ Park. Dis.* **2021**, *7*, 2. [CrossRef]
386. Wang, J.; Zhang, X.N.; Fang, J.N.; Hua, F.F.; Han, J.Y.; Yuan, Z.Q.; Xie, A.M. The mechanism behind activation of the Nod-like receptor family protein 3 inflammasome in Parkinson's disease. *Neural. Regen. Res.* **2022**, *17*, 898–904. [CrossRef]
387. Simola, N.; Morelli, M.; Carta, A.R. The 6-hydroxydopamine model of Parkinson's disease. *Neurotox. Res.* **2007**, *11*, 151–167. [CrossRef]
388. Si, X.L.; Fang, Y.J.; Li, L.F.; Gu, L.Y.; Yin, X.Z.; Tian, J.; Yan, Y.P.; Pu, J.L.; Zhang, B.R. From inflammasome to Parkinson's disease: Does the NLRP3 inflammasome facilitate exosome secretion and exosomal alpha-synuclein transmission in Parkinson's disease? *Exp. Neurol.* **2021**, *336*, 113525. [CrossRef]
389. Chen, J.; Mao, K.; Yu, H.; Wen, Y.; She, H.; Zhang, H.; Liu, L.; Li, M.; Li, W.; Zou, F. p38-TFEB pathways promote microglia activation through inhibiting CMA-mediated NLRP3 degradation in Parkinson's disease. *J. Neuroinflamm.* **2021**, *18*, 295. [CrossRef]
390. Panicker, N.; Kam, T.I.; Wang, H.; Neifert, S.; Chou, S.C.; Kumar, M.; Brahmachari, S.; Jhaldiyal, A.; Hinkle, J.T.; Akkentli, F.; et al. Neuronal NLRP3 is a parkin substrate that drives neurodegeneration in Parkinson's disease. *Neuron* **2022**, *110*, 2422–2437.e9. [CrossRef]

391. Gris, D.; Ye, Z.; Iocca, H.A.; Wen, H.; Craven, R.R.; Gris, P.; Huang, M.; Schneider, M.; Miller, S.D.; Ting, J.P. NLRP3 plays a critical role in the development of experimental autoimmune encephalomyelitis by mediating Th1 and Th17 responses. *J. Immunol.* **2010**, *185*, 974–981. [CrossRef]
392. Malhotra, S.; Río, J.; Urcelay, E.; Nurtdinov, R.; Bustamante, M.F.; Fernández, O.; Oliver, B.; Zettl, U.; Brassat, D.; Killestein, J.; et al. NLRP3 inflammasome is associated with the response to IFN-β in patients with multiple sclerosis. *Brain* **2015**, *138*, 644–652. [CrossRef]
393. Malhotra, S.; Costa, C.; Eixarch, H.; Keller, C.W.; Amman, L.; Martínez-Banaclocha, H.; Midaglia, L.; Sarró, E.; Machín-Díaz, I.; Villar, L.M.; et al. NLRP3 inflammasome as prognostic factor and therapeutic target in primary progressive multiple sclerosis patients. *Brain* **2020**, *143*, 1414–1430. [CrossRef]
394. Olcum, M.; Tastan, B.; Kiser, C.; Genc, S.; Genc, K. Microglial NLRP3 inflammasome activation in multiple sclerosis. *Adv. Protein Chem. Struct. Biol.* **2020**, *119*, 247–308. [CrossRef] [PubMed]
395. Cui, Y.; Yu, H.; Bu, Z.; Wen, L.; Yan, L.; Feng, J. Focus on the Role of the NLRP3 Inflammasome in Multiple Sclerosis: Pathogenesis, Diagnosis, and Therapeutics. *Front. Mol. Neurosci.* **2022**, *15*, 894298. [CrossRef] [PubMed]
396. Soares, J.L.; Oliveira, E.M.; Pontillo, A. Variants in NLRP3 and NLRC4 inflammasome associate with susceptibility and severity of multiple sclerosis. *Mult. Scler. Relat. Disord.* **2019**, *29*, 26–34. [CrossRef] [PubMed]
397. Vidmar, L.; Maver, A.; Drulović, J.; Sepčić, J.; Novaković, I.; Ristič, S.; Šega, S.; Peterlin, B. Multiple Sclerosis patients carry an increased burden of exceedingly rare genetic variants in the inflammasome regulatory genes. *Sci. Rep.* **2019**, *9*, 9171. [CrossRef]
398. Keane, R.W.; Dietrich, W.D.; de Rivero Vaccari, J.P. Inflammasome Proteins as Biomarkers of Multiple Sclerosis. *Front. Neurol.* **2018**, *9*, 135. [CrossRef] [PubMed]
399. Kadowaki, A.; Quintana, F.J. The NLRP3 inflammasome in progressive multiple sclerosis. *Brain* **2020**, *143*, 1286–1288. [CrossRef] [PubMed]
400. Farooqi, N.; Gran, B.; Constantinescu, C.S. Are current disease-modifying therapeutics in multiple sclerosis justified based on studies in experimental autoimmune encephalomyelitis? *J. Neurochem.* **2010**, *115*, 829–844. [CrossRef] [PubMed]
401. Inoue, M.; Williams, K.L.; Gunn, M.D.; Shinohara, M.L. NLRP3 inflammasome induces chemotactic immune cell migration to the CNS in experimental autoimmune encephalomyelitis. *Proc. Natl. Acad. Sci. USA* **2012**, *109*, 10480–10485. [CrossRef] [PubMed]
402. Khan, N.; Kuo, A.; Brockman, D.A.; Cooper, M.A.; Smith, M.T. Pharmacological inhibition of the NLRP3 inflammasome as a potential target for multiple sclerosis induced central neuropathic pain. *Inflammopharmacology* **2018**, *26*, 77–86. [CrossRef] [PubMed]
403. Bakhshi, S.; Shamsi, S. MCC950 in the treatment of NLRP3-mediated inflammatory diseases: Latest evidence and therapeutic outcomes. *Int. Immunopharmacol.* **2022**, *106*, 108595. [CrossRef]
404. Hou, B.; Zhang, Y.; Liang, P.; He, Y.; Peng, B.; Liu, W.; Han, S.; Yin, J.; He, X. Inhibition of the NLRP3-inflammasome prevents cognitive deficits in experimental autoimmune encephalomyelitis mice via the alteration of astrocyte phenotype. *Cell Death Dis.* **2020**, *11*, 377. [CrossRef]
405. Inoue, M.; Shinohara, M.L. The role of interferon-β in the treatment of multiple sclerosis and experimental autoimmune encephalomyelitis—In the perspective of inflammasomes. *Immunology* **2013**, *139*, 11–18. [CrossRef]
406. Gros-Louis, F.; Gaspar, C.; Rouleau, G.A. Genetics of familial and sporadic amyotrophic lateral sclerosis. *Biochim. Biophys. Acta* **2006**, *1762*, 956–972. [CrossRef]
407. Jaarsma, D.; Teuling, E.; Haasdijk, E.D.; De Zeeuw, C.I.; Hoogenraad, C.C. Neuron-specific expression of mutant superoxide dismutase is sufficient to induce amyotrophic lateral sclerosis in transgenic mice. *J. Neurosci.* **2008**, *28*, 2075–2088. [CrossRef]
408. Yamanaka, K.; Boillee, S.; Roberts, E.A.; Garcia, M.L.; McAlonis-Downes, M.; Mikse, O.R.; Cleveland, D.W.; Goldstein, L.S. Mutant SOD1 in cell types other than motor neurons and oligodendrocytes accelerates onset of disease in ALS mice. *Proc. Natl. Acad. Sci. USA* **2008**, *105*, 7594–7599. [CrossRef]
409. Mackenzie, I.R.; Rademakers, R.; Neumann, M. TDP-43 and FUS in amyotrophic lateral sclerosis and frontotemporal dementia. *Lancet Neurol.* **2010**, *9*, 995–1007. [CrossRef]
410. Arai, T.; Hasegawa, M.; Akiyama, H.; Ikeda, K.; Nonaka, T.; Mori, H.; Mann, D.; Tsuchiya, K.; Yoshida, M.; Hashizume, Y.; et al. TDP-43 is a component of ubiquitin-positive tau-negative inclusions in frontotemporal lobar degeneration and amyotrophic lateral sclerosis. *Biochem. Biophys. Res. Commun.* **2006**, *351*, 602–611. [CrossRef]
411. Neumann, M.; Sampathu, D.M.; Kwong, L.K.; Truax, A.C.; Micsenyi, M.C.; Chou, T.T.; Bruce, J.; Schuck, T.; Grossman, M.; Clark, C.M.; et al. Ubiquitinated TDP-43 in frontotemporal lobar degeneration and amyotrophic lateral sclerosis. *Science* **2006**, *314*, 130–133. [CrossRef]
412. Ezzi, S.A.; Urushitani, M.; Julien, J.P. Wild-type superoxide dismutase acquires binding and toxic properties of ALS-linked mutant forms through oxidation. *J. Neurochem.* **2007**, *102*, 170–178. [CrossRef]
413. Roberts, K.; Zeineddine, R.; Corcoran, L.; Li, W.; Campbell, I.L.; Yerbury, J.J. Extracellular aggregated Cu/Zn superoxide dismutase activates microglia to give a cytotoxic phenotype. *Glia* **2013**, *61*, 409–419. [CrossRef]
414. Zhao, W.; Beers, D.R.; Bell, S.; Wang, J.; Wen, S.; Baloh, R.H.; Appel, S.H. TDP-43 activates microglia through NF-κB and NLRP3 inflammasome. *Exp. Neurol.* **2015**, *273*, 24–35. [CrossRef] [PubMed]
415. Banerjee, P.; Elliott, E.; Rifai, O.M.; O'Shaughnessy, J.; McDade, K.; Abrahams, S.; Chandran, S.; Smith, C.; Gregory, J.M. NLRP3 inflammasome as a key molecular target underlying cognitive resilience in amyotrophic lateral sclerosis. *J. Pathol.* **2022**, *256*, 262–268. [CrossRef] [PubMed]

416. Kadhim, H.; Deltenre, P.; Martin, J.J.; Sébire, G. In-situ expression of Interleukin-18 and associated mediators in the human brain of sALS patients: Hypothesis for a role for immune-inflammatory mechanisms. *Med. Hypotheses* **2016**, *86*, 14–17. [CrossRef] [PubMed]
417. Gugliandolo, A.; Giacoppo, S.; Bramanti, P.; Mazzon, E. NLRP3 Inflammasome Activation in a Transgenic Amyotrophic Lateral Sclerosis Model. *Inflammation* **2018**, *41*, 93–103. [CrossRef] [PubMed]
418. Debye, B.; Schmülling, L.; Zhou, L.; Rune, G.; Beyer, C.; Johann, S. Neurodegeneration and NLRP3 inflammasome expression in the anterior thalamus of SOD1(G93A) ALS mice. *Brain Pathol.* **2018**, *28*, 14–27. [CrossRef]
419. Michaelson, N.; Facciponte, D.; Bradley, W.; Stommel, E. Cytokine expression levels in ALS: A potential link between inflammation and BMAA-triggered protein misfolding. *Cytokine Growth Factor Rev.* **2017**, *37*, 81–88. [CrossRef]
420. Van Schoor, E.; Ospitalieri, S.; Moonen, S.; Tomé, S.O.; Ronisz, A.; Ok, O.; Weishaupt, J.; Ludolph, A.C.; Van Damme, P.; Van Den Bosch, L.; et al. Increased pyroptosis activation in white matter microglia is associated with neuronal loss in ALS motor cortex. *Acta Neuropathol.* **2022**, *144*, 393–411. [CrossRef]
421. Stallings, N.R.; Puttaparthi, K.; Luther, C.M.; Burns, D.K.; Elliott, J.L. Progressive motor weakness in transgenic mice expressing human TDP-43. *Neurobiol. Dis.* **2010**, *40*, 404–414. [CrossRef]
422. Quek, H.; Cuní-López, C.; Stewart, R.; Colletti, T.; Notaro, A.; Nguyen, T.H.; Sun, Y.; Guo, C.C.; Lupton, M.K.; Roberts, T.L.; et al. ALS monocyte-derived microglia-like cells reveal cytoplasmic TDP-43 accumulation, DNA damage, and cell-specific impairment of phagocytosis associated with disease progression. *J. Neuroinflamm.* **2022**, *19*, 58. [CrossRef]
423. Kenney, C.; Powell, S.; Jankovic, J. Autopsy-proven Huntington's disease with 29 trinucleotide repeats. *Mov. Disord.* **2007**, *22*, 127–130. [CrossRef]
424. Fusco, F.R.; Paldino, E. Role of Phosphodiesterases in Huntington's Disease. *Adv. Neurobiol.* **2017**, *17*, 285–304. [CrossRef] [PubMed]
425. Paldino, E.; Fusco, F.R. Emerging Role of NLRP3 Inflammasome/Pyroptosis in Huntington's Disease. *Int. J. Mol. Sci.* **2022**, *23*, 8363. [CrossRef] [PubMed]
426. Menalled, L.B.; Chesselet, M.F. Mouse models of Huntington's disease. *Trends Pharmacol. Sci.* **2002**, *23*, 32–39. [CrossRef] [PubMed]
427. Paldino, E.; D'Angelo, V.; Sancesario, G.; Fusco, F.R. Pyroptotic cell death in the R6/2 mouse model of Huntington's disease: New insight on the inflammasome. *Cell Death Discov.* **2020**, *6*, 69. [CrossRef] [PubMed]
428. Paldino, E.; D'Angelo, V.; Laurenti, D.; Angeloni, C.; Sancesario, G.; Fusco, F.R. Modulation of Inflammasome and Pyroptosis by Olaparib, a PARP-1 Inhibitor.; in the R6/2 Mouse Model of Huntington's Disease. *Cells* **2020**, *9*, 2286. [CrossRef]
429. Chen, K.P.; Hua, K.F.; Tsai, F.T.; Lin, T.Y.; Cheng, C.Y.; Yang, D.I.; Hsu, H.T.; Ju, T.C. A selective inhibitor of the NLRP3 inflammasome as a potential therapeutic approach for neuroprotection in a transgenic mouse model of Huntington's disease. *J. Neuroinflamm.* **2022**, *19*, 56. [CrossRef]
430. Siew, J.J.; Chen, H.M.; Chen, H.Y.; Chen, H.L.; Chen, C.M.; Soong, B.W.; Wu, Y.R.; Chang, C.P.; Chan, Y.C.; Lin, C.H.; et al. Galectin-3 is required for the microglia-mediated brain inflammation in a model of Huntington's disease. *Nat. Commun.* **2019**, *10*, 3473. [CrossRef]
431. Barake, F.; Soza, A.; González, A. Galectins in the brain: Advances in neuroinflammation, neuroprotection and therapeutic opportunities. *Curr. Opin. Neurol.* **2020**, *33*, 381–390. [CrossRef]
432. Tricarico, P.M.; Caracciolo, I.; Crovella, S.; D'Agaro, P. Zika virus induces inflammasome activation in the glial cell line U87-MG. *Biochem. Biophys. Res. Commun.* **2017**, *492*, 597–602. [CrossRef]
433. Zheng, Y.; Liu, Q.; Wu, Y.; Ma, L.; Zhang, Z.; Liu, T.; Jin, S.; She, Y.; Li, Y.P.; Cui, J. Zika virus elicits inflammation to evade antiviral response by cleaving cGAS via NS1-caspase-1 axis. *EMBO J.* **2018**, *37*, e99347. [CrossRef]
434. Wang, W.; Li, G.; De, W.; Luo, Z.; Pan, P.; Tian, M.; Wang, Y.; Xiao, F.; Li, A.; Wu, K.; et al. Zika virus infection induces host inflammatory responses by facilitating NLRP3 inflammasome assembly and interleukin-1β secretion. *Nat. Commun.* **2018**, *9*, 106. [CrossRef]
435. Ramos, H.J.; Lanteri, M.C.; Blahnik, G.; Negash, A.; Suthar, M.S.; Brassil, M.M.; Sodhi, K.; Treuting, P.M.; Busch, M.P.; Norris, P.J.; et al. IL-1β signaling promotes CNS-intrinsic immune control of West Nile virus infection. *PLoS Pathog.* **2012**, *8*, e1003039. [CrossRef]
436. Yang, C.M.; Lin, C.C.; Lee, I.T.; Lin, Y.H.; Yang, C.M.; Chen, W.J.; Jou, M.J.; Hsiao, L.D. Japanese encephalitis virus induces matrix metalloproteinase-9 expression via a ROS/c-Src/PDGFR/PI3K/Akt/MAPKs-dependent AP-1 pathway in rat brain astrocytes. *J. Neuroinflamm.* **2012**, *9*, 12. [CrossRef]
437. He, W.; Zhao, Z.; Anees, A.; Li, Y.; Ashraf, U.; Chen, Z.; Song, Y.; Chen, H.; Cao, S.; Ye, J. p21-activated kinase 4 signaling promotes Japanese encephalitis virus-mediated inflammation in astrocytes. *Front. Cell. Infect. Microbiol.* **2017**, *7*, 271. [CrossRef]
438. Ashraf, U.; Ding, Z.; Deng, S.; Ye, J.; Cao, S.; Chen, Z. Pathogenicity and virulence of Japanese encephalitis virus: Neuroinflammation and neuronal cell damage. *Virulence* **2021**, *12*, 968–980. [CrossRef]
439. Burdo, T.H.; Lackner, A.; Williams, K.C. Monocyte/macrophages, and their role in HIV neuropathogenesis. *Immunol. Rev.* **2013**, *254*, 102–113. [CrossRef]
440. Walsh, J.G.; Reinke, S.N.; Mamik, M.K.; McKenzie, B.A.; Maingat, F.; Branton, W.G.; Broadhurst, D.I.; Power, C. Rapid inflammasome activation in microglia contributes to brain disease in HIV/AIDS. *Retrovirology* **2014**, *11*, 35. [CrossRef]

441. Breitinger, U.; Farag, N.S.; Sticht, H.; Breitinger, H.G. Viroporins: Structure, function, and their role in the life cycle of SARS-CoV-2. *Int. J. Biochem. Cell Biol.* **2022**, *145*, 106185. [CrossRef]
442. Poeck, H.; Bscheider, M.; Gross, O.; Finger, K.; Roth, S.; Rebsamen, M.; Hannesschläger, N.; Schlee, M.; Rothenfusser, S.; Barchet, W.; et al. Recognition of RNA virus by RIG-I results in activation of CARD9 and inflammasome signaling for interleukin 1 beta production. *Nat. Immunol.* **2010**, *11*, 63–69, Addendum in *Nat. Immunol.* **2014**, *15*, 109. [CrossRef]
443. Rajan, J.V.; Rodriguez, D.; Miao, E.A.; Aderem, A. The NLRP3 inflammasome detects encephalomyocarditis virus and vesicular stomatitis virus infection. *J. Virol.* **2011**, *85*, 4167–4172. [CrossRef] [PubMed]
444. Szabo, M.P.; Iba, M.; Nath, A.; Masliah, E.; Kim, C. Does SARS-CoV-2 affect neurodegenerative disorders? TLR2, a potential receptor for SARS-CoV-2 in the CNS. *Exp. Mol. Med.* **2022**, *54*, 447–454. [CrossRef] [PubMed]
445. Hung, E.C.; Chim, S.S.; Chan, P.K.; Tong, Y.K.; Ng, E.K.; Chiu, R.W.; Leung, C.B.; Sung, J.J.; Tam, J.S.; Lo, Y.M. Detection of SARS coronavirus RNA in the cerebrospinal fluid of a patient with severe acute respiratory syndrome. *Clin. Chem.* **2003**, *49*, 2108–2109. [CrossRef] [PubMed]
446. Sepehrinezhad, A.; Rezaeitalab, F.; Shahbazi, A.; Sahab-Negah, S. A Computational-Based Drug Repurposing Method Targeting SARS-CoV-2 and its Neurological Manifestations Genes and Signaling Pathways. *Bioinform. Biol. Insights.* **2021**, *15*, 11779322211026728. [CrossRef] [PubMed]
447. Chen, R.; Wang, K.; Yu, J.; Howard, D.; French, L.; Chen, Z.; Wen, C.; Xu, Z. The spatial and cell-type distribution of SARS-CoV-2 receptor ACE2 in the human and mouse brains. *Front. Neurol.* **2020**, *11*, 573095. [CrossRef] [PubMed]
448. Ribeiro, D.E.; Oliveira-Giacomelli, A.; Glaser, T.; Arnaud-Sampaio, V.F.; Andrejew, R.; Dieckmann, L.; Baranova, J.; Lameu, C.; Ratajczak, M.Z.; Ulrich, H. Hyperactivation of P2X7 receptors as a culprit of COVID-19 neuropathology. *Mol. Psychiatry* **2021**, *26*, 1044–1059. [CrossRef]
449. Sepehrinezhad, A.; Gorji, A.; Sahab Negah, S. SARS-CoV-2 may trigger inflammasome and pyroptosis in the central nervous system: A mechanistic view of neurotropism. *Inflammopharmacology* **2021**, *29*, 1049–1059. [CrossRef]
450. Helms, J.; Kremer, S.; Merdji, H.; Clere-Jehl, R.; Schenck, M.; Kummerlen, C.; Collange, O.; Boulay, C.; Fafi-Kremer, S.; Ohana, M.; et al. Neurologic features in severe SARS-CoV-2 infection. *N. Engl. J. Med.* **2020**, *382*, 2268–2270. [CrossRef]
451. Zhao, Y.; Li, W.; Lukiw, W. Ubiquity of the SARS-CoV-2 receptor ACE2 and upregulation in limbic regions of Alzheimer's disease brain. *Folia Neuropathol.* **2021**, *59*, 232–238. [CrossRef]
452. Ding, Q.; Shults, N.V.; Gychka, S.G.; Harris, B.T.; Suzuki, Y.J. Protein Expression of Angiotensin-Converting Enzyme 2 (ACE2) is Upregulated in Brains with Alzheimer's Disease. *Int. J. Mol. Sci.* **2021**, *22*, 1687. [CrossRef]
453. Theobald, S.J.; Simonis, A.; Georgomanolis, T.; Kreer, C.; Zehner, M.; Eisfeld, H.S.; Albert, M.C.; Chhen, J.; Motameny, S.; Erger, F.; et al. Long-lived macrophage reprogramming drives spike protein-mediated inflammasome activation in COVID-19. *EMBO Mol. Med.* **2021**, *13*, e14150. [CrossRef]
454. Xu, H.; Akinyemi, I.A.; Chitre, S.A.; Loeb, J.C.; Lednicky, J.A.; McIntosh, M.T.; Bhaduri-McIntosh, S. SARS-CoV-2 viroporin encoded by ORF3a triggers the NLRP3 inflammatory pathway. *Virology* **2022**, *568*, 13–22. [CrossRef]
455. Freeman, T.L.; Swartz, T.H. Targeting the NLRP3 Inflammasome in Severe COVID-19. *Front. Immunol.* **2020**, *11*, 1518. [CrossRef]
456. Fatima, S.; Zaidi, S.S.; Alsharidah, A.S.; Aljaser, F.S.; Banu, N. Possible Prophylactic Approach for SARS-CoV-2 Infection by Combination of Melatonin, Vitamin C and Zinc in Animals. *Front. Vet. Sci.* **2020**, *7*, 585789. [CrossRef]
457. Ding, H.G.; Deng, Y.Y.; Yang, R.Q.; Wang, Q.S.; Jiang, W.Q.; Han, Y.L.; Huang, L.Q.; Wen, M.Y.; Zhong, W.H.; Li, X.S.; et al. Hypercapnia induces IL-1β overproduction via activation of NLRP3 inflammasome: Implication in cognitive impairment in hypoxemic adult rats. *J. Neuroinflamm.* **2018**, *15*, 4. [CrossRef]
458. Heneka, M.T.; Golenbock, D.; Latz, E.; Morgan, D.; Brown, R. Immediate and long-term consequences of COVID-19 infections for the development of neurological disease. *Alzheimers Res. Ther.* **2020**, *12*, 69. [CrossRef]
459. Farheen, S.; Agrawal, S.; Zubair, S.; Agrawal, A.; Jamal, F.; Altaf, I.; Kashif Anwar, A.; Umair, S.M.; Owais, M. Pathophysiology of aging and immune-senescence: Possible correlates with comorbidity and mortality in middle-aged and old COVID-19 patients. *Front. Aging* **2021**, *2*, 748591. [CrossRef]
460. Flud, V.V.; Shcherbuk, Y.A.; Shcherbuk, A.Y.; Leonov, V.I.; Al-Sahli, O.A. Neurological complications and consequences of new coronavirus COVID-19 infection in elderly and old patients (literature review). *Adv. Gerontol.* **2022**, *35*, 231–242.
461. Fu, Y.W.; Xu, H.S.; Liu, S.J. COVID-19 and neurodegenerative diseases. *Eur. Rev. Med. Pharmacol. Sci.* **2022**, *26*, 4535–4544. [CrossRef]
462. Baazaoui, N.; Iqbal, K. COVID-19 and Neurodegenerative Diseases: Prion-Like Spread and Long-Term Consequences. *J. Alzheimers Dis.* **2022**, *88*, 399–416. [CrossRef]
463. Bernardini, A.; Gigli, G.L.; Janes, F.; Pellitteri, G.; Ciardi, C.; Fabris, M.; Valente, M. Creutzfeldt-Jakob disease after COVID-19: Infection-induced prion protein misfolding? A case report. *Prion* **2022**, *16*, 78–83. [CrossRef]
464. Tetz, G.; Tetz, V. Prion-like Domains in Spike Protein of SARS-CoV-2 Differ across Its Variants and Enable Changes in Affinity to ACE2. *Microorganisms* **2022**, *10*, 280. [CrossRef] [PubMed]
465. Wilson, L.; Stewart, W.; Dams-O'Connor, K.; Diaz-Arrastia, R.; Horton, L.; Menon, D.K.; Polinder, S. The chronic and evolving neurological consequences of traumatic brain injury. *Lancet Neurol.* **2017**, *16*, 813–825. [CrossRef] [PubMed]
466. Lee, J.H.; Kim, H.J.; Kim, J.U.; Yook, T.H.; Kim, K.H.; Lee, J.Y.; Yang, G. A Novel Treatment Strategy by Natural Products in NLRP3 Inflammasome-Mediated Neuroinflammation in Alzheimer's and Parkinson's Disease. *Int. J. Mol. Sci.* **2021**, *22*, 1324. [CrossRef] [PubMed]

467. Wu, A.G.; Zhou, X.G.; Qiao, G.; Yu, L.; Tang, Y.; Yan, L.; Qiu, W.Q.; Pan, R.; Yu, C.L.; Law, B.Y.; et al. Targeting microglial autophagic degradation in NLRP3 inflammasome-mediated neurodegenerative diseases. *Ageing Res. Rev.* **2021**, *65*, 101202. [CrossRef]
468. Freeman, L.; Guo, H.; David, C.N.; Brickey, W.J.; Jha, S.; Ting, J.P. NLR members NLRC4 and NLRP3 mediate sterile inflammasome activation in microglia and astrocytes. *J. Exp. Med.* **2017**, *214*, 1351–1370. [CrossRef]
469. McKenzie, B.A.; Mamik, M.K.; Saito, L.B.; Boghozian, R.; Monaco, M.C.; Major, E.O.; Lu, J.Q.; Branton, W.G.; Power, C. Caspase-1 inhibition prevents glial inflammasome activation and pyroptosis in models of multiple sclerosis. *Proc. Natl. Acad. Sci. USA* **2018**, *115*, E6065–E6074. [CrossRef]
470. Yap, J.K.Y.; Pickard, B.S.; Chan, E.W.L.; Gan, S.Y. The Role of Neuronal NLRP1 Inflammasome in Alzheimer's Disease: Bringing Neurons into the Neuroinflammation Game. *Mol. Neurobiol.* **2019**, *56*, 7741–7753. [CrossRef]
471. Nagyőszi, P.; Nyúl-Tóth, Á.; Fazakas, C.; Wilhelm, I.; Kozma, M.; Molnár, J.; Haskó, J.; Krizbai, I.A. Regulation of NOD-like receptors and inflammasome activation in cerebral endothelial cells. *J. Neurochem.* **2015**, *135*, 551–564. [CrossRef]
472. Feng, Y.S.; Tan, Z.X.; Wu, L.Y.; Dong, F.; Zhang, F. The involvement of NLRP3 inflammasome in the treatment of neurodegenerative diseases. *Biomed. Pharmacother.* **2021**, *38*, 111428. [CrossRef]
473. Lahooti, B.; Chhibber, T.; Bagchi, S.; Varahachalam, S.P.; Jayant, R.D. Therapeutic role of inflammasome inhibitors in neurodegenerative disorders. *Brain Behav. Immun.* **2021**, *91*, 771–783. [CrossRef]
474. Corcoran, S.E.; Halai, R.; Cooper, M.A. Pharmacological Inhibition of the Nod-Like Receptor Family Pyrin Domain Containing 3 Inflammasome with MCC950. *Pharmacol. Rev.* **2021**, *73*, 968–1000. [CrossRef]
475. Soriano-Teruel, P.M.; García-Laínez, G.; Marco-Salvador, M.; Pardo, J.; Arias, M.; DeFord, C.; Merfort, I.; Vicent, M.J.; Pelegrín, P.; Sancho, M.; et al. Identification of an ASC oligomerization inhibitor for the treatment of inflammatory diseases. *Cell Death Dis.* **2021**, *12*, 1155. [CrossRef]
476. De Souza, N. Model organisms: Mouse models challenged. *Nat. Methods* **2013**, *10*, 288. [CrossRef]
477. Pound, P.; Ritskes-Hoitinga, M. Is it possible to overcome issues of external validity in preclinical animal research? Why most animal models are bound to fail. *J. Transl. Med.* **2018**, *16*, 304. [CrossRef]
478. Gharib, W.H.; Robinson-Rechavi, M. When orthologs diverge between human and mouse. *Brief. Bioinform.* **2011**, *12*, 436–441. [CrossRef]
479. Seok, J.; Warren, H.S.; Cuenca, A.G.; Mindrinos, M.N.; Baker, H.V.; Xu, W.; Richards, D.R.; Mcdonald-Smith, G.P.; Gao, H.; Hennessy, L.; et al. Inflammation and Host Response to Injury, Large Scale Collaborative Research Program. Genomic responses in mouse models poorly mimic human inflammatory diseases. *Proc. Natl. Acad. Sci. USA* **2013**, *110*, 3507–3512. [CrossRef]
480. Greek, R.; Menache, A. Systematic reviews of animal models: Methodology versus epistemology. *Int. J. Med. Sci.* **2013**, *10*, 206–221. [CrossRef]
481. Akhtar, A. The flaws and human harms of animal experimentation. *Camb. Q. Health Ethics.* **2015**, *24*, 407–419. [CrossRef]
482. Balls, M. It's Time to Include Harm to Humans in Harm-Benefit Analysis—But How to Do It, that is the Question. *Altern. Lab. Anim.* **2021**, *49*, 182–196. [CrossRef]
483. Chiarini, A.; Gardenal, E.; Whitfield, J.F.; Chakravarthy, B.; Armato, U.; Dal Pra, I. Preventing the spread of Alzheimer's disease neuropathology: A role for calcilytics? *Curr. Pharm. Biotechnol.* **2015**, *16*, 696–706. [CrossRef]
484. Chiarini, A.; Armato, U.; Liu, D.; Dal Prà, I. Calcium-Sensing Receptor Antagonist NPS 2143 Restores Amyloid Precursor Protein Physiological Non-Amyloidogenic Processing in Aβ-Exposed Adult Human Astrocytes. *Sci. Rep.* **2017**, *7*, 1277. [CrossRef] [PubMed]
485. Van Dyck, C.H.; Swanson, C.J.; Aisen, P.; Bateman, R.J.; Chen, C.; Gee, M.; Kanekiyo, M.; Li, D.; Reyderman, L.; Cohen, S.; et al. Lecanemab in Early Alzheimer's Disease. *N. Engl. J. Med.* **2023**, *388*, 9–21. [CrossRef] [PubMed]

biomedicines

MDPI

Article

The Tryptophan-Kynurenine Metabolic System Is Suppressed in Cuprizone-Induced Model of Demyelination Simulating Progressive Multiple Sclerosis

Helga Polyák [1,2], Zsolt Galla [3], Nikolett Nánási [4], Edina Katalin Cseh [1], Cecília Rajda [1], Gábor Veres [5], Eleonóra Spekker [4], Ágnes Szabó [1,2], Péter Klivényi [1], Masaru Tanaka [4,†] and László Vécsei [1,4,*,†]

1 Department of Neurology, Albert Szent-Györgyi Medical School, University of Szeged, Semmelweis u. 6, H-6725 Szeged, Hungary; polyak.helga@med.u-szeged.hu (H.P.); cseh.edina@med.u-szeged.hu (E.K.C.); rajda.cecilia@med.u-szeged.hu (C.R.); szabo.agnes.4@med.u-szeged.hu (Á.S.); peter.klivenyi@med.u-szeged.hu (P.K.)
2 Doctoral School of Clinical Medicine, University of Szeged, Korányi fasor 6, H-6720 Szeged, Hungary
3 Department of Pediatrics, Albert Szent-Györgyi Faculty of Medicine, University of Szeged, H-6725 Szeged, Hungary; galla.zsolt@med.u-szeged.hu
4 Danube Neuroscience Research Laboratory, ELKH-SZTE Neuroscience Research Group, Eötvös Loránd Research Network, University of Szeged (ELKH-SZTE), Tisza Lajos krt. 113, H-6725 Szeged, Hungary; nannik1026@gmail.com (N.N.); spekker.eleonora@gmail.com (E.S.); tanaka.masaru.1@med.u-szeged.hu (M.T.)
5 Independent Researcher, H-6726 Szeged, Hungary; weresdzsi@gmail.com
* Correspondence: vecsei.laszlo@med.u-szeged.hu; Tel.: +36-62-545-351
† These authors contributed equally to this work.

Abstract: Progressive multiple sclerosis (MS) is a chronic disease with a unique pattern, which is histologically classified into the subpial type 3 lesions in the autopsy. The lesion is also homologous to that of cuprizone (CPZ) toxin-induced animal models of demyelination. Aberration of the tryptophan (TRP)-kynurenine (KYN) metabolic system has been observed in patients with MS; nevertheless, the KYN metabolite profile of progressive MS remains inconclusive. In this study, C57Bl/6J male mice were treated with 0.2% CPZ toxin for 5 weeks and then underwent 4 weeks of recovery. We measured the levels of serotonin, TRP, and KYN metabolites in the plasma and the brain samples of mice at weeks 1, 3, and 5 of demyelination, and at weeks 7 and 9 of remyelination periods by ultra-high-performance liquid chromatography with tandem mass spectrometry (UHPLC-MS/MS) after body weight measurement and immunohistochemical analysis to confirm the development of demyelination. The UHPLC-MS/MS measurements demonstrated a significant reduction of kynurenic acid, 3-hydoxykynurenine (3-HK), and xanthurenic acid in the plasma and a significant reduction of 3-HK, and anthranilic acid in the brain samples at week 5. Here, we show the profile of KYN metabolites in the CPZ-induced mouse model of demyelination. Thus, the KYN metabolite profile potentially serves as a biomarker of progressive MS and thus opens a new path toward planning personalized treatment, which is frequently obscured with immunologic components in MS deterioration.

Keywords: multiple sclerosis; PPMS; SPMS; tryptophan; kynurenine; cuprizone; demyelination; remyelination; animal model; translational

Citation: Polyák, H.; Galla, Z.; Nánási, N.; Cseh, E.K.; Rajda, C.; Veres, G.; Spekker, E.; Szabó, Á.; Klivényi, P.; Tanaka, M.; et al. The Tryptophan-Kynurenine Metabolic System Is Suppressed in Cuprizone-Induced Model of Demyelination Simulating Progressive Multiple Sclerosis. *Biomedicines* **2023**, *11*, 945. https://doi.org/10.3390/biomedicines11030945

Academic Editors: David R. Wallace and Víctor M. Rivera

Received: 17 January 2023
Revised: 7 March 2023
Accepted: 16 March 2023
Published: 20 March 2023

1. Introduction

Multiple sclerosis (MS) is an inflammatory, neurodegenerative, and immune-mediated disorder of the central nervous system (CNS), which is characterized by the appearance of inflammatory lesions of the white matter, axonal damage, loss of oligodendrocyte and axons, gliosis, demyelination, as well as neurodegeneration [1–5]. MS is still an incurable disease and is estimated to affect 2.8 million people worldwide [6]. It occurs most often in one of the most productive stages of life, i.e., in young adulthood, which makes the disease to be responsible for worsening the quality of the life not only for those affected but for society too [6,7]. The development of the disorder is the result of an interrelationship of various immune, genetic, epigenetic, and environmental factors [5]. The pathophysiology of the disease and the molecular and metabolic mechanisms underlying neuroaxonal damage are not yet fully elucidated, although oxidative stress greatly contributes to the progression of MS by inducing axonal and neuronal damage [5,8]. Furthermore, a growing body of research suggests that in addition to immune-mediated inflammatory responses, certain neurodegenerative processes, including glutamate excitotoxicity, mitochondrial dysfunction, and aforementioned oxidative stress also play a role in the pathogenesis and progression of the disease [8]. In most cases, MS starts with a relapsing–remitting (RRMS) course, often followed over the years by a relapse-independent deterioration in neurological function, known as the progressive phase of MS (SPMS) [9,10]. Progressive MS relates to gradual exacerbation in neurologic as well as psychiatric signs and symptoms. Cortical demyelinated lesions may be the pathological course leading to neural dysfunction, progression, and particularly cognitive impairments [11–17], which are clinical findings in patients with progressive MS. Extensive cortical demyelination has been observed in the frontal, temporal, insular, and cerebellar cortices, the cingulate gyrus, and the hippocampus [18,19].

Preclinical research makes a major contribution to prevention, diagnosis, and therapeutic management in clinical medicine, employing in vitro systems and animal models of diseases [20–24]. Experimental research collecting the data from animal studies and animal models of neurodegenerative and neuropsychiatric diseases has provided valuable clues to understanding the roles of endogenous peptides, hormones, and metabolites, searching for useful biomarkers and exploring novel targets for the treatment of diseases [25–29]. Thus, extensive preclinical, clinical, and computational studies are under extended research for their potential translationability and synthesizability in the search for novel preventive, diagnostic, and interventional measures for neurodegenerative diseases [30–34].

Various animal models, such as the experimental autoimmune encephalomyelitis, Theiler's murine encephalomyelitis virus, and cuprizone (CPZ) toxin model, have been applied for the investigation of the pathomechanism of MS [35,36]. The CPZ-induced model is an ideal tool to study the demyelination phase, concomitant pathological changes and processes, as well as remyelination, in the absence of a peripheral immune response and blood–brain barrier (BBB) disruption [35,36]. CPZ is a copper chelating agent that induces apoptosis in mature oligodendrocytes [4]. The chelating function may have a negative effect on copper cofactor-dependent mitochondrial enzymes in the respiratory chain [3]. Although the exact mechanism of the intoxication is not yet fully explained, it appears that mitochondrial dysfunction causes oxidative stress, which contributes to the apoptosis of oligodendrocytes and axonal damage, ultimately leading to demyelination [3,4,37]. During the early weeks of toxin treatment (acute phase), the activation of oligodendrocytosis, microglia, and astrocyte starts, and demyelination becomes apparent [36,38,39]. Glial activation facilities the removal of accumulating myelin debris and gliosis further exacerbates demyelination and oligodendrocytosis [36,38,40,41]. Significant demyelination and oligodendroglial cell death can be observed in the corpus callosum, the striatum, the cortex, the hippocampus, and the cerebellum, among others [36]. Based on the literature, the weight of the brain decreases after CPZ treatment, which reduction can be explained by the thinning of the corpus callosum and cortex [35]. Furthermore, the discontinuation of CPZ treatment leads to rapid regeneration, exemplified by remyelination [35,42,43].

The CPZ intoxication contributes to axonal and synaptic damage, leading to the development of functional impairment of the nervous system [36,44]. Furthermore, CPZ intoxication causes glutamate excitotoxicity and synaptotoxicity through the glutamate receptor subunits [36,45,46]. The ionotropic amino-3-hydroxy-5-methyl-4-isoxazolepropionic acid (AMPA) and kainate types of glutamate receptors, and *N*-methyl-*D*-aspartate (NMDA) receptors in the myelin sheaths expressed in the oligodendrocytes, contribute to the cell and myelin sheath swelling, vacuolation, and subsequent excitotoxic cell death [35,47]. CPZ treatment also causes decreased expression of AMPA receptors in hippocampal neurons and cortex [46], upregulation of the NR2A subunit of the NMDA receptor in the corpus callosum [45], and downregulation of the mGluR2 subunit of the metabotropic glutamate receptors [36,45].

More than 95% of the essential amino acid tryptophan (TRP) is degraded via the kynurenine (KYN) metabolic pathway (Figure 1) [48,49]. The first rate-determining step of the TRP-KYN metabolism is the conversion of TRP to L-KYN, catalyzed by TRP-2,3-dioxygenase (TDO) and indoleamine-2,3-dioxygenases (IDOs) [50,51]. TDO can be found in liver cells and, to a lesser extent in the brain, while the IDOs are expressed in the neurons, the astrocytes, and the microglia [51]. The KYN metabolites are formed via different subbranches. Kynurenine aminotransferases (KATs) catalyze the transamination of KYN to form KYNA, whereas kynureninase catalyzes KYN to form ANA. Those metabolites are further degraded to form 3-hydroxy-L-kynurenine (3-HK), xanthurenic acid (XA), 3-hydroxyanthranilic acid (3-HANA), picolinic acid (PICA), and quinolinic acid (QUIN) [8,50,52,53]. During TRP degradation, many KYN metabolites were shown to possess neuroactive properties, such as KYNA and PICA, whereas QUIN and 3-HANA were considered to have neurotoxic properties [52,54]. KYNA affects glutamatergic transmission at various glutamate receptor subunits, including NMDA, AMPA, or kainate receptors [55–57]. A certain concentration of KYNA has been considered neuroprotective, scavenging the insult of reactive oxygen species [48,58]. On the other hand, QUIN is a competitive agonist of the NMDA receptors, and it may cause glutamatergic excitotoxicity by Ca^{2+} influx, which produces neuronal cell death [52,59,60].

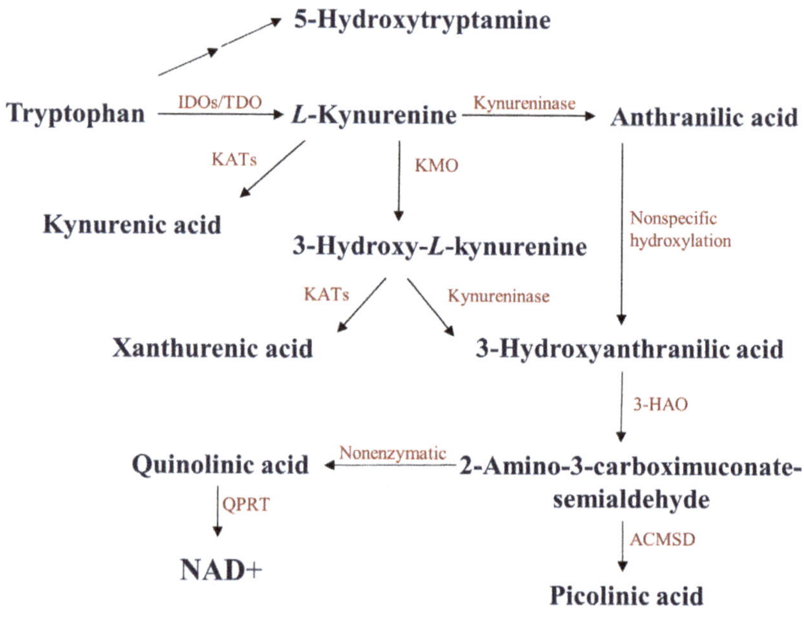

Figure 1. The kynurenine metabolic pathway of tryptophan degradation. 3-HAO: 3-hydroxyanthranilate oxidase; ACMSD: α-amino-β-carboxymuconate-semialdehyde-decarboxylase; IDOs: indoleamine 2,3-dioxygenases; TDO:/tryptophan 2,3-dioxygenase; KATs: kynurenine aminotransferases; KMO: kynurenine 3-monooxygenase; NAD^+: nicotinamide adenine dinucleotide: QPRT: quinolinate phosphoribosyltransferase.

Moreover, several studies have reported the alteration of TRP-KYN metabolites in neurodegenerative diseases and psychiatric disorders, including MS [48,52,61–64]. It was reported that KYNA levels were elevated in the relapse phase while decreased in the remission phase [65,66]. Meanwhile, reduced KYNA and PICA and elevated QUIN concentrations were reported in the cerebrospinal fluid of MS patients [48]. In addition, decreased KYNA levels were observed in the progressive phase of the disease [67]. A recent study has reported a decreased 3-HK level in the serum samples of MS patients [68].

Recently, we reported a significant decrease in KYNA levels in the cortex and the hippocampus, and the plasma during the demyelination period, which is normalized after remyelination in the CPZ-induced model of demyelination [69]. In this study, we measure the levels of serotonin, TRP, and eight TRP-KYN metabolites of the plasma and the brain samples of CPZ-intoxicated mice in the demyelination and remyelination periods in search of a TRP metabolite profile of progressive and recovery phases.

2. Materials and Methods

2.1. Animal Experiments and Sample Collection

In the experiment, eight-week-old C57BL/6J male mice were used (n = 160). The animals were bred and maintained under standard laboratory conditions with 12 h–12 h light/dark cycle at 24 ± 1 °C and 45–55% relative humidity in the Animal House of the Department of Neurology, University of Szeged. The investigations were in accordance with the Ethical Codex of Animal Experiments and were approved by the Ethics Committee of the Faculty of Medicine, University of Szeged, and the National Food Chain Safety Office with a permission number of XI/1101/2018. The experiment was performed as previously described in our study [69]. Briefly, the animals were housed in polycarbonate cages (530 cm^3 floor space) in groups of 5. Prior to the start of the experiment, all animals were acclimated to grounded standard rodent chow for 2 weeks, and animal's weight was monitored every other day.

The CPZ toxin was administered to half of the experimental animals (n = 80) for 5 weeks by a diet containing 0.2% CPZ (bis-cyclohexanone-oxaldihydrazone; Sigma-Aldrich) mixed into a grounded standard rodent chow with free access to water. For control group (CO), age and weight-matched animals were used (n = 80), which had rodent chow and free access to water. At the end of the first, third, and fifth weeks, 16 animals were randomly chosen from both CO and CPZ groups and terminated for further analysis. Thus, at the end of the demyelination phase, 96 animals were terminated (n = 96, 48 CPZ-treated, and 48 control animals). The surviving animals (n = 64, 32 CPZ treated, and 32 control animals) underwent the remyelination phase for 4 weeks and they were sacrificed at the end of the second and fourth weeks of the recovery phase (Figure 2).

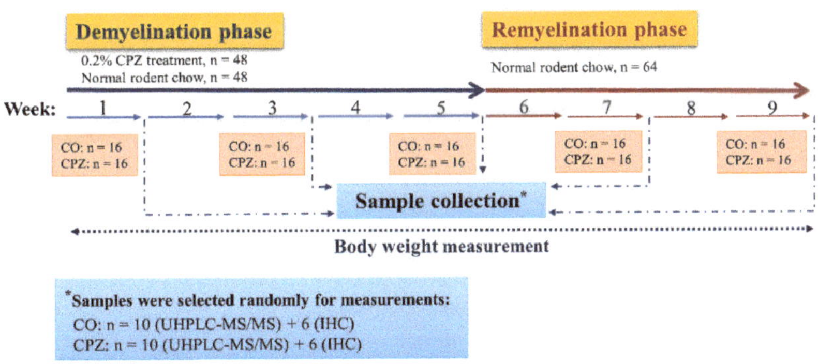

Figure 2. Timeline of the experimental procedure used. CO: control group; CPZ: cuprizone group; IHC: immunohistochemical studies; n: the number of animals; UHPLC-MS/MS: Ultra-high-performance liquid chromatography with tandem mass spectrometry; *: random sample selection was applied for both measurements in the CPZ-treated and the CO groups.

The animals were terminated according to our previous investigation [69]. The mice were anesthetized with intraperitoneal 4% chloral hydrate (10 mL/kg body weight). For the histological and immunohistochemical studies, mice (CPZ: $n = 30$, CO: $n = 30$) were perfused transcardially with artificial cerebrospinal fluid followed by 4% paraformaldehyde in 0.1 M phosphate buffer. Brain samples were dissected and postfixed in the same fixative overnight at 4 °C. Brains were embedded in paraffin, coronally sectioned in 8 μm thickness obtained from different regions (0.14, −0.22, −1.06, and −1.94 mm) according to the mouse brain atlas of Paxinos and Franklin (2001) and placed on gelatin-coated slides [70]. For bioanalytical measurements, the animals (CPZ: $n = 50$, CO: $n = 50$) were anesthetized and perfused as described above. Blood samples were taken from the left heart ventricle into Eppendorf tubes containing 5% disodium ethylenediaminetetraacetate dihydrate and plasma was separated by centrifugation (3500 rpm for 10 min at 4 °C). The brains were dissected into five different brain regions: the cerebellum, the brainstem, the striatum, the somatosensory cortex, and the hippocampus. All samples were placed on ice and stored at −80 °C. The samples were marked as groups of DEM, REM, or CO.

2.2. Luxol Fast Blue Crystal Violet Staining and Myelin Status Determination by Densitometric Analysis

Myelin damage was examined by luxol fast blue crystal violet staining. The brain slides were deparaffinized, rehydrated with 95% alcohol, and incubated in a 0.01% luxol fast blue solution overnight at 60 °C, after that the sections were differentiated in 0.05% lithium carbonate solution and counterstained with crystal violet. For measurements of the stained sections were taken using a Zeiss AxioImager M2 microscope, supplied with an AxioCam MRc Rev. 3 camera (Carl Zeiss Microscopy). Zeiss Zen 2.6 (blue edition)® image analysis software program was applied, which measured the mean intensity of different color channels on a scale from 0 to 65,536. In our case, the low-intensity value characterizes the higher myelin content in the control (CO) group as there was a higher tissue staining. On the other hand, the higher intensity measured in the CPZ-treated (DEM) group shows a lower rate of tissue staining resulting in a decreased myelin content. To determine the myelin content, we performed luxol fast blue crystal violet staining, then the corpus callosum was marked on each section based on the mouse brain atlas of Paxinos and Franklin (2001), and intensity measurement was used on this designated area in order to determine myelin content in the corpus callosum.

2.3. Ultra-High-Performance Liquid Chromatography with Tandem Mass Spectrometry Measurement

All reagents and chemicals were of analytical or liquid chromatography–mass spectrometry grade. TRP and its metabolites, and their deuterated forms: d4-SERO, d5-TRP, d4-KYN, d5-KYNA, d4-XA, d5-5-HIAA, d3-3-HANA, d4-PICA, and d3-QUIN were purchased from Toronto Research Chemicals (Toronto, ON, Canada). d3-3-HK was obtained from Buchem B. V. (Apeldoorn, The Netherlands). Acetonitrile (ACN) was provided by Molar Chemicals (Halásztelek, Hungary). Methanol (MeOH) was purchased from LGC Standards (Wesel, Germany). Formic acid (FA) and water were obtained from VWR Chemicals (Monroeville, PA, USA).

The preparation of the standards, internal standards (IS), and quality control (QC) solutions were necessary for the measurement, as well as the preparation of the animal plasma sample for analysis, which was based on the description published by Tömösi et al. [71]. As for the brain samples, after measuring the weight of the five different brain regions, we homogenized them (UP100H, Hielscher Ultrasound Technology, Germany; amplitude: 100%, cycle: 0.5) in 3× amount of ice-cooled LC-MS water (for example, 90 μL water was pipetted to 30.0 mg sample). After that, the same steps as for the plasma samples were performed, with the difference that the precipitation was carried out with 100% acetonitrile. Then, plasma samples and brain regions were measured according to the previously published methodology using UHPLC-MS/MS [72]. Multiple reaction monitoring (MRM) transition of picolinic acid was 124.0/106.0 using 75 V as declustering potential and 13 V as collision energy, retention time: 1.21 min.

2.4. Statistical Analysis

For the statistical analysis of body weight, two-way repeated-measures ANOVA was used. For the densitometric analysis of LFB statistical differences were determined by one-way analysis of variance (ANOVA), then depending on the variances of data, Sidak or Tamhane's T2 post hoc test was applied. Pairwise comparisons of group means were based on the estimated marginal means with Sidak or Tamhane's T2 post hoc test with adjustment for multiple comparisons. Group values were given as means ± SEM, analyses were performed in SPSS Statistics software (version 20.0 for Windows, SPSS Inc. IBM, Armork, NY, USA). Regarding the UHPLC-MS/MS measurements, all statistical analyses were performed with the help of the R software (R Development Core Team). After checking for its assumptions (checking for outliers, Shapiro and Levene tests), we performed two-way ANOVA with estimated marginal means post hoc tests to determine significance between treatment groups, measurement times, and their interaction. In case of the assumptions were not met, we applied the Sheirer–Ray–Hare test with Dunn test as post hoc. Type I errors from multiple comparisons were controlled with the Bonferroni method. We rejected null hypotheses when the corrected p level was < 0.05, and in such cases, the differences were considered significant.

3. Results

3.1. Investigation of Body Weight

The results of body weight measurement during the CPZ treatment and the recovery phase can be seen in the supplementary material (Figure S1).

3.2. Evaluation of Cuprizone Damage in the Demyelination and Remyelination Phases

The extent of myelin damage was examined by luxol fast blue crystal violet staining. A detailed description and figures of the immunohistochemical analyses and subsequent intensity measurements can be found in the supplementary material (Figures S2 and S3).

3.3. Ultra-High-Performance Liquid Chromatography with Tandem Mass Spectrometry Measurement of Kynurenine Metabolites

During our experiment, UHPLC-MS/MS bioanalytical measurements were performed to detect TRP and different KYN metabolites from both plasma and different brain region samples, including the striatum, cortex, hippocampus, cerebellum, and brainstem. These samples were collected on the 1st, 3rd, and 5th week of treatment, as well as on the 2nd and 4th week of the recovery phase. Due to the large amount of data obtained, only the significant changes were shown and published. Namely, in the case of plasma, we observed a significant difference in the levels of KYNA, 3-HK, and XA in terms of the CPZ treatment time and the degree of damage. Already, at the beginning of the CPZ treatment, a significant decrease in the level of these metabolites was observed, and differences persisted until the end of treatment. In the first half of the recovery phase, these concentration differences disappeared while the level of KYNA, 3-HK, and XA metabolites was in the same range in both groups until the end of remyelination (detailed in Figure 3). In addition to the mentioned metabolites, at the 5th week of treatment, there was also a difference in plasma TRP and ANA levels between CO and CPZ groups (Figure 4).

In the analysis of the brain regions, a significant difference was detected in the concentration of 3-HK, already in the first week of the CPZ treatment in the striatum. However, in the 3rd week of the CPZ intoxication, in addition to the striatum, a significant reduction in 3-HK concentration was also noticed in the cortex, the hippocampus, and the brainstem, which was still maintained in the 5th week of CPZ treatment and was even more pronounced in the cortex and the hippocampus (Figure 5).

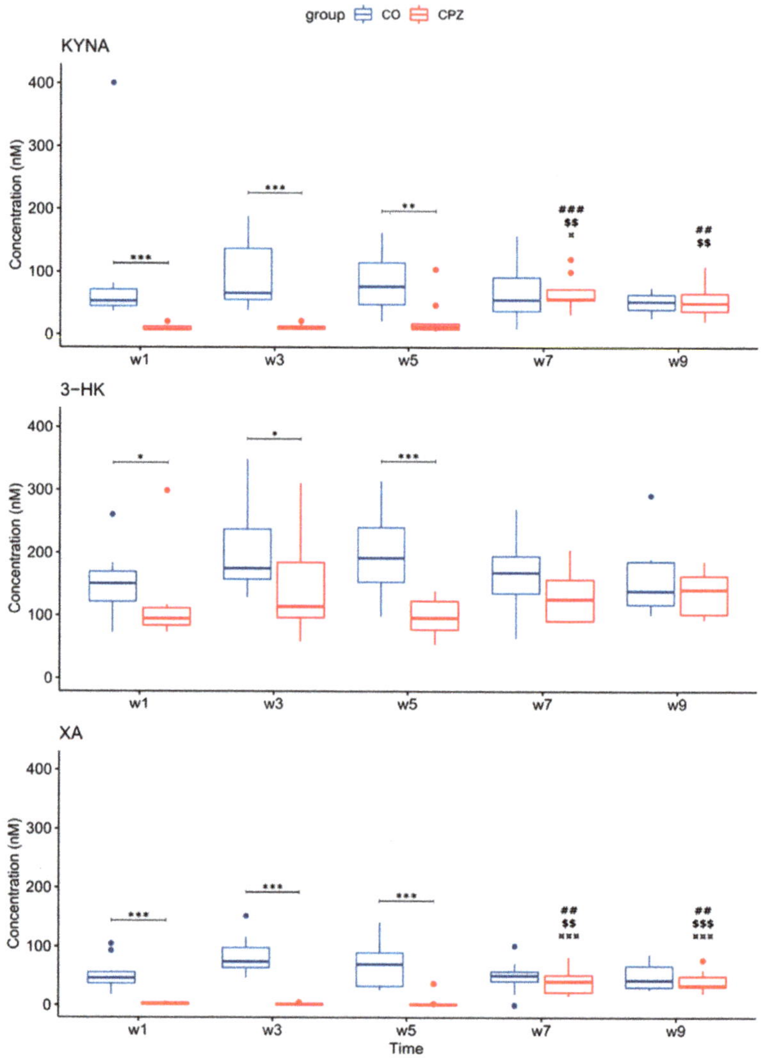

Figure 3. Changes in kynurenine metabolites during weeks of cuprizone treatment in the plasma of cuprizone-treated and control groups. In the first week of the CPZ treatment, a significant decrease in KYNA, 3-HK, and XA levels was observed in the CPZ group compared to the control, and differences remained until the end of the CPZ intoxication. In the first half of the recovery phase, these differences disappeared during remyelination between groups. 3-HK: 3-hydroxy-L-kynurenine; CO: control group; CPZ: cuprizone group; KYNA: kynurenic acid; XA: xanthurenic acid; w: week; *: $p < 0.05$: **: $p < 0.01$; ***: $p < 0.001$: ##: $p < 0.01$; ###: $p < 0.001$ compared to week 1; $$: $p < 0.01$; $$$: $p < 0.001$ compared to week 3; ¤: $p < 0.05$; ¤¤¤: $p < 0.001$ compared to week 5.

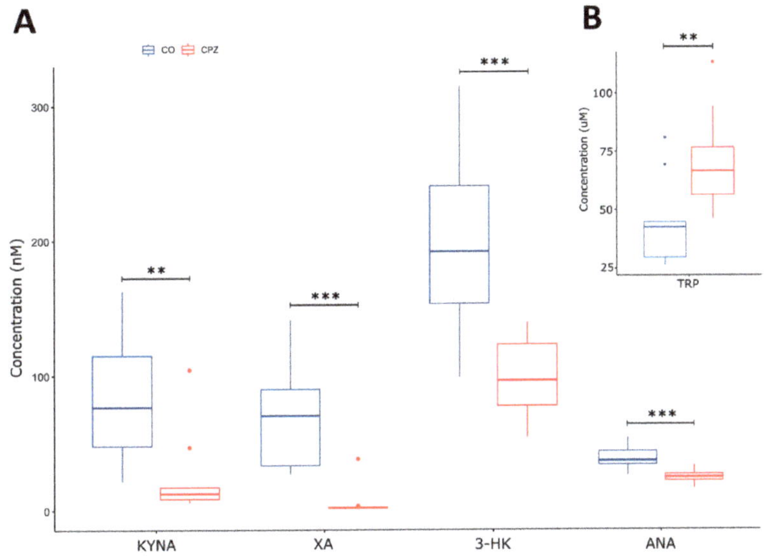

Figure 4. Alteration in plasma metabolite levels at week 5 of CPZ treatment. In addition to the KYNA, 3-HK, and XA (nM), we also observed a significant difference in the concentration of ANA (nM) (**A**) and TRP (μM) metabolites (**B**) by the 5th week of CPZ treatment between CO and CPZ groups. CO: control group; CPZ: cuprizone group: 3-HK: 3-hydroxy-L-kynurenine; ANA: anthranilic acid; KYNA: kynurenic acid; TRP: tryptophan; XA: xanthurenic acid; w: week; **: $p < 0.01$; ***: $p < 0.001$.

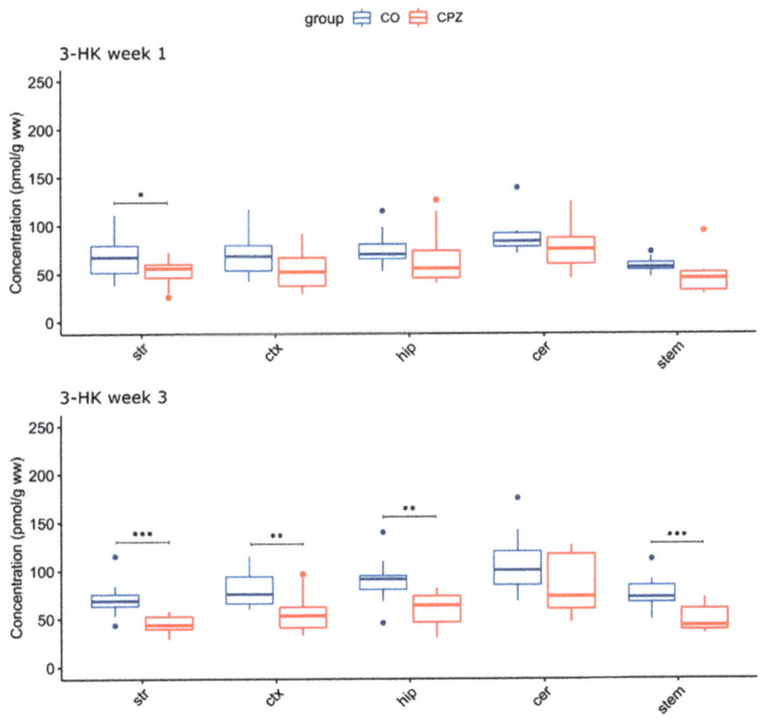

Figure 5. *Cont.*

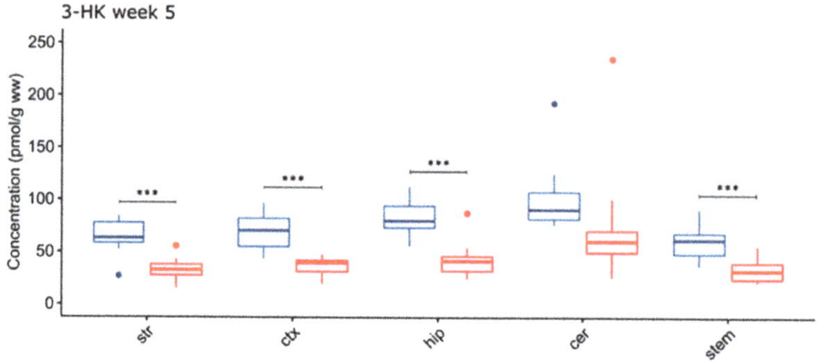

Figure 5. Decreased 3-hydroxy-L-kynurenine concentrations in certain brain regions in time progression in the cuprizone-treated group. A significant difference in 3-HK levels emerged between the groups as the damage caused by the CPZ treatment worsened. 3-HK: 3-hydroxy-L-kynurenine; CO: control group; cer: cerebellum: CPZ: cuprizone group: ctx: cortex; hip: hippocampus; stem: brainstem; str: striatum; ww: wet weight; *: $p < 0.05$; **: $p < 0.01$; ***: $p < 0.001$.

In addition, in the 3rd week of treatment, an increase in TRP concentration was observed in the cortex and hippocampus in the CPZ group. The elevated TRP level became even more apparent by the 5th week of CPZ intoxication, and it was also increased in the striatum and likewise in the cortex and the hippocampus (Figure 6).

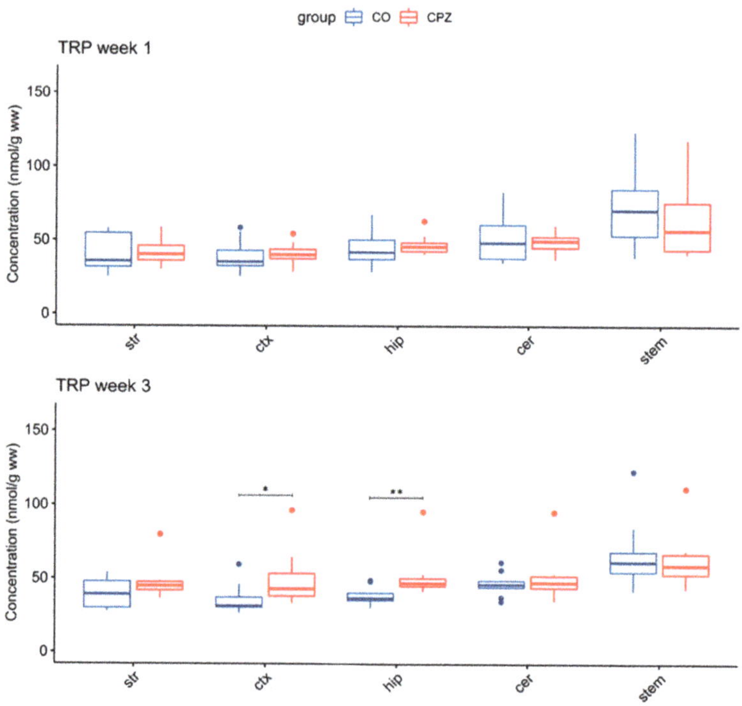

Figure 6. *Cont.*

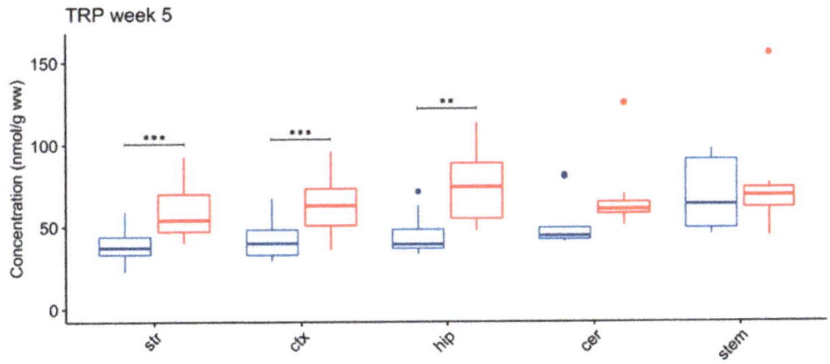

Figure 6. Elevated tryptophan concentrations in some brain regions in the cuprizone-treated group. In the 3rd week of CPZ treatment, the TRP levels were significantly increased in the cortex and the hippocampus. By the 5th week, the TRP concentration was also significantly increased in the striatum during the CPZ treatment. cer: cerebellum; CO: control group; CPZ: cuprizone group; ctx: cortex: hip: hippocampus; stem: brainstem: str: striatum; ww: wet weight; TRP tryptophan; *: $p < 0.05$; **: $p < 0.01$; ***: $p < 0.001$.

Moreover, in addition to the differences in 3-HK and TRP levels, there was also a significant decrease in ANA concentration by the 5th week of CPZ intoxication in the striatum, the cortex, the hippocampus, and the brainstem of the CPZ group compared to the CO (Figure 7).

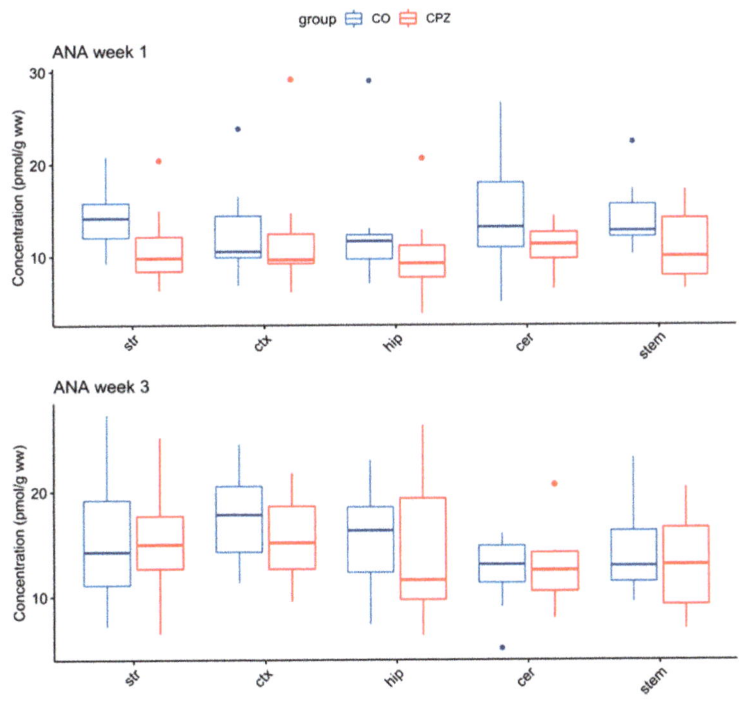

Figure 7. *Cont.*

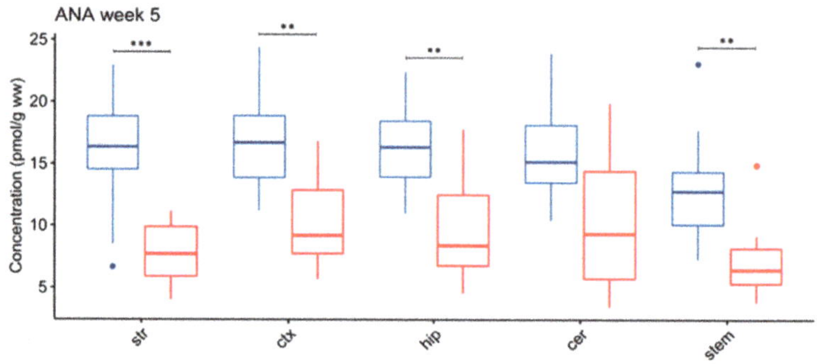

Figure 7. Change in concentration of ANA during cuprizone treatment. In the 5th week of CPZ treatment, a marked decrease in ANA concentrations in the striatum, the cortex, the hippocampus, and the brainstem in CPZ-treated group was observed, compared to the control group. ANA: anthranilic acid; cer: cerebellum; CO: control group; CPZ: cuprizone group; ctx: cortex; hip: hippocampus; stem: brainstem; str: striatum; ww: wet weight; **: $p < 0.01$; ***: $p < 0.001$.

4. Discussion

In this study, we examined the TRP-KYN metabolites in the CPZ-induced animal model of demyelination. The CPZ-treated model is a widely used model of MS, which is considered to be spared the participation of immunological components. Histologically, the model is characterized by oligodendrocyte apoptosis, demyelination, as well as microglia and macrophage activation in the absence of inflammatory processes, which provides an opportunity to investigate the progressive phase of MS in the preclinical model. To ensure the reliability of the CPZ-induced model, we monitored the changes in the body weight of the animals, and then we performed immunohistochemical analyses to determine the degree of demyelination in the corpus callosum.

Bioanalytical measurements were applied to investigate serotonin (5-HT), TRP, and KYN metabolites in the plasma and various brain regions, including areas affected by demyelination and oligodendrocytosis-induced damage [35,36]. This complex bioanalytical study was made to expand our knowledge obtained in the previous study regarding the measurement of four KP metabolites in the CPZ-induced model of MS [69].

In a series of our studies, we noticed a significant body weight decrease in the CPZ-treated animals a few days after the CPZ treatment. Upon the withdrawal of the CPZ toxin, the difference in the body weight starts decreasing and the CPZ-treated mice gained as much weight as that of the CO group by the end of the remyelination phase. Histologically, we observed myelin damage at the beginning of the CPZ treatment, which becomes more significant over time and progressed to extensive severe demyelination. These changes are in line with the previous data [35,36].

Collecting samples at a different time frames made it possible to monitor the progress and extent of the damage, as well as to analyze changes in the concentrations of various metabolites. At the end of the first week of CPZ treatment, immunohistochemical analyzes did not show any difference in myelin sheath between the CPZ and CO groups, as presumably, demyelination damage was not detectable [35,36,73]. By the end of the third week, significant myelin damages were observed in the CPZ toxin-treated group, including oligodendrocyte degeneration, and significant microglia and astrocyte activation processes were observed in this phase [36,73]. By the fifth week, we observed large and severe demyelination in the corpus callosum of the CZP group. The significant demyelination, the phagocytosis of the myelin sheath by microglia, extensive and severe axon damage, astrocytosis, and microgliosis are also supported by the literature data [35,36,73].

These observations have encouraged us to explore a metabolic change in TRP metabolism in response to CPZ treatment, particularly during the demyelination and remyelination phases. Thus, UHPLC-MS/MS analysis was applied to measure the concentrations of 5-HT,

TRP, and KYN metabolites. In the plasma samples, reduced levels of KYNA, XA, and 3-HK were measured as early as in the first week in the CPZ-treated group. The difference persisted toward the third and fifth weeks of CPZ treatment. Furthermore, by the fifth week, the differences in both TRP and ANA levels were observed in the CPZ-treated group. During the remyelination phase, the differences in those metabolites decrease to become insignificant and become none eventually.

In this study, by performing immunohistochemical analyses and plasma metabolite concentration measurements performed at different times of CPZ treatment and recovery phase, we wanted to investigate whether the concentration of KYN metabolites shows the differences already at the beginning of the intoxication and how the levels change during the CPZ treatment, whether they are consistent with the demyelination and remyelination processes in the periphery and the central nervous system. At the beginning of CPZ poisoning, oligodendrocytosis starts already, resulting in demyelination, and later micro-glia and macrophage activation occurs (for more details, see [36]). During the plasma examination, already in the first week of the CPZ treatment, we observed significant concentration differences for certain metabolites between the groups, which became even more evident by the 5th week, but by the 2nd week of recovery, the differences has disappeared, and it seems that the levels of KYN metabolites had normalized by remyelination process.

However, CPZ treatment alters normal liver function due to the megamitochodrium formation, and as a result, plasma amino acid levels also change during treatment [74]. In addition to our immunohistochemical analyses, we found differences in both weight and TRP-KYN metabolite levels as a result of CPZ treatment in the demyelination phase. However, in the recovery phase, these differences between the groups disappeared and it seems that parallel to body weights, metabolite concentrations also normalized relatively quickly in the remyelination process.

Similarly, we observed a significant decrease in 3-HK levels in the brain samples including the striatum, the cortex, the hippocampus, and the brainstem, and by the end of the fifth week, a marked decrease in both 3-HK and ANA levels was observed in the same brain regions. Interestingly, the TRP levels were elevated during CPZ treatment in the cortex and the hippocampus and by the fifth week in the striatum. The differences likewise disappeared in the remyelination phase upon CPZ withdrawal, suggesting the remedy of demyelination. Moreover, in the present study, we observed a significant 3-HK reduction both in the periphery and the brain tissues. Saraste and colleagues reported a reduced level of 3-HK in patients with MS, which is considered to be related to microglial activity [68]. Similarly, a significant microglia/macrophage activation has been described already in the first few weeks of treatment, which persists up to five weeks during the demyelination period in the CPZ animal model [35,36]. The dysregulation of TRP-KYN metabolism is significantly associated with neurodegenerative processes, especially via microglia activation [75–77]. Furthermore, the correlation between microglial activation and low 3-HK levels may be ascribed to the genetic variability of the enzymes, the locational difference in enzyme activity in the body, and the resultant difference in metabolite levels, employed in TRP-KYN metabolism [68,78]. The BBB also plays an important role in a locational difference in the metabolite concentrations: 3-HK and TRP cross the BBB efficiently, while KYNA and QUIN hardly cross it, while the BBB remains relatively intact during CPZ treatment [68,79,80]. Additionally, a reduced 3-HK plasma level was also observed in patients with major depressive disorder compared to the healthy controls [81]. A significant decrease in 3-HK levels was observed in the serum of schizophrenic patients [82]. Thus, these diseases may be associated with microglial activation [83–85].

Some KYN metabolites are neuroactive. Furthermore, the changes in the level of KYN metabolites can affect the activity of enzymes. CPZ intoxication may affect the enzyme functions in KYN metabolism. Indeed, it was shown that elevated copper concentration affects the function of the KAT enzymes in the periphery [86] and presumably thereby the level of KYNA. Probably, this may explain the decrease in KYNA levels in the periphery, as a result of CPZ treatment.

Based on studies, the copper–zinc superoxide dismutase cuproenzyme also shows reduced activity when treated with CPZ [87–89]. Copper, as a cofactor of various copper enzymes, among others superoxide dismutase [90], dopamine-β-hydroxylase [91], monoamine oxidase [92], the cytochrome c oxidase family [93], cytochrome c oxidase assembly protein [94,95]; plays a significant role in several cellular processes, and neurodegeneration may develop in the event of a disturbance in its homeostasis [35].

In contrast, 3-HK and QUIN are considered neurotoxic metabolites at least in certain concentrations and environments, inducing cell death by various excitotoxic processes [75,96,97]. Recently, an increasing amount of evidence highlighted that 3-HK is not always toxic, because 3-HK together with 3-HANA may contribute to the regulation of the redox balance of brain tissue and prevent further damage [75,98,99]. Therefore, 3-HK is a controversial metabolite, as it can function as a scavenger but promote oxidative damage, depending on the redox environment, pH, or cell type, among others [99]. As for XA, studies suggest that XA has antioxidant properties, as it can inhibit lipid peroxidation and prevent iron-induced NADP-isocitrate dehydrogenase inactivation [99–101]; furthermore, it can scavenge free radicals [99,102], bind the superoxide anion, and inhibit hematoxylin autoxidation (see the review [99]).

ANA has been attracting increasing attention as a neuroprotective agent. ANA has proven to be a strong radical scavenger by effectively binding the largest subset of free radicals [99,102]. Moreover, ANA affects respiratory parameters [103] and is a precursor of the synthesis of nonsteroidal anti-inflammatory drugs, [63,104]. Considering this, XA and ANA may play a major role as antioxidants in CPZ intoxication. Presumably, the elevated TRP levels of the CNS can be explained by the reduction of enzyme activities of downward KYN enzymes and/or a compensatory mechanism to ensure TRP availability due to CPZ intoxication. In a recent study, a significantly higher TRP level was observed among patients with MS compared to the control group [105].

The metabolite concentration differences observed in individual brain regions may be related to changes in the volume of certain regions.

CPZ-induced oligodendrocytosis is unequally distributed in the CNS. CPZ poisoning causes extensive oligodendrocytosis and severe demyelination, among others in the corpus callosum, cerebral cortex, hippocampus, and to a lesser extent in the cerebellum and brainstem. The reasons for the regional variability are not known, but it may be influenced by the uneven distribution of different oligodendrocyte subtypes in the CNS. Thus, it may happen that CPZ is much more toxic to some subtypes, while less to others. In addition, altered gene expression can affect the sensitivity of certain areas to injuries (for more details, see [36]).

Furthermore, based on studies, the dry mass of the brain irreversibly reduced after CPZ intoxication [106,107], which was also indicated by the thinning of the corpus callosum and cortex after CPZ treatment [35,108–110]. In our study, we found a significant difference in metabolite concentration in the brain regions, that were mentioned in the literature as severely demyelinated areas, including the cortex and hippocampus as well as the brainstem.

Moreover, based on the literature data, CPZ also causes damage to neurotransmitter homeostasis. CPZ exerts an inhibitory effect on glutamic acid decarboxylase, an increase in glutamate (GLU) level, and a decrease in the gamma-aminobutyric acid (GABA) [111]. Another study, on the other hand, described an increased GABA level during 3 weeks of CPZ treatment, in contrast to the reduced GABA level seen during 8 weeks of treatment, which may point to changes in neurotransmitter concentration over time during CPZ intoxication [35,112]. Furthermore, based on research, dopaminergic and noradrenergic synapses are also affected during CPZ treatment. Specifically, CPZ poisoning has an inhibitory effect on the functioning of dopamine hydroxylase and monoamine oxidase enzymes, which affect dopamine and norepinephrine concentrations (for more details, see [35]).

The authors acknowledge the limitations of this study and different outcomes depending on analytical methods. In the present study, UHPLC-MS/MS analysis did not re-produce the TRP and KYNA concentrations in the brain regions, which were measured by the HPLC method. The discrepancy may be ascribed to the difference in sample preparation (precipitation with perchloric acid vs. acetonitrile) and in detection methods (UV–VIS vs. UHPLC-MS/MS). Interestingly, the UHPLC-MS/MS method measures KYNA in a much narrower concentration range from brain samples, compared to measurements with fluorescent detectors [113]. Recently, more bioanalytical measurements were carried out by mass spectrometry analysis in the advantage of its outstanding selectivity, sensitivity, detection specificity, and reproducibility, compared to HPLC methods.

Many factors play a role in the examination and analysis of brain regions, such as the complexity of the regions, sample preparation and its limitations, and many additional different metabolites present in the CNS. Nevertheless, it is important to mention a problem that affects many and is becoming increasingly common, namely the reproducibility crisis including using different analytical methods. The scientists have drawn attention to, and focus on the dilemma that the repetition of studies is not necessarily reproducible in animal as well as in human studies [114].

Nonetheless, this is the first study comprehensively showing the concentrations of 5-HT, TRP, and KYN metabolites at different time points during CPZ treatment and the recovery phase, in the plasma and five different brain regions. The study has confirmed metabolic changes on both sides of the BBB during demyelination. In accordance with our previous study, here we showed the involvement of KYN metabolites in the CPZ-induced animal model of demyelination, which is analog to progressive MS in terms of some markers. In addition to complementing the previous data, this study has revealed the profile of KYN metabolites during progressive demyelination. Further studies might shed light on the mechanism behind the alteration of KYN metabolites and possible changes in enzyme activities of TRP-KYN metabolism.

Supplementary Materials: The following supporting information can be downloaded at: https://www.mdpi.com/article/10.3390/biomedicines11030945/s1, Figure S1: Alteration in body weight of the animals during the experiment; Figure S2: Luxol fast blue-crystal violet staining in the corpus callosum of the control and cuprizone-treated groups in the first, third, and fifth week of CPZ treatment (DEM), and in the seventh and ninth weeks of the experiment, which is the second and fourth weeks of the recovery phase (remyelination); Figure S3: Formation of corpus callosum demyelination by CPZ treatment in the CO and CPZ group.

Author Contributions: Conceptualization, H.P.; writing—original draft preparation and editing, H.P., Z.G., N.N., E.K.C., C.R., P.K., M.T. and L.V.; performed the experiments, analyzed, and interpreted the data; H.P., Z.G., N.N. and G.V.; visualization, E.S. and Á.S.; project administration, L.V.; funding acquisition, M.T. and L.V. All authors have read and agreed to the published version of the manuscript.

Funding: This research was funded by National Research, Development, and Innovation Office–NKFIH K138125, and ELKH-SZTE Neuroscience Research Group. Helga Polyák was supported by the ÚNKP-22-3-New National Excellence Program of the Ministry for Culture and Innovation from the source of the National Research, Development, and Innovation Fund; and EFOP 3.6.3-VEKOP-16-2017-00009.

Institutional Review Board Statement: The animal study protocol approved by the Protection of Animals in Research of the University of Szeged (Szeged, Hungary) and the Scientific Ethics Committee for Animal Research of the Protection of Animals Advisory Board (XI/1101/2018), and the protocol for animal care approved both by the Hungarian Health Committee (40/2013 (II.14.)) and the directive of the European Parliament (2010/63/EU).

Informed Consent Statement: Not applicable.

Data Availability Statement: The data presented in this study are available on request from the corresponding author.

Acknowledgments: We are grateful to Orsolya Ivánkovitsné Kiss for their valuable professional help. Graphical abstract was created with BioRender.com (accessed on 1 January 2023).

Conflicts of Interest: The authors declare no conflict of interest.

Abbreviations

ANA	anthranilic acid
AMPA	amino-3-hydroxy-5methyl-4 isoxazole propionic acid
BBB	blood–brain barrier
CO	control
CNS	central nervous system
CPZ	cuprizone
DEM	demyelination
3-HANA	3-hydroxyanthranilic acid
3-HK	3-hydroxy-L-kynurenine
HPLC	high-performance liquid chromatography
IDO	indoleamine 2,3- dioxygenase
KAT	kynurenine aminotransferase
KMO	kynurenine 3-monooxygenase
KP	kynurenine pathway
KYN	kynurenine
KYNA	kynurenic acid
UHPLC-MS/MS	Ultra-high-performance liquid chromatography with tandem mass spectrometry
LFB	luxol fast blue
MRM	multiple reaction monitoring
MS	multiple sclerosis
NAD+	nicotinamide adenine dinucleotide
NMDA	N-methyl-D-aspartate
ROS	reactive oxygen species
PICA	picolinic acid
REM	remyelination
RRMS	relapsing–remitting multiple sclerosis
5-HT	serotonin
TDO	tryptophan 2,3-dioxygenase
TRP	tryptophan
XA	xanthurenic acid
QUIN	quinolinic acid

References

1. Frohman, E.M.; Racke, M.K.; Raine, C.S. Multiple sclerosis—The plaque and its pathogenesis. *N. Engl. J. Med.* **2006**, *354*, 942–955. [CrossRef] [PubMed]
2. Karussis, D.; Karageorgiou, C.; Vaknin-Dembinsky, A.; Gowda-Kurkalli, B.; Gomori, J.M.; Kassis, I.; Bulte, J.W.; Petrou, P.; Ben-Hur, T.; Abramsky, O.; et al. Safety and immunological effects of mesenchymal stem cell transplantation in patients with multiple sclerosis and amyotrophic lateral sclerosis. *Arch. Neurol.* **2010**, *67*, 1187–1194. [CrossRef] [PubMed]
3. Kipp, M.; Nyamoya, S.; Hochstrasser, T.; Amor, S. Multiple sclerosis animal models: A clinical and histopathological perspective. *Brain Pathol.* **2017**, *27*, 123–137. [CrossRef] [PubMed]
4. Lassmann, H.; Bradl, M. Multiple sclerosis: Experimental models and reality. *Acta Neuropathol.* **2017**, *133*, 223–244. [CrossRef]
5. Shiri, E.; Pasbakhsh, P.; Borhani-Haghighi, M.; Alizadeh, Z.; Nekoonam, S.; Mojaverrostami, S.; Pirhajati Mahabadi, V.; Mehdi, A.; Zibara, K.; Kashani, I.R. Mesenchymal Stem Cells Ameliorate Cuprizone-Induced Demyelination by Targeting Oxidative Stress and Mitochondrial Dysfunction. *Cell Mol. Neurobiol.* **2021**, *41*, 1467–1481. [CrossRef]
6. Walton, C.; King, R.; Rechtman, L.; Kaye, W.; Leray, E.; Marrie, R.A.; Robertson, N.; La Rocca, N.; Uitdehaag, B.; van der Mei, I.; et al. Rising prevalence of multiple sclerosis worldwide: Insights from the Atlas of MS, third edition. *Mult. Scler.* **2020**, *26*, 1816–1821. [CrossRef]
7. Calandri, E.; Graziano, F.; Borghi, M.; Bonino, S. Young adults' adjustment to a recent diagnosis of multiple sclerosis: The role of identity satisfaction and self-efficacy. *Disabil. Health J.* **2019**, *12*, 72–78. [CrossRef]
8. Rajda, C.; Pukoli, D.; Bende, Z.; Majláth, Z.; Vécsei, L. Excitotoxins, Mitochondrial and Redox Disturbances in Multiple Sclerosis. *Int. J. Mol. Sci.* **2017**, *18*, 353. [CrossRef]

9. Cree, B.A.C.; Arnold, D.L.; Chataway, J.; Chitnis, T.; Fox, R.J.; Pozo Ramajo, A.; Murphy, N.; Lassmann, H. Secondary Progressive Multiple Sclerosis. *Neurology* **2021**, *97*, 378–388. [CrossRef]
10. Lassmann, H.; van Horssen, J.; Mahad, D. Progressive multiple sclerosis: Pathology and pathogenesis. *Nat. Rev. Neurol.* **2012**, *8*, 647–656. [CrossRef]
11. Battaglia, S.; Thayer, J.F. Functional interplay between central and autonomic nervous systems in human fear conditioning. *Trends Neurosci.* **2022**, *45*, 504–506. [CrossRef] [PubMed]
12. Battaglia, S.; Orsolini, S.; Borgomaneri, S.; Barbieri, R.; Diciotti, S.; di Pellegrino, G. Characterizing cardiac autonomic dynamics of fear learning in humans. *Psychophysiology* **2022**, *59*, e14122. [CrossRef] [PubMed]
13. Di Gregorio, F.; La Porta, F.; Petrone, V.; Battaglia, S.; Orlandi, S.; Ippolito, G.; Romei, V.; Piperno, R.; Lullini, G. Accuracy of EEG Biomarkers in the Detection of Clinical Outcome in Disorders of Consciousness after Severe Acquired Brain Injury: Preliminary Results of a Pilot Study Using a Machine Learning Approach. *Biomedicines* **2022**, *10*, 1897. [CrossRef] [PubMed]
14. Nyatega, C.O.; Qiang, L.; Adamu, M.J.; Kawuwa, H.B. Gray matter, white matter and cerebrospinal fluid abnormalities in Parkinson's disease: A voxel-based morphometry study. *Front. Psychiatry* **2022**, *13*, 1027907. [CrossRef]
15. Nyatega, C.O.; Qiang, L.; Adamu, M.J.; Younis, A.; Kawuwa, H.B. Altered Dynamic Functional Connectivity of Cuneus in Schizophrenia Patients: A Resting-State fMRI Study. *Appl. Sci.* **2021**, *11*, 11392. [CrossRef]
16. Orso, B.; Lorenzini, L.; Arnaldi, D.; Girtler, N.; Brugnolo, A.; Doglione, E.; Mattioli, P.; Biassoni, E.; Massa, F.; Peira, E.; et al. The Role of Hub and Spoke Regions in Theory of Mind in Early Alzheimer's Disease and Frontotemporal Dementia. *Biomedicines* **2022**, *10*, 544. [CrossRef]
17. Okanda Nyatega, C.; Qiang, L.; Jajere Adamu, M.; Bello Kawuwa, H. Altered striatal functional connectivity and structural dysconnectivity in individuals with bipolar disorder: A resting state magnetic resonance imaging study. *Front. Psychiatry* **2022**, *13*, 1054380. [CrossRef]
18. Battaglia, S.; Fabius, J.H.; Moravkova, K.; Fracasso, A.; Borgomaneri, S. The Neurobiological Correlates of Gaze Perception in Healthy Individuals and Neurologic Patients. *Biomedicines* **2022**, *10*, 627. [CrossRef]
19. Nyatega, C.O.; Qiang, L.; Jajere, M.A.; Kawuwa, H.B. Atypical Functional Connectivity of Limbic Network in Attention Deficit/Hyperactivity Disorder. *Clin. Schizophr. Relat. Psychoses* **2022**, *16*, 2. [CrossRef]
20. Datki, Z.; Sinka, R. Translational biomedicine-oriented exploratory research on bioactive rotifer-specific biopolymers. *Adv. Clin. Exp. Med.* **2022**, *31*, 931–935. [CrossRef]
21. Kwon, K.-M.; Lee, M.-J.; Chung, H.-S.; Pak, J.-H.; Jeon, C.-J. The Organization of Somatostatin-Immunoreactive Cells in the Visual Cortex of the Gerbil. *Biomedicines* **2022**, *10*, 92. [CrossRef]
22. Lieb, A.; Thaler, G.; Fogli, B.; Trovato, O.; Posch, M.A.; Kaserer, T.; Zangrandi, L. Functional Characterization of Spinocerebellar Ataxia Associated Dynorphin A Mutant Peptides. *Biomedicines* **2021**, *9*, 1882. [CrossRef] [PubMed]
23. Palotai, M.; Telegdy, G.; Tanaka, M.; Bagosi, Z.; Jászberényi, M. Neuropeptide AF induces anxiety-like and antidepressant-like behavior in mice. *Behav. Brain Res.* **2014**, *274*, 264–269. [CrossRef] [PubMed]
24. Simon, C.; Soga, T.; Ahemad, N.; Bhuvanendran, S.; Parhar, I. Kisspeptin-10 Rescues Cholinergic Differentiated SHSY-5Y Cells from α-Synuclein-Induced Toxicity In Vitro. *Int. J. Mol. Sci.* **2022**, *23*, 5193. [CrossRef] [PubMed]
25. Song, A.; Cho, G.-W.; Vijayakumar, K.A.; Moon, C.; Ang, M.J.; Kim, J.; Park, I.; Jang, C.H. Neuroprotective Effect of Valproic Acid on Salicylate-Induced Tinnitus. *Int. J. Mol. Sci.* **2021**, *23*, 23. [CrossRef] [PubMed]
26. Telegdy, G.; Tanaka, M.; Schally, A.V. Effects of the growth hormone-releasing hormone (GH-RH) antagonist on brain functions in mice. *Behav. Brain Res.* **2011**, *224*, 155–158. [CrossRef]
27. Tanaka, M.; Kádár, K.; Tóth, G.; Telegdy, G. Antidepressant-like effects of urocortin 3 fragments. *Brain Res. Bull.* **2011**, *84*, 414–418. [CrossRef]
28. Ibos, K.E.; Bodnár, É.; Bagosi, Z.; Bozsó, Z.; Tóth, G.; Szabó, G.; Csabafi, K. Kisspeptin-8 Induces Anxiety-like Behavior and Hypolocomotion by Activating the HPA Axis and Increasing GABA Release in the Nucleus Accumbens in Rats. *Biomedicines* **2021**, *9*, 112. [CrossRef]
29. Rákosi, K.; Masaru, T.; Zarándi, M.; Telegdy, G.; Tóth, G.K. Short analogs and mimetics of human urocortin 3 display antidepressant effects in vivo. *Peptides* **2014**, *62*, 59–66. [CrossRef]
30. Tanaka, M.; Szabó, Á.; Vécsei, L. Integrating Armchair, Bench, and Bedside Research for Behavioral Neurology and Neuropsychiatry: Editorial. *Biomedicines* **2022**, *10*, 2999. [CrossRef]
31. Younis, A.; Qiang, L.; Nyatega, C.O.; Adamu, M.J.; Kawuwa, H.B. Brain Tumor Analysis Using Deep Learning and VGG-16 Ensembling Learning Approaches. *Appl. Sci.* **2022**, *12*, 7282. [CrossRef]
32. Tanaka, M.; Vécsei, L. Editorial of Special Issue 'Dissecting Neurological and Neuropsychiatric Diseases: Neurodegeneration and Neuroprotection. *Int. J. Mol. Sci.* **2022**, *23*, 6991. [CrossRef] [PubMed]
33. Tanaka, M.; Vécsei, L. Editorial of Special Issue "Crosstalk between Depression, Anxiety, and Dementia: Comorbidity in Behavioral Neurology and Neuropsychiatry". *Biomedicines* **2021**, *9*, 517. [CrossRef] [PubMed]
34. Tanaka, M. Crosstalk between Depression, Anxiety, and Dementia: Comorbidity in Behavioral Neurology and Neuropsychiatry. *Biomedicines* **2022**. [CrossRef]
35. Praet, J.; Guglielmetti, C.; Berneman, Z.; Van der Linden, A.; Ponsaerts, P. Cellular and molecular neuropathology of the cuprizone mouse model: Clinical relevance for multiple sclerosis. *Neurosci. Biobehav. Rev.* **2014**, *47*, 485–505. [CrossRef]

36. Sen, M.K.; Mahns, D.A.; Coorssen, J.R.; Shortland, P.J. Behavioural phenotypes in the cuprizone model of central nervous system demyelination. *Neurosci. Biobehav. Rev.* **2019**, *107*, 23–46. [CrossRef]
37. Kalman, B.; Laitinen, K.; Komoly, S. The involvement of mitochondria in the pathogenesis of multiple sclerosis. *J. Neuroimmunol.* **2007**, *188*, 1–12. [CrossRef]
38. Gudi, V.; Gingele, S.; Skripuletz, T.; Stangel, M. Glial response during cuprizone-induced de- and remyelination in the CNS: Lessons learned. *Front. Cell Neurosci.* **2014**, *8*, 73. [CrossRef]
39. Hiremath, M.M.; Saito, Y.; Knapp, G.W.; Ting, J.P.-Y.; Suzuki, K.; Matsushima, G.K. Microglial/macrophage accumulation during cuprizone-induced demyelination in C57BL/6 mice. *J. Neuroimmunol.* **1998**, *92*, 38–49. [CrossRef]
40. Lampron, A.; Larochelle, A.; Laflamme, N.; Préfontaine, P.; Plante, M.-M.; Sánchez, M.G.; Yong, V.W.; Stys, P.K.; Tremblay, M.È.; Rivest, S. Inefficient clearance of myelin debris by microglia impairs remyelinating processes. *J. Exp. Med.* **2015**, *212*, 481–495. [CrossRef]
41. Skripuletz, T.; Hackstette, D.; Bauer, K.; Gudi, V.; Pul, R.; Voss, E.; Berger, K.; Kipp, M.; Baumgärtner, W.; Stangel, M. Astrocytes regulate myelin clearance through recruitment of microglia during cuprizone-induced demyelination. *Brain* **2013**, *136*, 147–167. [CrossRef]
42. Mason, J.L.; Jones, J.J.; Taniike, M.; Morell, P.; Suzuki, K.; Matsushima, G.K. Mature oligodendrocyte apoptosis precedes IGF-1 production and oligodendrocyte progenitor accumulation and differentiation during demyelination/remyelination. *J. Neurosci. Res.* **2000**, *61*, 251–262. [CrossRef] [PubMed]
43. Matsushima, G.K.; Morell, P. The Neurotoxicant, Cuprizone, as a Model to Study Demyelination and Remyelination in the Central Nervous System. *Brain Pathol.* **2001**, *11*, 107–116. [CrossRef] [PubMed]
44. Nyamoya, S.; Steinle, J.; Chrzanowski, U.; Kaye, J.; Schmitz, C.; Beyer, C.; Kipp, M. Laquinimod Supports Remyelination in Non-Supportive Environments. *Cells* **2019**, *8*, 1363. [CrossRef] [PubMed]
45. Azami Tameh, A.; Clarner, T.; Beyer, C.; Atlasi, M.A.; Hassanzadeh, G.; Naderian, H. Regional regulation of glutamate signaling during cuprizone-induced demyelination in the brain. *Ann. Anat.—Anat. Anz.* **2013**, *195*, 415–423. [CrossRef] [PubMed]
46. Dutta, R.; Chomyk, A.M.; Chang, A.; Ribaudo, M.V.; Deckard, S.A.; Doud, M.K.; Edberg, D.D.; Bai, B.; Li, M.; Baranzini, S.E.; et al. Hippocampal demyelination and memory dysfunction are associated with increased levels of the neuronal microRNA miR-124 and reduced AMPA receptors. *Ann. Neurol.* **2013**, *73*, 637–645. [CrossRef] [PubMed]
47. Dewar, D.; Underhill, S.M.; Goldberg, M.P. Oligodendrocytes and ischemic brain injury. *J. Cereb. Blood Flow Metab.* **2003**, *23*, 263–274. [CrossRef]
48. Vécsei, L.; Szalárdy, L.; Fülöp, F.; Toldi, J. Kynurenines in the CNS: Recent advances and new questions. *Nat. Rev. Drug Discov.* **2013**, *12*, 64–82. [CrossRef]
49. Morales-Puerto, N.; Giménez-Gómez, P.; Pérez-Hernández, M.; Abuin-Martínez, C.; Gil de Biedma-Elduayen, L.; Vidal, R.; Gutiérrez-López, M.D.; O'Shea, E.; Colado, M.I. Addiction and the kynurenine pathway: A new dancing couple? *Pharmacol. Ther.* **2021**, *223*, 107807. [CrossRef]
50. Badawy, A.A.-B. Kynurenine pathway and human systems. *Exp. Gerontol.* **2020**, *129*, 110770. [CrossRef]
51. Rajda, C.; Majláth, Z.; Pukoli, D.; Vécsei, L. Kynurenines and Multiple Sclerosis: The Dialogue between the Immune System and the Central Nervous System. *Int. J. Mol. Sci.* **2015**, *16*, 18270–18282. [CrossRef] [PubMed]
52. Biernacki, T.; Sandi, D.; Bencsik, K.; Vécsei, L. Kynurenines in the Pathogenesis of Multiple Sclerosis: Therapeutic Perspectives. *Cells* **2020**, *9*, 1564. [CrossRef] [PubMed]
53. Chen, L.-M.; Bao, C.-H.; Wu, Y.; Liang, S.-H.; Wang, D.; Wu, L.-Y.; Huang, Y.; Liu, H.R.; Wu, H.G. Tryptophan-kynurenine metabolism: A link between the gut and brain for depression in inflammatory bowel disease. *J. Neuroinflamm.* **2021**, *18*, 135. [CrossRef]
54. Taleb, O.; Maammar, M.; Klein, C.; Maitre, M.; Mensah-Nyagan, A.G. A Role for Xanthurenic Acid in the Control of Brain Dopaminergic Activity. *Int. J. Mol. Sci.* **2021**, *22*, 6974. [CrossRef]
55. Birch, P.J.; Grossman, C.J.; Hayes, A.G. Kynurenate and FG9041 have both competitive and non-competitive antagonist actions at excitatory amino acid receptors. *Eur. J. Pharmacol.* **1988**, *151*, 313–315. [CrossRef]
56. Kessler, M.; Terramani, T.; Lynch, G.; Baudry, M. A glycine site associated with N-methyl-D-aspartic acid receptors: Characterization and identification of a new class of antagonists. *J. Neurochem.* **1989**, *52*, 1319–1328. [CrossRef] [PubMed]
57. Zádori, D.; Klivényi, P.; Plangár, I.; Toldi, J.; Vécsei, L. Endogenous neuroprotection in chronic neurodegenerative disorders: With particular regard to the kynurenines. *J. Cell Mol. Med.* **2011**, *15*, 701–717. [CrossRef]
58. Bohár, Z.; Toldi, J.; Fülöp, F.; Vécsei, L. Changing the Face of Kynurenines and Neurotoxicity: Therapeutic Considerations. *Int. J. Mol. Sci.* **2015**, *16*, 9772–9793. [CrossRef]
59. Guillemin, G.J. Quinolinic acid, the inescapable neurotoxin. *FEBS J.* **2012**, *279*, 1356–1365. [CrossRef]
60. Sandi, D.; Fricska-Nagy, Z.; Bencsik, K.; Vécsei, L. Neurodegeneration in Multiple Sclerosis: Symptoms of Silent Progression, Biomarkers and Neuroprotective Therapy—Kynurenines Are Important Players. *Molecules* **2021**, *26*, 3423. [CrossRef]
61. Sandi, D.; Biernacki, T.; Szekeres, D.; Füvesi, J.; Kincses, Z.T.; Rózsa, C.; Mátyás, K.; Kása, K.; Matolcsi, J.; Zboznovits, D.; et al. Prevalence of cognitive impairment among Hungarian patients with relapsing-remitting multiple sclerosis and clinically isolated syndrome. *Mult. Scler. Relat. Disord.* **2017**, *17*, 57–62. [CrossRef]
62. Tanaka, M.; Tóth, F.; Polyák, H.; Szabó, Á.; Mándi, Y.; Vécsei, L. Immune Influencers in Action: Metabolites and Enzymes of the Tryptophan-Kynurenine Metabolic Pathway. *Biomedicines* **2021**, *9*, 734. [CrossRef] [PubMed]

63. Tanaka, M.; Spekker, E.; Szabó, Á.; Polyák, H.; Vécsei, L. Modelling the neurodevelopmental pathogenesis in neuropsychiatric disorders. Bioactive kynurenines and their analogues as neuroprotective agents—In celebration of 80th birthday of Professor Peter Riederer. *J. Neural. Transm.* **2022**, *129*, 627–642. [CrossRef] [PubMed]
64. Tanaka, M.; Toldi, J.; Vécsei, L. Exploring the Etiological Links behind Neurodegenerative Diseases: Inflammatory Cytokines and Bioactive Kynurenines. *Int. J. Mol. Sci.* **2020**, *21*, 2431. [CrossRef] [PubMed]
65. Rejdak, K.; Petzold, A.; Kocki, T.; Kurzepa, J.; Grieb, P.; Turski, W.A.; Stelmasiak, Z. Astrocytic activation in relation to inflammatory markers during clinical exacerbation of relapsing-remitting multiple sclerosis. *J. Neural Transm.* **2007**, *114*, 1011. [CrossRef] [PubMed]
66. Rejdak, K.; Bartosik-Psujek, H.; Dobosz, B.; Kocki, T.; Grieb, P.; Giovannoni, G.; Turski, W.A.; Stelmasiak, Z. Decreased level of kynurenic acid in cerebrospinal fluid of relapsing-onset multiple sclerosis patients. *Neurosci. Lett.* **2002**, *331*, 63–65. [CrossRef]
67. Lim, C.K.; Bilgin, A.; Lovejoy, D.B.; Tan, V.; Bustamante, S.; Taylor, B.V.; Bessede, A.; Brew, B.J.; Guillemin, G.J. Kynurenine pathway metabolomics predicts and provides mechanistic insight into multiple sclerosis progression. *Sci. Rep.* **2017**, *7*, 41473. [CrossRef]
68. Saraste, M.; Matilainen, M.; Rajda, C.; Galla, Z.; Sucksdorff, M.; Vécsei, L.; Airas, L. Association between microglial activation and serum kynurenine pathway metabolites in multiple sclerosis patients. *Mult. Scler. Relat. Disord.* **2022**, *59*, 103667. [CrossRef]
69. Polyák, H.; Cseh, E.K.; Bohár, Z.; Rajda, C.; Zádori, D.; Klivényi, P.; Toldi, J.; Vécsei, L. Cuprizone markedly decreases kynurenic acid levels in the rodent brain tissue and plasma. *Heliyon* **2021**, *7*, e06124. [CrossRef]
70. Acs, P.; Kipp, M.; Norkute, A.; Johann, S.; Clarner, T.; Braun, A.; Berente, Z.; Komoly, S.; Beyer, C. 17β-estradiol and progesterone prevent cuprizone provoked demyelination of corpus callosum in male mice. *Glia* **2009**, *57*, 807–814. [CrossRef]
71. Tömösi, F.; Kecskeméti, G.; Cseh, E.K.; Szabó, E.; Rajda, C.; Kormány, R.; Szabó, Z.; Vécsei, L.; Janáky, T. A validated UHPLC-MS method for tryptophan metabolites: Application in the diagnosis of multiple sclerosis. *J. Pharm. Biomed. Anal.* **2020**, *185*, 113246. [CrossRef] [PubMed]
72. Galla, Z.; Rajda, C.; Rácz, G.; Grecsó, N.; Baráth, Á.; Vécsei, L.; Bereczki, C.; Monostori, P. Simultaneous determination of 30 neurologically and metabolically important molecules: A sensitive and selective way to measure tyrosine and tryptophan pathway metabolites and other biomarkers in human serum and cerebrospinal fluid. *J. Chromatogr. A* **2021**, *1635*, 461775. [CrossRef] [PubMed]
73. Zhan, J.; Mann, T.; Joost, S.; Behrangi, N.; Frank, M.; Kipp, M. The Cuprizone Model: Dos and Do Nots. *Cells* **2020**, *9*, 843. [CrossRef] [PubMed]
74. Goldberg, J.; Daniel, M.; van Heuvel, Y.; Victor, M.; Beyer, C.; Clarner, T.; Kipp, M. Short-Term Cuprizone Feeding Induces Selective Amino Acid Deprivation with Concomitant Activation of an Integrated Stress Response in Oligodendrocytes. *Cell Mol. Neurobiol.* **2013**, *33*, 1087–1098. [CrossRef]
75. Capucciati, A.; Galliano, M.; Bubacco, L.; Zecca, L.; Casella, L.; Monzani, E.; Nicolis, S. Neuronal Proteins as Targets of 3-Hydroxykynurenine: Implications in Neurodegenerative Diseases. *ACS Chem. Neurosci.* **2019**, *10*, 3731–3739. [CrossRef]
76. Lovelace, M.D.; Varney, B.; Sundaram, G.; Lennon, M.J.; Lim, C.K.; Jacobs, K.; Guillemin, G.J.; Brew, B.J. Recent evidence for an expanded role of the kynurenine pathway of tryptophan metabolism in neurological diseases. *Neuropharmacology* **2017**, *112*, 373–388. [CrossRef]
77. Zinger, A.; Barcia, C.; Herrero, M.T.; Guillemin, G.J. The involvement of neuroinflammation and kynurenine pathway in Parkinson's disease. *Parkinsons. Dis.* **2011**, *2011*, 716859. [CrossRef]
78. Schwarcz, R.; Bruno, J.P.; Muchowski, P.J.; Wu, H.-Q. Kynurenines in the mammalian brain: When physiology meets pathology. *Nat. Rev. Neurosci.* **2012**, *13*, 465–477. [CrossRef]
79. Fukui, S.; Schwarcz, R.; Rapoport, S.I.; Takada, Y.; Smith, Q.R. Blood–Brain Barrier Transport of Kynurenines: Implications for Brain Synthesis and Metabolism. *J. Neurochem.* **1991**, *56*, 2007–2017. [CrossRef]
80. Kita, T.; Morrison, P.F.; Heyes, M.P.; Markey, S.P. Effects of systemic and central nervous system localized inflammation on the contributions of metabolic precursors to the L-kynurenine and quinolinic acid pools in brain. *J. Neurochem.* **2002**, *82*, 258–268. [CrossRef]
81. Cathomas, F.; Guetter, K.; Seifritz, E.; Klaus, F.; Kaiser, S. Quinolinic acid is associated with cognitive deficits in schizophrenia but not major depressive disorder. *Sci. Rep.* **2021**, *11*, 9992. [CrossRef] [PubMed]
82. Oxenkrug, G.; van der Hart, M.; Roeser, J.; Summergrad, P. Anthranilic Acid: A Potential Biomarker and Treatment Target for Schizophrenia. *Ann. Psychiatry Ment. Health* **2016**, *4*, 1059.
83. Bloomfield, P.S.; Selvaraj, S.; Veronese, M.; Rizzo, G.; Bertoldo, A.; Owen, D.R.; Bloomfield, M.A.; Bonoldi, I.; Kalk, N.; Turkheimer, F.; et al. Microglial Activity in People at Ultra High Risk of Psychosis and in Schizophrenia: An [(11)C]PBR28 PET Brain Imaging Study. *Am. J. Psychiatry* **2016**, *173*, 44–52. [CrossRef]
84. Brisch, R.; Wojtylak, S.; Saniotis, A.; Steiner, J.; Gos, T.; Kumaratilake, J.; Henneberg, M.; Wolf, R. The role of microglia in neuropsychiatric disorders and suicide. *Eur. Arch. Psychiatry Clin. Neurosci.* **2022**, *272*, 929–945. [CrossRef]
85. Cakmak, J.D.; Liu, L.; Poirier, S.E.; Schaefer, B.; Poolacherla, R.; Burhan, A.M.; Sabesan, P.; St Lawrence, K.; Théberge, J.; Hicks, J.W.; et al. The functional and structural associations of aberrant microglial activity in major depressive disorder. *J. Psychiatry Neurosci.* **2022**, *47*, E197–E208. [CrossRef] [PubMed]

86. El-Sewedy, S.M.; Abdel-Tawab, G.A.; El-Zoghby, S.M.; Zeitoun, R.; Mostafa, M.H.; Shalaby, S.M. Studies with tryptophan metabolites in vitro. effect of zinc, manganese, copper and cobalt ions on kynurenine hydrolase and kynurenine aminotransferase in normal mouse liver. *Biochem. Pharmacol.* **1974**, *23*, 2557–2565. [CrossRef]
87. Acs, P.; Selak, M.A.; Komoly, S.; Kalman, B. Distribution of oligodendrocyte loss and mitochondrial toxicity in the cuprizone-induced experimental demyelination model. *J. Neuroimmunol.* **2013**, *262*, 128–131. [CrossRef] [PubMed]
88. Zhang, Y.; Xu, H.; Jiang, W.; Xiao, L.; Yan, B.; He, J.; Wang, Y.; Bi, X.; Li, X.; Kong, J.; et al. Quetiapine alleviates the cuprizone-induced white matter pathology in the brain of C57BL/6 mouse. *Schizophr. Res.* **2008**, *106*, 182–191. [CrossRef] [PubMed]
89. Ljutakova, S.G.; Russanov, E.M. Differences in the in vivo effects of cuprizone on superoxide dismutase activity in rat liver cytosol and mitochondrial intermembrane space. *Acta Physiol. Pharmacol. Bulg.* **1985**, *11*, 56–61.
90. Fridovich, I. Superoxide Dismutases. *Annu. Rev. Biochem.* **1975**, *44*, 147–159. [CrossRef]
91. Blackburn, N.J.; Mason, H.S.; Knowles, P.F. Dopamine-β-hydroxylase: Evidence for binuclear copper sites. *Biochem. Biophys. Res. Commun.* **1980**, *95*, 1275–1281. [CrossRef] [PubMed]
92. Zhang, X.; McIntire, W.S. Cloning and sequencing of a copper-containing, topa quinone-containing monoamine oxidase from human placenta. *Gene* **1996**, *179*, 279–286. [CrossRef]
93. Horn, D.; Barrientos, A. Mitochondrial Copper Metabolism and Delivery to Cytochrome c Oxidase. *IUBMB Life* **2008**, *60*, 421–429. [CrossRef] [PubMed]
94. Takahashi, Y.; Kako, K.; Kashiwabara, S.; Takehara, A.; Inada, Y.; Arai, H.; Nakada, K.; Kodama, H.; Hayashi, J.; Baba, T.; et al. Mammalian Copper Chaperone Cox17p Has an Essential Role in Activation of Cytochrome c Oxidase and Embryonic Development. *Mol. Cell Biol.* **2002**, *22*, 7614–7621. [CrossRef]
95. Herring, N.R.; Konradi, C. Myelin, copper, and the cuprizone model of schizophrenia. *Front. Biosci.* **2011**, *3*, 23–40.
96. Lugo-Huitrón, R.; Ugalde Muñiz, P.; Pineda, B.; Pedraza-Chaverrí, J.; Ríos, C.; Pérez-de la Cruz, V. Quinolinic Acid: An Endogenous Neurotoxin with Multiple Targets. *Oxid. Med. Cell Longev.* **2013**, *2013*, 104024. [CrossRef]
97. Okuda, S.; Nishiyama, N.; Saito, H.; Katsuki, H. 3-Hydroxykynurenine, an Endogenous Oxidative Stress Generator, Causes Neuronal Cell Death with Apoptotic Features and Region Selectivity. *J. Neurochem.* **1998**, *70*, 299–307. [CrossRef]
98. Chobot, V.; Hadacek, F.; Weckwerth, W.; Kubicova, L. Iron chelation and redox chemistry of anthranilic acid and 3-hydroxyanthranilic acid: A comparison of two structurally related kynurenine pathway metabolites to obtain improved insights into their potential role in neurological disease development. *J. Organomet. Chem.* **2015**, *782*, 103–110. [CrossRef]
99. González Esquivel, D.; Ramírez-Ortega, D.; Pineda, B.; Castro, N.; Ríos, C.; Pérez de la Cruz, V. Kynurenine pathway metabolites and enzymes involved in redox reactions. *Neuropharmacology* **2017**, *112*, 331–345. [CrossRef]
100. Lima, V.L.A.; Dias, F.; Nunes, R.D.; Pereira, L.O.; Santos, T.S.R.; Chiarini, L.B.; Ramos, T.D.; Silva-Mendes, B.J.; Perales, J.; Valente, R.H.; et al. The Antioxidant Role of Xanthurenic Acid in the Aedes aegypti Midgut during Digestion of a Blood Meal. *PLoS ONE* **2012**, *7*, e38349. [CrossRef]
101. Murakami, K.; Ito, M.; Yoshino, M. Xanthurenic acid inhibits metal ion-induced lipid peroxidation and protects NADP-isocitrate dehydrogenase from oxidative inactivation. *J. Nutr. Sci. Vitaminol.* **2001**, *47*, 306–310. [CrossRef]
102. Pérez-González, A.; Alvarez-Idaboy, J.R.; Galano, A. Free-radical scavenging by tryptophan and its metabolites through electron transfer based processes. *J. Mol. Model* **2015**, *21*, 213. [CrossRef] [PubMed]
103. Baran, H.; Staniek, K.; Kepplinger, B.; Stur, J.; Draxler, M.; Nohl, H. Kynurenines and the respiratory parameters on rat heart mitochondria. *Life Sci.* **2003**, *72*, 1103–1115. [CrossRef] [PubMed]
104. Badawy, A.A.-B. Hypothesis kynurenic and quinolinic acids: The main players of the kynurenine pathway and opponents in inflammatory disease. *Med. Hypotheses* **2018**, *118*, 129–138. [CrossRef]
105. Negrotto, L.; Correale, J. Amino Acid Catabolism in Multiple Sclerosis Affects Immune Homeostasis. *J. Immunol.* **2017**, *198*, 1900–1909. [CrossRef] [PubMed]
106. Venturini, G. Enzymic Activities and Sodium, Potassium and Copper Concentrations in Mouse Brain and Liver after Cuprizone Treatment in Vivo. *J. Neurochem.* **1973**, *21*, 1147–1151. [CrossRef]
107. Wakabayashi, T.; Asano, M.; Kurono, C. Mechanism of the formation of megamitochondria induced by copper-chelating agents. II. Isolation and some properties of megamitochondria from the cuprizone-treated mouse liver. *Acta Pathol. Jpn.* **1975**, *25*, 39–49.
108. Song, S.-K.; Yoshino, J.; Le, T.Q.; Lin, S.-J.; Sun, S.-W.; Cross, A.H.; Armstrong, R.C. Demyelination increases radial diffusivity in corpus callosum of mouse brain. *Neuroimage* **2005**, *26*, 132–140. [CrossRef]
109. Fairless, A.H.; Dow, H.C.; Toledo, M.M.; Malkus, K.A.; Edelmann, M.; Li, H.; Talbot, K.; Arnold, S.E.; Abel, T.; Brodkin, E.S. Low Sociability is Associated with Reduced Size of the Corpus Callosum in the BALB/cJ Inbred Mouse Strain. *Brain Res.* **2008**, *1230*, 211–217. [CrossRef]
110. Parenti, R.; Cicirata, F.; Zappalà, A.; Catania, A.; La Delia, F.; Cicirata, V.; Tress, O.; Willecke, K. Dynamic expression of Cx47 in mouse brain development and in the cuprizone model of myelin plasticity. *Glia* **2010**, *58*, 1594–1609. [CrossRef]
111. Kesterson, J.W.; Carlton, W.W. Cuprizone toxicosis in mice—Attempts to antidote the toxicity. *Toxicol. Appl. Pharmacol.* **1972**, *22*, 6–13. [CrossRef] [PubMed]
112. Biancotti, J.C.; Kumar, S.; de Vellis, J. Activation of inflammatory response by a combination of growth factors in cuprizone-induced demyelinated brain leads to myelin repair. *Neurochem. Res.* **2008**, *33*, 2615–2628. [CrossRef] [PubMed]

113. Sadok, I.; Gamian, A.; Staniszewska, M.M. Chromatographic analysis of tryptophan metabolites. *J. Sep. Sci.* **2017**, *40*, 3020–3045. [CrossRef] [PubMed]
114. Baker, M. 1500 scientists lift the lid on reproducibility. *Nature* **2016**, *533*, 452–454. [CrossRef]

biomedicines

MDPI

Article

Human iPSC Modeling of Genetic Febrile Seizure Reveals Aberrant Molecular and Physiological Features Underlying an Impaired Neuronal Activity

Stefania Scalise [1], Clara Zannino [1], Valeria Lucchino [1], Michela Lo Conte [1], Luana Scaramuzzino [1], Pierangelo Cifelli [2], Tiziano D'Andrea [3], Katiuscia Martinello [4], Sergio Fucile [3,4], Eleonora Palma [3], Antonio Gambardella [5], Gabriele Ruffolo [3,6,*], Giovanni Cuda [1,*] and Elvira Immacolata Parrotta [5]

1 Department of Experimental and Clinical Medicine, University Magna Graecia of Catanzaro, Viale Europa, 88100 Catanzaro, Italy; stefania.scalise@unicz.it (S.S.); clara.zannino@unicz.it (C.Z.); valeria.lucchino@unicz.it (V.L.); michela.loconte@studenti.unicz.it (M.L.C.); scaramuzzino.luana@unicz.it (L.S.)
2 Department of Biotechnological and Applied Clinical Sciences (DISCAB), University of Aquila, 67100 Aquila, Italy; pierangelo.cifelli@univaq.it
3 Department of Physiology and Pharmacology, University of Rome, Sapienza, P.le Aldo Moro, 5, 00185 Rome, Italy; t.dandrea@uniroma1.it (T.D.); sergio.fucile@uniroma1.it (S.F.); eleonora.palma@uniroma1.it (E.P.)
4 IRCCS Neuromed, Via Atinense, 86077 Pozzilli, Italy; katiuscia.martinello@neuromed.it
5 Department of Medical and Surgical Sciences, University Magna Graecia of Catanzaro, Viale Europa, 88100 Catanzaro, Italy; a.gambardella@unicz.it (A.G.); parrotta@unicz.it (E.I.P.)
6 IRCCS San Raffaele Roma, Via della Pisana, 00163 Rome, Italy
* Correspondence: gabriele.ruffolo@uniroma1.it (G.R.); cuda@unicz.it (G.C.)

Citation: Scalise, S.; Zannino, C.; Lucchino, V.; Lo Conte, M.; Scaramuzzino, L.; Cifelli, P.; D'Andrea, T.; Martinello, K.; Fucile, S.; Palma, E.; et al. Human iPSC Modeling of Genetic Febrile Seizure Reveals Aberrant Molecular and Physiological Features Underlying an Impaired Neuronal Activity. *Biomedicines* **2022**, *10*, 1075. https://doi.org/10.3390/biomedicines10051075

Academic Editors: Masaru Tanaka and Eleonóra Spekker

Received: 13 April 2022
Accepted: 3 May 2022
Published: 5 May 2022

Publisher's Note: MDPI stays neutral with regard to jurisdictional claims in published maps and institutional affiliations.

Abstract: Mutations in *SCN1A* gene, encoding the voltage-gated sodium channel (VGSC) $Na_V1.1$, are widely recognized as a leading cause of genetic febrile seizures (FS), due to the decrease in the Na^+ current density, mainly affecting the inhibitory neuronal transmission. Here, we generated induced pluripotent stem cells (iPSCs)-derived neurons (idNs) from a patient belonging to a genetically well-characterized Italian family, carrying the c.434T > C mutation in *SCN1A* gene (hereafter $SCN1A^{M145T}$). A side-by-side comparison of diseased and healthy idNs revealed an overall maturation delay of $SCN1A^{M145T}$ cells. Membranes isolated from both diseased and control idNs were injected into *Xenopus* oocytes and both GABA and AMPA currents were successfully recorded. Patch-clamp measurements on idNs revealed depolarized action potential for $SCN1A^{M145T}$, suggesting a reduced excitability. Expression analyses of VGSCs and chloride co-transporters *NKCC1* and *KCC2* showed a cellular "dysmaturity" of mutated idNs, strengthened by the high expression of SCN3A, a more fetal-like VGSC isoform, and a high *NKCC1*/*KCC2* ratio, in mutated cells. Overall, we provide strong evidence for an intrinsic cellular immaturity, underscoring the role of mutant $Na_V1.1$ in the development of FS. Furthermore, our data are strengthening previous findings obtained using transfected cells and recordings on human slices, demonstrating that diseased idNs represent a powerful tool for personalized therapy and ex vivo drug screening for human epileptic disorders.

Keywords: febrile seizure; induced pluripotent stem cells; mesial temporal lobe epilepsy; voltage gated sodium channel $Na_V1.1$; disease model

1. Introduction

Febrile seizures (FS), i.e., seizures occurring during fever not due to a central nervous system (CNS) infection, are convulsive events commonly affecting children [1]. Retrospective studies have linked childhood FS to the development of hippocampal sclerosis (HS) and mesial temporal lobe epilepsy (MTLE) later in life, especially when in the presence of a family history of febrile convulsions [2]. MTLE with HS is a drug-resistant form of epilepsy and often requires patients to undergo temporal lobotomy to achieve seizure freedom [3]. Missense and nonsense mutations in the *SCN1A* gene, encoding the α-subunit of the $Na_V1.1$ VGSC, are associated in a wide range of epileptic disorders in which FS are involved, such as Dravet syndrome [4], generalized epilepsy with febrile seizures plus (GEFS+) [5], simple FS [6] and MTLE with HS [7]. Additionally, SCN1A mouse mutants exposed to recurrent early-life FS developed an increased risk of seizures susceptibility during adult life [8]. Although very useful, animal models fail in trying to faithfully recapitulate the mechanisms underlying human disease, since the patient-specific genetic background is not taken into account. For instance, remarkable differences between GEFS+ [9] and simple FS [6] clinical phenotypes do exist among patients carrying the same mutation. Limitations associated with animal models can be overcome by the generation of iPSCs from a patient's own cells [10–13]. In this study, we generated iPSCs from a patient carrying a missense mutation in the *SCN1A* gene. This mutation causes the substitution of a highly conserved methionine with a threonine in position 145 of the $Na_V1.1$ protein (M145T). This patient belongs to an Italian family of 13 individuals carrying the mutation and affected by FS [6]. In addition, this patient developed MTLE with HS during adolescence, showing severe recurrent drug-resistant seizures [14,15]. Then, neurosurgery became necessary at the age of 27 years to remove hippocampal sclerotic tissue and achieve a control of seizures [15]. Patch-clamp recordings in human cell lines, transfected with a plasmid carrying the *SCN1A*-M145T mutant gene, revealed a loss-of-function mutation leading to a 60% reduction in the Na+ current density and a positive shift of about 15 mV in the voltage-dependent activation of the channel [6]. Furthermore, electrophysiological experiments conducted on fresh hippocampal slices obtained from the same patient from which iPSCs were generated, showed a more depolarized action potential (AP) threshold and an impairment of GABAergic neurotransmission in interneurons [15], a hallmark of *SCN1A* mutations in epileptic phenotypes [16,17]. Notably, the latter was coupled to an increase in GABA current use-dependent desensitization in oocytes micro-transplanted with the same hippocampal tissue [15,18].

The pivotal involvement of inhibitory interneurons in epilepsy was also shown in studies based on iPSCs models of pathogenic *SCN1A* mutations [19,20]; others have instead demonstrated the involvement of both glutamatergic and GABAergic populations in the epileptic brain hyperexcitability [21]. In this study, we used patient-specific iPSCs-derived neurons (idNs) to investigate the molecular and electrophysiological mechanisms underlying the $SCN1A^{M145T}$ disease phenotype. Our results show a significant alteration in the development and maturation processes of $SCN1A^{M145T}$ idNs compared to their healthy control counterpart. Electrophysiological measurements conducted on single neurons during their development add further knowledge to this scenario, with findings that successfully recapitulate those previously recorded on hippocampal sclerotic tissue from the same patient. Taken together, our results strengthen the potential of iPSCs technology for a more comprehensive understanding of the complexity of epileptic-like human phenotypes.

2. Materials and Methods

2.1. Clinical Features and iPSCs Generation from a Patient with Missense Mutation in the SCN1A Gene

In this study, we generated iPSCs from a male subject who carried a missense mutation (c.434T > C in exon 3) in the *SCN1A* gene encoding for the α-subunit of the $Na_V1.1$ VGSC. The patient (referred as subject IV-3 in the pedigree described in [6,14]) belongs to a family in which thirteen members were affected by FS during childhood, all carrying the same c.434T > C missense mutation, which causes the substitution of a highly conserved methionine residue with a threonine within the S1 segment of the domain 1 in the $Na_V1.1$ channel (Figure 1A). The patient of this study experienced FS lasting up to 15 min, suggesting a clinical phenotype of complex FS, until the age of six. Seven years later, he started suffering from focal complex seizures, compatible with MTLE. The disease progressed, and the patient became refractory to antiepileptic drugs (AEDs). Neurological evaluation reported bilateral mesial temporal epileptiform spikes mostly localized (>70%) on the right side, while brain MRI evidenced significant sclerosis in the right hippocampus, requiring right temporal lobectomy to achieve seizure freedom (Colosimo et al., 2007). iPSCs were also generated from skin fibroblasts isolated from a healthy thirty-year-old male and were used as a control line in our experiments. The generation and characterization of SCN1A^{M145T} and healthy control iPSCs is described in [22] (line identified as UNIMGi001-A) and [23] (see line hiPSCs-3), respectively. iPSCs were cultured on Matrigel-coated dishes with mTeSR1 medium (StemCell Technologies, Vancouver, BC, Canada), in a humidified incubator at 37 °C at 5% CO_2. Cells were split every 4–5 days (80% confluence) with the use of Gentle Cell Dissociation Reagent (StemCell Technologies). Both cell lines were routinely tested for Mycoplasma with the Mycoplasma PCR Detection Kit (Applied Biological Materials, Richmond, BC, Canada).

2.2. Generation of iPSCs-Derived Neurons (idNs)

We coaxed iPSCs from both control and SCN1A^{M145T} to differentiate into neural stem cells (NSCs) using Gibco® PSC Neural Induction Medium (Thermo Fisher Scientific, Waltham, MA, USA), following the manufacturer's instructions. To obtain neurons, NSCs were then plated at a density of 5×10^4 cells/cm^2 on dishes coated with Poly-D-Lysine (molecular weight 30,000–70,000) plus Laminin (both from Merck, Darmstadt, Germany) and cultured in Neuronal Differentiation Medium (NDMC), composed of Neurobasal Medium, 1× B27 supplement, 1× Glutamax, 1× CultureOne™ Supplement, 200 μM ascorbic acid and 0.2% Penicillin/Streptomycin (all from Thermo Fisher Scientific). NDMC was supplemented with GDNF at 10 ng/mL and BDNF at 20 ng/mL (both from PeproTech, London, UK) at NSCs plating; the concentration of GDNF and BDNF was lowered to 5 ng/m and 10 ng/m, respectively, at the first medium change. Subsequently, NDMC medium was supplemented with BDNF only, used at 5 ng/mL during the second medium change and at 2.5 ng/mL during the whole culture period. Neurons were kept in culture 28–35 days until they reached full maturation for subsequent analysis.

2.3. RNA Extraction and qRT-PCR Analysis

Total RNA was obtained by phenol/chloroform extraction using TRIzol reagent (Thermo Fisher Scientific) and 1 μg RNA was retro-transcribed in cDNA using the High-Capacity cDNA Reverse Transcription Kit (Thermo Fisher Scientific). cDNA was used for relative quantitation of gene expression by qRT-PCR, using the SensiFAST SYBR Hi-ROX kit (Meridian Bioscience, Cincinnati, OH, USA). Gene expression levels were normalized to Glyceraldehyde 3-phosphate dehydrogenase (*GAPDH*) as a housekeeping gene. qRT-PCR was performed by QuantStudio™ 7 Pro Real-Time PCR System (Thermo Fisher Scientific). A list of primers used in this study is provided in Supplementary Table S1.

2.4. cDNA Sequencing

Primers based on the cDNA sequence of the *SCN1A* gene were designed to amplify the exon 3 in which the c.434T > C mutation is found. The amplification of the target region was obtained with the use of specific primer pairs (FW: 5′-ATTGAAAGACGCATTGCAGA-3′ and RV: 5′-TGTTCCTCCAAGGAAGCATT-3) and the following PCR program: 3 min at 95 °C, 30 cycles of 30 s at 95 °C, 30 s at 52 °C and 45 s at 72 °C, with a final extension at 72 °C for 5 min. Following gel electrophoresis to confirm the amplicon size (797 bp), PCR products were extracted from gel using the EZ-10 Spin Column DNA Gel Extraction Kit (Bio Basic Inc., Markham, ON, Canada) and Sanger sequenced (Eurofins Genomics, Ebersberg, Germany).

2.5. Western Blot Analysis

For total protein extraction, cells were harvested in ice-cold phosphate-buffered saline (PBS) and lysed in RIPA buffer (Merck) containing Halt™ Protease Inhibitor and Halt™ Phosphatase Inhibitor Cocktails (Thermo Fisher Scientific). Protein concentration was measured by Bradford assay. After denaturation for 10 min at 70 °C in Laemmli Sample Buffer, 70 μg of proteins were resolved in acrylamide/bisacrylamide precast gels Mini-PROTEAN TGX (Bio-Rad, Hercules, CA, USA) and transferred to nitrocellulose membrane. The membrane was incubated overnight at 4 °C with the following primary antibodies: anti-Na_V1.1 (rabbit polyclonal, 0.5 μg/mL, ab24820, Abcam, Cambridge, UK) and anti-TUBB3 (mouse monoclonal, 1:10,000, 480,011, Thermo Fisher Scientific). After washing, horseradish peroxidase conjugated secondary antibody anti-rabbit IgG and anti-mouse IgG (Jackson ImmunoResearch, Cambridge, UK) were added to the membrane and incubated for 1 h at room temperature. The protein bands on the membranes were detected by Clarity™ Western ECL Blotting Substrates (Bio-Rad) using the Alliance™ Q9-Atom (Uvitec, Cambridge, UK). Western blot bands were quantified using the Analyze Gels tool of Fiji Software [24].

2.6. Immunofluorescence

Immunofluorescence analysis was performed on poly-D-Lysine plus laminin-coated permanox chamber slides (Thermo Fisher Scientific). Cells were fixed in 4% (vol/vol) paraformaldehyde (PFA) and subjected to immunostaining with the following primary antibodies: anti-TUBB3 (mouse monoclonal, 1:250, 480011, Thermo Fisher Scientific), anti-MAP2 (mouse monoclonal, 1:1000, MA5-12826, and chicken polyclonal, 1:5000, PA1-10005, both from Thermo Fisher Scientific), anti-NEFH (rabbit polyclonal, 1:1000, ab8135, Abcam), anti-GAD1 (chicken polyclonal, 1:1000, AP31805PU-N, Origene, Rockville, MD, USA), anti-SST (mouse monoclonal, 1:200, Ma5-17182, Thermo Fisher Scientific), anti-CALB2 (rabbit polyclonal, 1:100, PA5-16681, Thermo Fisher Scientific) and anti-CALB1 (Rabbit monoclonal, 1:100, NB120-11427, Abcam), anti-Na_V1.1 (rabbit polyclonal, 1:100, ab24820 Abcam) and anti-vGLUT1 (mouse monoclonal, 1:100, sc-377425, Santa Cruz Biotechnology, Dallas, TX, USA). Incubation with primary antibodies was carried overnight at 4 °C. After washing with PBS, cells were incubated with AlexaFluor-594, or -488 conjugated secondary antibodies (all from Thermo Fisher Scientific) for 1 h at room temperature. Nuclei were stained with DAPI (4′,6-diamidino-2-phenylindole, Thermo Fisher Scientific) and mounted with Dako Fluorescent Mounting Medium (Agilent, Santa Clara, CA, USA). Images were acquired with a Leica microscopy system (DMi8), using LAS X (v.3.7.4.23463) software. For quantification of double positive cells (Figures 2D and 3D,E,F), neurons from two different differentiation experiments were manually counted using the multipoint tool in Fiji software in a blinded manner.

2.7. Patch-Clamp Recordings on idNs

Whole-cell patch clamp recordings were performed on idNs of WT and SCN1A^{M145T} mutant at day of differentiation 35 in 35 mm Petri-dishes. The measures were performed at 25 °C. The identification of neurons followed morphological criteria: highly birefringent cells with small diameter processes were selected, and 100% of the WT cells exhibited action potentials (APs). APs were recorded from neurons applying depolarizing current steps (4–50 pA, 1 s) using glass electrodes (3–4 MΩ) filled with (in mM): 140 KCl, 10 Hepes, 5 BAPTA, 2 Mg-ATP (pH 7.3, adjusted with KOH). Membrane potentials were acquired at 50 kHz and filtered at 3 kHz with an amplifier HEKA EPC 800 (HEKA Elektronik, Reutlingen, Germany) and analyzed off-line. During recordings, cells were continuously perfused using a gravity-driven perfusion system with the following external solution: 140 mM NaCl, 10 mM HEPES, 2.8 mM KCl, 2 mM $CaCl_2$, 2 mM $MgCl_2$, 10 mM glucose, (pH 7.3 adjusted with NaOH). Membrane potentials have been corrected for junction potential.

2.8. Membrane Preparation from idNs

We isolated cellular membranes from approximately 7×10^7 idNs of WT and SCN1A^{M145T}. The procedure was similar to that already described in [18] for human tissues. The cells were scraped at day of differentiation 35, spun down and subsequently homogenized in membrane buffer (200 mM glycine, 150 mM NaCl, 50 mM EGTA, 50 mM EDTA, and

300 mM sucrose; plus 10 µL/mL of protease inhibitors, P2714 (Sigma)—pH 9 adjusted with NaOH). Then, the vials were centrifuged for 15 min at 9500× *g*. Afterwards, the supernatant was centrifuged for 2 h at 100,000× *g* with an ultracentrifuge. Finally, the pellet was resuspended in glycine 5 mM and used directly or aliquoted and kept at −80 °C for later usage.

2.9. Xenopus Laevis Oocytes Injection and Voltage-Clamp Recordings

Xenopus oocytes were collected and injected as previously described in [18]. The animal protocols were approved by the Italian Ministry of Health (authorization no. 427/2020-PR). The electrophysiology experiments were carried out from 12 to 48 h after injection, with the technique of two-electrode voltage-clamp. The two microelectrodes were filled with 3M KCl. The oocytes were placed in a recording chamber (0.1 mL volume) and perfused continuously with oocyte Ringer solution (OR: NaCl 82.5 mM; KCl 2.5 mM; $CaCl_2$ 2.5 mM; $MgCl_2$ 1 mM; Hepes 5 mM, adjusted to pH 7.4 with NaOH) at room temperature (20–22 °C). The neurotransmitters (GABA or AMPA) were administered through a gravity driven multi-valve perfusion system (9–10 mL/min) controlled by a computer (Biologique RSC-200; Claix, France) to ensure the exact duration of each application. AMPA currents were recorded in presence of cyclothiazide (CTZ, 20 µM) in order to avoid receptor desensitization [18]. GABA, AMPA, CTZ, Bicuculline methochloride and NBQX were purchased from Tocris Bioscience (Bristol, UK) and dissolved in sterile water (GABA, AMPA and Bicuculline methochloride) or DMSO (CTZ, NBQX) before final dilution to the desired concentration in OR. The solutions containing DMSO were always used with a final DMSO concentration lower than 1:1000. GABA current rundown was defined as the decrease in the current peak amplitude after six 10 s applications of GABA at 40 s intervals, expressed as percentage of the first response [25].

2.10. Statistical Analysis

For molecular biology data, the number of biological replicates used in each experiment was indicated in the figure legends. Statistical analysis was performed using two-tailed *t*-test or multiple unpaired *t*-test with Welch correction in GraphPad Prism software, version 9.3.1. Data are represented as means of biological replicates ± SEM and *p*-values ≤ 0.05 were considered significant. For patch-clamp experiments, data sampling and analysis were performed using pClamp 10 software (Molecular devices, Sunnyvale, CA, USA). Statistical significance was assessed with ANOVA, unless otherwise stated.

The figures of this work were created with BioRender.com.

3. Results

3.1. Generation of $SCN1A^{M145T}$ and Control idNs

SCN1A loss-of-function mutations are reported to affect both GABAergic and glutamatergic neurons [21]. Therefore, we directed WT and $SCN1A^{M145T}$ iPSCs differentiation toward forebrain neurons. idNs presented a well-defined neuronal morphology (Figure 1B) and a high expression of neuronal marker genes *MAP2, NEFL, NEFM, SYP*, and *PSD95* with respect to their undifferentiated counterpart (iPSCs) (Figure 1C), while *ALDH1L1* and *OLIG2,* astrocytes- and oligodendrocytes-specific markers, respectively, are expressed at lower level compared to *MAP2* expression (Supplementary Figure S1). Moreover, immunofluorescence analysis showed that idNs co-express the pan-neuronal marker protein TUBB3, dendrite marker MAP2, and axonal marker NEFH (Figure 1D).

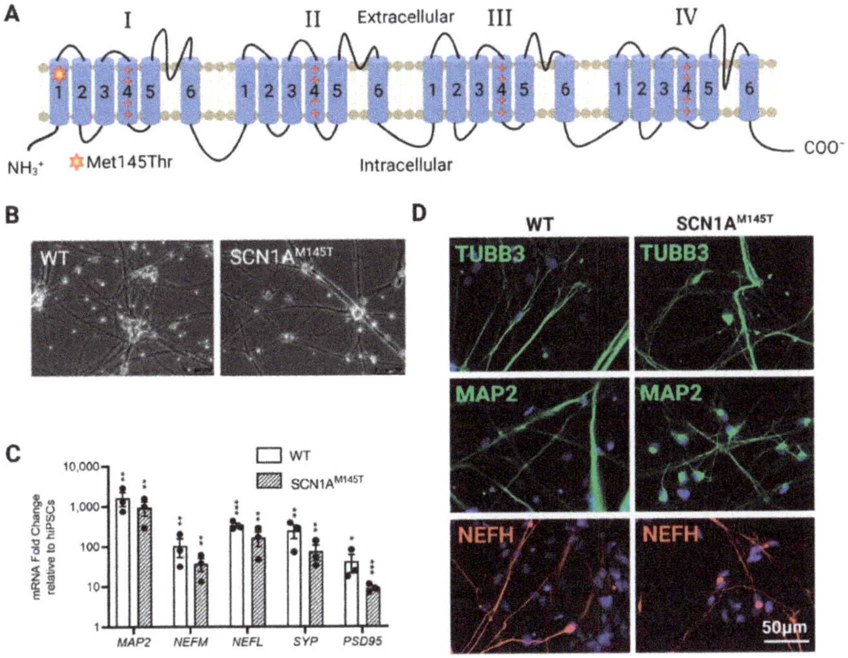

Figure 1. Characterization of idNs. (**A**) Representation of $Na_V1.1$ channel. The star in segment 1 of domain I shows the localization of the mutated aminoacid (Met145Thr). The relative missense mutation c.434T > C is found in the exon 3 of the translated sequence. (**B**) Bright-field images of idNs from WT and SCN1A^{M145T}-iPSCs (20× magnification). (**C**) Differentiated idNs show high expression levels of neuronal specific genes such as *MAP2, NEFM, NEFL, SYP* and *PSD95* compared to their undifferentiated counterparts (iPSCs). *GAPDH* was used as a housekeeping control. Data are presented as mean ± SEM of three biological replicates (black dots), * $p < 0.05$, ** $p < 0.01$, *** $p < 0.001$, *t*-test has been calculated vs. expression in iPSCs. (**D**) Immunostaining of neuronal markers TUBB3 (neurites marker), MAP2 (cell body and dendrites marker), and NEFH (axonal marker) in WT and SCN1A^{M145T} idNs. DAPI nuclear counterstain is shown in all images in blue (63× magnification).

Morphological analysis of cultured idNs revealed the presence of both bipolar (inhibitory) and pyramidal (excitatory) populations (Figure 2A). The expression of glutamate decarboxylase (*GAD2*, GABAergic marker) and vesicular glutamate transporter (v*GLUT2*, glutamatergic marker) by qRT-PCR analysis revealed a slightly higher expression of the GABAergic marker in idNs from WT and SCN1A^{M145T} (Figure 2B). In addition, by immunofluorescence analysis, we found that about 90% of MAP2^{+} neurons indeed co-express the GABAergic marker GAD1 (Figure 2C,D). On the other hand, the glutamatergic marker vGLUT1 could be slightly detected in few cells only (Supplementary Figure S2).

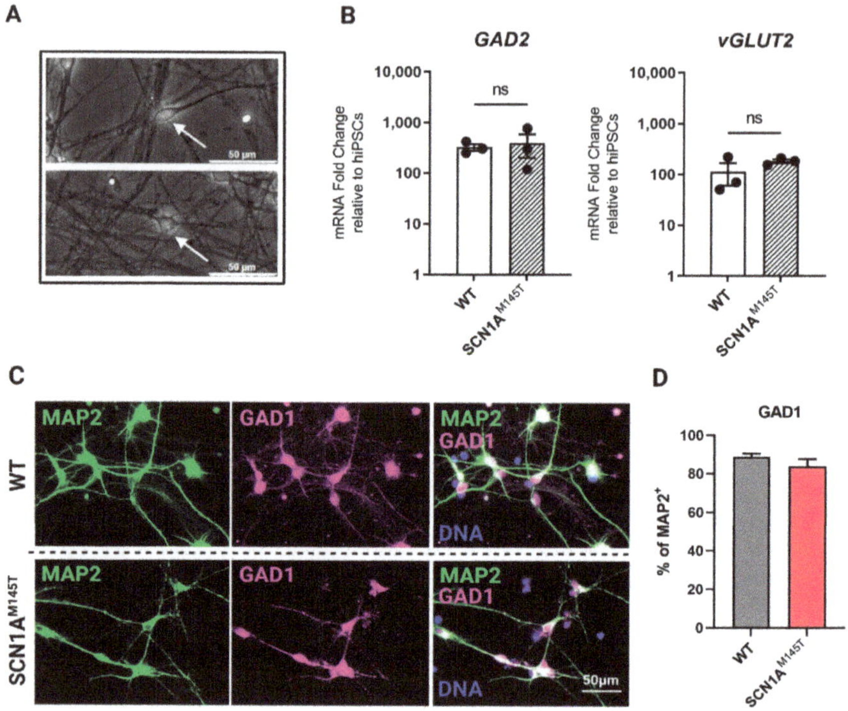

Figure 2. Expression of GABAergic and glutamatergic markers in idNs. (**A**) Generated idNs are composed of a mixed neuronal population containing neurons with both bipolar (upper panel, white arrow) and pyramidal (lower panel, white arrow) morphology. (**B**) Expression analysis of mRNAs relative to GABAergic (inhibitory) marker glutamate decarboxylase 2 (*GAD2*) and glutamatergic (excitatory) marker vesicular glutamate transporter 2 (*vGLUT2*) in idNs relative to undifferentiated iPSCs. *GAPDH* was used as a housekeeping gene. qRT-PCR analysis did not reveal significant differences in the expression of *GAD2* and *vGLUT2* between diseased and control idNs relatively to their undifferentiated iPSCs (*p*-value = non-significant (ns), *t*-test), even though a higher prevalence of GABAergic marker *GAD2* was detected, as shown by its higher fold change relative to iPSCs. Data are presented as mean ± SEM of three biological replicates (dots). (**C**) Representative immunofluorescence images of GABA synthesis enzyme GAD1 compared to neuronal marker MAP2 in idNs of WT (upper line images) and SCN1A^{M145T} (lower line images) cells (63× magnification). (**D**) The diagram shows that about 90% of MAP2^{+} cells co-express the GAD1 marker. For each cell line, at least 300 neurons were counted, and data are presented as mean ± SEM of two independent experiments.

Mature GABAergic interneurons express specific neuropeptides and calcium binding protein, thus we performed immunostaining of idNs at day 35 of differentiation using antibodies against somatostatin (SST) (Figure 3A), Calretinin (CALB2) (Figure 3B), Calbindin (CALB1) (Figure 3C), and parvalbumin (PV). Results of immunofluorescence analysis indicate that 10% of MAP2^{+} neurons expressed SST (Figures 3D and S3A), about 9% expressed CALB2 (Figures 3D and S3B), while CALB1 is slightly present (less than 1% of MAP2^{+}, Figures 3D and S3C). We could not detect PV positive neurons in idNs from WT and SCN1A^{M145T} in line with the observation that this protein is expressed later during neuronal differentiation of iPSCs [26].

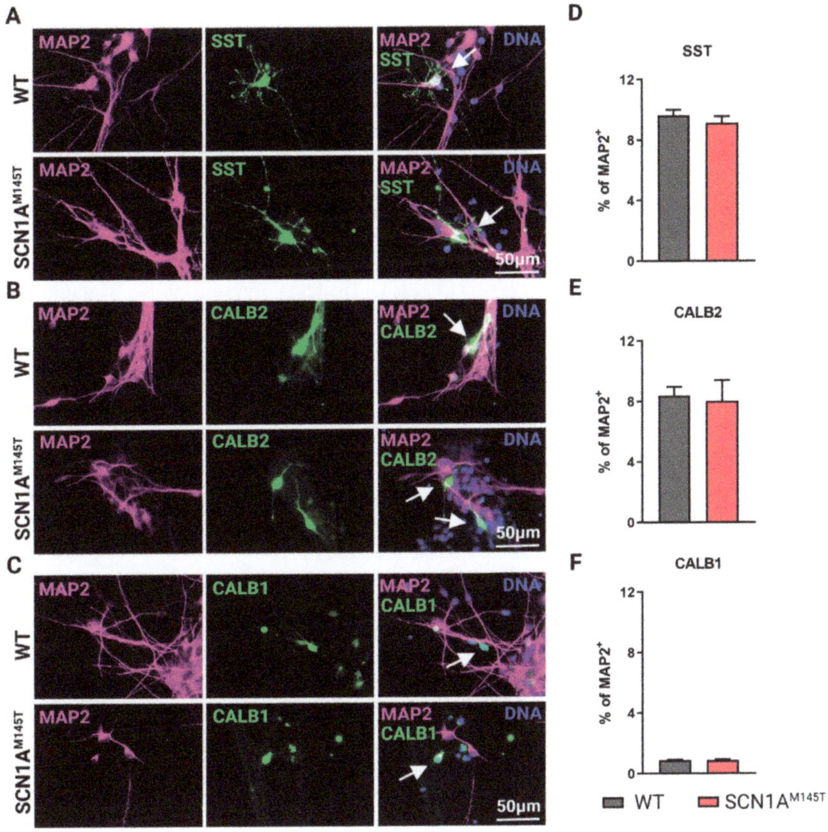

Figure 3. Types of interneurons generated from iPSCs. Immunofluorescence analysis of idNs stained with antibodies against interneuronal subtypes markers (**A**) somatostatin (SST), (**B**) calretinin (CALB2), and (**C**) calbindin (CALB1). In each group of images, WT cells are shown in the upper panel, while SCN1A^{M145T} idNs are shown in the lower panel. White arrows in the merged images indicate neurons expressing the interneuronal makers indicated (63× magnification). (**D–F**) Quantification of percentage of MAP2$^+$ neurons co-expressing interneuronal markers immunostained in panels (**A–C**). About 9–1% of idNs express SST and CALB2, while CALB1 is present in less than 1% percent of neurons. At least 200 cells were counted for each bar, and data are presented as mean ± SEM of two independent experiments.

3.2. Expression of *SCN1A* Gene and $Na_V1.1$ Protein in idNs

Given that the SCN1A^{M145T} patient is heterozygous for the mutation, we inquired whether both *SCN1A* alleles are transcribed in idNs. To this purpose, we used Sanger sequencing to analyze cDNA obtained by reverse transcription of *SCN1A* mRNA derived from idNs of both WT and the SCN1A^{M145T}. Interestingly, the results confirmed the heterozygous status in SCN1A^{M145T}-idNs, as demonstrated by the presence of a double peak at the mutation site (Figure 4A). Subsequently, we performed a side-by-side analysis of the expression of *SCN1A* mRNA together with others VGSCs known to be predominantly expressed in CNS, such as *SCN2A*, *SCN3A* and *SCN8A*, in idNs at different time points during differentiation (d0, which corresponds to the NSCs stage, d14, d21, and d28). Interestingly, we found that, although the expression of *SCN1A* increased in both WT- and SCN1A^{M145T} idNs over time in culture, the expression of *SCN1A* channel results lower, in all time-points tested, in patient idNs compared to control idNs (Figure 4B). We observed a similar trend of expression also for the *SCN2A*, which has the higher expression level among all the sodium channels analyzed (Supplementary Figure S4A). On the other hand, *SCN3A*, which is considered an embryonic isoform [27], progressively decreases during the differentiation period in WT idNs, while mutated idNs show an opposite expression trend of *SCN3A*, which results up-regulated over time in culture (Figure 4C). Lastly, the expression of *SCN8A* channel was low in both control and mutated idNs (Supplementary Figure S4B), in accordance with the notion that *SCN8A* is poorly expressed in developing neurons [27]. Based on the difference found in the expression of *SCN1A* at the mRNA level, we analyzed the expression levels of $Na_V1.1$ protein in total lysates from control and mutated idNs. Intriguingly, immunoblot analysis showed that $SCN1A^{M145T}$ neurons express significantly lower levels of $Na_V1.1$ protein compared to healthy neurons (Figure 4D,E). Moreover, our immunofluorescence data revealed a high expression of $Na_V1.1$ within the soma of $GAD1^+$ neuronal cells (Figure 4F), in line with the in vivo data showing that $Na_V1.1$ is primarily expressed in GABAergic neurons [17].

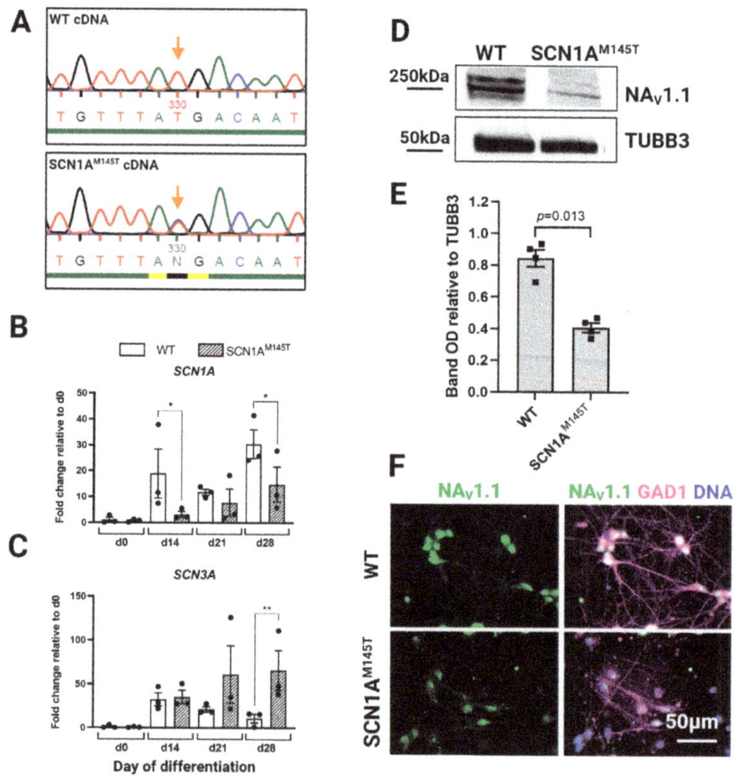

Figure 4. Expression of *SCN1A* gene and $Na_V1.1$ protein in idNs. (**A**) Sequencing of cDNA obtained by reverse transcription of *SCN1A* mRNA from WT- and $SCN1A^{M145T}$-idNs. In the $SCN1A^{M145T}$ cells, the double peak in the mutation site (indicated by the orange arrow) demonstrates that both alleles (one with the original nucleotide T and the other with the mutated one C) were transcribed. (**B**) qRT-PCR analysis of *SCN1A* gene in idNs at day of differentiation 0 (NSCs), d14, d21 and d28. $SCN1A^{M145T}$ cells showed a lower expression of SCN1A during differentiation compared to WT, although in both cell lines the expression increases following idNs maturation. (**C**) The expression of the embryonic isoform of sodium channel *SCN3A* increases with the progress of cell maturation in $SCN1A^{M145T}$ idNs, while it decreases in WT idNs as the cells become more differentiated. For both graphs, *GAPDH* was used as a housekeeping gene; data are mean ± SEM of three biological

replicates (black dots), * $p < 0.05$, ** $p < 0.01$, *t*-test has been calculated vs. WT at the same day of differentiation. (**D**) Western blot analysis of $Na_V1.1$ protein in lysates obtained from WT- and SCN1A^{M145T} idNs at day 35 of differentiation. Tubulin Beta 3 Class III (TUBB3) was used as loading control. (**E**) Quantification of $Na_V1.1$ Western blot bands in four biological replicates ($n = 4$, OD = relative optical density calculated as ($Na_V1.1$ optical density)/(TUBB3 optical density), *p*-value calculated using *t*-test). (**F**) Immunofluorescence of idNs showing that GABAergic neurons (GAD1 positive) express $Na_V1.1$, mainly in the cell body. Immunofluorescence data show a lower expression of the channel in the neurons differentiated from SCN1A^{M145T} patient in respect to those of WT (63× magnification).

3.3. *Expression of Chloride Cotransporters in idNs*

The functionality of GABA neurons during development is intimately correlated with the expression of the chloride co-transporters, the Na-K-2Cl cotransporter isoform 1 (NKCC1) and the K-Cl cotransporter isoform 2 (KCC2) [28] and a high NKCC1/KCC2 ratio indicates neuronal immaturity [29,30]. We analyzed the expression of the two chloride co-transporters transcripts in idNs from WT and SCN1A^{M145T} at day 0 of differentiation (neural stem cells), d14, d21, d28, and d60. In accordance with data reported in the human brain transcriptome database [31], we observed that *KCC2* undergoes strong physiological increase in idNs from WT during development, while its expression remained at low levels in idNs from SCN1A^{M145T} during the whole experimental culture period (Figure 5A). The expression of *NKCC1* was instead similar between the two groups during the first phases of neuronal development (from d0 to d28) but became significantly over-expressed at day 60 of differentiation in SCN1A^{M145T} idNs only (Figure 5B). Additionally, the *NKCC1*/*KCC2* mRNA expression ratio was higher in mutated neurons at all differentiation time points tested (Figure 5C). Altogether, these findings provide strong evidence that an imbalanced NKCC1/KCC2 expression shift occurs in idNs derived from the patient carrying the M145T mutation in the *SCN1A* gene, suggesting that this mutation may promote the persistence of an immature phenotype.

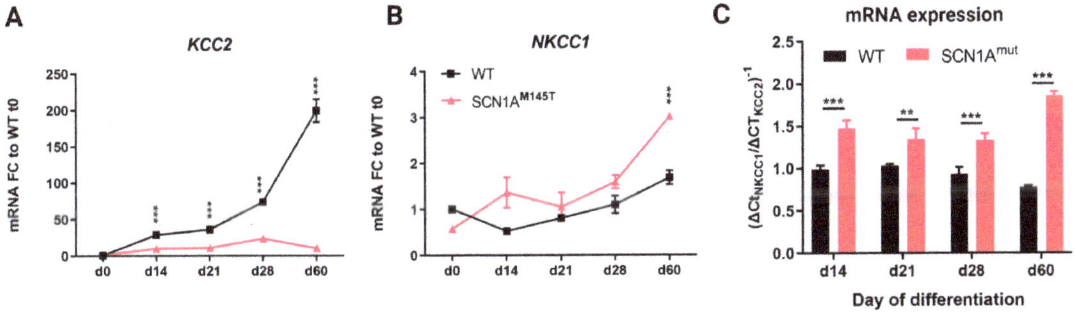

Figure 5. Expression of chloride co-transporters in idNs. (**A**,**B**) qRT-PCR analysis of chloride co-transporters *KCC2* and *NKCC1* in neurons from WT and SCN1A^{M145T} tested at day of differentiation 0 (NSCs), d14, d21, d28 and d60. (**C**) *NKCC1*/*KCC2* mRNA ratio in idNs at different days of differentiation. The ratio was calculated as the inverse of $\Delta Ct_{NKCC1}/\Delta Ct_{KCC2}$: ($\Delta Ct = Ct_{Gene\_Of\_Interest} - Ct_{GAPDH}$). For all graphs, the mean ± SEM of three biological replicates is shown; ** $p < 0.01$, *** $p < 0.001$, *t*-test has been calculated vs. WT at the same day of differentiation.

3.4. *Recording of GABA and AMPA Currents by Injection of idNs Membranes in Xenopus Oocytes*

Here, we recorded for the first time neurotransmitter-evoked currents from *Xenopus* oocytes injected with membranes obtained from idNs (Figure 6). First, we successfully evoked GABA and AMPA responses both from SCN1A^{M145T}-injected oocytes and control (WT) injected oocytes. We obtained responses that ranged from to 3.1 nA to 75.0 nA for

GABA (500 μM, 4 s, mean = 18.2 ± 2.5 nA; N = 44) and from 6.3 to 43.2 nA to for AMPA (20 μM, 10 s with a short 20 s pretreatment with CTZ 20 μM, mean = 16.4 ± 1.6; n = 27). In order to verify that currents we recorded were genuine, we blocked the evoked current with the respective specific blockers (Figure 6A,B). As expected, GABA currents (500 μM) were totally blocked by co-application of bicuculline, a competitive antagonist of $GABA_A$Rs, (100 μM; n = 6) and AMPA currents (20 μM) were totally blocked by the specific AMPA receptor blocker NBQX, a competitive antagonist of AMPARs (50 μM; n = 6). Both GABA (Figure 6A) and AMPA (Figure 6B) currents recovered their original amplitude after the washout of the blocker. In another set of experiments, we measured the GABA current rundown, a GABAergic dysfunction associated with drug-resistant epilepsy [25,32,33], from $SCN1A^{M145T}$-injected oocytes and WT-injected oocytes. Not surprisingly, we measured a value of current rundown in WT-injected oocytes that was similar to that recorded in previous studies [34] using cortical tissue samples from individuals without any neurological disorder (71.0 ± 2.7%; n = 8; Figure 6C). On the other hand, interestingly, we did not measure a significant increase in current rundown in $SCN1A^{M145T}$-injected oocytes (65.7 ± 3.2%; n = 10; Figure 6C).

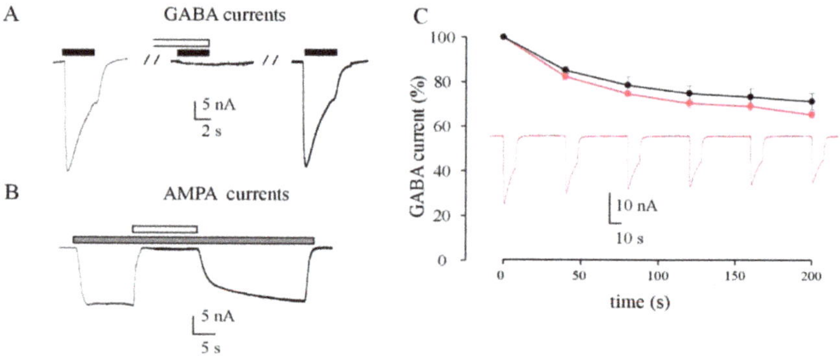

Figure 6. *Xenopus* oocytes injected with membranes from idNs incorporated functional neurotransmitter receptors. (**A**) Sample currents evoked by 500 μM GABA or (**B**) 20 μM AMPA on oocytes microinjected with membranes extracted from cultured idNs obtained from a patient carrying the M145T mutation of the *SCN1A* gene. (**A**) GABA currents were completely inhibited by a brief preincubation (30 s) with bicuculline (100 μM) and subsequently recovered following the washout of the inhibitor. (**B**) AMPA currents were completely inhibited by co-administration of NBQX (50 μM), and they recovered to the original amplitude once NBQX administration was interrupted. AMPA currents were recorded in presence of CTZ (20 μM). Black bars = GABA; gray bars = AMPA; white bars in (**A**) bicuculline; in (**B**) NBQX. (**C**) Time course of the GABA current rundown evoked by six consecutive GABA applications (500 μM, 10 s) interspaced by a 40 s washout, in oocytes injected with membranes from control (black dots; •) and M145T idNs (magenta; •; $p > 0.05$). The dots represent GABA currents expressed as a percentage of the first evoked response (• = 16.6 ± 1.0 nA, n = 8; • = 23.7 ± 1.1 nA, n = 10).

3.5. Patch-clamp recordings of WT and $SCN1A^{M145T}$ idNs

WT- and $SCN1A^{M145T}$ idNs at day of differentiation 35 were analyzed to compare their functional properties. Resting membrane potential and cell capacitance values were similar in WT and $SCN1A^{M145T}$: −48 ± 2 mV vs. −44 ± 2 mV, and 28 ± 3 pF vs. 35 ± 2 pF, respectively. All WT neurons tested fired APs upon current injection (Figure 7A, 15 out 15 cells), while responsive $SCN1A^{M145T}$ idNs were the 62% of the total (Figure 7A inset, 20 out 34 cells; p = 0.004, Fisher Exact test). Although at this stage AP threshold is not expected to be at the full maturation level, WT neurons exhibited a more hyperpolarized mean AP threshold value than $SCN1A^{M145T}$ (−37 ± 1 mV vs. −31 ± 1 mV, p = 0.003; Figure 7A–C, [35]). Furthermore, AP amplitudes were larger in WT than in $SCN1A^{M145T}$

idNs (57 ± 3 mV vs. 45 ± 2 mV, $p = 0.006$; Figure 7D), while no differences were observed in AP kinetics.

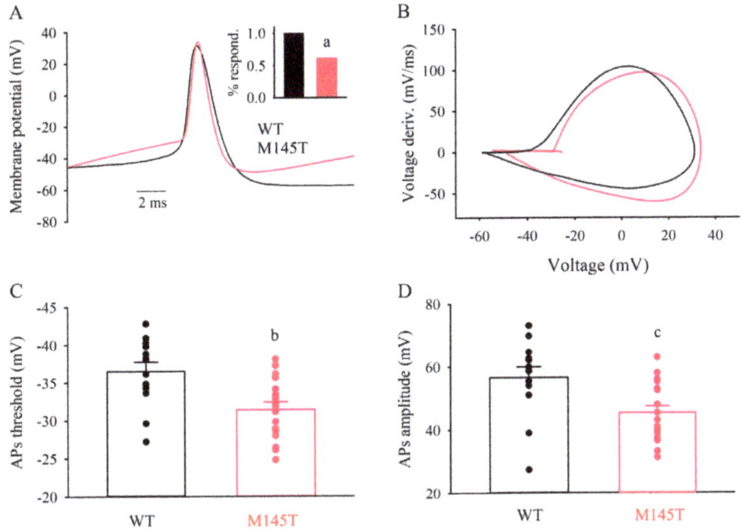

Figure 7. SCN1A^{M145T} idNs exhibit depolarized action potential (AP) threshold. (**A**) superimposed typical AP traces recorded from one WT and one M145T neuron (black and magenta traces, respectively). Inset: bar graphs representing the frequency of AP-firing cells (a, $p = 0.004$, Fisher Exact test). (**B**) superimposed phase-plane plot obtained from the two APs shown in (**A**). (**C**) bar graphs representing the mean AP threshold values averaged from 13 WT- and 19 SCN1A^{M145T} idNs, as indicated. Circles indicate the AP threshold of individual cells (b, $p = 0.003$). (**D**) bar graphs representing the mean AP amplitude values. Same cells as in (**C**) (c, $p = 0.006$).

4. Discussion

Na$_V$1.1 sodium channel, encoded by *SCN1A* gene, belongs to the family of VGSCs and allows the sodium influx from extracellular space into the cytosol during depolarization. Na$_V$1.1 is highly expressed in the CNS and peripheral nervous system (PNS) and mainly localizes in the cell bodies and proximal processes of neurons, where it is involved in the generation of action potential [36,37]. Mutations in the *SCN1A* gene are responsible for a plethora of diseases, collectively known as channelopathies affecting the entire nervous system [38,39]. In particular, pathogenic variants in *SCN1A*/Na$_V$1.1 are responsible for several epilepsy syndromes including Dravet Syndrome (DS), a severe childhood form of epilepsy characterized by beginning with complex FS. Genetically caused FS are associated with missense loss-of-function mutations in *SCN1A*. These mutations are often functionally linked to hypoexcitability of at least some type of γ-aminobutyric acid (GABA)ergic neurons, due to decrease in Na^+ current density [16,40–43], even if the exact mechanisms responsible for the disease are still unknown. Given that, additional approaches are necessary to unravel further physiopathological features that can be targeted by novel therapeutic strategies. In this study, we took advantage of the iPSCs technology to investigate the function underlying the clinical phenotype of a patient belonging to a well characterized Italian family with FS due to genetic defect in *SCN1A* gene [6]. Particularly, the patient described here harbors a c.434T > C missense mutation in the *SCN1A* gene (SCN1A^{M145T}), responsible for the substitution of a highly conserved methionine residue with a threonine within the S1 segment of the domain 1 in the Na$_V$1.1. Clinically, the patient was characterized by a complex pathophysiology with FS lasting up to 15 min [14] and developed MTLE with HS which in the end required neurosurgery. A similar condition affects a significant number of people suffering from drug-resistant epileptic seizures and, notwithstanding the recent advances that constantly improve the outcomes of surgical interventions [44], this invasive procedure is not yet completely free from complications [45]. This is mostly due to the failure of the available ASM which may effectively decrease frequency and severity of seizures without tackling the pathophysiological mechanisms, still partly unknown, that underlie their generation and recurrence [46]. It is for this reason that the search for new models and research approaches is currently one of the main topics in the field of drug-resistant epilepsy [47,48], since it may open new perspectives towards alternative therapeutic strategies.

Here, our main purpose was to build a comprehensive model of the disease by performing a side-by-side comparison of neurons differentiated from diseased (SCN1A^{M145T} idNs) and healthy control iPSCs (WT idNs). Our study allowed us to draw the following major conclusions: (i) SCN1A^{M145T} idNs show an overall immature phenotype, as demonstrated by the altered expression of the chloride co-transporters, *NKCC1* and *KCC2*, and VGSC isoforms; (ii) SCN1A^{M145T} idNs show a depolarized action potential threshold compared to the WT counterpart measured by patch-clamp, suggesting a reduced excitability; (iii) we were able, for the first time to our knowledge, to record AMPA and GABA currents from both SCN1A^{M145T}- and control-idNs membranes micro-transplanted into *Xenopus* oocytes. Concerning chloride co-transporters, we could detect an increase in *NKCC1/KCC2* ratio for SCN1A^{M145T} idNs, that well fits with an overall immaturity of diseased neurons [49]. Furthermore, this observation is in line with our previous study of a patient affected by Dravet syndrome [50]. Indeed, previous studies have reported that $GABA_A$ receptors (Rs) function is strongly dependent on chloride homeostasis ensured by the chloride co-transporters NKCC1 and KCC2 both in physiological and pathological conditions [49]. The action of NKCC1 was shown to prevail during the first phases of neuronal development, where it is involved in the depolarizing, or "less hyperpolarizing", current through $GABA_A$Rs [29] and later during development this equilibrium is shifted in favor of KCC2 [51]. Interestingly, our findings indicate that SCN1A^{M145T} shows a persistent increase in NKCC1/KCC2 mRNA ratio, which supports the hypothesis that other physiopathological mechanisms, beyond the complexity of sodium channel mutations, deserve further investigation for a comprehensive understanding of channelopathies complexity [52].

Additionally, we observed a high expression of *SCN3A*, the fetal isoform of VGSCs slightly detectable in mature cells [53], and a lower expression of *SCN1A* and *SCN2A* in SCN1A^{M145T} idNs over time in culture, while WT idNs display a VGSCs expression pattern that mirrors the changes observed during normal developmental and maturation processes [27]. The progressive up-regulation of *SCN3A* in diseased neurons may reflect a sort of compensatory effect, as previously reported in mice carrying loss-of-function mutations in *SCN1A* and where the increased $Na_V1.3$ expression was observed [16]. Overall, our findings, using the most common glial and neuronal markers, are in accordance with those by other authors providing evidence that differentiation of pluripotent stem cells mainly produce interneurons [21,54]. In addition, we found that SCN1A^{M145T} neurons express significantly lower levels of $Na_V1.1$ protein compared to healthy neurons, even if this protein co-localizes with GAD1. Therefore, we may hypothesize that the loss of function of the mutated protein expressed on interneurons is contributing to their decreased excitability, leading to a reduced GABA release on synaptic targets. This "interneurons hypothesis" [55] contributes, at least in part, to defining a pathophysiological substrate for the generation and recurrence of seizures in these patients [15,50]. Indeed, to further support this hypothesis, here we found a depolarized action potential threshold in SCN1A^{M145T} neurons compared to WT, although in both situations we could not record a fully developed AP threshold value as expected in these experimental conditions [35]. Interestingly, the patient from which we differentiated idNs was also suffering from drug-resistant MTLE, which prompted us to measure GABA current rundown, a GABAergic dysfunction which is a hallmark of this condition [25,32,56,57], in idNs membranes-injected oocytes. Indeed, we could not measure a significant increase in current rundown in SCN1A^{M145T}-injected oocytes, in contrast to what was observed measuring this electrophysiological parameter from surgically resected brain tissue of patients with drug-resistant MTLE, including the patient of this study [15,34]. This is an important point since functional impairment of GABAergic neurons and GABA current-rundown tightly correlate with MTLE phenotype. A reasonable explanation for this discrepancy could lie in the fact that GABA rundown might arise as a part of the cascade of pathological events eventually leading to generation and recurrence of seizures [58,59]. As such, it is unlikely to detect a high GABA rundown in iPSCs-derived neurons, since these cells have never undergone continuous insults such as seizures or hippocampal sclerosis occurrence. This result offers an additional and intriguing perspective for the analysis of our results. Indeed, previous studies support the idea that GABA current rundown emerges in the "chronic" stages of the epileptic disorder, after the first spontaneous seizure [56,58]. Unfortunately, this means that there would be scarce therapeutic opportunity to prevent the consolidation of this aberration of GABAergic synaptic transmission since patients usually require medical attention after the appearance of spontaneous seizures [60]. On the other hand, there are patients that clearly carry additional "risk factors" for developing epilepsy, such as genetic mutations. Here and in previous studies [61], we clearly hypothesize that the aforementioned channelopathies can induce other synaptic dysfunctions [50,62], thus an early therapeutic intervention may be possible in conditions where a definite risk factor can be identified. Moreover, we can hypothesize that preventing the appearance of synaptic dysfunctions may have an impact on the evolution of the disease [63,64]. Clearly, additional experiments will be designed to further develop this hypothesis. For instance, future studies using the methodologies described here will allow the evaluation of the effect of candidate pharmacological agents on key pathophysiological alterations, such as those reported above. Moreover, an interesting outlook would be the implementation of our methodology with innovative and dynamic techniques of cell culture [65–68]. Additional future investigations will focus on transcriptomic and proteomic profiling for a comprehensive understanding of how genes and proteins are expressed and interconnected in the complex disease phenotype. Overall, our results, albeit limited by the fact that the data are obtained from cells generated from a single patient, are characterized by a high robustness and contribute by shedding light on the molecular mechanisms responsible for this particular form of FS, opening new stimulating perspectives on the ex vivo precision medicine approaches for a better management of patients with FS and MTLE, and for the prevention of potential development of drug resistance.

5. Conclusions

We report the generation and characterization of idNs from a patient belonging to a genetically well-characterized Italian family, carrying the c.434T > C mutation in SCN1A gene, responsible for FS and MTLE. Notably, electrophysiological experiments mirror the profile recorded from hippocampal tissue resected from the same patient, strengthening the validity of iPSCs technology for disease modeling. Moreover, our functional data clearly show that this channelopathy induces additional synaptic dysfunctions that may be a consequence of seizures or hippocampal sclerosis which may be prevented by early and targeted interventions. Using a multidisciplinary approach, our results reveal an aberrant maturation and altered electrophysiological features in neurons derived from the $SCN1A^{M145T}$ patient and set the ground for future use of this approach for personalized medicine.

Supplementary Materials: The following supporting information can be downloaded at: https://www.mdpi.com/article/10.3390/biomedicines10051075/s1, Figure S1: expression of astrocyte and oligodendrocyte markers in idNs culture; Figure S2: expression of vGLUT1 in idNs; Figure S3: expression of interneuronal subtype markers in idNs; Figure S4: expression of SCN2A and SCN8A in idNs. Table S1: List of primers.

Author Contributions: Conceptualization, S.S., A.G., G.C. and E.I.P.; data curation, S.S., E.P., G.R. and E.I.P.; formal analysis, S.S., S.F., E.P., G.R. and E.I.P.; funding acquisition, E.P. and G.C.; Investigation, S.S., C.Z., P.C., T.D. and K.M.; project administration, E.I.P.; resources, E.P., A.G. and G.C.; software, P.C., T.D., K.M. and S.F.; supervision, G.C. and E.I.P.; validation, S.S., C.Z., V.L., M.L.C., L.S., P.C., T.D., K.M. and S.F.; visualization, A.G., G.R. and G.C.; writing—original draft, S.S., E.P., G.R., G.C. and E.I.P.; writing—review and editing, G.R., G.C. and E.I.P. All authors have read and agreed to the published version of the manuscript.

Funding: This research was funded by: Italian Ministry of University and Research, grant number PRIN 2017 #2017CH4RNP_01 (G.C.); Ateneo project grant number RG12117A8697DCF1 (E.P.); Italian Ministry of Health "Ricerca corrente" (G.R.). This project has received funding from the European Union's Horizon 2020 research and innovation programme under Grant Agreement No. 952455 (E.P., S.F., G.R.).

Institutional Review Board Statement: The study was conducted according to the guidelines of the Declaration of Helsinki, and approved by the Ethics Committee of the "Magna Graecia" University of Catanzaro and the *Azienda Ospedaliero—Universitaria* "Mater Domini" (Approval number: AOM92_2020).

Informed Consent Statement: Informed consent was obtained from all subjects involved in the study.

Data Availability Statement: All the data presented in this study are available from the corresponding author upon reasonable request.

Conflicts of Interest: The authors declare no conflict of interest.

References

1. Camfield, P.; Camfield, C. Incidence, Prevalence and Aetiology of Seizures and Epilepsy in Children. *Epileptic Disorders* **2015**, *17*, 117–123. [CrossRef] [PubMed]
2. Saghazadeh, A.; Mastrangelo, M.; Rezaei, N. Genetic Background of Febrile Seizures. *Rev. Neurosci.* **2014**, *25*, 129–161. [CrossRef] [PubMed]
3. de Tisi, J.; Bell, G.S.; Peacock, J.L.; McEvoy, A.W.; Harkness, W.F.; Sander, J.W.; Duncan, J.S. The Long-Term Outcome of Adult Epilepsy Surgery, Patterns of Seizure Remission, and Relapse: A Cohort Study. *Lancet* **2011**, *378*, 1388–1395. [CrossRef]
4. Marini, C.; Scheffer, I.E.; Nabbout, R.; Suls, A.; De Jonghe, P.; Zara, F.; Guerrini, R. The Genetics of Dravet Syndrome: Genetics of Dravet Syndrome. *Epilepsia* **2011**, *52*, 24–29. [CrossRef]
5. Escayg, A.; Heils, A.; MacDonald, B.T.; Haug, K.; Sander, T.; Meisler, M.H. A Novel SCN1A Mutation Associated with Generalized Epilepsy with Febrile Seizures Plus—And Prevalence of Variants in Patients with Epilepsy. *Am. J. Hum. Genet.* **2001**, *68*, 866–873. [CrossRef]
6. Mantegazza, M.; Gambardella, A.; Rusconi, R.; Schiavon, E.; Annesi, F.; Cassulini, R.R.; Labate, A.; Carrideo, S.; Chifari, R.; Canevini, M.P.; et al. Identification of an Nav1.1 Sodium Channel (SCN1A) Loss-of-Function Mutation Associated with Familial Simple Febrile Seizures. *Proc. Natl. Acad. Sci. USA* **2005**, *102*, 18177–18182. [CrossRef]
7. Kasperavičiūtė, D.; Catarino, C.B.; Matarin, M.; Leu, C.; Novy, J.; Tostevin, A.; Leal, B.; Hessel, E.V.S.; Hallmann, K.; Hildebrand, M.S.; et al. Epilepsy, Hippocampal Sclerosis and Febrile Seizures Linked by Common Genetic Variation around SCN1A. *Brain* **2013**, *136*, 3140–3150. [CrossRef]
8. Dutton, S.B.B.; Dutt, K.; Papale, L.A.; Helmers, S.; Goldin, A.L.; Escayg, A. Early-Life Febrile Seizures Worsen Adult Phenotypes in Scn1a Mutants. *Exp. Neurol.* **2017**, *293*, 159–171. [CrossRef]
9. Scheffer, I.E.; Zhang, Y.-H.; Jansen, F.E.; Dibbens, L. Dravet Syndrome or Genetic (Generalized) Epilepsy with Febrile Seizures Plus? *Brain Dev.* **2009**, *31*, 394–400. [CrossRef]
10. Orban, M.; Goedel, A.; Haas, J.; Sandrock-Lang, K.; Gärtner, F.; Jung, C.B.; Zieger, B.; Parrotta, E.; Kurnik, K.; Sinnecker, D.; et al. Functional Comparison of Induced Pluripotent Stem Cell- and Blood-Derived GPIIbIIIa Deficient Platelets. *PLoS ONE* **2015**, *10*, e0115978. [CrossRef]
11. Parrotta, E.I.; Scalise, S.; Scaramuzzino, L.; Cuda, G. Stem Cells: The Game Changers of Human Cardiac Disease Modelling and Regenerative Medicine. *Int. J. Mol. Sci.* **2019**, *20*, 5760. [CrossRef] [PubMed]
12. De Angelis, M.T.; Santamaria, G.; Parrotta, E.I.; Scalise, S.; Lo Conte, M.; Gasparini, S.; Ferlazzo, E.; Aguglia, U.; Ciampi, C.; Sgura, A.; et al. Establishment and Characterization of Induced Pluripotent Stem Cells (IPSCs) from Central Nervous System Lupus Erythematosus. *J. Cell. Mol. Med.* **2019**, *23*, 7382–7394. [CrossRef] [PubMed]
13. Lucchino, V.; Scaramuzzino, L.; Scalise, S.; Grillone, K.; Lo Conte, M.; Esposito, C.; Aguglia, U.; Ferlazzo, E.; Perrotti, N.; Malatesta, P.; et al. Generation of Human Induced Pluripotent Stem Cell Lines (UNIMGi003-A and UNIMGi004-A) from Two Italian Siblings Affected by Unverricht-Lundborg Disease. *Stem Cell Res.* **2021**, *53*, 102329. [CrossRef] [PubMed]

14. Colosimo, E.; Gambardella, A.; Mantegazza, M.; Labate, A.; Rusconi, R.; Schiavon, E.; Annesi, F.; Cassulini, R.R.; Carrideo, S.; Chifari, R.; et al. Electroclinical Features of a Family with Simple Febrile Seizures and Temporal Lobe Epilepsy Associated with SCN1A Loss-of-Function Mutation. *Epilepsia* **2007**, *48*, 1691–1696. [CrossRef]
15. Ruffolo, G.; Martinello, K.; Labate, A.; Cifelli, P.; Fucile, S.; Di Gennaro, G.; Quattrone, A.; Esposito, V.; Limatola, C.; Giangaspero, F.; et al. Modulation of GABAergic Dysfunction Due to SCN1A Mutation Linked to Hippocampal Sclerosis. *Ann. Clin. Transl. Neurol.* **2020**, *7*, 1726–1731. [CrossRef]
16. Yu, F.H.; Mantegazza, M.; Westenbroek, R.E.; Robbins, C.A.; Kalume, F.; Burton, K.A.; Spain, W.J.; McKnight, G.S.; Scheuer, T.; Catterall, W.A. Reduced Sodium Current in GABAergic Interneurons in a Mouse Model of Severe Myoclonic Epilepsy in Infancy. *Nat. Neurosci.* **2006**, *9*, 1142–1149. [CrossRef]
17. Cheah, C.S.; Yu, F.H.; Westenbroek, R.E.; Kalume, F.K.; Oakley, J.C.; Potter, G.B.; Rubenstein, J.L.; Catterall, W.A. Specific Deletion of NaV1.1 Sodium Channels in Inhibitory Interneurons Causes Seizures and Premature Death in a Mouse Model of Dravet Syndrome. *Proc. Natl. Acad. Sci. USA* **2012**, *109*, 14646–14651. [CrossRef]
18. Palma, E.; Esposito, V.; Mileo, A.M.; Di Gennaro, G.; Quarato, P.; Giangaspero, F.; Scoppetta, C.; Onorati, P.; Trettel, F.; Miledi, R.; et al. Expression of Human Epileptic Temporal Lobe Neurotransmitter Receptors in Xenopus Oocytes: An Innovative Approach to Study Epilepsy. *Proc. Natl. Acad. Sci. USA* **2002**, *99*, 15078–15083. [CrossRef]
19. Higurashi, N.; Uchida, T.; Lossin, C.; Misumi, Y.; Okada, Y.; Akamatsu, W.; Imaizumi, Y.; Zhang, B.; Nabeshima, K.; Mori, M.X.; et al. A Human Dravet Syndrome Model from Patient Induced Pluripotent Stem Cells. *Mol. Brain* **2013**, *6*, 19. [CrossRef]
20. Sun, Y.; Paşca, S.P.; Portmann, T.; Goold, C.; Worringer, K.A.; Guan, W.; Chan, K.C.; Gai, H.; Vogt, D.; Chen, Y.-J.J.; et al. A Deleterious Nav1.1 Mutation Selectively Impairs Telencephalic Inhibitory Neurons Derived from Dravet Syndrome Patients. *eLife* **2016**, *5*, e13073. [CrossRef]
21. Liu, Y.; Lopez-Santiago, L.F.; Yuan, Y.; Jones, J.M.; Zhang, H.; O'Malley, H.A.; Patino, G.A.; O'Brien, J.E.; Rusconi, R.; Gupta, A.; et al. Dravet Syndrome Patient-Derived Neurons Suggest a Novel Epilepsy Mechanism. *Ann. Neurol.* **2013**, *74*, 128–139. [CrossRef] [PubMed]
22. Scalise, S.; Scaramuzzino, L.; Lucchino, V.; Esposito, C.; Malatesta, P.; Grillone, K.; Perrotti, N.; Cuda, G.; Parrotta, E.I. Generation of IPSC Lines from Two Patients Affected by Febrile Seizure Due to Inherited Missense Mutation in SCN1A Gene. *Stem Cell Res.* **2020**, *49*, 102083. [CrossRef] [PubMed]
23. Parrotta, E.I.; Scalise, S.; Taverna, D.; De Angelis, M.T.; Sarro, G.; Gaspari, M.; Santamaria, G.; Cuda, G. Comprehensive Proteogenomic Analysis of Human Embryonic and Induced Pluripotent Stem Cells. *J. Cell. Mol. Med.* **2019**, *23*, 5440–5453. [CrossRef] [PubMed]
24. Schindelin, J.; Arganda-Carreras, I.; Frise, E.; Kaynig, V.; Longair, M.; Pietzsch, T.; Preibisch, S.; Rueden, C.; Saalfeld, S.; Schmid, B.; et al. Fiji: An Open-Source Platform for Biological-Image Analysis. *Nat. Methods* **2012**, *9*, 676–682. [CrossRef]
25. Ruffolo, G.; Di Bonaventura, C.; Cifelli, P.; Roseti, C.; Fattouch, J.; Morano, A.; Limatola, C.; Aronica, E.; Palma, E.; Giallonardo, A.T. A Novel Action of Lacosamide on GABAA Currents Sets the Ground for a Synergic Interaction with Levetiracetam in Treatment of Epilepsy. *Neurobiol. Dis.* **2018**, *115*, 59–68. [CrossRef] [PubMed]
26. Sun, A.X.; Yuan, Q.; Tan, S.; Xiao, Y.; Wang, D.; Khoo, A.T.T.; Sani, L.; Tran, H.-D.; Kim, P.; Chiew, Y.S.; et al. Direct Induction and Functional Maturation of Forebrain GABAergic Neurons from Human Pluripotent Stem Cells. *Cell Rep.* **2016**, *16*, 1942–1953. [CrossRef]
27. Heighway, J.; Sedo, A.; Garg, A.; Eldershaw, L.; Perreau, V.; Berecki, G.; Reid, C.A.; Petrou, S.; Maljevic, S. Sodium Channel Expression and Transcript Variation in the Developing Brain of Human, Rhesus Monkey, and Mouse. *Neurobiol. Dis.* **2022**, *164*, 105622. [CrossRef]
28. Rahmati, N.; Hoebeek, F.E.; Peter, S.; De Zeeuw, C.I. Chloride Homeostasis in Neurons With Special Emphasis on the Olivocerebellar System: Differential Roles for Transporters and Channels. *Front. Cell. Neurosci.* **2018**, *12*, 101. [CrossRef]
29. Blaesse, P.; Airaksinen, M.S.; Rivera, C.; Kaila, K. Cation-Chloride Cotransporters and Neuronal Function. *Neuron* **2009**, *61*, 820–838. [CrossRef]
30. Schulte, J.T.; Wierenga, C.J.; Bruining, H. Chloride Transporters and GABA Polarity in Developmental, Neurological and Psychiatric Conditions. *Neurosci. Biobehav. Rev.* **2018**, *90*, 260–271. [CrossRef]
31. Kang, H.J.; Kawasawa, Y.I.; Cheng, F.; Zhu, Y.; Xu, X.; Li, M.; Sousa, A.M.M.; Pletikos, M.; Meyer, K.A.; Sedmak, G.; et al. Spatio-Temporal Transcriptome of the Human Brain. *Nature* **2011**, *478*, 483–489. [CrossRef] [PubMed]
32. Li, G.; Yang, K.; Zheng, C.; Liu, Q.; Chang, Y.; Kerrigan, J.F.; Wu, J. Functional Rundown of Gamma-Aminobutyric AcidA Receptors in Human Hypothalamic Hamartomas. *Ann. Neurol.* **2011**, *69*, 664–672. [CrossRef] [PubMed]
33. Ragozzino, D.; Palma, E.; Di Angelantonio, S.; Amici, M.; Mascia, A.; Arcella, A.; Giangaspero, F.; Cantore, G.; Di Gennaro, G.; Manfredi, M.; et al. Rundown of GABA Type A Receptors Is a Dysfunction Associated with Human Drug-Resistant Mesial Temporal Lobe Epilepsy. *Proc. Natl. Acad. Sci. USA* **2005**, *102*, 15219–15223. [CrossRef] [PubMed]
34. Palma, E.; Roseti, C.; Maiolino, F.; Fucile, S.; Martinello, K.; Mazzuferi, M.; Aronica, E.; Manfredi, M.; Esposito, V.; Cantore, G.; et al. GABAA-Current Rundown of Temporal Lobe Epilepsy Is Associated with Repetitive Activation of GABAA "Phasic" Receptors. *Proc. Natl. Acad. Sci. USA* **2007**, *104*, 20944–20948. [CrossRef]
35. Tidball, A.M.; Lopez-Santiago, L.F.; Yuan, Y.; Glenn, T.W.; Margolis, J.L.; Clayton Walker, J.; Kilbane, E.G.; Miller, C.A.; Martina Bebin, E.; Scott Perry, M.; et al. Variant-Specific Changes in Persistent or Resurgent Sodium Current in SCN8A-Related Epilepsy Patient-Derived Neurons. *Brain* **2020**, *143*, 3025–3040. [CrossRef]

36. Whitaker, W.R.J.; Faull, R.L.M.; Waldvogel, H.J.; Plumpton, C.J.; Emson, P.C.; Clare, J.J. Comparative Distribution of Voltage-Gated Sodium Channel Proteins in Human Brain. *Mol. Brain Res.* **2001**, *88*, 37–53. [CrossRef]
37. Duflocq, A.; Le Bras, B.; Bullier, E.; Couraud, F.; Davenne, M. Nav1.1 Is Predominantly Expressed in Nodes of Ranvier and Axon Initial Segments. *Mol. Cell. Neurosci.* **2008**, *39*, 180–192. [CrossRef]
38. Scheffer, I.E.; Nabbout, R. SCN1A-Related Phenotypes: Epilepsy and Beyond. *Epilepsia* **2019**, *60*, S17–S24. [CrossRef]
39. Bartolini, E.; Campostrini, R.; Kiferle, L.; Pradella, S.; Rosati, E.; Chinthapalli, K.; Palumbo, P. Epilepsy and Brain Channelopathies from Infancy to Adulthood. *Neurol. Sci.* **2020**, *41*, 749–761. [CrossRef]
40. Tang, B.; Dutt, K.; Papale, L.; Rusconi, R.; Shankar, A.; Hunter, J.; Tufik, S.; Yu, F.H.; Catterall, W.A.; Mantegazza, M.; et al. A BAC Transgenic Mouse Model Reveals Neuron Subtype-Specific Effects of a Generalized Epilepsy with Febrile Seizures Plus (GEFS+) Mutation. *Neurobiol. Dis.* **2009**, *35*, 91–102. [CrossRef]
41. Mashimo, T.; Ohmori, I.; Ouchida, M.; Ohno, Y.; Tsurumi, T.; Miki, T.; Wakamori, M.; Ishihara, S.; Yoshida, T.; Takizawa, A.; et al. A Missense Mutation of the Gene Encoding Voltage-Dependent Sodium Channel (Nav1.1) Confers Susceptibility to Febrile Seizures in Rats. *J. Neurosci.* **2010**, *30*, 5744–5753. [CrossRef] [PubMed]
42. Hedrich, U.B.S.; Liautard, C.; Kirschenbaum, D.; Pofahl, M.; Lavigne, J.; Liu, Y.; Theiss, S.; Slotta, J.; Escayg, A.; Dihné, M.; et al. Impaired Action Potential Initiation in GABAergic Interneurons Causes Hyperexcitable Networks in an Epileptic Mouse Model Carrying a Human NaV1.1 Mutation. *J. Neurosci.* **2014**, *34*, 14874–14889. [CrossRef] [PubMed]
43. Das, A.; Zhu, B.; Xie, Y.; Zeng, L.; Pham, A.T.; Neumann, J.C.; Safrina, O.; Benavides, D.R.; MacGregor, G.R.; Schutte, S.S.; et al. Interneuron Dysfunction in a New Mouse Model of SCN1A GEFS+. *ENeuro* **2021**, *8*, ENEURO.0394-20.2021. [CrossRef] [PubMed]
44. Engel, J. The Current Place of Epilepsy Surgery. *Curr. Opin. Neurol.* **2018**, *31*, 192–197. [CrossRef] [PubMed]
45. Baxendale, S. The Cognitive Costs, Contraindications and Complications of Epilepsy Surgery in Adults. *Curr. Opin. Neurol.* **2020**, *33*, 207–212. [CrossRef]
46. Löscher, W.; Klitgaard, H.; Twyman, R.E.; Schmidt, D. New Avenues for Anti-Epileptic Drug Discovery and Development. *Nat. Rev. Drug Discov.* **2013**, *12*, 757–776. [CrossRef]
47. Löscher, W. Animal Models of Seizures and Epilepsy: Past, Present, and Future Role for the Discovery of Antiseizure Drugs. *Neurochem. Res.* **2017**, *42*, 1873–1888. [CrossRef]
48. Aguilar-Castillo, M.J.; Cabezudo-García, P.; Ciano-Petersen, N.L.; García-Martin, G.; Marín-Gracia, M.; Estivill-Torrús, G.; Serrano-Castro, P.J. Immune Mechanism of Epileptogenesis and Related Therapeutic Strategies. *Biomedicines* **2022**, *10*, 716. [CrossRef]
49. Kaila, K.; Price, T.J.; Payne, J.A.; Puskarjov, M.; Voipio, J. Cation-Chloride Cotransporters in Neuronal Development, Plasticity and Disease. *Nat. Rev. Neurosci.* **2014**, *15*, 637–654. [CrossRef]
50. Ruffolo, G.; Cifelli, P.; Roseti, C.; Thom, M.; van Vliet, E.A.; Limatola, C.; Aronica, E.; Palma, E. A Novel GABAergic Dysfunction in Human Dravet Syndrome. *Epilepsia* **2018**, *59*, 2106–2117. [CrossRef]
51. Ben-Ari, Y. The GABA Excitatory/Inhibitory Developmental Sequence: A Personal Journey. *Neuroscience* **2014**, *279*, 187–219. [CrossRef] [PubMed]
52. Musto, E.; Gardella, E.; Møller, R.S. Recent Advances in Treatment of Epilepsy-Related Sodium Channelopathies. *Eur. J. Paediatr. Neurol.* **2020**, *24*, 123–128. [CrossRef] [PubMed]
53. Smith, R.S.; Kenny, C.J.; Ganesh, V.; Jang, A.; Borges-Monroy, R.; Partlow, J.N.; Hill, R.S.; Shin, T.; Chen, A.Y.; Doan, R.N.; et al. Sodium Channel SCN3A (NaV1.3) Regulation of Human Cerebral Cortical Folding and Oral Motor Development. *Neuron* **2018**, *99*, 905–913.e7. [CrossRef] [PubMed]
54. Koch, P.; Opitz, T.; Steinbeck, J.A.; Ladewig, J.; Brüstle, O. A Rosette-Type, Self-Renewing Human ES Cell-Derived Neural Stem Cell with Potential for in Vitro Instruction and Synaptic Integration. *PNAS* **2009**, *106*, 3225–3230. [CrossRef]
55. Chopra, R.; Isom, L.L. Untangling the Dravet Syndrome Seizure Network: The Changing Face of a Rare Genetic Epilepsy: The Paradox of Dravet Syndrome. *Epilepsy Curr.* **2014**, *14*, 86–89. [CrossRef]
56. Cifelli, P.; Palma, E.; Roseti, C.; Verlengia, G.; Simonato, M. Changes in the Sensitivity of GABAA Current Rundown to Drug Treatments in a Model of Temporal Lobe Epilepsy. *Front. Cell. Neurosci.* **2013**, *7*, 108. [CrossRef]
57. Gambardella, A.; Labate, A.; Cifelli, P.; Ruffolo, G.; Mumoli, L.; Aronica, E.; Palma, E. Pharmacological Modulation in Mesial Temporal Lobe Epilepsy: Current Status and Future Perspectives. *Pharmacol. Res.* **2016**, *113*, 421–425. [CrossRef]
58. Mazzuferi, M.; Palma, E.; Martinello, K.; Maiolino, F.; Roseti, C.; Fucile, S.; Fabene, P.F.; Schio, F.; Pellitteri, M.; Sperk, G.; et al. Enhancement of GABAA-Current Run-down in the Hippocampus Occurs at the First Spontaneous Seizure in a Model of Temporal Lobe Epilepsy. *Proc. Natl. Acad. Sci. USA* **2010**, *107*, 3180–3185. [CrossRef]
59. Cifelli, P.; Di Angelantonio, S.; Alfano, V.; Morano, A.; De Felice, E.; Aronica, E.; Ruffolo, G.; Palma, E. Dissecting the Molecular Determinants of GABAA Receptors Current Rundown, a Hallmark of Refractory Human Epilepsy. *Brain Sci.* **2021**, *11*, 441. [CrossRef]
60. Pitkänen, A.; Löscher, W.; Vezzani, A.; Becker, A.J.; Simonato, M.; Lukasiuk, K.; Gröhn, O.; Bankstahl, J.P.; Friedman, A.; Aronica, E.; et al. Advances in the Development of Biomarkers for Epilepsy. *Lancet Neurol.* **2016**, *15*, 843–856. [CrossRef]
61. Stern, W.M.; Sander, J.W.; Rothwell, J.C.; Sisodiya, S.M. Impaired Intracortical Inhibition Demonstrated in Vivo in People with Dravet Syndrome. *Neurology* **2017**, *88*, 1659–1665. [CrossRef] [PubMed]

62. Ruffolo, G.; Iyer, A.; Cifelli, P.; Roseti, C.; Mühlebner, A.; van Scheppingen, J.; Scholl, T.; Hainfellner, J.A.; Feucht, M.; Krsek, P.; et al. Functional Aspects of Early Brain Development Are Preserved in Tuberous Sclerosis Complex (TSC) Epileptogenic Lesions. *Neurobiol. Dis.* **2016**, *95*, 93–101. [CrossRef] [PubMed]
63. Deidda, G.; Parrini, M.; Naskar, S.; Bozarth, I.F.; Contestabile, A.; Cancedda, L. Reversing Excitatory GABAAR Signaling Restores Synaptic Plasticity and Memory in a Mouse Model of Down Syndrome. *Nat. Med.* **2015**, *21*, 318–326. [CrossRef] [PubMed]
64. Tang, X.; Kim, J.; Zhou, L.; Wengert, E.; Zhang, L.; Wu, Z.; Carromeu, C.; Muotri, A.R.; Marchetto, M.C.N.; Gage, F.H.; et al. KCC2 Rescues Functional Deficits in Human Neurons Derived from Patients with Rett Syndrome. *Proc. Natl. Acad. Sci. USA* **2016**, *113*, 751–756. [CrossRef]
65. Vatine, G.D.; Barrile, R.; Workman, M.J.; Sances, S.; Barriga, B.K.; Rahnama, M.; Barthakur, S.; Kasendra, M.; Lucchesi, C.; Kerns, J.; et al. Human IPSC-Derived Blood-Brain Barrier Chips Enable Disease Modeling and Personalized Medicine Applications. *Cell Stem Cell* **2019**, *24*, 995–1005.e6. [CrossRef] [PubMed]
66. Pelkonen, A.; Mzezewa, R.; Sukki, L.; Ryynänen, T.; Kreutzer, J.; Hyvärinen, T.; Vinogradov, A.; Aarnos, L.; Lekkala, J.; Kallio, P.; et al. A Modular Brain-on-a-Chip for Modelling Epileptic Seizures with Functionally Connected Human Neuronal Networks. *Biosens. Bioelectron.* **2020**, *168*, 112553. [CrossRef]
67. Coluccio, M.L.; D'Attimo, M.A.; Cristiani, C.M.; Candeloro, P.; Parrotta, E.; Dattola, E.; Guzzi, F.; Cuda, G.; Lamanna, E.; Carbone, E.; et al. A Passive Microfluidic Device for Chemotaxis Studies. *Micromachines* **2019**, *10*, 551. [CrossRef]
68. Guzzi, F.; Candeloro, P.; Coluccio, M.L.; Cristiani, C.M.; Parrotta, E.I.; Scaramuzzino, L.; Scalise, S.; Dattola, E.; D'Attimo, M.A.; Cuda, G.; et al. A Disposable Passive Microfluidic Device for Cell Culturing. *Biosensors* **2020**, *10*, 18. [CrossRef]

biomedicines

MDPI

Article

The Secretome of Human Dental Pulp Stem Cells and Its Components GDF15 and HB-EGF Protect Amyotrophic Lateral Sclerosis Motoneurons against Death

Richard Younes [1,2,3], Youssef Issa [1], Nadia Jdaa [1], Batoul Chouaib [2,4], Véronique Brugioti [1], Désiré Challuau [1], Cédric Raoul [1], Frédérique Scamps [1], Frédéric Cuisinier [2,†] and Cécile Hilaire [1,*,†]

[1] INM, University of Montpellier, INSERM, 34295 Montpellier, France; richard.younes@inserm.fr (R.Y.); youssef.issa@inserm.fr (Y.I.); nadia.jdaa@hotmail.fr (N.J.); veronique.brugioti@inserm.fr (V.B.); desire.challuau@inserm.fr (D.C.); cedric.raoul@inserm.fr (C.R.); frederique.scamps@inserm.fr (F.S.)
[2] LBN, University of Montpellier, 34193 Montpellier, France; batoul_chouaib@hotmail.com (B.C.); frederic.cuisinier@umontpellier.fr (F.C.)
[3] Neuroscience Research Center, Faculty of Medical Sciences, Lebanese University, Beirut 6573, Lebanon
[4] Human Health Department, IRSN, SERAMED, LRMed, 92262 Fontenay-aux-Roses, France
* Correspondence: cecile.hilaire@umontpellier.fr
† These authors contributed equally to this work.

Abstract: Amyotrophic lateral sclerosis (ALS) is a fatal and incurable paralytic disorder caused by the progressive death of upper and lower motoneurons. Although numerous strategies have been developed to slow disease progression and improve life quality, to date only a few therapeutic treatments are available with still unsatisfactory therapeutic benefits. The secretome of dental pulp stem cells (DPSCs) contains numerous neurotrophic factors that could promote motoneuron survival. Accordingly, DPSCs confer neuroprotective benefits to the *SOD1*^{G93A} mouse model of ALS. However, the mode of action of DPSC secretome on motoneurons remains largely unknown. Here, we used conditioned medium of human DPSCs (DPSCs-CM) and assessed its effect on survival, axonal length, and electrical activity of cultured wildtype and *SOD1*^{G93A} motoneurons. To further understand the role of individual factors secreted by DPSCs and to circumvent the secretome variability bias, we focused on GDF15 and HB-EGF whose neuroprotective properties remain elusive in the ALS pathogenic context. DPSCs-CM rescues motoneurons from trophic factor deprivation-induced death, promotes axon outgrowth of wildtype but not *SOD1*^{G93A} mutant motoneurons, and has no impact on the spontaneous electrical activity of wildtype or mutant motoneurons. Both GDF15 and HB-EGF protect *SOD1*^{G93A} motoneurons against nitric oxide-induced death, but not against death induced by trophic factor deprivation. GDF15 and HB-EGF receptors were found to be expressed in the spinal cord, with a two-fold increase in expression for the GDF15 low-affinity receptor in *SOD1*^{G93A} mice. Therefore, the secretome of DPSCs appears as a new potential therapeutic candidate for ALS.

Keywords: amyotrophic lateral sclerosis; neuropathology; death; axon outgrowth; electrical activity; dental pulp stem cell; conditioned medium; neurotrophic factors; motoneuron; cell therapy

Citation: Younes, R.; Issa, Y.; Jdaa, N.; Chouaib, B.; Brugioti, V.; Challuau, D.; Raoul, C.; Scamps, F.; Cuisinier, F.; Hilaire, C. The Secretome of Human Dental Pulp Stem Cells and Its Components GDF15 and HB-EGF Protect Amyotrophic Lateral Sclerosis Motoneurons against Death. *Biomedicines* **2023**, *11*, 2152. https://doi.org/10.3390/biomedicines11082152

Academic Editor: Masaru Tanaka

Received: 31 May 2023
Revised: 27 July 2023
Accepted: 28 July 2023
Published: 30 July 2023

1. Introduction

Amyotrophic lateral sclerosis (ALS) is an incurable neurodegenerative disease that leads to the selective loss of upper and lower motoneurons resulting in progressive paralysis and death from respiratory failure usually within 3 years of diagnosis. It remains one of the most devastating neurodegenerative diseases to date, due to the lack of effective treatment.

Approximately 90% of cases are sporadic, with no known family history of the disease. Sporadic and inherited forms, referred to as familial, are mostly indistinguishable by clinical and pathological markers. ALS is primarily characterized by degeneration of motoneurons in the brain and spinal cord [1]. Numerous studies have established that cellular and molecular pathophysiological mechanisms act concomitantly or sequentially to lead to neuronal dysfunction and, ultimately, death [2]. These include abnormal aggregates of misfolded proteins, defects in axonal transport and RNA metabolism, mitochondrial dysfunction, oxidative stress, glutamate excitotoxicity, and neuroinflammation [3–7]. These different molecular pathways, as well as the different cell types, represent different therapeutic targets, thus underlining the complexity of developing effective therapeutic strategies.

More recently, echoing clinical findings, neuroimaging and neuropathological studies in humans have revealed a decrease in the volume of several subcortical gray nuclei, including the thalamus, hippocampus, amygdala, caudate nucleus, and nucleus accumbens [8]. Cerebellar dysfunction has also been reported [9]. The thalamus is affected, not only in its motor part but also in areas associated with learning and memory encoding, emotional, cognitive, and sensory processes [10]. As a result, this disease is now often regarded as part of the frontotemporal dementia spectrum, which is in line with the clinical picture, since a significant proportion of patients exhibit, in addition to motor pyramidal symptoms, extrapyramidal, cognitive, and sensory symptoms [10]. For the moment, the cellular and molecular mechanisms affecting these structures in ALS are not yet known. Studies in ALS mouse models also show that at presymptomatic stages of the disease, changes in the motor thalamus are already present, with neuronal death in motor nuclei and mild gliosis [11]. A cerebellar pathology is also present [9].

The identification of new areas and specific brain circuits affected by ALS can eventually help us to develop new, more complex preclinical models that better reflect the heterogeneity of the disease and thus improve clinical transfer, to develop new treatments [12,13]. Meanwhile, it may also help us to continue dissecting the various pathophysiological mechanisms leading to motoneuron death in known mouse models of the disease, and to decipher new molecular pathways to discover other potential therapeutic targets.

Despite our better pathophysiological identification of the disease, to date only two approved drugs, riluzole and edaravone, are used to treat ALS patients [7]. Riluzole has mainly an anti-glutamatergic effect [14,15] and edaravone is a free-radical scavenger that protects against oxidative stress [16–18]. Unfortunately, the clinical outcomes of these two treatments are still unsatisfactory for patients. These two approved drugs only modestly slow disease progression and extend the patient's life by few months. A phase III clinical trial has evaluated antisense oligonucleotides (ASOs) targeting SOD1. The administration of ASOs in patients led to a decrease in SOD1 levels in the cerebrospinal fluid and the neurofilament light chain biomarker in the plasma [19]. However, ASO delivery did not significantly change the revised ALS functional rating scale score, the main clinical endpoint.

Another strategy for developing new therapies is to focus on neurotrophic factors (NTFs). Indeed, NTFs are essential for the maintenance, survival, neurite outgrowth, and axonal regeneration of motoneurons during development and adulthood [20,21]. The pleiotropic characteristics of NTFs have made them attractive therapeutic candidates for ALS to restore neuromuscular synapses and promote motoneuron survival. Unfortunately, after thirty years of clinical trials based on the administration of recombinant NTFs, the clinical outcomes remain largely disappointing [22]. The choice of route of administration, passage through the blood–brain barrier, and the limited half-life of NTFs in the blood are parameters that obviously need to be considered for these recombinant proteins. The NTFs to be evaluated, which when delivered to the periphery can lead to undesirable effects, also raise questions about methods for targeting the central nervous system, such as that based on viral gene transfer [23]. Another consideration is that of using a combinatorial approach to NTFs to optimize therapeutic effects by targeting different cell populations. Therapies based on complex protein solutions remain interesting to explore, either directly as biomaterial mixtures of therapeutic factors or as a means of identifying new pro-survival factors.

Dental pulp stem cells (DPSCs) secrete numerous NTFs that give them potent neuroprotective properties [24–26]. Moreover, several studies have shown the potential of these cells or their secretome in the treatment of neurodegenerative diseases [27–30].

Growing evidence has demonstrated the potential of DPSCs for the treatment of various diseases and dysfunctions, including neurodegenerative diseases. DPSCs can be used either directly as differentiated neurons or via their secretome. A major interest in the use of DPSCs lies in their isolation from the human third molar (wisdom tooth), since human molars are easy to extract and are considered medical waste. DPSCs proliferate rapidly and have the ability to differentiate into various cell types. Moreover, their secretome contains growth and neurotrophic factors. This secretome or conditioned medium (CM) can replace cells which, due to their ability to multiply, in some cases can be a source of cancer [31,32]. The potential therapeutic value of DPSCs has been demonstrated in spinal cord injury [33], traumatic brain injury [34], stroke [35], cerebral ischemia [36], Alzheimer's disease [30], Parkinson's disease [27], aneurysmal subarachnoid hemorrhage [37], retinal lesions [38], and ALS [29].

For ALS, the pioneering study by Wang et al. showed that intraperitoneal injection of DPSC secretome in an ALS mouse model expressing the ALS-causing mutation G93A in superoxide dismutase 1 (SOD1^{G93A}) increased lifespan and the number of surviving motoneurons [29]. However, the mechanism of action of DPSC secretome on motoneurons remains largely unknown. Moreover, DPSC secretome is a complex medium containing many factors [39]. Its composition is difficult to control and largely depends on manufacturing methods and protocols [40]. Identifying individual components is therefore an important approach to reducing constraints on the use of this complex media.

In this study, we first studied the effects of DPSC secretome on the survival, axon outgrowth, and synaptically driven electrical activity of mouse primary motoneurons from wildtype and *SOD1^{G93A}* mice. We then evaluated the neuroprotective potential of two candidate molecules that we selected from our previous work [39], growth differentiation factor 15 (GDF15) and heparin-binding EGF-like growth factor (HB-EGF).

We show that the secretome of DPSCs promotes the survival of wildtype and *SOD1^{G93A}* motoneurons without altering their electrical activity. DPSC secretome also enhances axon outgrowth of wildtype but not *SOD1^{G93A}* motoneurons. Interestingly, both GDF15 and HB-EGF confer neuroprotection to SOD1^{G93A}-expressing motoneurons against nitric oxide (NO)-induced cytotoxicity. Altogether, our results propose new therapeutic perspectives to explore based on DPSC secretome content, GDF15, and HB-EGF.

2. Materials and Methods

2.1. Preparation of Conditioned Media

2.1.1. DPSC-Conditioned Media

DPSCs were obtained from the wisdom teeth of a 15-year-old patient. Informed consent was obtained from the patient after receiving approval by the local ethics committee. DPSCs were recovered as previously described [41,42]. Briefly, after being disinfected, the teeth were cut along the cementum–enamel junction using a diamond disc and were split in two parts. Pulps were then collected and incubated with 3 mg/mL of type I collagenase and 4 mg/mL dispase for 1 h. Digested pulps were filtered, centrifuged, and the pellet was resuspended and cultured in α-MEM (ThermoFisher Scientific, Waltham, MA, USA) with 1% penicillin–streptomycin, 10% fetal bovine serum, and 1 μg/mL of recombinant human basic fibroblast growth factor (R&D System, Minneapolis, MN, USA). The culture medium was changed after 24 h and then changed twice a week [39]. DPSCs were allowed to multiply until they reached 80–90% confluency and then split with 0.05% trypsin-EDTA (ThermoFisher Scientific, Waltham, MA, USA) for 3 min at 37 °C to enhance the colony. After the 3rd or 4th passage, when DPSCs reached 90% confluency, cells were washed twice with phosphate buffered saline (PBS) and culture medium was replaced with serum-free Neurobasal medium containing 50 μM L-glutamine, 2% B-27 supplement, and 1% penicillin–streptomycin (ThermoFisher Scientific, Waltham, MA, USA). After 48 h of culture, the medium was collected, centrifuged once at 450× g for 5 min and then centrifuged at 1800× g for 3 min to remove cell debris. Unless used fresh, the conditioned medium was stored at −80 °C [39].

2.1.2. Adipose Derived Stems Cells (ASCs) and Fibroblast-Conditioned Media

Human adipose tissue was provided by the Institute for Regenerative Medicine and Biotherapy (IRMB, Montpellier, France) and human skin fibrobasts were given by Dr. Vasiliki Kalatzis (INM, Inserm UMR1298, Montpellier, France). The conditioned medium for these two cell types was produced in the same way as for DPSCs and used at passage 4 for ASCs and passage 6 for fibroblasts.

2.2. Animals

All animal experiments were approved by the national ethics committee on animal experimentation, and were conducted in compliance with the European Community and national directives for the care and use of laboratory animals. B6.Cg-Tg(SOD1*G93A)1Gur/J (*SOD1*^{G93A}) mice and B6.Cg-Tg(Hlxb9-GFP)1Tmj/J (*Hb9::GFP*) mice were purchased from the Jackson laboratory and maintained on a C57BL/6J background under specific-pathogen-free conditions. They were housed in cages with a 12 h light/12 h dark cycle with food and water supplied ad libitum in our animal facility accredited by the French Ministry of Food, Agriculture and Fisheries (B-34 172 36-11 March 2010). Experiments were conducted in accordance with the Directives of the Council of the European Communities of November 24, 1986 (86/609/EEC) and the French Ethics Committee (approval A34-506). Thirty-seven wildtype mice and twenty-three SOD1^{G93A} mice were used for embryonic motoneuron primary cultures. Five wildtype and five SOD1^{G93A} 3-month-old mice were used for quantitative RT-PCR.

2.3. Motoneurons Immunopurification and Culture

Motoneurons were purified from the spinal cords of wildtype and *SOD1*^{G93A} embryos using 5.2% iodixanol density gradient centrifugation combined with anti-p75-based magnetic cell isolation (Miltenyi Biotec, Bergisch Gladbach, Germany, clone REA648) as we previously described [43,44]. Motoneurons were plated on poly-ornithine (3 µg/mL)/laminin (2 µg/mL)-coated 4-well plates in supplemented Neurobasal medium containing 2% (*v*/*v*) horse serum, 25 µM L-glutamate, 25 µM β-mercaptoethanol, 50 µM mM L-glutamine, 2% (vol/vol) B-27 supplement, and 0.5% penicillin–streptomycin (ThermoFisher Scientific, Waltham, MA, USA). When indicated, a cocktail of neurotrophic factors (0.1 ng/mL GDNF (Sigma-Aldrich, Saint-Louis, MO, USA, G1401), 1 ng/mL brain-derived neurotrophic factor (BDNF) (ImmunoTools, MGC34632), and 10 ng/mL CNTF (R&D Systems, Minneapolis, MN, USA, 557-NT/CF)) was added to the supplemented Neurobasal medium. Recombinant mouse GDF15 (R&D Systems, Minneapolis, MN, USA, 8944-GD) and HB-EGF (E4643, Sigma-Aldrich, Saint-Louis, MO, USA) were added at the time of seeding in basal supplemented Neurobasal medium. DETANONOate (Enzo Life Sciences, Farmingdale, NY, USA, ALX-430-014) was added after 24 h of culture.

For the electrophysiological recordings, motoneurons were isolated from the ventral spinal cord of *Hb9::GFP* and *SOD1*^{G93A} E12.5 embryos using iodixanol density gradient centrifugation [45].

2.4. Motoneurons Survival

Motoneurons were seeded at a density of 1250 cells/cm^2. Wildtype and *SOD1*^{G93A} immunopurified motoneurons were counted using a phase contrast microscope using morphological criteria. Motoneurons are considered as living cells if their axons are more than three times the length of the cell body, if those axons are not completely degraded, and if the cell body does not contain vacuoles [44,46]. Counting was performed on three or four separate samples, and the number of surviving motoneurons was determined after 24 h of culture. To allow comparison of values from different experiments, survival values were normalized relative to the value in the absence of neurotrophic factors.

2.5. Axon Length Measurements

For the quantification of axon length, motoneurons were seeded at a density of 750 cells/cm^2. At one day in vitro, motoneurons were processed for immunostaining with β-tubulin III antibodies. Images were acquired with a ZEISS AXIO Imager Z2 Apotome and axon length was determined using ImageJ software v1.53 and the NeuronJ plugin (National Institutes of Health, Bethesda, MD, USA) [47]. Total axon length was determined by measuring the length of the longest neurite with connected branches of β-tubulin III-positive motoneurons [46].

2.6. Electrophysiology

For electrophysiological analysis, motoneurons were seeded at a density of 1250 cells/cm^2. Spontaneous electrical activity was recorded at room temperature using the loose-patch electrophysiological technique with an Axopatch 200B amplifier (Molecular Devices, San José, CA, USA). The bathing solution contained 145 mM NaCl, 5 mM KCl, 10 mM D-glucose, 10 mM HEPES, 2 mM CaCl2, and 1.5 mM MgCl2, adjusted to pH 7.4 and 310 mosm. The electrode was filled with the same extracellular solution and to obtain a good resolution of extracellular recordings of the spontaneous activity, the contact with the cell membrane had a resistance in the range of 30–100 MΩ. Cultures were maintained at 37 °C, 7.5% CO_2. All recordings were conducted between 7 and 9 days of culture [45,48]. Motoneurons were identified according to morphological criteria (large size, multipolar with high dendritic complexity) or GFP expression [44].

2.7. Immunocytochemistry

Motoneurons were first fixed with 4% paraformaldehyde (PFA) in PBS that was directly added to the culture medium (1:1) for 10 min, and then fixed with 4% PFA in PBS for 15 min on ice. Cells were then washed with PBS and incubated for 1 h in blocking solution containing 4% bovine serum albumin, 4% donkey serum, and 0.1% Triton-X100 in PBS. Coverslips were then incubated overnight at +4 °C with rabbit anti-βIII tubulin (Sigma-Aldrich, Saint-Louis, MO, USA, T2200). Cells were washed 3 times for 10 min each with PBS, incubated with the appropriate fluorescent-conjugated secondary antibody (ThermoFisher Scientific, Waltham, MA, USA), washed, and mounted in Moviol solution [45]. Image acquisition was carried out on a Zeiss (Car Zeiss AG, Oberkochen, Germany) Axio Imager Z2-module Apotome 2.0. ImageJ (National Institutes of Health, Bethesda, MD, USA) was used for axon tracing.

2.8. Immunohistochemistry

As previously described [43] for the spinal cord, mice were anaesthetized and transcardially perfused with 4% PFA in PBS. The lumbar spinal cord was removed, post-fixed in 4% PFA for 2 h, incubated in 30% sucrose in PBS, embedded in OCT, flash-frozen, and cut at 18 µm thickness. The sections were then rinsed for 5 min with PBS and incubated for 2 h at room temperature in blocking solution (PBS, 5% donkey serum, 0.3% Triton-X100, 0.05% Tween-20). This was followed by overnight incubation at +4 °C with the following primary antibodies: rabbit anti-TGFβ-R1 (Sigma-Aldrich, Saint-Louis, MO, USA, SAB4502958, 1:200) and goat anti-choline acetyl transferase (ChAT) (Merck, Darmstadt, Germany, AB144P, 1:100). Sections were washed and incubated for 1 h at room temperature with appropriate AlexaFluor-conjugated secondary antibodies (ThermoFisher Scientifc, Waltham, MA, USA). Slides were mounted in Mowiol solution. Image acquisition was carried out on a Zeiss Axio imager Z2-module Apotome 2.0 (Car Zeiss AG, Oberkochen, Germany).

2.9. Quantitative Reverse Transcription Polymerase Chain Reaction

For the lumbar spinal cord, 90-day-old mice were deeply anaesthetized, perfused transcardially with PBS, and tissue was harvested in RNAprotect tissue reagent (Qiagen, Hilden, Germany). Lysis buffer was used to homogenize the tissue that was then passed through needles.

The lysates were mixed with an equal volume of 70% ethanol, and total mRNA was separated from other cellular components on RNeasy minispin columns. The eluted mRNA was quantified spectrophotometrically (Nanodrop). After removal of genomic DNA, reverse transcription was performed with 1 µg of mRNA using the Quantitect RT kit (Qiagen, Hilden, Germany). The collected cDNA was diluted to 100 ng in H_20 and stored at −20 °C until use. Quantitative PCR was performed on 10 ng of cDNA with SYBR green dye (Qiagen, Hilden, Germany) using the LightCycler 480 system (Roche Diagnostics, Basel, Switzerland). The primers used were as follows: *Egfr*, 5′-ATTAATCCCGGAGAGCCAGA-3′ and 5′-TGTGCCTTGGCAGACTTTCT-3′; *ErbB4*, 5′-TCCACTTTACCACAACACGCT-3′ and 5′-TCAAAGCCATGATCACCAGGA-3′; *Tgfbr1*, 5′-ATGTGGAAATGGAAGCCCAGA-3′ and 5′-ATGACAGTGCGGTTATGGCA-3′; *Tgfbr2* variant 1, 5′-TGTTGAGATTGCAGGATCTGG-3′ and 5′-TGGACAGTCTCACATCGCAA-3′; *Tgfbr2* variant 2, 5′-TTCCCAAGTCGGTTAACAGTGA-3′ and 5′-TTCTGGTTGTCGCAAGTGGA-3′; *Polymerase (RNA) II polypeptide J (Polr2j)*, 5′-ACCACACTCTGGGGAACATC-3′ and 5′-CTCGCTGATGAGGTCTGTGA-3′. The PCR conditions were 15 s at 94 °C, 20 s at 60 °C, 35 s at 72 °C for a total of 45 cycles. After PCR amplification, melting curve analysis was performed to verify PCR specificity. The level of the domestic gene *polymerase (RNA) II polypeptide J (Polr2J)* was used to normalize cDNA amounts. Ct was calculated as the difference between Ct values, determined with Equation (2)-Ct [48].

2.10. Statistical Analysis

Data are presented as means ± standard error of the mean (SEM) (Table S1 Supplemental Data). Statistical analysis was performed using Prism 8 (GraphPad, San Diego, CA, USA). One-way ANOVA was used for multiple comparisons followed by Tukey's *post hoc* test. For qRT-PCR experiments, the Mann–Whitney test was used. Statistical significance was accepted at the level of $p < 0.05$ (Supplementary Materials Table S2).

3. Results

3.1. DPSCs-CM Promotes the Survival of Motoneurons

We first aimed to determine whether DPSCs-CM can have an effect on motoneuron survival under deprivation of NTFs. For this purpose, we studied the effect of increasing concentrations of DPSCs-CM in the culture medium without the standard cocktail of GNDF, BDNF, and CNTF. These NTFs are known to promote optimal survival of embryonic motoneurons [44]. The efficacy of the different concentrations of DPSCs-CM was assessed 24 h after seeding and motoneuron survival was expressed relative to the control condition where motoneurons were cultured in the absence of NTFs. We did not observe any neurotrophic effect for 5 and 10% DPSCS-CM concentrations (Figure 1A). However, the survival rate was 2 to 2.5 times higher when motoneurons were cultured in the presence of 25, 50, and 75% DPSCs-CM. We did not find any significant difference in motoneuron survival between the conditions with 25, 50, and 75% concentrations or with the NTFs. However, when motoneurons were cultured only in the presence of DPSCs-CM (100%), a detrimental effect on survival was observed (Figure 1A). Our results indicate that DPSCs-CM rescue motoneurons from death induced by trophic deprivation.

The DPSCs are neural crest-derived mesenchymal stem cells originating from the ectoderm [49]. To determine whether the neurotrophic effect we observed was specific to DPSCs-CM, we analyzed the neurotrophic properties of CMs derived from human ASCs (ASCs-CM), which are mesenchymal stem cells originating from the mesoderm, and human fibroblasts (Fibro-CM). Following 24 h of culture, we did not observe any neurotrophic effect of ASCs-CM at 50 or 100% on motoneuron survival compared with the NTF-free medium (Figure 1B). Similarly, 50 and 100% Fibro-CM did not improve the survival of motoneurons compared with the NTF-free medium (Figure 1C). These results demonstrate that DPSCs-CM has selective neurotrophic properties for embryonic motoneurons.

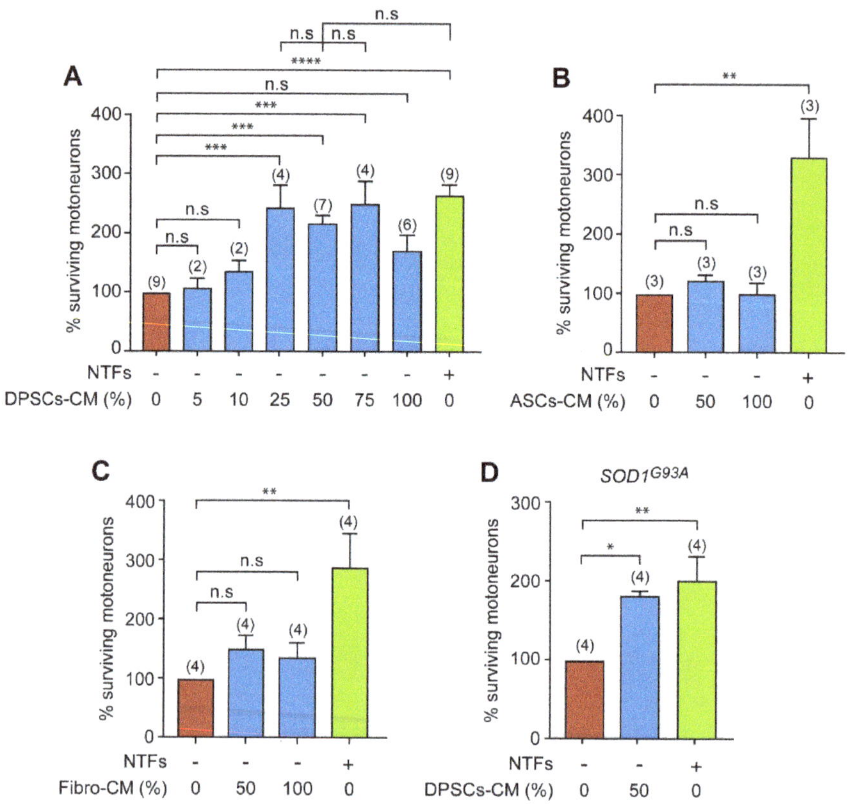

Figure 1. DPSCs-CM increases the survival of wildtype and *SOD1*^{G93A} motoneurons.the number of independent experiments (each conducted in triplicate or quadruplicate) is indicated in brackets. (**A**) Motoneurons were immunopurified from the spinal cord of E12.5 mouse embryos and cultured in the absence (-) of a cocktail of NTFs and with increasing concentration of DPSCs-CM. Motoneurons were also cultured in the presence of NTFs (+) only. The percentage of surviving motoneurons, expressed relative to the condition without any trophic support, was calculated after 24 h. Data are means ± SEM and the number of independent experiments (each conducted in triplicate or quadruplicate) is indicated in brackets. Note that for the low concentrations of DPSCs-CM (5% and 10%), which are not relevant for motoneuron survival, the experiments were repeated twice independently, each time in triplicate. One-way ANOVA followed by Tukey's post hoc test, * $p < 0.05$, ** $p < 0.01$, *** $p < 0.001$, **** $p < 0.0001$, n.s, non-significant. (**B**,**C**) Motoneurons were isolated and seeded in the absence of NTFs and with the same optimal dose previously described (50%), or only (100%) with ASCs-CM (**B**) or Fibro-CM (**C**). The number of surviving motoneurons was determined 24 h later and expressed relative to the survival in the absence of NTFs. All values are expressed as the means ± SEM of three (**B**) or four (**C**) independent experiments (triplicate or quadruplicate). One-way ANOVA followed by Tukey's test, ** $p < 0.01$, n.s, non-significant. (**D**) Mutant *SOD1*^{G93A} motoneurons were cultured with DPSCs-CM (at 50%) or NTFs. Motoneuron survival was determined 24 h later and expressed relative to the basal condition (without NTFs). Data represent the mean ± SEM of triplicates or quadruplicates of four independent experiments.

We next investigated whether DPSCs-CM could also improve the survival of motoneurons purified from the SOD1^{G93A} ALS mouse model. The survival of SOD1^{G93A} motoneurons in the absence of NTFs after 24 h of culture was similar to that of wildtype motoneurons (not shown), as already described [44]. We then analyzed the survival of

$SOD1^{G93A}$ motoneurons in the optimal concentration of DPSCs-CM that we previously determined (Figure 1A). In the presence of 50% DPSCs-CM, we observed a significant increase in the percentage of surviving ALS motoneurons compared with the negative control (Figure 1D). There were no significant differences between the survival of motoneurons cultured in the presence of 50% DPSCs-CM and those cultured in the presence of NTFs.

Therefore, our results indicate that DPSCs-CM can provide a robust neurotrophic support to motoneurons expressing ALS-causing SOD1 mutation.

3.2. DPSCs-CM Promotes Axon Outgrowth of Wildtype, but Not $SOD1^{G93A}$ Motoneurons

We next investigated whether DPSCs-CM could also influence neurite growth of wildtype and *$SOD1^{G93A}$* motoneurons. Motoneurons were immunopurified from wildtype and *$SOD1^{G93A}$* mouse embryos and immunostained with anti-βIII tubulin to trace the total length of the axon (including branchings) using the NeuronJ ImageJ plugin, as we previously described [47] (Figure 2A). We found that DPSCs-CM significantly increased the axon length of wildtype motoneurons cultured in the absence of NTFs (Figure 2B). Addition of DPSCs-CM to motoneuron culture elicited axon outgrowth as efficiently as the addition of the cocktail of NTFs. Interestingly, we observed that neither DPSCs-CM nor the cocktail of NTFs increased the total axon length of motoneurons (Figure 2C).

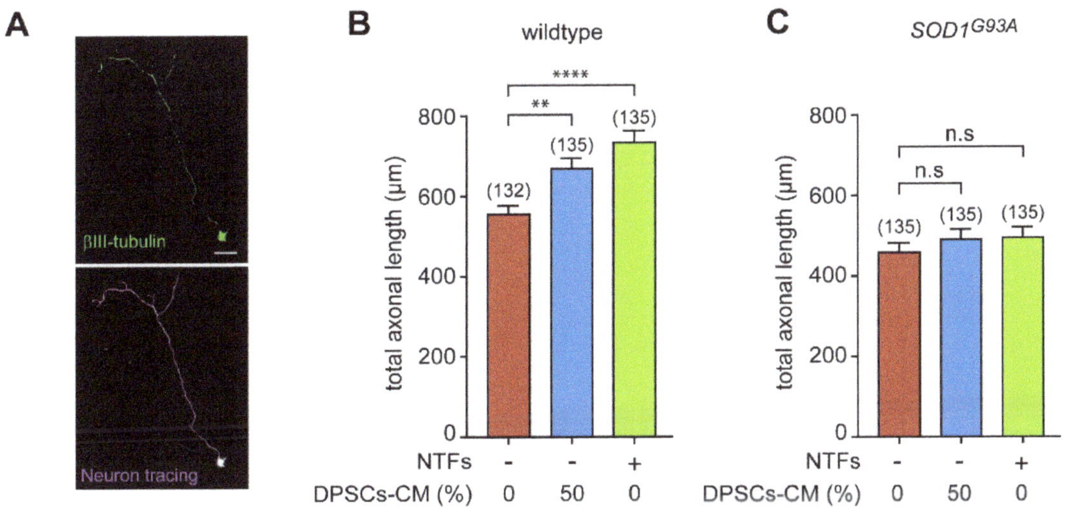

Figure 2. DPSCs-CM induces axon outgrowth of wildtype but not *$SOD1^{G93A}$* motoneurons. The number in brackets indicating the total number of motoneurons measured. (**A**) Representative immunostaining of a wildtype motoneuron with anti-βIII-tubulin antibody. The total axon length that also included collaterals of motoneurons (violet trace) was measured with ImageJ using the NeuronJ plugin. Scale bar = 20 µm. (**B**,**C**) Freshly purified motoneurons from E12.5 wildtype (**B**) or $SOD1^{G93A}$ (**C**) embryos were treated (or not) with DPSCs-CM at the optimal concentration of 50% and NTFs. Measurement of the total axon length was performed after 24 h of culture and values are expressed relative to the basal condition (in the absence of NTFs). Graphs represent the mean value ± SEM of three independent experiments, the number in brackets indicating the total number of motoneurons measured. Data were analyzed by one-way ANOVA followed by Tukey's post hoc test. ** $p < 0.01$, **** $p < 0.0001$, and n.s, non-significant.

Our results highlight that an uncoupling between axonal outgrowth and survival promoted by the neurotrophic support of DPSCs-CM or recombinant NTFs occurs when an ALS causal gene is expressed in motoneurons.

3.3. DPSCs-CM Does Not Modify the Synaptically Driven Activity of Wildtype and *SOD1^{G93A}* Motoneurons

To complete the range of functional properties of DPSCs-CM on motoneurons, we then focused on the spontaneous electrical activity that results from the synaptic network activity. We performed extracellular recordings of the spontaneous firing rate of both wildtype and *SOD1^{G93A}* neurons using the loose-patch technique. As we have previously shown, after 7 days in vitro (DIV), primary motoneurons exhibit adult-like intrinsic electrical activity and are spontaneously active due to a synaptic excitatory network [45,48] (Figure 3A). We did not observe any difference in the frequency of the spontaneous electrical activity of wildtype and *SOD1^{G93A}* neurons cultured in the presence or the absence of NTFs (Figure 3B,C). When we then cultured neurons in the presence of DPSCs-CM (50%), we did not find any significant change in the spontaneous spike frequency (approximately 2 Hz in average) of either wildtype or SOD1^{G93A}-expressing neurons (Figure 3B,C). Therefore, DPSC-CM does not modify the synaptically driven electrical activity of wildtype or *SOD1^{G93A}* neurons.

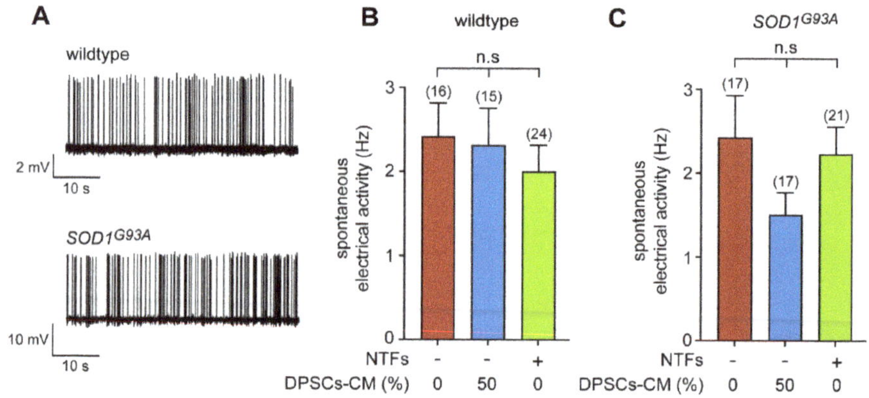

Figure 3. DPSCs-CM does not modify the spontaneous electrical activity of motoneurons. (**A**) Representative trace of recordings of spontaneous electrical activity of wildtype and *SOD1^{G93A}* neurons after 7 DIV in the presence of NTFs. (**B**,**C**) Motoneuron-enriched cultures were prepared from *Hb9::GFP* wildtype (**B**) and SOD1mutant mice (**C**) and cultured for 7 DIV in the absence of NTFs, with 50% DPSCs-CM or in the presence of trophic support. Spontaneous electrical activity was measured using the loose-patch technique and spike frequency calculated. The total number of recorded cells is indicated in brackets. Values are means ± SEM of three and four independent experiments for wildtype and *SOD1^{G93A}* conditions, respectively. Data were analyzed by one-way ANOVA followed by Tukey's post hoc test, n.s, non-significant.

3.4. GDF15 and HB-EGF Do Not Provide Any Neutrotrophic Support to Motoneurons

Among the wide palette of secreted proteins that we identified previously [39] we sought to focus on two proteins, GDF15 and HB-EGF, that have been described for their activity on neuronal survival, but whose trophic benefits for ALS motoneurons have not been evaluated [50,51]. Using a human growth factor antibody array we previously found that GDF15 and HB-EGF concentration amounted to 12 ± 12.8 pg/mL and 2 ± 4 pg/mL, respectively, in DPSCs-CM [39].

We first evaluated whether addition of recombinant GDF15 could promote the survival of wildtype motoneurons when cultured in the absence of NTFs. Motoneurons were treated with increasing concentrations of GDF15, based on what was previously described [51], and the percentage of surviving motoneurons determined after 24 h of culture. We found that GDF15 does not affect the survival of wildtype motoneurons (Figure 4A). When HB-EGF was evaluated for its prosurvival properties in motoneurons, using the optimal concentration previously described [50], we did not observe any neuroprotective benefits against NTF deprivation (Figure 4B).

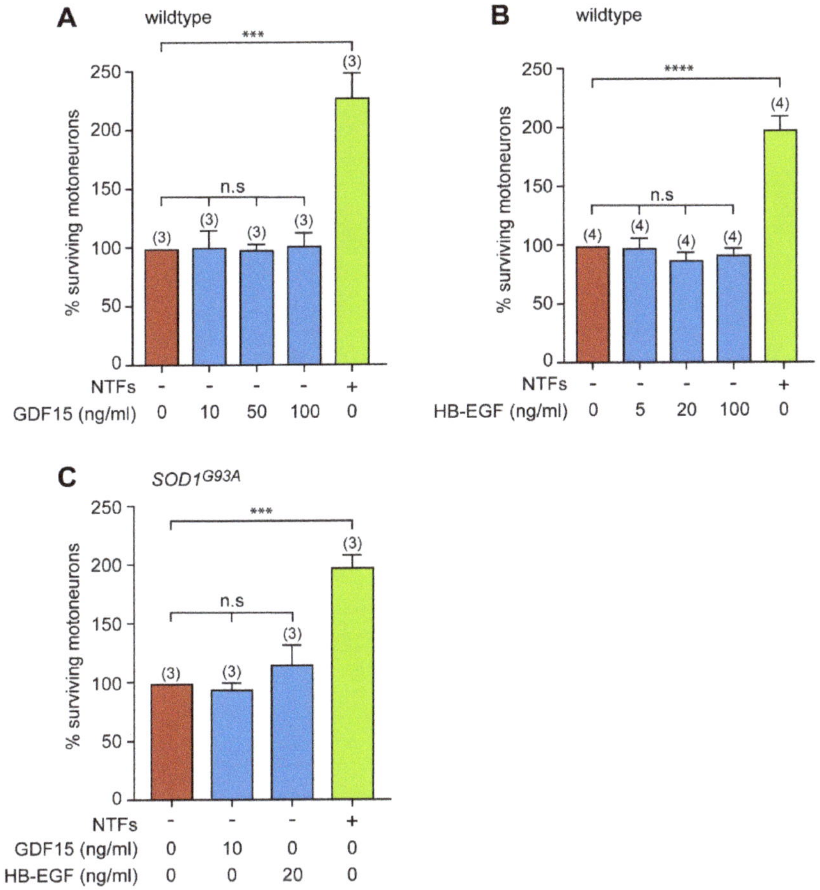

Figure 4. Recombinant GDF15 and HB-EGF do not provide trophic support to either wildtype or SOD1 mutant motoneurons. (**A**) Motoneurons were plated in basal conditions (without NTFs) and treated with indicated concentrations of recombinant GDF15 (or cultured with NTFs only). The survival of motoneurons was determined after 24 h. (**B**) Mouse motoneurons were treated with increasing concentrations (5, 20, and 100 ng/mL) of HB-EGF at the time of seeding. Twenty-four hours later, motoneuron survival was assessed. (**C**) Motoneurons immunopurified from $SOD1^{G93A}$ E12.5 embryos were incubated in the absence of NTFs with either GDF15 (10 ng/mL) or HB-EGF (20 ng/mL). The percentage of surviving $SOD1^{G93A}$ motoneurons was determined after 24h of treatment. Data represent the mean values ± SEM of triplicates of three (**A**,**C**) and four(**B**) independent experiments (number in brackets). Data were analyzed by one-way ANOVA followed by Tukey's post hoc test, *** $p < 0.001$, **** $p < 0.0001$, n.s, non-significant.

We then asked whether GDF15 or HB-EGF might promote the survival of $SOD1^{G93A}$ motoneurons placed in basal conditions. We observed that neither GDF15 (10 ng/mL) nor HB-EGF (20 ng/mL) saved motoneurons from death induced by the absence of NTFs (Figure 4C).

From our data, neither GDF15 nor HB-EGF are able to rescue motoneurons from death induced by neurotrophic factor deprivation.

3.5. GDF15 and HB-EGF Prevent Motoneuron Death from Oxidative Insult

We also investigated whether these two candidates could rescue motoneuron death from oxidative stress. It was shown that motoneurons expressing SOD1 mutants have an exacerbated susceptibility to NO exposure [44]. Motoneurons purified from wildtype and $SOD1^{G93A}$ mice were then cultured in the presence of NTFs, with GDF15 or HB-EGF, and exposed to the NO donor DETANONOate for 48 h. As previously described, DETANONOate induces death of mutant but not wildtype motoneurons (Figure 5A,B). Interestingly, we found that both GDF15 and HB-EGF rescued ALS motoneuron from death induced by NO (Figure 5B).

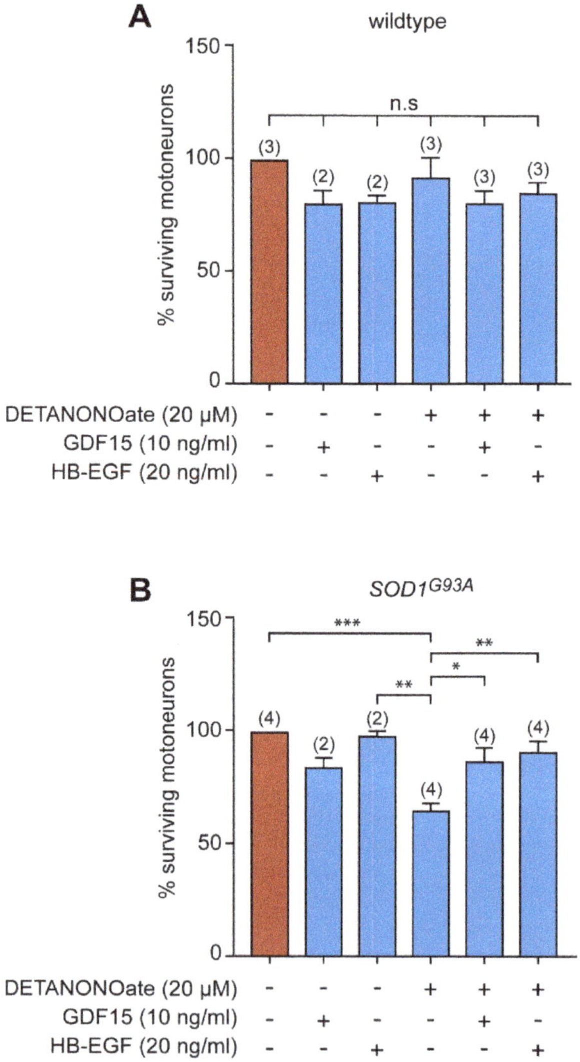

Figure 5. GDF15 and HB-EGF protect *$SOD1^{G93A}$* motoneurons from NO-induced death. the number of independent experiments (each performed in triplicate or quadruplicate) is indicated in brackets. (**A**) Wildtype motoneurons were cultured for 24 h in the presence of NTFs and incubated with the NO donor DETANONOate (20 µM) in combination with GDF15 (10 ng/mL) and HB-EGF (20 ng/mL).

The percentage of surviving motoneurons was determined 48 h later. (**B**) SOD1^{G93A}-expressing motoneurons were maintained in culture for 24 h and treated (or not) with DETANONOate (20 µM), GDF15 (10 ng/mL), or HB-EGF (20 ng/mL) for 48 h. The number of surviving motoneurons is expressed as a percentage of the number of motoneurons in the control condition (in the presence of NTFs only). Histograms show mean values ± SEM; the number of independent experiments (each performed in triplicate or quadruplicate) is indicated in brackets. One-way ANOVA followed by Tukey's post hoc test, * $p < 0.05$, ** $p < 0.01$, *** $p < 0.001$.

These data reveal a clear therapeutic potential of GDF15 and HB-EGF by rescuing motoneurons from death under pathological conditions.

3.6. Expression of GDF15 and HB-EGF in Adult Spinal Cord

To uncover the potential implication of HB-EGF and GDF15 signaling in ALS pathogenesis, we first analyzed the mRNA expression levels of their cognate receptors in the spinal cords of 3-month-old wildtype and SOD1^{G93A} mice using quantitative RT-PCR. At this presymptomatic stage, there is no substantial loss of motoneurons, but a stress response already takes place in the vulnerable population of motoneurons [52]. The transcript levels of the two HB-EGF receptors, Erbb4 and Egfr, were detected in the spinal cords of wildtype mice. The levels of *Egfr* remained unchanged in presymptomatic SOD1^{G93A} mice, while there was a significant two-fold decrease in *Erbb4* expression levels (Figure 6A). We found that the cognate high-affinity receptor of GDF15, Gfral, was not expressed in either the wildtype or SOD1^{G93A} spinal cord ($n = 3$), while the transcripts of the GDF15 low-affinity receptors, Tgfbr1 and Tgfbr2 (variants 1 and 2) were detected. Moreover, we evidenced that *Tgfbr1* levels were increased in the *SOD1^{G93A}* spinal cord (Figure 6A). Our results suggest that GDF15-TGFβ-R1 axis in motoneurons could be a therapeutic target for ALS. Therefore, we analyzed the expression profile in the spinal cords of wildtype and *SOD1^{G93A}* mice at 3 months of age. TGFβ-R1 was found to be expressed in nearly all ChAT-positive motoneurons in the spinal cords of both wildtype and *SOD1^{G93A}* mice (Figure 6B). In addition, we detected TGFβ-R1-expressing cells in the white matter that had the morphology of radial glia (not shown). Of note, TGFβ-R1 was also observed in glial cells reminiscent of microglia, consistent with the role of TGF-β1 signalling in microglial cells [53,54].

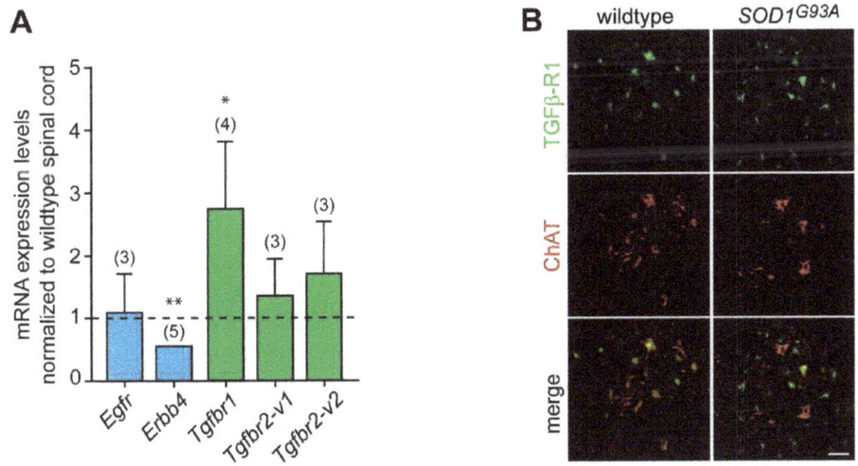

Figure 6. GDF15 and HB-EGF receptors are differentially expressed in the spinal cords of presymptomatic mice. (**A**) Quantitative RT-PCR was performed on total RNA isolated from the spinal cords of 3-month-old wildtype and SOD1^{G93A} mice. Transcript levels of *Egfr*, *Erbb4*, *Gfral*, *Tgfbr1*, and *Tgfbr2*

were expressed relative to polymerase (RNA) II polypeptide J (*Polr2J*) transcript. The expression levels of each transcript in the spinal cords of *SOD1*^{G93A} mice were normalized to those obtained in the spinal cords of wildtype mice (represented by the dashed line). Values are means ± SEM, the number of mice is indicated in brackets. Data were analyzed by Mann–Whitney test. * $p < 0.05$, ** $p < 0.01$ vs. wildtype spinal cord. (**B**) Lumbar spinal cord sections of wildtype and SOD1^{G93A} mice at 3 months of age were immunostained with antibodies against TGFβ-R1 (in green) and ChAT (in red). Scale bar, 50 μm.

These results demonstrate expression of GDF15 and HB-EGF receptors and their differential regulation in the spinal cord of ALS mice. This finding further supports a potential role of these factors in the pathophysiology of the peripheral motor system.

In summary, using primary culture of motoneurons, our study revealed the regenerative and death-protective effect of human dental pulp cell secretome under ALS conditions. Furthermore, GDF-15 and HB-EGF present in the secretome protect ALS motoneurons exclusively under oxidative stress conditions which confer disease-specific effects on these factors.

4. Discussion

Despite extensive research in both fundamental and clinical fields, ALS remains a disease with no effective treatment. The ongoing research on ALS disease mechanisms is therefore of high importance for discovering new therapies to enhance a patient's quality of life and substantially prolong their life expectancy. Since the pioneering work of Wang et al. on the ALS mouse model, the therapeutic benefits of the DPSC secretome appear as promising therapeutic means for ALS [29]. These cells are capable of secreting neurotrophic factors that are essential for neuronal survival and neurite growth, which make them promising for use in therapeutic procedures [24,25].

Our in vitro study has clarified to some extent the mechanism of the survival effect of DPSC secretome on ALS *SOD1*^{G93A} mice by directly influencing the survival of motoneurons. DPSCs-CM has a differential effect on axon outgrowth depending on the expression of SOD1 mutation, but does not change the synaptically driven electrical activity of wildtype or *SOD1*^{G93A} neurons. Moreover, GDF15 and HB-EGF, which we have previously found to be secreted by DPSCs-CM, have a neuroprotective effect on *SOD1*^{G93A} motoneurons only after exposure to oxidative stress.

DPSCs-CM is composed of many proteins including trophic factors [39,55,56]. Consistent with the effects on motoneuron survival that we have described here, the DPSCs-CM contains factors that are well known to promote neuronal survival and neurite growth during development and adulthood, such as NT-3 or VEGF [57,58]. Additional studies have also shown the expression of BDNF, CNTF, and GDNF by DPSCs, mainly at the transcriptional levels. Our CM preparation does not show a high representability of BDNF and GDNF, although CNTF was not evaluated in our previous study. It is therefore reasonable to propose that the additional effect of several factors allows an optimal survival of motoneurons, finally at a similar extent as that obtained with the combination of the NTFs we used. In addition to the survival effect, DPSCs-CM has no detrimental effect on the synaptic network, which was an important point to verify as hyperexcitability or hypoexcitability are hallmarks of ALS [59]. By rescuing motoneurons from injury, DPSCs could also preserve their functional role which further renders them suitable for in vivo applications.

The beneficial effect of the CM on the survival of wildtype motoneurons seems to be specific to DPSCs, as the CM from ASCs or skin fibroblasts did not promote motoneuron survival, as also observed with fibroblasts on trigeminal motoneurons [3]. Concerning ASCs, their secretome contains various growth and neurotrophic factors that were shown to confer neuroprotective benefits when administrated in SOD1 mutant mice [60,61]. This protective effect of ASCs in a mouse model of ALS is corroborated with their ability to rescue motoneurons from the neurotoxicity of astrocytes derived from ALS patients [62].

While DPSCs-CM promotes the survival of both wildtype and *SOD1*^{G93A} motoneuron, our work reveals that DPSCs-CM specifically induces axon outgrowth in wildtype motoneurons but not in *SOD1*^{G93A} motoneurons. This intriguing observation parallels recent work on early axonal transport defects, which are observed well before the onset of symptoms in ALS mice [63]. They show that BDNF is able to stimulate the anterograde transport of signaling endosomes in wildtype embryonic motoneurons, but not those expressing *SOD1*^{G93A}, due to increased expression of the truncated kinase-deficient form of TrkB and p75^{NTR} at the muscle, sciatic nerve, and Schwann cell levels [63]. Here, our study identified SOD1^{G93A}-associated cell-autonomous signaling defects in axonal growth that may involve, as suggested by the previous study [63], differential expression of receptors or their signaling components of the NTFs secreted by DPSCs.

As mentioned above, the DPSC secretome is a complex medium composed of a plethora of soluble factors [40]. Among them, we identified two candidates, GDF15 and HB-EGF, we thought would be of interest, based on the following considerations. Both proteins were shown to confer neuroprotection [50,51,64,65], they have not been studied in the context of ALS, and they were not detected in the secretome of ASCs [66]. We showed that GDF15 or HB-EGF do not increase the survival of wildtype or *SOD1*^{G93A}-expressing motoneurons. While the effect of HB-EGF on motoneuron survival has not been documented to our knowledge, HB-EGF has been reported to confer neuroprotection following ischemic injury [50,67]. This suggests that HB-EGF does not have an effect on naturally occurring developmental death but is more associated with degenerative processes in the adult as shown in an experimental model of Alzheimer's disease [68]. We previously demonstrated that only motoneurons that express ALS-linked mutant forms of SOD1 have increased susceptibility to NO, while the response to excitotoxicity or trophic factor deprivation is comparable to that of the wild type [44]. We found that exogenous HB-EGF can rescue SOD1^{G93A}-expressing motoneurons from NO-induced death, supporting the proposition that HB-EGF is a promising neuroprotective factor for adult-onset neurodegenerative disorders including AD and ALS.

Contrary to our results, GDF15 has been shown to contribute to neuronal survival, including in spinal motoneurons [51]. The different culture conditions, age of embryos (E12.5 vs. E13.5), and the isolation of motoneurons from the lumbar or whole spinal cord [51,69] might explain this discrepancy. However, it is worth noting that *gdf15*-deficient mice display a 19% loss of trigeminal and facial motoneurons, and 21% loss of lumbar motoneurons [51]. The discrepancy between our study performed on motoneurons purified from the whole spinal cord and the one carried out on motoneurons from the lumbar part of the spinal cord [51] suggests that GDF15 may indeed be beneficial for some motoneuron subpopulations, as already observed for hepatocyte growth factor [70].

Interestingly, while GDF15 and HB-EGF do not affect spinal motoneurons under conditions of NTF deprivation, they are able to protect *SOD1*^{G93A} motoneurons from NO-induced death. This suggests a protective effect under pathological conditions. In a model of spinal cord injury, GDF15 was found to inhibit the oxidative-stress-dependent ferroptotic death of neurons, and RNA-interference-mediated silencing of GDF15 reduced the locomotor recovery of mice [71]. In a mouse model of spinal muscular atrophy (SMA), a longitudinal transcriptomic study of vulnerable and resistant motoneuron pools showed that in SMA-resistant ocular motoneurons, *gdf15* was highly upregulated [72]. In addition, GDF15 was able to rescue human motoneurons derived from induced pluripotent stem cells, IPSCs, from degeneration. Jennings et al. recently reported an elevation of GDF15 in the serum of patients with Charcot–Marie–Tooth disease and a mouse model of this disorder, proposing that the induction of GDF15 is an adaptive response to stress to promote peripheral neuron regeneration [73].

To complete our study, we conducted a quantitative transcriptomic analysis of the receptors for HB-EGF, *Egfr* and *ErbB4*, and GDF15, *Gfral*, *Tgfbr1*, and *Tgfbr2* in the lumbar spinal cords of wildtype and *SOD1*^{G93A} mice. Transcripts of both HB-EGF receptors were expressed in the spinal cord. Consistent with a previous study [74], we observed down-regulation of *ErbB4* in the spinal cords of *SOD1*^{G93A} mice. ErbB4 is highly expressed at the membrane of motoneurons and at the time of disease onset, ErbB4 immunoreactivity is reduced in some type of motoneurons, followed by greater loss of ErbB4 in the remaining *SOD1*^{G93A} motoneurons at symptomatic stages [74]. Of note, a decreased ErbB4 immunoreactivity was also observed in motoneurons of patients with sporadic ALS and was associated with cytoplasmic aggregation of TDP-43 [75]. Therefore, strategies aimed to increase ErB4 receptor or related signaling pathways could have neuroprotective effects in ALS, as has been previously shown with the HGF receptor MET [76].

Transcripts of GFRAL, the high-affinity receptor of GDF15, were not detected in the spinal cord, which is consistent with expression exclusively in the hindbrain [77]. However, GDF15 can bind to its low-affinity receptor, TGFβ-R. TGFβ-R signaling has been well-studied as its ligand TGFβ plays significant role under inflammatory conditions, as observed in ALS [78]. Notably, microglial activation is a hallmark of ALS and therefore it was not surprising that TGFβ-R1 expression reveals the presence of microglia specifically in the SOD1^{G93A} spinal cord, which could explain the increase in its transcript expression. Interestingly, we also observed TGFβ-R1 expression in the soma of motoneurons, whose activation could protect them from oxidative stress as we demonstrated *in vitro*. Overall, these results indicate that GDF15 signaling by TGFβ-R1 may be a novel therapeutic strategy for the treatment of ALS at an early stage of the disease.

Our findings have important clinical implications for improving spinal motoneuron survival and in particular under oxidative stress, a condition encountered during the neurodegenerative process of ALS. Moreover, our previous work demonstrated the regenerative properties of DPSCs on sensory neurons [39] which are also affected during ALS progression in humans [79]. Altogether, these results should encourage continued development of therapeutic strategies. However, the effects of these factors on other neuronal structures that are also affected in ALS need also to be addressed.

Our study also has some limitations as it underestimates the properties of DPSCs-CM when considering the non-cell-autonomous components of motoneuron degeneration in ALS. The regenerative effects of DPSCs in ALS mice could indeed also be mediated by other cell types including glial cells, as shown with ASCs [62]. It would therefore be interesting to evaluate the neuroprotective effects of DPSCs-CM, GDF15, and HB-EGF on human motoneurons challenged with astrocytes, oligodendrocytes, or microglia derived from ALS patient IPSCs. Moreover, GDF15 and HB-EGF act on many other cell types which necessitates development of therapeutic strategies that specifically target the spinal cord or the motoneurons.

5. Conclusions

In conclusion, our study reveals the therapeutic potential of DPSC secretome in promoting motoneuron survival and also provides a first step in understanding the key components responsible for this therapeutic effect. The main areas of interest for future studies should include focus on other factors that have never been explored, to rescue motoneurons from death as well to assess new combinations of complementary molecules. We believe that such discoveries will drive the development of new strategies for specifically delivering active and cell-state-dependent cocktails of secretome-derived molecules for the treatment of ALS.

Supplementary Materials: The following supporting information can be downloaded at https://www.mdpi.com/article/10.3390/biomedicines11082152/s1, Table S1 Supplemental Data: statistical values; Table S2 Supplemental Data: statistical analysis.

Author Contributions: Conceptualization, C.H., F.S., C.R. and F.C.; Investigation, R.Y., Y.I., N.J., F.S., C.H., V.B. and D.C.; Formal analysis; C.H., F.S. and R.Y.; Resources, B.C. and F.C.; Supervision, C.H. and F.C.; Validation, C.H. and F.C.; Visualization, R.Y., C.R., C.H. and F.S.; Writing—original draft preparation, C.H., F.S., C.R. and R.Y. All authors have read and agreed to the published version of the manuscript.

Funding: This research was funded by the National Institute of Health and Medical Research (Inserm) and the Association pour la Recherche sur la SLA (ARSLA) (F.C and C.R). R.Y. is a recipient of a Lebanese University and CoEN Montpellier Center of Excellence in Neurodegeneration PhD fellowship. Y.I. is a recipient of an ARSLA PhD fellowship.

Institutional Review Board Statement: The use of animals was approved by the general direction of research and innovation of the Ministry of Higher Education, Research and Innovation, agreement APAFIS#6224-201672711103977.

Informed Consent Statement: Informed consent was obtained from the subject involved in the study.

Data Availability Statement: The data presented in this study are available on request from the corresponding author.

Acknowledgments: We are grateful of Vasiliki Kalatzis for her generous gift of human fibroblasts, Hayat Harati, Elias Estephan, and Youssef Fares for their administrative advice. We are grateful to the personnel of the MRI imaging facility, the RAM-Neuro INM animal core facility, and PCEA.

Conflicts of Interest: The authors declare no conflict of interest.

References

1. Taylor, J.P.; Brown, R.H.; Cleveland, D.W. Decoding ALS: From Genes to Mechanism. *Nature* **2016**, *539*, 197–206. [CrossRef] [PubMed]
2. Robberecht, W.; Philips, T. The Changing Scene of Amyotrophic Lateral Sclerosis. *Nat. Rev. Neurosci.* **2013**, *14*, 248–264. [CrossRef] [PubMed]
3. Gosset, P.; Camu, W.; Raoul, C.; Mezghrani, A. Prionoids in Amyotrophic Lateral Sclerosis. *Brain Commun.* **2022**, *4*, fcac145. [CrossRef] [PubMed]
4. Medinas, D.B.; Valenzuela, V.; Hetz, C. Proteostasis Disturbance in Amyotrophic Lateral Sclerosis. *Hum. Mol. Genet.* **2017**, *26*, R91–R104. [CrossRef]
5. Crabé, R.; Aimond, F.; Gosset, P.; Scamps, F.; Raoul, C. How Degeneration of Cells Surrounding Motoneurons Contributes to Amyotrophic Lateral Sclerosis. *Cells* **2020**, *9*, 2550. [CrossRef]
6. López-Pingarrón, L.; Almeida, H.; Soria-Aznar, M.; Reyes-Gonzales, M.C.; Terrón, M.P.; García, J.J. Role of Oxidative Stress on the Etiology and Pathophysiology of Amyotrophic Lateral Sclerosis (ALS) and Its Relation with the Enteric Nervous System. *CIMB* **2023**, *45*, 3315–3332. [CrossRef]
7. Rothstein, J.D. Edaravone: A New Drug Approved for ALS. *Cell* **2017**, *171*, 725. [CrossRef]
8. Finegan, E.; Hi Shing, S.L.; Chipika, R.H.; McKenna, M.C.; Doherty, M.A.; Hengeveld, J.C.; Vajda, A.; Donaghy, C.; McLaughlin, R.L.; Hutchinson, S.; et al. Thalamic, Hippocampal and Basal Ganglia Pathology in Primary Lateral Sclerosis and Amyotrophic Lateral Sclerosis: Evidence from Quantitative Imaging Data. *Data Brief* **2020**, *29*, 105115. [CrossRef]
9. Chipika, R.H.; Mulkerrin, G.; Pradat, P.-F.; Murad, A.; Ango, F.; Raoul, C.; Bede, P. Cerebellar Pathology in Motor Neuron Disease: Neuroplasticity and Neurodegeneration. *Neural Regen. Res.* **2022**, *17*, 2335–2341. [CrossRef]
10. Chipika, R.H.; Finegan, E.; Li Hi Shing, S.; McKenna, M.C.; Christidi, F.; Chang, K.M.; Doherty, M.A.; Hengeveld, J.C.; Vajda, A.; Pender, N.; et al. "Switchboard" Malfunction in Motor Neuron Diseases: Selective Pathology of Thalamic Nuclei in Amyotrophic Lateral Sclerosis and Primary Lateral Sclerosis. *Neuroi. Clin.* **2020**, *27*, 102300. [CrossRef]
11. Debye, B.; Schmülling, L.; Zhou, L.; Rune, G.; Beyer, C.; Johann, S. Neurodegeneration and NLRP3 Inflammasome Expression in the Anterior Thalamus of SOD1(G93A) ALS Mice. *Brain Pathol.* **2018**, *28*, 14–27. [CrossRef]
12. Fisher, E.M.C.; Greensmith, L.; Malaspina, A.; Fratta, P.; Hanna, M.G.; Schiavo, G.; Isaacs, A.M.; Orrell, R.W.; Cunningham, T.J.; Arozena, A.A. Opinion: More Mouse Models and More Translation Needed for ALS. *Mol. Neurodegener.* **2023**, *18*, 30. [CrossRef]
13. Bonifacino, T.; Zerbo, R.A.; Balbi, M.; Torazza, C.; Frumento, G.; Fedele, E.; Bonanno, G.; Milanese, M. Nearly 30 Years of Animal Models to Study Amyotrophic Lateral Sclerosis: A Historical Overview and Future Perspectives. *Int. J. Mol. Sci.* **2021**, *22*, 12236. [CrossRef]
14. Bruijn, L.I.; Miller, T.M.; Cleveland, D.W. Unraveling the Mechanisms Involved in Motor Neuron Degeneration in ALS. *Annu. Rev. Neurosci.* **2004**, *27*, 723–749. [CrossRef]
15. Lazarevic, V.; Yang, Y.; Ivanova, D.; Fejtova, A.; Svenningsson, P. Riluzole Attenuates the Efficacy of Glutamatergic Transmission by Interfering with the Size of the Readily Releasable Neurotransmitter Pool. *Neuropharmacology* **2018**, *143*, 38–48. [CrossRef]
16. Abe, S.; Kirima, K.; Tsuchiya, K.; Okamoto, M.; Hasegawa, T.; Houchi, H.; Yoshizumi, M.; Tamaki, T. The Reaction Rate of Edaravone (3-Methyl-1-Phenyl-2-Pyrazolin-5-One (MCI-186)) with Hydroxyl Radical. *Chem. Pharm. Bull. (Tokyo)* **2004**, *52*, 186–191. [CrossRef]
17. Fujisawa, A.; Yamamoto, Y. Edaravone, a Potent Free Radical Scavenger, Reacts with Peroxynitrite to Produce Predominantly 4-NO-Edaravone. *Redox Rep.* **2016**, *21*, 98–103. [CrossRef]
18. Takei, K.; Watanabe, K.; Yuki, S.; Akimoto, M.; Sakata, T.; Palumbo, J. Edaravone and Its Clinical Development for Amyotrophic Lateral Sclerosis. *Amyotroph. Lateral Scler. Front. Degener.* **2017**, *18*, 5–10. [CrossRef]

19. Miller, T.M.; Cudkowicz, M.E.; Genge, A.; Shaw, P.J.; Sobue, G.; Bucelli, R.C.; Chiò, A.; Van Damme, P.; Ludolph, A.C.; Glass, J.D.; et al. Trial of Antisense Oligonucleotide Tofersen for SOD1 ALS. *N. Engl. J. Med.* **2022**, *387*, 1099–1110. [CrossRef]
20. Kanning, K.C.; Kaplan, A.; Henderson, C.E. Motor Neuron Diversity in Development and Disease. *Annu. Rev. Neurosci.* **2010**, *33*, 409–440. [CrossRef]
21. Tovar-y-Romo, L.B.; Ramírez-Jarquín, U.N.; Lazo-Gómez, R.; Tapia, R. Trophic Factors as Modulators of Motor Neuron Physiology and Survival: Implications for ALS Therapy. *Front. Cell. Neurosci.* **2014**, *8*, 61. [CrossRef] [PubMed]
22. Bartus, R.T.; Johnson, E.M. Clinical Tests of Neurotrophic Factors for Human Neurodegenerative Diseases, Part 1: Where Have We Been and What Have We Learned? *Neurobiol. Dis.* **2017**, *97*, 156–168. [CrossRef] [PubMed]
23. Cappella, M.; Pradat, P.-F.; Querin, G.; Biferi, M.G. Beyond the Traditional Clinical Trials for Amyotrophic Lateral Sclerosis and The Future Impact of Gene Therapy. *J. Neuromuscul. Dis.* **2021**, *8*, 25–38. [CrossRef] [PubMed]
24. Nosrat, I.V.; Widenfalk, J.; Olson, L.; Nosrat, C.A. Dental Pulp Cells Produce Neurotrophic Factors, Interact with Trigeminal Neurons in Vitro, and Rescue Motoneurons after Spinal Cord Injury. *Dev. Biol.* **2001**, *238*, 120–132. [CrossRef]
25. Nosrat, I.V.; Smith, C.A.; Mullally, P.; Olson, L.; Nosrat, C.A. Dental Pulp Cells Provide Neurotrophic Support for Dopaminergic Neurons and Differentiate into Neurons in Vitro; Implications for Tissue Engineering and Repair in the Nervous System. *Eur. J. Neurosci.* **2004**, *19*, 2388–2398. [CrossRef] [PubMed]
26. Kolar, M.K.; Itte, V.N.; Kingham, P.J.; Novikov, L.N.; Wiberg, M.; Kelk, P. The Neurotrophic Effects of Different Human Dental Mesenchymal Stem Cells. *Sci. Rep.* **2017**, *7*, 12605. [CrossRef]
27. Simon, C.; Gan, Q.F.; Kathivaloo, P.; Mohamad, N.A.; Dhamodharan, J.; Krishnan, A.; Sengodan, B.; Palanimuthu, V.R.; Marimuthu, K.; Rajandas, H.; et al. Deciduous DPSCs Ameliorate MPTP-Mediated Neurotoxicity, Sensorimotor Coordination and Olfactory Function in Parkinsonian Mice. *Int. J. Mol. Sci.* **2019**, *20*, 568. [CrossRef]
28. Ueda, T.; Inden, M.; Ito, T.; Kurita, H.; Hozumi, I. Characteristics and Therapeutic Potential of Dental Pulp Stem Cells on Neurodegenerative Diseases. *Front. Neurosci.* **2020**, *14*, 407. [CrossRef]
29. Wang, J.; Zuzzio, K.; Walker, C.L. Systemic Dental Pulp Stem Cell Secretome Therapy in a Mouse Model of Amyotrophic Lateral Sclerosis. *Brain Sci.* **2019**, *9*, 165. [CrossRef]
30. Zhang, X.-M.; Ouyang, Y.-J.; Yu, B.-Q.; Li, W.; Yu, M.-Y.; Li, J.-Y.; Jiao, Z.-M.; Yang, D.; Li, N.; Shi, Y.; et al. Therapeutic Potential of Dental Pulp Stem Cell Transplantation in a Rat Model of Alzheimer's Disease. *Neural Regen. Res.* **2021**, *16*, 893–898. [CrossRef]
31. Sandonà, M.; Di Pietro, L.; Esposito, F.; Ventura, A.; Silini, A.R.; Parolini, O.; Saccone, V. Mesenchymal Stromal Cells and Their Secretome: New Therapeutic Perspectives for Skeletal Muscle Regeneration. *Front. Bioeng. Biotechnol.* **2021**, *9*, 652970. [CrossRef]
32. Frisbie, L.; Buckanovich, R.J.; Coffman, L. Carcinoma-Associated Mesenchymal Stem/Stromal Cells: Architects of the Pro-Tumorigenic Tumor Microenvironment. *Stem Cells* **2022**, *40*, 705–715. [CrossRef]
33. Luo, L.; Albashari, A.A.; Wang, X.; Jin, L.; Zhang, Y.; Zheng, L.; Xia, J.; Xu, H.; Zhao, Y.; Xiao, J.; et al. Effects of Transplanted Heparin-Poloxamer Hydrogel Combining Dental Pulp Stem Cells and BFGF on Spinal Cord Injury Repair. *Stem Cells Int.* **2018**, *2018*, 2398521. [CrossRef]
34. Király, M.; Kádár, K.; Horváthy, D.B.; Nardai, P.; Rácz, G.Z.; Lacza, Z.; Varga, G.; Gerber, G. Integration of Neuronally Predifferentiated Human Dental Pulp Stem Cells into Rat Brain In Vivo. *Neurochem. Int.* **2011**, *59*, 371–381. [CrossRef]
35. Winderlich, J.N.; Kremer, K.L.; Koblar, S.A. Adult Human Dental Pulp Stem Cells Promote Blood-Brain Barrier Permeability through Vascular Endothelial Growth Factor-a Expression. *J. Cereb. Blood Flow Metab.* **2016**, *36*, 1087–1097. [CrossRef]
36. Nito, C.; Sowa, K.; Nakajima, M.; Sakamoto, Y.; Suda, S.; Nishiyama, Y.; Nakamura-Takahashi, A.; Nitahara-Kasahara, Y.; Ueda, M.; Okada, T.; et al. Transplantation of Human Dental Pulp Stem Cells Ameliorates Brain Damage Following Acute Cerebral Ischemia. *Biomed. Pharmacother.* **2018**, *108*, 1005–1014. [CrossRef] [PubMed]
37. Chen, T.-F.; Chen, K.-W.; Chien, Y.; Lai, Y.-H.; Hsieh, S.-T.; Ma, H.-Y.; Wang, K.-C.; Shiau, C.-Y. Dental Pulp Stem Cell-Derived Factors Alleviate Subarachnoid Hemorrhage-Induced Neuroinflammation and Ischemic Neurological Deficits. *Int. J. Mol. Sci.* **2019**, *20*, 3747. [CrossRef] [PubMed]
38. Bray, A.F.; Cevallos, R.R.; Gazarian, K.; Lamas, M. Human Dental Pulp Stem Cells Respond to Cues from the Rat Retina and Differentiate to Express the Retinal Neuronal Marker Rhodopsin. *Neuroscience* **2014**, *280*, 142–155. [CrossRef] [PubMed]
39. Chouaib, B.; Collart-Dutilleul, P.-Y.; Blanc-Sylvestre, N.; Younes, R.; Gergely, C.; Raoul, C.; Scamps, F.; Cuisinier, F.; Romieu, O. Identification of Secreted Factors in Dental Pulp Cell-Conditioned Medium Optimized for Neuronal Growth. *Neurochem. Int.* **2021**, *144*, 104961. [CrossRef]
40. Chouaib, B.; Cuisinier, F.; Collart-Dutilleul, P.-Y. Dental Stem Cell-Conditioned Medium for Tissue Regeneration: Optimization of Production and Storage. *World J. Stem Cells* **2022**, *14*, 287–302. [CrossRef]
41. Collart-Dutilleul, P.-Y.; Secret, E.; Panayotov, I.; Deville De Périère, D.; Martín-Palma, R.J.; Torres-Costa, V.; Martin, M.; Gergely, C.; Durand, J.-O.; Cunin, F.; et al. Adhesion and Proliferation of Human Mesenchymal Stem Cells from Dental Pulp on Porous Silicon Scaffolds. *ACS Appl. Mater. Interfaces* **2014**, *6*, 1719–1728. [CrossRef]
42. Panayotov, I.V.; Collart-Dutilleul, P.-Y.; Salehi, H.; Martin, M.; Végh, A.; Yachouh, J.; Vladimirov, B.; Sipos, P.; Szalontai, B.; Gergely, C.; et al. Sprayed Cells and Polyelectrolyte Films for Biomaterial Functionalization: The Influence of Physical PLL-PGA Film Treatments on Dental Pulp Cell Behavior. *Macromol. Biosci.* **2014**, *14*, 1771–1782. [CrossRef]
43. Soulard, C.; Salsac, C.; Mouzat, K.; Hilaire, C.; Roussel, J.; Mezghrani, A.; Lumbroso, S.; Raoul, C.; Scamps, F. Spinal Motoneuron TMEM16F Acts at C-Boutons to Modulate Motor Resistance and Contributes to ALS Pathogenesis. *Cell Rep.* **2020**, *30*, 2581–2593.e7. [CrossRef]

44. Raoul, C.; Estévez, A.G.; Nishimune, H.; Cleveland, D.W.; deLapeyrière, O.; Henderson, C.E.; Haase, G.; Pettmann, B. Motoneuron Death Triggered by a Specific Pathway Downstream of Fas. Potentiation by ALS-Linked SOD1 Mutations. *Neuron* **2002**, *35*, 1067–1083. [CrossRef]
45. Benlefki, S.; Sanchez-Vicente, A.; Milla, V.; Lucas, O.; Soulard, C.; Younes, R.; Gergely, C.; Bowerman, M.; Raoul, C.; Scamps, F.; et al. Expression of ALS-Linked SOD1 Mutation in Motoneurons or Myotubes Induces Differential Effects on Neuromuscular Function In Vitro. *Neuroscience* **2020**, *435*, 33–43. [CrossRef]
46. Benlefki, S.; Younes, R.; Challuau, D.; Bernard-Marissal, N.; Hilaire, C.; Scamps, F.; Bowerman, M.; Kothary, R.; Schneider, B.L.; Raoul, C. Differential Effect of Fas Activation on Spinal Muscular Atrophy Motoneuron Death and Induction of Axonal Growth. *Cell. Mol. Biol.* 2023, *in press*.
47. Otsmane, B.; Moumen, A.; Aebischer, J.; Coque, E.; Sar, C.; Sunyach, C.; Salsac, C.; Valmier, J.; Salinas, S.; Bowerman, M.; et al. Somatic and Axonal LIGHT Signaling Elicit Degenerative and Regenerative Responses in Motoneurons, Respectively. *EMBO Rep.* **2014**, *15*, 540–547. [CrossRef]
48. Bowerman, M.; Salsac, C.; Bernard, V.; Soulard, C.; Dionne, A.; Coque, E.; Benlefki, S.; Hince, P.; Dion, P.A.; Butler-Browne, G.; et al. KCC3 Loss-of-Function Contributes to Andermann Syndrome by Inducing Activity-Dependent Neuromuscular Junction Defects. *Neurobiol. Dis.* **2017**, *106*, 35–48. [CrossRef]
49. Gancheva, M.R.; Kremer, K.L.; Gronthos, S.; Koblar, S.A. Using Dental Pulp Stem Cells for Stroke Therapy. *Front. Neurol.* **2019**, *10*, 422. [CrossRef]
50. Zhou, Y.; Besner, G.E. Heparin-Binding Epidermal Growth Factor-like Growth Factor Is a Potent Neurotrophic Factor for PC12 Cells. *Neurosignals* **2010**, *18*, 141–151. [CrossRef]
51. Strelau, J.; Strzelczyk, A.; Rusu, P.; Bendner, G.; Wiese, S.; Diella, F.; Altick, A.L.; von Bartheld, C.S.; Klein, R.; Sendtner, M.; et al. Progressive Postnatal Motoneuron Loss in Mice Lacking GDF-15. *J. Neurosci.* **2009**, *29*, 13640–13648. [CrossRef]
52. Saxena, S.; Cabuy, E.; Caroni, P. A Role for Motoneuron Subtype-Selective ER Stress in Disease Manifestations of FALS Mice. *Nat. Neurosci.* **2009**, *12*, 627–636. [CrossRef] [PubMed]
53. Butovsky, O.; Jedrychowski, M.P.; Moore, C.S.; Cialic, R.; Lanser, A.J.; Gabriely, G.; Koeglsperger, T.; Dake, B.; Wu, P.M.; Doykan, C.E.; et al. Identification of a Unique TGF-β-Dependent Molecular and Functional Signature in Microglia. *Nat. Neurosci.* **2014**, *17*, 131–143. [CrossRef] [PubMed]
54. Spittau, B.; Dokalis, N.; Prinz, M. The Role of TGFβ Signaling in Microglia Maturation and Activation. *Trends Immunol.* **2020**, *41*, 836–848. [CrossRef]
55. Gugliandolo, A.; Mazzon, E. Dental Mesenchymal Stem Cell Secretome: An Intriguing Approach for Neuroprotection and Neuroregeneration. *Int. J. Mol. Sci.* **2021**, *23*, 456. [CrossRef] [PubMed]
56. Mead, B.; Logan, A.; Berry, M.; Leadbeater, W.; Scheven, B.A. Dental Pulp Stem Cells, a Paracrine-Mediated Therapy for the Retina. *Neural Regen. Res.* **2014**, *9*, 577–578. [CrossRef]
57. Keefe, K.M.; Sheikh, I.S.; Smith, G.M. Targeting Neurotrophins to Specific Populations of Neurons: NGF, BDNF, and NT-3 and Their Relevance for Treatment of Spinal Cord Injury. *Int. J. Mol. Sci.* **2017**, *18*, 548. [CrossRef]
58. Calvo, P.M.; Hernández, R.G.; de la Cruz, R.R.; Pastor, A.M. Role of Vascular Endothelial Growth Factor as a Critical Neurotrophic Factor for the Survival and Physiology of Motoneurons. *Neural Regen. Res.* **2023**, *18*, 1691–1696. [CrossRef]
59. Scamps, F.; Aimond, F.; Hilaire, C.; Raoul, C. Synaptic Transmission and Motoneuron Excitability Defects in Amyotrophic Lateral Sclerosis. In *Amyotrophic Lateral Sclerosis*; Araki, T., Ed.; Exon Publications: Brisbane, Australia, 2021; ISBN 978-0-645-00177-8.
60. Fontanilla, C.V.; Gu, H.; Liu, Q.; Zhu, T.Z.; Zhou, C.; Johnstone, B.H.; March, K.L.; Pascuzzi, R.M.; Farlow, M.R.; Du, Y. Adipose-Derived Stem Cell Conditioned Media Extends Survival Time of a Mouse Model of Amyotrophic Lateral Sclerosis. *Sci. Rep.* **2015**, *5*, 16953. [CrossRef]
61. Walker, C.L.; Meadows, R.M.; Merfeld-Clauss, S.; Du, Y.; March, K.L.; Jones, K.J. Adipose-Derived Stem Cell Conditioned Medium Impacts Asymptomatic Peripheral Neuromuscular Denervation in the Mutant Superoxide Dismutase (G93A) Transgenic Mouse Model of Amyotrophic Lateral Sclerosis. *Restor. Neurol. Neurosci.* **2018**, *36*, 621–627. [CrossRef]
62. Ciervo, Y.; Gatto, N.; Allen, C.; Grierson, A.; Ferraiuolo, L.; Mead, R.J.; Shaw, P.J. Adipose-Derived Stem Cells Protect Motor Neurons and Reduce Glial Activation in Both in Vitro and in Vivo Models of ALS. *Mol. Ther. Methods Clin. Dev.* **2021**, *21*, 413–433. [CrossRef]
63. Tosolini, A.P.; Sleigh, J.N.; Surana, S.; Rhymes, E.R.; Cahalan, S.D.; Schiavo, G. BDNF-Dependent Modulation of Axonal Transport Is Selectively Impaired in ALS. *Acta Neuropathol. Commun.* **2022**, *10*, 121. [CrossRef]
64. Higashiyama, S.; Abraham, J.A.; Miller, J.; Fiddes, J.C.; Klagsbrun, M. A Heparin-Binding Growth Factor Secreted by Macrophage-like Cells That Is Related to EGF. *Science* **1991**, *251*, 936–939. [CrossRef]
65. Nakagawa, T.; Sasahara, M.; Hayase, Y.; Haneda, M.; Yasuda, H.; Kikkawa, R.; Higashiyama, S.; Hazama, F. Neuronal and Glial Expression of Heparin-Binding EGF-like Growth Factor in Central Nervous System of Prenatal and Early-Postnatal Rat. *Brain Res. Dev. Brain Res.* **1998**, *108*, 263–272. [CrossRef]
66. Chouaib, B. Dental Pulp Stem Cell-Conditioned Medium for Tissue Regeneration. Ph.D. Thesis, Montpellier University, Montpellier, France, 2020.
67. Opanashuk, L.A.; Mark, R.J.; Porter, J.; Damm, D.; Mattson, M.P.; Seroogy, K.B. Heparin-Binding Epidermal Growth Factor-like Growth Factor in Hippocampus: Modulation of Expression by Seizures and Anti-Excitotoxic Action. *J. Neurosci.* **1999**, *19*, 133–146. [CrossRef]

68. Maurya, S.K.; Mishra, J.; Abbas, S.; Bandyopadhyay, S. Cypermethrin Stimulates GSK3β-Dependent Aβ and p-Tau Proteins and Cognitive Loss in Young Rats: Reduced HB-EGF Signaling and Downstream Neuroinflammation as Critical Regulators. *Mol. Neurobiol.* **2016**, *53*, 968–982. [CrossRef]
69. Wiese, S.; Pei, G.; Karch, C.; Troppmair, J.; Holtmann, B.; Rapp, U.R.; Sendtner, M. Specific Function of B-Raf in Mediating Survival of Embryonic Motoneurons and Sensory Neurons. *Nat. Neurosci.* **2001**, *4*, 137–142. [CrossRef]
70. Lamballe, F.; Genestine, M.; Caruso, N.; Arce, V.; Richelme, S.; Helmbacher, F.; Maina, F. Pool-Specific Regulation of Motor Neuron Survival by Neurotrophic Support. *J. Neurosci.* **2011**, *31*, 11144–11158. [CrossRef]
71. Xia, M.; Zhang, Q.; Zhang, Y.; Li, R.; Zhao, T.; Chen, L.; Liu, Q.; Zheng, S.; Li, H.; Qian, Z.; et al. Growth Differentiation Factor 15 Regulates Oxidative Stress-Dependent Ferroptosis Post Spinal Cord Injury by Stabilizing the P62-Keap1-Nrf2 Signaling Pathway. *Front. Aging Neurosci.* **2022**, *14*, 905115. [CrossRef]
72. Nichterwitz, S.; Nijssen, J.; Storvall, H.; Schweingruber, C.; Comley, L.H.; Allodi, I.; van der Lee, M.; Deng, Q.; Sandberg, R.; Hedlund, E. LCM-Seq Reveals Unique Transcriptional Adaptation Mechanisms of Resistant Neurons and Identifies Protective Pathways in Spinal Muscular Atrophy. *Genome Res.* **2020**, *30*, 1083–1096. [CrossRef]
73. Jennings, M.J.; Kagiava, A.; Vendredy, L.; Spaulding, E.L.; Stavrou, M.; Hathazi, D.; Grüneboom, A.; De Winter, V.; Gess, B.; Schara, U.; et al. NCAM1 and GDF15 Are Biomarkers of Charcot-Marie-Tooth Disease in Patients and Mice. *Brain* **2022**, *145*, 3999–4015. [CrossRef] [PubMed]
74. Lasiene, J.; Komine, O.; Fujimori-Tonou, N.; Powers, B.; Endo, F.; Watanabe, S.; Shijie, J.; Ravits, J.; Horner, P.; Misawa, H.; et al. Neuregulin 1 Confers Neuroprotection in SOD1-Linked Amyotrophic Lateral Sclerosis Mice via Restoration of C-Boutons of Spinal Motor Neurons. *Acta Neuropathol. Commun.* **2016**, *4*, 15. [CrossRef] [PubMed]
75. Takahashi, Y.; Uchino, A.; Shioya, A.; Sano, T.; Matsumoto, C.; Numata-Uematsu, Y.; Nagano, S.; Araki, T.; Murayama, S.; Saito, Y. Altered Immunoreactivity of ErbB4, a Causative Gene Product for ALS19, in the Spinal Cord of Patients with Sporadic ALS. *Neuropathology* **2019**, *39*, 268–278. [CrossRef] [PubMed]
76. Genestine, M.; Caricati, E.; Fico, A.; Richelme, S.; Hassani, H.; Sunyach, C.; Lamballe, F.; Panzica, G.C.; Pettmann, B.; Helmbacher, F.; et al. Enhanced Neuronal Met Signalling Levels in ALS Mice Delay Disease Onset. *Cell Death Dis.* **2011**, *2*, e130. [CrossRef]
77. Rochette, L.; Zeller, M.; Cottin, Y.; Vergely, C. Insights Into Mechanisms of GDF15 and Receptor GFRAL: Therapeutic Targets. *Trends Endocrinol. Metab.* **2020**, *31*, 939–951. [CrossRef]
78. Galbiati, M.; Crippa, V.; Rusmini, P.; Cristofani, R.; Messi, E.; Piccolella, M.; Tedesco, B.; Ferrari, V.; Casarotto, E.; Chierichetti, M.; et al. Multiple Roles of Transforming Growth Factor Beta in Amyotrophic Lateral Sclerosis. *Int. J. Mol. Sci.* **2020**, *21*, 4291. [CrossRef]
79. Iglesias, C.; Sangari, S.; El Mendili, M.-M.; Benali, H.; Marchand-Pauvert, V.; Pradat, P.-F. Electrophysiological and Spinal Imaging Evidences for Sensory Dysfunction in Amyotrophic Lateral Sclerosis. *BMJ Open* **2015**, *5*, e007659. [CrossRef]

biomedicines

Review

Rehabilitation: Neurogenic Bone Loss after Spinal Cord Injury

Giovanna E. Leone [1,2], Donald C. Shields [2], Azizul Haque [1,2,3,*] and Narendra L. Banik [1,2,3,*]

1 Department of Microbiology and Immunology, Medical University of South Carolina, Charleston, SC 29425, USA; gleone@email.sc.edu
2 Department of Neurosurgery, Medical University of South Carolina, Charleston, SC 29425, USA; donshields@sbcglobal.net
3 Ralph H. Johnson Veterans Administration Medical Center, Charleston, SC 29401, USA
* Correspondence: haque@musc.edu (A.H.); baniknl@musc.edu (N.L.B.)

Abstract: Osteoporosis is a common skeletal disorder which can severely limit one's ability to complete daily tasks due to the increased risk of bone fractures, reducing quality of life. Spinal cord injury (SCI) can also result in osteoporosis and sarcopenia. Most individuals experience sarcopenia and osteoporosis due to advancing age; however, individuals with SCI experience more rapid and debilitating levels of muscle and bone loss due to neurogenic factors, musculoskeletal disuse, and cellular/molecular events. Thus, preserving and maintaining bone mass after SCI is crucial to decreasing the risk of fragility and fracture in vulnerable SCI populations. Recent studies have provided an improved understanding of the pathophysiology and risk factors related to musculoskeletal loss after SCI. Pharmacological and non-pharmacological therapies have also provided for the reduction in or elimination of neurogenic bone loss after SCI. This review article will discuss the pathophysiology and risk factors of muscle and bone loss after SCI, including the mechanisms that may lead to muscle and bone loss after SCI. This review will also focus on current and future pharmacological and non-pharmacological therapies for reducing or eliminating neurogenic bone loss following SCI.

Keywords: neurodegeneration; osteopenia; osteoporosis; sarcopenia; spinal cord injury

Citation: Leone, G.E.; Shields, D.C.; Haque, A.; Banik, N.L. Rehabilitation: Neurogenic Bone Loss after Spinal Cord Injury. *Biomedicines* **2023**, *11*, 2581. https://doi.org/10.3390/biomedicines11092581

Academic Editor: Marco Segatto

Received: 26 July 2023
Revised: 12 September 2023
Accepted: 18 September 2023
Published: 20 September 2023

1. Introduction

Spinal cord injury (SCI) is a severe neurological disorder that results from sudden and damaging impact to the spine and vertebrae [1,2]. SCI is one of the most commonly caused damages in vehicle injuries [3], but can also be caused by falls, athletic injuries, and various other reasons [4]. SCI impacts more than 10,000 individuals each year and poses a significant economic burden to the U.S [5]. SCI can be detrimental and life threatening, and while there are therapeutic modalities being studied, more research on how to mitigate the short- and long-term effects of SCI is still needed. The immediate impacts of SCI can vary and depend largely on the specific location and magnitude of the injury [1,6]. In general, the higher up the level of injury is to the spinal cord, the more severe the symptoms. Injuries to the spinal cord of any magnitude and location can have both localized and global effects on bone composition. The local effects include paralysis, reduced function in the lower body, and bone loss, most commonly in the femurs, tibias, fibulas, and pelvic bones. The global effects of SCI (i.e., neurogenic bone loss) include changes in neural signaling over time, which can lead to a disruption in bone remodeling throughout the body, not just in regions directly impacted by the SCI. The global effects of SCI may also include disruptions to bone vascularity, as there is a synergistic relationship between the skeletal and vascular systems. A decrease in bone vascularity and reduced neoangiogenesis can limit the healing capacity and progress of SCI rehabilitation modalities, and thus limit bone remodeling and repair [7]. People with a SCI are two to five times more likely to die prematurely than people without an SCI, and this carries substantial individual and societal costs. Short-term impacts often include gliosis, axonal damage, neuronal death, immobilization, and a loss of sensory and motor function, while long-term impacts include organ dysfunction, sarcopenia, osteopenia, bone fractures, and osteoporosis [1,4,8].

Demyelination and axonal degeneration are short-term but chronic outcomes of SCI, because they last for prolonged periods of time after the injury and are often irreversible [2,6]. Axonal degeneration occurs when the axons are lesioned, causing severe neuronal transmission deficits distal to the lesion site. This damage is furthered if the axon is lesioned in the central nervous system (CNS). Although there are potential therapeutic approaches to slowing axonal degeneration, this damage is usually permanent if the axonal lesion site is in the CNS [1,4]. Demyelination and a buildup of myelin debris are other immediate outcomes of SCI, which then lead to excessive levels of gliosis and glial scar formation [9,10]. These are just some of the immediate, short-term effects of SCI that come along with a multitude of long-term effects.

Many of the long-term outcomes of SCI are related to muscle and bone loss due to immobilization. Due to lack of physical activity and increased immobilization after one suffers from severe SCI, muscle and bone tissue severely decrease [11,12]. Osteoporosis is a common issue experienced after SCI and is defined as a skeletal disorder in which bone strength is compromised, leaving a person with a greater risk of fracture [13,14]. Individuals with osteoporosis experience large levels of osteopenia and are prone to fractures, which severely decrease quality of life and require substantial medical resources. Due to osteopenia after SCI, bone fractures are extremely common in individuals with SCI, because of their lower osteogenic load and increased bone demineralization [15–17]. The absolute causes of bone loss after SCI are not yet known; however, some of the possible causes are neurogenic factors, hormonal factors, and sarcopenia [15,18]. Immobility and disuse are other causes of osteopenia and sarcopenia in SCI patients due to the decrease in mechanical loading in the bone while one recovers from SCI. Sarcopenia, also known as muscle loss, has been linked to being a possible cause of osteopenia; however, more research is needed to evaluate the relationship between osteopenia and sarcopenia in SCI [11,19]. Diagnosis, prevention, and treatment for decreasing osteopenia and osteoporosis after SCI are critical to helping the thousands of individuals who suffer from SCI each year [15,17].

Therapies for reducing the negative outcomes of SCI are urgently needed. Although there has been promising research on therapies such as blocking 4-1BB and RANKL signaling [20,21], increasing Wnt signaling and calcium-regulated hormones [22,23], and loading of the bones and muscles [24], further research is still needed and there is research being conducted now on prospects for SCI treatments. The purpose of this review article is to discuss the pathophysiology of osteoporosis and determine the known treatments for bone loss and osteoporosis after SCI to reveal where more research needs to be conducted, as well as to cover the promising treatment options that are currently being studied.

2. Pathophysiology of Bone Loss after SCI

Individuals with complete paralysis after SCI show the most extensive bone loss and fracture risk [25,26]. Understanding the mechanisms that lead to bone loss and osteoporosis after SCI is important to determining how to slow bone loss after SCI. Common causes of bone loss after severe SCI are immobility and de-loading, which result in increased bone resorption and a decrease in osteoblast activity [4,27]. When one is immobile due to an injury, less stress is placed on the bones, leading to a direct response from other systems in the body, including the neurogenic and musculoskeletal systems [27]. Immobility has a direct effect on the musculoskeletal system, since it causes an increase in bone resorption and a decrease in osteoblast activity, resulting in osteopenia [4,27]. However, bone loss following SCI is believed to be distinct, as compared to the response to other disuse conditions in terms of both severity and mechanism. Although our focus is SCI, other factors secondary to SCI may also promote bone loss, including systemic hormonal changes, altered bone innervation, and impaired bone perfusion [26,28]. In an SCI study conducted on rats, significant bone loss was observed during a bone compartment analysis on the SCI animals compared to controls [11,29]. Overall, decreases in bone mineral content, trabecular structure, and bone mineral density were observed in all the SCI groups.

The next systems that immobilization and bone loss impact are the CNS, peripheral nervous system (PNS), and endocrine system. Bone cells have many nerve endings close to them, which greatly impact the CNS and PNS. Bone cells also connect the skeleton to the endocrine system through various receptors and neuromediators [27]. Skeletal loss may also promote sarcopenia and endocrine system dysfunction via multiple receptors and neuromediators, thus influencing the adipose tissue production of leptin and anorexigenics, which both affect bone remodeling [27,30]. Moreover, immobilization impacts skeletal vascularization, which is required for bone remodeling and osteoblast function. The resulting vasoconstriction further contributes to the muscular, endocrine, and nervous system impairments associated with osteoporosis in SCI patients.

The vascular system is a necessary contributor to osteogenesis after SCI. Neo-angiogenesis (i.e., the formation of new blood vessels) plays a crucial role in bone development after SCI, because it ensures that bone tissues are obtaining the necessary blood and oxygen supply to stimulate bone formation, maintenance, and repair [7]. Following SCI, individuals often experience disruptions to the circulatory system from mechanical trauma. Ischemia, hypoxia, and localized edema are potential secondary effects of SCI impacting the vascular system, thus impeding healing and rehabilitation [31]. The secondary effects of SCI on the vascular network not only potentially cause secondary injury and can further deteriorate bone and spinal cord tissue, but a reduced vascularity can also mitigate healing from SCI treatment [7,31]. Various SCI treatments, including cell transplantation, are ineffective if the local blood vessels are damaged, leading to a lack of oxygen and nutrients that the transplanted cells need for survival [31]. Pericytes and endothelial cells are important structures of the vascular system that play essential roles in angiogenesis; however, they cannot sustain and mediate angiogenesis to osteogenesis when there is damage to the blood vessels in the affected area [7,31]. Physical rehabilitation and therapeutic strategies, such as surgical anastomosis and exogenous pericyte cell transplantation, are available to help to stimulate angiogenesis after SCI [7]. Research is still limited on the effectiveness of therapy and rehabilitation for stimulating angiogenesis after SCI.

3. Disuse and Bone Loss after SCI

The disuse of physical activity and loading is a main cause of osteopenia, which can cause localized bone loss and bone fractures, which are most commonly fractures of the distal femur and proximal tibia [32]. Bone loss after disuse is caused mainly by skeletal and mechanical unloading, meaning there is no pressure put on the skeleton, so it gradually and continually weakens [33]. The loss of bone appears to primarily be a consequence of decreased osteoblastic activity and number, although an increase in osteoclastic activity cannot be excluded (Figure 1). In studies performed on animals, de-loading has been found to be a direct cause of osteoblast activity and bone resorption [34]. After SCI, there are also multiple factors that can contribute to a decrease in mechanical loading on the skeletal tissue. Physical exertion stimulates osteoblast activity, which increases bone tissue via the mineralization of the skeleton. A lack of physical activity and skeletal loading (in many SCI patients due to paresis) is related to osteopenia and resulting fractures, most commonly of the distal femur and proximal tibia [32–34].

A lack of physical activity can also cause a decrease in the body mass (musculature and adipose tissues) load on the musculoskeletal system, thus creating less stimulation for osteoblast activity. Abdelrahman et al. examined the changes in total bone mineral content (BMC) and bone mineral density (BMD) in ten concentric sectors at the 4% site using tomography scans. They also analyzed the regional changes in BMC and cortical BMD in thirty-six polar sectors at the 66% site using linear mixed-effects models. They showed that the total BMC (p = 0.001) significantly decreased with time at the 4% site. Interestingly, the absolute losses of BMC and cortical BMD were similar at the 66% site. In a rat model, the SCI-induced bone changes observed were not solely attributable to bone loss [35], but also to suppress bone growth, suggesting that decreased whole-bone mechanical properties could be the result of changes in the spatial distribution of bone.

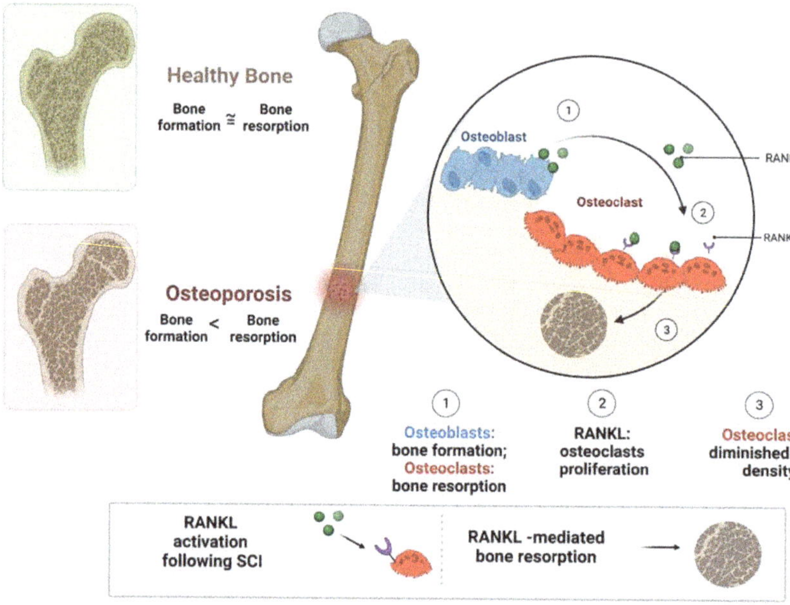

Figure 1. The pathophysiology of RANKL in bone resorption and osteoporosis after SCI. In healthy individuals, osteoblast/osteoclast activity provides for a healthy balance of bone formation and resorption. After SCI, RANKL increases osteoclastic activity, leading to increased bone resorption and osteoporosis. Figure created on Biorender.com.

4. Risk Factors in SCI Individuals

The incidence and prevalence of SCI and its related complications have been increasing, with the incidence rate being estimated at from 15 to 40 cases per million worldwide [36–38]. The specific risk factors associated with SCI include age, gender, lifestyle, body mass index, and physical health conditions. SCIs are most common in males, who make up 78% of new SCI injuries in the U.S. [4,39]. Certain age ranges are more highly associated with SCI prevalence, including post-menopausal women and males aged 18–21 [32,40]. Post-menopausal women are likely at a higher risk of SCI due to the combination of having a higher risk for falls and a decreasing bone density. Males aged 18–21 commonly suffer from SCI due to lifestyles and behaviors that are common causes of SCI, such as contact sports and high falls [41,42]. Studies have suggested that the mean age of the SCI patient in developed countries is higher compared to that in developing countries over the same time period. Possible reasons for this are the aging of the populations in developed countries and/or the larger male-to-female ratio in developing countries in relation to developed countries [38,43]. Thus, it is likely that the elderly SCI populations in developed countries are suffering from additional complications such as bone fracture.

In recent years, epidemiological studies from countries worldwide have focused on traumatic SCIs, since the information about non-traumatic SCIs is limited and their risk factors are variable [41,42]. Traumatic incidents that are common risks of SCIs include sporting accidents, traffic accidents, and high falls [41,42]. Overall, there has been no obvious breakthrough in the determination of risk factors and clinical treatment of SCI and its associated complications; therefore, the emphasis has been on the prevention of traumatic SCIs.

Beyond structural loading, multiple factors, including an increased age, increased time since SCI, and lower body mass index, may be contributory risk factors to SCI [44,45]. Likewise, post-menopausal bone loss may exacerbate the skeletal effects following an SCI. Moreover, bone mineral density measurements shortly after an SCI are informative predictors of osteoporosis in the 12-month period following an SCI [32,40]. The type of SCI is also an important indicator of who will be at a greater risk of bone fractures. Recent findings have suggested that individuals who suffer from motor-complete SCIs have a higher risk of skeletal fractures; moreover, those who consume alcohol post-SCI are at a greater risk for fractures [46,47].

5. Cellular and Molecular Events following SCI

5.1. 4-1BB Signaling after Acute SCI

The receptor 4-1BB (also known as CD137) is a costimulatory and inflammatory receptor that is expressed on activated T cells [48] and some nonimmune cells, such as endothelial cells, glial cells, and neurons [49,50]. 4-1BB ligand (4-1BBL, also known as CD137L) is highly expressed on macrophages and antigen-presenting cells and can receive and transmit reverse signals into cells by binding to its receptor, 4-1BB [49,51,52]. The expressions of 4-1BB and 4-1BBL are upregulated on neuronal and immune cells following injury, and 4-1BB/4-1BBL signaling contributes to the progression of inflammation by controlling the communication of peripheral nerve fibers with cutaneous immune cells. Thus, 4-1BB/4-1BBL signaling might be involved in the regulation of glial and neuronal interaction, controlling neuroinflammation in the CNS. However, the underlying mechanisms and precise role of 4-1BB/4-1BBL signaling in the interplay of peripheral sensory neurons with immune cells are still not clear. Studies have shown the role of 4-1BB in the skeletal system in terms of osteoclast and function [53,54]. Increased bone resorption and decreased bone formation have also been found in aged mice compared to young mice. However, very little information is available on whether high-level 4-1BB/4-1BBL expression in bone marrow is associated with bone loss.

Increasing evidence has suggested that bone loss following an SCI may be affected by tumor necrosis factor receptor 4-1BB signaling. Animal studies have demonstrated that older mice have higher levels of 4-1BB in their bone marrow and have also been found to have a significantly greater bone loss than younger mice with less 4-1BB [21,55]. Targeted anti-4-1BB signaling may prevent bone loss in individuals who have just experienced an SCI. Likewise, anti-4-1BB-directed therapies are effective in treating various neoplasms; however, the treatment must be targeted directly to the tumor to limit the toxicity to bone marrow [56].

5.2. RANKL Signaling after SCI

Bone resorption and osteoclast function are also related to the release of the receptor activator of nuclear factor kappa-B ligand (RANKL) after SCI [57,58]. When individuals experience immobilization due to SCI, RANKL can cause much of the bone loss they experience [57]. The binding of RANKL to its receptor RANK can trigger osteoclast precursors to differentiate into osteoclasts (Figure 1). This process mainly depends on RANKL–RANK signaling, which is temporally regulated by various adaptor proteins and kinases. RANK is expressed in bone marrow mesenchymal stem cells (BMSCs) and is decreased during osteogenic differentiation [59]. RANKL expression can be reduced by the increased secretion of lipid-modified signaling glycoprotein, Wnt, which also stimulates osteoblast function and new bone cell production. Unfortunately, after SCI, Wnt is typically reduced, while RANKL is increased [57]. In addition to Wnt, ellagic acid (EA) has been found to block the interaction between RANK and RANKL, which inhibits the RANKL pathways and suppresses osteoclast activity [60].

5.3. Wnt Signaling after SCI

The Wnt/Beta-catenin pathway has been implicated in neuronal development and regeneration [61]. The central nervous system also utilizes this pathway after SCI for the regeneration of bone and CNS tissue via DNA replication, mitotic recombination, collagen/fibrin organization, and cell development [61–63]. Wnt-3a demonstrates a neuroprotective effect, contributing to neuropathic pain remission and neuronal survival. In animal studies, SCI subjects whose Wnt signals were blocked recovered three weeks after the animals without Wnt signal inhibition [62,64]. Moreover, SCI-related bone loss is reduced in rodents with increased Wnt signaling, related, in part, to reduced osteoclastogenesis and osteoclast activation. Furthermore, the Wnt pathway causes the secretion of glycoproteins from myofibers and satellite cells, with resulting increased levels of beta catenin, a multifunctional protein that promotes cell proliferation and muscle regeneration [63,65,66].

6. Calcium-Regulated Hormones in Bone Loss after SCI

Calcium and vitamin D play roles in bone health and regeneration. Immobilization, aging, and musculoskeletal disuse impede the metabolisms of vitamin D and calcium [33,67,68]. Although some controversy remains, there are reasonable data showing evidence that individuals who are either on a low calcium intake and/or have a vitamin D deficiency suffer from limited gastrointestinal calcium absorption, and may have an increased risk of fracture [68–70]. Individuals with SCI are also known to have a higher prevalence of vitamin D deficiency than the healthy population [70,71]. Studies have suggested that a significant depression in the ionized serum calcium concentration may trigger a secondary increase in the parathyroid hormone (PTH) concentration, which may result in an increased bone turnover in SCI individuals [72–74]. Vitamin D deficiency and abnormal PTH levels are also common in both acute and chronic SCI. The PTH levels are significantly reduced in SCI due to the hypercalcemia that accompanies increased bone resorption [70]. Thus, low PTH may contribute to SCI-induced bone loss. Insulin-like growth factor 1 can also play a role in blood calcium level regulation and changes in PTH in SCI [67,72]. The suppression of these hormonal factors, along with low estrogen/testosterone levels, are associated with bone and muscle atrophy [67]. Of note, PTH is not reduced significantly immediately after SCI, but instead slowly decreases over time. Thus, osteopenia secondary to SCI may play a decisive role in PTH reduction [67].

Individuals with SCI often show bone loss below the level of injury, and sometimes, it can happen throughout the body [75]. A recent study showed the progression of bone loss in SCI mice, which can begin as early as one week following injury in the hind limbs [72]. The total bone mineral density (BMD) and the BMD in areas above the level of injury are not significantly affected until the chronic stages of the injury. This study suggests that chronic SCI may induce a global dysregulation of bone homeostasis. Another study tested and compared the time course of bone loss following SCI in rats with different severities [76]. In severe SCI, rapid bone loss was observed as early as 2–3 weeks, and this bone loss was significant by 8 weeks. Thus, investigating how a loss of PTH following SCI affects the bones may help to develop effective therapies.

7. Bone Density and Fractures after SCI

Bone loss after SCI leads to an increased risk of low-impact fractures and significantly increases the morbidity and mortality of SCI individuals. Even though many severe SCI individuals employ wheelchairs for mobilization, they are still at risk for low-impact fractures [17,32,77]. Osteoporotic fractures are associated with chronic and disabling pain and can markedly increase the chances of death, especially in individuals over the age of 70 [78]. Common distal femur/proximal tibia fractures further limit mobility and impede rehabilitation [27,57]. Fractures after SCI are less common in the first year after injury, but as osteopenia and osteoporosis worsen over time, fractures become increasingly common [40,79]. Therefore, patients who experience SCI can benefit from bone density measurements and preventative treatments soon after injury to prevent future skeletal fracture. Adipocytes also secrete a protein, adiponectin, which may be a predictor of osteopenia in SCI patients. Adiponectin appears to induce osteoclast activity and osteoclastogenesis [80,81].

Studies evaluating SCI patients have found an inverse relationship between adiponectin levels and bone mineral density following SCI [34,80]. Adiponectin has also been identified as a marker for elevated fracture risks. A recent study characterized the time courses of cancellous and cortical bone deficits in a clinically relevant rodent SCI contusion model to determine the mechanisms of skeletal deterioration after SCI [82]. The findings from this study are very important from a clinical perspective, given that fracture incidence is associated with mortality in this population [83]. Overall, the authors found that severe cancellous bone loss occurred at the distal femur and proximal tibia within 2 weeks of SCI and thereafter temporally delayed cortical bone deficits similar to biphasic bone loss in human SCI.

Hormonal imbalance can also contribute to bone fracture and osteoporosis [84,85]. Estrogen plays a protective role in bone health. When estrogen levels decrease, such as after menopause, the risk of osteoporosis and bone loss rises. While post-menopausal women are more prone to osteoporosis and an increased risk of fracture, older men are not immune to a weakening of their bones due to hormonal changes. As men age, their bone density decreases, making fractures more likely [86,87]. In men, the aromatase enzyme converts testosterone into estrogen, and a loss of testosterone can impact this process and lead to bone density loss. Thus, the risk factors of age, duration of SCI, and neurological deficit negatively influence BMD, leading to fracture and bone loss.

8. Therapeutic Strategies for Neurogenic Bone Loss after SCI

8.1. Pharmacological Therapy

Pharmacological therapies for the bone loss in SCI individuals have been relatively ineffective. While vitamin D supplementation is commonly used to restore the vitamin D levels in SCI individuals with a vitamin D deficiency, it has not been effective in preventing and restoring bone loss [88]. Thus, multiple pharmacological strategies may provide benefits for neurogenic bone loss after SCI. For example, ellagic acid (EA) has been found to bind to RANKL and downregulate osteoclast activity, although this endogenous compound may produce negative side effects at elevated concentrations [60,89,90]. Bisphosphonates and Denosumab have also been evaluated for their prevention of the loss of bone mass after SCI (Figure 2). Bisphosphonates act to slow bone loss by inhibiting bone resorption; these include Etidronate, Clodronate, Pamidronate, Tiludronate, and Alendronate [44,91,92]. Bisphosphonates used in SCI patients have been shown to reduce the risk of hip fractures (but not knee fractures) [29,93].

Despite some success, the effects of bisphosphonates have been inconsistent. Clodronate, Etidronate, and Tiludronate have been shown to yield increased bone mass in less than one year post injury (Figure 2), whereas Alendronate improved bone mass in more than one year after injury [44]. However, Pamidronate was not shown to improve bone mass in this study. In addition, the prolonged use of bisphosphonate therapy may produce adverse effects such as osteonecrosis of the jaw; thus, judicial administration is advised [94]. These therapies are currently available in oral or intravenous administrations, and single annualendroal bisphosphonate injections may be available for SCI patients in the future [29,94,95]. In a recent larger clinical trial on patients with chronic SCI, Teriparatide treatment was used, which resulted in a significant increase in spine BMD at 1 year and further improvements in the hip at 2 years [96,97]. Furthermore, Denosumab, a monoclonal antibody to RANKL, is FDA approved for osteoporosis treatment [98,99]. Denosumab prevents bone loss in SCI patients via the inhibition of osteoclast activity via the RANKL pathway, however, it must be frequently administered [93,100–102]. Denosumab thus reduces bone resorption and increases bone mineral density, reducing the risk of fractures.

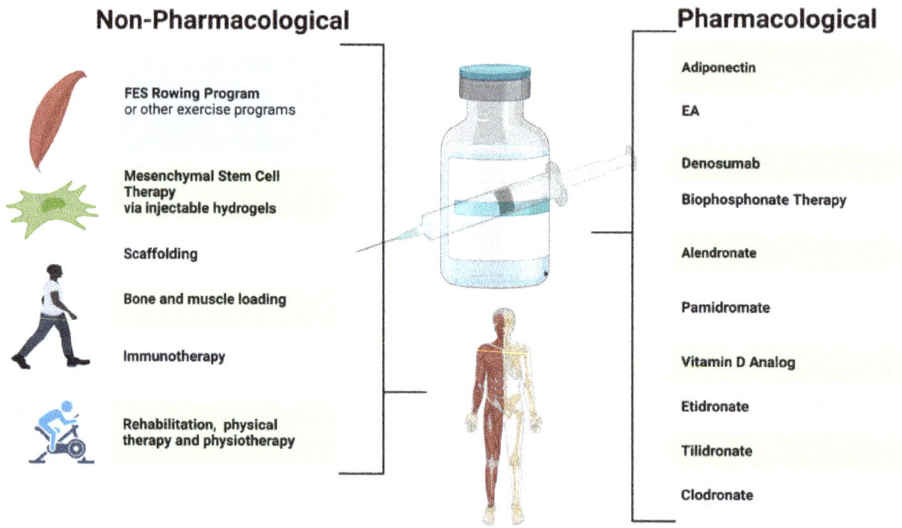

Figure 2. Rehabilitation Methods for Neurogenic Bone Loss After SCI. Figure created on https://www.biorender.com/.

8.2. Nonpharmacological Therapy

Pharmacological therapies to date are limited, as they do not provide a significant restoration of damaged spinal cord parenchyma. Therefore, non-pharmacological approaches, such as mesenchymal stem cell (MSC) therapy, physiotherapy, immunotherapy, injectable hydrogels, and stem cell secretome therapy, are under consideration [103,104]. MSCs from the bone marrow, umbilical cord, and/or adipose tissue may reduce inflammation and provide neuroprotective effects to prevent further injury to the spinal cord near the impact site. Injectable hydrogels, which facilitate MSC targeting, are also being studied [105].

These therapies, in conjunction with weight-bearing rehabilitation, may be increasingly employed to decrease osteopenia in patients with SCI [106,107]. Following SCI, a primary catalyst behind bone loss is the decrease in mechanical loading. When individuals with SCI cease weight-bearing activities, they face a heightened susceptibility to rapid bone resorption and osteocyte apoptosis, frequently leading to the development of osteoporosis. Engaging in any form of mechanical loading on the skeletal system, including compression, tension, torsion, or bending, will uphold bone density and promote bone mass recovery [108]. Therapies aimed at this axial loading encompass activities such as walking, jogging, and jumping. Rehabilitations that stimulate mechanical loading are practical, non-invasive, and economical methods for stimulating bone regeneration [109]. Rehabilitation improves mechanical loading by exposing tissues to a range of strains and forces, prompting osteocytes to sense stress and begin to stimulate regeneration [110]. Reciprocally, the subjection of mechanical loading on tissues from rehabilitation has been shown to be an effective therapy for tissue regeneration, which ultimately improves the bone's capacity for mechanical loading [109]. Rehabilitation also enhances mechanical loading by modifying and improving vascularization, thereby facilitating bone growth. Therapies with an increased musculoskeletal load have proven effective; however, this approach is limited in patients who are wheelchair-bound after SCI. Stand-up wheelchairs, standing frames, and suspended treadmills can provide useful alternatives [11,111,112]. Physical activity, which inherently stimulates the axial loading of the tibia, femur, and axial skeleton, may also promote bone density after SCI by improving bone vascularization and osteoblast activity [4,112].

Static loading and prone position muscle stimulation appear to be less effective techniques for the attenuation of bone loss after SCI [4,113]. Thus, functional electrical stimulation (FES) rowing following SCI has been evaluated. FES rowing employs cyclical exercise patterns coupled with electrical stimulation to simulate the functional motor patterns otherwise impaired by SCI. Rowing allows for paraparetic SCI patients to exercise in a sitting position (in some cases with a cycle ergometer), coordinating their upper body movements with the electrical stimulation of the lower body muscle groups to recreate the effects of full-body exercise [44,113,114]. In one trial, the bone loss in the distal femur and tibia appeared to be reduced in the majority of participants after 30 sessions; however, other results have suggested that bone loss is ameliorated with muscle electrical stimulation alone. Non-mechanical load-bearing exercises such as swimming and cycling are weaker therapies in terms of reducing bone loss; however, they have still been shown to be effective at maintaining muscle mass, which can indirectly reduce fracture risk. Further studies are therefore needed to determine how these therapies can be best implemented for SCI individuals who are wheelchair-bound. It is important to note that the extent of improvement in mechanical loading after SCI is highly dependent on individual aspects such as the severity of the injury. To yield the best results, rehabilitation should be started early, be consistent, and be tailored to individual needs and goals.

9. Conclusions

While SCI can lead to an irreversible loss of motor control and sensations below the level of trauma, the secondary consequences and complications associated with chronic SCI may be subjected to a devised repair strategy. SCI individuals experience a significant number of complications, including muscle wasting, osteopenia or osteoporosis, hormone dysregulation, cardiovascular problems, reduced angiogenesis, and immune deficiency. Although many of these complications appear soon after the injury, very little is known about the exact mechanism(s) underlying their development and progression overtime.

In general, SCI severely limits one's physical and functional capacity due to the many limitations caused after an injury. A significant comorbidity related to SCI is neurogenic bone loss, which predisposes these individuals to osteoporosis and fractures. To reduce the risks of long bone fractures after SCI, pharmacological approaches, including the administration of ellagic acid, Adiponectin, Denosumab, and bisphosphonates, are being evaluated. Non-pharmacological treatments further augment bone density; these include exercise therapies such as FES rowing, bone loading, physiotherapy, and mesenchymal stem cell therapy. The application of both types of therapeutic approaches must be appropriately tailored for individual SCI patients in relation to the time after injury, side-effects, and other patient-specific comorbidities. In addition, studies are needed to develop novel combination approaches and determine the most effective therapies and prevention methods for osteoporosis in people with SCI.

Author Contributions: G.E.L. designed, wrote the manuscript, drew the figures, and edited the manuscript. D.C.S. wrote and edited the manuscript. A.H. conceived, designed, wrote, and edited the manuscript. N.L.B. conceived, designed, and edited the manuscript. All authors have read and agreed to the published version of the manuscript.

Funding: This work was supported in part by funding from the Veterans Administration (1IOBX001262, 1I01 BX004269) and South Carolina State Spinal Cord Injury Research Fund (SCIRF-2015P-01, SCIRF-2015P-04, SCIRF-2015-I-01, SCIRF #2016 I-03, and SCIRF #2018 I-01). Contents do not necessarily represent the policy of the SCIRF and do not imply endorsement by the funding agency. Banik is the recipient of RCS Award (IK6BX005964) from the Department of Veterans Administration. This work was also supported in part by funding from the National Institutes of Health (1R21NS118393-01).

Institutional Review Board Statement: Not applicable.

Informed Consent Statement: Not applicable.

Data Availability Statement: The data used to support the findings of this manuscript are available from the corresponding authors upon reasonable written request after the publication.

Acknowledgments: We thank Citadel Military College in Charleston for supporting Giovanna Leone.

Conflicts of Interest: The authors declare no conflict of interest.

References

1. Alizadeh, A.; Dyck, S.M.; Karimi-Abdolrezaee, S. Traumatic spinal cord injury: An overview of pathophysiology, models and acute injury mechanisms. *Front. Neurol.* **2019**, *10*, 282. [CrossRef]
2. Ahuja, C.S.; Wilson, J.R.; Nori, S.; Kotter, M.; Druschel, C.; Curt, A.; Fehlings, M.G. Traumatic spinal cord injury. *Nat. Rev. Dis. Primers* **2017**, *3*, 17018. [CrossRef] [PubMed]
3. Chen, Y.; Tang, Y.; Vogel, L.; DeVivo, M. Causes of spinal cord injury. *Top. Spinal Cord Inj. Rehabil.* **2013**, *19*, 1–8. [CrossRef] [PubMed]
4. Shams, R.; Drasites, K.P.; Zaman, V.; Matzelle, D.; Shields, D.C.; Garner, D.P.; Sole, C.J.; Haque, A.; Banik, N.L. The Pathophysiology of Osteoporosis after Spinal Cord Injury. *Int. J. Mol. Sci.* **2021**, *22*, 3057. [CrossRef]
5. Chen, L.W.; Glinsky, J.V.; Islam, S.; Hossain, M.; Boswell-Ruys, C.L.; Kataria, C.; Redhead, J.; Xiong, Y.; Gollan, E.; Costa, P.D.; et al. The effects of 10,000 voluntary contractions over 8 weeks on the strength of very weak muscles in people with spinal cord injury: A randomised controlled trial. *Spinal Cord* **2020**, *58*, 857–864. [CrossRef]
6. Fouad, K.; Popovich, P.G.; Kopp, M.A.; Schwab, J.M. The neuroanatomical–functional paradox in spinal cord injury. *Nat. Rev. Neurol.* **2021**, *17*, 53–62. [CrossRef]
7. Zhu, S.; Bennett, S.; Kuek, V.; Xiang, C.; Xu, H.; Rosen, V.; Xu, J. Endothelial cells produce angiocrine factors to regulate bone and cartilage via versatile mechanisms. *Theranostics* **2020**, *10*, 5957–5965. [CrossRef] [PubMed]
8. Rodriguez, G.; Berri, M.; Lin, P.; Kamdar, N.; Mahmoudi, E.; Peterson, M.D. Musculoskeletal morbidity following spinal cord injury: A longitudinal cohort study of privately-insured beneficiaries. *Bone* **2021**, *142*, 115700. [CrossRef]
9. Wang, S.; Deng, J.; Fu, H.; Guo, Z.; Zhang, L.; Tang, P. Astrocytes directly clear myelin debris through endocytosis pathways and followed by excessive gliosis after spinal cord injury. *Biochem. Biophys. Res. Commun.* **2020**, *525*, 20–26. [CrossRef]
10. Yang, T.; Dai, Y.; Chen, G.; Cui, S. Dissecting the Dual Role of the Glial Scar and Scar-Forming Astrocytes in Spinal Cord Injury. *Front. Cell. Neurosci.* **2020**, *14*, 78. [CrossRef]
11. Lin, C.-Y.; Androjna, C.; Rozic, R.; Nguyen, B.T.; Parsons, B.; Midura, R.J.; Lee, Y.-S. Differential Adaptations of the Musculoskeletal System after Spinal Cord Contusion and Transection in Rats. *J. Neurotrauma* **2018**, *35*, 1737–1744. [CrossRef] [PubMed]
12. Marini, S.; Barone, G.; Masini, A.; Dallolio, L.; Bragonzoni, L.; Longobucco, Y.; Maffei, F. The Effect of Physical Activity on Bone Biomarkers in People with Osteoporosis: A Systematic Review. *Front. Endocrinol.* **2020**, *11*, 585689. [CrossRef] [PubMed]
13. Craven, B.C.; Cirnigliaro, C.M.; Carbone, L.D.; Tsang, P.; Morse, L.R. The Pathophysiology, Identification and Management of Fracture Risk, Sublesional Osteoporosis and Fracture among Adults with Spinal Cord Injury. *J. Pers. Med.* **2023**, *13*, 966. [CrossRef] [PubMed]
14. Varacallo, M.; Davis, D.D.; Pizzutillo, P. Osteoporosis in Spinal Cord Injuries. In *StatPearls*; Ineligible Companies: Treasure Island, FL, USA, 2023.
15. Edwards, W.B.; Schnitzer, T.J. Bone Imaging and Fracture Risk after Spinal Cord Injury. *Curr. Osteoporos. Rep.* **2015**, *13*, 310–317. [CrossRef] [PubMed]
16. Maïmoun, L.; Gelis, A.; Serrand, C.; Mura, T.; Humbert, L.; Boudousq, V.; de Santa-Barbara, P.; Laux, D.; Fattal, C.; Mariano-Goulart, D. Alteration of Volumetric Bone Mineral Density Parameters in Men with Spinal Cord Injury. *Calcif. Tissue Int.* **2023**, *113*, 304–316. [CrossRef]
17. Cirnigliaro, C.M.; La Fountaine, M.F.; Parrott, J.S.; Kirshblum, S.C.; Sauer, S.J.; Shapses, S.A.; McClure, I.A.; Bauman, W.A. Loss of lower extremity bone mineral density 1 year after denosumab is discontinued in persons with subacute spinal cord injury. *Osteoporos. Int.* **2023**, *34*, 741–748. [CrossRef]
18. Zhang, L.; Yin, Y.; Guo, J.; Jin, L.; Hou, Z. Chronic intermittent hypobaric hypoxia ameliorates osteoporosis after spinal cord injury through balancing osteoblast and osteoclast activities in rats. *Front. Endocrinol.* **2023**, *14*, 1035186. [CrossRef]
19. Ishimoto, R.; Mutsuzaki, H.; Shimizu, Y.; Kishimoto, H.; Takeuchi, R.; Hada, Y. Prevalence of Sarcopenic Obesity and Factors Influencing Body Composition in Persons with Spinal Cord Injury in Japan. *Nutrients* **2023**, *15*, 473. [CrossRef]
20. Bitra, A.; Doukov, T.; Croft, M.; Zajonc, D.M. Crystal structures of the human 4-1BB receptor bound to its ligand 4-1BBL reveal covalent receptor dimerization as a potential signaling amplifier. *J. Biol. Chem.* **2018**, *293*, 9958–9969. [CrossRef]
21. Wan, D.; Ai, S.; Ouyang, H.; Cheng, L. Activation of 4-1BB signaling in bone marrow stromal cells triggers bone loss via the p-38 MAPK-DKK1 axis in aged mice. *Exp. Mol. Med.* **2021**, *53*, 654–666. [CrossRef]
22. Cheng, P.; Liao, H.-Y.; Zhang, H.-H. The role of Wnt/mTOR signaling in spinal cord injury. *J. Clin. Orthop. Trauma* **2022**, *25*, 101760. [CrossRef]
23. Liu, J.; Xiao, Q.; Xiao, J.; Niu, C.; Li, Y.; Zhang, X.; Zhou, Z.; Shu, G.; Yin, G. Wnt/beta-catenin signalling: Function, biological mechanisms, and therapeutic opportunities. *Signal Transduct. Target. Ther.* **2022**, *7*, 3. [CrossRef] [PubMed]

24. Gifre, L.; Vidal, J.; Carrasco, J.L.; Filella, X.; Ruiz-Gaspà, S.; Muxi, A.; Portell, E.; Monegal, A.; Guañabens, N.; Peris, P. Effect of Recent Spinal Cord Injury on Wnt Signaling Antagonists (Sclerostin and Dkk-1) and Their Relationship with Bone Loss. A 12-Month Prospective Study. *J. Bone Miner. Res.* **2015**, *30*, 1014–1021. [CrossRef]
25. Sutor, T.W.; Kura, J.; Mattingly, A.J.; Otzel, D.M.; Yarrow, J.F. The Effects of Exercise and Activity-Based Physical Therapy on Bone after Spinal Cord Injury. *Int. J. Mol. Sci.* **2022**, *23*, 608. [CrossRef] [PubMed]
26. Yarrow, J.F.; Wnek, R.D.; Conover, C.F.; Reynolds, M.C.; Buckley, K.H.; Kura, J.R.; Sutor, T.W.; Otzel, D.M.; Mattingly, A.J.; Croft, S.; et al. Bone loss after severe spinal cord injury coincides with reduced bone formation and precedes bone blood flow deficits. *J. Appl. Physiol.* **2021**, *131*, 1288–1299. [CrossRef] [PubMed]
27. Morse, L.; Teng, Y.D.; Pham, L.; Newton, K.; Yu, D.; Liao, W.-L.; Kohler, T.; Müller, R.; Graves, D.; Stashenko, P.; et al. Spinal cord injury causes rapid osteoclastic resorption and growth plate abnormalities in growing rats (SCI-induced bone loss in growing rats). *Osteoporos. Int.* **2008**, *19*, 645–652. [CrossRef]
28. Le, B.; Ray, C.; Gonzalez, B.; Miskevics, S.; Weaver, F.M.; Priebe, M.; Carbone, L.D. Laboratory evaluation of secondary causes of bone loss in Veterans with spinal cord injury and disorders. *Osteoporos. Int.* **2019**, *30*, 2241–2248. [CrossRef]
29. Ma, Z.; Ma, M.; He, Y.; Sun, H.; Yang, B.; Dong, H.; Wang, Y. Bisphosphonates Alleviate Bone Loss in People with Acute Spinal Cord Injury: A Systematic Review and Meta-Analysis. *World Neurosurg.* **2023**, *170*, e584–e595. [CrossRef]
30. Ducher, G.; Courteix, D.; Même, S.; Magni, C.; Viala, J.; Benhamou, C. Bone geometry in response to long-term tennis playing and its relationship with muscle volume: A quantitative magnetic resonance imaging study in tennis players. *Bone* **2005**, *37*, 457–466. [CrossRef]
31. Zhu, S.; Chen, M.; Ying, Y.; Wu, Q.; Huang, Z.; Ni, W.; Wang, X.; Xu, H.; Bennett, S.; Xiao, J.; et al. Versatile subtypes of pericytes and their roles in spinal cord injury repair, bone development and repair. *Bone Res.* **2022**, *10*, 30. [CrossRef]
32. Zheng, X.; Qi, Y.; Zhou, H.; Kang, H.; Tong, Y.; Bi, L. Bone Mineral Density at the Distal Femur and Proximal Tibia and Related Factors During the First Year of Spinal Cord Injury. *Int. J. Gen. Med.* **2021**, *14*, 1121–1129. [CrossRef] [PubMed]
33. Alexandre, C.; Vico, L. Pathophysiology of bone loss in disuse osteoporosis. *Jt. Bone Spine* **2011**, *78*, 572–576. [CrossRef] [PubMed]
34. Tan, C.O.; Battaglino, R.A.; Doherty, A.L.; Gupta, R.; Lazzari, A.A.; Garshick, E.; Zafonte, R.; Morse, L.R. Adiponectin is associated with bone strength and fracture history in paralyzed men with spinal cord injury. *Osteoporos. Int.* **2014**, *25*, 2599–2607. [CrossRef] [PubMed]
35. Williams, J.A.; Huesa, C.; Turunen, M.J.; Oo, J.A.; Radzins, O.; Gardner, W.; Windmill, J.F.C.; Isaksson, H.; Tanner, K.E.; Riddell, J.S.; et al. Time course changes to structural, mechanical and material properties of bone in rats after complete spinal cord injury. *J. Musculoskelet. Neuronal Interact.* **2022**, *22*, 212–234.
36. Jackson, A.B.; Dijkers, M.; DeVivo, M.J.; Poczatek, R.B. A demographic profile of new traumatic spinal cord injuries: Change and stability over 30 years. *Arch. Phys. Med. Rehabil.* **2004**, *85*, 1740–1748. [CrossRef]
37. Dharnipragada, R.; Ahiarakwe, U.; Gupta, R.; Abdilahi, A.; Butterfield, J.; Naik, A.; Parr, A.; Morse, L.R. Pharmacologic and Nonpharmacologic Treatment Modalities for Bone Loss in SCI—Proposal for Combined Approach. *J. Clin. Densitom.* **2023**, *26*, 101359. [CrossRef]
38. Babaie, M.; Joulani, M.; Hameghavandi, M.H.R.; Asgardoon, M.H.; Nojomi, M.; O'Reilly, G.M.; Gholami, M.; Ghodsi, Z.; Rahimi-Movaghar, V. Risk of permanent medical impairment after road traffic crashes: A systematic review. *Chin. J. Traumatol.* **2022**, *22*, 115–118. [CrossRef]
39. Spinal cord injury facts and figures at a glance. *J. Spinal Cord Med.* **2013**, *36*, 170–171. [CrossRef]
40. Gifre, L.; Vidal, J.; Carrasco, J.L.; Muxi, A.; Portell, E.; Monegal, A.; Guañabens, N.; Peris, P. Risk factors for the development of osteoporosis after spinal cord injury. A 12-month follow-up study. *Osteoporos. Int.* **2015**, *26*, 2273–2280. [CrossRef]
41. New, P.W.; Simmonds, F.; Stevermuer, T. A population-based study comparing traumatic spinal cord injury and non-traumatic spinal cord injury using a national rehabilitation database. *Spinal Cord* **2011**, *49*, 397–403. [CrossRef]
42. Dalle, D.U.; Sriram, S.; Bandyopadhyay, S.; Egiz, A.; Kotecha, J.; Kanmounye, U.S.; Higginbotham, G.; Ooi, S.Z.Y.; Bankole, N.D.A. Management and Outcomes of Traumatic Pediatric Spinal Cord Injuries in Low- and Middle-Income Countries: A Scoping Review. *World Neurosurg.* **2022**, *165*, 180–187. [CrossRef] [PubMed]
43. Chiu, W.-T.; Lin, H.-C.; Lam, C.; Chu, S.-F.; Chiang, Y.-H.; Tsai, S.-H. Review Paper: Epidemiology of Traumatic Spinal Cord Injury: Comparisons Between Developed and Developing Countries. *Asia Pac. J. Public Health.* **2010**, *22*, 9–18. [CrossRef] [PubMed]
44. Ashe, M.; Craven, C.; Eng, J.; Krassioukov, A. Prevention and Treatment of Bone Loss After a Spinal Cord Injury: A Systematic Review. *Top. Spinal Cord Inj. Rehabil.* **2007**, *13*, 123–145. [CrossRef]
45. Alazzam, A.M.; Goldsmith, J.A.; Khalil, R.E.; Khan, M.R.; Gorgey, A.S. Denervation impacts muscle quality and knee bone mineral density after spinal cord injury. *Spinal Cord* **2023**, *61*, 276–284. [CrossRef]
46. Morse, L.R.; Battaglino, R.A.; Stolzmann, K.L.; Hallett, L.D.; Waddimba, A.; Gagnon, D.; Lazzari, A.A.; Garshick, E. Osteoporotic fractures and hospitalization risk in chronic spinal cord injury. *Osteoporos. Int.* **2009**, *20*, 385–392. [CrossRef] [PubMed]
47. Ibarra, A.; Alcántar-Garibay, O.; Incontri-Abraham, D. Spinal cord injury-induced cognitive impairment: A narrative review. *Neural Regen. Res.* **2022**, *17*, 2649–2654. [CrossRef]
48. Drenkard, D.; Becke, F.M.; Langstein, J.; Spruss, T.; Kunz-Schughart, L.A.; Tan, T.E.; Lim, Y.C.; Schwarz, H. CD137 is expressed on blood vessel walls at sites of inflammation and enhances monocyte migratory activity. *FASEB J.* **2006**, *21*, 456–463. [CrossRef]
49. Reali, C.; Curto, M.; Sogos, V.; Scintu, F.; Pauly, S.; Schwarz, H.; Gremo, F. Expression of CD137 and its ligand in human neurons, astrocytes, and microglia: Modulation by FGF-2. *J. Neurosci. Res.* **2003**, *74*, 67–73. [CrossRef]

50. Kim, C.-S.; Kim, J.G.; Lee, B.-J.; Choi, M.-S.; Choi, H.-S.; Kawada, T.; Lee, K.-U.; Yu, R. Deficiency for Costimulatory Receptor 4-1BB Protects Against Obesity-Induced Inflammation and Metabolic Disorders. *Diabetes* **2011**, *60*, 3159–3168. [CrossRef]
51. Shao, Z.; Schwarz, H. CD137 ligand, a member of the tumor necrosis factor family, regulates immune responses via reverse signal transduction. *J. Leukoc. Biol.* **2011**, *89*, 21–29. [CrossRef]
52. Tu, T.H.; Kim, C.-S.; Goto, T.; Kawada, T.; Kim, B.-S.; Yu, R. 4-1BB/4-1BBL Interaction Promotes Obesity-Induced Adipose Inflammation by Triggering Bidirectional Inflammatory Signaling in Adipocytes/Macrophages. *Mediat. Inflamm.* **2012**, *2012*, 972629. [CrossRef]
53. Saito, K.; Ohara, N.; Hotokezaka, H.; Fukumoto, S.; Yuasa, K.; Naito, M.; Fujiwara, T.; Nakayama, K. Infection-induced Up-regulation of the Costimulatory Molecule 4-1BB in Osteoblastic Cells and Its Inhibitory Effect on M-CSF/RANKL-induced in Vitro Osteoclastogenesis. *J. Biol. Chem.* **2004**, *279*, 13555–13563. [CrossRef] [PubMed]
54. Yang, J.; Park, O.J.; Lee, Y.J.; Jung, H.-M.; Woo, K.M.; Choi, Y. The 4-1BB ligand and 4-1BB expressed on osteoclast precursors enhance RANKL-induced osteoclastogenesis via bi-directional signaling. *Eur. J. Immunol.* **2008**, *38*, 1598–1609. [CrossRef] [PubMed]
55. Wakley, A.A.; Leeming, R.; Malon, J.; Arabatzis, T.J.; Koh, W.Y.; Cao, L. Contribution of CD137L to Sensory Hypersensitivity in a Murine Model of Neuropathic Pain. *eNeuro* **2018**, *5*, 218. [CrossRef] [PubMed]
56. Sanchez-Paulete, A.R.; Labiano, S.; Rodriguez-Ruiz, M.E.; Azpilikueta, A.; Etxeberria, I.; Bolaños, E.; Lang, V.; Rodriguez, M.; Aznar, M.A.; Jure-Kunkel, M.; et al. Deciphering CD137 (4-1BB) signaling in T-cell costimulation for translation into successful cancer immunotherapy. *Eur. J. Immunol.* **2016**, *46*, 513–522. [CrossRef]
57. Sun, L.; Pan, J.; Peng, Y.; Wu, Y.; Li, J.; Liu, X.; Qin, Y.; Bauman, W.A.; Cardozo, C.; Zaidi, M.; et al. Anabolic steroids reduce spinal cord injury-related bone loss in rats associated with increased Wnt signaling. *J. Spinal Cord Med.* **2013**, *36*, 616–622. [CrossRef]
58. Tian, Y.; Chen, J.; Yan, X.; Ren, D.; Liu, M.; Zhang, Q.; Zhang, Q.; Yuan, X. Overloaded Orthopedic Force Induces Condylar Subchondral Bone Absorption by Stimulating Rat Mesenchymal Stem Cells Differentiating into Osteoclasts via mTOR-Regulated RANKL/OPG Secretion in Osteoblasts. *Stem Cells Dev.* **2021**, *30*, 29–38. [CrossRef]
59. Cao, X. RANKL-RANK signaling regulates osteoblast differentiation and bone formation. *Bone Res.* **2018**, *6*, 35. [CrossRef]
60. Xu, H.; Chen, F.; Liu, T.; Xu, J.; Li, J.; Jiang, L.; Wang, X.; Sheng, J. Ellagic acid blocks RANKL–RANK interaction and suppresses RANKL-induced osteoclastogenesis by inhibiting RANK signaling pathways. *Chem. Biol. Interact.* **2020**, *331*, 109235. [CrossRef]
61. Gao, K.; Zhang, T.; Wang, F.; Lv, C. Therapeutic Potential of Wnt-3a in Neurological Recovery after Spinal Cord Injury. *Eur. Neurol.* **2019**, *81*, 197–204. [CrossRef]
62. Herman, P.E.; Papatheodorou, A.; Bryant, S.A.; Waterbury, C.K.M.; Herdy, J.R.; Arcese, A.A.; Buxbaum, J.D.; Smith, J.J.; Morgan, J.R.; Bloom, O. Highly conserved molecular pathways, including Wnt signaling, promote functional recovery from spinal cord injury in lampreys. *Sci. Rep.* **2018**, *8*, 742. [CrossRef] [PubMed]
63. Yin, H.; Price, F.; Rudnicki, M.A. Satellite Cells and the Muscle Stem Cell Niche. *Physiol. Rev.* **2013**, *93*, 23–67. [CrossRef] [PubMed]
64. Tang, Z.; Yang, C.; He, Z.; Deng, Z.; Li, X. Notoginsenoside R1 alleviates spinal cord injury through the miR-301a/KLF7 axis to activate Wnt/beta-catenin pathway. *Open Med.* **2022**, *17*, 741–755. [CrossRef] [PubMed]
65. Rudnicki, M.A.; Williams, B.O. Wnt signaling in bone and muscle. *Bone* **2015**, *80*, 60–66. [CrossRef]
66. Nedergaard, A.; Henriksen, K.; Karsdal, M.A.; Christiansen, C. Musculoskeletal ageing and primary prevention. *Best Pr. Res. Clin. Obstet. Gynaecol.* **2013**, *27*, 673–688. [CrossRef] [PubMed]
67. Kirk, B.; Zanker, J.; Duque, G. Osteosarcopenia: Epidemiology, diagnosis, and treatment-facts and numbers. *J. Cachexia Sarcopenia Muscle* **2020**, *11*, 609–618. [CrossRef]
68. LeBoff, M.S.; Chou, S.H.; Ratliff, K.A.; Cook, N.R.; Khurana, B.; Kim, E.; Cawthon, P.M.; Bauer, D.C.; Black, D.; Gallagher, J.C.; et al. Supplemental Vitamin D and Incident Fractures in Midlife and Older Adults. *N. Engl. J. Med.* **2022**, *387*, 299–309. [CrossRef] [PubMed]
69. Dawson-Hughes, B.; Harris, S.S.; Krall, E.A.; Dallal, G.E. Effect of Calcium and Vitamin D Supplementation on Bone Density in Men and Women 65 Years of Age or Older. *N. Engl. J. Med.* **1997**, *337*, 670–676. [CrossRef]
70. Bauman, W.A.; Zhang, R.-L.; Morrison, N.; Spungen, A.M. Acute Suppression of Bone Turnover with Calcium Infusion in Persons With Spinal Cord Injury. *J. Spinal Cord Med.* **2009**, *32*, 398–403. [CrossRef]
71. Flueck, J.L.; Perret, C. Vitamin D deficiency in individuals with a spinal cord injury: A literature review. *Spinal Cord* **2017**, *55*, 428–434. [CrossRef]
72. del Rivero, T.; Bethea, J.R. The effects of spinal cord injury on bone loss and dysregulation of the calcium/parathyroid hormone loop in mice. *Osteoporos. Sarcopenia* **2016**, *2*, 164–169. [CrossRef] [PubMed]
73. Ung, R.-V.; Lapointe, N.P.; Guertin, P.A. Early adaptive changes in chronic paraplegic mice: A model to study rapid health degradation after spinal cord injury. *Spinal Cord* **2008**, *46*, 176–180. [CrossRef] [PubMed]
74. Yalla, N.; Bobba, G.; Guo, G.; Stankiewicz, A.; Ostlund, R. Parathyroid hormone reference ranges in healthy individuals classified by vitamin D status. *J. Endocrinol. Investig.* **2019**, *42*, 1353–1360. [CrossRef] [PubMed]
75. Maïmoun, L.; Couret, I.; Mariano-Goulart, D.; Dupuy, A.M.; Micallef, J.-P.; Peruchon, E.; Ohanna, F.; Cristol, J.-P.; Rossi, M.; Leroux, J.-L. Changes in Osteoprotegerin/RANKL System, Bone Mineral Density, and Bone Biochemicals Markers in Patients with Recent Spinal Cord Injury. *Calcif. Tissue Int.* **2005**, *76*, 404–411. [CrossRef] [PubMed]
76. Voor, M.J.; Brown, E.H.; Xu, Q.; Waddell, S.W.; Burden, R.L.; Burke, D.A.; Magnuson, D.S.; Bramlett, H.M.; Dietrich, W.D.; Marcillo, A.; et al. Bone Loss following Spinal Cord Injury in a Rat Model. *J. Neurotrauma* **2012**, *29*, 1676–1682. [CrossRef]

77. Zehnder, Y.; Michel, D.; Knecht, H.; Perrelet, R.; Neto, I.; Kraenzlin, M.; Lippuner, K.; Lüthi, M.; Zäch, G. Long-term changes in bone metabolism, bone mineral density, quantitative ultrasound parameters, and fracture incidence after spinal cord injury: A cross-sectional observational study in 100 paraplegic men. *Osteoporos. Int.* **2004**, *15*, 180–189. [CrossRef]
78. Jang, E.J.; Lee, Y.-K.; Choi, H.J.; Ha, Y.-C.; Jang, S.; Shin, C.S.; Cho, N.H. Osteoporotic Fracture Risk Assessment Using Bone Mineral Density in Korean: A Community-based Cohort Study. *J. Bone Metab.* **2016**, *23*, 34–39. [CrossRef]
79. Dorado, M.T.F.; Merino, M.d.S.D.; Marco, D.G.; Boy, R.C.; Samper, B.B.; Dhier, L.M.; Bertol, C.L. Preventive treatment with alendronate of loss of bone mineral density in acute traumatic spinal cord injury. Randomized controlled clinical trial. *Spinal Cord* **2022**, *60*, 687–693. [CrossRef]
80. Naot, D.; Musson, D.S.; Cornish, J. The Activity of Adiponectin in Bone. *Calcif. Tissue Int.* **2017**, *100*, 486–499. [CrossRef]
81. Haugen, S.; He, J.; Sundaresan, A.; Stunes, A.K.; Aasarød, K.M.; Tiainen, H.; Syversen, U.; Skallerud, B.; Reseland, J.E. Adiponectin Reduces Bone Stiffness: Verified in a Three-Dimensional Artificial Human Bone Model In Vitro. *Front. Endocrinol.* **2018**, *9*, 236. [CrossRef]
82. Otzel, D.M.; Conover, C.F.; Ye, F.; Phillips, E.G.; Bassett, T.; Wnek, R.D.; Flores, M.; Catter, A.; Ghosh, P.; Balaez, A.; et al. Longitudinal Examination of Bone Loss in Male Rats After Moderate–Severe Contusion Spinal Cord Injury. *Calcif. Tissue Int.* **2019**, *104*, 79–91. [CrossRef] [PubMed]
83. Carbone, L.D.; Chin, A.S.; Burns, S.P.; Svircev, J.N.; Hoenig, H.; Heggeness, M.; Bailey, L.; Weaver, F. Mortality After Lower Extremity Fractures in Men with Spinal Cord Injury. *J. Bone Miner. Res.* **2014**, *29*, 432–439. [CrossRef] [PubMed]
84. Cheng, C.-H.; Chen, L.-R.; Chen, K.-H. Osteoporosis Due to Hormone Imbalance: An Overview of the Effects of Estrogen Deficiency and Glucocorticoid Overuse on Bone Turnover. *Int. J. Mol. Sci.* **2022**, *23*, 1376. [CrossRef] [PubMed]
85. Mills, E.G.; Yang, L.; Nielsen, M.F.; Kassem, M.; Dhillo, W.S.; Comninos, A.N. The Relationship Between Bone and Reproductive Hormones Beyond Estrogens and Androgens. *Endocr. Rev.* **2021**, *42*, 691–719. [CrossRef] [PubMed]
86. Wang, H.; Cheng, J.; Wei, D.; Wu, H.; Zhao, J. Causal relationships between sex hormone traits, lifestyle factors, and osteoporosis in men: A Mendelian randomization study. *PLoS ONE* **2022**, *17*, e0271898. [CrossRef] [PubMed]
87. Khosla, S.; Amin, S.; Orwoll, E. Osteoporosis in men. *Endocr. Rev.* **2008**, *29*, 441–464. [CrossRef]
88. Bauman, W.A.; Emmons, R.R.; Cirnigliaro, C.M.; Kirshblum, S.C.; Spungen, A.M. An effective oral vitamin D replacement therapy in persons with spinal cord injury. *J. Spinal Cord Med.* **2011**, *34*, 455–460. [CrossRef]
89. Tobeiha, M.; Moghadasian, M.H.; Amin, N.; Jafarnejad, S. RANKL/RANK/OPG Pathway: A Mechanism Involved in Exercise-Induced Bone Remodeling. *BioMed Res. Int.* **2020**, *2020*, 6910312. [CrossRef]
90. Liu, F.-L.; Chen, C.-L.; Lee, C.-C.; Wu, C.-C.; Hsu, T.-H.; Tsai, C.-Y.; Huang, H.-S.; Chang, D.-M. The Simultaneous Inhibitory Effect of Niclosamide on RANKL-Induced Osteoclast Formation and Osteoblast Differentiation. *Int. J. Med. Sci.* **2017**, *14*, 840–852. [CrossRef]
91. McDonald, C.L.; Lemme, N.J.; Testa, E.J.; Aaron, R.; Hartnett, D.A.; Cohen, E.M. Bisphosphonates in Total Joint Arthroplasty: A Review of Their Use and Complications. *Arthroplast. Today* **2022**, *14*, 133–139. [CrossRef]
92. Nardone, V.; D'Asta, F.; Brandi, M.L. Pharmacological management of osteogenesis. *Clinics* **2014**, *69*, 438–446. [CrossRef]
93. Bauman, W.A. Pharmacological approaches for bone health in persons with spinal cord injury. *Curr. Opin. Pharmacol.* **2021**, *60*, 346–359. [CrossRef] [PubMed]
94. Russell, R.G.G. Bisphosphonates: Mode of Action and Pharmacology. *Pediatrics* **2007**, *119* (Suppl. 2), S150–S162. [CrossRef] [PubMed]
95. Drake, M.T.; Clarke, B.L.; Khosla, S. Bisphosphonates: Mechanism of Action and Role in Clinical Practice. *Mayo Clin. Proc.* **2008**, *83*, 1032–1045. [CrossRef] [PubMed]
96. Liu, X.; Liu, M.; Turner, R.; Iwaniec, U.; Kim, H.; Halloran, B. Dried plum mitigates spinal cord injury-induced bone loss in mice. *JOR Spine* **2020**, *3*, e1113. [CrossRef] [PubMed]
97. Edwards, W.B.; Simonian, N.; Haider, I.T.; Anschel, A.S.; Chen, D.; Gordon, K.E.; Gregory, E.K.; Kim, K.H.; Parachuri, R.; Troy, K.L.; et al. Effects of Teriparatide and Vibration on Bone Mass and Bone Strength in People with Bone Loss and Spinal Cord Injury: A Randomized, Controlled Trial. *J. Bone Miner. Res.* **2018**, *33*, 1729–1740. [CrossRef]
98. Kostenuik, P.J.; Nguyen, H.Q.; McCabe, J.; Warmington, K.S.; Kurahara, C.; Sun, N.; Chen, C.; Li, L.; Cattley, R.C.; Van, G.; et al. Denosumab, a Fully Human Monoclonal Antibody to RANKL, Inhibits Bone Resorption and Increases BMD in Knock-In Mice That Express Chimeric (Murine/Human) RANKL*. *J. Bone Miner. Res.* **2009**, *24*, 182–195. [CrossRef]
99. Guo, Y.; Guo, T.; Di, Y.; Xu, W.; Hu, Z.; Xiao, Y.; Yu, H.; Hou, J. Pharmacokinetics, pharmacodynamics, safety and immunogenicity of recombinant, fully human anti-RANKL monoclonal antibody (MW031) versus denosumab in Chinese healthy subjects: A single-center, randomized, double-blind, single-dose, parallel-controlled trial. *Expert Opin. Biol. Ther.* **2023**, *23*, 705–715. [CrossRef]
100. Won, K.Y.; Kalil, R.K.; Kim, Y.W.; Park, Y.-K. RANK signalling in bone lesions with osteoclast-like giant cells. *Pathology* **2011**, *43*, 318–321. [CrossRef]
101. Miyagawa, K.; Ohata, Y.; Delgado-Calle, J.; Teramachi, J.; Zhou, H.; Dempster, D.D.; Subler, M.A.; Windle, J.J.; Chirgwin, J.M.; Roodman, G.D.; et al. Osteoclast-derived IGF1 is required for pagetic lesion formation in vivo. *JCI Insight* **2020**, *5*. [CrossRef]
102. Song, R.; Gu, J.; Liu, X.; Zhu, J.; Wang, Q.; Gao, Q.; Zhang, J.; Cheng, L.; Tong, X.; Qi, X.; et al. Inhibition of osteoclast bone resorption activity through osteoprotegerin-induced damage of the sealing zone. *Int. J. Mol. Med.* **2014**, *34*, 856–862. [CrossRef] [PubMed]

103. Liau, L.L.; Looi, Q.H.; Chia, W.C.; Subramaniam, T.; Ng, M.H.; Law, J.X. Treatment of spinal cord injury with mesenchymal stem cells. *Cell Biosci.* **2020**, *10*, 112. [CrossRef] [PubMed]
104. Abo-Aziza, F.A.; Zaki, A.K.A.; El-Maaty, A.M.A. Bone marrow-derived mesenchymal stem cell (BM-MSC): A tool of cell therapy in hydatid experimentally infected rats. *Cell Regen.* **2019**, *8*, 58–71. [CrossRef] [PubMed]
105. Boido, M.; Ghibaudi, M.; Gentile, P.; Favaro, E.; Fusaro, R.; Tonda-Turo, C. Chitosan-based hydrogel to support the paracrine activity of mesenchymal stem cells in spinal cord injury treatment. *Sci. Rep.* **2019**, *9*, 6402. [CrossRef]
106. Ning, Z.; Gu, P.; Zhang, J.; Cheung, C.W.; Lao, L.; Chen, H.; Zhang, Z.-J. Adiponectin regulates electroacupuncture-produced analgesic effects in association with a crosstalk between the peripheral circulation and the spinal cord. *Brain Behav. Immun.* **2021**, *99*, 43–52. [CrossRef]
107. Hook, M.A.; Falck, A.; Dundumulla, R.; Terminel, M.; Cunningham, R.; Sefiani, A.; Callaway, K.; Gaddy, D.; Geoffroy, C.G. Osteopenia in a Mouse Model of Spinal Cord Injury: Effects of Age, Sex and Motor Function. *Biology* **2022**, *11*, 189. [CrossRef]
108. Bergmann, P.; Body, J.J.; Boonen, S.; Boutsen, Y.; Devogelaer, J.P.; Goemaere, S.; Kaufman, J.; Reginster, J.Y.; Rozenberg, S. Loading and Skeletal Development and Maintenance. *J. Osteoporos.* **2010**, *2011*, 786752. [CrossRef]
109. Seo, B.R.; Mooney, D.J. Recent and Future Strategies of Mechanotherapy for Tissue Regenerative Rehabilitation. *ACS Biomater. Sci. Eng.* **2022**, *8*, 4639–4642. [CrossRef]
110. Takemura, Y.; Moriyama, Y.; Ayukawa, Y.; Kurata, K.; Rakhmatia, Y.D.; Koyano, K. Mechanical loading induced osteocyte apoptosis and connexin 43 expression in three-dimensional cell culture and dental implant model. *J. Biomed. Mater. Res. Part A* **2018**, *107*, 815–827. [CrossRef]
111. Harkema, S.J.; Ferreira, C.K.; Brand, R.J.v.D.; Krassioukov, A.V.; Jeffries, E.C.; Hoffman, S.M.; de Leon, R.; Dominguez, J.F.; Semerjian, T.Z.; Melgar, I.A.; et al. Improvements in Orthostatic Instability with Stand Locomotor Training in Individuals with Spinal Cord Injury. *J. Neurotrauma* **2008**, *25*, 1467–1475. [CrossRef]
112. Braaksma, J.M.; Vegter, R.J.; Leving, M.T.; van der Scheer, J.W.; Tepper, M.; Woldring, F.A.; van der Woude, L.H.; Houdijk, H.; de Groot, S. Handrim wheelchair propulsion technique in individuals with spinal cord injury with and without shoulder pain—A cross-sectional comparison. *Am. J. Phys. Med. Rehabil.* **2023**, *102*, 886–895. [CrossRef] [PubMed]
113. Lambach, R.L.; Stafford, N.E.; Kolesar, J.A.; Kiratli, B.J.; Creasey, G.H.; Gibbons, R.S.; Andrews, B.J.; Beaupre, G.S. Bone changes in the lower limbs from participation in an FES rowing exercise program implemented within two years after traumatic spinal cord injury. *J. Spinal Cord Med.* **2018**, *43*, 306–314. [CrossRef] [PubMed]
114. Bickel, C.S.; Yarar-Fisher, C.; Mahoney, E.T.; McCully, K.K. Neuromuscular Electrical Stimulation–Induced Resistance Training After SCI: A Review of the Dudley Protocol. *Top. Spinal Cord Inj. Rehabil.* **2015**, *21*, 294–302. [CrossRef] [PubMed]

biomedicines

MDPI

Systematic Review

Obliviate! Reviewing Neural Fundamentals of Intentional Forgetting from a Meta-Analytic Perspective

Olga Lucia Gamboa [1,2,*], Hu Chuan-Peng [3,4], Christian E. Salas [5] and Kenneth S. L. Yuen [4,6,*]

1 School of Psychology, University of Sydney, Sydney 2006, Australia
2 EQness, Sydney 2034, Australia
3 School of Psychology, Nanjing Normal University, Nanjing 210024, China; hcp4715@gmail.com
4 Leibniz Institute for Resilience Research, 55122 Mainz, Germany
5 Laboratory of Cognitive and Social Neuroscience, Faculty of Psychology, Diego Portales University, Santiago 7510457, Chile; salasriquelme@gmail.com
6 Neuroimaging Center (NIC), Focus Program Translational Neuroscience, Johannes Gutenberg University Medical Center, 55131 Mainz, Germany
* Correspondence: olga.gamboaarana@sydney.edu.au (O.L.G.); k.yuen@uni-mainz.de (K.S.L.Y.)

Abstract: Intentional forgetting (IF) is an important adaptive mechanism necessary for correct memory functioning, optimal psychological wellbeing, and appropriate daily performance. Due to its complexity, the neuropsychological processes that give birth to successful intentional forgetting are not yet clearly known. In this study, we used two different meta-analytic algorithms, Activation Likelihood Estimation (ALE) & Latent Dirichlet Allocation (LDA) to quantitatively assess the neural correlates of IF and to evaluate the degree of compatibility between the proposed neurobiological models and the existing brain imaging data. We found that IF involves the interaction of two networks, the main "core regions" consisting of a primarily right-lateralized frontal-parietal circuit that is activated irrespective of the paradigm used and sample characteristics and a second less constrained "supportive network" that involves frontal-hippocampal interactions when IF takes place. Additionally, our results support the validity of the inhibitory or thought suppression hypothesis. The presence of a neural signature of IF that is stable regardless of experimental paradigms is a promising finding that may open new venues for the development of effective clinical interventions.

Keywords: intentional forgetting; directed forgetting; fMRI; neuroimaging; meta-analysis; Activation Likelihood Estimation (ALE); Latent Dirichlet Allocation (LDA)

Citation: Gamboa, O.L.; Chuan-Peng, H.; Salas, C.E.; Yuen, K.S.L. Obliviate! Reviewing Neural Fundamentals of Intentional Forgetting from a Meta-Analytic Perspective. *Biomedicines* **2022**, *10*, 1555. https://doi.org/10.3390/biomedicines10071555

Academic Editor: Masaru Tanaka

Received: 4 June 2022
Accepted: 16 June 2022
Published: 29 June 2022

1. Introduction

Forgetting is an important adaptive mechanism essential for correct memory function. It helps regulate the content of memory storage in a way that only appropriate, relevant, and up-to-date information is kept [1,2]. The study of forgetting in animals has taken different forms, from behavioral measures like extinction of conditioned responses, pharmacological manipulations to block memory consolidations, to optogenetic manipulations of engram, or a mixture of all these techniques [3–6]. In human studies, memory extinction has been extensively studied in the domain of fear memory processes [5,7–11]. In parallel, a good amount of work has been devoted to study the failure of memory retention, i.e., incidental forgetting, and its neural correlates on declarative memory [12,13]. Both arms of study have revealed an overlapping brain network consisting of elevated activities in the ventromedial prefrontal cortex, anterior cingulate cortex, precuneus, coupled with the down regulation of the hippocampus, to support forgetting processes [7,8,12,13].

Incidental forgetting or extinction are both considered automatic processes. Intentional forgetting (IF), by contrast, represents an individual's active, volitional pursue to get rid of unwanted information [14]. It has its historical root in Freudian theory, known as suppression, and subsequently re-examined in a neurocognitive framework using neuroimaging techniques [15–17]. The relevance of intentional or motivated forgetting goes beyond mnemonic processes, as it is key for preserving good psychological health, supporting emotion regulation, structuring cognition, and facilitating behavioural flexibility [2]. Understanding the processes underlying intentional forgetting is of great value not only for cognitive scientists but for the medical community trying to develop optimized treatments directed to population suffering from disorders related to the inability to regulate intrusive thoughts. This understanding is even more important in face of the replication crisis of other memory manipulation techniques (e.g., memory extinction by reactivation, [18]). As such, we will solely focus on discussing IF in the current review.

1.1. Experimental Paradigms

Several experimental paradigms have been developed to study IF. They all follow the same principle: participants first learn some information, that later they will be instructed to either forget or remember [16,19,20]. The main difference between these paradigms lies in whether forgetting occurs at the encoding or retrieval phase.

In the think/no-think (TNT) paradigm, participants first go through a learning phase, studying cue-target pairs of items. In the critical phase (think/no-think task) only cue items are presented followed by an instruction to remember (think condition) or to suppress (no-think condition) the associated target. For the no-think condition, participants are instructed to fully avoid allowing the target to enter conscious awareness. Item pairs that are only shown in the learning phase but not the critical phase serve as the baseline condition. In the test phase, the cues from all three conditions (think, no-think and baseline) are shown and participants are asked to recall the correct target items [16,21]. Items in the no-think condition are recalled worse than in the other two conditions, while items in the think condition are recalled better than the ones in the baseline condition [16]. Essentially, the frequency of no-think operations and successful forgetting follows a dose-response relationship. Behavioural outcomes are explained via two theoretical accounts, the inhibitory hypothesis in which brain mechanisms related to inhibitory control are recruited by the no-think items [16] and the interference/substitution hypothesis suggesting that interference coming from information other than the no-think items, further aids forgetting [22,23].

The list–method directed forgetting paradigm, on the other hand, has a simpler experimental design. In the initial phase, participants are instructed to learn a list of items (list 1) to be tested later. Half of the participants are told to forget list 1 (forget condition), and then a second list of items (list 2) to be learned is presented to all participants. During the test phase, participants in both conditions are asked to recall both lists [24]. Participants in the forget condition perform worse at recalling list one, but recall list 2 better than participants in the remember condition [24]. Behavioural results have been explained through the retrieval inhibition hypothesis, in which the list to be remembered (list 2) interferes with the previously learned list (list 1) impairing its recall [20,25].

Finally, a variation of the list–method paradigm, is the item–method directed forgetting paradigm. Words are presented one by one to the participants, immediately followed by an instruction to either remember (R) or forget (F). In the test phase, participants are asked to recall all words regardless of the given instruction [26], displaying a better capacity to recall items to be remembered than items instructed to be forgotten [26,27]. In light of neuroimaging findings that show intentional forgetting as an active and complex mechanism [28–32], behavioural outcomes can be explained via the attentional inhibition-executive control hypothesis. Here, items to be forgotten experience an active inhibition that will remove them from working memory, limit their access to attentional resources and avoid future activations [33]. Meanwhile, the executive system actively regains processing resources boosting the rehearsal of items to be remembered [34–36].

1.2. Hypotheses of Brain Mechanisms: Thought Suppression and/or Substitution

Although the experimental paradigms mentioned above can successfully induce forgetting, there is not a clear understanding of the exact neuropsychological processes used to achieve IF. Researchers have also not yet reconciled to a cognitive framework explaining the underlying mechanisms supporting IF, therefore we see conceptually overlapping hypotheses are constantly being proposed. In our previous work we have qualitatively analyzed subjects' reported strategies employed during an item-method paradigm and found evidence of both active suppression and self-induced interference as predominant strategies to forget intentionally [29]. Similar hypotheses have been put forward by other research groups to explain how intentional forgetting occurs in the brain. First, the inhibitory or thought suppression hypothesis, which refers to a direct suppression of the unwanted memories, and second, the substitution or thought replacement hypothesis, a mechanism in which to-be-forgotten material is replaced by irrelevant content [29,37]. Experimental findings suggest that these two hypothesized processes are subserved by discrete neural circuitries: a fronto-hippocampal circuit that supports thought suppression/inhibition, and the ventral lateral prefrontal cortex (VLPFC) and the inferior frontal gyrus (IFG, labelled as caudal prefrontal cortex cPFC in the original paper) that supports thought substitution/replacement [38]. Attempts to test these two hypotheses have so far produced mixed results, as it has not been possible to replicate the differential patterns of neural activation comparing inhibition and thought substitution/replacement [39]. Therefore, it would be interesting to see if the two hypothesized patterns of neural activation can be observed in a meta-analysis of neuroimaging studies using the different IF paradigms.

1.3. Current Meta-Analysis

We have two main goals performing this meta-analysis: (1) to summarize and examine in a quantitative manner the neural correlates of intentional forgetting, (2) to establish to what extent the proposed neurobiological models (thought suppression and/or substitution) are supported by the reported data. We used two different meta-analytic algorithms, Activation Likelihood Estimation (ALE) & Latent Dirichlet Allocation (LDA) to provide complementary analyses on the convergence and divergence of brain activations reported in the literature [40]. The ALE algorithm is conventionally used for coordinate-based meta-analysis of neuroimaging results [41,42]. It identifies areas that exhibit a convergence of reported coordinates across experiments that is higher than expected under a random spatial association. While ALE analysis focuses on the convergence of activities across studies, complementary analysis using the Latent Dirichlet Allocation (LDA) algorithm can look into the divergence of neural circuitry underlying intentional forgetting. LDA is a data-driven Bayesian framework originally designed to perform automatic semantic extraction from a corpus of text. In recent years LDA and its variant, Author Topic Modelling (ATM), have been utilized to analyze neuroimaging data in order to reveal the latent cognitive network across experimental tasks or clinical conditions [40,43–46]. A combination of ALE and LDA is a novel approach that will strengthen our understanding of the mechanisms underlying intentional forgetting and may yield valuable information that can be very useful during the development of effective treatments for neuropsychiatric disorders related to intrusive thoughts and the inability to detach from unwanted memories.

2. Materials and Methods

2.1. Literature Search and Article Selection

Following the PRISMA 2009 flow diagram [47], we reported the literature search and the articles selection process as below (see Figure 1A). First, we performed an online-search to identify studies matching our scope in PubMed (date: May 2022), using the following syntax: intentional forgetting [Title/Abstract] OR motivated forgetting [Title/Abstract] OR instructed forgetting [Title/Abstract]) AND ((Magnetic Resonance Imaging) OR Directed forgetting [Title/Abstract]) AND ((Magnetic Resonance Imaging) OR (functional Magnetic Resonance Imaging) OR (Positron emission tomography)) filter English.

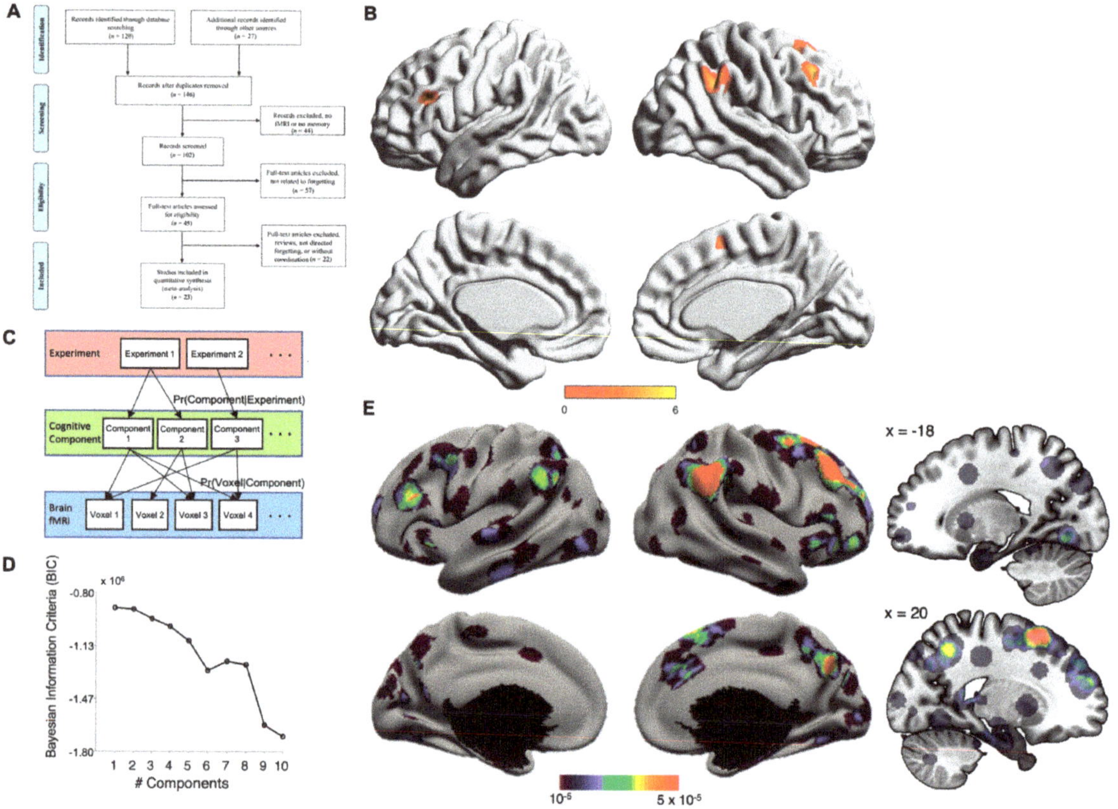

Figure 1. (**A**) PRISMA flow diagram; (**B**) results of ALE analysis; (**C**) Schematic diagram of LDA, (**D**) model selection of LDA results by BIC, and (**E**) LDA results (right hand side sagittal slides showing hippocampal & subcortical activations). Color bars at bottom of panel (**B**,**E**) represents the display threshold of the blobs presented.

2.2. Activation Likelihood Estimation (ALE) Analysis

Additional studies were identified through additional database (Google scholar) and the reference list obtained from the screened articles by the author (OLG, KY). After deleting the duplicated items, our search resulted in 147 studies for further screening. All studies were then screened according to our eligibility criteria: below): (1) studies that investigated intentional forgetting using fMRI and PET; (2) studies with healthy participants that were young to mid-aged adults, i.e., aged 18–45 years old. Studies focused on patients but reporting results from a healthy control group were included; (3) studies reporting whole-brain analysis (articles with results derived from only ROI analyses were excluded); (4) studies reporting standard reference frames such as MNI or Talairach; (5) if multiple papers used the same dataset, only one of these papers was included. Details of article selection is presented in Figure 1A. Conceptualization of this meta-analytoc review was pre-registered at the Open Science Foundation (https://osf.io/xaq5k, DOI: 10.17605/OSF.IO/XAQ5K, 2 June 2022).

We used the revised ALE algorithm for the coordinate-based meta-analysis of neuroimaging results [48,49]. This algorithm identifies areas that exhibit a convergence of reported coordinates across experiments that is higher than expected under a random spatial association. To account for the uncertainty associated with each activation cluster, ALE algorithm constructs 3D Gaussian probability distributions of activation likelihood based on each peak voxel. The Full-Width Half-Maximum (FWHM) of these Gaussian functions were determined based on the between-subject variance by the number of examined subjects per study so that foci with larger sample sizes can be modeled by "smaller" Gaussian distributions because they provide more reliable approximations of the "true" activation effect [48].

The probabilities of all foci reported in a given experiment were then combined for each voxel, resulting in a modeled activation (MA) map [41]. Taking the union across these MA maps yielded voxel-wise ALE scores that described the convergence of the results across experiments at each location of the brain. To distinguish "true" convergence among studies from random convergence (i.e., noise), we compared ALE scores to an empirical null distribution reflecting a random spatial association among experiments. Here, a random-effects inference was invoked, focusing on the inference on the above-chance convergence among studies rather than the clustering of foci within a particular study. Computationally, deriving this null-hypothesis involved sampling a voxel at random from each of the MA maps and taking the union of these values in the same manner as performed for the (spatially contingent) voxels in the true analysis, a process that can be solved analytically [41]. The p-value of the "true" ALE was then given by the proportion of equal or higher values obtained under the null-distribution. The resulting non-parametric p-values were then thresholded at the $p < 0.05$ (cluster-level corrected for multiple-comparison; cluster-forming threshold $p < 0.001$ at voxel level) [41]. All significant clusters were reported, and the volume, weighted center and locations, and Z-scores at the peaks within the regions are given.

2.3. Latent Dirichlet Allocation (LDA) Analysis

While ALE analysis focuses on the convergence of activities across studies, complementary analysis using the Latent Dirichlet Allocation (LDA) algorithm can look into the divergence of neural circuitry underlying intentional forgetting. LDA is a data-driven Bayesian framework originally designed to perform automatic semantic extraction from a corpus of text. In recent years LDA and its variant, Author Topic Modelling (ATM), have been utilized to analyze neuroimaging data in order to reveal the latent cognitive network across experimental tasks or clinical conditions [40,43–46,50]. In brief, LDA/ATM is data-driven Bayesian framework that estimates the latent cognitive component across observed voxel-wise activations (Figure 1C). As there are only two experimental paradigms available for neuroimaging studies of IF, applying ATM will tend to overfit the data with an extra layer of constrain. As such, LDA, i.e., equivalent to an ATM treating each individual study as having their own author, is more appropriate to model the data. Scripts for running LDA/ATM are based on the following Github despository (https://github.com/ThomasYeoLab/CBIG/tree/master/stable_projects/meta-analysis/Ngo2019_AuthorTopic, accessed on 27 March 2020). Conditional probabilities Pr(Voxel | Factor) and Pr(Factor | Study) are being estimated by the Collapsed Variational Bayesian (CVB) inference algorithm (with alpha = 100, eta = 0.01, 100 random seeds for each K), and model selection determining the number of optimal factors (K) is done by the Bayesian Information Criterion (BIC) (Figure 1D).

3. Results

Following standardized procedures, we performed keyword-based literature search, screening, for study inclusion into our meta-analysis (see Methods for details). In brief, we searched for functional neuroimaging studies with healthy participants that performs an IF task up to May 2022. As the study on the neural correlates of IF is relatively new, the majority of studies are conducted on young healthy adults so we focus on samples aged between 18–45 years old. The article selection resulted in 23 studies, with 466 subjects, 159 foci. While it is almost impossible to estimate the number of unpublished null findings that exist, a recent simulation study suggested that a minimum of 20 experiments, in combination with cluster-level correction, should provide adequate power and sensitivity to reveal a robust effect [51]. Our article selection should thus be considered representative of the subject matter. Figure 1A. presents the PRISMA flowchart on study selection (see also Table 1 for a list of the studies included). The ALE results revealed that four brain regions were convergently activated by directed forgetting > remembering contrast: right superior frontal gyrus (rSFG), right inferior parietal lobe (rIPL, including both supramarginal and angular gyri), bilateral middle frontal gyrus (MFG) (see Table 2, Figure 1B). The LDA analysis revealed a one-factor solution as the most optimal model (Figure 1D). The latent cognitive component revealed by this solution is highly similar to the ALE analysis, showing the involvement of bilateral middle frontal gyri, bilateral IPL, plus additional brain networks not revealed by ALE, involving bilateral hippocampal complex, precuneus, bilateral middle cingulum, primary visual cortex and cerebellum (Table 2, Figure 1E).

Table 1. List of fMRI studies included in the current meta-analysis.

Studies	n	Age	Software [1]	Paradigm [2]	Stimuli	Contrast
Anderson et al., 2004 [17]	24 (10F)	29–31	SPM99	T/NT	word pairs	suppression > recall
Bastin et al., 2012 [52]	17 (8F)	20–32	SPM5	DF	6-letter words	To be forget-forget > To be remember-forget
Benoit et al., 2012 [37]	18 (12F)	23.7	SPM8	T/NT	word pairs	suppression > recall
Benoit et al., 2015 [53]	16 (8F)	22	SPM8	T/NT	Picture	suppression > recall
Butler et al., 2010 [54]	14 (7F)	22.6	BV	T/NT	emotion pictures	NT > T (neutral)
Depue et al., 2007 [55]	16 (8F)	19–29	FSL	T/NT	face-picture pairs	Suppression > recall
Depue et al., 2016 [56]	21 (10F)	21.5	FSL	T/NT	neutral face pictures	Suppression > recall
Gagnepain et al., 2014 [57]	24 (11F)	22.	SPM8	T/NT	word-object pairs	Suppression > recall
Gagnepain et al., 2017 [58]	22 (8F)	18–35	SPM12	T/NT	face-scene pairs	NT > T
Gamboa et al., 2018 [28]	31 (15F)	27.5	SPM12	DF	vocal words	To be Forget > to be remember
Hanslmayr et al., 2012 [59]	22 (15F)	23.05	SPM5	DF	words	To be Forget > to be remember
Marchewka et al., 2016 [60]	18 (18F)	22.02	SPM12	DF	emotional pictures	TBF-F > TBR-F
Noreen et al., 2016 [38]	22 (18F)	18–29	SPM8	T/NT	word-autobiographic-memory pairs	no-think > think
Nowicka et al., 2011 [27]	16 (8F)	26.6	SPM8	DF	emotional pictures	TBF > TBR for neutral pictures
Reber et al., 2002 [61]	12 (9F)	20	NA	DF	faces	TBF > TBR
Rizio et al., 2013 [62]	24 (NA)	21.11	SPM8	DF	words	TBF > TBR
Sacchet et al., 2017 [63]	16 (8F)	31.7	AFNI	T/NT	word-pairs	no-think > think
Wang et al., 2019 [64]	20 (10F)	23.6	SPM 12	DF	pictures (scene, faces, objects)	TBF > TBR

Table 1. *Cont.*

Studies	n	Age	Software [1]	Paradigm [2]	Stimuli	Contrast
Wierzba et al., 2018 [29]	24 (24F)	24.6	SPM12	DF	neutral/affective words	TBF > TBR
Wylie et al., 2008 [35]	11 (6F)	26	AFNI	DF	word pairs	TBF > TBR
Yang, T. et al., 2016 [30]	21 (13F)	22.19	SPM8	DF	word pairs	TBF > TBR (neutral words)
Yang, W. et al., 2013 [31]	25 (14F)	30	SPM8	DF	word pairs	TBF > TBR (neutral words)
Yang, W. et al., 2016 [65]	32 (10F)	30	SPM8	DF	word pairs	TBF > TBR

[1] SPM = Statistical Parametric Mapping (The Wellcome Center for Human Neuroimaging, UCL Queens Square Institute of Neurology, London, UK); BV = Brain Voyager (Brain Innovation, Inc., The Netherlands); FSL = FMIRB Software Library (FMRIB, Oxford, UK); AFNI = Analysis of Functional NeuroImages (National Institute of Mental Health, USA); [2] DF = Directed forgetting; T/NT = think/no-think; TBF = To be forget; TBR = To be remember; TBF-F = To be forget and forget; TBR-F = To be remember but forget; NT > T = No think > Think.

Table 2. ALE and LDA results.

Cluster	Coordinates (MNI) X	Y	Z	Number of Voxels	L/R	Anatomical Structure
ALE						
1	16	16	60	221	R	Superior Frontal Gyrus
2	58	−46	36	212	R	Inferior Parietal Lobe
3	42	24	44	160	R	Middle Frontal Gyrus
4	−42	28	24	117	L	Middle Frontal Gyrus
LDA						
1	−45	15	1	4091	L	Inferior Frontal Gyrus
2	**−27**	**45**	**21**	**724**	**L**	**Middle Frontal Gyrus**
3	**37**	**27**	**41**	**16,556**	**R**	**Middle Frontal Gyrus**
4	−21	51	−3	536	L	Orbitofrontal Gyrus
5	−43	−1	45	2122	L	Precentral Gyrus
6	−15	−3	45	498	L	Middle Cingulum
7	21	−39	43	515	R	Middle Cingulum
8	57	−37	13	1104	R	Superior Temporal Gyrus
9	−55	−37	−17	2934	L	Middle Temporal Gyrus
10	69	−25	−17	1012	R	Middle Temporal Gyrus
11	53	−25	−33	1050	R	Inferior Temporal Gyrus
12	−11	−15	−23	552	L	Hippocampus
13	25	−25	−17	1500	R	Parahippocampal Gyrus
14	−55	−59	39	1182	L	Inferior Parietal Lobe
15	**57**	**47**	**41**	**13,717**	**R**	**Inferior Parietal Lobe**
16	−7	−39	63	536	L	Precuneus
17	−43	−79	−5	931	L	Inferior Occipital Gyrus
18	49	−87	−3	398	R	Infeiror Occipital Gyrus
19	−15	−75	−7	1332	L	Lingual Gyrus
20	25	−71	−9	4213	R	Lingual Gyrus
21	13	−93	3	1039	R	Calcarine Gyrus
22	−33	−59	−23	552	L	Cerebellum
23	27	−79	−35	498	R	Cerebellum

Note: LDA results in BOLD overlap with ALE activations.

4. Discussion

During this meta-analysis we examined the neuroimaging literature on intentional forgetting, as a means to get a better understanding of brain structures supporting such an important mechanism. With two different methods (ALE and LDA) we tested the convergence and divergence of underlying neural circuitry that supports IF. Comparing the two resultant activation maps, we found strikingly similar patterns of activation foci in right superior frontal gyrus, bilateral middle frontal gyri, and right inferior parietal lobe. Additional brain network consisting of the hippocampal complex and surrounding temporal areas, middle cingulum, precuneus, primary visual cortex, and cerebellum was revealed with LDA. ALE searches for convergence of neural activation hotspots observed across the selected studies, irrespective of the paradigm used and sample characteristics. Therefore, our results from ALE can be considered the core brain areas supporting IF that generalized across experimental paradigms and studies. Alternatively, LDA search for divergent, latent neural activities that varies in individual studies. Results from LDA can be considered a network of co-activated brain regions that varies in activation depending on the task and stage when IF happened.

4.1. Core IF Brain Regions

The converging neural clusters, right superior frontal gyrus (rSFG), right inferior parietal lobe (supramarginal gyrus/angular gyrus included), and bilateral middle frontal gyri (rMFG), are shown by both meta-analytic analyses to be correlates of intentional forgetting. Each of these areas has shown to have an active participation in tasks involving attentional control and inhibition. For instance, the right SFG has been associated with inhibitory control guided by "top-down" processes [66] and cognitive update of memory representations [67]. Its engagement in cognitive functions such as memory or attention may be related to its anatomical and functional connections with relevant frontal regions such as MFG [68].

The MFG has been considered to be an important contributor during retrieval processes. According to hemispherical specializations, attentional and response selection mechanisms have been attributed to the left MFG, while monitoring processes have been linked to right MFG activation [69,70]. Additional to this, it is thought that the right MFG leads attentional control processes by reorienting attention from the external to the internal environment [71] and by flexibly adjusting exogenous and endogenous attention according to the task at hand [72]. Notably, studies with patients suffering right frontal lobe injury support the idea of the right frontal regions as key areas during modulation of attentional processes [71,73] and memory retrieval. Particularly in these studies, these patients were unable to regulate rehearsal and retrieval processes [71].

Meanwhile, the right SMG/AG part of the inferior parietal lobe is a brain region instrumental in two of the main components of attention, alertness and focus on a task and attentional shift to respond to novel, salient information [73]. Being part of the ventral posterior parietal cortex (VPC) and of the ventral fronto-parietal attentional system (comprised of the ventral frontal cortex: middle and inferior frontal gyri, the inferior parietal lobe: supramarginal and angular gyri, and the right temporoparietal junction (TPJ), this region is thought to moderate bottom-up attention [72,74]. And interestingly, its degree of activation has been directly linked to encoding failure [56,75,76].

4.2. Supportive IF Network

In addition to the core IF brain regions shown by the ALE analysis, LDA further revealed a divergent group of brain areas that co-activate to support IF. These loosely defined network consist of hippocampal complex and surrounding temporal areas, middle cingulum, precuneus, primary visual cortex and cerebellum. Of particular interest among these brain regions is the role of hippocampal complex, including hippocampus and parahippocampal cortices, in intentional forgetting. Depue and colleagues in addition to the task-based activation reported (and included in our meta-analysishad used functional connectivity during the think-no-think paradigm, as well as fractional anisotropy to provide empirical support for the functional and structural connections between rMFG and hippocampus during forgetting processes. They found that functional communication between the rMFG and hippocampus is supported by the integrity of the cingulum bundle.

And that increased integrity of the anatomical pathway was a predictor of the functional connectivity between these two regions during intentional forgetting. Finally, they reported that both structural and functional connections mediated behavior, arguing that there is an ongoing elemental interplay between, brain structure, brain function, and behavior [77]. The functional coupling between hippocampus and rMFG was further demonstrated by Schmitz et al. who used Magnetic Resonance Spectroscopy (MRS) to investigate GABAergic neurotransmission in the hippocampus. They showed that GABAergic inhibition predicts functional coupling between rMFG and hippocampus that is enhanced during retrieval suppression in the think-no-think paradigm [78]. These results, together with the idea that during intentional forgetting frontal regions fulfill an important role as a cognitive control system that modulates parietal activity (in charge of attentional processes) [31], and, other brain structures (involved in mnemonic processes) such as the medial temporal lobe (MTL) [31,38], could be strong indicators of the cooperation between attentional and inhibitory systems to support to act of intentionally forgetting.

Other regions part of this "network" such as the temporal gyrus, orbitofrontal gyrus, and cerebellum, to name some, have been less studied in the context of IF probably because of their lack of direct involvement in the inhibition process. However, their implication may be associated with functions subserving forgetting. For example, the orbital frontal gyrus and temporal gyrus are known to interact with parietal and frontal areas to assist attentional switching an important mechanism in IF [79,80]. Similarly, the cerebellum, a region that is undeniably less known for its role in cognitive functions, has been found to be of great importance in attention [81]. In a clinical study, Gottwald et al. (2003), found that patients with cerebellar damage had difficulties performing a shifting-attention task [82]. The fact that the performance was poor but not eliminated, was interpreted as an indication of the role of the cerebellum as a center of preparation and optimization of higher cognitive functions such as attentional processes [54,83], which have been reported to be necessary for successful forgetting [29].

4.3. *Thought Suppression and/or Substitution?*

Benoit et al. [38] hypothesized two neurocognitive processes supporting IF: *direct suppression* mediated by an increase level of neural activation in DLPFC, coupled with attenuated activations in the hippocampus, or *thought substitution* mediated by increased activations in both right IFG (labelled as cPFC in the original paper) and VLPFC. In our current meta-analyses, the observed neural activations (either core or distributed IF brain regions) do coincide with the ROIs specified in Benoit et al.'s hypotheses.

The core IF regions we identified and discussed above, involving the frontal-parietal circuit, is strongly implicated in cognitive inhibitory processes. Additionally, our LDA results showed frontal-hippocampal involvement in IF and this frontal-hippocampal network resembles the direct suppression processes. It is important to note that the frontal-hippocampal network proposed by Benoit et al. was identified by means of functional connectivity analysis, and these findings are not included in the current meta-analysis (Table 1). Therefore, the observed ALE and LDA activation patterns can be treated as independent verification of Beniot et al.'s hypothesis.

Upon careful scrutiny of the more extensive LDA findings and the distributed IF network, we do observe IFG/cPFC and VLPFC involvement during IF (Figure 2). This finding provides some hints for the existence of thought substitution processes during IF. Nevertheless, it should be cautioned that a meta-analysis like the current one has no way to access individual's strategy used during IF, but only relies on the observed patterns of neural activation to make a reverse inference on the cognitive processes involved. Linking back to the qualitative analysis on the forgetting strategy in our previous study [29], it is very likely that both inhibition/suppression and thought substitution processes co-exist, or even work in tandem to enhance IF. Further experimental studies should test in detail the differential contribution of IF strategy to achieve successful forgetting.

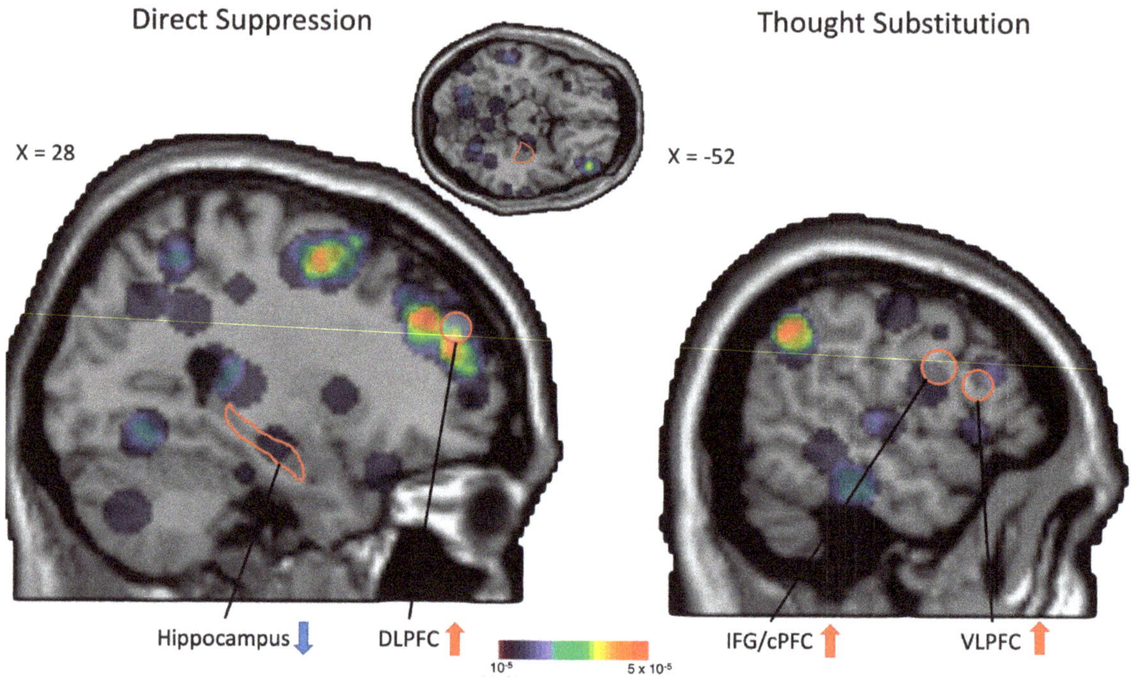

Figure 2. Distinct neural systems for direct suppression and thought substitution, as proposed by Benoit et al. (2012). They proposed that direct suppression involves recruitment of DLPFC and disengagement of hippocampus while thought substitution recruits IFG (caudal PFC) and VLPFC (ROIs in red enclosures, red upward arrows represent hypothetized engagement of brain regions whereas blue downward arrows represent hypothetized disengagement of brain regions). Results from our LDA analysis was overlaid on the MNI anatomical template. We observed distributed activities in all ROIs being mentioned. Here we treat direct suppression and inhibition are interchangeable constructs.

4.4. Applicability of the Findings

Research in the field of forgetting has provided relevant knowledge about how (mostly) the (healthy) brain deals with unwanted information. These results are of great importance for psychiatric disorders where individuals are constantly challenged by involuntary intrusions of unwanted memories. Knowing that purposely trying to forget an unwanted memory, triggers a cascade of mechanisms that leads to obstruction of memory representation and limits its future recovery, may be taken into account to develop new strategies that will help maintain unwanted memories out of awareness.

However, experiments have been mainly performed on healthy participants, and as such, results may not be fully applicable to people with disorders involving problematic thoughts (post-traumatic stress disorder, obsessive-compulsive disorder, depression, etc.), since, resistance or suppression of unwanted memories in a related clinical population may have detrimental outcomes for emotional and mental health [84]. For instance, experiments conducted on patients with anxiety, have shown that using suppression as a coping strategy, reduced the involuntary appearance of anxious thoughts. However, this effect was temporary, and a rebound effect was observed 7 days after the experimental session [85].

In general, studies performed in natural settings have shown that a repressive coping strategy significantly favors the appearance of traumatic memories [49,50], indicating that a more adaptive method to manage intrusive disturbing thoughts in a clinical population is to work with them rather than suppress them [52,85]. Yet, it has been reported that, lack of intentional inhibition of unwanted material results in unsuccessful forgetting [29].

This may imply that to appropriately reduce the strength and appearance of intrusive distressing memories, a certain level of intentional suppression is required, besides the regulation of their cognitive and emotional response [17]. Thus, the formulation of new strategies to regulate intrusive thoughts may benefit from developing methods directed to find the right amount of suppression for each individual combined with techniques such as transcranial magnetic stimulation and cognitive behavioral interventions involving the training of non-judgmental awareness of the disturbing thoughts.

Author Contributions: Conceptualization, O.L.G. and K.S.L.Y.; methodology, K.S.L.Y. and H.C.-P.; formal analysis, K.S.L.Y. and H.C.-P.; data curation, O.L.G., H.C.-P., C.E.S. and K.S.L.Y.; writing—original draft preparation, O.L.G.; and writing—review and editing, O.L.G., H.C.-P., C.E.S. and K.S.L.Y. All authors have read and agreed to the published version of the manuscript.

Funding: This research received no external funding.

Institutional Review Board Statement: Not applicable.

Informed Consent Statement: Not applicable.

Data Availability Statement: Documentations of our literature search, screening and exclusion can be found in https://osf.io/ph278/.

Conflicts of Interest: The authors declare no conflict of interest.

References

1. Bjork, R. Retrieval inhibition as an adaptive mechanism in human memory. In *Varieties of Memory and Consciousness: Essays in Honour of Endel Tulving*; Roediger, H.L., III, Craik, F., Eds.; Erlbaum: Hillsdale, NJ, USA, 1989; pp. 309–330.
2. Norby, S. Why Forget? On the Adaptive Value of Memory Loss. *Perspect. Psychol. Sci. J. Assoc. Psychol. Sci.* **2015**, *10*, 551–578. [CrossRef] [PubMed]
3. Nader, K.; Schafe, G.; Le Doux, J. Fear memories require protein synthesis in the amygdala for reconsolidation after retrieval. *Nature* **2000**, *406*, 722–726. [CrossRef] [PubMed]
4. Ryan, T.J.; Frankland, P.W. Forgetting as a form of adaptive engram cell plasticity. *Nat. Rev. Neurosci.* **2022**, *23*, 173–186. [CrossRef] [PubMed]
5. Haaker, J.; Gaburro, S.; Sah, A.; Gartmann, N.; Lonsdorf, T.B.; Meier, K.; Singewald, N.; Pape, H.C.; Morellini, F.; Kalisch, R. Single dose of L-dopa makes extinction memories context-independent and prevents the return of fear. *Proc. Natl. Acad. Sci. USA* **2018**, *110*, E2428–E2436. [CrossRef]
6. Salinas-Hernandez, X.I.; Vogel, P.; Betz, S.; Kalisch, R.; Sigurdsson, T.; Duvarci, S. Dopamine neurons drive fear extinction learning by signaling the omission of expected aversive outcomes. *eLife* **2018**, *7*, e38818. [CrossRef]
7. Kalisch, R.; Korenfeld, E.; Stephan, K.E.; Weiskopf, N.; Seymour, B.; Dolan, R.J. Context-dependent human extinction memory is mediated by a ventromedial prefrontal and hippocampal network. *J. Neurosci. Off. J. Soc. Neurosci.* **2016**, *26*, 9503–9511. [CrossRef]
8. Lonsdorf, T.B.; Haaker, J.; Kalisch, R. Long-term expression of human contextual fear and extinction memories involves amygdala, hippocampus and ventromedial prefrontal cortex: A reinstatement study in two independent samples. *Soc. Cogn. Affect. Neurosci.* **2014**, *9*, 1973–1983. [CrossRef]
9. Gerlicher, A.; Tüscher, O.; Kalisch, R. Dopamine-dependent prefrontal reactivations explain long-term benefit of fear extinction. *Nat. Commun.* **2018**, *9*, 4294. [CrossRef]
10. Thiele, M.; Yuen, K.; Gerlicher, A.; Kalisch, R. A ventral striatal prediction error signal in human fear extinction learning. *NeuroImage* **2021**, *229*, 117709. [CrossRef]
11. Kalisch, R.; Gerlicher, A.; Duvarci, S. A Dopaminergic Basis for Fear Extinction. *Trends Cogn. Sci.* **2019**, *23*, 274–277. [CrossRef]
12. Wagner, A.D.; Schacter, D.L.; Rotte, M.; Koutstaal, W.; Maril, A.; Dale, A.M.; Rosen, B.R.; Buckner, R.L. Building memories: Remembering and forgetting of verbal experiences as predicted by brain activity. *Science* **1998**, *281*, 1188–1191. [CrossRef]
13. Kim, H. Neural activity that predicts subsequent memory and forgetting: A meta-analysis of 74 fMRI studies. *Neuroimage* **2011**, *54*, 2446–2461. [CrossRef] [PubMed]
14. Golding, J.M.; Long, D.L. There's more to intentional forgetting than directed forgetting: An integrative review. In *Intentional Forgetting: Interdisciplinary Approaches*; Lawrence Erlbaum Associates Publishers: Mahwah, NJ, USA, 1998; pp. 59–102.
15. Freud, S. Resistance and suppression. In *A General Introduction to Psychoanalysis*; Horace Liveright: New York, NY, USA, 1920; pp. 248–261.
16. Anderson, M.C.; Green, C. Suppressing unwanted memories by executive control. *Nature* **2001**, *410*, 366–369. [CrossRef] [PubMed]
17. Anderson, M.C.; Ochsner, K.N.; Kuhl, B.; Cooper, J.; Robertson, E.; Gabrieli, S.W.; Glover, G.H.; Gabrieli, J.D.E. Neural systems underlying the suppression of unwanted memories. *Science* **2004**, *303*, 232–235. [CrossRef] [PubMed]
18. Chalkia, A.; Schroyens, N.; Leng, L.; Vanhasbroeck, N.; Zenses, A.-K.; Van Oudenhove, L.; Beckers, T. No persistent attenuation of fear memories in humans: A registered replication of the reactivation-extinction effect. *Cortex* **2020**, *129*, 496–509. [CrossRef]

19. Basden, B.H.; Basden, D.R. Directed forgetting: Further comparisons of the item and list methods. *Memory* **1996**, *4*, 633–653. [CrossRef]
20. MacLeod, C.M. The item and list methods of directed forgetting: Test differences and the role of demand characteristics. *Psychon. Bull. Rev.* **1999**, *6*, 123–129. [CrossRef]
21. Depue, B.E.; Banich, M.T.; Curran, T. Suppression of emotional and nonemotional content in memory: Effects of repetition on cognitive control. *Psychol. Sci.* **2006**, *17*, 441–447. [CrossRef]
22. Bulevich, J.B.; Roediger, H.L.; Balota, D.A.; Butler, A.C. Failures to find suppression of episodic memories in the think/no-think paradigm. *Mem. Cogn.* **2006**, *34*, 1569–1577. [CrossRef]
23. Hertel, P.T.; Calcaterra, G. Intentional forgetting benefits from thought substitution. *Psychon. Bull. Rev.* **2005**, *12*, 484–489. [CrossRef]
24. Geiselman, R.E.; Bjork, R.A.; Fishman, D.L. Disrupted retrieval in directed forgetting: A link with posthypnotic amnesia. *J. Exp. Psychol. Gen.* **1983**, *112*, 58–72. [CrossRef] [PubMed]
25. Basden, B.H.; Basden, D.R.; Gargano, G.J. Directed forgetting in implicit and explicit memory tests: A comparison of methods. *J. Exp. Psychol. Learn. Mem. Cogn.* **1993**, *19*, 603–616. [CrossRef]
26. MacLeod, C. Directed forgetting affects both direct and indirect tests of memory. *J. Exp. Psychol. Learn. Mem. Cogn.* **1989**, *15*, 13–21. [CrossRef]
27. Lee, Y.S. Cognitive load hypothesis of item-method directed forgetting. *Q. J. Exp. Psychol.* **2012**, *65*, 1110–1122. [CrossRef] [PubMed]
28. Nowicka, A.; Marchewka, A.; Jednoróg, K.; Tacikowski, P.; Brechmann, A. Forgetting of emotional information is hard: An fMRI study of directed forgetting. *Cereb. Cortex* **2011**, *21*, 539–549. [CrossRef] [PubMed]
29. Gamboa, O.L.; Yuen, K.S.L.; von Wegner, F.; Behrens, M.; Steinmetz, H. The challenge of forgetting: Neurobiological mechanisms of auditory directed forgetting. *Hum. Brain Mapp.* **2018**, *39*, 249–263. [CrossRef]
30. Wierzba, M.; Riegel, M.; Wypych, M.; Jednoróg, K.; Grabowska, A.; Marchewka, A. Cognitive control over memory–individual differences in memory performance for emotional and neutral material. *Sci. Rep.* **2018**, *8*, 3808. [CrossRef]
31. Yang, T.; Lei, X.; Anderson, M. Decreased inhibitory control of negative information in directed forgetting. *Int. J. Psychophysiol.* **2016**, *100*, 44–51. [CrossRef]
32. Yang, W.; Liu, P.; Cui, Q.; Wei, D.; Li, W.; Qiu, J.; Zhang, Q. Directed forgetting of negative self-referential information is difficult: An FMRI study. *PLoS ONE* **2013**, *8*, e75190. [CrossRef]
33. Zacks, R.T.; Radvansky, G.; Hasher, L. Studies of directed forgetting in older adults. *J. Exp. Psychol. Learn. Mem. Cogn.* **1996**, *22*, 143–156. [CrossRef]
34. Hourihan, K.L.; Taylor, T.L. Cease remembering: Executive control processes in directed forgetting. *J. Exp. Psychol. Hum. Percept. Perform.* **2006**, *32*, 1354–1365. [CrossRef] [PubMed]
35. Taylor, T.L. Inhibition of return following instructions to remember and forget. *Q. J. Exp. Psychol. A* **2005**, *58*, 613–629. [CrossRef] [PubMed]
36. Wylie, G.R.; Foxe, J.J.; Taylor, T.L. Forgetting as an active process: An FMRI investigation of item-method-directed forgetting. *Cereb. Cortex* **2008**, *18*, 670–682. [CrossRef] [PubMed]
37. Gamboa, O.L.; Garcia-Campayo, J.; Müller, T.; von Wegner, F. Suppress to Forget: The Effect of a Mindfulness-Based Strategy during an Emotional Item-Directed Forgetting Paradigm. *Front. Psychol.* **2017**, *8*, 432. [CrossRef]
38. Benoit, R.G.; Anderson, M.C. Opposing mechanisms support the voluntary forgetting of unwanted memories. *Neuron* **2012**, *76*, 450–460. [CrossRef]
39. Noreen, S.; O'Connor, A.R.; MacLeod, M.D. Neural Correlates of Direct and Indirect Suppression of Autobiographical Memories. *Front. Psychol.* **2016**, *7*, 379. [CrossRef]
40. Ngo, G.H.; Eickhoff, S.B.; Nguyen, M.; Sevinc, G.; Fox, P.T.; Spreng, R.N.; Yeo, B.T. Beyond consensus: Embracing heterogeneity in curated neuroimaging meta-analysis. *Neuroimage* **2019**, *200*, 142–158. [CrossRef]
41. Eickhoff, S.B.; Laird, A.; Grefkes, C.; Wang, L.; Zilles, K.; Fox, P.T. Coordinate-based activation likelihood estimation meta-analysis of neuroimaging data: A random-effects approach based on empirical estimates of spatial uncertainty. *Hum. Brain Mapp.* **2009**, *30*, 2907–2926. [CrossRef]
42. Turkeltaub, P.E.; Eden, G.F.; Jones, K.M.; Zeffiro, T.A. Meta-analysis of the functional neuroanatomy of single-word reading: Method and validation. *Neuroimage* **2002**, *16*, 765–780. [CrossRef]
43. Yeo, B.T.T.; Krienen, F.M.; Eickhoff, S.B.; Yaakub, S.N.; Fox, P.T.; Buckner, R.L.; Asplund, C.; Chee, M. Functional specialization and flexibility in human association cortex. *Cereb. Cortex* **2015**, *25*, 3654–3672. [CrossRef]
44. Zhang, X.; Mormino, E.C.; Sun, N.; Sperling, R.A.; Sabuncu, M.R.; Yeo, B.T.T.; Weiner, M.W.; Aisen, P.; Petersen, R.; Jack, C.R.; et al. Bayesian model reveals latent atrophy factors with dissociable cognitive trajectories in Alzheimer's disease. *Proc. Natl. Acad. Sci. USA* **2016**, *113*, E6535–E6544. [CrossRef]
45. Sun, N.; Mormino, E.C.; Chen, J.; Sabuncu, M.R.; Yeo, B.T.; Alzheimer's Disease Neuroimaging Initiative. Multi-modal latent factor exploration of atrophy, cognitive and Tau heterogeneity in Alzheimer's Disease. *Neuroimage* **2019**, *201*, 116043. [CrossRef] [PubMed]

46. Groot, C.; Yeo, B.T.; Vogel, J.W.; Zhang, X.; Sun, N.; Mormino, E.C.; Pijnenburg, Y.A.; Miller, B.L.; Rosen, H.J.; La Joie, R.; et al. Latent atrophy factors related to phenotypical variants of posterior cortical atrophy. *Neurology* **2020**, *95*, e1672–e1685. [CrossRef] [PubMed]
47. Moher, D.; Liberati, A.; Tetzlaff, J.; Altman, D.G.; The PRISMA Group. Preferred reporting items for systematic reviews and meta-analyses: The PRISMA statement. *Int. J. Surg.* **2010**, *8*, 336–341. [CrossRef] [PubMed]
48. Turkeltaub, P.E.; Eickhoff, S.B.; Laird, A.; Fox, M.; Wiener, M.; Fox, P. Minimizing within-experiment and within-group effects in activation likelihood estimation meta-analyses. *Hum. Brain Mapp.* **2012**, *33*, 1–13. [CrossRef]
49. Eickhoff, S.B.; Bzdok, D.; Laird, A.; Kurth, F.; Fox, P.T. Activation likelihood estimation meta-analysis revisited. *Neuroimage* **2012**, *59*, 2349–2361. [CrossRef] [PubMed]
50. Zhang, H.; Qiu, B.; Giles, C.L.; Foley, H.C.; Yen, J. An LDA-based community structure discovery approach for large-scale social networks. In Proceedings of the 2007 IEEE Intelligence and Security Informatics, Brunswick, NJ, USA, 23–24 May 2007; pp. 200–207.
51. Eickhoff, S.B.; Nichols, T.E.; Laird, A.R.; Hoffstaedter, F.; Amunts, K.; Fox, P.T.; Bzdok, D.; Eickhoff, C.R. Behavior, sensitivity, and power of activation likelihood estimation characterized by massive empirical simulation. *Neuroimage* **2016**, *137*, 70–85. [CrossRef]
52. Hu, S.; Ide, J.S.; Zhang, S.; Li, C.-S.R. The Right Superior Frontal Gyrus and Individual Variation in Proactive Control of Impulsive Response. *J. Neurosci.* **2016**, *36*, 12688–12696. [CrossRef]
53. Bastin, C.; Feyers, D.; Majerus, S.; Balteau, E.; Degueldre, C.; Luxen, A.; Maquet, P.; Salmon, E.; Collette, F. The neural substracts of memory suppression: A fMRI exploration of directed forgetting. *PLoS ONE* **2012**, *7*, e29905. [CrossRef]
54. Benoit, R.G.; Hulbert, J.C.; Huddleston, E.; Anderson, M.C. Adaptive top-down suppression of hippocampal activity and the purging of intrusive memories from consciousness. *J. Cog. Neurosci.* **2015**, *27*, 96–111. [CrossRef]
55. Butler, A.J.; James, K.H. The neural correlates of attempting to suppress negative versus neutral memories. *Cogn. Affect. Behav. Neurosci.* **2010**, *10*, 182–194. [CrossRef] [PubMed]
56. Depue, B.E.; Curran, T.; Banich, M.T. Prefrontal regions orchestrate suppression of emotional memories via a two-phase process. *Science* **2007**, *317*, 215–219. [CrossRef] [PubMed]
57. Depue, B.E.; Orr, J.M.; Smolker, H.R.; Naaz, F.; Banich, M.T. The organization of right prefrontal networks reveals common mechanisms of inhibitory regulation across cognitive, emotional, and motor processes. *Cereb. Cortex* **2016**, *26*, 1634–1646. [CrossRef] [PubMed]
58. Gagnepain, P.; Henson, R.N.; Anderson, M.C. Suppressing unwanted memories reduces their unconscious influence via targeted cortical inhibition. *Proc. Natl. Acad. Sci. USA* **2014**, *111*, E1310–E1319. [CrossRef]
59. Gagnepain, P.; Hulbert, J.; Anderson, M.C. Parallel regulation of memory and emotional supports the suppression of intrusive memories. *J. Neurosci.* **2017**, *37*, 6423–6441. [CrossRef]
60. Hanslmayr, S.; Volberg, G.; Wimber, M.; Oehler, N.; Staudigl, T.; Hartmann, T.; Raabe, M.; Greenlee, M.W.; Bäuml, K.-H.T. Prefrontally driven downregulation of neural synchrony mediates goal-directed forgetting. *J. Neurosci.* **2012**, *32*, 14742–14751. [CrossRef]
61. Marchewka, A.; Wypych, M.; Michałowski, J.; Sińczuk, M.; Wordecha, M.; Jednoróg, K.; Nowicka, A. What is the effect of basic emotions on directed forgetting? Investigating the role of basic emotions in memory. *Front. Hum. Neurosci.* **2016**, *10*, 378. [CrossRef]
62. Reber, P.J.; Siwiec, R.M.; Gitleman, D.R.; Parrish, T.; Mesulam, M.-M.; Paller, K. Neural correlates of successful encoding identifies using functional magnetic resonance imaging. *J. Neurosci.* **2002**, *22*, 9541–9548. [CrossRef]
63. Rizio, A.; Dennis, N.A. The neural correlates of cognitive control: Successful remembering and intentional forgetting. *J. Cogn. Neurosci.* **2013**, *25*, 297–312. [CrossRef]
64. Sacchet, M.D.; Levy, B.J.; Hamilton, J.P.; Maksimovskiy, A.; Hertel, P.T.; Joormann, J.; Anderson, M.; Wagner, A.D.; Gotlib, I.H. Cognitive and neural consequences of memory suppression in major depressive disorder. *Cogn. Affect. Behav. Neurosci.* **2017**, *17*, 77–93. [CrossRef]
65. Wang, T.H.; Placek, K.; Lewis-Peacock, J.A. More is less: Increased processing of unwanted memories facilitates forgetting. *J. Neurosci.* **2019**, *39*, 3551–3560. [CrossRef] [PubMed]
66. Yang, W.; Chen, Q.; Liu, P.; Cheng, H.; Cui, Q.; Wei, D.; Zhang, Q.; Qiu, J. Abnormal brain activation during directed forgetting of negative memory in depressed patients. *J. Affect. Disord.* **2016**, *190*, 880–888. [CrossRef] [PubMed]
67. Tanaka, S.; Honda, M.; Sadato, N. Modality-specific cognitive function of medial and lateral human Brodmann area 6. *J. Neurosci.* **2005**, *25*, 496–501. [CrossRef] [PubMed]
68. Li, W.; Qin, W.; Liu, H.; Fan, L.; Wang, J.; Jiang, T.; Yu, C. Subregions of the human superior frontal gyrus and their connections. *Neuroimage* **2013**, *78*, 46–58. [CrossRef] [PubMed]
69. Rajah, M.N.; Languay, R.; Grady, C.L. Age-related changes in right middle frontal gyrus volume correlate with altered episodic retrieval activity. *J. Neurosci.* **2011**, *31*, 17941–17954. [CrossRef]
70. Rajah, M.N.; Ames, B.; D'Esposito, M. Prefrontal contributions to domain-general executive control processes during temporal context retrieval. *Neuropsychologia* **2008**, *46*, 1088–1103. [CrossRef]
71. Japee, S.; Holiday, K.; Satyshur, M.D.; Mukai, I.; Ungerleider, L.G. A role of right middle frontal gyrus in reorienting of attention: A case study. *Front. Syst. Neurosci.* **2015**, *9*, 23. [CrossRef]

72. Corbetta, M.; Patel, G.; Shulman, G.L. The reorienting system of the human brain: From environment to theory of mind. *Neuron* **2008**, *58*, 306–324. [CrossRef]
73. Conway, M.A.; Fthenaki, A. Disruption of inhibitory control of memory following lesions to the frontal and temporal lobes. *Cortex* **2003**, *39*, 667–686. [CrossRef]
74. Singh-Curry, V.; Husain, M. The functional role of the inferior parietal lobe in the dorsal and ventral stream dichotomy. *Neuropsychologia* **2009**, *47*, 1434–1448. [CrossRef]
75. Uncapher, M.R.; Wagner, A.D. Posterior parietal cortex and episodic encoding: Insights from fMRI subsequent memory effects and dual-attention theory. *Neurobiol. Learn. Mem.* **2009**, *91*, 139–154. [CrossRef] [PubMed]
76. Cabeza, R. Role of parietal regions in episodic memory retrieval: The dual attentional processes hypothesis. *Neuropsychologia* **2008**, *46*, 1813–1827. [CrossRef]
77. Daselaar, S.M.; Prince, S.E.; Cabeza, R. When less means more: Deactivations during encoding that predict subsequent memory. *Neuroimage* **2004**, *23*, 921–927. [CrossRef] [PubMed]
78. Schmitz, T.W.; Correia, M.M.; Ferreira, C.; Prescot, A.P.; Anderson, M.C. Hippocampal GABA enables inhibitory control over unwanted thoughts. *Nat. Commun.* **2017**, *8*, 1311. [CrossRef]
79. Kim, C.; Johnson, N.F.; Cilles, S.E.; Gold, B.T. Common and distinct mechanisms of cognitive flexibility in prefrontal cortex. *J. Neurosci.* **2011**, *31*, 4771–4779. [CrossRef] [PubMed]
80. Kim, C.; Cilles, S.E.; Johnson, N.F.; Gold, B.T. Domain general and domain preferential brain regions associated with different types of task switching: A Meta-analysis. *Hum. Brain Mapp.* **2012**, *33*, 130–142. [CrossRef]
81. Yoo, K.; Rosenberg, M.D.; Kwon, Y.H.; Lin, Q.; Avery, E.W.; Sheinost, D.; Constable, R.T.; Chun, M.M. A brain-based general measure of attentions. *Nat. Hum. Behav.* **2022**, *6*, 782–795. [CrossRef]
82. Gottwald, B.; Mihajlovic, Z.; Wilde, B.; Mehdorn, H.M. Does the cerebellum contribute to specific aspect of attention? *Neuropsychologia* **2003**, *11*, 1452–1460. [CrossRef]
83. Courchesne, E.; Allen, G. Prediction and preparation, fundamental functions of the cerebellum. *Learn. Mem.* **1997**, *4*, 1–35. [CrossRef]
84. Geraerts, E.; Merckelbach, H.; Jelicic, M.; Smeets, E. Long term consequences of suppression of intrusive anxious thoughts and repressive coping. *Behav. Res. Ther.* **2006**, *44*, 1451–1460. [CrossRef]
85. Dalgleish, T.; Hauer, B.; Kuyken, W. The mental regulation of autobiographical recollection in the aftermath of trauma. *Curr. Dir. Psychol. Sci.* **2008**, *17*, 259–263. [CrossRef]

biomedicines

MDPI

Review

The Neurobiological Correlates of Gaze Perception in Healthy Individuals and Neurologic Patients

Simone Battaglia [1,*], Jasper H. Fabius [2], Katarina Moravkova [2], Alessio Fracasso [2,†] and Sara Borgomaneri [1,3,*,†]

1 Centro Studi e Ricerche in Neuroscienze Cognitive, Dipartimento di Psicologia, Alma Mater Studiorum-Università di Bologna, 47521 Cesena, Italy
2 Institute of Neuroscience and Psychology, University of Glasgow, Glasgow G128QB, UK; j.h.fabius@uu.nl (J.H.F.); katarina.moravkova@glasgow.ac.uk (K.M.); alessio.fracasso@glasgow.ac.uk (A.F.)
3 IRCCS Fondazione Santa Lucia, 00179 Rome, Italy
* Correspondence: simone.battaglia@unibo.it (S.B.); sara.borgomaneri@unibo.it (S.B.)
† These authors contributed equally to this work.

Abstract: The ability to adaptively follow conspecific eye movements is crucial for establishing shared attention and survival. Indeed, in humans, interacting with the gaze direction of others causes the reflexive orienting of attention and the faster object detection of the signaled spatial location. The behavioral evidence of this phenomenon is called gaze-cueing. Although this effect can be conceived as automatic and reflexive, gaze-cueing is often susceptible to context. In fact, gaze-cueing was shown to interact with other factors that characterize facial stimulus, such as the kind of cue that induces attention orienting (i.e., gaze or non-symbolic cues) or the emotional expression conveyed by the gaze cues. Here, we address neuroimaging evidence, investigating the neural bases of gaze-cueing and the perception of gaze direction and how contextual factors interact with the gaze shift of attention. Evidence from neuroimaging, as well as the fields of non-invasive brain stimulation and neurologic patients, highlights the involvement of the amygdala and the superior temporal lobe (especially the superior temporal sulcus (STS)) in gaze perception. However, in this review, we also emphasized the discrepancies of the attempts to characterize the distinct functional roles of the regions in the processing of gaze. Finally, we conclude by presenting the notion of invariant representation and underline its value as a conceptual framework for the future characterization of the perceptual processing of gaze within the STS.

Keywords: gaze perception; gaze-cueing; neuroimaging; superior temporal sulcus; amygdala; neurologic patients; neurocognitive mechanisms; non-invasive brain stimulation

Citation: Battaglia, S.; Fabius, J.H.; Moravkova, K.; Fracasso, A.; Borgomaneri, S. The Neurobiological Correlates of Gaze Perception in Healthy Individuals and Neurologic Patients. *Biomedicines* **2022**, *10*, 627. https://doi.org/10.3390/biomedicines10030627

Academic Editors: Eleonóra Spekker and Masaru Tanaka

Received: 1 February 2022
Accepted: 5 March 2022
Published: 9 March 2022

1. Introduction

In any social situation, the direction of one's gaze serves as a fundamental method of communication through which individuals exchange information. Gaze can express relevant information about the mental states of others, support social control, regulate turn-taking, guide attention and communicate intimacy [1–4]. Shifting the attention to the direction in which another person's attention is oriented is associated with the gaze-cueing of attention [5–7]. The direct gaze of others means that attention is directed to the observer, while averted gaze implies that the attention of the other is directed to the environment; consequently, averted gaze may also cause the observer to make reflexive shifts of attention toward the environment [8]. The behavioral index of such a joint shift of attention is called the gaze-cueing effect [9] in which human observers have faster saccadic or manual reaction times (RT) to objects appearing at the gaze-congruent locations compared with objects presented in gaze-incongruent locations [9–11].

Gaze-cueing is considered an involuntary and reflexive effect (see [12,13] for a review), and it is assumed to occur rapidly due to the pivotal role of gaze as a social and biological trigger [14]. However, gaze-cueing is not an encapsulated effect, as it seems to be permeable to contextual influences carried by the face [15], such as facial expression [16–19] and social (e.g., familiarity) or perceptual (e.g., dominance) characteristics [10,20,21]. It is crucial to mention that within a neuroimaging context, it is possible to investigate the neural bases of gaze-cueing (i.e., the fMRI contrast involving averted and direct gaze observation, in the presence of a concomitant behavioral task to measure the gaze-cueing effect) and the neural correlates of gaze direction (i.e., the fMRI contrast between averted and direct gaze observation in the absence of a concomitant behavioral task to measure the gaze-cueing effect). Similarly to *gaze-cueing* in behavioral studies, the perception of *gaze direction* and contextual information can interact [22–24]: when emotion and gaze direction are congruent (e.g., an angry expression with a direct gaze), the perception of that emotion is facilitated, but when emotional expression and gaze direction are incongruent (e.g., an angry expression with an averted gaze), emotion perception is impaired.

From a neuroimaging perspective, there are still unanswered questions about gaze-cueing and gaze direction: (i) What are the neural correlates that mediate gaze-cueing and the perception of gaze direction? (ii) Are the neural bases of gaze-cueing and the perception of gaze direction affected by contextual factors, such as emotional expressions?

Potential neural candidates for the visual processing of others' gaze that could allow for interactions with social factors (such as emotional expressions) are the superior temporal sulcus (STS) and the amygdala (see Figure 1).

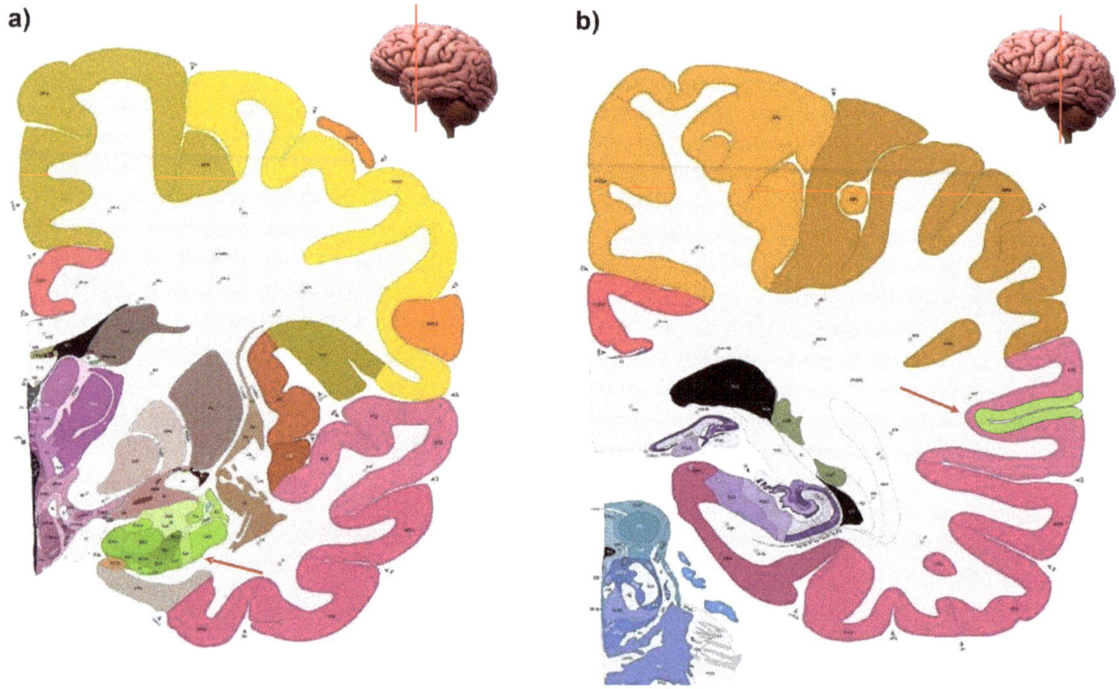

Figure 1. Coronal section of the brain in which the amygdala (panel **a**) and the supe-rior temporal sulcus (panel **b**) are colored in green. The red arrows are oriented towards the neuroanatomical structure of the amygdala and the superior temporal sulcus respectively. The representation of the coronal section of the brain in the present figure was generated free of charge using the Allen Brain Atlas (https://atlas.brain-map.org/, accessed on 24 February 2022).

The STS is part of a distributed neural system for facial information processing, together with the inferior occipital gyri and fusiform gyri [25]. The amygdala is responsive to emotional facial expressions [26–28]; thus, it is a likely candidate for the contextual modulation of face processing. It was reported that the STS is required for the encoding of gaze information [8,29,30]. Notably, the role of the amygdala in gaze processing was reported by several neuroimaging studies [8,31]. This may indicate that the amygdala has a broader role in attentional processes besides its role in emotion processing. In particular, it was argued that the amygdala is activated when the gaze signals a possible threat, as expressed by angry and fearful faces [32–35]. However, despite their mutual involvement in gaze processing, the amygdala and STS respond differently to emotion-related faces and to the concomitant presence of gaze and emotional expression. It is worthwhile to understand the neural bases of gaze direction and gaze-cueing and which contextual factors, such as facial emotional expressions, can modulate the subsequent brain activations. In turn, this helps with understanding the underlying processes causing impairments in social cognition and social functioning that are associated with various psychiatric, neurological and neurodegenerative illnesses [36–40].

Here, we investigated the close interaction between gaze perception and other changeable features of the face (e.g., emotional expressions), along with the neural systems underlying gaze-cueing and the processing of gaze direction. Although other areas were found to be implicated in gaze processing (e.g., fusiform gyrus and even frontal areas), we focused our review on the role of the STS and the amygdala since our main goal was to shed light on their roles in integrating gaze perception and emotional expressions. We concluded our review by showing data of brain-damaged patients and non-invasive brain stimulation evidence, highlighting the crucial roles of the STS and amygdala in gaze perception.

2. Neural Correlates of Gaze Direction: The Pivotal Role of the STS

The close interaction between eye movement, gaze perception and other changeable features of the face, such as emotional expressions, was demonstrated using neuroimaging and brain stimulation techniques, as well as in studies of brain-damaged patients. In this section, we discuss recent neuroimaging evidence investigating the visual processing of gaze-cueing and gaze direction. As mentioned in the introduction, likely neural candidates for visual processing of others' gaze direction that could allow for interactions with social factors are the superior temporal sulcus (STS) and the amygdala.

Neuroimaging studies showed that amygdala activation is not receptive to gaze direction per se [41], but instead is implicated in the emotional processing of gaze information [42,43]. This interpretation can explain why the amygdala shows differential activations when comparing the observation of direct versus averted sight [44]: some studies using neutral facial expressions reported that the amygdala was more active for direct than averted gaze [45,46], where this can be explained by the higher saliency of the direct gaze, which can be considered as a frequent trigger for social interaction, while others, using a social task (i.e., attribute hostile or friendly intentions to actors with a direct or averted gaze), found the opposite [47,48], indicating that the averted gaze may be more informative in an emotional context. Finally, other researchers reported no evidence of amygdala activity in response to gaze direction [49].

In terms of a functional difference between the STS and amygdala, it was suggested that the attentional shifting toward a joint object is a critical function of the STS, while the amygdala may appear to be required for combining emotional expression and gaze direction [32,50,51] rather than in the perception of gaze direction per se [41]. Supporting the role of STS in gaze perception, a seminal fMRI study [52] employed a one-back task in which participants attended selectively to the gaze direction or the identity of each face (match-to-sample task). Results revealed that the perception of gaze direction was mediated by regions in the STS, and it was distinct from the inferior occipital and fusiform gyri, which are dedicated to the perception of face identity. These results are compatible with the concept proposed by Haxby and colleagues [52] (also see [53]), indicating the posterior superior temporal sulcus (pSTS) as one of the three occipitotemporal regions attributed to visual face analysis (specifically, gaze perception), together with the inferior occipital gyrus (IOG) [54] and the lateral fusiform gyrus (lFG) [55]. The STS was also highlighted using a paradigm where participants were adjusted to gaze direction [56]. These authors measured blood oxygen level dependence (BOLD) by means of fMRI adaptation to gaze redirected to the left or right relative to the test faces with straight gazes. The authors found BOLD response suppression in the anterior STS and inferior parietal lobule (IPL), especially for faces with gaze averted to the adaptor's side. In 2003, Pelphrey and co-workers [57] observed shorter hemodynamic responses when participants viewed faces gazing toward objects (in this case, a checkerboard in the visual field) compared with faces gazing toward empty space. This pattern of responses was seen in the right STS, intraparietal sulcus and fusiform gyrus, supporting the assumption that the context in which an eye movement occurs affects brain activity associated with gaze shift perception [58,59].

An interesting study by Carlin and colleagues [60] helped to further clarify the role of the STS in response to faces with different gaze directions. These authors showed that the (bilateral) anterior part of the STS is implicated in the processing of gaze direction. Crucially, they demonstrated that gaze direction sensitivity in the anterior STS was independent of head orientation. These findings indicate the involvement of this area in the view-invariant representation of another's social gaze direction, suggesting a hierarchical processing stream for gaze perception with increasing invariance to gaze-independent features (head orientation) from the posterior to anterior STS [61].

Together, the studies mentioned here revealed that the STS has a pivotal role in processing the direction of gaze compared to the amygdala, at least when neutral facial expressions are presented.

3. Difference between Gaze Cues and Other Symbolic Cues

Multiple neuroimaging studies compared the neuroanatomical correlates of joint attention conveyed by gaze shifts to the correlates of attentional shifts induced by other symbolic cues without using face stimuli. These studies addressed the important question of the specificity of the STS to facial stimuli in the context of gaze perception. Previous studies investigated two alternatives: (i) Is the STS specialized for processing social attention (such as gaze cues)? (ii) Is the STS more broadly processing any symbolic cue (social or not)? In the following section, we summarize papers that investigated the specificity (or lack thereof) of the STS in the social orienting network. We identified recent studies comparing gaze and symbolic cues in fMRI data [42,62–72]. We first summarize their findings and then synthesize them. In agreement with the idea that the STS may be sensitive to intentions expressed by eye gaze [57,73], Hooker and colleagues [42] showed a specificity of the STS in response to gaze cues. The authors showed that gaze and arrow cues recruit distinct brain regions, with gaze cues preferentially activating occipito-temporal regions and arrow cues enhancing the activation of occipito-parietal regions. Importantly, relative to arrow cues, gaze cues showed increased activation in the human FFG and the STS. Furthermore, Lockhofen and colleagues [62] investigated the differences in the activation patterns associated with incongruent gaze and arrow cues in a gaze-cueing task, with a primary focus on brain regions that are involved in the processing of social information (i.e., the STS and fusiform gyrus). Their behavioral data showed that congruent stimuli produce a faster reaction time and that this improvement is more significant with non-social cues, while their imaging data were in line with other reports showing that the STS is more sensitive to gaze shifts than to directional arrow cues [42,63]. This study additionally showed that, unlike arrow cues, the STS did not reveal increased connectivity with the amygdala for gaze, but the connectivity was enhanced between the fusiform gyrus and the amygdala. Similar results were observed in 2008 [64] in which the authors, focusing on specific ROIs in the STS and IPS, indicated the role of the STS in processing eye gaze in a relevant social context. The authors found that gaze cues were associated with bilateral activations in the pSTS region and the cuneus (also see [74]). Therefore, higher STS activation may indicate gaze cues' social relevance over non-social cues.

However, in line with the idea of a lack of differences between symbolic and gaze cues in the STS, Sato and co-workers [75] used fMRI to examine participants' brain activity while they watched directional and non-directional stimuli (including eyes, hands and arrows), which may act as cues for a subsequent target object that could appear in a congruent or incongruent location. Behaviorally, no gaze cue was detected (i.e., the number of accurate responses or RTs did not significantly differ between the stimulus type or direction), and only eyes and arrow cues triggered activity in the right STS, while the 'hands' contrast was only significant using a more liberal statistical threshold. Modest amygdala activation was found specifically in response to eye direction. These results show that the amygdala might be involved in the processing of eye direction, while the STS region is considered to be active for both gaze-related functions and joint attention processes, regardless of the kind of cue that signals the shift of attention.

In addition, in 2014, Callejas and colleagues [66] tackled the question of STS specificity in gaze-cueing, probing the existence of a distinctive mechanism of attention for social cues by comparing the brain activation for social (i.e., computer-generated realistic photos of male/female faces with visible neck and shoulders) and symbolic cues (i.e., arrows), separately evaluating orienting attention in space and reorienting attention for object detection. Results showed no interaction effect between cue validity and cue type, and no evidence was discovered linking face-selective regions (i.e., bilateral fusiform face area (FFA), bilateral occipital face area (OFA) and right posterior superior temporal sulcus (pSTS)) to a hypothetical, specialized mechanism for orienting attention in response to gaze cues reflected by BOLD amplitudes. Interestingly, functional connectivity analysis revealed increased connectivity between face-selective regions and the right intra-parietal sulcus (IPS), pSTS, the amygdala and the inferior frontal gyrus (IFG) during gaze-cueing. The enhanced connectivity suggests that the attentional networks obtain gaze information from face-selective regions, which acquire them from gaze signals. In line with Callejas' findings, Greene and colleagues [67] showed the specific recruitment of extra-striate regions in gaze-cueing (not including the STS or the amygdala). There are also reports showing that symbolic cues selectively activate the STS rather than gaze cues, hence going against the idea of the specificity of the human STS in the social orienting network. Additionally, Engell and colleagues [68] further investigated whether attentional orienting due to the gaze-direction perception and symbolic directional cues (e.g., arrows) are processed by the same brain areas. Behaviorally, both arrow and gaze cues led to faster reaction times during valid trials than during invalid trials, similar to previous findings [76,77]. Importantly, invalid gaze/arrow cues produced a significantly stronger response than valid cues in the right lateral temporal cortex, including the pSTS/temporoparietal junction (TPJ) and the right inferior parietal lobe. In fact, the arrow cues were the main factor that determined this effect. Moreover, despite similar behavioral effects and neural activations, gaze and arrow stimuli triggered the attentional systems differently: invalid arrow cues increased the hemodynamic response in the ventral frontoparietal attention network (specifically, the right TPJ and IFG) compared to valid arrow cues. Gaze cues did not display this difference. Similarly, Hietanen and co-workers [69] reported a difference in the activation of the middle temporal gyrus (anatomically close to the STS) for non-symbolic and gaze cues, with stronger activations for non-symbolic cues, although a gaze generates a significantly larger cueing effect than arrow signals. In a similar study to that conducted by Engell and colleagues [68], Joseph and colleagues [70] found similar behavioral data; however, in contrast to Engell and colleagues' findings, they found that the contrast between invalid and valid gaze cues engages the ventral attention system, specifically, the TPJ and inferior parietal cortical nodes extending to the STS, when participants redirected their attention following the presentation of invalid gaze cues, but not after invalid arrow cues. From these data, the authors concluded that gaze direction engages the attention networks more robustly, suggesting that attentional differences between social and non-social stimuli might be quantitative rather than qualitative, contrary to existing theories, which suggest that the improved processing of social stimuli requires a specific network of brain areas [78].

Moreover, Uono and colleagues [71] performed an MEG source reconstruction analysis to examine the temporal patterns of the neural activation during attentional shifts induced by gazing and arrows. The results revealed that after 200 ms, the STS and the inferior frontal gyrus were activated only when directional cues were presented, suggesting that the brain mechanisms behind attentional shifts elicited by gazing and arrows have similar spatial and temporal distributions. The 350–400 ms time window showed different neural activations in response to gaze and arrow stimuli; however, such activation did not include the STS. Since the difference in attentional orienting between gaze and arrow cues might be distinguished only when the cues are part of a rich setting, Zhao and co-workers [72] compared gaze and arrow cues pairing social gazing and social voice or arrow and tone (congruent condition) or combining social gazing and tone or arrow and social voice (incongruent condition). The results showed that invalid conditions modulated the activity of the ventral frontoparietal network (i.e., left TPJ and IFG), while no activation differences in the cortical areas associated with attentional orienting (i.e., the STS and TPJ) were found in the contextual elaboration between gaze and arrow cues.

To conclude, while the STS is sensitive to gaze perception [25], its role in gaze-cueing is not clear. While several authors reported a specific role of the STS in processing gaze cues compared to non-symbolic cues [42,62,63,70], there is enough non-converging evidence regarding the specificity of the STS to be cautious about deriving definitive conclusions [68,69,72,75]. Indeed, it is important to mention that, even behaviorally, most of the aforementioned studies failed to show a higher cueing effect (i.e., faster reaction time for congruent relative to incongruent cues) for gaze cues compared to non-symbolic cues [62,66,68,75]. Thus, it is possible to speculate that this behavioral lack of advantage for a gaze versus non-symbolic cue is also reflected at the neural level. From the existing literature, it is not possible to disambiguate whether the human STS has a specific role in attention orienting to social stimuli or whether it encodes the visual input and passes information to a supra-modal attentional system.

4. Conjoint Processing of Emotional Expressions and Gaze Direction in the Amygdala and STS?

During social communication, emotional expression is useful for communicating the inner state of others and predicting their potential actions. When changes in emotional expression are associated with gaze shifts, the social cues of the other reveal information that guides individuals' behavior toward or away from other stimuli in the environment. Gaze cues play a significant role when conveying fear, where the meaning of the emotion is ambiguous until the source of the emotional change is determined (i.e., to identify the location of the threat source). The interaction between gaze perception and emotional expression is associated with amygdala and STS activation, which is crucial regarding both facial expression and gaze perception [33,65,79]. However, the neural basis of gaze and emotional cue integration during the attentional orienting process is still unknown. In the following section, we review recent studies investigating the conjoint processing of emotional expression and gaze direction, as well as aspects of the visual presentation of stimuli that can affect the processing of facial emotions (see Table 1) [32,33,42,50,65,79–86].

Although it was not the authors' primary focus of investigation, Hooker and colleagues [42] employed angry and happy faces in a gaze-cueing paradigm. They found that angry faces generated greater activation in the STS region than happy facial expressions. The activation was observed in an area extending from the dorsal STS to the IPS. The fact that STS activity is modulated by facial expressions strengthens the view that this area is receptive to gaze cues related to social interaction [87]. When the task was to detect an infrequent gaze direction change, increased selective amygdala activity was found. To investigate whether the amygdala is involved in the conjoint process of gaze direction encoding and facial expression processing, Adams and colleagues [32] combined the presentation of fearful/angry faces with direct or averted gaze. The authors found that the left amygdala responded strongly to angry averted and fearful directed gaze conditions, whose threat to the subject was difficult to determine. This evidence highlights the role of this neural structure in recognizing not only the presence of a facially conveyed threat but also the ambiguity associated with it. Interestingly, amygdala responses were significantly greater when anger was expressed with averted gaze, where it was only shown by cultural in-group members [80]. In 2004, Sato and colleagues [50] performed a similar experiment to Adams and co-workers. They recorded the BOLD signal while participants were presented with angry/neutral expressions facing or looking away from them. The authors focused their analysis on the amygdala and found that the amygdala was more active for angry expressions looking toward the subjects than those looking away from them (in contrast with Adam's results). These results jointly corroborate the concept that the amygdala shows high sensitivity to the emotional expression of a facial stimulus, whereas its response to gaze direction is less clear (however, see Section 4.2 on 'Transient or sustained presentation of visual stimuli' for a potential account of amygdala response to gaze direction).

Table 1. Summary of neuroimaging findings in studies with emotional facial expressions in different gaze direction paradigms.

Study	Emotional Facial Expression	fMRI Contrast	Paradigm	Coordinates (MNI)	Main Findings
Hooker et al. (2003)	Happy–angry	Angry > happy	Gaze-cueing	Right STS: 42 −57 26 Left STS: −35 −67 31	STS for angry faces > happy faces
Adams et al. (2003)	Fear–angry	Angry averted/fearful direct > angry direct/fearful averted	Passive observation	Left amygdala: −15 0 −18	Amygdala > angry averted and fearful directed gaze
Sato et al. (2004)	Angry–neutral	Expression × face direction	Gender task	Left amygdala: −22 −9 −16	Amygdala > angry expressions looking toward
Engell and Haxby (2007)	Neutral direct–averted gaze anger, disgust, fear and surprise with direct gaze	Emotions > neutral Averted > direct	Match identity task	Right STS (emotion): 52 −48 8 Right STS (gaze): 36 −54 15	STS emotions > neutral faces STS averted > direct gaze faces
Hadjikhani et al. (2008)	Fear	Fearful averted > fearful direct	Passive observation	Right STS (gaze): 48 −54 14 Amygdala (emotion): −12 −11 26	Amygdala and the STS fearful averted gaze > fearful direct gaze
N'Diaye et al. (2009)	Fear–angry (intensity morphing)	Emotions > neutral Averted > direct	Emotional intensity rating	Left amygdala (emotion): −24 +3 −18 Right amygdala (emotion): +24 +6 −15 Left pSTS (gaze): −57 −42 +9 Right pSTS (gaze): +45 −66 +12	Amygdala fearful averted gaze > directed gaze Amygdala angry directed gaze > averted gaze STS averted gaze > direct gaze
Ewbank et al. (2010)	Fear–angry–neutral	Angry direct > neutral direct Angry direct > angry averted Fearful averted > neutral averted	Gender task	Right amygdala (gaze): 22 −6 −12 Right amygdala (emotion × gaze): 24 −4 −12	Amygdala high-anxious angry direct gaze > angry averted gaze Highly anxious > fearful direct gaze and averted gaze
Straube et al. (2010)	Happy–angry–neutral	Averted > direct Valence × gaze	Gender task	Left amygdala (gaze): −27 −9 −14 Right amygdala (gaze): 16 −6 −11 Right STS (valence × gaze): 54 −49 16	Amygdala averted gaze > direct gaze STS averted emotional > averted neutral

Table 1. *Cont.*

Study	Emotional Facial Expression	fMRI Contrast	Paradigm	Coordinates (MNI)	Main Findings
Sato et al. (2010)	Happy–angry (static and dynamic)	Dynamic > static Direct > averted	Gender task	Left amygdala: −18 −10 −10 Right STS: 58 −46 8	Amygdala dynamic happy and angry directed gaze > averted gaze STS dynamic > static
Krämer et al. (2014)	Happy–angry	European angry averted > Asian angry averted	Valence rating	Right amygdala: 28 −8 16	Amygdala > angry averted gaze in ingroup
Ziaei et al. (2017)	Happy–angry–neutral	Expression with averted > all other conditions	Emotion recognition task	Left STS: 56 6 2	Stronger connectivity of the STS and mPFC with bilateral amygdala for angry averted gaze > direct gaze
Schobert et al. (2018)	Happy–angry (gaze shift and speech motion)	Emotional (happy and angry) > non-emotional (gaze and speech) Eyes (angry and gaze) > mouth (happy and speech)	Identity detection task	Left pSTS: −50 −48 15 Right pSTS: 50 −47 13	pSTS gaze-happy and gaze-angry > speech-happy and speech-angry
Kätsyri et al. (2020)	Angry–fear–neutral	Directed > averted + angry > fear	Circle detection in catch trials	pSTS: 65 −54 3	Amygdala direct gaze > averted gaze Amygdala threatening > neutral STS gaze direction × emotional expression

The conjoint processing expressions and gaze direction were explicitly studied also in the STS. An experiment was carried out by Engell and Haxby [81] that aimed at comparing the activation evoked within the STS by different emotional expressions and gaze directions. Regarding this aim, the authors showed participants with neutral faces with a direct or averted gaze, or emotionally expressive faces (e.g., anger, disgust, fear and surprise) with a direct gaze. The results showed a greater STS response when viewing emotional facial expressions as compared to neutral faces and that perceiving faces with an averted as compared to a direct gaze led to greater right STS activation. Unfortunately, the authors did not investigate possible differences across emotions in the activation of the STS. However, a further inspection of the responses within the right STS showed how expression and averted gaze activated distinct, though overlapping, cortical areas. These data are in line with clinical evidence that prosopagnosia patients can show an impaired perception of facial expressions without necessarily suffering from gaze perception deficits [88]. In 2008, Hadjikhani and co-workers [33] performed an fMRI study aiming at investigating whether fearful facial expressions with gaze averted toward a potential threat in the environment elicited more activation in brain areas involved in adaptive action planning. Fearful faces with an averted gaze altered the activity in the amygdala and STS more than fearful faces with a direct gaze. These data suggest that the gaze direction triggers the combined process of facial expression with the information given by the gaze direction. These results are in contrast with the study of Adams and colleagues [32] in which fearful faces with direct gaze were found to activate the amygdala. To further investigate the interaction between emotion and gaze as a way to shed light on the existing contrasting results [32,33,50], N'Diaye and co-workers [79] employed a controlled set of virtual animated faces, which could express emotions dynamically and with different intensities and gaze shifts directed toward or away from the participant. Their results showed that gaze direction modulated neural responses to parametrically manipulated fearful and angry expressions (i.e., by manipulating the intensity of these expressions using a morphing procedure) only when they were ambiguous (in the middle of the fearful–neutral continuum). In line with previous imaging studies, these results showed greater amygdala activation for fearful faces when their gaze was oriented away from rather than toward the observer [33,89], as well as stronger amygdala activation for angry faces looking toward rather than away from the observer [50,89]. Interestingly, the present data seem to suggest that a possible explanation for the contradictory results may be the effect of the emotion intensity on gaze-related effects. Critically, the STS was more active when faces were looking away from the observer relative to direct gaze trials (as in [81]). However, no evidence for the role of the STS in the combined perception of emotion and gaze was reported.

Furthermore, to investigate whether individual differences in anxiety levels may explain the abovementioned inconsistent findings, Ewbank and colleagues [82] examined the effect of gaze direction on amygdala activation for the facial expressions of anger, fear and neutral expressions, controlling for differences in individual anxiety levels. They found that angry faces evoked greater responses in highly anxious participants compared to angry faces with an averted gaze, while the response to fearful faces increased with anxiety for both direct and averted gaze conditions. Recent studies untangled the pathophysiology underlying anxiety disorders: experimental and preclinical evidence revealed that anxiety disorders are determined by the abnormal neural processing of threat-related stimuli, which is mediated by the cyclic AMP (cAMP)–protein kinase A (PKA) pathway [90–92]. Additionally, the downregulation of the regulatory subunit of PKA within the amygdala can lead to an augmentation of anxiety-like behavior, resulting in changes in amygdala activation in response to a gaze [93]. These findings confirm the amygdala's role in encoding the perceived threat level of faces, as well as the more general assumption that the amygdala encodes the relevance or significance of a stimulus to the observer [94]. These results suggest that individual anxiety levels may explain inconsistent findings from previous studies [95]. Straube and colleagues [83] investigated whether the amygdala is activated by a gaze in response to threat relevance or facial expression by testing its activation during the observation of neutral, happy and angry faces with either a direct or averted gaze. An averted versus direct gaze enhanced amygdala reactivity, regardless of the emotional facial expression, suggesting that the amygdala has a role in gaze processing, despite the valence of the facial expression (as suggested in an earlier study in monkeys [96]). On the other hand, STS displayed an increased activation due to averted emotional versus averted neutral faces, even though this difference was not seen during a direct gaze, in line with the idea that emotional averted faces might be considered more strongly than neutral averted faces because they may have a greater influence on proper behavior in (ambiguous) social interactions.

A potential confounding factor in the abovementioned studies is the use of static stimuli, which may have concealed the role of the amygdala. Therefore, Sato and colleagues [65] showed dynamic and static faces with angry or happy expressions. When dynamic stimuli were used, happy and angry faces with a direct gaze toward the subject caused higher amygdala activity than those looking away. Instead, when static stimuli were presented, a direct gaze had no effect on amygdala activation; only a trend for decreased amygdala activity was observed in response to facial expressions directed toward the subjects than for those directed away. In the same study, the STS was found to be more active in the contrast between dynamic and static stimuli, but not sensitive to the perception of emotion or gaze, in line with its role in perceiving biological motion [65].

In another fMRI study using an emotion recognition task [84], the authors investigated the neural networks underlying emotion recognition when a gaze is at play. Faces represented happiness, anger or a neutral expression with their gaze shifted toward their left or their right. The results showed the stronger connectivity of the STS and mPFC with both amygdalas while viewing an angry face with averted gaze than with a direct gaze, suggesting that gaze perception is not mediated by an encapsulated module, but emotional content can modulate the neural correlates of gaze perception. The idea of the involvement of STS in gaze processing was supported by a 7 T high-resolution fMRI experiment [85], which found a well-defined topographical subdivision within the STS, where the posterior and superior areas are related to gaze activity, while the middle and inferior regions are related to emotion. Finally, a recent study [86] replicated previous findings by demonstrating that the amygdala is activated by threatening emotional facial expressions and that it is highly responsive to direct gaze, but only in real faces and not in computer-generated faces. In addition, only the STS, and not the amygdala, responded differentially to gaze direction according to facial expression.

Overall, these results indicate a complex pattern of amygdala responses to gaze direction, which depends on individual differences, the type of emotion expressed by the face and gaze direction (see Table 1 for more details). Most of the aforementioned studies tested the role of the amygdala during averted and direct gaze only when threatening facial expressions were presented, preventing the possibility of investigating possible differences in emotion. However, in most of these studies, the elicited activity in the amygdala revealed a strong association between gazing and facial expression, suggesting its role in integrating the two sources of information, although not always in the same way. On the other hand, the human STS tends to be less responsive to the emotional expression of a facial stimulus but shows high sensitivity to gaze direction.

4.1. *The Time-Course of Processing of Emotional Expressions and Gaze Direction in the Amygdala and STS*

Magnetoencephalography (MEG) was also used to investigate the conjoint processing of emotion and gaze direction [97]. Lachat and colleagues [98] used MEG to assess whether combining averted gaze and a fearful facial expression could induce a differential early effect of attentional orienting response, in line with several EEG results that showed that fearful expressions were able to modulate early brain components (P1, N170 and N190) [99–102]. Lachat and collaborators found that averted gaze associated with a fearful expression led to an early attentional orienting effect only for congruent stimuli following fearful gaze cues, which was reflected at the neural level with the activation of the left superior parietal and left lateral middle occipital regions. No gaze-cueing effects were observed for the objects following happy gaze cues, and no evidence of the involvement of STS was reported [98]. However, these findings should be considered cautiously due to the limitations in the source localization of the MEG signal.

A subsequent MEG study [103] focused on the time course of the role of the amygdala in emotion and gaze perception, demonstrating that its activity was enhanced at an early stage for fearful compared to neutral faces (between 130 and 170 ms), while an effect of gaze direction in fearful faces was found at a later stage (between 190 and 350 ms). Moreover, a combined EEG and fMRI study [104] demonstrated that at earlier stages (170 ms), the amygdala processes the emotional information separately, while directional cues (gaze direction with pointing gesture) were integrated at approximately 190 ms in the parietal and supplementary motor cortices and not in the emotional system.

4.2. *Transient or Sustained Presentation of Visual Stimuli*

Speed processing represents a still unexplored element of threat perception. Most of the aforementioned studies [33,50,79] employed rapid presentation of visual stimuli, promoting visual input via the magnocellular pathway. This pathway is involved in immediate and quick orienting responses to danger [105]. Instead, Adams and colleagues [32] used prolonged (sustained) presentation of threatening faces, which favors the processing of visual information via the parvocellular system. Therefore, in a new study, the same authors [106] varied the presentation speed of fear displays. Interestingly, they found that responses of the amygdala to averted versus direct gaze in fearful expressions changed according to the presentation time: direct fearful gaze leads to increased left amygdala activity compared to averted fearful gaze when using longer presentation durations, replicating their original pattern of findings [32], while with a shorter presentation, a greater response was found for averted compared to direct fearful gaze (as in [33]).

These results hint at the presence of a fast and reflexive amygdala activation linked to obvious threat and a subsequent reflective activation linked to ambiguous threat (see also a subsequent study investigating sex-related differences in amygdala activation [107]). These findings were confirmed and expanded by a subsequent study from the same authors [97], which also found a higher phase locking to averted-gaze fear in the early response of the right amygdala, similarly to the initial connectivity between the left amygdala and orbitofrontal cortex. This finding suggests the rapid deployment of visual resources to congruent threat cues, and thus supports the right amygdala's role in the immediate recognition of emotionally relevant stimuli.

In the same study, the pSTS showed a different pattern of activation, with stronger responses for the short presentation of averted-gaze fear and the longer presentation of direct-gaze fear. The pSTS was involved in the late stages of longer stimuli presentations, highlighting its potential role as a constituent of the 'social brain' [108]. The 'social brain' is conceived as a core area in the visual integration of social cues [109], particularly when deducing the social meaning from both expressions and eye gazing [110], making it a likely candidate to solve the ambiguity of an incongruent threat cue during longer presentations.

4.3. Gaze Direction and Emotional Expression: Low-Level Visual Features

It is still unknown whether previously observed effects of gaze direction on amygdala activation depended to some degree on low-level features in the stimulus display (e.g., increase in visible eye sclera in the averted gaze conditions). Therefore, Hardee and colleagues [34] performed an fMRI study monitoring amygdala activation while participants observed fearful, gaze-shifting, happy and neutral eyes. They observed greater activity in the left amygdala only for the fearful eye observation. However, a condition combining fear and gaze was absent here since this issue was beyond the scope of the present work. Subsequently, Sauer and co-workers [111] included head direction manipulation in order to control for possible low-level confounding factors. Their data showed that faces with an averted gaze enhance amygdala activation more than faces with a direct gaze, despite the facial expression (angry, happy or neutral) or head orientation, suggesting that the amygdala plays an important role in attention processes and that averted gaze serves as a signal of socially relevant information. These findings support the study by Gamer and Büchel [112], which showed that amygdala activation was significantly enhanced when fearful faces were presented with the mouth aligned to fixation; therefore, when eye movements were directed to the eye region, this highlighted this area's function in reflexive gaze initiations in response to salient stimuli rather than being driven by simple low-level features in a stimulus display.

5. Subliminal vs. Supraliminal Cues

It is possible to trigger attentional orienting with eye gaze information even when such information is subliminal and participants are not aware of it, as demonstrated by multiple studies [113–116]. These studies propose that subliminal and supraliminal gaze cues can automatically foster attentional orienting and may, therefore, share the same neural circuit [117,118]. To address this issue, in 2016, Sato and colleagues [119] assessed brain activity using event-related fMRI while participants watched averted or straight-ahead gaze cues, displayed in supraliminal or subliminal conditions. The gaze cue was presented for 13 ms and then covered by a mask in the subliminal condition, while in the supraliminal condition, the gaze cue was shown for 200 ms without a mask. In both supraliminal and subliminal conditions, a gaze-cueing effect was reported (i.e., reaction times were faster when gaze cues were congruent with object locations than when they were incongruent), consistent with previous findings [120]. Widespread cortical activity was observed for both the supraliminal and subliminal presentation of stimuli with averted eyes compared to straight eyes. The middle temporal gyrus (including the superior temporal sulcus), inferior parietal lobule, anterior cingulate cortex, superior and middle frontal gyri and superior and middle frontal gyri were the most active regions. These findings suggest that the temporo-parieto-frontal attentional network is implicated in both conscious and unconscious gaze-triggered attentional shifts.

6. Brain-Damaged Patients and Transcranial Magnetic Stimulation (TMS) Evidence

Recent growing evidence suggests that non-invasive brain stimulation techniques (NIBS) can be used to induce neuroplasticity in order to modulate cognition and behavior [121,122], unveiling the critical role of specific neural circuits [123,124]. Therefore, NIBS might be applied to modulate low-level mechanisms that mediate, for example, attentional orienting to gaze cues, highlighting the functional role of brain areas (i.e., the amygdala and STS) in the gaze-cueing effect (see Figure 2) [21,125–127].

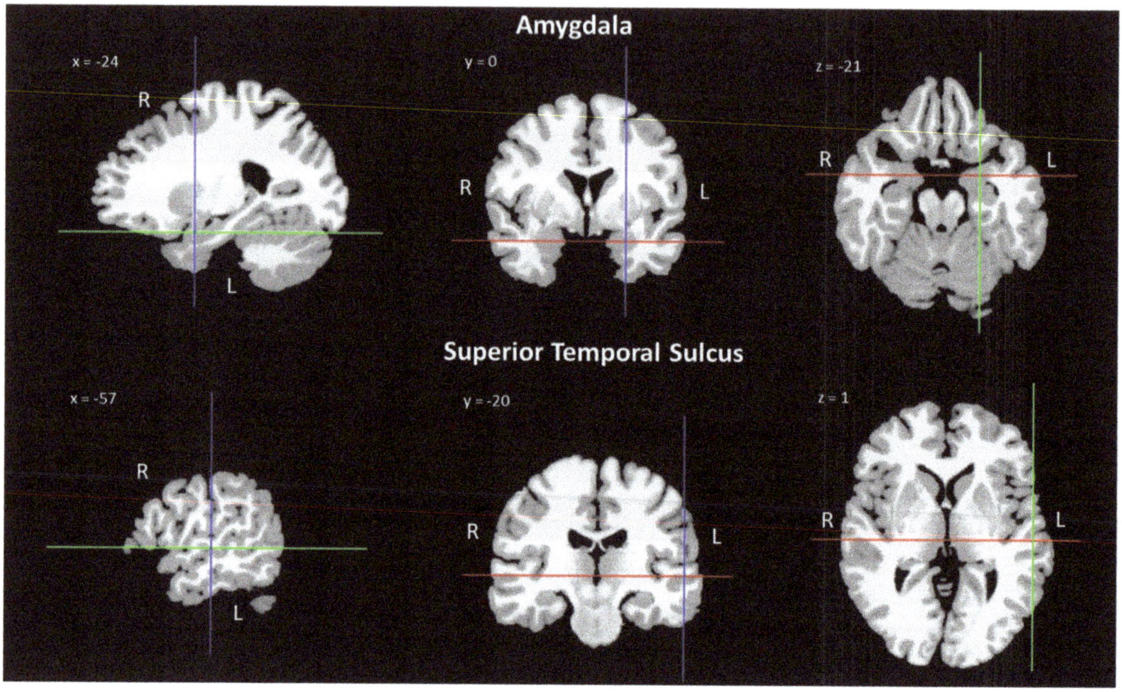

Figure 2. Neuroanatomic MRI images of the amygdala and superior temporal sulcus (STS) with MNI coordinates. The slices of the brain in the different axes of the present figure were generated free of charge using Bioimage Suite Web (https://bioimagesuiteweb.github.io/webapp/mni2tal.html, accessed on 24 February 2022).

Direct evidence for the roles of the STS and amygdala in gaze processing can also be derived from brain-damaged patients. Two studies published in 2006 [128,129] reported the case of a patient with a critical deficit in gaze perception following a stroke in the right STS. Similarly, findings were reported in non-human primates with temporal ablation in which STS damage resulted in an impaired recognition of different gaze angles [88]. For instance, a reversible deactivation (i.e., muscimol injections) of the pSTS in monkeys suppressed gaze-cueing behavior [130]. In addition to STS lesions, it was also reported that amygdala lesions impair attention to eye information and averted gaze-cueing effects [8,26,31,131–133]. Spezio and colleagues [133] demonstrated that amygdala lesions induce longer gazing at the mouth region compared to the eye region during a real conversation, suggesting that the amygdala is involved in recognizing and processing important social cues and directing attention accordingly [8,31]. Finally, an investigation on the interaction involving emotion and gaze processing in amygdala patients [132] demonstrated that a unilateral lesion in the right amygdala affects the interaction of gaze and expression during emotion perception. Thus, this finding adds support to recent theories that the human amygdala might be essential for encoding the self-relevance of emotional events. In particular, amygdala patients showed deficiencies in the effect of direct gaze on the perception of angry expression, implying that gazing plays a greater role in the interpretation of anger.

Although many studies have focused on the critical role of the STS in emotion perception [134–136], a limited number of human studies used non-invasive transcranial magnetic stimulation (TMS) to target pSTS and influence gaze processing [137,138]. Saitovitch and colleagues [138] administered inhibitory repetitive TMS (i.e., continuous theta-burst stimulation) over the right STS and found significantly fewer fixations toward someone else's gaze while viewing social scenes, thus suggesting a top-down effect of the STS on social attention (deploying attentional resources to the eyes and lips of the actors in the presented social scenes). In another TMS study [137], the authors applied single-pulse TMS to investigate the critical role of STS in a gaze direction task. The authors found that TMS applied over the STS 100 or 200 ms after the presentation of a second matching face stimulus (the same/different to the first face gaze direction) hampered the perception of gaze shift. This demonstrates the critical and causal function of the STS at early latencies in extrapolating gaze-shift information. Indeed, results showed that TMS impaired responses exclusively in trials in which there was a gaze shift between the first and the second presentation of faces in pairs (i.e., from straight to averted gaze or vice versa), but TMS had no effect on responses during trials with static gaze positions (straight–straight and averted–averted pairs). These findings suggest that TMS effects during gaze processing were notably content specific, primarily affecting the perception of gaze shifts instead of static eye positions. Together, the lesion and TMS studies suggest a critical and causal role of the STS at early latencies in the process of extrapolating gaze direction information.

7. Conclusions

In this review, we addressed the neural bases of the perceptual processing of the direction of another's eye gaze and the gaze-cued attentional orienting process. In addition, we reviewed modulation by contextual factors on the neural bases of gaze direction and gaze-cueing, such as the kind of cue that induced attention orienting (i.e., gazing or non-symbolic cues) or the emotional expression conveyed by the gaze cues. We reported many inconsistencies between the reviewed findings, which prompt further research in this area. Such inconsistencies may be explained by the different methods employed, such as task instruction (i.e., passive observation vs. gaze discrimination), type of emotional expressions (i.e., fear vs. anger as negative stimuli), kind of facial stimuli (i.e., static vs. dynamic) and scanning parameters. All these methodological differences represent a limit when results from different studies need to be compared. For example, N'Diaye and colleagues [79] demonstrated that the emotional intensity of the presented facial expression is able to interact with the gaze effect. Thus, the absence of systematic manipulation of the emotional intensity may represent a limit in the experimental design. Another example was reported by Zhao and colleagues [72], who proposed that attentional orienting differs between eye gaze and arrow cues when these cues are influenced through contextual processing. Similarly, the context seems to be able to modulate the activity of STS. Thus, choosing to add contextual information or not may dramatically impact both the behavioral and neuroimaging results. However, from the reviewed studies, we extracted the following key take-home messages.

First, we concluded that although several authors reported a specific function of the STS in processing gaze cues compared to non-symbolic cues [42,62,64,70], multiple studies questioned the specificity of the STS for the perception of gaze cues, preventing us from ascertaining with confidence whether the human STS is specialized for attentional orienting to social stimuli [68,69,72,75].

Second, when emotional expressions are presented, the STS seems to have a pivotal role in gaze perception and to be less sensitive to emotions, while the amygdala seems to play an important role in the interaction between gaze and emotional expressions, especially when negative expressions are presented [32,33,42,79]. This negative advantage is not surprising since it was demonstrated that faces expressing fear and anger are subject to a distinct processing advantage, which leads to them being analyzed quickly by the brain (~120 ms) akin to the way in which faces are structurally encoded [139–141], and evoking fast autonomic responses [142–145]. Studies measuring RTs have also revealed how faces expressing negative emotions enhance rapid and involuntary attention to the stimulus [146,147]. Moreover, negative expressions rapidly impact the corticospinal excitability, as measured with motor evoked potentials [148–154]. However, despite the clear importance of interpreting emotional expression and their direction of gaze for social communication, only a few studies reported increased STS activation for emotional expression (also positive) and an averted versus direct gaze [81].

Finally, few studies have found the amygdala to be activated for the averted gaze, regardless of facial expression [83], suggesting that STS possibly codes individual gaze directions [56], while the amygdala could play a specific part in perceiving or monitoring eye contact [45,46], possibly reflecting an emotional response to being looked at [155]. More effort is required to disclose the interplay between socially relevant stimuli and the information conveyed by gaze cues.

Together, the available neuroimaging results on the contextual (i.e., head orientation) modulation of gaze, especially the proposed increased invariance to gaze-independent characteristics from posterior to anterior STS [60], paves the way for questions such as whether and how social variables could potentially be processed in the framework of gaze perception. Carlin and Calder [156] suggested that gaze processing requires a large network of brain regions that includes anterior and posterior sections of the STS, the lateral parietal cortex and the medial prefrontal cortex. In particular, they suggested that the anterior STS may be important in the perceptual processing of gaze because it discriminates between different averted gaze directions in an approach that is invariant to head view and physical image features [60]. The preferential role of the anterior STS in view-invariant representations of gazing reveals a hierarchical processing stream for gaze perception, with greater invariance to gaze-independent features (e.g., head orientation) from the posterior to anterior STS. Indeed, the lateral parietal cortex, and the posterior part of the STS in particular, may contribute to gaze-cued attentional orienting, which is influenced by physical stimulus features [64]. Carlin's findings raise an interesting finding: on the one hand, emotional context, social status and familiarity conveyed by physical visual features could be filtered out in the posterior part of the STS before reaching an invariant representation that is useful for gaze perception in the anterior STS. On the other hand, some of these contextual variables (i.e., emotional context, social status and familiarity) could instead become an integral part of the invariant representation and be encoded in the high-order (supra-modal) processing of gaze-cueing that occurs in the anterior STS; however, future studies are needed to address this point. Taken together, previous imaging studies converge in suggesting a role of STS in attentional orienting, even when gaze information is subliminally presented, which is less sensitive to emotions, while the amygdala seems to play a role in jointly elaborating gaze and emotional expressions. These results are supported by brain-damaged patients and neuromodulation studies, which reported deficits in gaze perception and gaze-cueing behavior after lesions or the temporal inactivation of the STS. Similarly, amygdala lesions impair attention to eye information and averted gaze-cueing effects. Moreover, the unilateral lesion in the right amygdala affects the interaction of gaze and expression during emotion perception. However, the heterogeneity of the stimuli and paradigms employed indicate that the exact role of the two structures when multiple stimuli need to be processed is yet to be elucidated.

Author Contributions: Conceptualization, S.B. (Simone Battaglia), A.F. and S.B. (Sara Borgomaneri); Methodology, S.B. (Simone Battaglia), A.F. and S.B. (Sara Borgomaneri); Writing—Original Draft, S.B. (Simone Battaglia), A.F. and S.B. (Sara Borgomaneri); Writing—Review and Editing, S.B. (Simone Battaglia), J.H.F., K.M., A.F. and S.B. (Sara Borgomaneri); Visualization, S.B. (Simone Battaglia); Project Administration, S.B. (Simone Battaglia), A.F. and S.B. (Sara Borgomaneri); Funding Acquisition, A.F. and S.B. (Sara Borgomaneri). All authors have read and agreed to the published version of the manuscript.

Funding: This work was supported by grants awarded to A.F. and S.B. (Sara Borgomaneri). A.F. is supported by a grant from the Biotechnology and Biology research council (BBSRC, grant number: BB/S006605/1) and the Bial Foundation, Bial Foundation Grants Programme 2020/21 (A-29315). S. Borgomaneri is supported by Ministero della Salute, Italy (GR-2018-12365733).

Institutional Review Board Statement: Not applicable.

Informed Consent Statement: Not applicable.

Data Availability Statement: Not applicable.

Acknowledgments: The authors thank Gianluigi Serio and Chiara Di Fazio for their help in searching the relevant literature.

Conflicts of Interest: The authors declare that the research was conducted in the absence of any commercial or financial relationships that could be construed as a potential conflict of interest.

References

1. Argyle, M.; Cook, M. *Gaze and Mutual Gaze*; Cambridge U Press: Cambridge, UK, 1976.
2. Emery, N.J. The Eyes Have It: The Neuroethology, Function and Evolution of Social Gaze. *Neurosci. Biobehav. Rev.* **2000**, *24*, 581–604. [CrossRef]
3. Kleinke, C.L. Gaze and Eye Contact. A Research Review. *Psychol. Bull.* **1986**, *100*, 78–100. [CrossRef] [PubMed]
4. Leopold, D.; Rhodes, G. A Comparative View of Face Perception. *J. Comp. Psychol.* **2010**, *124*, 233–251. [CrossRef]
5. Chen, Y.; Zhao, Y.; Song, H.; Guan, L.; Wu, X. The Neural Basis of Intergroup Threat Effect on Social Attention. *Sci. Rep.* **2017**, *7*, 41062. [CrossRef] [PubMed]
6. Bayliss, A.P.; Tipper, S.P. Predictive Gaze Cues and Personality Judgments: Should Eye Trust You? *Psychol. Sci.* **2006**, *17*, 514–520. [CrossRef] [PubMed]
7. Deaner, R.O.; Platt, M.L. Reflexive Social Attention in Monkeys and Humans. *Curr. Biol.* **2003**, *13*, 1609–1613. [CrossRef] [PubMed]
8. Okada, T.; Sato, W.; Kubota, Y.; Usui, K.; Inoue, Y.; Murai, T.; Hayashi, T.; Toichi, M. Involvement of Medial Temporal Structures in Reflexive Attentional Shift by Gaze. *Soc. Cogn. Affect. Neurosci.* **2008**, *3*, 80–88. [CrossRef]
9. Driver, J.; Davis, G.; Ricciardelli, P.; Kidd, P.; Maxwell, E.; Baron-Cohen, S. Gaze Perception Triggers Reflexive Visuospatial Orienting. *Vis. Cogn.* **1999**, *6*, 509–540. [CrossRef]
10. Deaner, R.O.; Shepherd, S.V.; Platt, M.L. Familiarity Accentuates Gaze Cuing in Women but Not Men. *Biol. Lett.* **2007**, *3*, 64–67. [CrossRef]
11. Langton, S.R.H.; Bruce, V. Reflexive Visual Orienting in Response to the Social Attention of Others. *Vis. Cogn.* **1999**, *6*, 541–567. [CrossRef]
12. Frischen, A.; Bayliss, A.P.; Tipper, S.P. Gaze Cueing of Attention Visual Attention, Social Cognition, and Individual Differences. *Psychol. Bull.* **2007**, *133*, 694–724. [CrossRef] [PubMed]
13. Hadders-Algra, M. Human Face and Gaze Perception Is Highly Context Specific and Involves Bottom-up and Top-down Neural Processing. *Neurosci. Biobehav. Rev.* **2022**, *132*, 304–323. [CrossRef] [PubMed]
14. Friesen, C.K.; Kingstone, A. The Eyes Have It! Reflexive Orienting Is Triggered by Nonpredictive Gaze. *Psychon. Bull. Rev.* **1998**, *5*, 490–495. [CrossRef]
15. Dalmaso, M.; Castelli, L.; Galfano, G. Social Modulators of Gaze-Mediated Orienting of Attention: A Review Social Modulators of Gaze-Mediated Orienting of Attention: A Review. *Psychon. Bull. Rev.* **2020**, *27*, 833–855. [CrossRef]
16. Mathews, A.; Fox, E.; Yiend, J.; Calder, A. The Face of Fear: Effects of Eye Gaze and Emotion on Visual Attention. *Vis. Cogn.* **2003**, *10*, 823–835. [CrossRef]
17. Tipples, J. Fear and Fearfulness Potentiate Automatic Orienting to Eye Gaze. *Cogn. Emot.* **2006**, *20*, 309–320. [CrossRef]
18. Pecchinenda, A.; Petrucci, M. Emotion Unchained: Facial Expression Modulates Gaze Cueing under Cognitive Load. *PLoS ONE* **2016**, *11*, e0168111. [CrossRef]
19. McKay, K.; Grainger, S.; Coundouris, S.; Skorich, D.; Phillips, L.; Henry, J. Social Orienting of Attention: A Meta-Analytic Review of the Gaze-Cueing Effect. *J. Vis.* **2021**, *21*, 1977. [CrossRef]
20. Jones, B.C.; DeBruine, L.M.; Main, J.C.; Little, A.C.; Welling, L.L.M.; Feinberg, D.R.; Tiddeman, B.R. Facial Cues of Dominance Modulate the Short-Term Gaze-Cuing Effect in Human Observers. *Proc. Biol. Sci.* **2010**, *277*, 617–624. [CrossRef]

21. Masulli, P.; Galazka, M.; Eberhard, D.; Gillberg, C.; Billstedt, E.; Hadjikhani, N.; Andersen, T.S. Data-Driven Analysis of Gaze Patterns in Face Perception: Methodological and Clinical Contributions. *Sci. Direct* **2021**, *7*, 9–23. [CrossRef]
22. Adams, R.B.; Kleck, R.E. Effects of Direct and Averted Gaze on the Perception of Facially Communicated Emotion. *Emotion* **2005**, *5*, 3–11. [CrossRef] [PubMed]
23. Lebert, A.; Chaby, L.; Guillin, A.; Chekroun, S.; Vergilino-Perez, D. Are You "Gazing" at Me? How Others' Gaze Direction and Facial Expression Influence Gaze Perception and Postural Control. *Front. Psychol.* **2021**, *12*, 1–14. [CrossRef] [PubMed]
24. Narganes, C.; Ana, P.; Juan, B.C.; Andrea, L. Explicit vs. Implicit Spatial Processing in Arrow vs. Eye-Gaze Spatial Congruency Effects. *Psychol. Res.* **2022**, 0123456789. [CrossRef] [PubMed]
25. Hoffman, E.A.; Haxby, J.V. Distinct Representations of Eye Gaze and Identity in the Distributed Human Neural System for Face Perception. *Nat. Neurosci.* **2000**, *3*, 80–84. [CrossRef]
26. Adolphs, R.; Spezio, M. Role of the Amygdala in Processing Visual Social Stimuli. *Prog. Brain Res.* **2006**, *156*, 363–378. [CrossRef]
27. Wang, S.; Yu, R.; Tyszka, J.M.; Zhen, S.; Kovach, C.; Sun, S.; Huang, Y.; Hurlemann, R.; Ross, I.B.; Chung, J.M.; et al. The Human Amygdala Parametrically Encodes the Intensity of Specific Facial Emotions and Their Categorical Ambiguity. *Nat. Commun.* **2017**, *8*, 14821. [CrossRef]
28. Diano, M.; Tamietto, M.; Celeghin, A.; Weiskrantz, L.; Tatu, M.K.; Bagnis, A.; Duca, S.; Geminiani, G.; Cauda, F.; Costa, T. Dynamic Changes in Amygdala Psychophysiological Connectivity Reveal Distinct Neural Networks for Facial Expressions of Basic Emotions. *Sci. Rep.* **2017**, *7*, 45260. [CrossRef]
29. Pelphrey, K.A.; Viola, R.J.; McCarthy, G. When Strangers Pass: Processing of Mutual and Averted Social Gaze in the Superior Temporal Sulcus. *Psychol. Sci.* **2004**, *15*, 598–603. [CrossRef]
30. George, N.; Conty, L. Facing the Gaze of Others. *Neurophysiol. Clin.* **2008**, *38*, 197–207. [CrossRef]
31. Akiyama, T.; Kato, M.; Muramatsu, T.; Umeda, S.; Saito, F.; Kashima, H. Unilateral Amygdala Lesions Hamper Attentional Orienting Triggered by Gaze Direction. *Cereb. Cortex* **2007**, *17*, 2593–2600. [CrossRef]
32. Adams, R.B.; Gordon, H.L.; Baird, A.A.; Ambady, N.; Kleck, R.E. Effects of Gaze on Amygdala Sensitivity to Anger and Fear Faces. *Science* **2003**, *300*, 1536. [CrossRef] [PubMed]
33. Hadjikhani, N.; Hoge, R.; Snyder, J.; de Gelder, B. Pointing with the Eyes: The Role of Gaze in Communicating Danger. *Brain Cogn.* **2008**, *68*, 1–8. [CrossRef] [PubMed]
34. Hardee, J.E.; Thompson, J.C.; Puce, A. The Left Amygdala Knows Fear: Laterality in the Amygdala Response to Fearful Eyes. *Soc. Cogn. Affect. Neurosci.* **2008**, *3*, 47–54. [CrossRef] [PubMed]
35. Fichtenholtz, H.M.; Hopfinger, J.B.; Graham, R.; Detwiler, J.M.; LaBar, K.S. Event-Related Potentials Reveal Temporal Staging of Dynamic Facial Expression and Gaze Shift Effects on Attentional Orienting. *Soc. Neurosci.* **2009**, *4*, 317–331. [CrossRef]
36. Kennedy, D.P.; Adolphs, R. The Social Brain in Psychiatric and Neurological Disorders. *Trends Cogn. Sci.* **2012**, *16*, 559–572. [CrossRef]
37. Yu, R.L.; Wu, R.M. Social Brain Dysfunctions in Patients with Parkinson's Disease: A Review of Theory of Mind Studies. *Transl. Neurodegener.* **2013**, *2*, 7. [CrossRef]
38. Battaglia, S.; Serio, G.; Scarpazza, C.; D'Ausilio, A.; Borgomaneri, S. Frozen in (e)Motion: How Reactive Motor Inhibition Is Influenced by the Emotional Content of Stimuli in Healthy and Psychiatric Populations. *Behav. Res. Ther.* **2021**, *146*, 103963. [CrossRef]
39. Spekker, E.; Tanaka, M.; Szabó, Á.; Vécsei, L. Neurogenic Inflammation: The Participant in Migraine and Recent Advancements in Translational Research. *Biomedicines* **2022**, *10*, 76. [CrossRef]
40. Tanaka, M.; Vécsei, L. Editorial of Special Issue "Crosstalk between Depression, Anxiety, and Dementia: Comorbidity in Behavioral Neurology and Neuropsychiatry". *Biomedicines* **2021**, *9*, 517. [CrossRef]
41. Mormann, F.; Niediek, J.; Tudusciuc, O.; Quesada, C.M.; Coenen, V.; Elger, C.; Adolphs, R. Neurons in the Human Amygdala Encode Face Identity but Not Gaze Direction. *Nat. Neurosci.* **2016**, *18*, 1568–1570. [CrossRef]
42. Hooker, C.I.; Paller, K.A.; Gitelman, D.R.; Parrish, T.B.; Mesulam, M.-M.; Reber, P.J. Brain Networks for Analyzing Eye Gaze. *Brain Res. Cogn. Brain Res.* **2003**, *17*, 406–418. [CrossRef]
43. Graham, R.; Labar, K.S. Face Processing and Social Attention. *Neuropsychologia* **2013**, *50*, 553–566. [CrossRef] [PubMed]
44. Itier, R.J.; Batty, M. Neural Bases of Eye and Gaze Processing: The Core of Social Cognition. *Neurosci. Biobehav. Rev.* **2009**, *33*, 843–863. [CrossRef] [PubMed]
45. George, N.; Driver, J.; Dolan, R.J. Seen Gaze-Direction Modulates Fusiform Activity and Its Coupling with Other Brain Areas during Face Processing. *Neuroimage* **2001**, *13*, 1102–1112. [CrossRef]
46. Kawashima, R.; Sugiura, M.; Kato, T.; Nakamura, A.; Hatano, K.; Ito, K.; Fukuda, H.; Kojima, S.; Nakamura, K. The Human Amygdala Plays an Important Role in Gaze Monitoring: A PET Study. *Brain* **1999**, *122*, 779–783. [CrossRef]
47. Wicker, B.; Perrett, D.I.; Baron-Cohen, S.; Decety, J. Being the Target of Another's Emotion: A PET Study. *Neuropsychologia* **2003**, *41*, 139–146. [CrossRef]
48. Senju, A.; Johnson, M.H. The Eye Contact Effect: Mechanisms and Development. *Trends Cogn. Sci.* **2009**, *13*, 127–134. [CrossRef]
49. Pageler, N.M.; Menon, V.; Merin, N.M.; Eliez, S.; Brown, W.E.; Reiss, A.L. Effect of Head Orientation on Gaze Processing in Fusiform Gyrus and Superior Temporal Sulcus. *Neuroimage* **2003**, *20*, 318–329. [CrossRef]
50. Sato, W.; Yoshikawa, S.; Kochiyama, T.; Matsumura, M. The Amygdala Processes the Emotional Significance of Facial Expressions: An FMRI Investigation Using the Interaction between Expression and Face Direction. *Neuroimage* **2004**, *22*, 1006–1013. [CrossRef]

51. Pfeiffer, U.J.; Vogeley, K.; Schilbach, L. From Gaze Cueing to Dual Eye-Tracking: Novel Approaches to Investigate the Neural Correlates of Gaze in Social Interaction. *Neurosci. Biobehav. Rev.* **2013**, *37*, 2516–2528. [CrossRef]
52. Haxby, J.V.; Hoffman, E.A.E.; Gobbini, M.M.I. The Distributed Human Neural System for Face Perception. *Trends Cogn. Sci.* **2000**, *4*, 223–233. [CrossRef]
53. Nummenmaa, L.; Passamonti, L.; Rowe, J.; Engell, A.D.; Calder, A.J. Connectivity Analysis Reveals a Cortical Network for Eye Gaze Perception. *Cereb. Cortex* **2010**, *20*, 1780–1787. [CrossRef]
54. Chen, Y.; Byrne, P.; Yan, X.; Henriques, D.Y.P.; Crawford, J.D. Allocentric versus Egocentric Representation of Remembered Reach Targets in Human Cortex. *J. Neurosci.* **2014**, *34*, 12515–12526. [CrossRef] [PubMed]
55. Kanwisher, N.; McDermott, J.; Chun, M.M. The Fusiform Face Area: A Module in Human Extrastriate Cortex Specialized for Face Perception. *J. Neurosci.* **1997**, *17*, 4302–4311. [CrossRef] [PubMed]
56. Calder, A.J.; Beaver, J.D.; Winston, J.S.; Dolan, R.J.; Jenkins, R.; Eger, E.; Henson, R.N.A.A. Separate Coding of Different Gaze Directions in the Superior Temporal Sulcus and Inferior Parietal Lobule. *Curr. Biol.* **2007**, *17*, 20–25. [CrossRef] [PubMed]
57. Pelphrey, K.A.; Singerman, J.D.; Allison, T.; McCarthy, G. Brain Activation Evoked by Perception of Gaze Shifts: The Influence of Context. *Neuropsychologia* **2003**, *41*, 156–170. [CrossRef]
58. Puce, A.; Allison, T.; Bentin, S.; Gore, J.C.; Mccarthy, G. Temporal Cortex Activation in Humans Viewing Eye and Mouth Movements. *J. Neurosci.* **1998**, *18*, 2188–2199. [CrossRef]
59. Mcgettigan, C.; Jasmin, K.; Eisner, F.; Agnew, Z.K.; Josephs, O.J.; Calder, A.J.; Jessop, R.; Lawson, R.P.; Spielmann, M.; Scott, S.K. Neuropsychologia You Talkin' to Me? Communicative Talker Gaze Activates Left-Lateralized Superior Temporal Cortex during Perception of Degraded Speech. *Neuropsychologia* **2017**, *100*, 51–63. [CrossRef]
60. Carlin, J.D.; Calder, A.J.; Kriegeskorte, N.; Nili, H.; Rowe, J.B. A Head View-Invariant Representation of Gaze Direction in Anterior Superior Temporal Sulcus. *Curr. Biol.* **2011**, *21*, 1817–1821. [CrossRef]
61. Carlin, J.D.; Rowe, J.B.; Kriegeskorte, N.; Thompson, R.; Calder, A.J. Direction-Sensitive Codes for Observed Head Turns in Human Superior Temporal Sulcus. *Cereb. Cortex* **2012**, 735–744. [CrossRef]
62. Lockhofen, D.E.L.; Gruppe, H.; Ruprecht, C.; Gallhofer, B.; Sammer, G. Hemodynamic Response Pattern of Spatial Cueing Is Different for Social and Symbolic Cues. *Front. Hum. Neurosci.* **2014**, *8*, 912. [CrossRef] [PubMed]
63. Kingstone, A.; Tipper, C.; Ristic, J.; Ngan, E. The Eyes Have It!: An FMRI Investigation. *Brain Cogn.* **2004**, *55*, 269–271. [CrossRef] [PubMed]
64. Materna, S.; Dicke, P.W.; Thier, P. Dissociable Roles of the Superior Temporal Sulcus and the Intraparietal Sulcus in Joint Attention: A Functional Magnetic Resonance Imaging Study. *J. Cogn. Neurosci.* **2008**, *20*, 108–119. [CrossRef] [PubMed]
65. Sato, W.; Kochiyama, T.; Uono, S.; Yoshikawa, S. Amygdala Integrates Emotional Expression and Gaze Direction in Response to Dynamic Facial Expressions. *Neuroimage* **2010**, *50*, 1658–1665. [CrossRef]
66. Callejas, A.; Shulman, G.L.; Corbetta, M. Dorsal and Ventral Attention Systems Underlie Social and Symbolic Cueing. *J. Cogn. Neurosci.* **2014**, *26*, 63–80. [CrossRef]
67. Greene, D.J.; Mooshagian, E.; Kaplan, J.T.; Zaidel, E.; Iacoboni, M. The Neural Correlates of Social Attention: Automatic Orienting to Social and Nonsocial Cues. *Psychol. Res.* **2009**, *73*, 499–511. [CrossRef]
68. Engell, A.D.; Nummenmaa, L.; Oosterhof, N.N.; Henson, R.N.; Haxby, J.V.; Calder, A.J. Differential Activation of Frontoparietal Attention Networks by Social and Symbolic Spatial Cues. *Soc. Cogn. Affect. Neurosci.* **2010**, *5*, 432–440. [CrossRef]
69. Hietanen, J.K.; Nummenmaa, L.; Nyman, M.J.; Parkkola, R.; Hämäläinen, H. Automatic Attention Orienting by Social and Symbolic Cues Activates Different Neural Networks: An FMRI Study. *Neuroimage* **2006**, *33*, 406–413. [CrossRef]
70. Joseph, R.M.; Fricker, Z.; Keehn, B. Activation of Frontoparietal Attention Networks by Non-Predictive Gaze and Arrow Cues. *Soc. Cogn. Affect. Neurosci.* **2015**, *10*, 294–301. [CrossRef]
71. Uono, S.; Sato, W.; Kochiyama, T. Commonalities and Differences in the Spatiotemporal Neural Dynamics Associated with Automatic Attentional Shifts Induced by Gaze and Arrows. *Neurosci. Res.* **2014**, *87*, 56–65. [CrossRef]
72. Zhao, S.; Li, C.; Uono, S.; Yoshimura, S.; Toichi, M. Human Cortical Activity Evoked by Contextual Processing in Attentional Orienting. *Sci. Rep.* **2017**, *7*, 2962. [CrossRef] [PubMed]
73. Pelphrey, K.A.; Morris, J.P.; McCarthy, G. Grasping the Intentions of Others: The Perceived Intentionality of an Action Influences Activity in the Superior Temporal Sulcus during Social Perception. *J. Cogn. Neurosci.* **2004**, *16*, 1706–1716. [CrossRef] [PubMed]
74. Marquardt, K.; Ramezanpour, H.; Dicke, P.W.; Thier, P. Following Eye Gaze Activates a Patch in the Posterior Temporal Cortex That Is Not Part of the Human "Face Patch" System. *eNeuro* **2017**, *4*, ENEURO.0317-16.2017. [CrossRef]
75. Sato, W.; Kochiyama, T.; Uono, S.; Yoshikawa, S. Commonalities in the Neural Mechanisms Underlying Automatic Attentional Shifts by Gaze, Gestures, and Symbols. *Neuroimage* **2009**, *45*, 984–992. [CrossRef] [PubMed]
76. Ristic, J.; Friesen, C.K.; Kingstone, A. Are Eyes Special? It Depends on How You Look at It. *Psychon. Bull. Rev.* **2002**, *9*, 507–513. [CrossRef]
77. Tipples, J. Eye Gaze Is Not Unique: Automatic Orienting in Response to Uninformative Arrows. *Psychon. Bull. Rev.* **2002**, *9*, 314–318. [CrossRef]
78. Tipper, C.M.; Handy, T.C.; Giesbrecht, B.; Kingstone, A. Brain Responses to Biological Relevance. *J. Cogn. Neurosci.* **2008**, *20*, 879–891. [CrossRef]
79. N'Diaye, K.; Sander, D.; Vuilleumier, P. Self-Relevance Processing in the Human Amygdala: Gaze Direction, Facial Expression, and Emotion Intensity. *Emotion* **2009**, *9*, 798–806. [CrossRef]

80. Krämer, K.; Bente, G.; Kuzmanovic, B.; Barisic, I.; Pfeiffer, U.J.; Georgescu, A.L.; Vogeley, K. Neural Correlates of Emotion Perception Depending on Culture and Gaze Direction. *Cult. Brain* **2014**, *2*, 27–51. [CrossRef]
81. Engell, A.D.; Haxby, J.V. Facial Expression and Gaze-Direction in Human Superior Temporal Sulcus. *Neuropsychologia* **2007**, *45*, 3234–3241. [CrossRef]
82. Ewbank, M.P.; Fox, E.; Calder, A.J. The Interaction between Gaze and Facial Expression in the Amygdala and Extended Amygdala Is Modulated by Anxiety. *Front. Hum. Neurosci.* **2010**, *4*, 56. [CrossRef] [PubMed]
83. Straube, T.; Langohr, B.; Schmidt, S.; Mentzel, H.J.; Miltner, W.H.R. Increased Amygdala Activation to Averted versus Direct Gaze in Humans Is Independent of Valence of Facial Expression. *Neuroimage* **2010**, *49*, 2680–2686. [CrossRef] [PubMed]
84. Ziaei, M.; Ebner, N.C.; Burianová, H. Functional Brain Networks Involved in Gaze and Emotional Processing. *Eur. J. Neurosci.* **2017**, *45*, 312–320. [CrossRef]
85. Schobert, A.-K.; Corradi-Dell'Acqua, C.; Frühholz, S.; van der Zwaag, W.; Vuilleumier, P. Functional Organization of Face Processing in the Human Superior Temporal Sulcus: A 7T High-Resolution FMRI Study. *Soc. Cogn. Affect. Neurosci.* **2018**, *13*, 102–113. [CrossRef] [PubMed]
86. Kätsyri, J.; de Gelder, B.; de Borst, A.W. Amygdala Responds to Direct Gaze in Real but Not in Computer-Generated Faces. *Neuroimage* **2020**, *204*, 116216. [CrossRef]
87. Allison, T.; Puce, A.; McCarthy, G. Social Perception from Visual Cues: Role of the STS Region. *Trends Cogn. Sci.* **2000**, *4*, 267–278. [CrossRef]
88. Campbell, R.; Heywood, C.A.; Cowey, A.; Regard, M.; Landis, T. Sensitivity to Eye Gaze in Prosopagnosic Patients and Monkeys with Superior Temporal Sulcus Ablation. *Neuropsychologia* **1990**, *28*, 1123–1142. [CrossRef]
89. Boll, S.; Gamer, M.; Kalisch, R.; Büchel, C. Processing of Facial Expressions and Their Significance for the Observer in Subregions of the Human Amygdala. *Neuroimage* **2011**, *56*, 299–306. [CrossRef]
90. Balogh, L.; Tanaka, M.; Török, N.; Taguchi, S. Crosstalk between Existential Phenomenological Psychotherapy and Neurological Sciences in Mood and Anxiety Disorders. *Biomedicines* **2021**, *9*, 340. [CrossRef]
91. Tanaka, M.; Török, N.; Fanni, T.; Szab, Á. Co-Players in Chronic Pain: Neuroinflammation and the Tryptophan-Kynurenine Metabolic Pathway. *Biomedicines* **2021**, *9*, 897. [CrossRef]
92. Tanaka, M.; Fanni, T.; Poly, H.; Szab, Á.; Yvette, M. Immune Influencers in Action: Metabolites and Enzymes of the Tryptophan-Kynurenine Metabolic Pathway. *Biomedicines* **2021**, *9*, 734. [CrossRef] [PubMed]
93. Keil, M.F.; Briassoulis, G.; Stratakis, C.A.; Wu, T.J. Protein Kinase A and Anxiety-Related Behaviors: A Mini-Review. *Front. Endocrinol.* **2016**, *7*, 83. [CrossRef] [PubMed]
94. Sander, D.; Grafman, J.; Zalla, T. The Human Amygdala: An Evolved System for Relevance Detection. *Rev. Neurosci.* **2003**, *14*, 303–316. [CrossRef] [PubMed]
95. Talipski, L.A.; Bell, E.; Goodhew, S.C.; Dawel, A.; Edwards, M. Examining the Effects of Social Anxiety and Other Individual Differences on Gaze-Directed Attentional Shifts. *Q. J. Exp. Psychol.* **2021**, *74*, 771–785. [CrossRef]
96. Hoffman, K.L.; Gothard, K.M.; Schmid, M.C.; Logothetis, N.K. Facial-Expression and Gaze-Selective Responses in the Monkey Amygdala. *Curr. Biol.* **2007**, *17*, 766–772. [CrossRef]
97. Cushing, C.A.; Im, H.Y.; Adams, R.B.; Ward, N.; Albohn, D.N.; Steiner, T.G.; Kveraga, K. Neurodynamics and Connectivity during Facial Fear Perception: The Role of Threat Exposure and Signal Congruity. *Sci. Rep.* **2018**, *8*, 2776. [CrossRef]
98. Lachat, F.; Farroni, T.; George, N. Watch out! Magnetoencephalographic Evidence for Early Modulation of Attention Orienting by Fearful Gaze Cueing. *PLoS ONE* **2012**, *7*, e50499. [CrossRef]
99. Pourtois, G.; Thut, G.; Grave de Peralta, R.; Michel, C.; Vuilleumier, P. Two Electrophysiological Stages of Spatial Orienting towards Fearful Faces: Early Temporo-Parietal Activation Preceding Gain Control in Extrastriate Visual Cortex. *Neuroimage* **2005**, *26*, 149–163. [CrossRef]
100. Righart, R.; de Gelder, B. Context Influences Early Perceptual Analysis of Faces-an Electrophysiological Study. *Cereb. Cortex* **2006**, *16*, 1249–1257. [CrossRef]
101. Williams, L.M.; Palmer, D.; Liddell, B.J.; Song, L.; Gordon, E. The "when" and "Where" of Perceiving Signals of Threat versus Non-Threat. *Neuroimage* **2006**, *31*, 458–467. [CrossRef]
102. Borhani, K.; Borgomaneri, S.; Làdavas, E.; Bertini, C. The Effect of Alexithymia on Early Visual Processing of Emotional Body Postures. *Biol. Psychol.* **2016**, *115*, 1–8. [CrossRef] [PubMed]
103. Dumas, T.; Dubal, S.; Attal, Y.; Chupin, M.; Jouvent, R.; Morel, S.; George, N. MEG Evidence for Dynamic Amygdala Modulations by Gaze and Facial Emotions. *PLoS ONE* **2013**, *8*, e74145. [CrossRef]
104. Conty, L.; Dezecache, G.; Hugueville, L.; Grèzes, J. Early Binding of Gaze, Gesture, and Emotion: Neural Time Course and Correlates. *J. Neurosci.* **2012**, *32*, 4531–4539. [CrossRef] [PubMed]
105. Vuilleumier, P.; Armony, J.L.; Driver, J.; Dolan, R.J. Distinct Spatial Frequency Sensitivities for Processing Faces and Emotional Expressions. *Nat. Neurosci.* **2003**, *6*, 624–631. [CrossRef] [PubMed]
106. Adams, R.B.; Franklin, R.G.; Kveraga, K.; Ambady, N.; Kleck, R.E.; Whalen, P.J.; Hadjikhani, N.; Nelson, A.J. Amygdala Responses to Averted vs Direct Gaze Fear Vary as a Function of Presentation Speed. *Soc. Cogn. Affect. Neurosci.* **2012**, *7*, 568–577. [CrossRef] [PubMed]
107. Im, H.Y.; Adams, R.B.; Cushing, C.A.; Boshyan, J.; Ward, N.; Kveraga, K. Sex-Related Differences in Behavioral and Amygdalar Responses to Compound Facial Threat Cues. *Hum. Brain Mapp.* **2018**, *39*, 2725–2741. [CrossRef]

108. Lahnakoski, J.M.; Glerean, E.; Salmi, J.; Jääskeläinen, I.P.; Sams, M.; Hari, R.; Nummenmaa, L. Naturalistic FMRI Mapping Reveals Superior Temporal Sulcus as the Hub for the Distributed Brain Network for Social Perception. *Front. Hum. Neurosci.* **2012**, *6*, 233. [CrossRef]
109. Adolphs, R. The Social Brain: Neural Basis of Social Knowledge. *Annu. Rev. Psychol.* **2009**, *60*, 693–716. [CrossRef]
110. Adams, R.B.; Rule, N.O.; Franklin, R.G.; Wang, E.; Stevenson, M.T.; Yoshikawa, S.; Nomura, M.; Sato, W.; Kveraga, K.; Ambady, N. Cross-Cultural Reading the Mind in the Eyes: An FMRI Investigation. *J. Cogn. Neurosci.* **2010**, *22*, 97–108. [CrossRef]
111. Sauer, A.; Mothes-Lasch, M.; Miltner, W.H.R.; Straube, T. Effects of Gaze Direction, Head Orientation and Valence of Facial Expression on Amygdala Activity. *Soc. Cogn. Affect. Neurosci.* **2014**, *9*, 1246–1252. [CrossRef]
112. Gamer, M.; Büchel, C. Amygdala Activation Predicts Gaze toward Fearful Eyes. *J. Neurosci.* **2009**, *29*, 9123–9126. [CrossRef] [PubMed]
113. Mitsuda, T.; Masaki, S. Subliminal Gaze Cues Increase Preference Levels for Items in the Gaze Direction. *Cogn. Emot.* **2018**, *32*, 1146–1151. [CrossRef] [PubMed]
114. Sato, W.; Okada, T.; Toichi, M. Attentional Shift by Gaze Is Triggered without Awareness. *Exp. Brain Res.* **2007**, *183*, 87–94. [CrossRef] [PubMed]
115. Tokunaga, S.; Miyatani, M. The Effects of Facial Expressions on Gaze-Triggered Attention without Awareness. *Jpn. J. Cogn. Psychol.* **2010**, *8*, 53–61. [CrossRef]
116. Xu, S.; Zhang, S.; Geng, H. Gaze-Induced Joint Attention Persists under High Perceptual Load and Does Not Depend on Awareness. *Vis. Res.* **2011**, *51*, 2048–2056. [CrossRef]
117. Luo, Y.; Zhang, S.; Tao, R.; Geng, H. The Power of Subliminal and Supraliminal Eye Contact on Social Decision Making: An Individual-Difference Perspective. *Conscious. Cogn.* **2016**, *40*, 131–140. [CrossRef]
118. Schütz, C.; Güldenpenning, I.; Koester, D.; Schack, T. Social Cues Can Impact Complex Behavior Unconsciously. *Sci. Rep.* **2020**, *10*, 21017. [CrossRef]
119. Sato, W.; Kochiyama, T.; Uono, S.; Toichi, M. Neural Mechanisms Underlying Conscious and Unconscious Attentional Shifts Triggered by Eye Gaze. *Neuroimage* **2016**, *124*, 118–126. [CrossRef]
120. Bailey, P.E.; Slessor, G.; Rendell, P.G.; Bennetts, R.J.; Campbell, A.; Ruffman, T. Age Differences in Conscious versus Subconscious Social Perception: The Influence of Face Age and Valence on Gaze Following. *Psychol. Aging* **2014**, *29*, 491–502. [CrossRef]
121. Borgomaneri, S.; Serio, G.; Battaglia, S. Please, Don't Do It! Fifteen Years of Progress of Non-Invasive Brain Stimulation in Action Inhibition. *Cortex* **2020**, *132*, 404–422. [CrossRef]
122. Huang, Y.Z.; Lu, M.K.; Antal, A.; Classen, J.; Nitsche, M.; Ziemann, U.; Ridding, M.; Hamada, M.; Ugawa, Y.; Jaberzadeh, S.; et al. Plasticity Induced by Non-Invasive Transcranial Brain Stimulation: A Position Paper. *Clin. Neurophysiol.* **2017**, *128*, 2318–2329. [CrossRef] [PubMed]
123. Vosskuhl, J.; Strüber, D.; Herrmann, C.S. Non-Invasive Brain Stimulation: A Paradigm Shift in Understanding Brain Oscillations. *Front. Hum. Neurosci.* **2018**, *12*, 1–19. [CrossRef] [PubMed]
124. Miniussi, C.; Harris, J.A.; Ruzzoli, M. Modelling Non-Invasive Brain Stimulation in Cognitive Neuroscience. *Neurosci. Biobehav. Rev.* **2013**, *37*, 1702–1712. [CrossRef] [PubMed]
125. Zhang, L.; Xuan, R.; Chen, Q.; Zhao, Q.; Shi, Z.; Du, J.; Zhu, C.; Yu, F.; Ji, G.; Wang, K. High-definition transcranial direct current stimulation modulates eye gaze on emotional faces in college students with alexithymia: An eye-tracking study. *Prog. Neuro-Psychopharmacol. Biol. Psychiatry* **2022**, *116*, 110521. [CrossRef]
126. Betti, S.; Zani, G.; Granziol, U.; Guerra, S.; Castiello, U.; Sartori, L.; Shepherd, S.V. Look at Me: Early Gaze Engagement Enhances Corticospinal Excitability During Action Observation. *Front. Psychol.* **2018**, *9*, 1–13. [CrossRef]
127. Ito, T.; Kamiue, M.; Kihara, T.; Ishimaru, Y.; Kimura, D.; Tsubahara, A. Visual Attention and Motion Visibility Modulate Motor Resonance during Observation of Human Walking in Different Manners. *Brain Sci.* **2021**, *11*, 679. [CrossRef]
128. Akiyama, T.; Kato, M.; Muramatsu, T.; Saito, F.; Nakachi, R.; Kashima, H. A Deficit in Discriminating Gaze Direction in a Case with Right Superior Temporal Gyrus Lesion. *Neuropsychologia* **2006**, *44*, 161–170. [CrossRef]
129. Akiyama, T.; Kato, M.; Muramatsu, T.; Saito, F.; Umeda, S.; Kashima, H. Gaze but Not Arrows: A Dissociative Impairment after Right Superior Temporal Gyrus Damage. *Neuropsychologia* **2006**, *44*, 1804–1810. [CrossRef]
130. Roy, A.; Shepherd, S.V.; Platt, M.L. Reversible Inactivation of PSTS Suppresses Social Gaze Following in the Macaque (Macaca Mulatta). *Soc. Cogn. Affect. Neurosci.* **2014**, *9*, 209–217. [CrossRef]
131. Young, A.W.; Aggleton, J.P.; Hellawell, D.J.; Johnson, M.; Broks, P.; Hanley, J.R. Face Processing Impairments after Amygdalotomy. *Brain* **1995**, *118*, 15–24. [CrossRef]
132. Cristinzio, C.; N'Diaye, K.; Seeck, M.; Vuilleumier, P.; Sander, D. Integration of Gaze Direction and Facial Expression in Patients with Unilateral Amygdala Damage. *Brain* **2010**, *133*, 248–261. [CrossRef] [PubMed]
133. Spezio, M.L.; Huang, P.Y.S.; Castelli, F.; Adolphs, R. Amygdala Damage Impairs Eye Contact during Conversations with Real People. *J. Neurosci.* **2007**, *27*, 3994–3997. [CrossRef] [PubMed]
134. Sliwinska, M.W.; Pitcher, D. TMS Demonstrates That Both Right and Left Superior Temporal Sulci Are Important for Facial Expression Recognition. *Neuroimage* **2018**, *183*, 394–400. [CrossRef] [PubMed]
135. Paracampo, R.; Pirruccio, M.; Costa, M.; Borgomaneri, S.; Avenanti, A. Visual, Sensorimotor and Cognitive Routes to Understanding Others' Enjoyment: An Individual Differences RTMS Approach to Empathic Accuracy. *Neuropsychologia* **2018**, *116*, 86–98. [CrossRef] [PubMed]

136. Pitcher, D. Facial Expression Recognition Takes Longer in the Posterior Superior Temporal Sulcus than in the Occipital Face Area. *J. Neurosci.* **2014**, *34*, 9173–9177. [CrossRef] [PubMed]
137. Pourtois, G.; Sander, D.; Andres, M.; Grandjean, D.; Reveret, L.; Olivier, E.; Vuilleumier, P. Dissociable Roles of the Human Somatosensory and Superior Temporal Cortices for Processing Social Face Signals. *Eur. J. Neurosci.* **2004**, *20*, 3507–3515. [CrossRef] [PubMed]
138. Saitovitch, A.; Popa, T.; Lemaitre, H.; Rechtman, E.; Lamy, J.-C.; Grévent, D.; Calmon, R.; Meunier, S.; Brunelle, F.; Samson, Y.; et al. Tuning Eye-Gaze Perception by Transitory STS Inhibition. *Cereb. Cortex* **2016**, *26*, 2823–2831. [CrossRef]
139. Kawasaki, H.; Kaufman, O.; Damasio, H.; Damasio, A.R.; Granner, M.; Bakken, H.; Hori, T.; Howard, M.A.; Adolphs, R. Single-Neuron Responses to Emotional Visual Stimuli Recorded in Human Ventral Prefrontal Cortex. *Nat. Neurosci.* **2001**, *4*, 15–16. [CrossRef]
140. Eimer, M.; Holmes, A. An ERP Study on the Time Course of Emotional Face Processing. *Neuroreport* **2002**, *13*, 427–431. [CrossRef]
141. Eimer, M.; Kiss, M.; Holmes, A. Links between Rapid ERP Responses to Fearful Faces and Conscious Awareness. *J. Neuropsychol.* **2008**, *2*, 165–181. [CrossRef]
142. Ellena, G.; Battaglia, S.; Làdavas, E. The Spatial Effect of Fearful Faces in the Autonomic Response. *Exp. Brain Res.* **2020**, *238*, 2009–2018. [CrossRef] [PubMed]
143. Öhman, A.; Soares, J.J. "Unconscious Anxiety": Phobic Responses to Masked Stimuli. *J. Abnorm. Psychol.* **1994**, *103*, 231–240. [CrossRef] [PubMed]
144. Globisch, J.; Hamm, A.O.; Esteves, F.; Öhman, A. Fear Appears Fast: Temporal Course of Startle Reflex Potentiation in Animal Fearful Subjects. *Psychophysiology* **1999**, *36*, 66–75. [CrossRef] [PubMed]
145. Candini, M.; Battaglia, S.; Benassi, M.; Pellegrino, G.; Frassinetti, F. The Physiological Correlates of Interpersonal Space. *Sci. Rep.* **2021**, *11*, 2611. [CrossRef]
146. Öhman, A.; Lundqvist, D.; Esteves, F. The Face in the Crowd Revisited: A Threat Advantage with Schematic Stimuli. *J. Pers. Soc. Psychol.* **2001**, *80*, 381–396. [CrossRef]
147. Holmes, A.; Vuilleumier, P.; Eimer, M. The Processing of Emotional Facial Expression Is Gated by Spatial Attention: Evidence from Event-Related Brain Potentials. *Brain Res. Cogn. Brain Res.* **2003**, *16*, 174–184. [CrossRef]
148. Borgomaneri, S.; Vitale, F.; Avenanti, A. Early Changes in Corticospinal Excitability When Seeing Fearful Body Expressions. *Sci. Rep.* **2015**, *5*, 14122. [CrossRef]
149. Borgomaneri, S.; Vitale, F.; Gazzola, V.; Avenanti, A. Seeing Fearful Body Language Rapidly Freezes the Observer's Motor Cortex. *Cortex* **2015**, *65*, 232–245. [CrossRef]
150. Vicario, C.M.; Rafal, R.D.; Borgomaneri, S.; Paracampo, R.; Kritikos, A.; Avenanti, A. Pictures of Disgusting Foods and Disgusted Facial Expressions Suppress the Tongue Motor Cortex. *Soc. Cogn. Affect. Neurosci.* **2017**, *12*, 352–362. [CrossRef]
151. Borgomaneri, S.; Vitale, F.; Avenanti, A. Behavioral Inhibition System Sensitivity Enhances Motor Cortex Suppression When Watching Fearful Body Expressions. *Brain Struct. Funct.* **2017**, *222*, 3267–3282. [CrossRef]
152. Borgomaneri, S.; Vitale, F.; Avenanti, A. Early Motor Reactivity to Observed Human Body Postures Is Affected by Body Expression, Not Gender. *Neuropsychologia* **2020**, *146*, 107541. [CrossRef] [PubMed]
153. Borgomaneri, S.; Gazzola, V.; Avenanti, A. Temporal Dynamics of Motor Cortex Excitability during Perception of Natural Emotional Scenes. *Soc. Cogn. Affect. Neurosci.* **2014**, *9*, 1451–1457. [CrossRef] [PubMed]
154. Borgomaneri, S.; Vitale, F.; Battaglia, S.; Avenanti, A. Early Right Motor Cortex Response to Happy and Fearful Facial Expressions: A TMS Motor-Evoked Potential Study. *Brain Sci.* **2021**, *11*, 1203. [CrossRef] [PubMed]
155. Nummenmaa, L.; Calder, A.J. Neural Mechanisms of Social Attention. *Trends Cogn. Sci.* **2009**, *13*, 135–143. [CrossRef] [PubMed]
156. Carlin, J.D.; Calder, A.J. The Neural Basis of Eye Gaze Processing. *Curr. Opin. Neurobiol.* **2013**, *23*, 450–455. [CrossRef] [PubMed]

biomedicines

MDPI

Article

The Serum Brain-Derived Neurotrophic Factor Increases in Serotonin Reuptake Inhibitor Responders Patients with First-Episode, Drug-Naïve Major Depression

Reiji Yoshimura *, Naomichi Okamoto, Enkmurun Chibaatar, Tomoya Natsuyama and Atsuko Ikenouchi

Department of Psychiatry, University of Occupational and Environmental Health, Iseigaoka, Yahatanishi-ku, Kitakyushu 8078555, Fukuoka, Japan
* Correspondence: yoshi621@med.uoeh-u.ac.jp; Tel.: +81-936917253; Fax: +81-936924894

Abstract: Brain-derived neurotrophic factor (BDNF) is a growth factor synthesized in the cell bodies of neurons and glia, which affects neuronal maturation, the survival of nervous system, and synaptic plasticity. BDNF play an important role in the pathophysiology of major depression (MD). The serum BDNF levels changed over time, or with the improvement in depressive symptoms. However, the change of serum BDNF during pharmacotherapy remains obscure in MDD. In particular, the changes in serum BDNF associated with pharmacotherapy have not yet been fully elucidated. The present study aimed to compare the changes in serum BDNF concentrations in first-episode, drug-naive patients with MD treated with antidepressants between treatment-response and treatment-nonresponse groups. The study included 35 inpatients and outpatients composed of 15 males and 20 females aged 36.7 ± 6.8 years at the Department of Psychiatry of our University Hospital. All patients met the DSM-5 diagnostic criteria for MD. The antidepressants administered included paroxetine, duloxetine, and escitalopram. Severity of depressive state was assessed using the 17-item HAMD before and 8 weeks after drug administration. Responders were defined as those whose total HAMD scores at 8 weeks had decreased by 50% or more compared to those before drug administration, while non-responders were those whose total HAMD scores had decreased by less than 50%. Here we showed that serum BDNF levels were not significantly different at any point between the two groups. The responder group, but not the non-responder group, showed statistically significant changes in serum BDNF 0 and serum BDNF 8. The results suggest that the changes of serum BDNF might differ between the two groups. The measurement of serum BDNF has the potential to be a useful predictor of pharmacotherapy in patients with first-episode, drug-naïve MD.

Keywords: brain-derived neurotrophic factor; serum; trajectory; major depression; first-episode; drug-naïve; Diagnostic and Statistical Manual of Mental Disorders Fifth Edition; Hamilton Rating Scale for Depression; paroxetine; escitalopram; duloxetine

Citation: Yoshimura, R.; Okamoto, N.; Chibaatar, E.; Natsuyama, T.; Ikenouchi, A. The Serum Brain-Derived Neurotrophic Factor Increases in Serotonin Reuptake Inhibitor Responders Patients with First-Episode, Drug-Naïve Major Depression. *Biomedicines* **2023**, *11*, 584. https://doi.org/10.3390/biomedicines11020584

Academic Editor: Masaru Tanaka

Received: 11 January 2023
Revised: 13 February 2023
Accepted: 14 February 2023
Published: 16 February 2023

1. Introduction

Major depression (MD) is common, with almost one in five people experiencing one episode at some point in their lifetime [1]. The symptoms of MD are related to structural and neurochemical deficits in the corticolimbic brain regions. The behavioral symptoms of MD are extensive, covering the emotional, motivational, cognitive, and physiological domains, and include anhedonia, aberrant reward-associated perception, and memory alterations [2]. The precise pathophysiology of MD, however, remains unknown. The understanding of the pathophysiology of MD has progressed, but the single model and its mechanisms cannot explain the entirety of MD. Monoamines [3], the abnormality of the hypothalamic-pituitary-adrenal axis [4], inflammation [5], neuroplasticity, neurogenesis, neurocircuit [6], or structural change including in the prefronto-striatal-limbic and fronto-parietal network, which is associated with cognitive impairment, [7–9] or reduced resilience to mitochondrial impairment [10] might be involved in the pathophysiology of MD.

BDNF plays a major role in neuronal growth and survival, functions as a neurotransmitter modulator, and contributes to neuronal plasticity. BDNF stimulates and regulates the growth of new neurons from neural stem cells (i.e., neurogenesis); BDNF has been linked to synaptic re-modeling, being able to both induce and be induced by long-term potentiation [11,12]. The human BDNF gene consists of 11 exons and 9 promoters [13]. Both BDNF proteins and mRNA have been detected in various brain regions, including the olfactory bulb, cerebral cortex, hippocampus, basal forebrain, mesencephalon, hypothalamus, brainstem, and spinal cord [14–19]. Therefore, it has been suggested that abnormalities in BDNF in the brain are associated with the pathogenesis of MD [20–26]. It has been reported that the BDNF val66met polymorphism affects the activity-dependent secretion of BDNF and human memory and hippocampal function. In short, BDNF gene val/met polymorphism is associated with human memory and hippocampal function, and val/met polymorphism exerts these effects by impacting intracellular trafficking and the activity-dependent secretion of BDNF [27]. Furthermore, the BDNF gene Val66Met polymorphism is associated with vulnerability to MD [28,29].

Several meta-analyses have shown that serum and plasma levels of BDNF were significantly decreased in patients with MD compared to healthy controls [30–35]. Similarly, in our previous work, we also reported that serum BDNF concentrations were significantly lower in untreated patients with MD than in healthy controls [34,35]. A meta-analysis showed that various antidepressant treatments increase serum and plasma BDNF concentrations in patients with MD [34,36–39]. We also previously found that paroxetine and milnacipran significantly increased serum BDNF concentrations after 8 weeks in untreated patients with MD in the treatment response group [36,37]. Although there have been reports of increased serum and plasma BDNF levels after antidepressant treatment in patients with MD [21,22], the findings of the relationship between changes in serum or plasma BDNF levels and response to pharmacotherapy, however controversial, cannot be fully elucidated. Response to duloxetine was associated with a higher baseline serum BDNF level and greater reduction of the Hamilton Rating Scale for Depression (HAMD) scores for MD [40]. The absence of an early increase of serum BDNF levels in conjunction with early non-response to antidepressants can be a highly specific peripheral biomarker predictive for treatment failure of pharmacotherapy in MD [41]. Alternatively, the combination of an early increase (day 7) of plasma BDNF with early reduction of the HAMD scores could be a useful predictive marker for pharmaco-treatment in MD [42]. A decrease of serum BDNF levels at week 2 of selective serotonin reuptake inhibitor (SSRI) treatment might be associated with later SSRI response in adolescents with MD [43]. Pretreatment serum BDNF levels have been reported to be correlated with antidepressant responses, and responders to treatment improvement in severity of MD had higher pretreatment serum BDNF levels than did non-responders [40,43]. In short, the results of the time course of serum/plasma BDNF levels and response to antidepressants were not consistent and remain obscure. Moreover, there are no reports of detailed observations of the time course of serum or plasma BDNF levels and the response to antidepressants in MD patients.

Considering the above results and evidence, we sought to compare the changes of serum BDNF concentrations between the treatment-response and treatment-nonresponse groups in first-episode, drug-naive patients with MD treated with antidepressants. In other words, the aim of this study is to identify the changes in serum BDNF as an indicator of drug responsiveness.

2. Patients and Methods

2.1. Study Population

This study included 35 inpatients and outpatients, composed of 15 males and 20 females with a mean age of 36.7 $\pm$ 6.8 years, at the Department of Psychiatry, Occupational and Environmental Medicine Hospital. The detailed demographics of the patients are described in Table 1. All patients met the Diagnostic and Statistical Manual of Mental

Disorders Fifth Edition (DSM-5) [44] diagnostic criteria for MD and were first-episode and drug-naive.

Table 1. Background of treatment response group and treatment cost response group.

	Responder (n = 20)	Non-Responder (n = 15)
Age, year	37.8 (7.21)	35.3 (6.67)
Sex (male), (%)	8 (40.0%)	7 (46.6%)
Smokers, (%)	4 (20%)	4 (26%)
History of abuse, (%)	0 (0%)	0 (0%)
Comorbidity, (%)	0 (0%)	0 (0%)
Duration of illness, weeks	8.41 (2.31)	8.80 (2.47)
HAMD 0 (points)	23.3 (2.87)	23.0 (2.42)
HAMD 8 (points)	9.45 (2.03)	15.4 (2.26)

Data are expressed as mean (standard deviation).

2.2. Treatment and Assessment of Depression

The antidepressants administered were paroxetine in 15 patients (average dose: 32.0 ± 9.7 mg/day), duloxetine in 11 patients (average dose: 49.0 ± 9.9 mg/day), and escitalopram in 9 patients (average dose: 18.8 ± 3.1 mg/day) (Table 2). Severity of depressive state was assessed using the 17-item Hamilton Depression Rating Scale (HAMD) [45] before and 8 weeks after drug administration. Responders were defined as those whose total HAMD scores at 8 weeks had decreased by 50% or more compared to those before drug administration, while non-responders were those whose total HAMD scores had decreased by less than 50%.

Table 2. Drugs and daily dose of the patients.

Drug	#	Min. Dose (mg/day)	Max. Dose (mg/day)	Mean (SD) (mg/day)
Escitalopram	9	10	20	18.8 (3.1)
Paroxetine	11	20	40	49.0 (9.9)
Duloxetine	15	40	60	32.0 (9.7)

2.3. Blood Collection and Measurement of Serum BDNF

Blood samples were collected at 8–10 a.m. before drug administration; after 2, 4, and 8 weeks; before meals; and after a 30-min rest (Figure 1). The serum was separated, and the serum BDNF concentration was measured by enzyme-linked immunosorbent assay (ELISA) according to a previously reported method [46]. In brief, 96-well microplates were coated with anti-BDNF monoclonal antibodies and incubated at 4 °C for 18 h. The plates were incubated in a blocking buffer for 1 h at room temperature. The samples were diluted with an assay buffer by 100-times and BDNF standards were maintained at room temperature under conditions of horizontal shaking for 2 h, followed by washing with the appropriate washing buffer. The plates were incubated with antihuman BDNF polyclonal antibodies at room temperature for 2 h and washed with the washing buffer. The plates were then incubated with anti-IgY antibody conjugated to horseradish peroxidase for 1 h at room temperature and incubated in peroxidase substrate and tetramethylbenzidine solution to induce a color reaction. The reaction was stopped with 1 mol/L hydrochloric acid. The absorbance at 450 nm was measured with an Emax automated microplate reader. Measurements were performed in duplicates. The standard curve was linear from 0.5 ng/mL to 50 ng/mL, and the detection limit was 5 ng/mL. The cross-reactivity to related neurotrophies (NT-3, NT-4, NGF) was less than 3%. Intra- and inter-assay coefficients of variation were about 5% and 7%, respectively. The recovery rate of the exogenous added BDNF in the measured serum samples was more than 95%. All measurements were performed in triplicate, and the average value was used as the measured value.

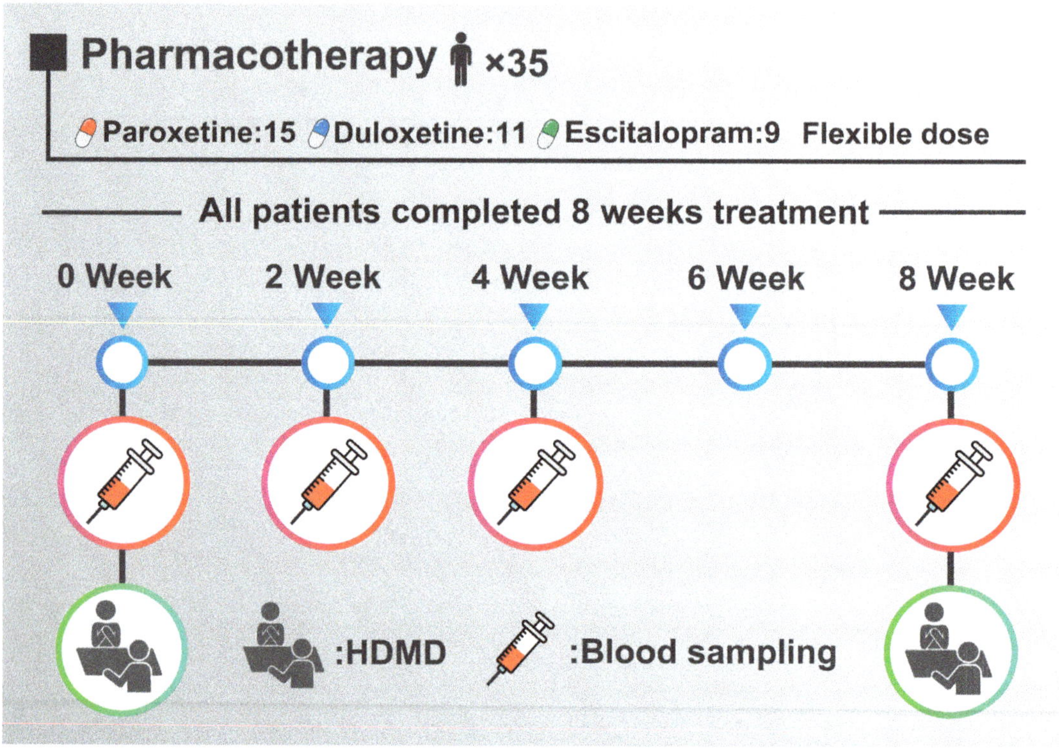

Figure 1. Study protocol.

2.4. Statistical Analysis

All statistical analyses were performed using EZR 1.60 and STATA 17.0. *p*-values were calculated using Welch's *t*-test for the difference in BDNF levels between the two groups at 0, 2, 4, and 8 weeks. Adjusted *p*-values were calculated using a multiple regression analysis to adjust for age and sex.

Using a mixed-effects model, changes in BDNF levels at 2, 4, and 8 weeks were compared to baseline values (0 week) in the response and non-response groups, respectively. Interactions between responders and non-responders were calculated using a mixed-effects model. Interactions were evaluated for the difference in the slope of BDNF values between the two groups compared to baseline values (0 week) at 2, 4, and 8 weeks.

Three mixed-effects models were created: a response group, a non-response group, and both groups combined for the interaction, with age and sex substituted as covariates. The intercept was set to random effects and time (i.e., 0, 2, 4, and 8 weeks) was treated as a nominal variable. In the mixed effects model, an ANOVA (global test) was conducted to confirm the null hypothesis that all means and changes are equal at the four time points. The validity of the mixed effects model was confirmed by using histograms of the normality of the residuals. All tests were two-tailed, and a *p*-value < 0.05 was considered statistically significant. Demographic data were expressed as mean (standard deviation) or percentage, and statistical values obtained were expressed as mean (standard error).

2.5. Ethical Statement

The study protocol was approved by the Ethics Committee of the University of Occupational and Environmental Health (Approved date: August 2018, Approved#: UOEHCRB21-057). Written informed consent was obtained from all participants.

3. Results

Patient Demographics and BDNF Levels

There were no significant differences in sex, age, and HAMD scores before drug administration between the responder and non-responder groups (Table 1). Serum BDNF levels at 0, 2, 4, and 8 weeks were not significantly different between the responder and non-responder groups (Table 3). The responder group showed statistically significant changes in serum BDNF 0 (baseline) and serum BDNF 8. In contrast, the non-responder group showed no statistically significant changes from baseline (Tables 4 and 5, Figures 2 and 3). There was a statistically significant interaction between serum BDNF 0 (baseline) and serum BDNF 8 between the responder and non-responder groups (Table 4).

Table 3. Serum BDNF concentrations in treatment responders and non-responders.

	Responder	Non-Responder	*p*-Value	Adjusted *p*-Value
BDNF 0	10.2 (1.98)	11.1(2.25)	0.24	0.24
BDNF 2	10.1 (1.90)	10.9 (2.21)	0.24	0.24
BDNF 4	10.5 (1.84)	11.0 (2.18)	0.45	0.52
BDNF 8	11.1 (2.22)	11.1 (2.30)	0.98	0.97

Data are expressed as mean (standard deviation). The *p*-value was calculated by Welch's *t*-test. Adjusted *p*-values were calculated using multiple regression analysis adjusted for age and sex.

Table 4. Changes in serum BDNF (Change from BDNF 0).

	Estimate	95% CI	Standard Error	t-Value	Adjusted *p*-Value
Responder					
BDNF 2	−0.145	−0.520, 0.230	0.193	−0.750	0.45
BDNF 4	0.225	−0.150, 0.600	0.193	1.164	0.24
BDNF 8	0.845	0.469, 1.220	0.193	4.371	<0.001 *
Non-responder					
BDNF 2	−0.160	−0.406, 0.086	0.127	−1.255	0.21
BDNF 4	−0.113	−0.360, −0.133	0.127	−0.889	0.37
BDNF 8	−0.006	−0.253, 0.240	0.127	−0.052	0.95

ANOVA (global test's *p*-value for responder group in mixed effect model < 0.001 *). ANOVA (global test's *p*-value for non-responder group in mixed effect model = 0.51). Adjusted *p*-values were calculated using a mixed effects model adjusted for age and sex. CI means confidence interval.

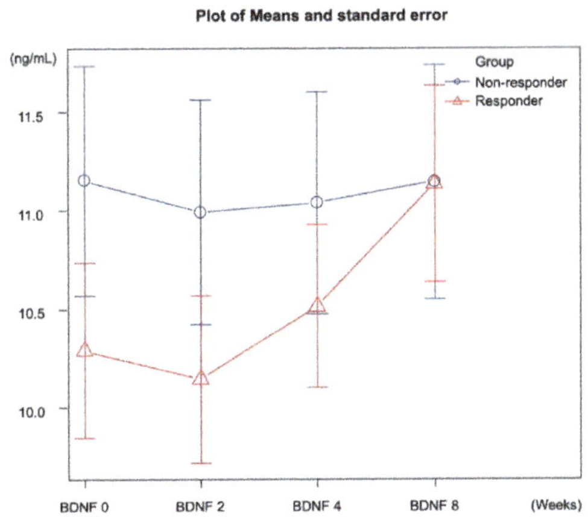

Figure 2. Changes in serum BDNF levels in the responders and non-responders. Data was expressed as mean and standard error.

Table 5. Interaction analysis between the two groups (Change from BDNF 0).

	Estimate	95% CI	Standard Error	t-Value	Adjusted *p*-Value
Group BDNF 2	0.015	−0.464, 0.494	0.249	0.060	0.95
Group BDNF 4	0.338	−0.140, 0.817	0.249	1.356	0.17
Group BDNF 8	0.851	0.372, 1.330	0.249	3.413	<0.001 *

ANOVA (global test's *p*-value for interaction analysis in mixed effect model = 0.0025 *). The adjusted *p*-value was calculated by a mixed effects model adjusted for age and sex. CI means confidence interval.

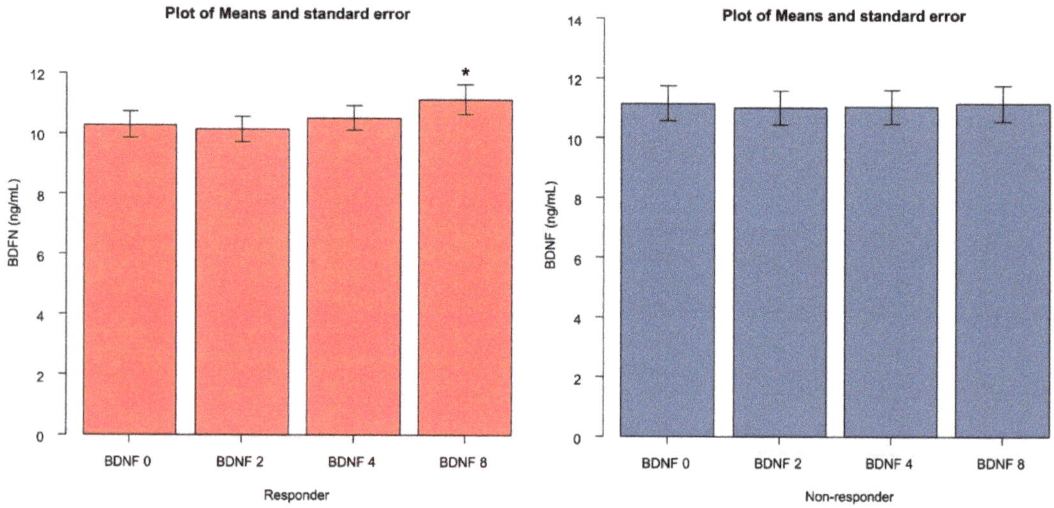

Figure 3. Changes in serum BDNF levels in the responders (red column) and non-responders (blue column). Data was expressed as mean and standard error. ANOVA (global test's *p*-value for interaction analysis in mixed effect model; * $p = 0.0025$.

4. Discussion

The results of the current study showed that serum BDNF concentrations increased significantly after 8 weeks in the paroxetine, escitalopram, and duloxetine response groups, but not after 2 or 4 weeks. However, serum BDNF concentrations did not increase at any point in time in the non-response group. We also found that serum BDNF levels at week 0 (baseline) did not statistically differ between the responder group and the non-responder group.

Previous studies have reported that serum BDNF concentrations increased during antidepressant treatment [33,34,47,48]. However, the duration of the response and the type of drug used varied. In particular, selective serotonin receptor inhibitors (SSRIs) and serotonin-norepinephrine receptor inhibitors (SNRIs) have been shown to increase serum BDNF levels after 8 weeks of treatment [34,47,49]. The current study examined serum BDNF concentrations after 8 weeks of paroxetine or milnacipran (an SNRI) treatment and found that serum BDNF concentrations after 8 weeks significantly increased in the group of patients who had responded to paroxetine or milnacipran treatment, while serum BDNF concentrations after 4 weeks were unchanged with treatment with both antidepressants. In contrast, there was no change in serum BDNF concentrations in the group of non-responders. In the second set of 35 untreated patients, the serum BDNF concentration after 8 weeks of treatment with paroxetine, duloxetine, or escitalopram was significantly increased, while the serum BDNF concentration after 2 or 4 weeks of treatment was unchanged in the response group. Taken together, 8 weeks may be necessary for serotonin reuptake inhibitors to increase serum BDNF in the responder group. Our current study demonstrated that mirtazapine, a different class of SSRI or SNRI, could increase serum BDNF at week 4 in the responder group [50]. Thus, mirtazapine might have a rapid effect on the increase in serum BDNF. In a study of sertraline, venlafaxine, and escitalopram, elevation in the serum BDNF level was observed at 5 weeks and 6 months post-dose in the sertraline group and at 6 months post-dose in the venlafaxine group, whereas no change was observed in the non-responder group at any point in time. In contrast, the escitalopram group showed no increase in serum BDNF after 6 months [51]. Taking these findings into account, the changes in serum or plasma BDNF seem to be inconsistent. The one important reason for this inconsistence might be a heterogeneity of MD.

A decrease in serum BDNF levels in the early phase of SSRI treatment may be associated with a later SSRI response in adolescents with MD [39]. A previous report demonstrated that responders to treatment ($\geq$50% improvement in depression ratings) had higher pre-treatment BDNF levels than did non-responders [52,53]. On the other hand, we found that baseline serum BDNF did not differ between the groups. A possible explanation is that baseline serum BDNF levels vary between patients (Figure 4a,b). If we confirm the result using a larger sample, serum BDNF levels the in non-responder group might be higher than in the responder group. In short, this might be a type II error. A study reported that plasma BDNF was not significantly changed after 1–2 days of single ketamine administration compared to placebo, which does not support the hypothesis that ketamine treatment increases BDNF plasma levels in patients with MD [54]. Another report demonstrated that BDNF was significantly elevated for only 1 week following the first ketamine infusion in those classified as responders [43]. No correlations were found between plasma BDNF levels and response to venlafaxine and paroxetine treatment at week 10 in patients with MD [55]. Treatment with venlafaxine for 4 weeks decreased serum BDNF levels, whereas treatment with mirtazapine for 4 weeks increased serum BDNF levels in patients with MD [56]. Treatment with mirtazapine for 12 weeks increased serum BDNF levels, which is associated with its response [57]. Based on these findings, the relationship between antidepressants, duration of treatment, and treatment response is also inconsistent and complicated.

We previously reported that early changes in serum BDNF levels (from week 0 and week 4) did not predict the response to treatment with SSRIs [58]. A recent systematic review and network meta-analysis found a significant effect of antidepressants on increased BDNF levels [standardized mean difference (SMD) = 0.62; 95% confidence interval (CI) = 0.31–0.94, Z = 3.92, $p < 0.0001$] [39]. Increases in BDNF levels over time were also associated with significant decreases in HAMD scores (SMD = 2.78, 95% CI = 2.31–3.26, Z = 11.57, $p < 0.00001$). The review also reported that SNRIs showed higher effect sizes than SSRIs (0.92 vs. 0.68). In addition, four antidepressants were analyzed separately for their role in increasing BDNF levels. Among these, only sertraline showed a significant increase in BDNF levels after treatment (SMD = 0.53, 95% CI = 0.13–0.93, Z = 2.62, $p = 0.009$), while venlafaxine, paroxetine, and escitalopram did not. In any case, a follow-up study with a large sample and a uniform protocol is needed regarding antidepressant treatment and changes in serum or plasma BDNF in MD.

Furthermore, it has been reported that electroconvulsive therapy (ECT) also could alter serum BDNF levels in MD patients [59–63], but other reports did not produce the same findings [64,65]. Repetitive transcranial magnetic stimulation (rTMS) also increases serum BDNF levels [66–70]. Taken together, the BDNF pathway is a common pathway for antidepressants, ECT, and rTMS to improve depressive symptoms, which is no longer controversial.

In the current study, antidepressants generally had a significant effect on the increase in serum BDNF levels after 8 weeks. Our prospective study of serum BDNF concentrations in a relatively small number of patients demonstrated that antidepressant-responsive patients had the first significant increase after 8 weeks of treatment, while non-responders showed no change at any point in time. These results are consistent with those of Zhou et al. [39] and did not contradict our previous report demonstrating that early changes in serum BDNF levels (from week 0 and week 4) did not predict the response to treatment with SSRIs [57].

Another systematic review and meta-analysis reported that peripheral measurements of BDNF are inadequate predictors of treatment response in treatment-refractory MD patients [69]. In our previous reports [71–73], catecholamine metabolites were altered after 4 weeks in the antidepressant response group, whereas BDNF was altered after 8 weeks in the antidepressant response group in the present results. These results suggest that changes in blood catecholamine metabolites precede changes in blood BDNF. Based on these findings, serum BDNF may be a candidate as a predictive factor for treatment response; however, it is difficult to predict the treatment of MD simply from BDNF changes alone. Combining BDNF changes with other biomarkers including microRNA [74], DNA methylation [75], proteomic markers [76], genetic information [77], and imaging findings [78] may help to more accurately predict treatment response and prognosis.

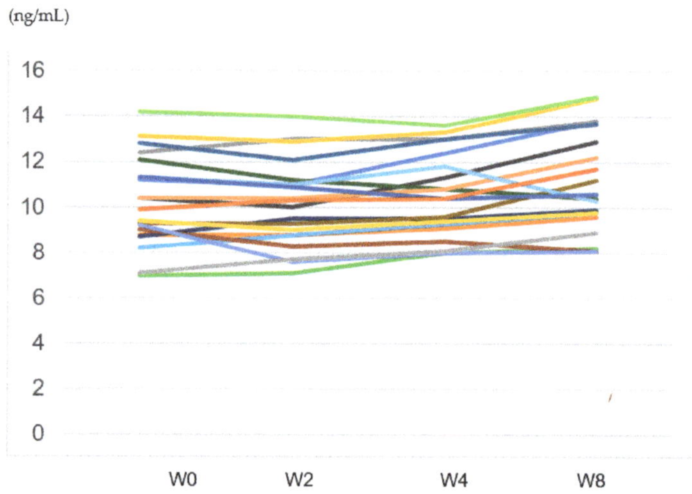

(**a**)

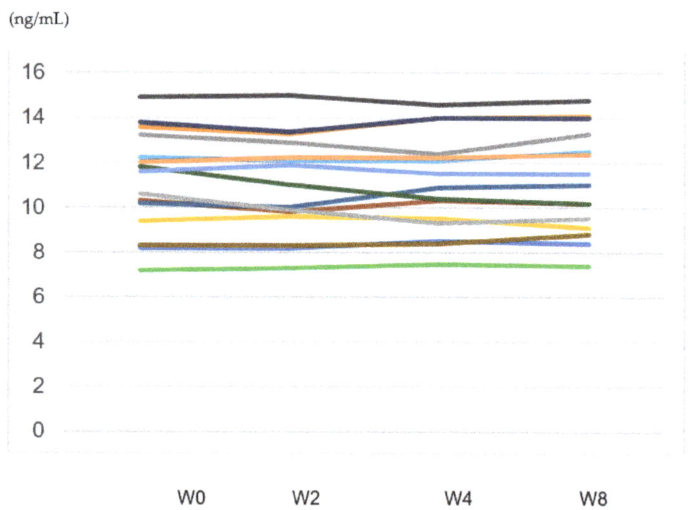

(**b**)

Figure 4. (**a**) Trajectory of serum BDNF levels in each patient with responders. (**b**) Trajectory of serum BDNF levels in each patient with responders.

This study has several important limitations. The first is the flexible dose design, with the type and dosage of antidepressants left to the discretion of the attending physician. Second, the number of patients was small (n = 35), and we did not perform a power analysis to determine the sample number. In addition, basal levels vary between patients, and the levels in the same patients at different points in time is much more consistent within the same patient (see Table 4), which might lead to the result that baseline serum BDNF levels did not differ in both groups. If we collected larger samples of serum BDNF levels in the non-responders, they might have been higher than in the responder group. Alternatively, the trajectory of serum BDNF changes might be important, but not the basal levels. In any case, further studies are required to explain this point. Third, there was no placebo group. Another limitation is the short clinical course of the patients, who were followed up for only 8 weeks after antidepressant administration. We are now undertaking a large-scale study including a placebo group, with a longer duration follow-up study (considering the above problems) to overcome these limitations. We hope to eventually establish a system that more accurately predicts antidepressant responsiveness in MD patients using not only serum BDNF data, but also technologies such as genomics, epigenomics, transcriptomics, proteomics, metabolomics, and connectomes [79].

5. Conclusions

In patients with first-episode and drug-naive MD treated with antidepressants, serum BDNF concentrations in the treatment response group increased significantly only after 8 weeks but not after 2 or 4 weeks of treatment. In contrast, no change in the serum BDNF concentration was observed in the non-responder group at any point in time. The difference of the changes of serum BDNF levels between the responders and the non-responders to antidepressants might be complicated, and must be further elucidated for each antidepressant. We believe that combined serum BDNF changes with other information including genomics, epigenomics, transcriptomics, proteomics, metabolomics, and connectomes could become an accurate method of prediction for drug response in MD patients. We urgently need to establish such a system for clinical practice.

Author Contributions: R.Y., N.O. and E.C. designed the study and prepared the manuscript. N.O., T.N. and A.I. conducted the literature review and prepared the tables. N.O. participated in the analysis. All authors have read and agreed to the published version of the manuscript.

Funding: This research received no external funding.

Institutional Review Board Statement: The study was conducted in accordance with the Declaration of Helsinki and was approved by the Clinical Research Ethics Committee of the University of Occupational and Environmental Health, Japan.

Informed Consent Statement: Informed consent was obtained from all the subjects involved in the study.

Data Availability Statement: The data presented in this study is available upon request from the corresponding author.

Conflicts of Interest: The authors declare no conflict of interest.

References

1. Malhi, G.S.; Mann, J.J. Depression. *Lancet* **2018**, *392*, 2299–2312. [CrossRef] [PubMed]
2. Filatova, E.; Shadrina, M.; Slominsky, P. Major Depression: One Brain, One Disease, One Set of Intertwined Processes. *Cells* **2021**, *10*, 1283. [CrossRef]
3. Delgado, P.L. Depression: The case for a monoamine deficiency. *J. Clin. Psychiatry* **2000**, *61*, 7–11.
4. Almeida, F.; Pinna, G.; Barros, H. The Role of HPA Axis and Allopregnanolone on the Neurobiology of Major Depressive Disorders and PTSD. *Int. J. Mol. Sci.* **2021**, *22*, 5495. [CrossRef]
5. Suneson, K.; Lindahl, J.; Hårsmar, S.C.; Söderberg, G.; Lindqvist, D. Inflammatory Depression—Mechanisms and Non-Pharmacological Interventions. *Int. J. Mol. Sci.* **2021**, *22*, 1640. [CrossRef]
6. Kraus, C.; Castrén, E.; Kasper, S.; Lanzenberger, R. Serotonin and neuroplasticity–Links between molecular, functional and structural pathophysiology in depression. *Neurosci. Biobehav. Rev.* **2017**, *77*, 317–326. [CrossRef]

7. Zheng, R.; Zhang, Y.; Yang, Z.; Han, S.; Cheng, J. Reduced Brain Gray Matter Volume in Patients With First-Episode Major Depressive Disorder: A Quantitative Meta-Analysis. *Front. Psychiatry* **2021**, *12*, 671348. [CrossRef] [PubMed]
8. Ippolito, G.; Bertaccini, R.; Tarasi, L.; Di Gregorio, F.; Trajkovic, J.; Battaglia, S.; Romei, V. The Role of Alpha Oscillations among the Main Neuropsychiatric Disorders in the Adult and Developing Human Brain: Evidence from the Last 10 Years of Research. *Biomedicines* **2022**, *10*, 3189. [CrossRef]
9. Chen, T.; Chen, Z.; Gong, Q. White Matter-Based Structural Brain Network of Major Depression. *Adv. Exp. Med. Biol.* **2021**, *1305*, 35–55. [CrossRef] [PubMed]
10. Tanaka, M.; Szabó, Á.; Spekker, E.; Polyák, H.; Tóth, F.; Vécsei, L. Mitochondrial Impairment: A Common Motif in Neuropsychiatric Presentation? The Link to the Tryptophan–Kynurenine Metabolic System. *Cells* **2022**, *11*, 2607. [CrossRef]
11. Kovalchuk, Y.; Hanse, E.; Kafitz, K.W.; Konnerth, A. Postsynaptic Induction of BDNF-Mediated Long-Term Potentiation. *Science* **2002**, *295*, 1729–1734. [CrossRef]
12. Dragunow, M.; Beilharz, E.; Mason, B.; Lawlor, P.; Abraham, W.; Gluckman, P. Brain-derived neurotrophic factor expression after long-term potentiation. *Neurosci. Lett.* **1993**, *160*, 232–236. [CrossRef] [PubMed]
13. Pruunsild, P.; Kazantseva, A.; Aid, T.; Palm, K.; Timmusk, T. Dissecting the human BDNF locus: Bidirectional transcription, complex splicing, and multiple promoters. *Genomics* **2007**, *90*, 397–406. [CrossRef]
14. Bathina, S.; Das, U.N. Brain-derived neurotrophic factor and its clinical implications. *Arch. Med. Sci.* **2015**, *11*, 1164–1178. [CrossRef] [PubMed]
15. Leal, G.; Comprido, D.; Duarte, C.B. BDNF-induced local protein synthesis and synaptic plasticity. *Neuropharmacology* **2014**, *76 Pt C*, 639–656. [CrossRef]
16. Wang, C.S.; Kavalali, E.T.; Monteggia, L.M. BDNF signaling in context: From synaptic regulation to psychiatric disorders. *Cell* **2021**, *185*, 62–76. [CrossRef] [PubMed]
17. Song, M.; Martinowich, K.; Lee, F.S. BDNF at the synapse: Why location matters. *Mol. Psychiatry* **2017**, *22*, 1370–1375. [CrossRef] [PubMed]
18. Kowiański, P.; Lietzau, G.; Czuba, E.; Waśkow, M.; Steliga, A.; Moryś, J. BDNF: A Key Factor with Multipotent Impact on Brain Signaling and Synaptic Plasticity. *Cell. Mol. Neurobiol.* **2018**, *38*, 579–593. [CrossRef]
19. Colucci-D'Amato, L.; Speranza, L.; Volpicelli, F. Neurotrophic Factor BDNF, Physiological Functions and Therapeutic Potential in Depression, Neurodegeneration and Brain Cancer. *Int. J. Mol. Sci.* **2020**, *21*, 7777. [CrossRef]
20. Caviedes, A.; Lafourcade, C.; Soto, C.; Wyneken, U. BDNF/NF-κB Signaling in the Neurobiology of Depression. *Curr. Pharm. Des.* **2017**, *23*, 3154–3163. [CrossRef]
21. Levy, M.J.F.; Boulle, F.; Steinbusch, H.W.; van den Hove, D.; Kenis, G.; Lanfumey, L. Neurotrophic factors and neuroplasticity pathways in the pathophysiology and treatment of depression. *Psychopharmacology* **2018**, *235*, 2195–2220. [CrossRef]
22. Lin, C.-C.; Huang, T.-L. Brain-derived neurotrophic factor and mental disorders. *Biomed. J.* **2020**, *43*, 134–142. [CrossRef] [PubMed]
23. Yuluğ, B.; Ozan, E.; Gönül, A.S.; Kilic, E. Brain-derived neurotrophic factor, stress and depression: A minireview. *Brain Res. Bull.* **2009**, *78*, 267–269. [CrossRef] [PubMed]
24. Fujii, R.; Watanabe, K.; Okamoto, N.; Natsuyama, T.; Tesen, H.; Igata, R.; Konishi, Y.; Ikenouchi, A.; Kakeda, S.; Yoshimura, R. Hippocampal Volume and Plasma Brain-Derived Neurotrophic Factor Levels in Patients With Depression and Healthy Controls. *Front. Mol. Neurosci.* **2022**, *6*, 857293. [CrossRef]
25. Halbach, O.V.B.U.; Halbach, V.V.B.U. BDNF effects on dendritic spine morphology and hippocampal function. *Cell Tissue Res.* **2018**, *373*, 729–741. [CrossRef] [PubMed]
26. Erbay, L.G.; Karlidag, R.; Oruc, M.; Cigremis, Y.; Celbis, O. Association of BDNF/TrkB and ngf/trka levels in postmortem brain with major depression and suicide. *Psychiatr. Danub.* **2021**, *33*, 491–498. [CrossRef]
27. Egan, M.F.; Kojima, M.; Callicott, J.H.; Goldberg, T.E.; Kolachana, B.S.; Bertolino, A.; Zaitsev, E.; Gold, B.; Goldman, D.; Dean, M.; et al. The BDNF val66met Polymorphism Affects Activity-Dependent Secretion of BDNF and Human Memory and Hippocampal Function. *Cell* **2003**, *112*, 257–269. [CrossRef]
28. Youssef, M.M.; Underwood, M.D.; Huang, Y.-Y.; Hsiung, S.-C.; Liu, Y.; Simpson, N.R.; Bakalian, M.J.; Rosoklija, G.B.; Dwork, A.J.; Arango, V.; et al. Association of BDNF Val66Met Polymorphism and Brain BDNF Levels with Major Depression and Suicide. *Int. J. Neuropsychopharmacol.* **2018**, *21*, 528–538. [CrossRef]
29. Hosang, G.M.; Shiles, C.; Tansey, K.E.; McGuffin, P.; Uher, R. Interaction between stress and the BDNFVal66Met polymorphism in depression: A systematic review and meta-analysis. *BMC Med.* **2014**, *12*, 7. [CrossRef] [PubMed]
30. Sen, S.; Duman, R.; Sanacora, G. Serum Brain-Derived Neurotrophic Factor, Depression, and Antidepressant Medications: Meta-Analyses and Implications. *Biol. Psychiatry* **2008**, *64*, 527–532. [CrossRef]
31. Bocchio-Chiavetto, L.; Bagnardi, V.; Zanardini, R.; Molteni, R.; Nielsen, M.G.; Placentino, A.; Giovannini, C.; Rillosi, L.; Ventriglia, M.; Riva, M.A.; et al. Serum and plasma BDNF levels in major depression: A replication study and meta-analyses. *World J. Biol. Psychiatry* **2010**, *11*, 763–773. [CrossRef]
32. Kishi, T.; Yoshimura, R.; Ikuta, T.; Iwata, N. Brain-Derived Neurotrophic Factor and Major Depressive Disorder: Evidence from Meta-Analyses. *Front. Psychiatry* **2018**, *8*, 308. [CrossRef]
33. Polyakova, M.; Stuke, K.; Schuemberg, K.; Mueller, K.; Schoenknecht, P.; Schroeter, M.L. BDNF as a biomarker for successful treatment of mood disorders: A systematic & quantitative meta-analysis. *J. Affect. Disord.* **2015**, *174*, 432–440. [CrossRef]

34. Shi, Y.; Luan, D.; Song, R.; Zhang, Z. Value of peripheral neurotrophin levels for the diagnosis of depression and response to treatment: A systematic review and meta-analysis. *Eur. Neuropsychopharmacol.* **2020**, *41*, 40–51. [CrossRef] [PubMed]
35. Molendijk, M.; Spinhoven, P.; Polak, M.; Bus, B.A.A.; Penninx, B.W.J.H.; Elzinga, B.M. Serum BDNF concentrations as peripheral manifestations of depression: Evidence from a systematic review and meta-analyses on 179 associations (N = 9484). *Mol. Psychiatry* **2014**, *19*, 791–800. [CrossRef]
36. Yoshimura, R.; Mitoma, M.; Sugita, A.; Hori, H.; Okamoto, T.; Umene, W.; Ueda, N.; Nakamura, J. Effects of paroxetine or milnacipran on serum brain-derived neurotrophic factor in depressed patients. *Prog. Neuro-Psychopharmacol. Biol. Psychiatry* **2007**, *31*, 1034–1037. [CrossRef]
37. Yoshimura, R.; Kishi, T.; Hori, H.; Atake, K.; Katsuki, A.; Nakano-Umene, W.; Ikenouchi-Sugita, A.; Iwata, N.; Nakamura, J. Serum proBDNF/BDNF and response to fluvoxamine in drug-naïve first-episode major depressive disorder patients. *Ann. Gen. Psychiatry* **2014**, *13*, 19. [CrossRef] [PubMed]
38. Naveen, G.H.; Varambally, S.; Thirthalli, J.; Rao, M.; Christopher, R.; Gangadhar, B.N. Serum cortisol and BDNF in patients with major depression-effect of yoga. *Int. Rev. Psychiatry* **2016**, *28*, 273–278. [CrossRef]
39. Zhou, C.; Zhong, J.; Zou, B.; Fang, L.; Chen, J.; Deng, X.; Zhang, L.; Zhao, X.; Qu, Z.; Lei, Y.; et al. Meta-analyses of comparative efficacy of antidepressant medications on peripheral BDNF concentration in patients with depression. *PLoS ONE* **2017**, *12*, e0172270. [CrossRef] [PubMed]
40. Mikoteit, T.; Beck, J.; Eckert, A.; Hemmeter, U.; Brand, S.; Bischof, R.; Holsboer-Trachsler, E.; Delini-Stula, A. High baseline BDNF serum levels and early psychopathological improvement are predictive of treatment outcome in major depression. *Psychopharmacology* **2014**, *231*, 2955–2965. [CrossRef]
41. Tadić, A.; Wagner, S.; Gorbulev, S.; Dahmen, N.; Hiemke, C.; Braus, D.F.; Lieb, K. Peripheral blood and neuropsychological markers for the onset of action of antidepressant drugs in patients with Major Depressive Disorder. *BMC Psychiatry* **2011**, *11*, 16. [CrossRef] [PubMed]
42. Dreimüller, N.; Schlicht, K.F.; Wagner, S.; Peetz, D.; Borysenko, L.; Hiemke, C.; Lieb, K.; Tadić, A. Early reactions of brain-derived neurotrophic factor in plasma (pBDNF) and outcome to acute antidepressant treatment in patients with Major Depression. *Neuropharmacology* **2012**, *62*, 264–269. [CrossRef]
43. Wolkowitz, O.M.; Wolf, J.; Shelly, W.; Rosser, R.; Burke, H.M.; Lerner, G.K.; Reus, V.I.; Nelson, J.C.; Epel, E.S.; Mellon, S.H. Serum BDNF levels before treatment predict SSRI response in depression. *Prog. Neuro-Psychopharmacol. Biol. Psychiatry* **2011**, *35*, 1623–1630. [CrossRef] [PubMed]
44. American Psychiatric Association. *Diagnostic and Statistical Manual of Mental Disorders 5*; American Psychiatric Publishing: Washington, DC, USA, 2013.
45. Hamilton, M. A rating scale for depression. *J. Neurol. Neurosurg. Psychiatry* **1960**, *23*, 56–62. [CrossRef]
46. Yoshimura, R.; Kishi, T.; Atake, K.; Katsuki, A.; Iwata, N. Serum Brain-Derived Neurotrophic Factor, and Plasma Catecholamine Metabolites in People with Major Depression: Preliminary Cross-Sectional Study. *Front. Psychiatry* **2018**, *9*, 52. [CrossRef] [PubMed]
47. Shimizu, E.; Hashimoto, K.; Okamura, N.; Koike, K.; Komatsu, N.; Kumakiri, C.; Nakazato, M.; Watanabe, H.; Shinoda, N.; Okada, S.; et al. Alterations of serum levels of brain-derived neurotrophic factor (BDNF) in depressed patients with or without antidepressants. *Biol. Psychiatry* **2003**, *54*, 70–75. [CrossRef]
48. Brunoni, A.R.; Lopes, M.; Fregni, F. A systematic review and meta-analysis of clinical studies on major depression and BDNF levels: Implications for the role of neuroplasticity in depression. *Int. J. Neuropsychopharmacol.* **2008**, *11*, 1169–1180. [CrossRef]
49. Gonul, A.S.; Akdeniz, F.; Taneli, F.; Donat, O.; Eker, Ç.; Vahip, S. Effect of treatment on serum brain–derived neurotrophic factor levels in depressed patients. *Eur. Arch. Psychiatry Clin. Neurosci.* **2005**, *255*, 381–386. [CrossRef]
50. Katsuki, A.; Yoshimura, R.; Kishi, T.; Hori, H.; Umene-Nakano, W.; Ikenouchi-Sugita, A.; Hayashi, K.; Atake, K.; Iwata, N.; Nakamura, J. Serum levels of brain-derived neurotrophic factor (BDNF), BDNF gene Val66Met polymorphism, or plasma catecholamine metabolites, and response to mirtazapine in Japanese patients with major depressive disorder (MDD). *CNS Spectrums* **2012**, *17*, 155–163. [CrossRef]
51. Matrisciano, F.; Bonaccorso, S.; Ricciardi, A.; Scaccianoce, S.; Panaccione, I.; Wang, L.; Ruberto, A.; Tatarelli, R.; Nicoletti, F.; Girardi, P.; et al. Changes in BDNF serum levels in patients with major depression disorder (MDD) after 6 months treatment with sertraline, escitalopram, or venlafaxine. *J. Psychiatr. Res.* **2009**, *43*, 247–254. [CrossRef]
52. Lee, J.; Lee, K.H.; Kim, S.H.; Han, J.Y.; Hong, S.-B.; Cho, S.-C.; Kim, J.-W.; Brent, D. Early changes of serum BDNF and SSRI response in adolescents with major depressive disorder. *J. Affect. Disord.* **2020**, *265*, 325–332. [CrossRef] [PubMed]
53. Jiang, H.; Veldman, E.R.; Tiger, M.; Ekman, C.-J.; Lundberg, J.; Svenningsson, P. Plasma Levels of Brain-Derived Neurotrophic Factor and S100B in Relation to Antidepressant Response to Ketamine. *Front. Neurosci.* **2021**, *15*, 698633. [CrossRef]
54. Allen, A.; Naughton, M.; Dowling, J.; Walsh, A.; Ismail, F.; Shorten, G.; Scott, L.; McLoughlin, D.; Cryan, J.; Dinan, T.; et al. Serum BDNF as a peripheral biomarker of treatment-resistant depression and the rapid antidepressant response: A comparison of ketamine and ECT. *J. Affect. Disord.* **2015**, *186*, 306–311. [CrossRef]
55. Carboni, L.; McCarthy, D.J.; Delafont, B.; Filosi, M.; Ivanchenko, E.; Ratti, E.; Learned, S.M.; Alexander, R.; Domenici, E. Biomarkers for response in major depression: Comparing paroxetine and venlafaxine from two randomised placebo-controlled clinical studies. *Transl. Psychiatry* **2019**, *9*, 182. [CrossRef] [PubMed]

56. Deuschle, M.; Gilles, M.; Scharnholz, B.; Lederbogen, F.; Lang, U.E.; Hellweg, R. Changes of Serum Concentrations of Brain-Derived Neurotrophic Factor (BDNF) during Treatment with Venlafaxine and Mirtazapine: Role of Medication and Response to Treatment. *Pharmacopsychiatry* **2015**, *48*, 292–293. [CrossRef] [PubMed]
57. Gupta, R.; Gupta, K.; Tripathi, A.; Bhatia, M.; Gupta, L.K. Effect of Mirtazapine Treatment on Serum Levels of Brain-Derived Neurotrophic Factor and Tumor Necrosis Factor-α in Patients of Major Depressive Disorder with Severe Depression. *Pharmacology* **2016**, *97*, 184–188. [CrossRef]
58. Yoshimura, R.; Kishi, T.; Hori, H.; Katsuki, A.; Sugita-Ikenouchi, A.; Umene-Nakano, W.; Atake, K.; Iwata, N.; Nakamura, J. Serum Levels of Brain-Derived Neurotrophic Factor at 4 Weeks and Response to Treatment with SSRIs. *Psychiatry Investig.* **2014**, *11*, 84–88. [CrossRef]
59. Vanicek, T.; Kranz, G.S.; Vyssoki, B.; Komorowski, A.; Fugger, G.; Höflich, A.; Micskei, Z.; Milovic, S.; Lanzenberger, R.; Eckert, A.; et al. Repetitive enhancement of serumBDNFsubsequent to continuationECT. *Acta Psychiatr. Scand.* **2019**, *140*, 426–434. [CrossRef]
60. Vanicek, T.; Kranz, G.S.; Vyssoki, B.; Fugger, G.; Komorowski, A.; Höflich, A.; Saumer, G.; Milovic, S.; Lanzenberger, R.; Eckert, A.; et al. Acute and subsequent continuation electroconvulsive therapy elevates serum BDNF levels in patients with major depression. *Brain Stimul.* **2019**, *12*, 1041–1050. [CrossRef]
61. Bilgen, A.E.; Zincir, S.B.; Zincir, S.; Özdemir, B.; Ak, M.; Aydemir, E.; Şener, I. Effects of electroconvulsive therapy on serum levels of brain-derived neurotrophic factor and nerve growth factor in treatment resistant major depression. *Brain Res. Bull.* **2014**, *104*, 82–87. [CrossRef]
62. Mindt, S.; Neumaier, M.; Hellweg, R.; Sartorius, A.; Kranaster, L. Brain-Derived Neurotrophic Factor in the Cerebrospinal Fluid Increases During Electroconvulsive Therapy in Patients with Depression. *J. ECT* **2020**, *36*, 193–197. [CrossRef] [PubMed]
63. Okamoto, T.; Yoshimura, R.; Ikenouchi-Sugita, A.; Hori, H.; Umene-Nakano, W.; Inoue, Y.; Ueda, N.; Nakamura, J. Efficacy of electroconvulsive therapy is associated with changing blood levels of homovanillic acid and brain-derived neurotrophic factor (BDNF) in refractory depressed patients: A pilot study. *Prog. Neuro-Psychopharmacol. Biol. Psychiatry* **2008**, *32*, 1185–1190. [CrossRef] [PubMed]
64. Fernandes, B.; Gama, C.S.; Massuda, R.; Torres, M.; Camargo, D.; Kunz, M.; Belmonte-De-Abreu, P.S.; Kapczinski, F.; Fleck, M.P.D.A.; Lobato, M.I. Serum brain-derived neurotrophic factor (BDNF) is not associated with response to electroconvulsive therapy (ECT): A pilot study in drug resistant depressed patients. *Neurosci. Lett.* **2009**, *453*, 195–198. [CrossRef]
65. Rapinesi, C.; Kotzalidis, G.D.; Curto, M.; Serata, D.; Ferri, V.R.; Scatena, P.; Carbonetti, P.; Napoletano, F.; Miele, J.; Scaccianoce, S.; et al. Electroconvulsive therapy improves clinical manifestations of treatment-resistant depression without changing serum BDNF levels. *Psychiatry Res.* **2015**, *227*, 171–178. [CrossRef] [PubMed]
66. Valiuliene, G.; Valiulis, V.; Dapsys, K.; Vitkeviciene, A.; Gerulskis, G.; Navakauskiene, R.; Germanavicius, A. Brain stimulation effects on serum BDNF, VEGF, and TNFα in treatment-resistant psychiatric disorders. *Eur. J. Neurosci.* **2021**, *53*, 3791–3802. [CrossRef] [PubMed]
67. Zhao, X.; Li, Y.; Tian, Q.; Zhu, B.; Zhao, Z. Repetitive transcranial magnetic stimulation increases serum brain-derived neurotrophic factor and decreases interleukin-1β and tumor necrosis factor-α in elderly patients with refractory depression. *J. Int. Med. Res.* **2019**, *47*, 1848–1855. [CrossRef]
68. Yukimasa, T.; Yoshimura, R.; Tamagawa, A.; Uozumi, T.; Shinkai, K.; Ueda, N.; Tsuji, S.; Nakamura, J. High-Frequency Repetitive Transcranial Magnetic Stimulation Improves Refractory Depression by Influencing Catecholamine and Brain-Derived Neurotrophic Factors. *Pharmacopsychiatry* **2006**, *39*, 52–59. [CrossRef]
69. Peng, S.; Li, W.; Lv, L.; Zhang, Z.; Zhan, X. BDNF as a biomarker in diagnosis and evaluation of treatment for schizophrenia and depression. *Discov. Med.* **2018**, *26*, 127–136.
70. Meshkat, S.; Alnefeesi, Y.; Jawad, M.Y.; Di Vincenzo, J.D.; Rodrigues, N.B.; Ceban, F.; Lui, L.M.; McIntyre, R.S.; Rosenblat, J.D. Brain-Derived Neurotrophic Factor (BDNF) as a biomarker of treatment response in patients with Treatment Resistant Depression (TRD): A systematic review & meta-analysis. *Psychiatry Res.* **2022**, *317*, 114857. [CrossRef]
71. Ueda, N.; Yoshimura, R.; Shinkai, K.; Nakamura, J. Plasma Levels of Catecholamine Metabolites Predict the Response to Sulpiride or Fluvoxamine in Major Depression. *Pharmacopsychiatry* **2002**, *35*, 175–181. [CrossRef]
72. Yoshimura, R.; Nakamura, J.; Shinkai, K.; Ueda, N. Clinical response to antidepressant treatment and 3-methoxy-4-hydroxyphenylglycol levels: Mini review. *Prog. Neuro-Psychopharmacol. Biol. Psychiatry* **2004**, *28*, 611–616. [CrossRef]
73. Atake, K.; Yoshimura, R.; Hori, H.; Katsuki, A.; Ikenouchi-Sugita, A.; Umene-Nakano, W.; Nakamura, J. Duloxetine, a Selective Noradrenaline Reuptake Inhibitor, Increased Plasma Levels of 3-Methoxy-4-hydroxyphenylglycol but Not Homovanillic Acid in Patients with Major Depressive Disorder. *Clin. Psychopharmacol. Neurosci.* **2014**, *12*, 37–40. [CrossRef]
74. Lopez, J.P.; Fiori, L.M.; Cruceanu, C.; Lin, R.; Labonte, B.; Cates, H.M.; Heller, E.A.; Vialou, V.; Ku, S.M.; Gerald, C.; et al. MicroRNAs 146a/b-5 and 425-3p and 24-3p are markers of antidepressant response and regulate MAPK/Wnt-system genes. *Nat. Commun.* **2017**, *8*, 15497. [CrossRef] [PubMed]
75. Lisoway, A.J.; Zai, C.C.; Tiwari, A.K.; Kennedy, J.L. DNA methylation and clinical response to antidepressant medication in major depressive disorder: A review and recommendations. *Neurosci. Lett.* **2017**, *669*, 14–23. [CrossRef] [PubMed]
76. Silva-Costa, L.C.; Carlson, P.T.; Guest, P.C.; de Almeida, V.; Martins-De-Souza, D. Proteomic Markers for Depression. *Adv. Exp. Med. Biol.* **2019**, *1118*, 191–206. [CrossRef]

77. Fabbri, C.; Hosák, L.; Mössner, R.R.; Giegling, I.; Mandelli, L.L.; Bellivier, F.; Claes, S.; Da Collier, D.A.; Corrales, A.A.; DeLisi, L.E.; et al. Consensus paper of the WFSBP Task Force on Genetics: Genetics, epigenetics and gene expression markers of major depressive disorder and antidepressant response. *World J. Biol. Psychiatry* **2017**, *18*, 5–28. [CrossRef] [PubMed]
78. Han, K.-M.; Ham, B.-J.; Kim, Y.-K. Development of Neuroimaging-Based Biomarkers in Major Depression. *Adv. Exp. Med. Biol.* **2021**, *1305*, 85–99. [CrossRef] [PubMed]
79. Gadad, B.S.; Jha, M.K.; Czysz, A.; Furman, J.L.; Mayes, T.L.; Emslie, M.P.; Trivedi, M.H. Peripheral biomarkers of major depression and antidepressant treatment response: Current knowledge and future outlooks. *J. Affect. Disord.* **2018**, *233*, 3–14. [CrossRef]

biomedicines

MDPI

Article

Unlocking the Secrets: Exploring the Biochemical Correlates of Suicidal Thoughts and Behaviors in Adults with Autism Spectrum Conditions

Ivan Mirko Cremone [1], Benedetta Nardi [1], Giulia Amatori [1], Lionella Palego [2], Dario Baroni [2], Danila Casagrande [1], Enrico Massimetti [3], Laura Betti [2], Gino Giannaccini [2], Liliana Dell'Osso [1] and Barbara Carpita [1,*]

1 Department of Clinical and Experimental Medicine, University of Pisa, via Roma 67, 56126 Pisa, Italy; ivan.cremone@gmail.com (I.M.C.); benedetta.nardi@live.it (B.N.); giulia.amatori@libero.it (G.A.); danilacasagrande@yahoo.it (D.C.); liliana.dellosso@unipi.it (L.D.)

2 Department of Pharmacy, University of Pisa, 56126 Pisa, Italy; lionella.palego@unipi.it (L.P.); dariobaroni@ymail.com (D.B.); laura.betti@unipi.it (L.B.); gino.giannaccini@unipi.it (G.G.)

3 ASST Bergamo Ovest, SSD Psychiatric Diagnosis and Treatment Service, 24047 Treviglio, Italy; e.massimetti@yahoo.it

* Correspondence: barbara.carpita@unipi.it; Tel.: +39-3911105675

Abstract: Involving 1 million people a year, suicide represents one of the major topics of psychiatric research. Despite the focus in recent years on neurobiological underpinnings, understanding and predicting suicide remains a challenge. Many sociodemographical risk factors and prognostic markers have been proposed but they have poor predictive accuracy. Biomarkers can provide essential information acting as predictive indicators, providing proof of treatment response and proposing potential targets while offering more assurance than psychological measures. In this framework, the aim of this study is to open the way in this field and evaluate the correlation between blood levels of serotonin, brain derived neurotrophic factor, tryptophan and its metabolites, IL-6 and homocysteine levels and suicidality. Blood samples were taken from 24 adults with autism, their first-degree relatives, and 24 controls. Biochemical parameters were measured with enzyme-linked immunosorbent assays. Suicidality was measured through selected items of the MOODS-SR. Here we confirm the link between suicidality and autism and provide more evidence regarding the association of suicidality with increased homocysteine (0.278) and IL-6 (0.487) levels and decreased tryptophan (−0.132) and kynurenic acid (−0.253) ones. Our results suggest a possible transnosographic association between these biochemical parameters and increased suicide risk.

Keywords: autism spectrum disorder; brain derived neurotrophic factor; biomarkers; homocysteine; interleukin-6; inflammation; kynurenine pathway; serotonin; suicidality; suicidal behavior; tryptophan

Citation: Cremone, I.M.; Nardi, B.; Amatori, G.; Palego, L.; Baroni, D.; Casagrande, D.; Massimetti, E.; Betti, L.; Giannaccini, G.; Dell'Osso, L.; et al. Unlocking the Secrets: Exploring the Biochemical Correlates of Suicidal Thoughts and Behaviors in Adults with Autism Spectrum Conditions. *Biomedicines* **2023**, *11*, 1600. https://doi.org/10.3390/biomedicines11061600

Academic Editors: Masaru Tanaka and Eleonóra Spekker

Received: 27 April 2023
Revised: 27 May 2023
Accepted: 28 May 2023
Published: 31 May 2023

1. Introduction

According to the World Health Organization, approximately 1 million individuals worldwide commit suicide each year [1]. Suicide represents one of the major topics of psychiatric research; however, its definition varies in the literature, and there is still an open debate over standardizing its nomenclature, particularly for elements regarding the intentionality [2,3]. A commonly used definition for suicide is a "fatal self-inflicted self-destructive act with explicit or inferred intent to die" [4], but that only represents a phenomenon that lies at the extreme end of a continuum of actions that are collectively referred to as suicidal behaviors, which also includes suicidal ideation and attempted suicide [3,5]. In particular, suicidal ideation is described as contemplating or planning to commit suicide, while suicide attempts are non-lethal, self-directed injurious behavior with an intent to die [6]. Despite there being a great amount of studies in recent years focusing on the possible neurobiological underpinnings of suicidal ideation and behaviors, to date, understanding and predicting suicidal attempts remains a challenge. Although a significant number of clinical, psychosocial, sociodemographical risk factors and prognostic markers have been proposed [7], they typically have poor predictive accuracy [8] and are frequently unmodifiable [9]. Moreover, the prediction of suicidal attempts relies on subjective indicators, such as patient accounts of thoughts, behavior, and family history [10–12]. For all these reasons, the literature stresses the need for more objective indicators in order to facilitate the development of effective techniques for prevention and intervention, as well as the improvement in the ability of prediction, risk screening, and evaluation of suicidality [9]. Furthermore, a better understanding of the neurobiological processes of suicidal thoughts and behaviors are also needed to provide unambiguous knowledge of the pathophysiology of suicide and find new treatment targets and strategies [13].

In this context, biomarkers can offer crucial information by serving as predictive indicators, providing proof of treatment response and proposing potential targets for new treatment approaches [6]. Biomarkers, which are broadly referred to as objective indicators of a biological state or condition, may be quantified in a variety of ways, such as changes in protein expression, epigenetic markers, and metabolomic alterations that can be found in both the central nervous system (CNS) and peripheral nervous system (PNS) [14]. In a clinical setting, biomarkers for suicidality (both ideation and attempts) may offer more assurance regarding the diagnosis or the course of therapy rather than psychological measures alone [6].

The recent literature has highlighted the role of neuroinflammation as a key player in the pathophysiology of suicidality, including both suicidal ideation and behaviors [15–19]; this evidence is in line with items of research that have consistently observed central and peripheral chronic inflammation in many psychiatric disorders [20]. In fact, regardless of their age or gender, individuals who manifested suicidal thoughts and attempts are reported to have higher levels of inflammation and oxidative stress markers in both the central nervous system (CNS) and peripheral tissues [21–23]. In addition, the fact that the stress response is significantly correlated with the immune system strongly supports the theory that investigating inflammation could be a promising path to grasp a better understanding of suicidality [13,24,25]. However, to date, results concerning the role that different mediators of inflammation play in the etiopathology of suicide are still mixed [26,27]. Numerous items of research have emphasized the link between suicidality and inflammatory states, highlighting the presence of higher levels of proinflammatory cytokines in the blood, the cerebrospinal fluid (CSF), and post-mortem brain [13,28–35] of suicide attempters, indicating that cytokine activation may affect vulnerable people's tendency to suicide. In particular, among the cytokines, the most proposed potential markers of suicidality are interleukin-1 beta (IL-1β) and interleukin-6 (IL-6), which have been repeatedly shown to be elevated in the blood and the CSF of suicidal patients [17,21,36–40]. IL-6 is a member of the proinflammatory cytokine family that not only stimulates the production of several proteins involved in acute inflammation processes, but also plays a significant role in human cell differentiation and proliferation [41]. Moreover, IL-6 has been linked to the physiological homeostasis of neural tissue and to the pathogenesis of many diseases with significant neuropathological changes [42], while other studies have suggested that increased amounts of IL-6 may be able to modify the behavioral output from the brain and may influence neuronal plasticity, neurogenesis, and neurotransmission, even though the precise mechanism has not yet been clarified [43–45]. Additionally, IL-6 has been linked to mental illnesses such as anxiety, despair, and suicide ideation [46], and numerous studies have historically reported a correlation between IL-6 levels and suicidality both in adults and adolescents [47–49].

Numerous items of research have emphasized how the stimulation of the kynurenine (KYN) pathway of tryptophan (TRP) degradation, caused by neuroinflammation, may lead to an imbalance of the pathway metabolites, some of which may directly influence the manifestation of suicidal symptoms via the regulation of glutamate neurotransmission [50]. Moreover, the stimulation of the kynurenine (KYN) pathway leads to a reduction of TRP that ultimately causes a depletion of serotonin (5-HT) and melatonin [51], which is not only associated with an increase in aggressive and impulsive behaviors [52] but is also reported to be one of the major findings that differentiated biomarkers of suicide from those of other psychiatric disorders [53,54]. Additionally, kynurenic acid (KYNA)'s enhanced metabolism into the neurotoxic NMDA receptor agonist, quinolinic acid (QUIN), may lead to an overactivation of the glutamatergic system, which in turn affect brain-derived neurotrophic factor (BDNF) synthesis, with an eventual impact on neuroplasticity and cognition [55]. Regardless of the presence of co-occurring mood disorders, higher levels of QUIN were found in the cerebrospinal fluid (CSF) of patients who attempted suicide and those who had suicidal thoughts [51,56] and were also correlated to a higher score on the Suicide Intent Scale [51,57].

On the other hand, as reported by multiple twin and family studies, genetics may be one of the variables influencing the risk of suicide [58]. In particular, the genetic component for suicide also appears to be shared with the inheritance of psychiatric disorders [59,60], and a recent genome-wide association study revealed an intriguing correlation between suicide attempts and depressive symptoms, neuroticism, schizophrenia, insomnia, and major depressive disorder [61,62]. However, it is still difficult to pinpoint the genetic basis for suicide behaviors. To date, a variety of single nucleotide polymorphisms have been identified as being associated with suicide in a genome-wide association study with polygenic risk scores [63–65], but the heritability estimated from these common variants only ranges between 4% and 4.6% [62,66], thus suggesting a complex biological background and a potential multifactorial origin behind these behaviors.

On the other hand, results from clinical and psychopathologic studies have stressed how one of the neuropsychiatric conditions most frequently associated with increased suicidal risk is the autism spectrum disorder (ASD) or also subthreshold autistic traits (AT) [67–70]. ASD is a neurodevelopmental disorder characterized by severe and pervasive impairment in reciprocal socialization, qualitative impairment in communication, and repetitive or odd behavior. Recently, it has been determined that autism is a disorder of genetic [71] and neurological origin with defects in the coordinated functioning of various brain regions [72,73]. Many studies have reported a lower synchronization of the activation (or functional connectivity) between frontal and posterior brain regions in ASD subjects in a wide variety of tasks [74–81], including language comprehension tasks [82]. Moreover, an aberrant trajectory of brain development is often reported, which is consistent with autism being a developmental condition. Many researchers have reported cortico–cortical connection abnormalities in ASD subjects, possibly caused by abnormalities in any of numerous brain development pathways [76,82–88], and defection in the pruning of synapses during later stages of neural development [89,90]. These abnormalities in maturational processes are in line the observation, during the early stages of development, of larger brain sizes in autistic patients [91,92], especially in the frontal cortex [93], and loss of cerebellar Purkinje cells, which is correlated with the severity of social impairments [94]. Additionally, many authors have highlighted increased brain size and white matter volumetric abnormalities [95], particularly in the frontal lobes, in young children with autism [92,93], impaired white matter connectivity, that can be assumed to be a neural substrate for socio-emotional dysfunction in ASD, specifically for social signal interpretation in social interaction [96–98], and reduction in the amount of white matter over adolescence and adulthood [83,99,100]. Subthreshold autistic traits have been firstly identified among first degree relatives of ASD probands, where they are known under the name of Broad autism phenotype (BAP), a label that stresses the possible common genetic underpinnings between different presentations of the autism spectrum [101]. However, recent literature has highlighted how AT seem to be continuously distributed from the clinical to general population, being particularly frequent among psychiatric patients with other kinds of psychiatric disorders [65,102–109], leading to the formulation of a neurodevelopemental hypothesis for psychiatric disorders [110]. Indeed, a growing body of research is suggesting that not only ASD subjects share an increased risk of suicidal thoughts and behaviors [95,111–118], but also that the presence of elevated AT in non-autistic populations represent a confirmed risk factor for lifetime suicidality [66,119–121]. However, to the best of our knowledge, no study has specifically investigated the biochemical correlates of suicidal ideation and behaviors among adult subjects in the autism spectrum.

In this framework, the aim of this study was to evaluate how the blood levels of 5-HT, BDNF, TRP, and its metabolites of the KYN pathway (specifically, KYN, KYNA, and QUIN), as well as the levels of IL-6, and of homocysteine (HCY) (as a potential marker of altered trans-sulfuration and transmethylation pathways, eventually leading to impaired redox balance and DNA methylation) may be associated with suicidality in a sample of adult ASD patients, their first-degree relatives (BAP), and healthy controls (HC).

2. Materials and Methods

For the study, a sample of 24 adults with autism, their first-degree relatives, and 24 controls was recruited. A blood sample was taken from each subject and all biochemical parameters were measured with enzyme-linked immunosorbent assays. Suicidality was measured through selected items of the MOODS-SR questionnaire.

2.1. Recruitment Procedures

ASD subjects (ASD group) were recruited among out-patient and in-patients followed at the Psychiatric Department of the Azienda Ospedaliera Universitaria Pisana (AOUP), University of Pisa. In order to participate in the study, the patient group must have received an ASD diagnosis, be aged at least 18 and not be aged below 65 years, and have no or only milder intellectual impairment and/or language developmental alteration. During the recruitment procedures, for each patient the participation of a parent or a sibling was requested: the relatives' group was labeled as the "BAP group". Being unable to complete the evaluations due to a linguistic or intellectual disability, having a diagnosis of schizophrenia, a substance use disorder, a neurodegenerative disease, or any other relevant medical or neurological disorder were the main exclusion criteria for both groups. Additionally, BAP subjects were excluded if they had been given a DSM-5 diagnosis of ASD or another neurodevelopmental disorder and if they were younger than 18 years old. The HC group was recruited on a voluntary basis. Exclusion criteria for this group were the same used for the ASD group, with the exception that subjects in the CTL groups were also excluded if they received a diagnosis of a psychiatric disorder according to DSM-5 criteria.

Psychometric measures and a structured clinical interview were used to evaluate each individual. A blood sample was taken from each participant in order to conduct the biochemical assessment. All subjects received clear information about the study and had the opportunity to ask questions before providing a written informed consent. The present research was led in accordance with the declaration of Helsinki, and all procedures were approved by the local ethical committee.

2.2. Psychometric Instruments

All subjects were clinically evaluated by trained psychiatrists, and the ASD diagnosis was confirmed according to DSM-5 criteria. The Structured Clinical Interview for DSM-5 disorders (SCID-5) was employed for evaluating the eventual presence of other comorbid psychiatric conditions. Mood symptoms and suicidality scores were evaluated by means of the Mood Spectrum Questionnaire (MOODS-SR).

2.2.1. The Structured Clinical Interview for DSM-5 Disorders (SCID-5)

The SCID-5 is the gold standard structured clinical interview for investigating the presence of major psychiatric disorders according to DSM-5 [122]. It must be administered by trained mental health professionals. It is composed of 10 independent modules; the sequence of questions follows the order of the related diagnostic manual (DSM-5), and the different items of each module guide the interviewer through the evaluation of the presence of symptoms that may satisfy the diagnostic criteria.

2.2.2. The Mood Spectrum Questionnaire (MOODS-SR)

The MOODS-SR is a questionnaire designed to assess a wide range of temperamental characteristics and mood symptoms throughout the course of a lifetime. It has 160 items and is divided into Manic Component, Depressive Component, and Rhithmicity domains. There are two possible answers for each item: "Yes" and "No." The instrument's Cronbach's alphas for the Italian translation ranged from 0.72 to 0.92. The MOODS-SR was used in earlier studies to assess suicidality (with subsections for suicidal ideation and behaviors) [67,70,123], through the instrument's items 102 to 107.

2.3. *Biochemical Evaluations*

Biochemical analyses were conducted on the blood samples of all participants, which were collected in the mornings after at least 12 h of fasting in K3EDTA vacutainer tubes or in clot activator tubes for serum separation. Platelet rich plasma (PRP) was separated from other blood elements by centrifuging the samples at 150× *g* for 15 min, at room temperature (RT). Subsequently, platelet poor plasma (PPP) and platelet pellets were obtained by transferring the PRP aliquots in falcon tubes and centrifuging them again at 1500× *g* for 15 min, at RT. In order to obtain serum aliquots, the samples in clot activator tubes were centrifuged only one time at 1500× *g* for 15 min, at RT. Finally, the obtained PRP and serum aliquots were transferred to high-quality, low-binding protein Eppendorf Safe-Lock test tubes, while platelet pellets remained in Falcon tubes. All these samples were maintained at −80 °C until the day of the assay. All biochemical parameters were measured with enzyme-linked immunosorbent assays (ELISA). IL-6 and BDNF levels were measured in the PPP with sandwich ELISA kits (Picokine IL-6 assay, Boster Biological Technology, Pleasanton, CA, USA and Biosensis, mature BDNF RapidTM, Thebarton, Australia) featuring a 96-well microplate. All measures were performed in duplicate. The ELISA kits featured a primary monoclonal anti-IL-6 or anti-BDNF antibody, a second biotinylated antibody, and a streptavidin-biotin complex. Levels of HCY, 5-HT, TRP, and KYN pathway metabolites (KYN, QUIN, KYNA) were measured through indirect competitive ELISA kits (ImmuSmol, Bordeaux, France). All these parameters were measured in PPP, with the exception of KYNA, which, according to the kit instruction, was measured in serum. For 5-HT, intra-platelet levels were also measured. These kinds of ELISA kits featured microplates pre-coated with quotes of the same analyte object of investigation; in order to perform the competitive essay, the analyte as well as a first specific antibody were added to each well. Subsequently, for the detection reaction, a second a second biotinylated antibody linked to horseradish peroxidase (HRP) was added, followed by the HRP substrate, 3,3′,5,5′-tetramethylbenzidine (TMB). Regarding HCY, due to the risk of matrix effects if measured directly, according to the kit instruction PPP aliquots were incubated with an enzyme reaction mixture containing the S-adenosyl-L-homocysteine hydrolase enzyme and its substrate adenosine/dithiothreitol (DTT) before the assay, thus transforming HCY to S-adenosyl-homocysteine. Regarding the measurement of 5-HT intra-platelet levels (ng/mL), in order to avoid eventual biases associated with individual differences in platelet count, we proceeded according to Bradford's method [124], normalizing the obtained concentrations for total proteins (mg/mL). The normalized concentrations were reported as ng/mg protein.

2.4. Statistical Analysis

We performed an ANOVA analysis of variance in order to compare MOOD-SR total and domain scores, as well as biochemical parameter mean levels among groups, followed by Bonferroni post-hoc tests. Pearson correlation coefficients were calculated for measuring the correlations between biochemical parameters and scores reported on psychometric scales. In order to evaluate eventual biochemical parameters statistically predictive of suicidality in the sample, a linear regression analysis was performed with the suicide sub-scale mean scores as the dependent variable and with mean levels of biochemical parameters as independent variables. All the analyses were performed using SPSS, version 24 (IBM Corp., Armonk, NY, USA, 2016). For biochemical assays, calibration curves and the regression analyses were performed by means of GraphPad Prism (Version 7.0, San Diego, CA, USA). See Figure 1.

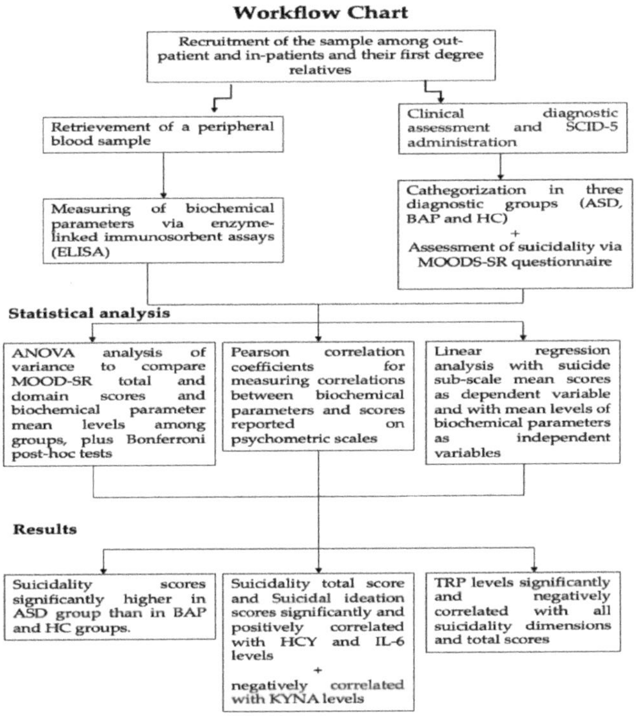

Figure 1. Summarizing flowchart.

3. Results

Comparison of socio-demographic variables, as well as other clinical and psychopathological features of the sample, are described elsewhere [125–127]. Considering MOODS-SR scores, the ANOVA analysis showed significant differences among groups on MOODS Total (F = 33.62; dF = 2; $p < 0.001$) as well as on MOODS manic component (F = 18.13; dF = 2; $p < 0.001$), depressive component (F = 35.39; dF = 2; $p < 0.001$), and Rhythmicity (F = 12.74; dF = 2; $p < 0.001$). According to Bonferroni post-hoc tests, the ASD group reported higher MOODS-SR totals and domain scores than the BAP group, which in turn reported significantly higher scores than the HC group (see Table 1).

Table 1. Comparison of MOODS-SR scores among groups.

MOODS-SR	ASD (n = 24) (Mean ± SD)	BAP (n = 24) (Mean ± SD)	CTL (n = 24) (Mean ± SD)	F	df	*p*
Manic component	22.77 ± 12.20	14.32 ± 9.32	6.12 ± 5.73	18.13	2	<0.001 *
Depressive component	34.32 ± 11.84	16.23 ± 13.73	7.79 ± 5.79	35.39	2	<0.001 *
Rhythmicity	13.00 ± 5.35	9.24 ± 5.36	5.54 ± 4.30	12.74	2	<0.001 *
MOODS total	70.09 ± 23.87	40.62 ± 24.97	19.46 ± 12.68	33.62	2	<0.001 *

* ASD > B AP > CTL; $p < 0.05$ (Bonferroni post-hoc test).

Suicidality scores were also significantly different among groups considering total score (F = 14.81; dF = 2; $p < 0.001$), suicidal ideation score (F = 12.22; dF = 2; $p < 0.001$), and suicidal behavior score (F = 8.62; dF = 2; $p < 0.001$). In particular, according to Bonferroni post-hoc tests, all suicidality scores were significantly higher in the ASD group than in the BAP and HC groups, without significant differences between the latter two (see Table 2).

Table 2. Comparison of suicidality scores among groups.

Suicidality Sub-Scale	ASD (n = 24) (Mean ± SD)	BAP (n = 24) (Mean ± SD)	CTL (n = 24) (Mean ± SD)	F	df	*p*
Suicidal ideation	1.87 ± 1.73	0.54 ± 1.25	0.12 ± 0.61	12.22	2	<0.001 *
Suicidal behavior	0.50 ± 0.83	0.00 ± 0.00	0.00 ± 0.00	8.62	2	<0.001 *
Suicidality total	2.37 ± 2.24	0.54 ± 1.25	0.12 ± 0.61	14.81	2	<0.001 *

* ASD > BAP, CTL; $p < 0.05$ (Bonferroni post-hoc test).

Results from Pearson correlation coefficient analysis between biochemical parameters and MOODS-SR scores showed significant positive correlations of IL-6 levels and MOODS total and domain scores, with the exception of the MOODS-SR manic component. Significant positive correlations were also reported between HCY levels and MOODS-SR depressive component scores. Moreover, significant negative correlations were reported between all MOODS-SR domain scores and both KYNA and TRP levels (see Table 3).

Table 3. Correlations between MOODS-SR scores and biochemical parameters levels.

	Manic Component	Depressive Component	Rhythmicity	MOODS Total
IL-6 pg/mL	0.233	0.428 **	0.538 **	0.422 *
HCY μM	0.119	0.283 *	0.208	0.232
5-HT (PPP) ng/mL	−0.137	−0.188	−0.134	−0.178
5-HT (intra-platelet) ng/mg prot	−0.093	−0.136	−0.064	−0.120
KYN ng/mL	−0.218	−0.099	−0.052	−0.146
KYNA ng/mL	−0.250 *	−0.398 **	−0.315 *	−0.357 **
QUIN ng/mL	−0.069	−0.111	−0.100	−0.104
TRP μM	−0.269 *	−0.259 *	−0.277 *	−0.294 *
BDNF ng/mL	−0.021	0.017	−0.113	−0.022

* $p < 0.05$; ** $p < 0.01$.

Regarding correlations with suicidality scores, we found that suicidality total score and suicidal ideation scores were significantly and positively correlated with both HCY and IL-6 levels, while they were negatively correlated with KYNA concentrations. Finally, TRP levels were significantly and negatively correlated with all suicidality dimensions and total scores (see Table 4 and Figure 2).

Table 4. Correlations between suicidality scores and biochemical parameters levels.

	Suicidal Ideation	**Suicidal Behavior**	**Suicidality Total**
IL-6 pg/mL	0.476 **	0.289	0.487 **
HCY μM	0.297 *	0.105	0.278 *
5-HT (PPP) ng/mL	−0.084	−0.124	−0.105
5-HT (intra-platelet) ng/mg prot	−0.152	0.044	−0.111
KYN ng/mL	−0.077	−0.215	−0.127
KYNA ng/mL	−0.267 *	−0.116	−0.253 *
QUIN ng/mL	−0.152	−0.159	−0.171
TRP μM	−0.276 *	−0.294 *	−0.132 **
BDNF ng/mL	0.055	0.114	0.079

* $p < 0.05$; ** $p < 0.01$.

According to the linear regression analysis, HCY levels were identified among all the biochemical parameters as significant positive predictors of higher suicidality score (see Table 5).

Table 5. Linear regression analysis with suicidality score as dependent variable and biochemical parameters levels as independent variables.

	b (SE)	**Beta**	**t**	***p***	**C.I.$_{95\%}$**
Constant	−0.74 (0.55)		−1.36	0.183	−1.85; −0.37
HCY μM	0.18 (0.05)	0.54	3.79	0.001	0.08; 0.28

R2 = 0.290, corrected R2 = 0.270. F = 14.33 (p = 0.001). Excluded variables: KYNA, BDNF, IL-6, 5-HT (PPP), 5-HT (intra-platelet), TRP, KYN, QUIN.

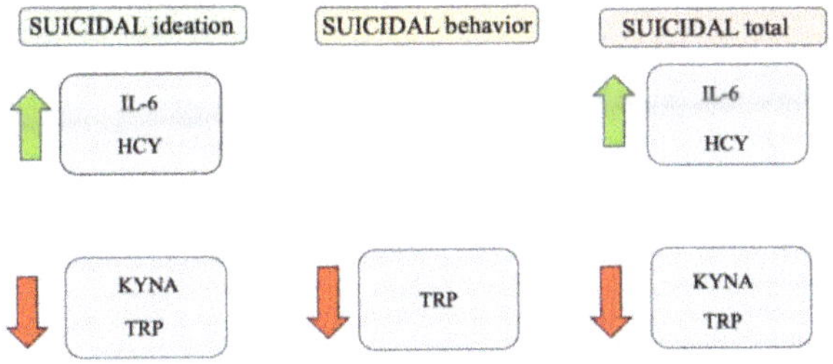

Figure 2. Schematic representation of associations between suicidality and biochemical variables.

Our results are in line with the previous literature suggesting an involvement of inflammation in the development and maintenance of various psychiatric disorders, in this case suicidal behaviors.

4. Discussion

Our results highlighted higher suicidality scores in the ASD group compared to the BAP and HC groups, while no differences were reported between the latter. Concerning the biochemical parameters, a significantly negative correlation between KYNA, TRP, and all MOODS-SR domain scores, as well as a significant negative correlation between the same parameters and suicidality scores, was highlighted. Contrarywise, we found a positive correlation between the proinflammatory cytokine IL-6 and MOODS-SR total and domain scores, as well as a significant positive correlation between IL-6 and suicidality and a positive correlation between HCY and the MOODS-SR depressive component, as well as with suicidality scores.

This evidence is in line with previous items of research that have reported an increased suicidality risk in ASD populations [58,95,111–121]. The association between ASD and suicidality has long been discussed, and a number of factors contributing to the correlation have been identified. According to the "interpersonal theory of suicide", the social difficulties and their consequences (such as loneliness [128,129], peer victimization [130,131], adoption of camouflaging behaviors [132–134]), and the resulting lack of meaningful social connections, may increase the risk of contemplating suicide. A recent study that investigated the relationship between AT and suicidality highlighted how the deficits in non-verbal communication and the restricted interest and ruminations were significantly and positively correlated with suicidality scores [67]. However, our findings, reporting similar levels of suicidality between BAP and HC groups, partially differ from those of another study that also highlighted increased suicidality scores mong subjects with subthreshold AT. This difference may be explained by the more limited sample size of the current study, and/or by the different methods of selection of the subthreshold AT traits/BAP group [70].

Results from the evaluation of the biochemical parameters highlighted a significantly negative correlation between KYNA, TRP, and all MOODS-SR domain scores, as well as a significant negative correlation between the same parameters and suicidality scores. KYNA is a metabolite of the TRP metabolic system, is robustly synthesized in the endothelium, and its serum levels correlate with homocysteine [135]. These evidences are in line with the strong and widespread correlation between the KYN pathway and depression that has been reported in the scientific literature, starting from its first appearance in The Lancet in 1969 with the name of the "serotonin hypothesis" [136], which claimed that the increment of TRP-KYN pathway metabolism, caused by the activation of the hepatic TRP-pyrrolase (TDO) due to elevated steroid levels, had a great impact on various neurological functions [137,138]. These statements have been confirmed and widened by the growing number of studies that are reporting an alteration of the TRP-KYN metabolism in a wide range of illnesses, including neoplastic, immunologic, neurological, and psychiatric disorders [18,139–141]. Subsequently, various studies have evaluated the levels of the KYN pathways in patients suffering from mood disorders. In particular, a recent meta-analysis including 22 studies reported that patients suffering from depression showed reduced levels of KYNA and KYN and increased levels of QUIN [142]. The first evidence of a link between the dysregulation of the KYN pathway and suicidality emerged many years later, in a study that highlighted elevated plasma KYN levels in depressed patients who had attempted suicide compared to patient who never attempted [143]. However, in agreement with our findings, opposite results came from more recent studies. In particular, recent research reported a 40% decrease in plasma TRP levels in suicidal adolescents with major depression (MDD), compared to non-suicidal individuals with MDD and HC [144], which is in agreement with many other studies that have linked lower peripheral TRP levels to suicide tendencies [145–147] in vulnerable subjects. Similarly, other studies described significantly lower levels of KYNA in the peripheral blood of depressed patients with a history of suicidal thoughts [148] and of suicidal patients with schizophrenia [149] compared to the non-suicidal ones. In this framework, two potential mechanisms through which the conversion of TRP to KYN may affect depression and suicidality have been suggested [19]. One theory holds that 5-HT production is decreased as a result of TRP depletion brought on by cytokine activation of

IDO [150]. This proposed route is plausible given that hypofunction of the 5-HT system has been linked to recurrent depression and suicide attempts separately [151]. The alternative theory speculates that an increase in KYN metabolites may have a more direct impact on the brain [152]. It is interesting to note that KYNA can also affect the tone of the glutamatergic [153], dopaminergic [154,155], and cholinergic [156] systems, due to its actions as an inhibitor of the α7 nicotinic acetylcholine receptor [157], as well as alter the ability of contextual learning [158], plausibly contributing to executive functioning deficits linked to suicidality [159,160] (see Figure 3). Moreover, KYN metabolites are known to exhibit a wide range of bioactive properties such as oxidant, antioxidant, anti-inflammatory, neurotoxin, neuroprotectant, and/or immunomodulating activity. Their actions depend on the concentration and the cellular environment, and their metabolic system functions under complex positive and negative feedback loops [161]. However, to date, the evidence on the action of KYN metabolites are conflicting, and there is no reached consensus on them.

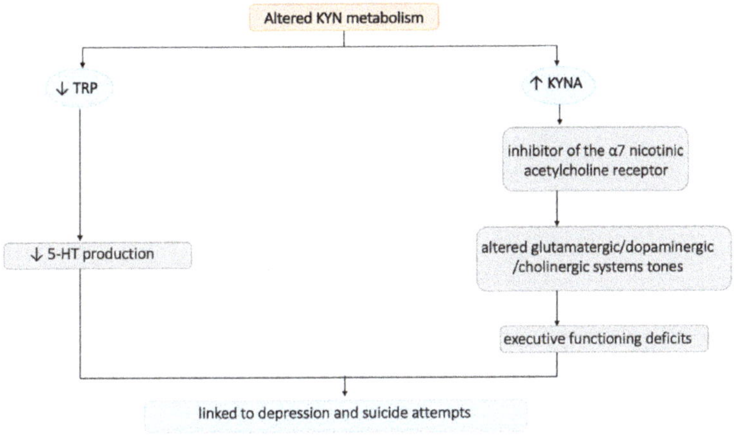

Figure 3. Summarizing figure on the altered KYN metabolism's contribution to suicidality.

Moreover, our results highlighted a positive correlation between the proinflammatory cytokine IL-6 and MOODS-SR total and domain scores, as well as a significant positive correlation between IL-6 and suicidality. The correlation between IL-6 and the MOOD-SR scores is in line with the growing evidence of alterations in IL-6 levels in patients with mood disorders, especially MDD. In particular, according to three meta-analyses, subjects with MDD have higher serum/plasma IL-6 levels than those without depression [162–164]. Many items of research have also focused on typifying the depression, reporting significantly higher levels of IL-6 in melancholic depression [165–167] and atypical depression [168,169], both characterized by a severe alteration in the circadian rhythmicity pattern, thus validating our results and highlighting a positive correlation between IL-6 and rhythmicity scores. The correlation between IL-6 and suicidality is in line with the branch of scientific literature reporting higher levels of proinflammatory cytokines in the blood, the CSF, and the brain of subjects with a different kind of SB [13,26–33], leading to the suggestion that cytokine stimulation may be involved in the development of suicidal thoughts and behaviors among vulnerable subjects. As a matter of fact, many studies have described higher quantities of IL-6 in the serum and CSF of patients exhibiting suicidal tendencies [47] as well as in children with suicidal tendencies prior to pharmaceutical treatment [49]. Increased levels of IL-6 have been linked to both current and past suicidal tendencies [48], confirming our results (See Figure 4).

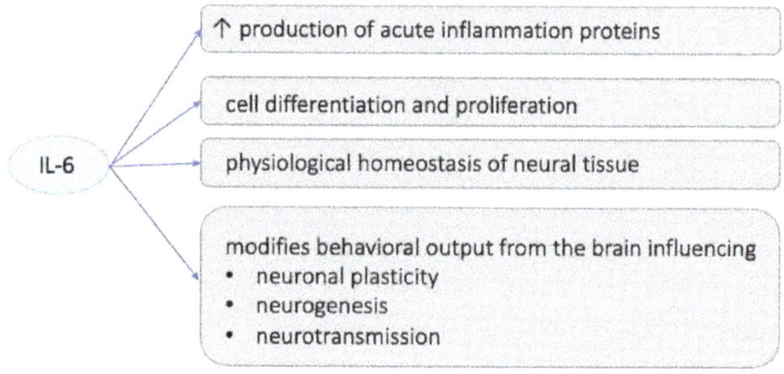

Figure 4. Summarizing figure on the altered IL-6 level's contribution to suicidality.

Lastly, our analysis reported a positive correlation between HCY and MOODS-SR depressive component as well as with suicidality scores. Our results are in line with previous research, which reported a substantial correlation between HCY and depression [170–173] but also identified HCY plasma levels as risk factors for the development of depression [174,175]. This evidence has been recently supported by a meta-analysis that included 46 observational studies [176]. In addition, HCY has been recently linked to suicide risks among patients with depression [177]; this result is in line with our data, which suggested a possible link between HCY levels and suicidality also among subjects in the autism spectrum. One of the possible mechanisms through which HCY levels may be involved in mood balance was hypothesized to lie in the excitatory function of HCY, which leads to an increase in the glutammatergic neurotransmission and, as a result, a calcium influx that has neurotoxic effects, ultimately leading to a greater instability of the affective symptomatology [178,179]. Moreover, a few community-based investigations highlighted a correlation between HCY concentrations and the severity and course of depression symptoms [180–183], also linking it with an increase in hostility and aggression and even by psychotic symptoms [184–187]. All this evidence seem to support the hypothesis arising from our results of the existence of a correlation between plasma levels of HCY and suicidality (See Figure 5).

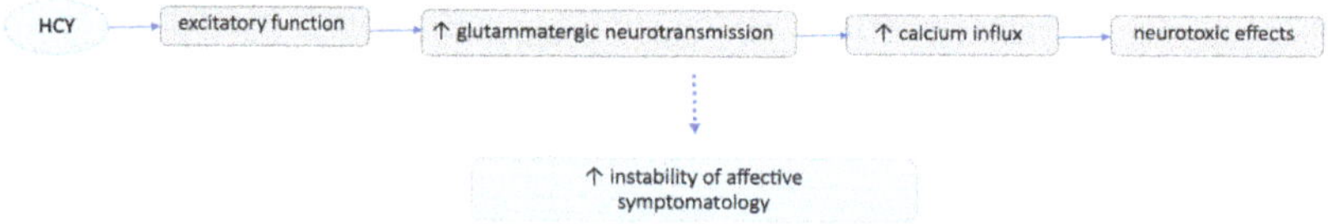

Figure 5. Summarizing figure on the altered HCY level's contribution to suicidality.

In this framework, research on microbiota has recently gained increasing importance in the field of psychiatric disorder etiopathogenesis. In fact, a rising body of evidence suggests that gut flora and depression are intimately related, and that gut microorganisms may interact with the brain via peripheral inflammation [188–192]. Moreover, intestinal and mucosal layer thinning increases microbiota transmission to the brain and may result in chronic, low-grade inflammation, which is seen in many psychiatric disorders [193]. Furthermore, microbiome composition was examined in the context of reported sleep issues because chronic fatigue has been linked to dysbiosis [194] and insomnia is considered a risk factor for suicidal behaviors [195,196]. Additionally, it has been proven that BDNF in the amygdala and hippocampus, as well as other peripheral and central nervous system events, such as cytokine production, short chain fatty acid release, and microglial maturation and activation, are all influenced by microbe–brain interactions [197,198].

Interestingly, the link between the gut microbiota and inflammation and immunological response has been the subject of extensive research, and numerous studies have demonstrated that the balance between pro- and anti-inflammatory responses in the gut is directly influenced by the gut flora [199]. The so formulated leaky-gut hypothesis describes how the connection between the gut microbiota, CNS, and the periphery leads to a vicious loop that activates inflammatory responses [200], and one of the proposed pathways for the induction of depressive symptoms by pro-inflammatory cytokines include the activation of the enzyme indoleamine 2,3-dioxygenase (IDO), which promotes the metabolic conversion of TRP into KYN [201]. On the other hand, altered KYN/TRP ratio and plasma levels of KYN have been positively linked to the severity of depressive symptoms, the latter also being correlated with suicidality [202], thus suggesting a possible role of microbiota as a marker of suicidal behaviors.

In this framework, our findings suggest, in line with previous studies, that we reconsider the importance of inflammation in the pathogenesis of mental illnesses, in particular the involvement of HCY, the TRP metabolism, and the interleukins system. On the other hand, further research should broadly investigate those alteration in all its pathways and branches. Further research investigating biochemical correlates should move forward acknowledging the potential existence of intertwined relationships between various systems and metabolic pathways, as well as between central and peripheral systems, in shaping suicidal behaviors in all their spectrum of manifestations. Our results also suggest the utility of metabolomics/proteomics approaches in this field of research, which could take into account multiple variables and metabolic signatures. Expanding the research in this area while learning more about the physiological correlates of suicide behaviors may increase the ability to identify useful risk factors and ultimately reveal potential novel therapy targets.

5. Limits and Further Directions

Some limitations should be considered in this study. Firstly, the results' applicability is limited by the small sample size. Additionally, the cross-sectional design of the study did not allow making inferences about possible temporal or causal relationships between the variables. Moreover, there was no information on the dietary status of the subjects, which may have a big impact on TRP levels and metabolism. In addition, the instruments used in the study were self-reported, and as a consequence subjects may have over- or underestimated their symptoms. Finally, this study lacks an investigation of the microbiota activity, which, considering that the bioactivity of the microbiota directly influences serotonin, KYN, and TRP levels, could be of relevance in this context.

Globally, a further in-depth understanding of the pathophysiological mechanisms underlying suicidal thoughts and behaviors is still required, and a special focus should be dedicated to research potential mediators and regulators of the inflammatory response that increase susceptibility or resilience to suicide [34], aiming to develop efficient panels of various biomarkers that could open to a new approach to treatment of suicide with novel treatment targets [187].

Literature regarding the possible therapeutic use of these correlates is still scant. Meanwhile, the possible involvement of glutammatergic routes as well as inflammatory processes in the pathophysiology of psychiatric disorders may eventually promote alternative targets for pharmacotherapy, eventually with a modulatory action on inflammation and glummate pathways [188]. However, further studies are needed in this field.

6. Conclusions

In conclusion, although preliminary, our findings confirm the link between suicidality and ASD and provides more evidence regarding the association of suicidality with increased HCY and IL-6 levels as well as with decreased TRP and KYNA ones. These biochemical findings, previously reported by studies among patients with other kinds of disorders, are confirmed herein also among patients in the autism spectrum, further stressing their association with increased suicide risk. Our results also suggest the utility of metabolomics/proteomics approaches in this field of research, which could take into account multiple variables and metabolic signatures. This highlights the need not only to reconsider the importance of inflammation in the pathogenesis of mental illnesses, in particular the involvement of HCY, the TRP metabolism, and the interleukins system, but also to spread the investigation considering these alterations in different metabolic pathways and branches. Future studies should take into account the possible role that metabolic pathways and systems, as well as the interactions between the central nervous system and peripheral systems, may have in influencing suicidal behavior in all of its forms. The ability to identify meaningful risk variables may improve with further studies in this field, also deepening our understanding regarding the pathophysiological correlates of suicide behaviors, which may ultimately offer possible innovative therapeutic targets.

Author Contributions: Conceptualization, L.D., I.M.C., B.C., G.G. and L.B.; methodology, I.M.C., B.C., L.D., G.G., L.B. and L.P.; formal analysis, I.M.C. and B.C.; investigation, B.C., I.M.C., B.N., G.A., D.B., E.M., D.C. and L.P.; resources, L.D. and G.G.; writing—original draft preparation, B.N., B.C. and I.M.C.; writing—review and editing, I.M.C. and B.C.; supervision, I.M.C., L.D., G.G. and B.C. All authors have read and agreed to the published version of the manuscript.

Funding: This research received no external funding.

Institutional Review Board Statement: The study was conducted in accordance with the Declaration of Helsinki and approved by the Ethics Committee of the Azienda Ospedaliero-Universitaria of Pisa (protocol code id14295, date of approval 17 January 2023).

Informed Consent Statement: Informed consent was obtained from all subjects involved in the study.

Data Availability Statement: All data generated or analyzed during this study are included in this published article.

Conflicts of Interest: The authors declare no conflict of interest.

References

1. World Health Organization. *Suicide in the World: Global Health Estimates*; World Health Organization: Geneva, Switzerland, 2019.
2. O'Carroll, P.W.; Berman, A.L.; Maris, R.W.; Moscicki, E.K.; Tanney, B.L.; Silverman, M.M. Beyond the Tower of Babel: A nomenclature for suicidology. *Suicide Life Threat. Behav.* **1996**, *26*, 237–252. [CrossRef] [PubMed]
3. Malafosse, A. Genetics of suicidal behavior. *Am. J. Med. Genet. C. Semin. Med. Genet.* **2005**, *133*, 1–2. [CrossRef] [PubMed]
4. Goldsmith, S.K.; Pellmar, T.C.; Kleinman, A.M.; Bunney, W.E. *Reducing Suicide: A National Imperative*; The National Academies Press: Washington, DC, USA, 2002.
5. Turecki, G.; Ernst, C.; Jollant, F.; Labonté, B.; Mechawar, N. The neurodevelopmental origins of suicidal behavior. *Trends Neurosci.* **2012**, *35*, 14–23. [CrossRef] [PubMed]
6. Johnston, J.N.; Campbell, D.; Caruncho, H.J.; Henter, I.D.; Ballard, E.D.; A Zarate, C. Suicide Biomarkers to Predict Risk, Classify Diagnostic Subtypes, and Identify Novel Therapeutic Targets: 5 Years of Promising Research. *Int. J. Neuropsychopharmacol.* **2022**, *25*, 197–214. [CrossRef] [PubMed]
7. Hawton, K.; i Comabella, C.C.; Haw, C.; Saunders, K. Risk factors for suicide in individuals with depression: A systematic review. *J. Affect. Disord.* **2013**, *147*, 17–28. [CrossRef]
8. May, A.M.; Klonsky, E.D.; Klein, D.N. Predicting future suicide attempts among depressed suicide ideators: A 10-year longitudinal study. *J. Psychiatr. Res.* **2012**, *46*, 946–952. [CrossRef]
9. Bokor, J.; Sutori, S.; Torok, D.; Gal, Z.; Eszlari, N.; Gyorik, D.; Baksa, D.; Petschner, P.; Serafini, G.; Pompili, M.; et al. Inflamed Mind: Multiple Genetic Variants of IL6 Influence Suicide Risk Phenotypes in Interaction with Early and Recent Adversities in a Linkage Disequilibrium-Based Clumping Analysis. *Front. Psychiatry* **2021**, *12*, 746206. [CrossRef]
10. Davis, A.T.; Schrueder, C. The prediction of suicide. *Med. J. Aust.* **1990**, *153*, 552–554. [CrossRef]
11. Blasco-Fontecilla, H.; Lopez-Castroman, J.; Giner, L.; Baca-García, E.; Oquendo, M.A. Predicting Suicidal Behavior: Are We Really that Far Along? Comment on Discovery and Validation of Blood Biomarkers for Suicidality. *Curr. Psychiatry Rep.* **2013**, *15*, 424. [CrossRef]
12. Smith, E.G.; Kim, H.M.; Ganoczy, D.; Stano, C.; Pfeiffer, P.N.; Valenstein, M. Suicide Risk Assessment Received Prior to Suicide Death by Veterans Health Administration Patients with a History of Depression. *J. Clin. Psychiatry* **2013**, *74*, 226–232. [CrossRef]
13. Ganança, L.; Oquendo, M.A.; Tyrka, A.R.; Cisneros-Trujillo, S.; Mann, J.J.; Sublette, M.E. The role of cytokines in the pathophysiology of suicidal behavior. *Psychoneuroendocrinology* **2016**, *63*, 296–310. [CrossRef] [PubMed]

14. Niciu, M.J.; Mathews, D.C.; Ionescu, D.F.; Richards, E.M.; Furey, M.L.; Yuan, P.; Nugent, A.C.; Henter, I.D.; Machado-Vieira, R.; Zarate, C.A., Jr. Biomarkers in mood disorders research: Developing new and improved therapeutics. *Rev. Psiquiatr. Clin.* **2014**, *41*, 131–134. [CrossRef]
15. Bryleva, E.Y.; Brundin, L. Kynurenine pathway metabolites and suicidality. *Neuropharmacology* **2017**, *112*, 324–330. [CrossRef]
16. Orsolini, L.; Latini, R.; Pompili, M.; Serafini, G.; Volpe, U.; Vellante, F.; Fornaro, M.; Valchera, A.; Tomasetti, C.; Fraticelli, S.; et al. Understanding the Complex of Suicide in Depression: From Research to Clinics. *Psychiatry Investig.* **2020**, *17*, 207–221. [CrossRef] [PubMed]
17. Amitai, M.; Taler, M.; Ben-Baruch, R.; Lebow, M.; Rotkopf, R.; Apter, A.; Fennig, S.; Weizman, A.; Chen, A. Increased circulatory IL-6 during 8-week fluoxetine treatment is a risk factor for suicidal behaviors in youth. *Brain Behav. Immun.* **2020**, *87*, 301–308. [CrossRef] [PubMed]
18. Tanaka, M.; Szabó, Á.; Spekker, E.; Polyák, H.; Tóth, F.; Vécsei, L. Mitochondrial Impairment: A Common Motif in Neuropsychiatric Presentation? The Link to the Tryptophan–Kynurenine Metabolic System. *Cells* **2022**, *11*, 2607. [CrossRef] [PubMed]
19. Tanaka, M.; Tóth, F.; Polyák, H.; Szabó, Á.; Mándi, Y.; Vécsei, L. Immune Influencers in Action: Metabolites and Enzymes of the Tryptophan-Kynurenine Metabolic Pathway. *Biomedicines* **2021**, *9*, 734. [CrossRef]
20. Bauer, M.E.; Teixeira, A.L. Inflammation in psychiatric disorders: What comes first? *Ann. N. Y. Acad. Sci.* **2018**, *1437*, 57–67. [CrossRef]
21. Wichers, M.C.; Koek, G.H.; Robaeys, G.; Verkerk, R.; Scharpé, S.; Maes, M. IDO and interferon-α-induced depressive symptoms: A shift in hypothesis from tryptophan depletion to neurotoxicity. *Mol. Psychiatry* **2005**, *10*, 538–544. [CrossRef]
22. Salas-Magaña, M.; Tovilla-Zárate, C.A.; González-Castro, T.B.; Juárez-Rojop, I.E.; López-Narváez, M.L.; Rodríguez-Pérez, J.M.; Bello, J.R. Decrease in brain-derived neurotrophic factor at plasma level but not in serum concentrations in suicide behavior: A systematic review and meta-analysis. *Brain Behav.* **2017**, *7*, e00706. [CrossRef]
23. Carpita, B.; Betti, L.; Palego, L.; Bartolommei, N.; Chico, L.; Pasquali, L.; Siciliano, G.; Monzani, F.; Franchi, R.; Rogani, S.; et al. Plasma redox and inflammatory patterns during major depressive episodes: A cross-sectional investigation in elderly patients with mood disorders. *CNS Spectr.* **2020**, *26*, 416–426. [CrossRef]
24. Marazziti, D.; Abelli, M.; Baroni, S.; Carpita, B.; Piccinni, A.; Dell'Osso, L. Recent findings on the pathophysiology of social anxiety disorder. *Clin. Neuropsych.* **2014**, *11*, 91–100.
25. Rengasamy, M.; Zhong, Y.; Marsland, A.; Chen, K.; Douaihy, A.; Brent, D.; Melhem, N.M. Signaling networks in inflammatory pathways and risk for suicidal behavior. *Brain Behav. Immun. Health* **2020**, *7*, 100122. [CrossRef] [PubMed]
26. Hong, H.; Kim, B.S.; Im, H.-I. Pathophysiological Role of Neuroinflammation in Neurodegenerative Diseases and Psychiatric Disorders. *Int. Neurourol. J.* **2016**, *20*, S2–S7. [CrossRef]
27. Wang, A.K.; Miller, B.J. Meta-analysis of Cerebrospinal Fluid Cytokine and Tryptophan Catabolite Alterations in Psychiatric Patients: Comparisons Between Schizophrenia, Bipolar Disorder, and Depression. *Schizophr. Bull.* **2018**, *44*, 75–83. [CrossRef]
28. Lindqvist, D.; Janelidze, S.; Hagell, P.; Erhardt, S.; Samuelsson, M.; Minthon, L.; Hansson, O.; Björkqvist, M.; Träskman-Bendz, L.; Brundin, L. Interleukin-6 Is Elevated in the Cerebrospinal Fluid of Suicide Attempters and Related to Symptom Severity. *Biol. Psychiatry* **2009**, *66*, 287–292. [CrossRef]
29. Janelidze, S.; Mattei, D.; Westrin, A.; Träskman-Bendz, L.; Brundin, L. Cytokine levels in the blood may distinguish suicide attempters from depressed patients. *Brain Behav. Immun.* **2011**, *25*, 335–339. [CrossRef] [PubMed]
30. Lindqvist, D.; Janelidze, S.; Erhardt, S.; Träskman-Bendz, L.; Engström, G.; Brundin, L. CSF biomarkers in suicide attempters-a principal component analysis. *Acta Psychiatr. Scand.* **2011**, *124*, 52–61. [CrossRef]
31. Martinez, J.M.; Garakani, A.; Yehuda, R.; Gorman, J.M. Proinflammatory and "resiliency" proteins in the CSF of patients with major depression. *Depress. Anxiety* **2011**, *29*, 32–38. [CrossRef]
32. Pandey, G.N.; Rizavi, H.S.; Ren, X.; Fareed, J.; Hoppensteadt, D.A.; Roberts, R.C.; Conley, R.R.; Dwivedi, Y. Proinflammatory cytokines in the prefrontal cortex of teenage suicide victims. *J. Psychiatr. Res.* **2012**, *46*, 57–63. [CrossRef]
33. Hoyo-Becerra, C.; Huebener, A.; Trippler, M.; Lutterbeck, M.; Liu, Z.J.; Truebner, K.; Bajanowski, T.; Gerken, G.; Hermann, D.M.; Schlaak, J.F. Concomitant Interferon Alpha Stimulation and TLR3 Activation Induces Neuronal Expression of Depression-Related Genes That Are Elevated in the Brain of Suicidal Persons. *PLoS ONE* **2013**, *8*, e83149. [CrossRef]
34. O'Donovan, A.; Rush, G.; Hoatam, G.; Hughes, B.M.; McCrohan, A.; Kelleher, C.; O'Farrelly, C.; Malone, K.M. Suicidal ideation is associated with elevated inflammation in patients with major depressive disorder. *Depress. Anxiety* **2013**, *30*, 307–314. [CrossRef] [PubMed]
35. Pandey, G.N.; Rizavi, H.S.; Zhang, H.; Bhaumik, R.; Ren, X. Abnormal protein and mRNA expression of inflammatory cytokines in the prefrontal cortex of depressed individuals who died by suicide. *J. Psychiatry Neurosci.* **2018**, *43*, 376–385. [CrossRef]
36. Serafini, G.; Parisi, V.M.; Aguglia, A.; Amerio, A.; Sampogna, G.; Fiorillo, A.; Pompili, M.; Amore, M. A Specific Inflammatory Profile Underlying Suicide Risk? Systematic Review of the Main Literature Findings. *Int. J. Environ. Res. Public. Health* **2020**, *17*, 2393. [CrossRef]
37. Miná, V.A.; Lacerda-Pinheiro, S.F.; Maia, L.C.; Pinheiro, R.F., Jr.; Meireles, C.B.; de Souza, S.I.; Reis, A.O.; Bianco, B.; Rolim, M.L. The influence of inflammatory cytokines in physiopathology of suicidal behavior. *J. Affect. Disord.* **2015**, *172*, 219–230. [CrossRef]
38. Niculescu, A.B.; Levey, D.; Phalen, P.; Le-Niculescu, H.; Dainton-Howard, H.; Jain, N.; Belanger, E.; James, A.; George, S.; Weber, H.; et al. Understanding and predicting suicidality using a combined genomic and clinical risk assessment approach. *Mol. Psychiatry* **2015**, *20*, 1266–1285. [CrossRef]

39. Brundin, L.; Sellgren, C.M.; Lim, C.K.; Grit, J.; Pålsson, E.; Landén, M.; Samuelsson, M.; Lundgren, K.; Brundin, P.; Fuchs, D.; et al. An enzyme in the kynurenine pathway that governs vulnerability to suicidal behavior by regulating excitotoxicity and neuroinflammation. *Transl. Psychiatry* **2016**, *6*, e865. [CrossRef]
40. Niculescu, A.B.; Le-Niculescu, H.; Levey, D.F.; Phalen, P.L.; Dainton, H.L.; Roseberry, K.; Niculescu, E.M.; Niezer, J.O.; Williams, A.; Graham, D.L.; et al. Precision medicine for suicidality: From universality to subtypes and personalization. *Mol. Psychiatry* **2017**, *22*, 1250–1273. [CrossRef] [PubMed]
41. Uciechowski, P.; Dempke, W.C. Interleukin-6: A Masterplayer in the Cytokine Network. *Oncology* **2020**, *98*, 131–137. [CrossRef]
42. Rothaug, M.; Becker-Pauly, C.; Rose-John, S. The role of interleukin-6 signaling in nervous tissue. *Biochim. Biophys. Acta (BBA) Mol. Cell Res.* **2016**, *1863*, 1218–1227. [CrossRef] [PubMed]
43. Monje, M.L.; Toda, H.; Palmer, T.D. Inflammatory Blockade Restores Adult Hippocampal Neurogenesis. *Science* **2003**, *302*, 1760–1765. [CrossRef]
44. Khairova, R.A.; Machado-Vieira, R.; Du, J.; Manji, H.K. A potential role for pro-inflammatory cytokines in regulating synaptic plasticity in major depressive disorder. *Int. J. Neuropsychopharmacol.* **2009**, *12*, 561–578. [CrossRef] [PubMed]
45. Roohi, E.; Jaafari, N.; Hashemian, F. On inflammatory hypothesis of depression: What is the role of IL-6 in the middle of the chaos? *J. Neuroinflammation* **2021**, *18*, 45. [CrossRef] [PubMed]
46. Hodes, G.E.; Ménard, C.; Russo, S.J. Integrating Interleukin-6 into depression diagnosis and treatment. *Neurobiol. Stress* **2016**, *4*, 15–22. [CrossRef]
47. González-Castro, T.B.; Tovilla-Zárate, C.A.; López-Narváez, M.L.; Genis-Mendoza, A.D.; Juárez-Rojop, I.E. Interleukin-6 Levels in Serum, Plasma, and Cerebral Spinal Fluid in Individuals with Suicide Behavior: Systematic Review and Meta-Analysis with Meta-Regression. *J. Interf. Cytokine Res.* **2021**, *41*, 258–267. [CrossRef]
48. Fernández-Sevillano, J.; González-Ortega, I.; MacDowell, K.; Zorrilla, I.; López, M.P.; Courtet, P.; Gabilondo, A.; Martínez-Cengotitabengoa, M.; Leza, J.C.; Sáiz, P.; et al. Inflammation biomarkers in suicide attempts and their relation to abuse, global functioning and cognition. *World J. Biol. Psychiatry* **2021**, *23*, 307–317. [CrossRef] [PubMed]
49. Amitai, M.; Taler, M.; Lebow, M.; Ben-Baruch, R.; Apter, A.; Fennig, S.; Weizman, A.; Chen, A. An increase in IL-6 levels at 6-month follow-up visit is associated with SSRI-emergent suicidality in high-risk children and adolescents treated with fluoxetine. *Eur. Neuropsychopharmacol.* **2020**, *40*, 61–69. [CrossRef]
50. Bryleva, E.Y.; Brundin, L. Suicidality and Activation of the Kynurenine Pathway of Tryptophan Metabolism. *Curr. Top Behav. Neurosci.* **2016**, *31*, 269–284. [CrossRef]
51. Erhardt, S.; Lim, C.K.; Linderholm, K.R.; Janelidze, S.; Lindqvist, D.; Samuelsson, M.; Lundberg, K.; Postolache, T.T.; Träskman-Bendz, L.; Guillemin, G.J.; et al. Connecting inflammation with glutamate agonism in suicidality. *Neuropsychopharmacology* **2012**, *38*, 743–752. [CrossRef]
52. Pandey, G.N.; Dwivedi, Y. Peripheral Biomarkers for Suicide. In *The Neurobiological Basis of Suicide*; Dwivedi, Y., Ed.; CRC Press: Boca Raton, FL, USA; Taylor & Francis: Millton, UK, 2012.
53. Arango, V.; Underwood, M.D.; Mann, J. Chapter 35 Serotonin brain circuits involved in major depression and suicide. *Prog. Brain Res.* **2002**, *136*, 443–453. [CrossRef]
54. Azmitia, E.C. Evolution of Serotonin: Sunlight to Suicide. In *Handbook of Behavioral Neuroscience*; Elsevier: Amsterdam, The Netherlands, 2020; Volume 31, pp. 3–22.
55. Pandey, G.N. Biological basis of suicide and suicidal behavior. *Bipolar Disord.* **2013**, *15*, 524–541. [CrossRef] [PubMed]
56. Hardingham, G.E.; Fukunaga, Y.; Bading, H. Extrasynaptic NMDARs oppose synaptic NMDARs by triggering CREB shut-off and cell death pathways. *Nat. Neurosci.* **2002**, *5*, 405–414. [CrossRef]
57. Sudol, K.; Mann, J.J. Biomarkers of Suicide Attempt Behavior: Towards a Biological Model of Risk. *Curr. Psychiatry Rep.* **2017**, *19*, 31. [CrossRef]
58. Lybech, L.K.M.; Calabró, M.; Briuglia, S.; Drago, A.; Crisafulli, C. Suicide Related Phenotypes in a Bipolar Sample: Genetic Underpinnings. *Genes* **2021**, *12*, 1482. [CrossRef] [PubMed]
59. Voracek, M.; Loibl, L.M. Genetics of suicide: A systematic review of twin studies. *Wien. Klin. Wochenschr.* **2007**, *119*, 463–475. [CrossRef]
60. Brent, D.A.; Mann, J.J. Family genetic studies, suicide, and suicidal behavior. *Am. J. Med. Genet. Part C Semin. Med. Genet.* **2005**, *133*, 13–24. [CrossRef] [PubMed]
61. Mullins, N.; Bigdeli, T.B.; Børglum, A.D.; Coleman, J.R.; Demontis, D.; Mehta, D.; Power, R.A.; Ripke, S.; Stahl, E.A.; Starnawska, A.; et al. GWAS of Suicide Attempt in Psychiatric Disorders and Association with Major Depression Polygenic Risk Scores. *Am. J. Psychiatry* **2019**, *176*, 651–660. [CrossRef]
62. Ruderfer, D.M.; Walsh, C.G.; Aguirre, M.W.; Tanigawa, Y.; Ribeiro, J.D.; Franklin, J.C.; Rivas, M.A. Significant shared heritability underlies suicide attempt and clinically predicted probability of attempting suicide. *Mol. Psychiatry* **2019**, *25*, 2422–2430. [CrossRef]
63. Stein, M.B.; Ware, E.B.; Mitchell, C.; Chen, C.-Y.; Borja, S.; Cai, T.; Dempsey, C.L.; Fullerton, C.S.; Gelernter, J.; Heeringa, S.G.; et al. Genomewide association studies of suicide attempts in US soldiers. *Am. J. Med. Genet. Part B Neuropsychiatr. Genet.* **2017**, *174*, 786–797. [CrossRef] [PubMed]

64. Mullins, N.; Perroud, N.; Uher, R.; Butler, A.W.; Cohen-Woods, S.; Rivera, M.; Malki, K.; Euesden, J.; Power, R.A.; Tansey, K.E.; et al. Genetic relationships between suicide attempts, suicidal ideation and major psychiatric disorders: A genome-wide association and polygenic scoring study. *Am. J. Med. Genet. Part B Neuropsychiatr. Genet.* **2014**, *165*, 428–437. [CrossRef]
65. Galfalvy, H.; Haghighi, F.; Hodgkinson, C.; Goldman, D.; Oquendo, M.A.; Burke, A.; Huang, Y.-Y.; Giegling, I.; Rujescu, D.; Bureau, A.; et al. A genome-wide association study of suicidal behavior. *Am. J. Med. Genet. Part B Neuropsychiatr. Genet.* **2015**, *168*, 557–563. [CrossRef]
66. Erlangsen, A.; Appadurai, V.; Wang, Y.; Turecki, G.; Mors, O.; Werge, T.; Mortensen, P.B.; Starnawska, A.; Børglum, A.D.; Schork, A.; et al. Genetics of suicide attempts in individuals with and without mental disorders: A population-based genome-wide association study. *Mol. Psychiatry* **2018**, *25*, 2410–2421. [CrossRef]
67. Dell'Osso, L.; Cremone, I.M.; Amatori, G.; Cappelli, A.; Cuomo, A.; Barlati, S.; Massimetti, G.; Vita, A.; Fagiolini, A.; Carmassi, C.; et al. Investigating the Relationship between Autistic Traits, Ruminative Thinking, and Suicidality in a Clinical Sample of Subjects with Bipolar Disorder and Borderline Personality Disorder. *Brain Sci.* **2021**, *11*, 621. [CrossRef]
68. Pelton, M.K.; Cassidy, S.A. Are autistic traits associated with suicidality? A test of the interpersonal-psychological theory of suicide in a non-clinical young adult sample. *Autism Res.* **2017**, *10*, 1891–1904. [CrossRef]
69. Chen, Y.; Chen, Y.; Gau, S.S. Suicidality in Children with Elevated Autistic Traits. *Autism Res.* **2020**, *13*, 1811–1821. [CrossRef] [PubMed]
70. Dell'Osso, L.; Carpita, B.; Muti, D.; Morelli, V.; Salarpi, G.; Salerni, A.; Scotto, J.; Massimetti, G.; Gesi, C.; Ballerio, M.; et al. Mood symptoms and suicidality across the autism spectrum. *Compr. Psychiatry* **2019**, *91*, 34–38. [CrossRef] [PubMed]
71. Thomas, S.D.; Jha, N.K.; Ojha, S.; Sadek, B. mTOR Signaling Disruption and Its Association with the Development of Autism Spectrum Disorder. *Molecules* **2023**, *28*, 1889. [CrossRef]
72. Sultan, S. Translating neuroimaging changes to neuro-endophenotypes of autistic spectrum disorder: A narrative review. *Egypt. J. Neurol. Psychiatry Neurosurg.* **2022**, *58*, 139. [CrossRef]
73. Just, M.A.; Keller, T.A.; Malave, V.L.; Kana, R.K.; Varma, S. Autism as a neural systems disorder: A theory of frontal-posterior underconnectivity. *Neurosci. Biobehav. Rev.* **2012**, *36*, 1292–1313. [CrossRef] [PubMed]
74. Just, M.; Minshew, N.; Williams, D.; Cherkassky, V.; Kana, R.; Keller, T.; Damarla, S. Cortical underconnectivity coupled with preserved visuospatial cognition in autism: Evidence from an fMRI study of an embedded figures task. *Autism Res.* **2010**, *5*, 273–279. [CrossRef]
75. Kana, R.K.; Keller, T.A.; Minshew, N.J.; Just, M.A. Inhibitory Control in High-Functioning Autism: Decreased Activation and Underconnectivity in Inhibition Networks. *Biol. Psychiatry* **2007**, *62*, 198–206. [CrossRef]
76. Kana, R.K.; Keller, T.A.; Cherkassky, V.L.; Minshew, N.J.; Just, M.A. Sentence comprehension in autism: Thinking in pictures with decreased functional connectivity. *Brain* **2006**, *129*, 2484–2493. [CrossRef]
77. Kana, R.K.; Keller, T.A.; Cherkassky, V.L.; Minshew, N.J.; Just, M. Atypical frontal-posterior synchronization of Theory of Mind regions in autism during mental state attribution. *Soc. Neurosci.* **2009**, *4*, 135–152. [CrossRef]
78. Koshino, H.; Carpenter, P.A.; Minshew, N.J.; Cherkassky, V.L.; Keller, T.A.; Just, M.A. Functional connectivity in an fMRI working memory task in high-functioning autism. *NeuroImage* **2005**, *24*, 810–821. [CrossRef]
79. Koshino, H.; Kana, R.K.; Keller, T.A.; Cherkassky, V.L.; Minshew, N.J.; Just, M.A. fMRI Investigation of Working Memory for Faces in Autism: Visual Coding and Underconnectivity with Frontal Areas. *Cereb. Cortex* **2008**, *18*, 289–300. [CrossRef]
80. Mason, R.A.; Williams, D.L.; Kana, R.K.; Minshew, N.; Just, M.A. Theory of Mind disruption and recruitment of the right hemisphere during narrative comprehension in autism. *Neuropsychologia* **2008**, *46*, 269–280. [CrossRef] [PubMed]
81. Mizuno, A.; Liu, Y.; Williams, D.L.; Keller, T.A.; Minshew, N.J.; Just, M. The neural basis of deictic shifting in linguistic perspective-taking in high-functioning autism. *Brain* **2011**, *134*, 2422–2435. [CrossRef]
82. Just, M.A.; Cherkassky, V.L.; Keller, T.A.; Minshew, N.J. Cortical activation and synchronization during sentence comprehension in high-functioning autism: Evidence of underconnectivity. *Brain* **2004**, *127*, 1811–1821. [CrossRef] [PubMed]
83. Bailey, A.; Luthert, P.; Dean, A.; Harding, B.; Janota, I.; Montgomery, M.; Rutter, M.; Lantos, P. A clinicopathological study of autism. *Brain* **1998**, *121*, 889–905. [CrossRef]
84. Bauman, M.; Kemper, T.L. Histoanatomic observations of the brain in early infantile autism. *Neurology* **1985**, *35*, 866–874. [CrossRef] [PubMed]
85. Courchesne, E.; Karns, C.M.; Davis, H.R.; Ziccardi, R.; Carper, R.A.; Tigue, Z.D.; Chisum, H.J.; Moses, P.; Pierce, K.; Lord, C.; et al. Unusual brain growth patterns in early life in patients with autistic disorder: An MRI study. *Neurology* **2001**, *57*, 245–254. [CrossRef]
86. Courchesne, E.; Carper, R.; Akshoomoff, N. Evidence of Brain Overgrowth in the First Year of Life in Autism. *JAMA* **2003**, *290*, 337–344. [CrossRef] [PubMed]
87. Hazlett, H.C.; Poe, M.; Gerig, G.; Smith, R.G.; Provenzale, J.; Ross, A.; Gilmore, J.; Piven, J. Magnetic Resonance Imaging and Head Circumference Study of Brain Size in Autism. *Arch. Gen. Psychiatry* **2005**, *62*, 1366–1376. [CrossRef]
88. Castelli, F.; Frith, C.; Happé, F.; Frith, U. Autism, Asperger syndrome and brain mechanisms for the attribution of mental states to animated shapes. *Brain* **2002**, *125*, 1839–1849. [CrossRef]
89. Frith, C. What Do Imaging Studies Tell Us About the Neural Basis of Autism? In *Autism: Neural Basis and Treatment Possibilities: Novartis Foundation Symposium*; Bock, G., Goode, J., Eds.; John Wiley & Sons: Chichester, UK, 2003; Volume 251, pp. 149–176.

90. Schultz, R.T.; Klin, A. Genetics of Childhood Disorders: XLIII. Autism, Part 2: Neural Foundations. *J. Am. Acad. Child. Adolesc. Psychiatry* **2002**, *41*, 1259–1262. [CrossRef] [PubMed]
91. Aylward, E.H.; Minshew, N.J.; Field, K.; Sparks, B.F.; Singh, N. Effects of age on brain volume and head circumference in autism. *Neurology* **2002**, *59*, 175–183. [CrossRef]
92. Piven, J.; Arndt, S.; Bailey, J.; Havercamp, S.; Andreasen, N.C.; Palmer, P. An MRI study of brain size in autism. *Am. J. Psychiatry* **1995**, *152*, 1145–1149. [CrossRef]
93. Carper, R.A.; Courchesne, E. Localized enlargement of the frontal cortex in early autism. *Biol. Psychiatry* **2005**, *57*, 126–133. [CrossRef] [PubMed]
94. Thabault, M.; Turpin, V.; Maisterrena, A.; Jaber, M.; Egloff, M.; Galvan, L. Cerebellar and Striatal Implications in Autism Spectrum Disorders: From Clinical Observations to Animal Models. *Int. J. Mol. Sci.* **2022**, *23*, 2294. [CrossRef]
95. Khadem-Reza, Z.K.; Zare, H. Evaluation of brain structure abnormalities in children with autism spectrum disorder (ASD) using structural magnetic resonance imaging. *Egypt. J. Neurol. Psychiatry Neurosurg.* **2022**, *58*, 135. [CrossRef]
96. Carper, R.A.; Moses, P.; Tigue, Z.D.; Courchesne, E. Cerebral Lobes in Autism: Early Hyperplasia and Abnormal Age Effects. *Neuroimage* **2002**, *16*, 1038–1051. [CrossRef]
97. Herbert, M.R.; Ziegler, D.A.; Makris, N.; Filipek, P.A.; Kemper, T.L.; Normandin, J.J.; Sanders, H.A.; Kennedy, D.N.; Caviness, V.S. Localization of white matter volume increase in autism and developmental language disorder. *Ann. Neurol.* **2004**, *55*, 530–540. [CrossRef] [PubMed]
98. Billeci, L.; Calderoni, S.; Conti, E.; Gesi, C.; Carmassi, C.; Dell'Osso, L.; Cioni, G.; Muratori, F.; Guzzetta, A. The Broad Autism (Endo)Phenotype: Neurostructural and Neurofunctional Correlates in Parents of Individuals with Autism Spectrum Disorders. *Front. Neurosci.* **2016**, *10*, 346. [CrossRef] [PubMed]
99. Courchesne, E.; Redcay, E.; Kennedy, D.P. The autistic brain: Birth through adulthood. *Curr. Opin. Neurol.* **2004**, *17*, 489–496. [CrossRef] [PubMed]
100. Waiter, G.D.; Williams, J.H.; Murray, A.; Gilchrist, A.; Perrett, D.; Whiten, A. Structural white matter deficits in high-functioning individuals with autistic spectrum disorder: A voxel-based investigation. *Neuroimage* **2005**, *24*, 455–461. [CrossRef]
101. Carpita, B.; Carmassi, C.; Calderoni, S.; Muti, D.; Muscarella, A.; Massimetti, G.; Cremone, I.M.; Gesi, C.; Conti, E.; Muratori, F.; et al. The broad autism phenotype in real-life: Clinical and functional correlates of autism spectrum symptoms and rumination among parents of patients with autism spectrum disorder. *CNS Spectr.* **2020**, *25*, 765–773. [CrossRef]
102. Carpita, B.; Muti, D.; Muscarella, A.; Dell'Oste, V.; Diadema, E.; Massimetti, G.; Signorelli, M.S.; Fusar Poli, L.; Gesi, C.; Aguglia, E.; et al. Sex Differences in the Relationship between PTSD Spectrum Symptoms and Autistic Traits in a Sample of University Students. *Clin. Pract. Epidemiol. Ment. Health* **2019**, *15*, 110–119. [CrossRef]
103. Carpita, B.; Cremone, I.M.; Amatori, G.; Cappelli, A.; Salerni, A.; Massimetti, G.; Borgioli, D.; Carmassi, C.; Massai, R.; Dell'osso, L. Investigating the relationship between orthorexia nervosa and autistic traits in a university population. *CNS Spectr.* **2022**, *27*, 613–620. [CrossRef]
104. Dell'Osso, L.; Carpita, B.; Bertelloni, C.A.; Diadema, E.; Barberi, F.M.; Gesi, C.; Carmassi, C. Subthreshold autism spectrum in bipolar disorder: Prevalence and clinical correlates. *Psychiatry Res.* **2019**, *281*, 112605. [CrossRef]
105. Dell'Osso, L.; Cremone, I.M.; Carpita, B.; Fagiolini, A.; Massimetti, G.; Bossini, L.; Vita, A.; Barlati, S.; Carmassi, C.; Gesi, C. Correlates of autistic traits among patients with borderline personality disorder. *Compr. Psychiatry* **2018**, *83*, 7–11. [CrossRef]
106. Dell'Osso, L.; Carpita, B.; Cremone, I.M.; Gesi, C.; D'Eermo, A.; De Iorio, G.; Massimetti, G.; Aguglia, E.; Bucci, P.; Carpiniello, B.; et al. Autism spectrum in patients with schizophrenia: Correlations with real-life functioning, resilience, and coping styles. *CNS Spectr.* **2021**, *12*, 457–467. [CrossRef] [PubMed]
107. Dell'Osso, L.; Carmassi, C.; Cremone, I.M.; Muti, D.; Salerni, A.; Barberi, F.M.; Massimetti, E.; Gesi, C.; Politi, P.; Aguglia, E.; et al. Defining the Optimal Threshold Scores for Adult Autism Subthreshold Spectrum (AdAS Spectrum) in Clinical and General Population. *Clin. Pract. Epidemiol. Ment. Health* **2020**, *16*, 204–211. [CrossRef] [PubMed]
108. Dell'Osso, L.; Conversano, C.; Corsi, M.; Bertelloni, C.A.; Cremone, I.M.; Carpita, B.; Carbone, M.G.; Gesi, C.; Carmassi, C. Polysubstance and Behavioral Addictions in a Patient with Bipolar Disorder: Role of Lifetime Subthreshold Autism Spectrum. *Case Rep. Psychiatry* **2018**, *2018*, 1547975. [CrossRef] [PubMed]
109. Dell'Osso, L.; Bertelloni, C.A.; Di Paolo, M.; Avella, M.T.; Carpita, B.; Gori, F.; Pompili, M.; Carmassi, C. Problematic Internet Use in University Students Attending Three Superior Graduate Schools in Italy: Is Autism Spectrum Related to Suicide Risk? *Int. J. Environ. Res. Public Health* **2019**, *16*, 1098. [CrossRef] [PubMed]
110. Dell'Osso, L.; Lorenzi, P.; Carpita, B. The neurodevelopmental continuum towards a neurodevelopmental gradient hypothesis. *J. Psychopathol.* **2019**, *25*, 179–182.
111. Cassidy, S.; Rodgers, J. Understanding and prevention of suicide in autism. *Lancet Psychiatry* **2017**, *4*, e11. [CrossRef]
112. Hedley, D.; Uljarević, M. Systematic Review of Suicide in Autism Spectrum Disorder: Current Trends and Implications. *Curr. Dev. Disord. Rep.* **2018**, *5*, 65–76. [CrossRef]
113. Hirvikoski, T.; Mittendorfer-Rutz, E.; Boman, M.; Larsson, H.; Lichtenstein, P.; Bölte, S. Premature mortality in autism spectrum disorder. *Br. J. Psychiatry* **2016**, *208*, 232–238. [CrossRef]
114. Hwang, Y.I.; Srasuebkul, P.; Foley, K.; Arnold, S.; Trollor, J.N. Mortality and cause of death of Australians on the autism spectrum. *Autism Res.* **2019**, *12*, 806–815. [CrossRef]

115. Kirby, A.V.; Bakian, A.V.; Zhang, Y.; Bilder, D.A.; Keeshin, B.R.; Coon, H. A 20-year study of suicide death in a statewide autism population. *Autism Res.* **2019**, *12*, 658–666. [CrossRef]
116. Mayes, S.D.; Gorman, A.A.; Hillwig-Garcia, J.; Syed, E. Suicide ideation and attempts in children with autism. *Res. Autism Spectr. Disord.* **2013**, *7*, 109–119. [CrossRef]
117. Paquette-Smith, M.; Weiss, J.; Lunsky, Y. History of Suicide Attempts in Adults with Asperger Syndrome. *Crisis* **2014**, *35*, 273–277. [CrossRef]
118. Segers, M.; Rawana, J. What Do We Know About Suicidality in Autism Spectrum Disorders? A Systematic Review. *Autism Res.* **2014**, *7*, 507–521. [CrossRef] [PubMed]
119. Upthegrove, R.; Abu-Akel, A.; Chisholm, K.; Lin, A.; Zahid, S.; Pelton, M.; Apperly, I.; Hansen, P.C.; Wood, S.J. Autism and psychosis: Clinical implications for depression and suicide. *Schizophr. Res.* **2018**, *195*, 80–85. [CrossRef]
120. Cassidy, S.; Bradley, L.; Shaw, R.; Baron-Cohen, S. Risk markers for suicidality in autistic adults. *Mol. Autism* **2018**, *9*, 42. [CrossRef] [PubMed]
121. Richards, G.; Kenny, R.; Griffiths, S.; Allison, C.; Mosse, D.; Holt, R.; O'connor, R.C.; Cassidy, S.; Baron-Cohen, S. Autistic traits in adults who have attempted suicide. *Mol. Autism* **2019**, *10*, 26. [CrossRef]
122. First, M.B.; Williams, J.B.; Karg, R.S.; Spitzer, R.L. *SCID-5-CV: Structured Clinical Interview for DSM-5 Disorders, Clinician Version*; American Psychiatric Association: Arlington, VA, USA, 2015.
123. Dell'Osso, L.; Cremone, I.M.; Carpita, B.; Dell'Oste, V.; Muti, D.; Massimetti, G.; Barlati, S.; Vita, A.; Fagiolini, A.; Carmassi, C.; et al. Rumination, posttraumatic stress disorder, and mood symptoms in borderline personality disorder. *Neuropsychiatr. Dis. Treat.* **2019**, *15*, 1231–1238. [CrossRef]
124. Bradford, M.M. A rapid and sensitive method for the quantitation of microgram quantities of protein utilizing the principle of protein-dye binding. *Anal. Biochem.* **1976**, *72*, 248–254. [CrossRef]
125. Carpita, B.; Nardi, B.; Palego, L.; Cremone, I.M.; Massimetti, G.; Carmassi, C.; Betti, L.; Giannaccini, G.; Dell'osso, L. Kynurenine pathway and autism spectrum phenotypes: An investigation among adults with autism spectrum disorder and their first-degree relatives. *CNS Spectr.* **2022**, *28*, 374–385. [CrossRef]
126. Carpita, B.; Stagnari, R.; Palego, L.; Baroni, D.; Massimetti, G.; Nardi, B.; Cremone, I.M.; Betti, L.; Giannaccini, G.; Dell'Osso, L. Circulating levels of 5-HT and BDNF in Adults with Autism Spectrum Conditions: An Investigation in a Sample of Subjects with Autism Spectrum Disorder, their first-degree Relatives and Controls. *Curr. Med. Chem.* **2023**. *ahead of print*. [CrossRef]
127. Carpita, B.; Massoni, L.; Battaglini, S.; Palego, L.; Cremone, I.M.; Massimetti, G.; Betti, L.; Giannaccini, G.; Dell'osso, L. IL-6, homocysteine, and autism spectrum phenotypes: An investigation among adults with autism spectrum disorder and their first-degree relatives. *CNS Spectr.* **2023**, 1–9. [CrossRef]
128. Hedley, D.; Uljarević, M.; Foley, K.-R.; Richdale, A.; Trollor, J. Risk and protective factors underlying depression and suicidal ideation in Autism Spectrum Disorder. *Depress. Anxiety* **2018**, *35*, 648–657. [CrossRef]
129. Jobe, L.E.; Williams White, S. Loneliness, social relationships, and a broader autism phenotype in college students. *Pers. Individ Differ.* **2007**, *42*, 1479–1489. [CrossRef]
130. Holt, M.K.; Vivolo-Kantor, A.M.; Polanin, J.R.; Holland, K.M.; DeGue, S.; Matjasko, J.L.; Wolfe, M.; Reid, G. Bullying and Suicidal Ideation and Behaviors: A Meta-Analysis. *Pediatrics* **2015**, *135*, e496–e509. [CrossRef]
131. Klomek, A.B.; Marrocco, F.; Kleinman, M.; Schonfeld, I.S.; Gould, M.S. Peer Victimization, Depression, and Suicidiality in Adolescents. *Suicide Life-Threat. Behav.* **2008**, *38*, 166–180. [CrossRef] [PubMed]
132. Rynkiewicz, A.; Schuller, B.; Marchi, E.; Piana, S.; Camurri, A.; Lassalle, A.; Baron-Cohen, S. An investigation of the 'female camouflage effect' in autism using a computerized ADOS-2 and a test of sex/gender differences. *Mol. Autism* **2016**, *7*, 10. [CrossRef] [PubMed]
133. Cassidy, S.A.; Gould, K.; Townsend, E.; Pelton, M.; Robertson, A.E.; Rodgers, J. Is Camouflaging Autistic Traits Associated with Suicidal Thoughts and Behaviours? Expanding the Interpersonal Psychological Theory of Suicide in an Undergraduate Student Sample. *J. Autism Dev. Disord.* **2020**, *50*, 3638–3648. [CrossRef]
134. Cremone, I.M.; Carpita, B.; Nardi, B.; Casagrande, D.; Stagnari, R.; Amatori, G.; Dell'osso, L. Measuring Social Camouflaging in Individuals with High Functioning Autism: A Literature Review. *Brain Sci.* **2023**, *13*, 469. [CrossRef] [PubMed]
135. Martos, D.; Tuka, B.; Tanaka, M.; Vécsei, L.; Telegdy, G. Memory Enhancement with Kynurenic Acid and Its Mechanisms in Neurotransmission. *Biomedicines* **2022**, *10*, 849. [CrossRef]
136. Chen, L.-M.; Bao, C.-H.; Wu, Y.; Liang, S.-H.; Di Wang, D.; Wu, L.-Y.; Huang, Y.; Liu, H.-R.; Wu, H.-G. Tryptophan-kynurenine metabolism: A link between the gut and brain for depression in inflammatory bowel disease. *J. Neuroinflammation* **2021**, *18*, 135. [CrossRef]
137. Lapin, I.; Oxenkrug, G. Intensification of the central serotoninergic processes as a possible determinant of the thymoleptic effect. *Lancet* **1969**, *293*, 132–136. [CrossRef] [PubMed]
138. Oxenkrug, G.F. Tryptophan-kynurenine metabolism as a common mediator of genetic and environmental impacts in major depressive disorder: The serotonin hypothesis revisited 40 years later. *Isr. J. Psychiatry Relat. Sci.* **2010**, *47*, 56–63. [PubMed]
139. Tanaka, M.; Vécsei, L. Monitoring the kynurenine system: Concentrations, ratios or what else? *Adv. Clin. Exp. Med.* **2021**, *30*, 775–778. [CrossRef] [PubMed]
140. Marrugo-Ramírez, J.; Rodríguez-Núñez, M.; Marco, M.-P.; Mir, M.; Samitier, J. Kynurenic Acid Electrochemical Immunosensor: Blood-Based Diagnosis of Alzheimer's Disease. *Biosensors* **2021**, *11*, 20. [CrossRef] [PubMed]

141. Tanaka, M.; Toldi, J.; Vécsei, L. Exploring the Etiological Links behind Neurodegenerative Diseases: Inflammatory Cytokines and Bioactive Kynurenines. *Int. J. Mol. Sci.* **2020**, *21*, 2431. [CrossRef]
142. Ogyu, K.; Kubo, K.; Noda, Y.; Iwata, Y.; Tsugawa, S.; Omura, Y.; Wada, M.; Tarumi, R.; Plitman, E.; Moriguchi, S.; et al. Kynurenine pathway in depression: A systematic review and meta-analysis. *Neurosci. Biobehav. Rev.* **2018**, *90*, 16–25. [CrossRef]
143. Sublette, M.E.; Galfalvy, H.C.; Fuchs, D.; Lapidus, M.; Grunebaum, M.F.; Oquendo, M.A.; Mann, J.J.; Postolache, T.T. Plasma kynurenine levels are elevated in suicide attempters with major depressive disorder. *Brain Behav. Immun.* **2011**, *25*, 1272–1278. [CrossRef]
144. Bradley, K.A.; Case, J.A.; Khan, O.; Ricart, T.; Hanna, A.; Alonso, C.M.; Gabbay, V. The role of the kynurenine pathway in suicidality in adolescent major depressive disorder. *Psychiatry Res.* **2015**, *227*, 206–212. [CrossRef]
145. Almeida-Montes, L.G.; Valles-Sanchez, V.; Moreno-Aguilar, J.; A Chavez-Balderas, R.; A García-Marín, J.; Sotres, J.F.C.; Hheinze-Martin, G. Relation of serum cholesterol, lipid, serotonin and tryptophan levels to severity of depression and to suicide attempts. *J. Psychiatry Neurosci.* **2000**, *25*, 371–377.
146. Clark, D.B. Serum tryptophan ratio and suicidal behavior in adolescents: A prospective study. *Psychiatry Res.* **2003**, *119*, 199–204. [CrossRef]
147. Pfeffer, C.R.; McBride, P.; Anderson, G.M.; Kakuma, T.; Fensterheim, L.; Khait, V. Peripheral serotonin measures in prepubertal psychiatric inpatients and normal children: Associations with suicidal behavior and its risk factors. *Biol. Psychiatry* **1998**, *44*, 568–577. [CrossRef] [PubMed]
148. Myint, A.-M.; Kim, Y.K.; Verkerk, R.; Scharpé, S.; Steinbusch, H.; Leonard, B. Kynurenine pathway in major depression: Evidence of impaired neuroprotection. *J. Affect. Disord.* **2007**, *98*, 143–151. [CrossRef] [PubMed]
149. Carlborg, A.; Jokinen, J.; Jönsson, E.G.; Erhardt, S.; Nordström, P. CSF kynurenic acid and suicide risk in schizophrenia spectrum psychosis. *Psychiatry Res.* **2013**, *205*, 165–167. [CrossRef] [PubMed]
150. Maes, M.; Meltzer, H.Y.; Scharpè, S.; Bosmans, E.; Suy, E.; De Meester, I.; Calabrese, J.; Cosyns, P. Relationships between lower plasma L-tryptophan levels and immune-inflammatory variables in depression. *Psychiatry Res.* **1993**, *49*, 151–165. [CrossRef] [PubMed]
151. Mann, J. Neurobiology of suicidal behaviour. *Nat. Rev. Neurosci.* **2003**, *4*, 819–828. [CrossRef]
152. Müller, N.; Schwarz, M.J. The immune-mediated alteration of serotonin and glutamate: Towards an integrated view of depression. *Mol. Psychiatry* **2007**, *12*, 988–1000. [CrossRef]
153. Amori, L.; Wu, H.-Q.; Marinozzi, M.; Pellicciari, R.; Guidetti, P.; Schwarcz, R. Specific inhibition of kynurenate synthesis enhances extracellular dopamine levels in the rodent striatum. *Neuroscience* **2009**, *159*, 196–203. [CrossRef]
154. Poeggeler, B.; Rassoulpour, A.; Wu, H.-Q.; Guidetti, P.; Roberts, R.; Schwarcz, R. Dopamine receptor activation reveals a novel, kynurenate-sensitive component of striatal N-methyl-d-aspartate neurotoxicity. *Neuroscience* **2007**, *148*, 188–197. [CrossRef]
155. Schwarcz, R.; Pellicciari, R. Manipulation of brain kynurenines: Glial targets, neuronal effects, and clinical opportunities. *J. Pharmacol. Exp. Ther.* **2002**, *303*, 1–10. [CrossRef]
156. Hilmas, C.; Pereira, E.F.R.; Alkondon, M.; Rassoulpour, A.; Schwarcz, R.; Albuquerque, E.X. The Brain Metabolite Kynurenic Acid Inhibits $\alpha 7$ Nicotinic Receptor Activity and Increases Non-$\alpha 7$ Nicotinic Receptor Expression: Physiopathological Implications. *J. Neurosci.* **2001**, *21*, 7463–7473. [CrossRef]
157. Chess, A.C.; Simoni, M.K.; Alling, T.E.; Bucci, D.J. Elevations of Endogenous Kynurenic Acid Produce Spatial Working Memory Deficits. *Schizophr. Bull.* **2007**, *33*, 797–804. [CrossRef] [PubMed]
158. Keilp, J.G.; Sackeim, H.A.; Brodsky, B.S.; Oquendo, M.A.; Malone, K.M.; Mann, J.J. Neuropsychological Dysfunction in Depressed Suicide Attempters. *Am. J. Psychiatry* **2001**, *158*, 735–741. [CrossRef]
159. Marzuk, P.M.; Hartwell, N.; Leon, A.C.; Portera, L. Executive functioning in depressed patients with suicidal ideation. *Acta Psychiatr. Scand.* **2005**, *112*, 294–301. [CrossRef] [PubMed]
160. Dowlati, Y.; Herrmann, N.; Swardfager, W.; Liu, H.; Sham, L.; Reim, E.K.; Lanctôt, K.L. A Meta-Analysis of Cytokines in Major Depression. *Biol. Psychiatry* **2010**, *67*, 446–457. [CrossRef] [PubMed]
161. Ortega, D.R.; Muñiz, P.E.U.; Ayala, T.B.; Cervantes, G.I.V.; Huitrón, R.L.; Pineda, B.; Esquivel, D.F.G.; de la Cruz, G.P.; Chaverrí, J.P.; Chapul, L.S.; et al. On the Antioxidant Properties of L-Kynurenine: An Efficient ROS Scavenger and Enhancer of Rat Brain Antioxidant Defense. *Antioxidants* **2022**, *11*, 31. [CrossRef] [PubMed]
162. Howren, M.B.; Lamkin, D.M.; Suls, J. Associations of Depression with C-Reactive Protein, IL-1, and IL-6: A Meta-Analysis. *Psychosom. Med.* **2009**, *71*, 171–186. [CrossRef]
163. Liu, Y.; Ho, R.C.-M.; Mak, A. Interleukin (IL)-6, tumour necrosis factor alpha (TNF-α) and soluble interleukin-2 receptors (sIL-2R) are elevated in patients with major depressive disorder: A meta-analysis and meta-regression. *J. Affect. Disord.* **2012**, *139*, 230–239. [CrossRef]
164. Maes, M.; Scharpé, S.; Meltzer, H.Y.; Bosmans, E.; Suy, E.; Calabrese, J.; Cosyns, P. Relationships between interleukin-6 activity, acute phase proteins, and function of the hypothalamic-pituitary-adrenal axis in severe depression. *Psychiatry Res.* **1993**, *49*, 11–27. [CrossRef]
165. Rush, G.; O'donovan, A.; Nagle, L.; Conway, C.; McCrohan, A.; O'farrelly, C.; Lucey, J.V.; Malone, K.M. Alteration of immune markers in a group of melancholic depressed patients and their response to electroconvulsive therapy. *J. Affect. Disord.* **2016**, *205*, 60–68. [CrossRef]

166. Yang, C.; Tiemessen, K.M.; Bosker, F.J.; Wardenaar, K.J.; Lie, J.; Schoevers, R.A. Interleukin, tumor necrosis factor-α and C-reactive protein profiles in melancholic and non-melancholic depression: A systematic review. *J. Psychosom. Res.* **2018**, *111*, 58–68. [CrossRef]
167. E Thase, M. Recognition and diagnosis of atypical depression. *J. Clin. Psychiatry* **2007**, *68*, 11–16.
168. Rudolf, S.; Greggersen, W.; Kahl, K.G.; Hüppe, M.; Schweiger, U. Elevated IL-6 levels in patients with atypical depression but not in patients with typical depression. *Psychiatry Res.* **2014**, *217*, 34–38. [CrossRef]
169. Morris, M.S.; Fava, M.; Jacques, P.F.; Selhub, J.; Rosenberg, I.H. Depression and Folate Status in the US Population. *Psychother. Psychosom.* **2003**, *72*, 80–87. [CrossRef] [PubMed]
170. Fava, M.; Borus, J.S.; E Alpert, J.; A Nierenberg, A.; Rosenbaum, J.F.; Bottiglieri, T. Folate, vitamin B12, and homocysteine in major depressive disorder. *Am. J. Psychiatry* **1997**, *154*, 426–428. [CrossRef] [PubMed]
171. Bjelland, I.; Tell, G.S.; Vollset, S.E.; Refsum, H.; Ueland, P.M. Folate, Vitamin B12, Homocysteine, and the MTHFR 677C→T Polymorphism in Anxiety and Depression. *Arch. Gen. Psychiatry* **2003**, *60*, 618–626. [CrossRef]
172. Bottiglieri, T.; Laundy, M.; Crellin, R.; Toone, B.K.; Carney, M.W.P.; Reynolds, E.H. Homocysteine, folate, methylation, and monoamine metabolism in depression. *J. Neurol. Neurosurg. Psychiatry* **2000**, *69*, 228–232. [CrossRef] [PubMed]
173. Almeida, O.P.; Lautenschlager, N.; Flicker, L.; Leedman, P.; Vasikaran, S.; Gelavis, A.; Ludlow, J. Association Between Homocysteine, Depression, and Cognitive Function in Community-Dwelling Older Women from Australia. *J. Am. Geriatr. Soc.* **2004**, *52*, 327–328. [CrossRef] [PubMed]
174. Refsum, H.; Nurk, E.; Smith, A.D.; Ueland, P.M.; Gjesdal, C.G.; Bjelland, I.; Tverdal, A.; Tell, G.S.; Nygård, O.; Vollset, S.E. The Hordaland Homocysteine Study: A Community-Based Study of Homocysteine, Its Determinants, and Associations with Disease. *J. Nutr.* **2006**, *136*, 1731–1740. [CrossRef]
175. Moradi, F.; Lotfi, K.; Armin, M.; Clark, C.C.; Askari, G.; Rouhani, M.H. The association between serum homocysteine and depression: A systematic review and meta-analysis of observational studies. *Eur. J. Clin. Investig.* **2021**, *51*, e13486. [CrossRef]
176. Kim, J.-M.; Kang, H.-J.; Kim, J.-W.; Choi, W.; Lee, J.-Y.; Kim, S.-W.; Shin, I.-S.; Kim, M.-G.; Chun, B.J.; Stewart, R. Multiple serum biomarkers for predicting suicidal behaviours in depressive patients receiving pharmacotherapy. *Psychol. Med.* **2022**, *17*, 1–10. [CrossRef]
177. Ho, P.I.; Ortiz, D.; Rogers, E.; Shea, T.B. Multiple aspects of homocysteine neurotoxicity: Glutamate excitotoxicity, kinase hyperactivation and DNA damage. *J. Neurosci. Res.* **2002**, *70*, 694–702. [CrossRef] [PubMed]
178. Sawada, S.; Yamamoto, C. Gamma-D-glutamylglycine and cis-2,3-piperidine dicarboxylate as antagonists of excitatory amino acids in the hippocampus. *Exp. Brain Res.* **1984**, *55*, 351–358. [CrossRef]
179. Sachdev, P.S.; Parslow, R.A.; Lux, O.; Salonikas, C.; Wen, W.; Naidoo, D.; Christensen, H.; Jorm, A.F. Relationship of homocysteine, folic acid and vitamin B_{12} with depression in a middle-aged community sample. *Psychol. Med.* **2005**, *35*, 529–538. [CrossRef] [PubMed]
180. Nanri, A.; Mizoue, T.; Matsushita, Y.; Sasaki, S.; Ohta, M.; Sato, M.; Mishima, N. Serum folate and homocysteine and depressive symptoms among Japanese men and women. *Eur. J. Clin. Nutr.* **2010**, *64*, 289–296. [CrossRef]
181. Beydoun, M.A.; Shroff, M.R.; Beydoun, H.A.; Zonderman, A.B. Serum Folate, Vitamin B-12, and Homocysteine and Their Association with Depressive Symptoms among U.S. Adults. *Psychosom. Med.* **2010**, *72*, 862–873. [CrossRef]
182. Elstgeest, L.E.M.; A Brouwer, I.; Penninx, B.W.H.; van Schoor, N.M.; Visser, M. Vitamin B12, homocysteine and depressive symptoms: A longitudinal study among older adults. *Eur. J. Clin. Nutr.* **2017**, *71*, 468–475. [CrossRef]
183. Fraguas, R.; Papakostas, G.I.; Mischoulon, D.; Bottiglieri, T.; Alpert, J.; Fava, M. Anger Attacks in Major Depressive Disorder and Serum Levels of Homocysteine. *Biol. Psychiatry* **2006**, *60*, 270–274. [CrossRef] [PubMed]
184. Hapuarachchi, J.R.; Chalmers, A.H.; Winefield, A.; Blake-Mortimer, J.S. Changes in Clinically Relevant Metabolites with Psychological Stress Parameters. *Behav. Med.* **2003**, *29*, 52–59. [CrossRef]
185. Stoney, C.M.; Engebretson, T.O. Plasma homocysteine concentrations are positively associated with hostility and anger. *Life Sci.* **2000**, *66*, 2267–2275. [CrossRef]
186. Levine, J.; Sela, B.-A.; Osher, Y.; Belmaker, R. High homocysteine serum levels in young male schizophrenia and bipolar patients and in an animal model. *Prog. Neuro Psychopharmacol. Biol. Psychiatry* **2005**, *29*, 1181–1191. [CrossRef]
187. Brundin, L.; Erhardt, S.; Bryleva, E.Y.; Achtyes, E.D.; Postolache, T.T. The role of inflammation in suicidal behaviour. *Acta Psychiatr. Scand.* **2015**, *132*, 192–203. [CrossRef] [PubMed]
188. Carpita, B.; Marazziti, D.; Palego, L.; Giannaccini, G.; Betti, L.; Dell'Osso, L. Microbiota, Immune System and Autism Spectrum Disorders: An Integrative Model towards Novel Treatment Options. *Curr. Med. Chem.* **2020**, *27*, 5119–5136. [CrossRef] [PubMed]
189. Farzi, A.; Hassan, A.; Zenz, G.; Holzer, P. Diabesity and mood disorders: Multiple links through the microbiota-gut-brain axis. *Mol. Asp. Med.* **2018**, *66*, 80–93. [CrossRef] [PubMed]
190. Sarkar, S.R.; Banerjee, S. Gut microbiota in neurodegenerative disorders. *J. Neuroimmunol.* **2019**, *328*, 98–104. [CrossRef]
191. Hsiao, E.Y.; McBride, S.W.; Hsien, S.; Sharon, G.; Hyde, E.R.; McCue, T.; Codelli, J.A.; Chow, J.; Reisman, S.E.; Petrosino, J.F.; et al. Microbiota Modulate Behavioral and Physiological Abnormalities Associated with Neurodevelopmental Disorders. *Cell* **2013**, *155*, 1451–1463. [CrossRef]
192. Simpson, C.A.; Diaz-Arteche, C.; Eliby, D.; Schwartz, O.S.; Simmons, J.G.; Cowan, C.S. The gut microbiota in anxiety and depression—A systematic review. *Clin. Psychol. Rev.* **2021**, *83*, 101943. [CrossRef]

193. Kelly, J.R.; Kennedy, P.J.; Cryan, J.F.; Dinan, T.G.; Clarke, G.; Hyland, N.P. Breaking down the barriers: The gut microbiome, intestinal permeability and stress-related psychiatric disorders. *Front. Cell Neurosci.* **2015**, *9*, 392. [CrossRef]
194. Safadi, J.M.; Quinton, A.M.G.; Lennox, B.R.; Burnet, P.W.J.; Minichino, A. Gut dysbiosis in severe mental illness and chronic fatigue: A novel trans-diagnostic construct? A systematic review and meta-analysis. *Mol. Psychiatry* **2021**, *27*, 141–153. [CrossRef]
195. McCall, W.V.; Black, C.G. The Link Between Suicide and Insomnia: Theoretical Mechanisms. *Curr. Psychiatry Rep.* **2013**, *15*, 389. [CrossRef]
196. McCall, W.V.; Benca, R.M.; Rosenquist, P.B.; Youssef, N.A.; McCloud, L.; Newman, J.C.; Case, D.; Rumble, M.E.; Szabo, S.T.; Phillips, M.; et al. Reducing Suicidal Ideation Through Insomnia Treatment (REST-IT): A Randomized Clinical Trial. *Am. J. Psychiatry* **2019**, *176*, 957–965. [CrossRef]
197. Rea, K.; Dinan, T.G.; Cryan, J.F. The microbiome: A key regulator of stress and neuroinflammation. *Neurobiol. Stress* **2016**, *4*, 23–33. [CrossRef] [PubMed]
198. Heijtz, R.D.; Wang, S.; Anuar, F.; Qian, Y.; Björkholm, B.; Samuelsson, A.; Hibberd, M.L.; Forssberg, H.; Pettersson, S. Normal gut microbiota modulates brain development and behavior. *Proc. Natl. Acad. Sci. USA* **2011**, *108*, 3047–3052. [CrossRef] [PubMed]
199. Bi, C.; Guo, S.; Hu, S.; Chen, J.; Ye, M.; Liu, Z. The microbiota–gut–brain axis and its modulation in the therapy of depression: Comparison of efficacy of conventional drugs and traditional Chinese medicine approaches. *Pharmacol. Res.* **2022**, *183*, 106372. [CrossRef] [PubMed]
200. He, X.; Hu, Y.; Liu, W.; Zhu, G.; Zhang, R.; You, J.; Shao, Y.; Li, Y.; Zhang, Z.; Cui, J.; et al. Deciphering the Effective Constituents and Mechanisms of Portulaca oleracea L. for Treating NASH via Integrating Bioinformatics Analysis and Experimental Pharmacology. *Front. Pharmacol.* **2022**, *12*, 818227. [CrossRef]
201. Tsuchiyagaito, A.; Smith, J.L.; El-Sabbagh, N.; Zotev, V.; Misaki, M.; Al Zoubi, O.; Teague, T.K.; Paulus, M.P.; Bodurka, J.; Savitz, J. Real-time fMRI neurofeedback amygdala training may influence kynurenine pathway metabolism in major depressive disorder. *NeuroImage Clin.* **2021**, *29*, 102559. [CrossRef] [PubMed]
202. Hunt, C.; e Cordeiro, T.M.; Suchting, R.; de Dios, C.; Leal, V.A.C.; Soares, J.C.; Dantzer, R.; Teixeira, A.L.; Selvaraj, S. Effect of immune activation on the kynurenine pathway and depression symptoms—A systematic review and meta-analysis. *Neurosci. Biobehav. Rev.* **2020**, *118*, 514–523. [CrossRef]

biomedicines

MDPI

Article

Breaking Barriers: Artificial Intelligence Interpreting the Interplay between Mental Illness and Pain as Defined by the International Association for the Study of Pain

Franciele Parolini [1,2,3,*], Márcio Goethel [2,3], Klaus Becker [2,3,*], Cristofthe Fernandes [4], Ricardo J. Fernandes [2,3], Ulysses F. Ervilha [2,5], Rubim Santos [1] and João Paulo Vilas-Boas [2,3]

1 Center for Rehabilitation Research (CIR), School of Health, Polytechnic Institute of Porto, Rua Dr. António Bernardino de Almeida, 400, 4200-072 Porto, Portugal; rss@ess.ipp.pt
2 Porto Biomechanics Laboratory, University of Porto, 4200-450 Porto, Portugal; gbiomech@fade.up.pt (M.G.); ricfer@fade.up.pt (R.J.F.); ulyervil@usp.br (U.F.E.); jpvb@fade.up.pt (J.P.V.-B.)
3 Center of Research, Education, Innovation and Intervention in Sport, Faculty of Sport, University of Porto, 4200-450 Porto, Portugal
4 Faculty of Psychology and Educational Sciences of the University of Porto, 4099-002 Porto, Portugal; up201802555@up.pt
5 Laboratory of Physical Activity Sciences, School of Arts, Sciences and Humanities, University of São Paulo, São Paulo 03828-000, Brazil
* Correspondence: fcsp@ess.ipp.pt (F.P.); up201803272@fade.up.pt (K.B.)

Abstract: Low back pain is one of the main causes of motor disabilities and psychological stress, with the painful process encompassing sensory and affective components. Noxious stimuli originate on the periphery; however, the stimuli are recombined in the brain and therefore processed differently due to the emotional environment. To better understand this process, our objective was to develop a mathematical representation of the International Association for the Study of Pain (IASP) model of pain, covering the multidimensional representation of this phenomenon. Data from the Oswestry disability index; the short form of the depression, anxiety, and stress scale; and pain catastrophizing daily questionnaires were collected through online completion, available from 8 June 2022, to 8 April 2023 (1021 cases). Using the information collected, an artificial neural network structure was trained (based on anomaly detection methods) to identify the patterns that emerge from the relationship between the variables. The developed model proved to be robust and able to show the patterns and the relationship between the variables, and it allowed for differentiating the groups with altered patterns in the context of low back pain. The distinct groups all behave according to the main finding that psychological and pain events are directly associated. We conclude that our proposal is effective as it is able to test and confirm the definition of the IASP for the study of pain. Here we show that the fiscal and mental dimensions of pain are directly associated, meaning that mental illness can be an enhancer of pain episodes and functionality.

Keywords: lower back pain; mental illness; pain catastrophizing; artificial intelligence (AI); motor disability; pain perception; affective components; central nervous system; emotions; peripheral nervous system; sensation

Citation: Parolini, F.; Goethel, M.; Becker, K.; Fernandes, C.; Fernandes, R.J.; Ervilha, U.F.; Santos, R.; Vilas-Boas, J.P. Breaking Barriers: Artificial Intelligence Interpreting the Interplay between Mental Illness and Pain as Defined by the International Association for the Study of Pain. *Biomedicines* **2023**, *11*, 2042. https://doi.org/10.3390/biomedicines11072042

Academic Editor: Masaru Tanaka

Received: 5 June 2023
Revised: 12 July 2023
Accepted: 15 July 2023
Published: 20 July 2023

1. Introduction

In several countries, low back pain is referred to as the major source of musculoskeletal complaints, with a high impact on health and the economy due to the limitations and disability it imposes on individuals and their need for healthcare and absenteeism from work [1]. The process that leads to the perception of pain is a complex succession of peripheral and central neural system activities, including modulation at different levels. Despite the essentially peripheral origin of low back pain, its cause is not identified in 85% of the cases [1,2]. The pain process encompasses sensory, cognitive, and affective components [3], with this last component including feelings of annoyance, sadness, anxiety, and depression in response to a harmful stimulus [4]. Brain activity in patients with low back pain for two months showed activation in the insular cortex, thalamus, anterior cingulate cortex, and prefrontal cortex.

For over a decade, it has been noted that those who suffer from long-term lower back pain exhibit activity in certain areas of the brain, specifically the perigenual anterior cingulate and medial prefrontal cortexes, as well as the amygdala [3,4]. This implies that as acute pain becomes chronic, there may be a shift towards emotional pathways instead of just sensory ones. Furthermore, the experience of pain, as well as anxiety and depression, is often intertwined with the idea of suffering [5]. The association of pain with the subjects' psychological state has been investigated [6–8], and there seems to be a consensus that psychological and social factors are fused in biopsychosocial processes that characterize chronic pain [9].

In recent years, there has been a growing recognition of the close relationship between pain and mental illness [10,11]. It has become increasingly clear that mental health conditions, such as depression, anxiety, and stress, can significantly influence the experience and perception of pain [10]. Conversely, chronic pain can also have a profound impact on mental well-being, leading to increased levels of psychological distress and impairment [6]. Recent studies elucidate those different types of chronic pain conditions such as fibromyalgia and low back pain and chronic pain conditions from underlying medical conditions such as post-trauma, neuropathic, and musculoskeletal pain have distinct pathogenic pathways [12,13]. So perhaps chronic back pain is best understood in the framework of pain perception, including cognitive, emotional, and social components; therefore, the association of mental health and pain perception appears to be a logical association [13,14]. Understanding and addressing this intricate relationship between mental illness and pain is crucial for providing comprehensive and effective care to individuals experiencing pain.

As of 2020, the meaning of pain has been redefined by the International Association for the Study of Pain (IASP). According to the new definition, pain is an unpleasant sensory and emotional experience associated with or similar to that associated with actual or potential tissue damage [10]. Those who have extensive knowledge in pain-related fields, including clinical and fundamental science, decided on the model by examining existing definitions and annotations and deciding whether they still apply or need modification. Although it seems to be very well accepted in the community, a global multivariate model can provide more robust support for what is currently the most accepted definition. If such a model considers, with the respective weights, the interaction of a set of variables involved, this multivariate phenomenon will certainly be better understood, and consequently, more accurate and adequate diagnostic and therapeutic tools will be developed.

To gain a deeper understanding of the complex interplay between mental illness and low back pain, researchers have turned to mathematical modeling and artificial intelligence as powerful tools [15–18]. Machine learning and deep learning algorithms offer the ability to analyze small and large numbers of data and discover hidden patterns and associations that may not be evident through traditional statistical approaches [19,20]. One of the main advantages of using artificial intelligence to study the relationship between mental ill-health and low back pain is its ability to capture and analyze multiple dimensions of pain [21]. Traditional research methods often focus only on the physiological aspects of pain, such as measuring pain intensity or identifying biomarkers. While these aspects are undeniably important, they provide only a partial understanding of the pain experience [18,22].

Mathematical modeling can provide a more comprehensive representation of pain by integrating functional, psychological, and emotional factors into the analysis, and artificial intelligence algorithms allow researchers to analyze complex and heterogeneous data and can help identify patterns and relationships between variables, determining the relationship between low back pain and its interaction with mental illness [23–26]. Pain is a subjective experience, the evaluation of which depends largely on self-reported measures. These measures often include questionnaires, surveys, and diaries to capture people's perceptions, emotions, and behaviors related to pain. AI algorithms can process and analyze these data sources, generating meaningful insights and identifying patterns that help understand how pain and mental illness relate to and affect an individual's quality of life [27].

In order to improve understanding of the link between low back pain and psychological conditions, and to aid in better assessment and decision-making by healthcare professionals, artificial intelligence has shown great efficacy [20]. Artificial intelligence algorithms can analyze behavioral, language, and emotional functional patterns captured in digital data, such as text messages, social media, or electronic health records, and identify indicators of emotional pain and distress [28]. This information can be used to develop low back pain tracking tools and continuous monitoring for more timely and individualized interventions.

Therefore, the aim of the current study was two-fold: (i) to develop a mathematical representation based on a multivariate model to elucidate the relation between low back pain and biopsychosocial aspects and (ii) to identify subpopulations that present deviations from the pattern that emerged. We hypothesize that it is possible to test the IASP concept of pain through a mathematical representation (evidencing its coherence) and that there is a strong relationship between mental health and the way the subject copes with the experience of pain and its functional consequences.

2. Materials and Methods

Details of the study design are presented, including methodological approaches that we use to analyze the complex interactions of low back pain phenomena and try to understand the underlying patterns and relationships, with the auxilium of mathematical modeling and the algorithms of artificial intelligence [28–30]. The methodological design of our study allows us to provide a comprehensive view of the research process and aims to ensure the validity and reliability of the results obtained. With targeted methods, we intend to expand our knowledge in this field, advancing our understanding of the interaction between low back pain and mental illness [20].

This was a cross-sectional observational study approved by the ethics committee of the School of Health of the Polytechnique of Porto (CE0092B), and the study objectives and procedures were developed and conducted in accordance with the guidelines of the Declaration of Helsinki. Volunteers consented to participate in the study through their informed consent form. The sample consisted of 1.021 young adults (73% females), aged between 18 and 35 years (24.68 ± 1.5 years, height 167.9 ± 0.1 m, and weight 65.8 ± 3.5 kg). The exclusion criteria were <18 years old, >35 years old, or not completing the survey. The research involved the Center for Rehabilitation and Research (CIR) of the Higher School of Health of the Polytechnic of Porto and the Laboratory of Biomechanics of the University of Porto (LABIOMEP).

2.1. Data Collection

The survey focusing on the relation of low back pain with psychological variables in young adults was created with Lime Survey version 3.28.56 + 230404, an online survey application software written in pre-processed Python text. Data were collected through online auto-completion on the Lime platform in the period from 8 June 2022, to 8 April 2023. The link to access the survey was disseminated through the institutional emails of the Polytechnic of Porto and the University of Porto to the entire academic population and also in social networks. Participants provided information related to gender, mass, age, height, sociodemographic information, the existence of medical diagnosis of psychiatric disorder, and the frequency of episodes of low back pain in six weeks.

2.2. Instruments

The Oswestry disability index I [31] was used in the survey as a specific instrument that measures the impact of back pain on daily living activities (particularly regarding pain intensity, lifting weights, social interaction, sitting, standing, traveling, sex life, sleeping, walking, and caring). It is composed of 10 questions with 6 alternatives (each ranging in scores from 0 to 5). The first question assesses the intensity of pain, while the others score the pain impact on daily activities (such as personal care, lifting weights, walking, sitting, standing, sleeping, social activities, and mobility). The total score is multiplied by the number of questions answered and then multiplied by 5, with the result expressed as a percentage ([score ÷ (number of questions answered × 5)] × 100). The scores are classified as minimal, moderate, and severe disabilities (0–20, 21–40, and 41–60%, respectively); disabled (61–80%); and bedridden (81–100%).

The short form of the depression, anxiety, and stress scale [32] was also used (including 21 items) and was designed to assess depression, anxiety, and stress domains (each one being represented by 7 items). Participants rated each item on a 0 ("did not apply to me at all") to 3 ("applied to me very much or most of the time") scale. Each domain is represented by a subscale score (the sum of the item responses for that subscale multiplied by two to be comparable with the original 42-item depression, anxiety, and stress scale). This instrument was previously validated and considered reliable [32], with a high score representing worse depression, anxiety, or stress. Cut points for normal, mild, moderate, severe, and extremely severe score classification, based on population norms, are provided in its manual. Classification symptoms are rated as 0–10 (normal), 11–18 (mild), 19–26 (moderate), 27–34 (severe), and 35–42 (extremely severe) for stress; 0–6 (normal), 7–9 (mild), 10- 14 (moderate), 15–19 (severe), and 20–42 (extremely severe) for anxiety; and 0–9 (normal), 10–12 (mild), 13–20 (moderate), 21–17 (severe), and 28–42 (extremely) severe for depression.

Pain catastrophizing daily [33] is a questionnaire with 14 points that aims to assess disasters in the last 24 h, whose items were also rated by our participants on a scale of 0 ("never") to 4 ("always"). The total score was calculated as the sum of the item responses (range 0–56), with higher scores representing greater catastrophizing of pain. The use of the daily catastrophe questionnaire may lead to greater analytical accuracy in research, health tools and platforms, and studies of psychosocial diaries that seek to understand the adaptive mechanisms of pain.

2.3. Anomaly Detection Structure

Anomaly detection refers to the problem of finding data patterns that do not conform to the expected behavior [23]. In the current research, a dataset of 1.021 volunteers was used to model the relationship patterns between the low back pain-related variables. An artificial neural network structure with two hidden layers was trained, with each of the hidden layers including tangent hyperbolic transfer and a logistic sigmoid with 20 neurons, and fully connected (Figure 1). The input layer was composed of socioanthropometric dimension-related variables (age, sex, body mass, height, and body mass index) and data from the Oswestry disability index I [31]; depression, anxiety, and stress scale [32]; and pain catastrophizing daily [33] questionnaire scores. The output layer contained the same information but with a randomized subject order. The output space consisted of a "1" or "2" binary classification, indicating "no change" and "change" in the general functional profile (respectively). The learning algorithm used was Bayesian regularization. The dataset was randomly split into 80% of samples for training and 20% for testing.

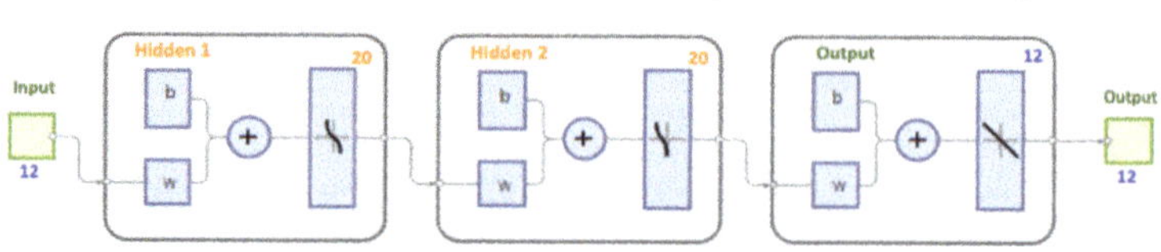

Figure 1. Used artificial neural network structure.

After 727 epochs, a mean square error performance value of 0.001 was obtained. The accuracy achieved after training equals R = 0.9903, 0.9625, and 0.9846 in training, in the test, and for all samples (respectively). Then, data of all subjects were simulated using the model obtained, and the estimates were compared with the real data through a single linear regression, where the target was the dependent variable and the output was the independent variable.

Subsequently, three subgroups were created, determined by the position of the R in relation to the 25th and 75th percentiles (the first formed by subjects with values < 25th percentile; the second, from 25th to 75th percentiles; and the third, >75th percentile). Since data did not show a normal distribution, the between-group comparison was performing using the Kruskal–Wallis test (with the pairwise comparison conducted using the Mann–Whitney U test adjusted with the Bonferroni correction).

3. Results

The model seems to capture some interesting differences between the groups (Figure 2), showing a relationship between the variables of number of lower back pain events in a 6-week period (p = 0.001), medical diagnosis of lower back pathology (p = 0.002), ODI (p = 0.001), age (p = 0.030), and anthropometric data and correlated with the psychological variables, stress (p = 0.001), anxiety (p = 0.001), depression (p = 0.001) and catastrophizing in the last 24 h in episodes of low back pain (p = 0.001). The results are expressed as the multiplication factor (MF) of each condition that is multiplied by the constant value (as mean) of each variable.

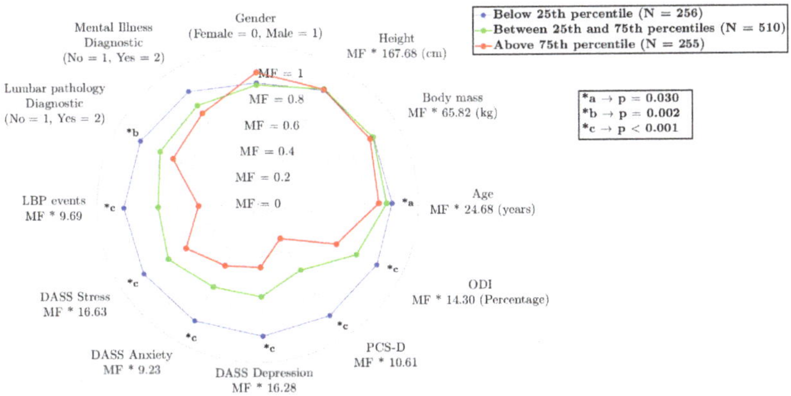

Figure 2. Comparison of different variables among the three subgroups, with the variable names being followed by the value to be multiplied by the multiplication factor. Legend: stress (DASS Stress), anxiety (DASS Anxiety), depression (DASS Depression) scale short; pain catastrophizing daily (PCS-D); and Oswestry disability index (ODI).

Statistical Analysis

Table 1 shows the difference between groups and effect size regarding each variable, followed by median and interquartile range values. The GPower 3.1.7 software (University of Kiel, Kiel, Germany) was used to calculate the effect size (ES) and determine the power of analysis using the Mann–Whitney U, followed by Cohen's d criterion (small: >0.2; moderate: >0.50; large: >0.80) [34]. No differences were found between groups regarding body mass, height, gender, or mental illness diagnosis. Lumbar pathology was higher in group 1 than in group 3 (p < 0.001) and in group 2 than in group 3 (p = 0.039), and low back pain events presented a similar behavior, i.e., group 1 > 2 (p = 0.001) and group 1 > 3 (p = 0.003). The psychological variables differed between groups, with stress being higher in group 1 than in group 2 and in group 1 than in group 3 (both for a p < 0.001); anxiety being higher in group 1 than in group 3 (p < 0.001), in group 1 than in group 2 (p < 0.001), and in group 2 than in group 3 (p = 0.031); depression displaying higher values in group 2 than in group 3 (p = 0.019), in group 1 than in group 3 (p < 0.001), and in group 1 than in group 2 ($p < 0.001$); and pain catastrophizing daily showing the results of group 1 > 2 and group 1 > 3 (both for a $p < 0.001$) due to its epistemological proximity. Given that psychological variables are factors that can exacerbate pain, the higher Oswestry disability index I values in group 1 than in 3 ($p = 0.050$), showing a mild difficulty in lumbar functionality, are not surprising.

Table 1. Comparison of different variables among the three subgroups. Legend: Cohen's d test value; stress (DASS Stress), anxiety (DASS Anxiety), depression (DASS Depression) scale short; pain catastrophizing daily (PCS-D); Oswestry disability index (ODI); group 1 (G1); group 2 (G2); group 3 (G3), * has binary values described in the results session, *p*-value, and Cohen's d test value.

Variables	Comparison		*p*	*d*
Age	G1–22 (8)	G2	0.322	1.494
		G3	0.000	0.212
	G2–22 (8)	G1	0.322	1.494
		G3	0.008	0.179
	G3–21 (5)	G1	0.000	0.212
		G2	0.008	0.179
Body mass	G1–62 (21)	G2	0.012	1.503
		G3	0.001	0.301
	G2–63 (16)	G1	0.012	1.503
		G3	0.437	0.191
	G3–60 (12)	G1	0.001	0.301
		G2	0.437	0.191
Height	G1–167 (11)	G2	0.801	1.403
		G3	1.000	0.177
	G2–165 (12)	G1	0.801	1.403
		G3	1.000	0.054
	G3–165 (12)	G1	1.000	0.177
		G2	1.000	0.054
Gender	G1 *	G2	1.000	0.041
		G3	1.000	0.138
	G2 *	G1	1.000	0.041
		G3	1.000	0.034
	G3 *	G1	1.000	0.138
		G2	1.000	0.034
Mental illness diagnosis	G1 *	G2	0.054	0.111
		G3	0.060	0.158
	G2 *	G1	0.054	0.111
		G3	1.000	0.033
	G3 *	G1	0.060	0.158
		G2	1.000	0.033
Lumbar pathology diagnosis	G1 *	G2	0.041	1.421
		G3	0.000	0.200
	G2 *	G1	0.041	1.421
		G3	0.142	0.136
	G3 *	G1	0.000	0.200
		G2	0.142	0.136

Table 1. *Cont.*

Variables	Comparison		*p*	*d*
LBP events	G1–4 (9)	G2	0.000	1.469
		G3	0.000	0.260
	G2–4 (4)	G1	0.000	1.469
		G3	0.028	0.280
	G3–4 (4)	G1	0.000	0.260
		G2	0.028	0.280
DASS—stress	G1–14 (16)	G2	0.000	1.686
		G3	0.000	0.517
	G2–10 (10)	G1	0.000	1.686
		G3	0.015	0.424
	G3–10 (8)	G1	0.000	0.517
		G2	0.015	0.424
DASS—anxiety	G1–8 (12)	G2	0.000	1.677
		G3	0.000	0.506
	G2–4 (6)	G1	0.000	1.677
		G3	0.021	0.413
	G3–2 (8)	G1	0.000	0.506
		G2	0.021	0.413
DASS—depression	G1–12 (20)	G2	0.000	1.757
		G3	0.000	0.596
	G2–6 (10)	G1	0.000	1.757
		G3	0.001	0.510
	G3–6 (8)	G1	0.000	0.596
		G2	0.001	0.510
PCS-D	G1–2 (15)	G2	0.000	1.523
		G3	0.000	0.326
	G2–1 (8)	G1	0.000	1.523
		G3	0.000	0.218
	G3–1 (5)	G1	0.000	0.326
		G2	0.000	0.218
ODI	G1–10 (18)	G2	0.015	1.416
		G3	0.000	0.193
	G2–10 (16)	G1	0.015	1.416
		G3	0.036	0.072
	G3–8 (10)	G1	0.000	0.193
		G2	0.036	0.072

These findings provide valuable information about the factors that contribute to low back pain in young adults and emphasize the importance of considering physiological and psychological aspects in understanding and managing this condition.

4. Discussion

Pain and mental illness together should be part of an integrated treatment approach. It should involve a multi-professional team, with a combination of physical interventions, such as exercise, physical therapy, medication to manage pain, and psychological interventions, to address the mental status and improve the functional status [22]. Therefore, research in this area, with the aid of multivariate models, is of great importance, as it allows the identification of risk and protection factors associated with pain and mental illness. These include genetic, environmental, psychosocial, and behavioral factors that may influence the development of these conditions. Understanding these factors enables the implementation of more effective preventive strategies and the development of targeted interventions, playing an important role in reducing the stigma associated with these conditions [35–37].

The mathematical modeling we used in our study can lead to advances in the delivery of care from all areas of healthcare. Using effective screening artificial intelligence algorithms, unusual patterns in the frequency, intensity, or duration of low back pain over time can be identified, which is useful for identifying episodes of severe acute pain or significant changes in pain patterns. This can be applied to identify specific activities, postures, or movements that lead to a significant increase in pain [29,38]. This information can help identify behaviors or situations that should be avoided or changed to improve pain management, and thus identify triggers associated with low back pain episodes and their physical and mental functions.

We found evidence of a relationship between the repetition of traumatic events and physical and mental functioning, particularly stress, anxiety, and ultimately depression. According to the literature and the data obtained in this study, the repeated experience of pain can have a significant impact on a person's daily functioning and can also increase the risk of developing or worsening depressive symptoms [30,39]. Recurrent or persistent pain can limit a person's ability to carry out daily activities, such as work, exercise, socializing, and self-care. In the case of persistent pain, it can affect sleep, energy, mood, and quality of life, leading to symptoms of depression [20,31]. Mental and emotional health plays a significant role in the experience and perception of pain, and addressing these aspects can bring substantial benefits to patients [26,38]; thus, this study has significant potential by exploring the direct relationship between musculoskeletal pain and mental ill-health.

Considering that pain is an unpleasant sensory and emotional experience associated with (or resembling) actual/potential tissue damage, there should be quantifiable emotional variables that allow transcribing it into a mathematical model. Moreover, due to the sensory–motor nature of this phenomenon, movement measures or scores should be included in the model. Data from human movement biomechanical variables are commonly heterogeneous and form a large volume of information, making it difficult to treat them using inferential statistics. However, advanced analytical techniques used to evaluate informative data features and model underlying relationships that cannot be treated with traditional statistics can increase the research quality [29,40]. For a more global understanding of low back pain multivariate phenomena, widely used artificial intelligence tools [41] should be employed. Aiming to mathematically represent the IASP [10] definition of pain using an artificial neural network approach, based on the current study results, we advocate that it is possible to mathematically model and represent it.

The mathematical model that we have presented processed information from 1.021 volunteers allowing us to assess the linear and nonlinear relationships between variables that construct the phenomenon. It showed a very robust final performance and identified the subpopulations that presented deviations from the pattern in the context of low back pain and biopsychosocial aspects [29,41]. The relationships between the variables that emerged from this model can be seen in the group profiles. An interesting fact in the group < 25th percentile is that the lumbar pathology diagnosis is closely linked to the depression, anxiety, and stress scale-related variables [32]; pain catastrophizing daily [33]; and low back pain events, promoting a slight functional incapacity of the individual. It seems that this functional incapacity makes it difficult for individuals to carry out their usual activities [24], eventually leading to social isolation and having a major negative effect on individual well-being.

The current study results show an interdependence of variables, meaning that, for example, our oldest group also has a higher prevalence of diagnosis of lumbar problems and low back pain flares, as well as scoring worse on depression, anxiety, and stress scale and pain catastrophizing daily and Oswestry disability index I surveys. However, our data cannot give a good explanation about the underlying mechanism, i.e., if the low back pain flares lead to worse psychological variables or if the psychological impairment leads to perception and aggravation of the pain (leading to seeking medical diagnosis).

The relationship between low back pain, psychological distress, and mild functional disability observed by us is in line with previous data that identified high levels of pain intensity associated with poor psychological and physiological capacity and high levels of anxiety and depression [42]. Based on the current study results and on the literature, it is possible that the mental disorder in low back pain may be a predictor of reduced functionality [32,43] and to hypothesize that individuals with a medical diagnosis of lower back pathology have a higher number of lower back pain episodes over a six-week period and higher levels of pain catastrophizing.

Our data are in line with a study with 84 patients with rotator cuff tears that were evaluated for the presence of differences in pain, function, and/or psychological distress associated with pain and analyzed for the association between psychological distress with shoulder pain and function during adjustment for cuff tear severity [43]. Results demonstrated that baseline psychological distress is related to patients' pain and shoulder function more than the diagnosis of rotator cuff tears, suggesting that the size and severity of the lesion are not fully related to symptoms (e.g., pain and functional limitation) but rather to psychological distress [43,44]. Anxiety and avoidance can cause an inflated sense of pain [45,46], while fear of pain influences short-term pleasure seeking [25] due to pain's catastrophic aftermath [47,48]. These behavioral patterns are not connected to the disease at hand.

Based on these statements, a study in mice examines whether long-term associations with remembering fear stored in neural engrams in the prefrontal cortex can determine how painful episodes evolve into later-life painful experiences [49]. It was evidenced that long-term fear memory is associated with pathological changes in nociceptive sensitivity following tissue injury, a key feature of pathological pain disorders and known to be regulated by the cortex [50]. Pain and fear are independent behavioral states that are interrelated [46,51], with fear acutely potentiating the perception of pain [49] that is fundamental to survival. It was concluded that a painful experience could encode a memory of fear (that will be stored in a discrete and specific cohort of prefrontal cortical neurons). This will be subjected to reactivation after exposure to a new painful stimulus in future life events, and as a result, it will produce an intensification of pain perception [50,52].

According to the approach mentioned above and the data from our study, it can be underlined that the catastrophizing of pain leads to excessive fear of pain, and the associative long-term memory of fear induced by previous exposure to pain may also be a critical predisposing factor for pain chronicity [51,52]. Thus, the fear of pain can provoke avoidance of motion behaviors and exacerbate pain in the long term, implying an increase in the functional disability of the individual. It is important to address that the relationship between pain, functionality, and depression is bidirectional.

This study has some limitations. Data from self-completion questionnaires rely on the accuracy and honesty of participants' responses. However, these responses may be subject to self-report bias, where participants may provide inaccurate or biased responses. This may occur due to memory problems, lack of understanding of the questions, or desire to please the researcher or hide certain information, besides not having a face-to-face and objective verification of the data provided by the participants. We took these limitations into consideration when constructing and applying the survey and interpreting the study results. We understood the possible sources of bias, which helped us to assess the validity and reliability of the results obtained. In addition, we combined different methods of complementary analysis which allowed us to strengthen the conclusion of our study.

5. Conclusions

In view of the above, we conclude that it is possible to validate and confirm the IASP definition of pain through mathematical modeling. The identified subpopulations showed a direct relationship between pain and mental illness, with these two inducing greater disabilities. Even if these results may help to improve the understanding of mental illness as a possible enhancer of pain episodes and functionality, future studies evaluating other variables, like the level of physical activity and the sedentary behavior of the subjects, are required to better understand the mentioned association.

Author Contributions: Conceptualization, F.P., K.B., U.F.E., M.G., R.S. and J.P.V.-B.; methodology, F.P., M.G., C.F., R.J.F. and K.B.; formal analysis, F.P., K.B., U.F.E., C.F., M.G., R.J.F., R.S. and J.P.V.-B.; resources, F.P., K.B., M.G., R.J.F. and R.S.; writing—original draft preparation, F.P., M.G., K.B., U.F.E., C.F., R.J.F., J.P.V.-B. and R.S.; writing—review and editing, F.P., U.F.E., R.J.F., C.F., K.B., J.P.V.-B., M.G. and R.S.; supervision, F.P., K.B., R.S. and M.G.; project administration, F.P., K.B., M.G., U.F.E., C.F., R.J.F. and R.S. All authors have read and agreed to the published version of the manuscript.

Funding: This research was funded by the Rehabilitation Research Center-Foundation for Science and Technology (FCT) through R&D Units funding UI/BD/151415/2021.

Institutional Review Board Statement: The study was conducted in accordance with the guidelines of the Declaration of Helsinki and approved by the Ethics Committee School of Health of the Polytechnic of Porto (CE0092B).

Informed Consent Statement: Informed consent was obtained from all the participants involved in the study.

Data Availability Statement: The data presented in this study are available on request from the corresponding author. The data are not publicly available due to privacy.

Conflicts of Interest: The authors declare no conflict of interest.

References

1. Vos, T.; Flaxman, A.D.; Naghavi, M.; Lozano, R.; Michaud, C.; Ezzati, M.; Shibuya, K.; Salomon, J.A.; Abdalla, S.; Aboyans, V.; et al. Years lived with disability (YLDs) for 1160 sequelae of 289 diseases and injuries 1990–2010: A systematic analysis for the Global Burden of Disease Study. *Lancet* **2012**, *380*, 2163–2196. [CrossRef] [PubMed]
2. Maher, C.; Underwood, M.; Buchbinder, R. Non-specific low back pain. *Lancet* **2017**, *389*, 736–747. [CrossRef] [PubMed]
3. Lumley, M.A.; Schubiner, H.; Lockhart, N.A.; Kidwell, K.M.; Harte, S.E.; Clauw, D.J.; Williams, D.A. Emotional awareness and expression therapy, cognitive behavioral therapy, and education for fibromyalgia: A cluster-randomized controlled trial. *Pain* **2017**, *158*, 2354–2363. [CrossRef]
4. Paulus, M.P.; Stein, M.B. Interoception in anxiety and depression. *Brain Struct. Funct.* **2010**, *214*, 451–463. [CrossRef]
5. Mescouto, K.; Olson, R.E.; Hodges, P.W.; Setchell, V. A critical review of the biopsychosocial model of low back pain care: Time for a new approach? *Disabil. Rehabil.* **2022**, *44*, 3270–3284. [CrossRef] [PubMed]
6. Ashar, Y.K.; Gordon, A.; Schubiner, H.; Uipi, C.; Knight, K.; Anderson, Z.; Carlisle, J.; Polisky, L.; Geuter, S.; Flood, T.F.; et al. Effect of Pain Reprocessing Therapy vs. Placebo and Usual Care for Patients with Chronic Back Pain: A Randomized Clinical Trial. *JAMA Psychiatry* **2022**, *79*, 13–23. [CrossRef] [PubMed]
7. Bushnell, M.; Čeko, M.; Low, L. Cognitive and emotional control of pain and its disruption in chronic pain. *Nat. Rev. Neurosci.* **2013**, *14*, 502–511. [CrossRef]
8. Edwards, R.R.; Dworkin, R.H.; Sullivan, M.D.; Turk, D.C.; Wasan, A.D. The Role of Psychosocial Processes in the Development and Maintenance of Chronic Pain. *J. Pain* **2016**, *17*, T70–T92. [CrossRef]
9. Gutiérrez, L.; Écija, C.; Catalá, P.; Peñacoba, C. Sedentary Behavior and Pain after Physical Activity in Women with Fibromyalgia—The Influence of Pain-Avoidance Goals and Catastrophizing. *Biomedicines* **2023**, *11*, 154.
10. Raja, S.N.; Carr, D.B.; Cohen, M.; Finnerup, N.B.; Flor, H.; Gibson, S.; Keefe, V.; Mogil, J.S.; Ringkamp, M.; Sluka, V.; et al. The revised International Association for the Study of Pain definition of pain: Concepts, challenges, and compromises. *Pain* **2020**, *161*, 1976–1982. [CrossRef]
11. Lovelace, M.D.; Varney, B.; Sundaram, G.; Franco, N.F.; Ng, M.L.; Pai, S.; Lim, C.K.; Guillemin, G.J.; Brew, B.J. Current Evidence for a Role of the Kynurenine Pathway of Tryptophan Metabolism in Multiple Sclerosis. *Front. Immunol.* **2016**, *7*, 246. [CrossRef] [PubMed]
12. Ong, W.Y.; Stohler, C.S.; Herr, D.R. Role of the Prefrontal Cortex in Pain Processing. *Mol. Neurobiol.* **2019**, *56*, 1137–1166. [CrossRef] [PubMed]
13. Tanaka, M.; Török, N.; Tóth, F.; Szabó, Á.; Vécsei, L. Co-Players in Chronic Pain: Neuroinflammation and the Tryptophan-Kynurenine Metabolic Pathway. *Biomedicines* **2021**, *9*, 897. [CrossRef]
14. Okafor, C.; Levin, M.J.; Boadi, P.; Cook, C.; George, S.; Klifto, C.; Anakwenze, O. Pain associated psychological distress is more strongly associated with shoulder pain and function than tear severity in patients undergoing rotator cuff repair. *JSES Int.* **2023**, *7*, 544–549. [CrossRef] [PubMed]

15. Gatchel, R.J.; Peng, Y.B.; Peters, M.L.; Fuchs, P.N.; Turk, D.C. The biopsychosocial approach to chronic pain: Scientific advances and future directions. *Psychol. Bull.* **2007**, *133*, 581–624. [CrossRef]
16. Dworkin, R.H.; Turk, D.C.; McDermott, M.P.; Peirce-Sandner, S.; Burke, L.B.; Cowan, P.; Farrar, J.T.; Hertz, S.; Raja, S.N.; Rappaport, B.A.; et al. Interpreting the clinical importance of group differences in chronic pain clinical trials: IMMPACT recommendations. *Pain* **2009**, *146*, 238–244. [CrossRef]
17. Wager, T.D.; Atlas, L.Y.; Lindquist, M.A.; Roy, M.; Woo, C.W.; Kross, E. An fMRI-based neurologic signature of physical pain. *N. Engl. J. Med.* **2013**, *15*, 1388–1397. [CrossRef]
18. Rajkomar, A.; Dean, J.; Kohane, I. Machine learning in medicine. *N. Engl. J. Med.* **2019**, *14*, 1347–1358. [CrossRef]
19. Smith, A.B.; Jones, C.D.; Johnson, L.M. Investigating the relationship between mental illness and pain using artificial intelligence: A systematic review. *J. Pain Res.* **2021**, *14*, 2385–2397.
20. Baron, R.; Binder, A.; Wasner, G. Neuropathic pain: Diagnosis, pathophysiological mechanisms, and treatment. *Lancet Neurol.* **2010**, *9*, 807–819. [CrossRef]
21. Bair, M.J.; Robinson, R.L.; Katon, W.; Kroenke, K. Depression and pain comorbidity: A literature review. *Arch. Intern. Med.* **2003**, *163*, 2433–2445. [CrossRef]
22. Sun, L.; Wang, Z.; Zhang, Y.; Wang, G. A Feature-Trajectory-Smoothed High-Speed Model for Video Anomaly Detection. *Sensors* **2023**, *23*, 1612. [CrossRef] [PubMed]
23. Silva, M.C.; Fassa, A.G.; Valle, N.C.J. Chronic low back pain in a Southern Brazilian adult population: Prevalence and associated factors. *Cad. Saude Publica* **2004**, *112*, 214–220.
24. Malhotra, P.; Ramakrishnan, A.; Anand, G.; Vig, L.; Agarwal, P.; Shroff, G. LSTM-based encoder-decoder for multi-sensor anomaly detection. *arXiv* **2016**, arXiv:1607.00148.
25. Samariya, D.; Ma, D.; Aryal, S.; Zhao, X. Detection and explanation of anomalies in healthcare data. *Health Inf. Sci. Syst.* **2023**, *11*, 20. [CrossRef]
26. Nagireddi, J.N.; Vyas, A.K.; Sanapati, M.R.; Soin, A.; Manchikanti, L. The Analysis of Pain Research through the Lens of Artificial Intelligence and Machine Learning. *Pain Physician* **2022**, *25*, 211–243.
27. Taherdoost, H.; Madanchian, M. Artificial Intelligence and Sentiment Analysis: A Review in Competitive Research. *Computers* **2023**, *12*, 37. [CrossRef]
28. Goethel, M.F.; Gonçalves, M.; Brietzke, C.; Cardozo, A.C.; Vilas-Boas, J.P.; Ervilha, U.F. A global view on how local muscular fatigue affects human performance. *Proc. Natl. Acad. Sci. USA* **2020**, *117*, 19866–19872. [CrossRef]
29. Hooten, W.M. Chronic Pain and Mental Health Disorders: Shared Neural Mechanisms, Epidemiology, and Treatment. *Mayo Clin. Proc.* **2016**, *91*, 955–970. [CrossRef]
30. Li, Y.; Cheng, H. Application of Anomaly Detection in Medical Data: A Review. *Sensors* **2021**, *9*, 7364–7380.
31. Davidson, M.; Keating, J.L. A comparison of five low back disability questionnaires: Reliability and responsiveness. *Phys. Ther.* **2002**, *82*, 8–24. [CrossRef] [PubMed]
32. Marijanović, I.; Kraljević, M.; Buhovac, T.; Cerić, T.; Mekić Abazović, A.; Alidžanović, J.; Gojković, Z.; Sokolović, E. Use of the Depression, Anxiety and Stress Scale (DASS-21) Questionnaire to Assess Levels of Depression, Anxiety, and Stress in Healthcare and Administrative Staff in 5 Oncology Institutions in Bosnia and Herzegovina during the 2020 COVID-19 Pandemic. *Med. Sci. Monit.* **2021**, *27*, 81–93. [CrossRef]
33. Darnall, B.D.; Sturgeon, J.A.; Cook, K.F.; Taub, C.J.; Roy, A.; Burns, J.W.; Sullivan, M.; Macke, S.C. Development and Validation of a Daily Pain Catastrophizing Scale. *J. Pain* **2017**, *18*, 1139–1149. [CrossRef] [PubMed]
34. Cohen, J. *Statistical Power Analysis for the Behavioral Sciences*; Lawrence Erlbaum Associates: Mahwah, NJ, USA, 1988; Volume 2.
35. Doan, L.; Manders, T.; Wang, J. Neuroplasticity underlying the comorbidity of pain and depression. *Neural Plast.* **2015**, *215*, 504691. [CrossRef] [PubMed]
36. Nickinson, R.S.; Board, T.N.; Kay, P.R. Post-operative anxiety and depression levels in orthopaedic surgery: A study of 56 patients undergoing hip or knee arthroplasty. *J. Eval. Clin.* **2009**, *15*, 307–310. [CrossRef] [PubMed]
37. Singh, V.; Kumar, A.; Gupta, S. Mental Health Prevention and Promotion-A Narrative Review. *Front. Psychiatry* **2022**, *13*, 898–909. [CrossRef]
38. Quartana, P.J.; Campbell, C.M.; Edwards, R.R. Pain catastrophizing: A critical review. *Expert Rev. Neurother.* **2009**, *9*, 745–758. [CrossRef]
39. Kalyan, K.; Jakhia, B.; Lele, R.D.; Joshi, M.; Chowdhary, A. Artificial neural network application in the diagnosis of disease conditions with liver ultrasound images. *Adv. Bioinform.* **2014**, *2014*, 708279. [CrossRef]
40. Halilaj, E.; Rajagopal, A.; Fiterau, M.; Hicks, J.L.; Hastie, T.J.; Delp, S.L. Machine learning in human movement biomechanics: Best practices, common pitfalls, and new opportunities. *J. Biomech.* **2018**, *81*, 1–11. [CrossRef]
41. Lillefjell, M.; Krokstad, S.; Espnes, G.A. Prediction of function in daily life following multidisciplinary rehabilitation for individuals with chronic musculoskeletal pain; a prospective study. *BMC Musculoskelet. Disord.* **2007**, *8*, 65. [CrossRef]
42. Popescu, V.G.; Burdea, G.C.; Bouzit, M.; Hentz, V.R. A virtual-reality-based telerehabilitation system with force feedback. *Technol. Biomed.* **2000**, *4*, 45–51. [CrossRef] [PubMed]
43. Park, I.; Lee, H.J.; Kim, S.K.; Park, M.S.; Kim, Y.-S. Factors Related to Preoperative Shoulder Pain in Patients with Atraumatic Painful Rotator Cuff Tears. *Clin. Shoulder Elb.* **2019**, *22*, 128–134. [CrossRef]

44. Crombez, C.; Eccleston, C.; Van Damme, S.; Vlaeyen, J.W.; Karoly, P. Fear-avoidance model of chronic pain: The next generation. *Clin. J. Pain* **2012**, *28*, 475–483. [CrossRef] [PubMed]
45. Sanz-Baños, Y.; Pastor, M.Á.; Velasco, L.; López-Roig, S.; Peñacoba, C.; Lledo, A.; Rodríguez, C. To walk or not to walk: Insights from a qualitative description study with women suffering from fibromyalgia. *Rheumatol. Int.* **2016**, *36*, 1135–1143. [CrossRef] [PubMed]
46. Écija, C.; Luque-Reca, C.; Suso-Ribera, C.; Catala, P.; Peñacoba, C. Associations of Cognitive Fusion and Pain Catastrophizing with Fibromyalgia Impact through Fatigue, Pain Severity, and Depression: An Exploratory Study Using Structural Equation Modeling. *J. Clin. Med.* **2020**, *9*, 1763. [CrossRef]
47. Pastor-Mira, M.A.; López-Roig, S.; Martínez-Zaragoza, F.; León, E.; Abad, E.; Lledó, A.; Peñacoba, C. Goal preferences, affect, activity patterns and health outcomes in women with fibromyalgia. *Front. Psychol.* **2019**, *10*, 1912. [CrossRef]
48. Vlaeyen, J.W.S.; Linton, S.J. Fear-avoidance and its consequences in chronic musculoskeletal pain: A state of the art. *Pain* **2000**, *85*, 317–332. [CrossRef]
49. Kuner, R.; Kuner, T. Cellular Circuits in the Brain and Their Modulation in Acute and Chronic Pain. *Physiol. Rev.* **2021**, *101*, 213–258. [CrossRef]
50. Baliki, M.N.; Apkarian, A.V. Nociception, Pain, Negative Moods, and Behavior Selection. *Neuron* **2015**, *87*, 74–91. [CrossRef]
51. Stegemann, A.; Liu, S.; Retana Romero, O.A.; Oswald, M.J.; Han, Y.; Beretta, C.A.; Gan, Z.; Tan, L.L.; Wisden, W.; Gräff, J.; et al. Prefrontal engrams of long-term fear memory perpetuate pain perception. *Nat. Neurosci.* **2022**, *26*, 820–829. [CrossRef]
52. Olango, W.M.; Finn, D.P. Neurobiology of stress-induced hyperalgesia. *Behav. Neurobiol. Chronic Pain* **2014**, *20*, 251–280.

MDPI
St. Alban-Anlage 66
4052 Basel
Switzerland
www.mdpi.com

Biomedicines Editorial Office
E-mail: biomedicines@mdpi.com
www.mdpi.com/journal/biomedicines

www.ingramcontent.com/pod-product-compliance
Lightning Source LLC
Chambersburg PA
CBHW061544220926
51850CB00005B/243
* 9 7 8 3 7 2 5 8 1 2 6 7 7 *